Innovations and Future Pathways for Sustainable and Resilient Civil Engineering

Sérgio António Neves Lousada
https://orcid.org/0000-0002-8429-2164
University of Madeira, Portugal

IGI Global
Scientific Publishing
Publishing Tomorrow's Research Today

Vice President of Editorial — Melissa Wagner
Director of Acquisitions — Mikaela Felty
Director of Book Development — Jocelynn Hessler
Production Manager — Mike Brehm
Cover Design — Jose Rosado

Published in the United States of America by
IGI Global Scientific Publishing
701 East Chocolate Avenue
Hershey, PA, 17033, USA
Tel: 717-533-8845 | Fax: 717-533-7115
Website: https://www.igi-global.com E-mail: cust@igi-global.com

Library of Congress Cataloging-in-Publication Data

Names: Lousada, Sérgio, 1981- editor
Title: Innovations and future pathways for sustainable and resilient civil engineering / edited by Sérgio António Neves Lousada.
Description: Hershey, PA : IGI Global Scientific Publishing, [2027] | Includes bibliographical references and index. | Summary: "This book presents a comprehensive and forward-thinking overview of the innovations, strategies, and transformative approaches that are shaping the future of sustainable and resilient civil engineering. Its primary objective is to bridge the gap between theory and practice by highlighting how technological advancements, climate adaptation strategies, and interdisciplinary collaboration are redefining infrastructure development in response to current and emerging global challenges"-- Provided by publisher.
Identifiers: LCCN 2025033253 (print) | LCCN 2025033254 (ebook) | ISBN 9798337361673 hardcover | ISBN 9798337361680 paperback | ISBN 9798337361697 ebook
Subjects: LCSH: Sustainable engineering | Smart materials | Civil engineering--Environmental aspects
Classification: LCC TA163 .I528 2026 (print) | LCC TA163 (ebook)
LC record available at https://lccn.loc.gov/2025033253
LC ebook record available at https://lccn.loc.gov/2025033254

British Cataloguing in Publication Data
A Cataloguing in Publication record for this book is available from the British Library.

Table of Contents

Preface xiii

Chapter 1
A Comprehensive Review of SCMs, LC^3, Alkali-Activated Materials, Recycled Aggregates, and Self-Healing Technologies 1
Parveen Kumar, Department of Civil Engineering, Maharishi Markandeshwar Engineering College, Maharishi Markandeshwar (Deemed to be) University, Mullana, India
Chadetrik Rout, Department of Civil Engineering, Maharishi Markandeshwar Engineering College, Maharishi Markandeshwar (Deemed to be) University, Mullana, India

Chapter 2
Circular Economy in Construction: Innovative Practices for Sustainable Material Use 39
Arun Murugesan, PSG Institute of Technology and Applied Research, India
Nidhya Rathinavel, PSG Institute of Technology and Applied Research, India
Rahul Balaji, PSG Institute of Technology and Appiled Research, India

Chapter 3
Innovative Practices That Reduce Material Consumption Through Recycling, Reuse, and Sustainable Construction Techniques Aligned With Circular Economy 69
Jyoti Rani, Parul University, India
Shashiranjan Kumar Yadav, Lovely Professional University, India
Sakthivel Santhanam, Presidency University, India

Chapter 4
Determination of Concrete Strength Used in Building Performance Analyses by Destructive and Non-Destructive Methods 107
Osman Uyanik, Faculty of Engineering and Natural Sciences, Süleyman Demirel University, Turkey
Nevbahar Ekin, Suleyman Demirel University, Turkey
Nurten Ayten Uyanik, Vocational School of Technical Sciences, Isparta University of Applied Sciences, Turkey
Ziya Öncü, Isparta University of Applied Sciences, Turkey

Chapter 5
Thermal Stability and Corrosion Resistance of AISI 904L Stainless Steel for Industrial Applications 149
Vaiavel Madeshwaren, Dhanalakshmi Srinivasan College of Engineering, Coimbatore, India
R. S. Gandhimathi, Annapoorana Engineering College, Salem, India
Bondala Ramakrishna, Aditya Institute of Technology and Management, Tekkali, India

Chapter 6
From Mold to Marvel: Engineering Excellence in Precast Systems..... 169
Ruchira Srivastava, Vollert India Pvt. Ltd., India
Krishan Mohan Sharma, Vollert India Pvt. Ltd., India

Chapter 7
Bridging the Gap: Aligning Precast Concrete Quality Management With ISO 9001:2015 Requirements 195
Ruchira Srivastava, Vollert India Pvt. Ltd., India
Krishan Mohan Sharma, Vollert India Pvt. Ltd, India

Chapter 8
Smart Structures for Smart Cities: The IoT-Driven Future of Precast Concrete 223
Ruchira Srivastava, Vollert India Pvt. Ltd., India
Krishan Mohan Sharma, Vollert India Pvt. Ltd, India

Chapter 9
Multi-Criteria Optimization for Multi-Hazard Resilience: Sustainable Seismic Retrofitting of Green Reinforced Concrete Structures 247
Ali Biglarifadafan, University of Golestan, Iran
Sérgio António Neves Lousada, Universidade da Madeira, Portugal

Chapter 10
Planning and Implementation of Hydrologically Sensitive Drainage Systems That Address Flooding and Water Quality in Urban Environments 333
Harshit K. Sheladiya, vertical-earth, USA

Chapter 11
Digital and Nature-Based Innovations for Climate-Resilient Urban Infrastructure: Smart Technologies and Sustainable Design 361
KM Ashifa, Istanbul Gelisim University, Turkey
R. Vettriselvan, Academy of Maritime Education and Training, Chennai, India
Vijay Singh Khunga, Saraswathi Institute of Medical Sciences, Hapur, India
V. Savitha, Department of Electrical Engineering, Saveetha Engineering College, Chennai, India
Sripriya Ranganathan, Department of Electrical Engineering, Saveetha Engineering College, Chennai, India

Chapter 12
Integrating Renewable Energy Systems in Civil Infrastructure: Thermo-Mechanical Innovations for Net-Zero Buildings... 385
Shalom Akhai, Maharishi Markandeshwar Engineering College, Maharishi Markandeshwar (Deemed to be) University, Mullana, India
Mahapara Abbass, Maharishi Markandeshwar Engineering College, Maharishi Markandeshwar (Deemed to be) University, Mullana, India

Chapter 13
Smart Aesthetics of Sustainable Construction as a Synthesis of Biomimicry, Emotion, and Analytics 433
Oleksandra Humenna, Kyiv National Economic University Named After Vadym Hetman, Ukraine
Hanna Melehanych, Uzhhorod National University, Ukraine
Maryna Chaikovska, Odesa I.I. Mechnikov National University, Ukraine

Compilation of References 467

About the Contributors 555

Index 563

Detailed Table of Contents

Preface xiii

Chapter 1
A Comprehensive Review of SCMs, LC^3, Alkali-Activated Materials, Recycled Aggregates, and Self-Healing Technologies 1
Parveen Kumar, Department of Civil Engineering, Maharishi Markandeshwar Engineering College, Maharishi Markandeshwar (Deemed to be) University, Mullana, India
Chadetrik Rout, Department of Civil Engineering, Maharishi Markandeshwar Engineering College, Maharishi Markandeshwar (Deemed to be) University, Mullana, India

The shift toward low-carbon and climate-resilient infrastructure has accelerated the development of advanced cementitious materials and reinforcement systems. This review summarizes recent progress in supplementary cementitious materials (SCMs), limestone–calcined clay cement (LC^3), alkali-activated binders, CO_2-cured concretes, and recycled-aggregate concretes, focusing on their contributions to clinker reduction, durability enhancement, and circular resource use. Performance is analysed across major exposure conditions, including chloride ingress, sulphate environments, carbonation, and freeze–thaw cycles. Waste-derived additives such as glass powder, crumb rubber, and recycled plastics are assessed for their functional benefits and engineering limitations. The paper also reviews self-healing technologies—capsule-based, crystalline, and microbial—that improve crack sealing, reduce permeability, and delay corrosion. Additionally, the role of FRP reinforcement in providing corrosion-free alternatives for aggressive environments is examined.

Chapter 2
Circular Economy in Construction: Innovative Practices for Sustainable Material Use 39
Arun Murugesan, PSG Institute of Technology and Applied Research, India
Nidhya Rathinavel, PSG Institute of Technology and Applied Research, India
Rahul Balaji, PSG Institute of Technology and Appiled Research, India

The construction industry uses around 40% between the consumed materials we use and 30% of the waste we produce globally. In general, the industry holds a linear model that reinforces negative impacts on the environment. The circular economy (CE)

aims to use waste as an opportunity for materials reuse and ecosystem regeneration. This chapter reviews CE strategies that have potential to transition the construction industry, such as modular construction, urban mining, and design for deconstruction. The study employed a mixed-methods approach through literature reviews, case studies, and interviews with industry stakeholders. It advocates for the use of tools such as the Building Information Model (BIM), digital twins, and material passports in the context of CE; aligns with the United Nations (UN) Sustainable Development Goals (SDGs), and presents a framework for the construction industry to become positive regenerative and resource-efficient built environment.

Chapter 3
Innovative Practices That Reduce Material Consumption Through Recycling, Reuse, and Sustainable Construction Techniques Aligned With Circular Economy 69
Jyoti Rani, Parul University, India
Shashiranjan Kumar Yadav, Lovely Professional University, India
Sakthivel Santhanam, Presidency University, India

The construction industry, a major consumer of natural resources and contributor to waste and carbon emissions, must shift from a linear "take-make-dispose" model to a circular economy for sustainable growth. This chapter explores innovative practices focusing on material efficiency, recycling, reuse, and sustainable design. Key strategies include adaptive building reuse, use of recycled and renewable materials, modular construction, and closed-loop supply chains. Through case studies and emerging technologies, it shows how circular principles reduce environmental impact and extend building lifecycles. The chapter also emphasizes supportive policies, collaborations, and research to advance circular construction globally toward a regenerative, resilient future.

Chapter 4
Determination of Concrete Strength Used in Building Performance Analyses by Destructive and Non-Destructive Methods 107
Osman Uyanik, Faculty of Engineering and Natural Sciences, Süleyman Demirel University, Turkey
Nevbahar Ekin, Suleyman Demirel University, Turkey
Nurten Ayten Uyanik, Vocational School of Technical Sciences, Isparta University of Applied Sciences, Turkey
Ziya Öncü, Isparta University of Applied Sciences, Turkey

Assessing the structural performance of concrete is vital for monitoring structures and planning interventions, especially after earthquakes. Accurately determining concrete strength ensures safety and cost-effective maintenance. Mechanical properties such as compressive strength, elasticity, and porosity are measured by destructive, semi-destructive, and non-destructive testing (NDT). Destructive testing provides reliable results but can damage the structure and is costly. NDT methods are faster and less invasive but provide estimated values. To improve accuracy, they must be calibrated with selective destructive, semi-destructive or non-destructive testing. Combined assessment methods, such as the SonReb (ultrasonic and rebound hammer) method, which combines both techniques, increase reliability. Integrating destructive and NDT approaches allows for safer, more economical assessments and better-informed engineering decisions.

Chapter 5
Thermal Stability and Corrosion Resistance of AISI 904L Stainless Steel for Industrial Applications 149
Vaiavel Madeshwaren, Dhanalakshmi Srinivasan College of Engineering, Coimbatore, India
R. S. Gandhimathi, Annapoorana Engineering College, Salem, India
Bondala Ramakrishna, Aditya Institute of Technology and Management, Tekkali, India

AISI 904L stainless steel is widely recognized for its superior corrosion resistance and excellent thermal stability, making it a preferred choice for demanding industrial environments. This study evaluates the material's corrosion behavior, high-temperature performance, and microstructural stability. Optical microscopy and scanning electron microscopy (SEM) were employed to analyze the microstructure. TGA results indicate strong oxidation resistance, with only 0–5% weight loss at 800°C. Mechanical tests reveal high resistance to deformation and significant stress retention at elevated temperatures, highlighting its capacity to withstand high strain rates. Corrosion testing in acidic environments shows improved resistance with increasing temperature, underscoring the alloy's resilience in harsh conditions. Supplementary analyses via XRD, EDX, and SEM confirm the alloy's structural integrity up to 1000°C. These findings affirm AISI 904L's suitability for applications that demand durability, thermal endurance, and long service life in aggressive industrial settings.

Chapter 6
From Mold to Marvel: Engineering Excellence in Precast Systems.................. 169
Ruchira Srivastava, Vollert India Pvt. Ltd., India
Krishan Mohan Sharma, Vollert India Pvt. Ltd., India

A recent analysis of the building sector reveals a low degree of automation in both manufacturing and construction, as well as extremely low material usage efficiency. There has been a clear shift towards automated manufacturing processes, while the sector also struggles with a shortage of competent people in mechanical engineering and construction. In India, the majority of building projects are completed using the traditional cast-in-place technique. However, India continues to have a massive housing need. Consequently, the construction work needs to be done considerably more quickly. This cannot be accomplished using traditional building techniques. With precast concrete for building, it is conceivable. When compared to normal concrete, precast concrete offers additional benefits. Thus, a variety of literature is examined, and this study provides a review of them all.

Chapter 7
Bridging the Gap: Aligning Precast Concrete Quality Management With ISO 9001:2015 Requirements 195
Ruchira Srivastava, Vollert India Pvt. Ltd., India
Krishan Mohan Sharma, Vollert India Pvt. Ltd, India

Customer satisfaction, regulatory compliance, and consistent quality are becoming important to the precast concrete sector. The internal quality control systems of many precast producers may not be entirely in line with the process-based, continuous improvement tenets of ISO 9001:2015 since they frequently function in silos. The discrepancies between the ISO 9001:2015 framework and standard precast concrete quality management procedures are investigated in this article. Process mapping, document analysis, and industry case studies are used to pinpoint crucial areas that need improvement, including stakeholder focus, risk-based thinking, and performance measurement. Design verification, material traceability, in-process inspections, and post-installation assessments are all integrated into a single quality management system as part of the study's detailed alignment roadmap specifically designed for precast operations. Precast producers can increase their operational effectiveness, lower their failure rates, and become more competitive in the market by using this alignment strategy.

Chapter 8
Smart Structures for Smart Cities: The IoT-Driven Future of Precast Concrete 223
Ruchira Srivastava, Vollert India Pvt. Ltd., India
Krishan Mohan Sharma, Vollert India Pvt. Ltd, India

In recent years, the precast concrete system has expanded significantly. Benefits including improved quality, cost savings, and construction process agility are to blame for this expansion. But uncertainty and One issue with this beneficial method is interdependencies throughout management procedures. Considering these difficulties, this paper suggests utilizing IoT technologies to make things better. There are several advantages to precast construction, including increased output, a better working environment, and a notable decrease in waste. The timely supply of precast components (PCs) is crucial to the successful completion of building projects, primarily because precast prefabrication operations are typically recognized as vital activities. Current precast production scheduling models, however, must adapt to the needs of dynamic situations where uncertainty is a problem for building projects. Building faster and more sustainably is a challenge for the construction sector.

Chapter 9
Multi-Criteria Optimization for Multi-Hazard Resilience: Sustainable Seismic Retrofitting of Green Reinforced Concrete Structures 247
Ali Biglarifadafan, University of Golestan, Iran
Sérgio António Neves Lousada, Universidade da Madeira, Portugal

This chapter proposes an Integrated Performance-Based Rehabilitation Framework that couples low-carbon, bio-based concrete retrofit solutions with a Multi-Criteria Decision-Making (MCDM) optimization scheme for existing reinforced-concrete structures exposed to compound hazards (e.g., earthquakes combined with flooding, heat, or wildfire). It argues that life-safety–focused retrofit practice is insufficient to manage trade-offs among resilience, lifecycle cost, carbon footprint, and implementability. The workflow integrates: (1) condition assessment and structural health monitoring; (2) stochastic degradation and service-life prediction; (3) performance-based evaluation of retrofit alternatives; (4) sustainable intervention design using bio-based materials; and (5) lifecycle management and decision support. A multi-objective formulation combines LCCA, LCA and probabilistic risk assessment, strengthened with robust decision methods under deep uncertainty, to identify retrofit designs that co-optimize safety and sustainability.

Chapter 10
Planning and Implementation of Hydrologically Sensitive Drainage Systems That Address Flooding and Water Quality in Urban Environments 333
Harshit K. Sheladiya, vertical-earth, USA

This study looks at how well SUDS work to cut down on flood risks and improve water quality in big US cities. The research used a Comparative Case Analysis. It pulled together performance numbers and opinions from stakeholders in five key cities: Philadelphia, Washington, D.C., Portland, Seattle, and New York City. The findings prove that SUDS dramatically lowers the number of floods. Across all five cities, flood frequency dropped by about 56% on average. Relevant statistics confirm that this success mostly comes from the systems controlling the amount of water runoff (R2 = 0.85). At the same time, SUDS methods greatly improved water quality. They were excellent at removing sediments (up to 73%) and heavy metals (up to 67%), which shows they successfully serve a dual purpose. Although experts had great faith in SUDS (about 75% to 80% saw them positively), the main obstacles to using them everywhere are still high upfront costs, the burden of long-term upkeep, and weak governance systems. The results emphasize that SUDS are vital for making cities stronger and more adaptable.

Chapter 11
Digital and Nature-Based Innovations for Climate-Resilient Urban Infrastructure: Smart Technologies and Sustainable Design 361
KM Ashifa, Istanbul Gelisim University, Turkey
R. Vettriselvan, Academy of Maritime Education and Training, Chennai, India
Vijay Singh Khunga, Saraswathi Institute of Medical Sciences, Hapur, India
V. Savitha, Department of Electrical Engineering, Saveetha Engineering College, Chennai, India
Sripriya Ranganathan, Department of Electrical Engineering, Saveetha Engineering College, Chennai, India

The challenges of climate change, rapid urbanisation and environmental degradation are increasing the risks on the urban infrastructure systems, and changes beyond the traditional civil engineering practices are necessary. This chapter examines how the integrating digital innovations with the nature-based solutions can enhance climate-resilient urban infrastructure as a whole. Through an analytical and application-oriented perspective, this chapter demonstrates how smart technologies such as digital twins, artificial intelligence, and real-time monitoring strengthen the performance, measurability, and adaptive management of nature-based solutions. The chapter creates a system as a whole linking technology, ecology and governance and makes the spaces of application and policy implications, recommending possibilities of having resilient, sustainable and inclusive developments of urban infrastructure.

Chapter 12
Integrating Renewable Energy Systems in Civil Infrastructure: Thermo-Mechanical Innovations for Net-Zero Buildings.. 385
Shalom Akhai, Maharishi Markandeshwar Engineering College, Maharishi Markandeshwar (Deemed to be) University, Mullana, India
Mahapara Abbass, Maharishi Markandeshwar Engineering College, Maharishi Markandeshwar (Deemed to be) University, Mullana, India

Modern infrastructure is undergoing a major transformation to meet sustainability and climate goals. This chapter explores how renewable energy systems and thermo-mechanical innovations enable Net-Zero Energy Buildings (NZEBs) and net-zero performance at district scale. It links envelope design, HVAC strategies, energy recovery, and storage with onsite renewables such as PV, PV-T, energy piles, and ambient loops. Key performance metrics—including energy use intensity, peak demand, and carbon intensity—are addressed alongside durability, safety, and techno-economic evaluation. Practical pathways are illustrated for composite, hot-humid, and cold climates, highlighting innovations like radiant systems, desiccant dehumidification, model predictive control, and life-cycle assessment. The discussion emphasizes integrated workflows, barrier-solution maps, and emerging technologies such as digital twins, low-GWP heat pumps, and advanced thermal storage. By combining efficiency, renewables, and resilience, the chapter outlines how civil infrastructure can achieve scalable net-zero design.

Chapter 13
Smart Aesthetics of Sustainable Construction as a Synthesis of Biomimicry, Emotion, and Analytics 433
Oleksandra Humenna, Kyiv National Economic University Named After Vadym Hetman, Ukraine
Hanna Melehanych, Uzhhorod National University, Ukraine
Maryna Chaikovska, Odesa I.I. Mechnikov National University, Ukraine

This chapter introduces an integrated model of smart aesthetics as a foundation for sustainable transformation in the construction sector. Responding to ecological challenges, post-war recovery needs, and rising expectations for human-centered design, the model combines three complementary components: biomimicry, Kansei engineering, and the Analytic Hierarchy Process. Biomimicry provides naturally optimized solutions for energy efficiency and climate resilience, while Kansei engineering enhances psychological comfort, emotional safety, and user well-being. AHP ensures balanced decision-making by evaluating ecological, social, cultural, and economic criteria within reconstruction projects. The chapter emphasizes that sustainable development requires not only engineering efficiency but also emotional and cultural intelligence embedded in the built environment. The proposed framework demonstrates practical value for schools, hospitals, residential areas, and public spaces, offering a pathway toward environmentally responsible and emotionally supportive urban environments.

Compilation of References 467

About the Contributors 555

Index 563

Preface

INTRODUCTION

The field of civil engineering is undergoing a profound transformation driven by the urgent need to develop sustainable, resilient, and future-ready infrastructure. In response to escalating environmental challenges, increasing climate-related risks, and rapid technological advancement, civil engineering is evolving beyond traditional practices toward more adaptive, integrated, and innovation-driven approaches. Today's professionals are increasingly adopting interdisciplinary methodologies and advanced tools that are reshaping how infrastructure systems are conceived, designed, constructed, and managed. This shift reflects a broader commitment to long-term sustainability, climate resilience, and responsible resource utilization, as well as a recognition of the need to support inclusive, low-impact, and intelligent built environments.

This edited volume, *Innovations and Future Pathways for Sustainable and Resilient Civil Engineering*, aims to provide a comprehensive and forward-looking examination of the innovations, strategies, and transformative approaches that are shaping the future of the discipline. Its central objective is to bridge the gap between theoretical advancements and practical applications by demonstrating how technological innovation, climate adaptation strategies, and interdisciplinary collaboration are collectively redefining infrastructure development in the context of pressing global challenges.

By bringing together contributions from researchers, academics, and industry practitioners, this book highlights recent developments across key areas such as digital construction technologies, smart infrastructure systems, green and sustainable design practices, circular economy approaches in construction, and nature-based solutions. These contributions collectively offer a diverse and rigorous body of knowledge that reflects the dynamic evolution of civil engineering and encourages new directions in research, education, and professional practice.

This volume serves both as a reference work and a source of inspiration for understanding how civil engineering can actively contribute to achieving global sustainability goals while enhancing urban and environmental resilience. It presents applied examples, critical analyses, and innovative methodologies that expand the conceptual and practical boundaries of the discipline in the twenty-first century. In doing so, it underscores the importance of integrating environmental, social, and economic dimensions into infrastructure planning and development.

The intended audience for this book is broad yet specialized, encompassing academicians, researchers, postgraduate students, policymakers, civil engineers, urban planners, architects, environmental scientists, and professionals engaged in infrastructure development and sustainability-focused practice. By fostering a deeper understanding of the emerging challenges and opportunities within civil engineering, this volume aims to support both scholarly inquiry and practical application. Ultimately, it seeks to encourage dialogue, collaboration, and informed decision-making across disciplines and sectors, contributing meaningfully to the advancement of sustainable and resilient infrastructure systems worldwide.

ORGANIZATION OF THE BOOK

Chapter 1 presents a comprehensive review of advanced cementitious materials and emerging reinforcement technologies that are redefining sustainable concrete systems. It examines supplementary cementitious materials (SCMs), limestone–calcined clay cement (LC^3), alkali-activated binders, CO_2-cured concretes, and recycled-aggregate concretes as key strategies for reducing clinker dependency and enhancing durability across aggressive exposure conditions. The chapter further evaluates waste-derived additives such as glass powder, crumb rubber, and recycled plastics, highlighting both their functional contributions and engineering limitations. In addition, it explores self-healing technologies—including capsule-based, crystalline, and microbial systems—as well as fiber-reinforced polymer (FRP) reinforcement, emphasizing their role in improving crack control, permeability reduction, and corrosion resistance. Collectively, this chapter establishes a material-level foundation for low-carbon and resilient infrastructure development.

Chapter 2 examines the transition of the construction sector from a linear to a circular economy model, emphasizing innovative practices that promote sustainable material use and resource efficiency. It highlights strategies such as modular construction, urban mining, and design for deconstruction, which enable the reuse and regeneration of building materials throughout their lifecycle. The chapter also underscores the integration of digital tools, including Building Information Modelling (BIM), digital twins, and material passports, as critical enablers of circular construction systems. Through a mixed-methods approach and alignment with the United Nations Sustainable Development Goals, the discussion provides a structured framework for transforming construction practices into regenerative, resource-efficient systems that minimize environmental impact while enhancing long-term sustainability.

Chapter 3 focuses on practical innovations that reduce material consumption in the construction industry through recycling, reuse, and sustainable building techniques aligned with circular economy principles. It explores adaptive reuse of existing structures, the incorporation of recycled and renewable materials, modular construction approaches, and closed-loop supply chains as key mechanisms for improving material efficiency. Drawing on case studies and emerging technologies, the chapter demonstrates how these strategies extend building lifecycles while significantly reducing environmental burdens. It further emphasizes the importance of supportive policy frameworks, interdisciplinary collaboration, and continued research to accelerate the global transition toward regenerative and resilient construction systems.

Chapter 4 addresses the critical importance of accurately determining concrete strength for structural performance assessment, particularly in post-disaster and maintenance contexts. It compares destructive, semi-destructive, and non-destructive testing (NDT) methods used to evaluate key mechanical properties such as compressive strength, elasticity, and porosity. While destructive testing remains highly reliable, its invasive nature limits applicability, whereas NDT methods offer faster and less intrusive alternatives that require careful calibration. The chapter highlights integrated approaches such as the SonReb method, which combines ultrasonic pulse velocity and rebound hammer testing to enhance accuracy. It concludes that combining multiple assessment techniques enables safer, more cost-effective, and more reliable engineering decisions in structural evaluation.

Chapter 5 investigates the thermal stability and corrosion resistance of AISI 904L stainless steel for use in demanding industrial environments. The chapter evaluates microstructural stability, mechanical performance, and corrosion behavior under high-temperature and acidic conditions using techniques such as optical microscopy, SEM, XRD, and TGA. Findings indicate excellent oxidation resistance, minimal weight loss at elevated temperatures, and strong mechanical integrity under stress. The alloy demonstrates superior corrosion resistance that improves with temperature, confirming its suitability for aggressive operational environments. Overall, the study reinforces AISI 904L as a high-performance material capable of ensuring durability and long service life in extreme industrial applications.

Chapter 6 explores the growing significance of precast concrete systems as an efficient alternative to traditional cast-in-place construction methods. It highlights the limitations of conventional construction practices in meeting rising housing demands and emphasizes the advantages of precast systems, including improved quality control, reduced material waste, and accelerated construction timelines. The chapter also discusses the increasing shift toward automation in manufacturing processes alongside challenges such as workforce shortages in technical and engineering domains. Through a synthesis of existing literature, it positions precast concrete as a critical solution for achieving faster, more efficient, and scalable infrastructure development.

Chapter 7 examines the alignment of precast concrete quality management systems with ISO 9001:2015 standards, emphasizing the importance of consistency, compliance, and continuous improvement. It identifies gaps between conventional precast industry practices and the process-based, risk-oriented framework of ISO standards, particularly in areas such as stakeholder engagement, performance measurement, and documentation systems. The chapter proposes a structured roadmap integrating design verification, material traceability, in-process inspection, and post-installation evaluation into a unified quality management system. This alignment is presented as a pathway to improved operational efficiency, reduced failure rates, and enhanced competitiveness within the precast construction sector.

Chapter 8 investigates the integration of Internet of Things (IoT) technologies into precast concrete systems to enhance efficiency, coordination, and production reliability. It highlights how uncertainty and interdependencies in traditional precast workflows can be mitigated through real-time monitoring, data-driven scheduling, and smart logistics systems. The chapter emphasizes the importance of timely component delivery and adaptive production planning in achieving efficient construction outcomes. By incorporating IoT-enabled solutions, the precast industry can improve productivity, reduce waste, and address the challenges of dynamic construction environments, thereby supporting faster and more sustainable infrastructure delivery.

Chapter 9 presents an integrated performance-based rehabilitation framework for enhancing the resilience of reinforced concrete structures exposed to multiple hazards. It combines low-carbon bio-based retrofit materials with multi-criteria decision-making (MCDM) techniques to optimize structural safety, sustainability, and lifecycle performance. The framework incorporates condition assessment, stochastic degradation modeling, performance evaluation, retrofit design, and lifecycle analysis under uncertainty. By integrating life-cycle cost assessment, environmental impact analysis, and probabilistic risk evaluation, the chapter demonstrates how retrofit strategies can be optimized to balance safety, resilience, and sustainability in complex hazard environments.

Chapter 10 evaluates the effectiveness of Sustainable Urban Drainage Systems (SUDS) in mitigating urban flooding and improving water quality across major U.S. cities. Based on comparative case analysis, the chapter reports significant reductions in flood frequency and notable improvements in pollutant removal, particularly sediments and heavy metals. Despite strong stakeholder support, challenges such as high initial costs, maintenance demands, and governance limitations remain barriers to widespread implementation. The findings emphasize SUDS as a critical component of climate-resilient urban infrastructure, capable of delivering dual benefits in flood mitigation and environmental enhancement.

Chapter 11 explores the integration of digital technologies and nature-based solutions to enhance climate resilience in urban infrastructure systems. It highlights how tools such as artificial intelligence, digital twins, and real-time monitoring can improve the performance, adaptability, and governance of ecological infrastructure. The chapter presents a holistic framework that connects technological innovation with ecological design and policy structures to support sustainable and inclusive urban development. By demonstrating the synergistic potential of digital and nature-based approaches, it underscores their role in strengthening urban resilience under increasing climate pressures.

Chapter 12 investigates the integration of renewable energy systems and thermo-mechanical innovations in achieving net-zero energy performance in buildings and infrastructure systems. It examines the relationship between building envelope design, HVAC optimization, energy recovery systems, and on-site renewable technologies such as photovoltaics and energy storage systems. The chapter also evaluates performance metrics including energy use intensity and carbon emissions, alongside techno-economic and lifecycle considerations. Through climate-specific design strategies and emerging technologies such as digital twins and advanced thermal storage, it outlines practical pathways toward scalable net-zero infrastructure development.

Chapter 13 introduces a smart aesthetics framework for sustainable construction that integrates biomimicry, Kansei engineering, and analytic hierarchy processes to support human-centered and environmentally responsive design. The chapter emphasizes the importance of combining ecological efficiency with emotional and cultural intelligence in the built environment. Biomimicry contributes nature-inspired design solutions, while Kansei engineering enhances user well-being and psychological comfort. The analytic hierarchy process enables balanced decision-making across environmental, social, cultural, and economic dimensions. Together, these components offer a comprehensive model for creating sustainable, resilient, and emotionally supportive urban spaces, particularly in contexts of reconstruction and urban renewal.

CONCLUSION

The collective contributions presented in *Innovations and Future Pathways for Sustainable and Resilient Civil Engineering* reflect a decisive shift in the discipline toward more adaptive, intelligent, and sustainability-driven paradigms. Across the chapters, a consistent narrative emerges: civil engineering is no longer defined solely by structural performance or construction efficiency, but by its capacity to respond to complex environmental, social, and technological challenges through integrated and forward-looking solutions. From advanced low-carbon materials and circular construction strategies to smart infrastructure systems, renewable energy integration, and resilience-based design frameworks, this volume demonstrates the breadth of innovation currently reshaping the built environment.

A key strength of this book lies in its interdisciplinary orientation. The convergence of material science, digital technologies, environmental systems, and human-centered design reflects the increasingly interconnected nature of modern infrastructure development. Whether addressing climate-resilient materials, IoT-enabled precast systems, sustainable drainage solutions, or multi-hazard structural retrofitting, the chapters collectively emphasize the importance of systems thinking and lifecycle-based approaches. These perspectives are essential for transitioning from reactive engineering practices to proactive, predictive, and regenerative infrastructure models.

Equally significant is the emphasis placed on sustainability as a multidimensional objective that extends beyond environmental performance to include economic viability, social inclusivity, and cultural relevance. The integration of circular economy principles, net-zero design strategies, and smart aesthetics underscores the need to balance technical innovation with broader societal values. In doing so, this volume reinforces the idea that resilient infrastructure must not only endure physical stresses but also support human well-being and equitable development.

As the global community continues to confront the escalating impacts of climate change, urbanization, and resource constraints, the role of civil engineering becomes increasingly pivotal. This book serves as both a reflection of current advancements and a guide for future inquiry, encouraging researchers, practitioners, and policymakers to adopt more holistic, data-informed, and sustainability-centered approaches. It is our hope that the insights presented within these pages will inspire continued innovation, foster interdisciplinary collaboration, and contribute meaningfully to the development of resilient infrastructure systems that can sustain future generations.

Chapter 1
A Comprehensive Review of SCMs, LC³, Alkali-Activated Materials, Recycled Aggregates, and Self-Healing Technologies

Parveen Kumar
http://orcid.org/0000-0003-3530-0700
Department of Civil Engineering, Maharishi Markandeshwar Engineering College, Maharishi Markandeshwar (Deemed to be) University, Mullana, India

Chadetrik Rout
Department of Civil Engineering, Maharishi Markandeshwar Engineering College, Maharishi Markandeshwar (Deemed to be) University, Mullana, India

ABSTRACT

The shift toward low-carbon and climate-resilient infrastructure has accelerated the development of advanced cementitious materials and reinforcement systems. This review summarizes recent progress in supplementary cementitious materials (SCMs), limestone–calcined clay cement (LC^3), alkali-activated binders, CO_2-cured concretes, and recycled-aggregate concretes, focusing on their contributions to clinker reduction, durability enhancement, and circular resource use. Performance is analysed across major exposure conditions, including chloride ingress, sulphate environments, carbonation, and freeze–thaw cycles. Waste-derived additives such as glass powder, crumb rubber, and recycled plastics are assessed for their func-

DOI: 10.4018/979-8-3373-6167-3.ch001

tional benefits and engineering limitations. The paper also reviews self-healing technologies—capsule-based, crystalline, and microbial—that improve crack sealing, reduce permeability, and delay corrosion. Additionally, the role of FRP reinforcement in providing corrosion-free alternatives for aggressive environments is examined.

1. INTRODUCTION

A major transformation in construction material selection is underway, driven by the dual imperatives of achieving net-zero greenhouse gas emissions and enhancing infrastructure resilience to climate-related challenges (Omar, 2025). Despite the global predominance of Portland-cement concrete, steel, and asphalt, their production processes and degradation during use present notable obstacles to both environmental sustainability and long-term durability (Abbass et al., 2024). The substantial embodied carbon from cement and steel manufacturing, combined with vulnerabilities to heat, moisture, chloride penetration, freeze–thaw cycles, and acidification, can reduce service life and elevate costs through increased maintenance and premature repairs. In response, building material choices are being rigorously reassessed to minimize life-cycle emissions while upholding critical standards of safety, reliability, and performance.

A new generation of sustainable, climate-resilient materials has emerged (Motia et al., 2024). Low-clinker cementitious systems like high-volume SCMs, LC^3, alkali-activated binders, and CO_2-cured concretes reduce embodied CO_2 without compromising durability or performance (Arthur et al., 2025). Approaches such as recycled aggregates, waste-derived additives, and self-healing concrete further improve circularity and longevity, while FRP reinforcement prevents corrosion. These innovations can cut life-cycle emissions by 30–50% per element and strengthen resistance against chloride diffusion, sulphate attack, cracking, and fatigue (Daehn et al., 2021).

The evaluation and adoption of these materials have increasingly been framed through performance-based design (PBD) and life cycle thinking (Mainini et al., 2024). In lieu of prescriptive composition limits, PBD specifies explicit service-life criteria—strength, stiffness retention, crack-width control, permeability, and corrosion initiation times—so that unconventional binders and reinforcements can be qualified on measurable outcomes. Life-cycle assessment (LCA) and life-cycle costing (LCC) have been integrated with these performance targets to quantify trade-offs and to avoid burden shifting (e.g., early-age performance penalties or added curing energy that negate carbon savings). Standardised disclosure through Environmental Product Declarations (EPDs) has been encouraged to support transparent, scenario-aligned comparisons across products; however, practical comparability has been shown to depend sensitively on functional unit alignment, system boundaries, module coverage, and treatment of biogenic carbon—factors that, if inconsistent, can obscure genuine differences among alternatives.

Despite clear progress, several implementation challenges have remained. Alternative binders may require tailored admixture chemistry and curing regimes; familiar durability tests (e.g., rapid chloride permeability) can yield non-comparable results across disparate pore solutions and should be complemented by diffusion-centric metrics; and early-age behaviour must be managed to avert unintended cracking. Circular constituents can introduce stiffness and strength penalties that necessitate mix redesign (e.g., pre-saturation and quality enhancement of recycled aggregates, chemical or mineral treatments, and particle-size optimization for waste glass to mitigate alkali–silica reaction). Self-healing strategies are constrained by crack-width thresholds (H. T. Hu et al., 2024), trigger conditions (moisture or CO_2 availability), and the one-time nature of capsule systems, whereas FRP reinforcement, though corrosion-free, demands attention to linear-elastic failure modes, service-temperature limits, and creep-rupture design. These constraints underscore the need for qualification protocols that couple material characterisation with structural-scale demonstration and code-compliant acceptance pathways (Akhai, 2024).

Concurrently, policy and procurement mechanisms have begun to accelerate diffusion. "Buy Clean" requirements (S. Kumar, 2022), EPD-based submittals, mixture-level global-warming-potential (GWP) caps, and resilience-weighted bid evaluations have been adopted to internalize life cycle and climate risk into project delivery. Major suppliers now offer low-carbon product lines (e.g., PLC/LC^3 concretes, CO_2-cured units), transportation agencies have expanded provisions for GFRP reinforcement in chloride-exposed decks, and municipal pilots have validated performance of self-healing admixtures and circular aggregates in representative exposures. As codes and standards evolve toward performance orientation—supported by emerging guidance for alkali-activated binders, LC^3 systems, and FRP reinforcement—the conditions for mainstreaming these materials are being established. Against this backdrop, the present work synthesizes state-of-the-art advances since 2020, interrogates mechanical and durability performance under multi-hazard exposures, and articulates an implementation roadmap that aligns design practice, specification language, and procurement incentives with the objectives of net-zero and climate-resilient infrastructure. In parallel, life cycle assessment (LCA) is being coupled with design decisions to favor materials with lower environmental impacts (Najjar et al., 2022). Tools like Environmental Product Declarations (EPDs) allow transparent comparison of materials, as shown in Table 1.

Table 1. EPDs for transparent comparison of construction materials

Study	Materials & scope	What they compared (FU & boundary)	Key findings on transparency/comparability
(Gelowitz & McArthur, 2017)	Multiple construction product EPDs (Type III)	Reviewed FU, system boundaries, data quality across EPDs	Large variability in data quality and PCR alignment; inconsistent units and module coverage frequently preclude direct comparison.
(Moré et al., 2022)	Broad set of construction EPDs	Meta-assessment of completeness & comparability	Only ~5% of EPDs were readily comparable due to divergent scenarios, missing modules, and inconsistent methods.
(Rasmussen et al., 2021)	81 EPDs for CLT, glulam, LVL, timber	EN 15804 EPDs; various scenarios	Even within one material class, results vary with plant, geography, and scenario choices; transparent reporting of modules and biogenic carbon is decisive.
(Oh et al., 2023)	RC vs CLT vs TCC floor slabs	Component LCAs (project-level comparatives)	Demonstrates that aligned FU (e.g., 1 m^2 slab meeting same performance) and common boundaries reveal clear impact differences; misaligned assumptions can flip rankings.
(Aragón & Alberti, 2024)	Digital EPD datasets	Machine-interpretability & scenario definitions	Digital EPDs still lack consistent schemas for scenarios and unique IDs, limiting automated, transparent comparisons across materials.
(Konradsen et al., 2024)	Method rules within EN 15804+A2	Treatment of biogenic carbon & modules	Biogenic carbon handling differs unless full life cycle is included; comparing A1–A3-only EPDs with cradle-to-grave can mislead.
(Asdrubali et al., 2025)	Insulation, bricks, concrete, cement, steel, stones	EPD-based GWP distributions across categories	Quantifies variance drivers (geography, production route, recycled content); shows EPDs can support material screening when PCRs and modules are harmonized.

Alongside carbon accounting, life cycle costing (LCC) captures total ownership costs (including maintenance, repairs, and downtime). A material that is more expensive upfront but significantly extends the time to the first repair or avoids failure can be justified while considering the net present cost and risk.

Although numerous studies report on individual low-carbon binders (SCMs, LC^3, AAMs, CO_2-cured systems) (Dhiman et al., 2025) and circular constituents (RAC, waste-derived additives), existing reviews are largely fragmented either material-specific, durability-specific, or focused only on embodied carbon (Abbass & Akhai, 2025b). There is limited work that systematically compares these systems under common exposure classes (chloride, sulphate, carbonation, freeze–thaw, acid) while also integrating mechanical performance, long-term durability, and life-cycle metrics (LCA/LCC). Furthermore, the literature seldom connects these material-level findings to performance-based qualification, EPD transparency, and procurement/policy instruments that are now driving "Buy Clean" and resilience-orientated specifications (Abbass & Akhai, 2025a). This paper leaves practitioners without a consolidated framework to select and implement low-carbon, high-durability solutions at infrastructure scale. This chapter fills in the gaps by giving a unified, exposure-class-based overview of low-carbon and high-durability concrete systems. These systems include SCM blends, LC^3, AAMs, CO_2-cured concretes, RAC, waste-derived additives, self-healing technologies, and FRP-reinforced systems. It combines data on mechanical and durability performance with GWP ranges, diffusion-based durability indicators, and indicative LCC trade-offs. This information is shown in comparative matrices and summary tables that can be used directly in design and specification. In parallel, it links these technical insights to performance-based design, EPD practice, and emerging policy tools, thereby offering a practical roadmap for deploying net-zero-aligned, climate-resilient materials in modern infrastructure.

1.2 SCMs and Limestone-Calcined Clay Cements (LC^3)

The most widely implemented approach to low-carbon concrete is partial substitution of Portland cement with supplementary cementitious materials (SCMs) such as fly ash (coal combustion ash), ground granulated blast-furnace slag (GGBS), silica fume, and natural pozzolans as illustrated in Figure 1. These SCMs are typically industrial by-products or natural volcanic materials with cementitious or pozzolanic properties. Replacing a portion of OPC clinker with SCMs directly reduces the CO_2 emitted per unit of cement (Antunes et al., 2021). Indeed, SCM blending is noted as *"the most common and cost-effective way to reduce the environmental impact of concrete,"* and the use of GGBS or fly ash has become routine in sustainable mix designs. GGBS can substitute up to 75% of OPC in mixes with little effect on 28-day strength, while fly ash can often replace 30–40% without affecting structural performance (Liu et al., 2022). Such high-volume SCM use can roughly halve the embodied CO_2 per cubic meter of concrete. In practice there are limits – beyond certain replacement percentages, early strength development may slow, and specific durability aspects must be confirmed. Nonetheless, many standards (e.g. EN 197 in Europe, ASTM C595 in the US) already cover common blended cements (Portland-slag, Portland-fly ash, etc.) and permit SCM use in specifications, which has smoothed their adoption in real projects.

Figure 1. a) fly ash and b) ground granulated blast-furnace slag

a) b)

A newer class of blended cement garnering attention is limestone calcined clay cement (LC^3) – a ternary blend typically containing roughly 50% clinker, 30% calcined clay (metakaolin-rich clay), ~15% limestone, plus a small gypsum addition (Zhang et al., 2025). The synergy between calcined clay (metakaolin) and finely ground limestone allows very high clinker replacement through complementary reactions. LC^3 can reduce CO_2 emissions by about 30–40% compared to pure OPC. Remarkably, despite having only around half the clinker content of OPC, LC^3 achieves 28-day strengths comparable to or even exceeding OPC, enabled by the pozzolanic reaction of calcined clay and the filler and chemical effects of limestone (which together generate additional calcium aluminosilicate hydrates). LC^3 also exhibits excellent durability: it shows greater resistance to sulfate attack and carbonation than OPC, and its denser hydrate matrix leads to significantly lower porosity and much reduced chloride diffusion. Field trials in multiple countries (e.g. Cuba, India, Switzerland) have demonstrated that LC^3 is feasible in real applications. These pilots indicate LC^3 concrete may have slightly lower 3-day strength due to reduced clinker, but by 7–28 days it meets or exceeds OPC strength and continues to gain strength over time. The high-rise built with mass timber shows that large projects using this low-carbon material are possible, as shown in Table 2.

Table 2. Matrix of exposure classes

Material family	Chloride (marine / de-icing)	Sulfate soils / waters	Carbonation (interior / urban)	Freeze–thaw (with de-icers)	Acid / industrial
OPC / PLC (CEM I/IL)	⚠ Baseline; ensure low w/b and cover. *Tests:* NT BUILD 492, ASTM C1556, RCPT (ASTM C1202, interpret with care).	⚠ Moderate; use Type V or sulfate-resistant cements where needed. *Test:* ASTM C1012.	⚠ Moderate carbonation; adequate cover/ curing needed. *Tests:* RILEM CPC-18/EN 13295.	With proper air-entrainment. *Tests:* ASTM C666, ASTM C457 (air-void).	Poor; OPC vulnerable to strong acids. *Test:* ASTM C267.
SCM blends (slag/fly ash/silica fume)	Slag + silica fume reduce Cl^- ingress; fly ash improves later-age. *Tests:* NT BUILD 492, C1556.	Slag excels; low C_3A fly ash blends good. *Test:* C1012.	⚠ Fly ash can raise carbonation unless cured well; silica fume mitigates. *Tests:* CPC-18.	Good with air system; later-age strength supports F-T. *Tests:* C666, C457.	⚠ Better than OPC but still limited; not for strong acids. *Test:* C267.
LC^3 (limestone–calcined clay)	Very low diffusion; good pore refinement. *Tests:* NT BUILD 492, C1556.	Good resistance (low permeability). *Test:* C1012.	⚠ Slightly higher carbonation rate than OPC if under-cured; ensure curing/ cover. *Tests:* CPC-18.	With air-entrainment; similar to PLC. *Tests:* C666, C457.	⚠ Limited; not for strong mineral acids. *Test:* C267.
Alkali-activated binders / geopolymers	⚠ Slag-rich AAMs: ; fly-ash-only mixes: ⚠ (curing critical). *Tests:* migration/ diffusion (avoid RCPT), bulk resistivity.	Generally good sulfate/ acid resistance (slag-rich best). *Test:* C1012.	Often low carbonation-induced permeability, but carbonation of pore solution can occur—check steel passivity. *Tests:* CPC-18 + half-cell.	⚠ Mix-dependent; verify air-void and curing. *Tests:* C666, mass loss.	Superior to OPC for many acids; still vulnerable to strong acids/ solvents. *Test:* C267.
CO_2-cured concretes (precast)	Lower Cl^- ingress if mix is optimised; ensure downstream curing for uncarbonated cores. *Tests:* NT BUILD 492/C1556.	⚠ Similar to PLC; verify permeability. *Test:* C1012.	⚠ Surface carbonation intentional; verify rebar cover and pH profile. *Tests:* CPC-18, phenolphthalein.	For precast units with air system. *Tests:* C666.	⚠ Limited acid resistance; verify by C267.
RAC (recycled-aggregate concrete)	⚠ Higher permeability unless RCA is treated; lower w/b + SCMs advised. *Tests:* NT BUILD 492/C1556.	⚠ Use treated RCA; prove via C1012.	⚠ Tends to carbonate faster; ensure curing/ cover. *Tests:* CPC-18.	⚠ Achieve proper air system; verify freeze–thaw. *Tests:* C666, C457.	Not recommended for strong acids.

continued on following page

Table 2. Continued

Material family	Chloride (marine / de-icing)	Sulfate soils / waters	Carbonation (interior / urban)	Freeze–thaw (with de-icers)	Acid / industrial
Rubberized concrete	For primary structural elements (strength loss); use only where toughness/damping sought. *Tests:* diffusion + strength.	----	⚠ Non-structural only; check aging.	Excellent F-T and impact resistance in non-structural uses.	----
Glass-powder SCM (fine)	Good when finely ground (pozzolanic).	Comparable to other SCM concretes.	⚠ Watch carbonation if under-cured.	With air system.	⚠ Limited.
FRP-reinforced RC (system level)	No steel corrosion; check serviceability (deflection/crack width). *Tests:* ASTM A944 bond (adapted), resistivity/diffusion for matrix.	No bar corrosion; matrix governs.	Rebar immune; matrix carbonation still relevant to concrete serviceability.	Bar immune; ensure concrete F-T performance.	Rebar immune; matrix durability drives performance.

Symbols:- acceptable/advantage, ⚠ conditional/needs control, poor/avoid.

Given LC^3's promise, efforts are underway to scale up its production. The required raw materials (suitable clays and limestone) are abundant in many regions, and LC^3 can be produced in existing cement plants with minor modifications (clay must be calcined around 700–800 °C, much lower than clinker's ~1450 °C) (Zhao & Zhang, 2023). Thus, clinker substitution with LC^3 can be ramped up relatively quickly without massive new investments, making it a promising near-term solution for low-carbon construction. Some national standards are beginning to incorporate LC^3 India has considered adding a Portland–calcined clay cement category). Practitioners specifying LC^3 should account for slight differences from conventional cement: LC^3 may require optimized superplasticizers and careful curing (the fine particles can increase water demand), and at very high clinker replacement levels the early-age strength is lower – so formwork removal times or curing durations might need adjustment. Overall, SCM blends and LC^3 represent a mature, largely "drop-in" strategy for cutting the carbon footprint of cement, backed by decades of field experience with fly ash/slag blends and growing real-world validation for LC^3.

1.3 Alkali-Activated Binders and CO_2-Cured Concretes

A more radical alternative to Portland clinker is alkali-activated materials (AAMs), such as geopolymers, which use aluminosilicate powders like fly ash or slag combined with alkaline solutions to create cementitious binders. These "geopolymer concretes" can cut CO_2 emissions by up to 85% compared to OPC (Adnan & Anas, 2025), especially when local industrial by-products are used. These materials repurpose waste, eliminate the need for limestone calcination, and maintain robust performance—certain formulations achieve strengths exceeding 40 MPa within 24 hours and exhibit substantial chemical resistance. Nevertheless, alkali-activated materials (AAMs) encounter several limitations: numerous mixes require high-temperature curing that is unsuitable for on-site applications, some compositions involve caustic chemicals that present safety concerns, and comprehensive long-term durability data remains limited. New developments include dry "one-part" activators and hybrid cement systems with reduced ordinary Portland cement, as standards are being set by organizations like ASTM and ACI. CO_2-cured concrete uses captured industrial CO_2 to form stable carbonate compounds within the material, as seen in technologies like Carbon Cure™, which inject CO_2 after casting to boost strength and speed production. The method sequesters CO_2 permanently, but emission reductions depend on optimizing mix designs—strength improvements should decrease cement use or eliminate traditional curing methods. LCA studies stress that adjusting mixes is vital for real carbon savings. While CO_2-cured concretes are generally more durable, they require thorough carbonation or moist curing to avoid shrinkage or cracking. This technique works best for controlled precast applications where conditions can be precisely managed.

1.4 Design and Durability Considerations

Applying new types of cement systems in real-world settings calls for close monitoring of durability, compatibility, and adherence to building codes. Important factors to consider include:

1.4.1 Binder Selection for Exposure: Different low-carbon binders excel in different environments, so the binder should be matched to the exposure class of the structure. High-volume slag or fly ash concretes generate less heat of hydration (useful for massive pours to reduce thermal cracking) and have improved sulfate resistance, but very high fly ash content may carbonate faster or set slowly in cold weather. Geopolymer concretes offer excellent acid and sulfate resistance (suitable for sewer infrastructure or industrial floors), but their performance under sustained high temperatures or fire is still being studied (Wong, 2022). LC^3 cement shows markedly lower chloride diffusivity – one study found an LC^3 concrete had roughly four times longer time to corrosion initiation in marine exposure than an OPC concrete (Sirivivatnanon et al., 2023). In contrast, a pure alkali-activated fly ash concrete might allow faster chloride ingress unless supplemented with slag or cured very well. Thus, designers should select binders appropriate for the exposure: e.g. use LC^3 or slag cement in marine/coastal structures for superior chloride resistance;

use moderate fly ash or limestone blends for large pours to minimize thermal cracking; consider geopolymers for highly acidic or sulfate-rich environments. It is essential to verify that any alternative-binder concrete meets relevant durability metrics (permeability, diffusion coefficient, electrical resistivity, etc.) required by codes for the given exposure class.

1.4.2 Cracking and Shrinkage: Low-CO_2 binders can alter concrete's cracking tendencies. High-SCM concretes often gain strength more slowly and have a lower early-age modulus, which can actually reduce early-age thermal and shrinkage cracking because less heat is generated, and stresses develop more gradually. Alkali-activated concretes have distinctive shrinkage patterns; alkali-activated slag concrete often shows greater autogenous shrinkage unless well cured (Z. Li et al., 2021), while some geopolymer concretes have drying shrinkage similar to OPC. Standard test methods for shrinkage or cracking risk may need adaptation for these materials (accelerated carbonation tests, for instance, can exaggerate shrinkage in some AAMs due to their different pore chemistry). In practice, providing adequate curing is critical (e.g. use curing membranes, and in the case of CO_2-cured concrete, ensure any uncarbonated cement hydrates subsequently), and employing crack control measures (shrinkage reinforcement or fibers) is prudent as with conventional concrete.

1.4.3 Admixture Compatibility: Novel binders may interact differently with common chemical admixtures (Bondarenko et al., 2025). For instance, high-carbon fly ash can adsorb air-entraining agents, making it challenging to achieve proper air void content for freeze–thaw durability in cold climates. LC^3 has generally been found compatible with standard polycarboxylate superplasticizers, though dosage or admixture formulations may require slight adjustments. Geopolymer mixes often cannot use certain admixtures designed for OPC (e.g. gypsum as a set regulator, or typical calcium-based accelerators) and instead may require admixtures specifically formulated for high-alkalinity systems. Specifications should therefore require trial batches to verify that properties like workability, set time, air content, and other fresh concrete characteristics are acceptable with the proposed binder–admixture combination.

1.4.4 Structural Design Parameters: Design codes may not yet explicitly list material properties for these new binders, but performance-based provisions typically allow their use given proper documentation and demonstration of equivalence. Engineers might need to establish certain properties – e.g. creep coefficient, drying shrinkage, tensile splitting strength development – through testing or literature if the code does not provide values for a given geopolymer or LC^3 concrete. Current research shows that LC^3 concrete has structural properties similar to OPC concrete. Steel rebar bond strength in LC^3 matches that of OPC, and LC^3 offers lower corrosion rates under chloride exposure due to refined pores (Wang et al., 2025). Projects without specific code provisions may use technical qualification or variance by demonstrating performance. Notably, standards bodies are moving to accommodate these materials: as mentioned, RILEM and ASTM are developing guidelines for alkali-activated binders, and ACI's report on alternate cements provides interim recommendations.

1.4.5 Quality Control and Testing: Field quality control procedures for alternative cements may require distinct assessments and criteria compared to those applied to ordinary Portland cement (OPC). Standard specifications such as ASTM C150, which define the chemical requirements for OPC, are not applicable to geopolymer binders; thus, quality assurance may focus on verifying activator concentrations or performing strength evaluations on trial mixes. Additionally, conventional concrete durability tests often necessitate reinterpretation. For instance, the rapid chloride permeability test (ASTM C1202), which measures electrical charge transfer, may yield non-comparable results for high alumina geopolymer concretes due to differences in pore solution chemistry and conductivity. Alternative approaches, such as surface resistivity or diffusion-based tests, are recommended to evaluate chloride ingress in these materials. When implementing an unfamiliar binder at scale, it is prudent to conduct mock-up pours or pilot sections to observe setting, finishing, and curing characteristics under actual field conditions, enabling informed adjustments as required as shown in Table 3.

Table 3. Typical benefit & trade-off ranges

Material family	GWP vs OPC (%)	28-day strength (MPa)	Apparent Cl^- diffusion D_app ($\times 10^{-12}$ m^2/s)	90-day drying shrinkage (με)	Creep coeff. φ(∞)	Special curing/QA	Typical cost delta
OPC / PLC	0 to −10	30–60	6–12	600–800	2.0–2.5	Standard moist curing; air control for F-T	Baseline
SCM blends (slag/fly ash/silica fume)	−20 to −50	30–70 (later-age gain)	2–6 (slag + SF lower end)	500–750	1.8–2.4	Extended moist curing; set/early strength monitoring	−5% to +5%
LC³	−30 to −40	35–65	1.5–4	500–700	1.8–2.3	Good curing; admixture re-tuning	−5% to +10%
AAM / geopolymers	−50 to −85 (feedstock-dependent)	35–70	1–5 (slag-rich) / 4–8 (FA-rich)	500–800 (mix-dependent)	1.6–2.3	Thermal/curing control; activator QA; avoid RCPT-only	+5% to +20%
CO_2-cured concretes (precast)	−10 to −25 (with mix optimisation)	40–70	2–6	500–700	1.8–2.3	Controlled carbonation + post-curing; strength-based cement reduction	0% to +10%
RAC (untreated RCA)	−2 to −8 (aggregate share)	25–50 (↓ with high RCA)	6–15	650–900	2.2–2.8	Pre-saturate RCA; tight w/b; SCMs	−5% to 0%

continued on following page

Table 3. Continued

Material family	GWP vs OPC (%)	28-day strength (MPa)	Apparent Cl^- diffusion D_app ($\times 10^{-12}$ m^2/s)	90-day drying shrinkage (με)	Creep coeff. φ(∞)	Special curing/QA	Typical cost delta
RAC (treated RCA: carbonation/nano-silica/slurry)	−2 to −8	30–60	3–8	600–800	2.0–2.5	Document treatment QA; absorption limits	0% to +10%
Rubberised concrete (5–20% fine repl.)	−2 to −6	15–40 (↓)	8–18	700–1000	2.3–3.0	Surface-treated rubber; define non-structural use	0% to +10%
Glass-powder SCM (10–20%)	−10 to −25	30–60	3–7	500–750	1.8–2.4	Verify fineness/ reactivity; ASR risk control	−5% to +5%
Plastic-modified concrete	−2 to −6	10–40 (↓)	10–20	700–1000	2.5–3.0	Lightweight/non-structural; fire/ performance checks	0% to +5%
FRP-reinforced RC (system)	N/A (concrete governs)	As per matrix	Matrix-dependent	Matrix-dependent	Matrix-dependent	Service stress ≤0.3–0.35 f_fu (GFRP); fire protection; crack-width control	+50–200% rebar cost (offset by lifecycle)

1.5 Circular Materials & Recycled Constituents

Moving toward a circular economy in construction involves substituting virgin raw materials with recycled or waste-derived constituents wherever possible. Both concrete and asphalt – which together consume vast quantities of aggregate and binder globally – have seen extensive research on incorporating recycled aggregates and diverse waste materials (plastic, glass, rubber, etc.) to reduce the use of new natural resources and divert waste from landfills. These strategies conserve aggregates and reduce waste disposal burdens, and in some cases can even improve certain material properties. However, they must be engineered carefully to avoid undesirable performance trade-offs. This section covers two main areas: (1) Recycled Aggregate Concrete (RAC), where crushed concrete or other recycled aggregate replaces some of the natural aggregate in new concrete, including impacts on performance and methods to improve RAC quality; and (2) waste-derived additives in concrete and asphalt (such as ground glass, scrap tire rubber, and recycled plastics) and how they influence material performance.

1.5.1 Recycled Aggregates: Performance and Pretreatment

Recycled Aggregate Concrete (RAC) refers to concrete in which a portion of the natural coarse and/or fine aggregates is replaced with recycled concrete aggregate

(RCA) obtained from crushed demolition concrete (and sometimes masonry). Using RCA directly addresses the massive construction and demolition (C&D) waste stream – on the order of 12 billion tons annually worldwide – and reduces the need to quarry new aggregate. Numerous studies and real projects show that RAC can perform well, although typically there is some reduction in mechanical properties if high replacement levels of RCA are used. A recent review quantified that replacing natural aggregate with recycled aggregate tends to reduce compressive strength by anywhere from a few percent up to ~40%, depending on the replacement rate and RCA quality (Lu, 2024). Tensile and flexural strengths likewise decline (often the tensile strength drops proportionally more than compressive strength) (KUMAR et al., 2025; P. Kumar & Walia, 2025), and the elastic modulus is particularly affected – roughly a 20–30% reduction in Young's modulus has been observed with 50% or more RCA, reflecting that RCA particles (which include residual mortar attached) are more porous and less stiff than natural stone (P. Kumar et al., 2025). As a rough guideline, using 50% coarse RCA might lead to about 15–30% lower 28-day compressive strength and noticeably higher creep and shrinkage, due to the higher water absorption and micro-cracking associated with RCA's attached old mortar.

Despite these performance gaps, many can be mitigated through mix design adjustments or by treating the RCA before use. The principal issue is the old cement paste adhering to RCA, which increases water absorption and acts as a weaker inclusion in the aggregate. Several pretreatment methods have been developed to improve RCA quality:

- **Mechanical removal of old paste:** Techniques like additional crushing, milling (e.g. ball milling), or even ultrasonic cleaning can knock off some of the adhered mortar, yielding a cleaner and more angular recycled aggregate. Autogenous grinding (tumbling RCA pieces against each other) has shown significant removal of loose mortar. Completely removing all residual mortar is often impractical or uneconomical, but partial removal can improve the aggregate's quality.
- **Strengthening the attached mortar:** Rather than removing it, another strategy is to harden or densify the old paste on the RCA. *Carbonation treatment* is one effective method: exposing RCA to CO_2 (often in a controlled chamber) converts some of the calcium hydroxide in the old cement paste into calcium carbonate, which hardens the residual mortar and reduces its porosity. Similarly, *polymer impregnation* can strengthen RCA surfaces (e.g. soaking RCA in a monomer like styrene and then polymerizing it in situ or using a resin to fill pores). Another approach is depositing fine minerals: for example, immersing RCA in a colloidal nano-silica solution or a cement slurry and then curing it so that micro-pores in the old mortar get filled and "healed."

These methods greatly enhance RAC strength and durability; properly treated concrete with 100% RCA can achieve compressive strength comparable to conventional concrete. Treated-RCA concretes also show lower chloride permeability and improved freeze–thaw resistance, mitigating concerns about RAC's permeability and frost vulnerability.

- **Combination approaches:** Researchers have tested pretreatment sequences—such as acid pre-soaks with lime or pozzolan slurries, or thermal cracking with mechanical abrasion—for better results, though each added step raises costs and complexity. A recent review found milling and autogenous grinding most effective for paste removal, while CO_2 carbonation, polymer impregnation, and nano-silica treatments significantly improved RAC properties.

The effects of such treatments can be notable. One study achieved about 95% crack sealing and ~90% strength recovery in a pre-damaged recycled aggregate concrete by applying a microbial calcium carbonate precipitation treatment to the cracks – indicating that even a weaker RAC can be substantially healed and strengthened with the right treatment (X. Hu et al., 2024). Another study found that treating RCA with silane coupling agents (to improve the bond at the interface) increased RAC compressive strength by 10–20% compared to untreated RCA (Huang et al., 2023), and further coating the RCA with a cementitious slurry led to an additional large strength increase (in some cases resulting in over double the strength of the untreated-RCA concrete). While such percentage gains are relative to a low baseline, they demonstrate that treated-RCA concrete can approach the strength of normal concrete. Correspondingly, durability is enhanced – researchers have measured reduced chloride permeability and improved freeze–thaw resistance in concretes made with treated RCA.

From a specification standpoint, many standards now allow a certain percentage of recycled aggregate in concrete, especially for non-structural or lower-stress applications. Often about 20–30% RCA content can be used without major modifications to the mix or design, and higher percentages are permitted for low-risk uses (such as road base, fill, or mass concrete in foundations). Some agencies specify using a slightly higher cement content or a higher strength class when high RCA contents are used, to offset potential strength loss. It is also recognized that replacing coarse aggregate is easier than replacing fine aggregate – using recycled sand can hurt workability and strength more than using recycled coarse aggregate, so RCA is more commonly applied to the coarse fraction. A simple practical measure always recommended is pre-saturating RCA before batching. Because RCA has higher water absorption, pre-wetting it to a saturated-surface-dry (SSD) condition prevents it from absorbing mix water and altering the effective water–cement ratio. For example, one study showed that using pre-saturated recycled coarse aggregate yielded concrete with 28-day compressive strength around 49 MPa and flexural strength 5.8 MPa, essentially equal to a control mix with natural aggregate (Sadrolodabaee et al., 2025), whereas using the same RCA dry led to variable results due to the extra water uptake. Modern mix designs that include high-range water reducers (superplasticizers) can also help accommodate RCA by maintaining workability at lower water contents, and incorporating SCMs can densify the cement paste matrix to mitigate RAC's higher porosity.

Even in structural members, moderate RCA replacement levels can be used with careful mix proportioning and perhaps slight design adjustments. The main drawbacks – reduced strength and stiffness – are well documented (e.g. a 100% RCA mix might have on the order of 15–30% lower compressive strength and substantially lower modulus, meaning designs must account for greater deflections and perhaps higher creep). Engineers should account for a lower elastic modulus in serviceability calculations and possibly higher creep and shrinkage when using high RCA content. From a sustainability perspective, replacing aggregate does reduce environmental impacts and saves natural resources. Interestingly, however, the direct carbon emissions savings from using recycled aggregate are relatively small compared to binder-related savings: replacing all the aggregate in concrete might cut only a few percent of the concrete's CO_2 footprint, since aggregate production is much less carbon-intensive than cement production. One analysis noted that 100% aggregate replacement reduced a concrete's embodied carbon by at most ~2%. Therefore, the motivation for RAC is often more about material conservation and waste reduction than about achieving large carbon reductions. Nevertheless, in a circular economy context, using 100% recycled aggregate where feasible – for example, constructing a new concrete structure using crushed concrete from a demolished predecessor – closes the material loop and contributes to sustainability by reusing resources that would otherwise become waste.

1.5.2 Waste-Derived Additives in Concrete and Asphalt (Glass, Rubber, Plastics)

Another avenue for incorporating waste materials is to use non-traditional additives or modifiers derived from waste streams. These can partially replace fine aggregate or act as supplementary fillers/fibers in concrete and similarly can modify asphalt binders or mixtures. Common examples include waste glass, scrap tire rubber, and recycled plastics. Each of these has distinct effects on material properties:

1.5.2.1 Waste Glass: Finely ground waste glass can serve as a pozzolanic SCM (sometimes referred to as glass powder). Glass is high in silica, and if ground sufficiently fine (e.g. particles <100 μm), it can react with calcium hydroxide in cement similar to fly ash, contributing to later-age strength and improving durability. Studies show that replacing about 10–20% of cement with ground glass powder can enhance long-term strength and reduce permeability (Nassar et al., 2021). For example, 20% glass powder in a mortar mix slightly increased 28-day strength compared to a control, indicating pozzolanic activity that continues beyond 28 days. However, using waste glass as a coarse aggregate (cullet) without fine grinding is problematic because the amorphous silica in glass can trigger the alkali–silica reaction (ASR) if the glass is reactive – a deleterious expansive reaction in concrete. To use glass safely, one must ensure either that the glass particles are sufficiently small to be consumed in benign pozzolanic reactions or take measures to mitigate ASR (such as using SCMs that tie up alkalis or adding lithium admixtures to suppress reactivity). Generally, only finely ground glass is recommended as a cementitious replacement; using larger glass pieces as aggregate is risky unless specialty low-alkali glass is used or the concrete alkalinity is carefully controlled. The type and composition of glass matter as well (ordinary soda-lime bottle glass is highly reactive in concrete; borosilicate glass is less reactive but is a less common waste). In practice, some precast products (paver blocks, tiles) and certain non-structural concretes incorporate roughly 10–15% ground glass powder as cement replacement, achieving waste utilization with slight durability gains due to pore refinement. Overall, glass powder can improve later-age strength and reduce chloride ingress provided the glass is ground fine enough. Using coarser glass sand, by contrast, tends to decrease strength and durability and can induce ASR if the glass is reactive, so particle fineness and proper mix design are key.

1.5.2.2 Crumb Rubber (Tire Rubber): Scrap tires can be processed into crumb rubber of various sizes (from fine powders up to small chips). In concrete, crumb rubber is typically used as a partial replacement for sand or fine aggregate (and occasionally as small coarse aggregate pieces). The primary effect of adding rubber to concrete is to increase ductility, toughness, and impact resistance, but at a significant cost to strength. Rubber particles are much softer than mineral aggregate and bond only weakly to the cement paste (even with surface treatments, the rubber–cement bond is limited). As a result, adding rubber causes a marked reduction in compressive and tensile strength, roughly proportional to the volume of aggregate replaced by rubber. Empirical data show that replacing just 5–10% of total aggregate volume with rubber can reduce compressive strength by on the order of 10–20% and splitting tensile strength by 30–50% (Karunarathna et al., 2021). Higher rubber contents lead to drastic drops: for instance, around 50% aggregate replacement with rubber often reduces compressive strength by more than half. Rubberized concrete is more deformable and fails less abruptly than traditional concrete. Rubber particles act as shock absorbers, enhancing impact resistance and ductility. This type of concrete offers better fatigue performance, vibration damping, reduced density, and improved thermal and acoustic insulation due to added air voids.

There are some durability benefits to adding rubber: for example, rubber particles can compress under freezing, so rubberized concrete often has improved freeze–thaw resistance – the rubber provides internal "air" voids that relieve pressure when water freezes. Rubber can also enhance energy absorption under blast or shock loading. However, durability concerns include the weak interfacial zone around rubber particles and the hydrophobic nature of untreated rubber, which can create voids or act as crack initiation points. Various surface treatments for rubber have been researched to improve its bond with cement paste: for instance, etching the rubber in NaOH solution or using silane coupling agents to make the rubber surface more hydrophilic and chemically bondable. Treating crumb rubber with NaOH or coating it with a silane has been shown to partly restore strength – e.g. silane-treated rubber mortar had ~15% higher compressive strength than an untreated one. Coating rubber particles with a thin layer of cement slurry before mixing (so they come pre-coated with a cementitious layer) has yielded even larger strength gains (in some studies, 50–150% higher strength compared to untreated rubber concrete, though the baseline was low). Essentially, these treatments help rubber particles carry load more effectively by improving their adhesion to the surrounding matrix.

Given the large strength penalties, rubberized concrete is generally used only in non-structural or secondary applications where high strength is not critical, but enhanced ductility, impact absorption, or reduced weight might be desired. Examples include lightweight concrete wall panels, architectural features, low-load pavement layers, shock-absorbing pads, or tunnel linings where vibration damping is useful. Even at low dosages, a bit of rubber could serve as a ductility-enhancing additive in structural concrete (some researchers suggest that a few percent rubber might improve post-crack behavior in beams or columns), but current design codes do not recognize any strength contribution from rubber, and significant rubber content would require re-evaluation of safety factors. One notable research avenue is ultra-high-performance rubberized concrete (UHP-RuC) – by combining rubber with steel fibers and optimized granular packing, some experimental mixes have achieved >100 MPa compressive strength while incorporating rubber, although these mixes are complex and not yet practical for routine use.

It is worth noting that the largest scale use of tire rubber in civil infrastructure is not in concrete but in asphalt pavements. Crumb Rubber-Modified Asphalt (CRMA) is widely employed in road paving: ground tire rubber is either blended into hot bitumen (the "wet" process, where the rubber swells and dissolves into the binder, modifying its properties) or added into the asphalt mix as part of the aggregate (the "dry" process). CRMA typically shows improved performance relative to conventional asphalt. The rubber increases the binder's viscosity and elasticity, which enhances resistance to rutting (permanent deformation in hot weather) and to fatigue cracking. It can also improve the binder's resistance to aging and oxidation and provides some noise reduction for traffic due to damping. Many highway agencies report longer-lasting pavements with crumb rubber. Studies show that rubber-modified asphalt can have higher stiffness and rut resistance at high temperatures and often better low-temperature flexibility (reducing thermal cracking), although careful formulation is needed – if not blended properly, rubber can sometimes introduce more air voids or slightly reduce rut resistance initially, but modern processes and additives usually mitigate these issues. On balance, CRMA is considered a superior alternative to unmodified asphalt in many situations, offering both performance gains and the sustainability benefit of consuming waste tires. The limitations are modest, mainly a higher material cost and some increased handling complexity (rubber can affect binder storage stability if not engineered well). Nonetheless, many U.S. states, European countries, and China routinely use crumb rubber in road surfacing, particularly for heavy-duty highways or noise-reducing surface layers. Life-cycle assessments of rubberized asphalt suggest a net sustainability win: the extended pavement life and reduced maintenance needs tend to outweigh the slightly higher initial emissions, and crucially it puts to productive use a waste material (scrap tires) that is otherwise an environmental nuisance.

1.5.2.3 Waste Plastics: There is growing interest in using waste plastics (like polyethylene, polypropylene, etc.) in both concrete and asphalt as well. In concrete, recycled plastics can be used as fibers or as lightweight aggregate replacements (shredded plastic particles). As micro-fibers recycled plastic fibers (for example, fibers made from recycled PET bottles) can help control plastic shrinkage cracking and provide some post-crack toughness, though they do not match the stiffness or strength contribution of steel fibers. As aggregate replacement, plastic – being much lighter and softer than mineral aggregate – tends to reduce concrete strength even more significantly than RCA or rubber does, so plastic aggregate is usually limited to low fractions of the mix. Research on using plastic aggregate in concrete shows mixed results: there can be slight improvements in ductility or impact resistance, but generally a notable strength drop and potential bonding issues because the plastic does not adhere well to cement. One clear upside is reduced density: concrete containing plastic aggregate can be considerably lighter, which might be useful for non-structural elements or partitions where weight saving is important. In terms of durability, most common plastics are fairly inert in concrete's alkaline environment at ambient temperatures, so chemical compatibility is usually fine – however, if a concrete with plastic aggregate is subjected to a fire, the plastic could melt or burn, potentially causing internal voids or toxic fumes, which is a serious consideration for structural use. Thus, plastic in concrete is currently more applicable to lightweight non-structural components or as secondary fiber reinforcement for shrinkage control.

In asphalt, waste plastic modifiers have gained attention as a way to stiffen and strengthen the binder similarly to commercial polymer modifiers (like SBS, styrene-butadiene-styrene). Adding certain melted waste plastics (such as shredded HDPE or PET) into the bitumen can increase its stiffness and high-temperature performance. Studies – including trials by companies in the UK and elsewhere – have found that incorporating on the order of 6–8% waste plastic by weight of the bitumen can significantly raise Marshall Stability and stiffness (often a 30–50% increase) (Hridoy et al., 2025), which translates to better rutting resistance on hot pavements. Essentially, the plastic can act like the polymers used in premium modified asphalts, improving the binder's high-temperature grade. However, concerns have been raised about low-temperature performance (some plastics could make the binder too stiff or brittle in cold temperatures, increasing the risk of thermal cracking) and about potential microplastic shedding – if the plastic does not fully melt and integrate into the binder, there's a risk that traffic wear could release tiny plastic particles from the pavement. Recent evidence suggests that when processed correctly (ensuring plastics fully melt and compatibilized with the bitumen), the risk of microplastic release is low and the performance gains are real. Several pilot road projects – in India, the UK, the Netherlands, etc. – have been built using waste plastic-modified asphalt and show promising durability. In India, thousands of kilometers of low-volume rural roads have been constructed using a simple "dry process" of sprinkling shredded waste polyethylene onto hot aggregates before bitumen mixing; these roads reportedly have improved resistance to rutting and moisture damage, and some Indian states have even mandated this method for certain categories of roads. Standardization for plastic-modified asphalts is still being developed, and performance can vary with plastic type (polypropylene vs polyethylene, for instance), but the trend is toward incorporating moderate amounts of waste plastic in asphalt where it can enhance performance while consuming waste.

Overall, waste-derived additives can contribute to sustainability by recycling problematic wastes into infrastructure materials. Their use must generally be justified by a specific functional benefit – for example, adding rubber to gain impact absorption and ductility, adding plastic to stiffen asphalt and extend pavement life, or adding glass powder for pozzolanic reaction – because these wastes often introduce property trade-offs (usually a reduction in strength when added to concrete).

1.6 Self-Healing and High-Durability Systems

Employing self-healing materials presents a significant advancement in prolonging infrastructure lifespan, as these systems can autonomously address cracks or damage, which in turn extends service life and reduces the frequency of maintenance. Within the field of concrete technology, extensive research has focused on self-healing solutions, utilizing processes such as mineral crystal formation or polymer release to effectively seal cracks upon their emergence. This section examines the main approaches to self-healing concrete: (1) autonomic healing via encapsulated chemical agents, (2) crystalline admixtures that precipitate minerals in cracks, and (3) microbial-induced healing (bio-concrete). We also discuss how these techniques improve durability by limiting permeability and mitigating rebar corrosion.

1.6.1 Capsule-Based, Crystalline, and Microbial Healing Mechanisms

1.6.1.1 Autonomic healing with capsules: One strategy is embedding small capsules or vascular networks containing a healing agent within the concrete matrix. When a crack propagates and intersects a capsule, the capsule ruptures, releasing the agent into the crack as illustrated in Figure 2. Common healing agents include epoxy or other polymer resins, cyanoacrylates (fast-setting adhesives), or sodium silicate solutions that can polymerize or crystallize to bind the crack flanks. Researchers have used microcapsules (on the order of 100 μm) as well as larger glass capillary tubes embedded in concrete. Upon cracking, the released resin or chemical ideally fills and seals the crack and can even bond the cracked faces back together. This approach has shown high efficiency in sealing cracks and restoring functionality, especially in laboratory tests on mortar specimens. For example, one study demonstrated that an encapsulated healing agent reduced water permeability by up to 95% after cracking (Soda et al., 2022), compared to an equivalent cracked sample without healing. Some recovery of mechanical strength is possible if the healing agent bonds well, though typically the focus is on restoring watertightness rather than full structural capacity. Key challenges for capsule-based healing are ensuring reliable capsule rupture at the right time and location (the capsules must break when and where cracks occur) and having sufficient agent volume to fill the cracks. New developments include nano-encapsulated healing agents and multiple capsule types that trigger sequentially, allowing repeated healing events in the same location if cracks re-form. A recent review highlighted the versatility of microcapsule-based self-repair and its sustainability benefits (Jiang et al., 2024) (by reducing the frequency of repairs). However, scaling up is challenging – a significant volume fraction of capsules may be needed to guarantee coverage of potential crack planes in a large structural concrete element, and the long-term stability of unbroken capsules in concrete's highly alkaline pore solution must be ensured so they don't degrade prematurely. Field trials are in progress; for instance, some bridge deck panels in Europe have been installed with encapsulated healing agents to test their performance in real-world conditions.

Figure 2. Microcapsule self-healing process in concrete (Song et al., 2013)

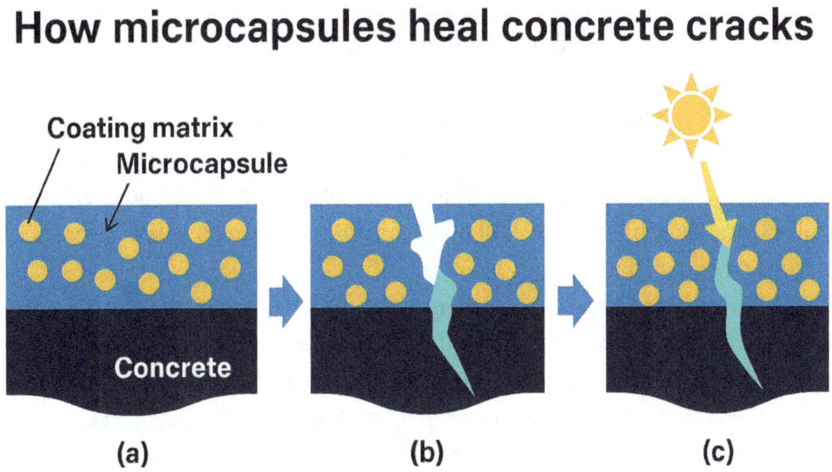

1.6.1.2 Crystalline admixtures: These are proprietary permeability-reducing admixtures added to concrete to enable self-sealing of cracks. They typically consist of reactive silica compounds and other active chemicals that remain dormant in the concrete until water penetrates through cracks, whereupon they react (often with calcium hydroxide or other minerals) to form insoluble crystals that block the pathways. In essence, they give concrete an inherent ability to self-seal microcracks (up to a certain width, often around 0.2–0.3 mm) whenever those cracks allow water ingress. One common formulation produces needle-like calcium silicate or calcium alumino-sulfate crystals that grow when water and un-hydrated cement particles at crack faces come into contact with the admixture chemicals. These crystalline products gradually plug pores and cracks, significantly reducing permeability. Studies show that crystalline admixtures can substantially cut water absorption and permeability (Tibbetts et al., 2021); combining a crystalline admixture with fly ash in concrete was observed to enhance late formation of additional C–S–H gel and produce new crystalline deposits in pores over time, thereby tightening the concrete. Such admixtures can even marginally increase long-term strength due to continued hydration and pore filling. In practice, many real-world waterproofing applications (basements, tunnels, water tanks) now use crystalline admixtures as an alternative to external membranes – the concrete itself becomes capable of self-sealing fine cracks. They are easy to use (typically added as a powder or liquid during batching like any admixture). While they do not "heal" cracks in the sense of restoring tensile capacity, they effectively turn a potentially leaking crack into a non-leaking crack when moisture is present. This is extremely valuable for durability, since it prevents the ingress of chlorides, sulfate, or other aggressive agents through microcracks, thereby protecting the reinforcement. In fact, by reducing crack permeability and connectivity, self-healing admixtures directly slow down reinforcement corrosion processes (since they limit water and oxygen access to the steel). Some building standards now recognize these crystalline technologies – for example, certain provisions allow them to replace traditional waterproof membranes in specific situations, given proof of performance.

1.6.1.3 Microbial (biogenic) healing: Perhaps the most fascinating approach uses bacteria to induce mineral precipitation that seals cracks – often termed *bio-concrete*. This method mixes calcite-producing bacteria (typically spore-forming Bacillus species that can survive in concrete's alkaline environment) and a suitable organic nutrient (such as calcium lactate) into the concrete, often by encapsulating the bacteria and nutrients in lightweight aggregates or gel beads to ensure their viability during mixing and curing. When cracks eventually form and water and air ingress occurs through the cracks, the dormant bacterial spores become active, consume the nutrient, and precipitate calcium carbonate (limestone) which gradually fills and plugs the crack. Laboratory studies have achieved up to ~95% crack sealing and over 90% compressive strength recovery in cracked mortar specimens using bacterial healing (Nair et al., 2023). The long-term viability of spores is essential; bacteria must stay dormant yet alive for years until concrete cracks. Research has found ways to protect spores, such as encasing them in silica gel or foam. While challenges like spore longevity and high costs remain, recent breakthroughs show Bacillus spores can survive over four years in dry concrete and activate when cracks appear. Microbes then precipitate $CaCO_3$, bonding the crack sides together. In marine or continuously wet environments, bacterial concrete has shown exceptional crack sealing and improved durability – one report cited about 95% crack closure in a marine exposure, with corresponding enhancements in resistance to chloride and sulfate ingress. There have been pilot implementations of microbial healing, such as sections of sewer pipes and irrigation canals treated with bacterial admixtures. Currently, the cost of these specialty admixtures is higher than conventional ones, but the hope is that if they can replace or greatly reduce maintenance (for instance, eliminating the need for epoxy injection of cracks or other repair interventions), they may become economical on a life-cycle basis. Importantly, microbial healing is very environmentally friendly – it leverages a natural biochemical process (biomineralization) to improve concrete durability, avoiding the use of harmful chemicals.

1.6.1.4 Comparing the methods: Encapsulated polymer-based healing can potentially seal larger cracks (up to a few millimeters wide, if enough healing agent is present) and does so almost immediately upon crack formation (as soon as capsules rupture and release resin). In contrast, crystalline and microbial methods are mostly effective on micro-cracks (typically up to ~0.2–0.5 mm width) and may take days or weeks of favorable conditions (continuous moisture for crystals to grow or bacteria to precipitate minerals) to fully seal the cracks. Capsules generally have a one-time effect – once a capsule's content is consumed, that specific location cannot heal again unless additional capsules are nearby – whereas crystalline admixtures and bacterial spores remain present in the concrete and can potentially respond to multiple cracking events over the structure's life (as long as nutrients or reactive compounds remain available for successive cycles of healing). Hybrid approaches are being explored to capture the benefits of both: for example, using a combination of microcapsules containing a quick-acting polymer resin and an embedded nutrient/bacteria system, so that an immediate crack seal is achieved by the resin and then longer-term biological healing can occur if needed.

1.6.2 Crack Control, Permeability Reduction, and Corrosion Mitigation

The overarching benefit of self-healing and high-durability concrete systems is tighter crack control and reduced permeability, which directly translates to better protection of reinforcement and extended service life. Cracks are the primary gateways for aggressive agents (like chloride ions from deicing salts or seawater, sulfate from soil or groundwater, or CO_2 from the atmosphere) to penetrate concrete and reach steel reinforcement. If cracks can be sealed promptly (or prevented from widening by design measures such as fiber reinforcement, as discussed in the next section), the ingress of corrosive species is dramatically curtailed. For instance, even a hairline 0.1 mm crack can allow chlorides to reach rebar much faster than in uncracked concrete – but if a crystalline admixture or microbial agent seals that crack when moisture appears, the diffusivity through that path drops by orders of magnitude. Studies have shown that self-healed concretes have significantly lower water absorption and chloride permeability after cracking compared to ordinary cracked concrete – in some experiments, cracked specimens that underwent self-healing were restored nearly to the watertightness of uncracked concrete. This means that corrosion initiation in the rebar is delayed or prevented, since water and chlorides cannot easily reach the steel. Additionally, some healing processes raise the local pH in the crack (certain bacteria produce alkaline byproducts, and some chemical healing agents are themselves alkaline), which can help re-passivate any steel surface that was exposed when the crack formed.

Another aspect is maintaining structural integrity and stiffness. While self-healing generally does not fully restore the original tensile strength across a large crack (especially not for wide cracks, or if the healed material is weaker than the original concrete), it can restore some stiffness and limit further crack propagation. A healed crack does not reopen as easily under cyclic or variable loads because the healing precipitate or polymer bears part of the load across the crack. In fatigue scenarios, this can improve performance: one study found that concrete beams with self-healing agents had longer fatigue life than control beams, attributing this to cracks not growing freely after each load cycle thanks to intermittent healing sealing them (De Belie et al., 2018). Moreover, experiments indicate that cracks up to about 0.3 mm can often be effectively closed by these methods. If the structure is designed such that typical service-load cracks remain below that width (for example, by using fibers or sufficient reinforcement to control crack size), then the self-healing mechanisms can manage and seal those cracks autogenously.

Importantly, corrosion mitigation has been demonstrated as a result of crack sealing: healed cracks exhibit dramatically lower corrosion rates on reinforcing bars than equivalent cracks left unhealed, because the formation of a tight seal removes the continuous moisture and ion path needed for a sustained corrosion cell, and limits oxygen and water access to the steel. Some crystalline admixtures also incorporate corrosion-inhibiting compounds which further protect the steel once the crack is sealed. By keeping cracks tiny or sealed, the service life of reinforced concrete can be significantly extended – potentially doubling the time to first major repair in aggressive environments by preventing early onset of rebar corrosion.

To leverage these benefits, new design paradigms are being considered. One is a “damage management” approach – accepting that micro-cracks will occur under service loads or due to shrinkage, but ensuring the material can automatically manage and mitigate that damage before it becomes critical. This could lead to drastically reduced maintenance demands. For example, a bridge deck made with self-healing concrete might largely eliminate the need for periodic crack sealing or epoxy injection over decades of service, because it autonomously maintains its watertightness. Infrastructure owners see this as a path to enhanced resilience with minimal intervention. However, practical deployment requires quantifying healing performance for inclusion in specifications. Standards organizations are developing test methods (e.g. repeated water permeability tests or crack re-opening cycles) to measure self-healing efficiency in a reproducible way. Project specifications might soon include performance criteria such as: “Concrete shall demonstrate at least an 80% reduction in water permeability within 28 days of crack formation (for a 0.3 mm crack), as verified by standardized test X.” As research continues and standard metrics are established, such requirements will enable engineers to specify – and contractors to reliably deliver – self-healing concrete functionality.

1.7 CFRP/GFRP Reinforcement and Strengthening (Corrosion-Free Composites)

Fiber-reinforced polymer (FRP) composites are materials consisting of high-strength fibers (commonly carbon, glass, aramid, or basalt) embedded in a polymer matrix (typically epoxy, vinyl ester, or polyester resin). FRPs have become invaluable in civil engineering in two main roles: **(a)** external strengthening of existing structural members, and **(b)**internal reinforcement (rebars or tendons) in new construction. Their attractive properties include very high tensile strength (especially carbon fibers, which can have tensile strengths on the order of 2–5 times that of steel), low density (about one-fourth the density of steel or less), and crucially, immunity to corrosion. Their downsides are a linear elastic (brittle) stress–strain behavior, relatively lower elastic modulus in some cases (GFRP's Young's modulus is ~45 GPa, roughly one-fifth of steel's 200 GPa; CFRP's modulus is higher, ~140–160 GPa, approaching steel's), and susceptibility of the polymer matrix to high temperature (most resins soften or degrade around 100–250 °C, so fire protection is needed for FRP components in fire-exposed structures).

1.7.1 External FRP strengthening: This refers to bonding FRP laminates or wrapping FRP fabric around structural elements to increase their load capacity or ductility. It has been widely applied since the 1990s for retrofitting and repair. For example, bonding carbon fiber strips with epoxy to the tension face of a concrete beam can restore or enhance its flexural capacity – useful if a building is being re-purposed for higher loads or a bridge beam has deteriorated and needs more strength. Guidelines like ACI 440.2R cover the design of such FRP strengthening systems. Similarly, wrapping a circular or rectangular concrete column with carbon or glass fiber fabric provides confinement pressure when the column tries to dilate under load, greatly improving the column's ductility and shear strength – this is vital for seismic retrofitting of older columns that lack sufficient transverse reinforcement. FRP jacketing of columns has been used extensively to upgrade the seismic performance of bridge piers and building columns. External FRP strengthening is now a mature practice: thousands of bridges and buildings worldwide have been upgraded using bonded CFRP plates or GFRP wraps. The benefits are significant: the FRP reinforcement is outside the concrete and does not corrode, so even in very harsh environments the strengthening system itself remains durable (the epoxy adhesives can be vulnerable to UV or moisture, but protective coatings can mitigate that). FRP plate bonding can often double the flexural strength of a beam or increase the shear capacity dramatically without any significant increase in member size or weight (Aprile et al., 2001). Compared to traditional steel plate bonding or enlargement techniques, FRP is much lighter and easier to install – often the work can be done with minimal equipment and without heavy cranes. One must ensure, however, that the bond between FRP and concrete is protected (if water gets between the FRP and concrete, freeze–thaw or other effects could deboned it over time), and designs must account for the linear elastic (non-ductile) failure mode of FRP. Typically, designs impose conservative limits so that failure is governed by something like yielding of existing steel reinforcement or crushing of concrete rather than rupture of the FRP, thereby maintaining some ductility. For columns, FRP jackets prevent concrete from spalling and delay buckling of longitudinal steel bars, increasing strength and allowing more deformation. For example, wrapping deficient bridge columns with CFRP has been shown to increase their ultimate drift (deformation) capacity by a factor of 2–3, effectively upgrading their seismic performance to modern standards without adding bulky external jackets.

1.7.2 FRP rebar and tendons: Using composite bars in new structures is particularly attractive for structures in corrosive environments – such as marine structures, bridge decks exposed to deicing salts, and chemical plants – where steel rebar would corrode quickly. GFRP rebar (glass fiber-reinforced polymer rods), usually with a surface sand coating or helical ribbing to ensure bond with concrete, has been commercialized since the 1990s. It has been used in many bridge decks (especially in Canada and parts of the U.S.) specifically to eliminate rebar corrosion problems. Because steel rebar corrosion can lead to serious deterioration (spalling, loss of bond, reduced cross-section) within just 20–30 years in a salt-exposed bridge deck, replacing steel with GFRP – which does not rust at all – can dramatically extend the service life. GFRP manufacturers often cite that their rebar can provide 100+ year service life in aggressive environments with minimal maintenance. Indeed, some of the first highway bridges built with all-GFRP rebar (for example, a bridge in Ohio in 1996, and the Morandi Bridge deck in Italy in 2001) have shown minimal signs of degradation in the composite bars after two decades, whereas an equivalent steel-reinforced deck might have needed major repairs by that age. Many recent bridges specify GFRP at least for the top mat of deck reinforcement, since that is the zone most vulnerable to salt ingress – using GFRP there prevents the primary cause of deck deterioration (top steel corrosion) and thus can greatly extend life.

1.7.3 CFRP tendons (prestressing strands) have also been used in select projects. CFRP tendons offer ultra-high strength (carbon fibers can have tensile strengths above 3000 MPa, much higher than steel prestressing strand ~1860 MPa) (Lucena et al., 2024) and essentially no creep or relaxation under sustained load. They also don't corrode, so prestressed concrete elements can be made more durable. However, CFRP tendons are very expensive compared to steel and require special anchorage systems (since they are linear elastic and can't deform to lock into wedges like steel strands). They have been used in some innovative projects, such as segmental concrete bridges in Germany and Japan, to ensure a very long-lasting prestressing system that won't suffer tendon corrosion. Designing with FRP bars or tendons requires acknowledging their linear elastic behavior: they have no yield plateau, so the design is governed by an ultimate strength limit state where either the FRP will rupture, or the concrete will crush. Usually it's desirable that concrete crushing (a compressive failure) occurs before FRP rupture to give warning, which is achieved by limiting the FRP stress/strain in design (e.g., designing such that FRP reaches only a certain strain well below its rupture strain at ultimate loads). FRP bars also have lower transverse strength and no ductility in bending, meaning bent FRP bars (hooks or stirrup bends) are weak points – as a result, standard practice is to avoid bending FRP bars and use mechanical anchorages or straight development lengths in lieu of hooks. Projects often design GFRP bars with headed ends or special anchor devices rather than rely on bent hooks, since glass fibers can break at a bent corner.

One distinct advantage of FRP reinforcement is that, because there is no concern about corrosion, the required concrete cover can be based solely on considerations like fire protection or bond development, not on providing a certain thickness to slow down corrosion penetration. This could allow thinner structural elements or more efficient cross-sections (though in practice building codes still specify minimum covers for reasons like fire resistance and general durability beyond corrosion). Additionally, FRP rebar is non-magnetic, which is useful for special structures where magnetic fields must be minimized (such as MRI rooms in hospitals or certain laboratories where steel rebar's magnetic field would be problematic).

Carbon FRP rebar (CFRP bars) has even higher tensile strength and higher elastic modulus than GFRP, but they are extremely costly, so GFRP (cheaper but lower stiffness) is far more commonly used for general infrastructure. Basalt FRP (BFRP) rebar is an emerging option that sits between glass and carbon in terms of cost and mechanical properties – BFRP bars have modulus and strength somewhat higher than GFRP and appear to have good durability in alkaline environments – they may become a competitive alternative in the future.

From a resilience perspective, corrosion-free reinforcement is a game-changer for structures in aggressive climates. A coastal bridge deck reinforced entirely with GFRP could be expected to last 75–100 years with minimal maintenance, whereas a conventional steel-reinforced deck might require major rehabilitation or even replacement after 25–40 years due to rebar corrosion damage. For example, the first vehicular bridge using all-GFRP rebar (built in 1996 in Missouri, USA) has shown essentially no deterioration after ~25 years in service, whereas a nearby similar bridge with steel rebar exhibited significant corrosion by that age. Many new bridges now at least use GFRP for the upper mat of deck steel or in barrier rails, to protect those most exposed zones.

In seismic design, one consideration is that steel reinforcement's ductility (its ability to yield and deform plastically) is a useful feature for energy dissipation. FRP rebar does not yield – it remains elastic until rupture – so an FRP-reinforced concrete frame would behave differently in a severe earthquake, with less hysteretic damping and a more elastic response, relying on concrete cracking and closing (or eventually FRP rupture) for energy dissipation. Because of this, current practice with FRP in seismic structures is often to use it selectively: for example, use traditional steel rebar in critical plastic hinge regions of a frame to provide ductility and energy dissipation, but use FRP in portions of the structure where ductility is less critical but durability is (such as in the lengths of members outside the hinge zones, or in secondary members). Research is exploring hybrid rebar that combines steel and FRP – for instance, a steel core surrounded by an FRP outer layer – to achieve both ductility and corrosion resistance, but these products are not yet widely available.

FRP's long-term durability is excellent in terms of corrosion (it simply does not corrode at all), but other issues must be managed: GFRP can exhibit creep rupture under sustained loads (the glass fiber and polymer matrix can slowly creep, and long-term strength under load is lower, so design codes include a creep reduction factor or stress limits for GFRP), and all FRPs lose strength at elevated temperatures or under direct UV exposure if not protected (resins can degrade). However, when properly embedded in concrete (providing a stable, alkaline, UV-shielded environment) and not overstressed, FRP rebar is expected to easily meet 100-year service life targets. Agencies such as the U.S. Federal Highway Administration (FHWA) and the Canadian Standards Association (CSA) have developed comprehensive design guides for FRP rebar (e.g. ACI 440.1R in the US and CSA S6 in Canada), so lack of code acceptance is no longer the primary barrier – the main barrier is cost. FRP bars typically cost several times more than steel for the same diameter, but when life-cycle costs are considered (accounting for greatly reduced maintenance and longer life) they can be economically justified in harsh environments. Additionally, the lower weight of FRP (roughly one-quarter the weight of steel) can reduce shipping costs and make construction handling easier and faster.

FRP composites are also being used to create entire structural profiles for specialized uses – for instance, pultruded FRP beams and deck panels for pedestrian bridges or even vehicular bridge decks. These take advantage of FRP's high strength-to-weight ratio to produce very lightweight structures that are quick to install, and of course their complete immunity to corrosion. These applications are currently niche due to high material cost and different design methodologies (designing an all-FRP structure requires careful analysis of buckling, creep, connection details, etc.), but they demonstrate the potential of FRP in load-bearing roles beyond just augmenting concrete or steel structures.

1.8 Bio-Based & Natural Materials

While many innovations in construction materials involve high-tech industrial processes, there is also a renaissance of bio-based and natural materials being used as sustainable alternatives. Timber, bamboo, straw, hemp, and other plant-derived materials are renewable, have low embodied carbon, and even store carbon absorbed during growth. Modern engineering has transformed these traditionally "low-tech" materials into high-performance products like mass timber (e.g. CLT – cross-laminated timber panels) and engineered bamboo lumber

1.8.1 Mass Timber & CLT: Mass timber refers to large, engineered wood components (glue-laminated beams, CLT panels, laminated veneer lumber or LVL, etc.) that enable wood to be used for mid- and even high-rise construction. Cross-Laminated Timber (CLT) consists of layers of dimensional lumber boards glued in alternating grain orientations to form stiff, plate-like panels (commonly 3, 5, or 7 layers thick). CLT panels can serve as floors, walls, and shear diaphragms in buildings, functioning analogously to concrete slabs or shear walls but at a fraction of the weight. Mass timber is appealing because it is both strong and lightweight. Wood has a high strength-to-weight ratio, certain engineered bamboo products have specific tensile strengths exceeding that of steel, and wood in general is roughly 5 times lighter than concrete for a given load-bearing capacity. It is also renewable (forests can regrow), and wood products store carbon that the trees absorbed from the atmosphere (effectively sequestering it until the wood decays or burns at end of life. Similarly, other analyses report on the order of 50–75% lower embodied emissions for wood buildings versus comparable concrete or steel buildings (Tozan et al., 2025).

Mass timber has been proven viable for tall buildings: numerous examples of 8- to 18-story wood buildings now exist worldwide, thanks to innovations in structural systems and connection design to address wood's unique properties (anisotropy, creep, etc.). Glue-laminated (glulam) timber beams and columns can replace steel or concrete members in many cases, and CLT panels can replace concrete floor slabs for moderate spans. Beyond sustainability, mass timber often enables faster construction due to a high degree of prefabrication – many components are manufactured off-site with precision CNC machining and simply assembled on site. The 18-story Brock Commons tower's timber structure was erected in only 70 days (J. Li & Zhao, 2025), since the CLT panels and glulam columns were fabricated beforehand and then rapidly lifted into place, resulting in a quieter and less disruptive construction process. This speed and precision are another advantage of mass timber construction.

1.8.2 Bamboo Composites: Bamboo, often called "nature's carbon fiber," has very high tensile strength and a lightweight hollow tubular form. It has been used traditionally in construction (such as bamboo scaffolding and vernacular housing in

Asia), but modern engineering techniques are elevating its performance by converting bamboo into engineered composites. Two forms are common: glue-laminated bamboo (layers of bamboo strips glued together, similar to glulam timber) and bamboo scrimber (bamboo culms are crushed or shredded into fiber bundles, then combined with resin and pressed into dense blocks or planks). The resulting engineered bamboo lumber has high and relatively isotropic strength – for instance, tests show tensile strengths on the order of 100–300 MPa and modulus around 15–20 GPa for some bamboo composite products (J. Hu et al., 2025). In terms of specific strength (strength per unit weight), certain bamboo composites can exceed that of steel. Engineers have begun using laminated bamboo beams and CLT-like bamboo panels in small buildings and pedestrian bridges.

Another approach using bamboo is as a substitute for steel rebar in concrete, utilizing whole bamboo poles or split sections as reinforcement. While conceptually attractive given bamboo's high tensile strength and renewable nature, raw bamboo in concrete faces challenges: it can rot when embedded in moist alkaline concrete unless properly treated, it has a different thermal expansion coefficient than concrete which can cause interface stresses, and achieving a reliable bond is difficult (bamboo's smooth surface and water-repellent silica skin hinder bond). Research continues on treating bamboo for use as rebar (e.g. coatings to make it waterproof and roughening or sand-coating to improve bond), but so far, the more promising route is engineered bamboo products where variability is reduced, and durability is enhanced through resins or treatments. Bamboo grows extremely fast (it can be harvested on a 3–5-year cycle, versus 20–50 years for most structural timbers), making it a highly renewable resource. It also sequesters carbon rapidly during growth. A 2021 study indicated that structures made with modern engineered bamboo can have environmental footprints comparable to or better than timber structures, with the added benefit that bamboo is abundant in tropical and subtropical regions where conventional structural wood can be scarce. Mechanically, engineered bamboo products have properties akin to hardwood timber or even mild steel in tension (high tensile strength, moderate stiffness).

1.9 Limitations

The review has several important limitations. It is largely material-centric, with limited treatment of whole-structure and hybrid-system behaviour (for example, timber–concrete–FRP composites). Many performance conclusions for LC^3, AAMs, CO_2-cured concretes, RAC and self-healing systems rest on short- to medium-term laboratory or pilot studies; robust long-term field data under combined mechanical, chemical and climatic stressors are scarce. Reported GWP, LCA and LCC values are sensitive to regional datasets, functional units and boundary choices, so methodological inconsistencies across EPDs and case studies persist. Evidence for circular additives (waste glass, rubber, plastics) remains fragmented and often small-scale, limiting clear guidance for structural use and code acceptance. Finally, AI/ML applications for mix design and durability prediction are promising but still nascent, with few validated models or open benchmark datasets.

1.10 Future scope

Future work should prioritize long-term, in-situ monitoring of low-clinker binders, CO_2-cured products, RAC and self-healing concretes under multi-hazard exposures to establish realistic service-life performance. Parallel to monitoring, standardized, diffusion-focused durability tests and reproducible self-healing metrics should be developed for adoption in codes and performance-based specifications. Research must also link transport properties and cracking behaviour to probabilistic service-life models and to harmonized LCA/LCC frameworks so that durability gains can be directly translated into lifecycle cost and carbon savings.

On the methodology and tools side, building open AI/ML datasets and validated predictive frameworks will accelerate data-driven mix optimization (minimizing GWP while meeting targets for diffusion, strength and shrinkage) and enable digital-twin approaches for maintenance forecasting. Finally, targeted research on hybrid net-zero systems (mass timber–concrete composites, FRP-reinforced low-clinker concretes) should integrate structural, fire and durability design, and should be coupled with pilot procurement experiments and policy pilots (GWP caps, performance-based procurement, digital EPD pilots) to bridge the gap between laboratory innovation and mainstream implementation.

1.11 Conclusion

1. Significant embodied-carbon cuts are achievable via clinker reduction (SCMs, LC^3) and CO_2-curing, without sacrificing strength or serviceability when mixes are performance-designed.
2. Alkali-activated binders offer large carbon savings and chemical durability but require application-specific curing, safety, and standardization pathways.
3. Circular constituents are viable when RCA is pretreated (e.g., carbonation, slurry/nano-silica) and when fine fractions are limited or engineered; pre-saturation and SCMs help recover stiffness and durability.
4. Waste glass is best used as finely ground pozzolan; coarser cullet demands ASR-mitigation. Rubber imparts toughness and impact resistance but reduces strength, suiting non-structural and specialized uses.

5. Self-healing (capsule, crystalline, microbial) reliably seals micro-cracks, lowers permeability, and delays corrosion initiation, extending time to first major repair.
6. FRP reinforcement eliminates rebar corrosion in aggressive exposures and is mature for decks, barriers, and marine elements; design must address linear-elastic behaviour, creep rupture (GFRP), and fire protection.
7. Performance-based specifications, coupled with diffusion-centric durability metrics (not solely electrical indices), provide fair qualification across unconventional binders.
8. LCA/LCC and EPD-based procurement are effective when functional units, boundaries, and modules are harmonized; otherwise, comparisons can be misleading.
9. Early-age behaviour, admixture compatibility, and tailored curing are critical control points for alternative binders and CO_2-cured systems.
10. Policy levers (GWP caps, EPD submittals, resilience-weighted bids) and evolving codes are creating clear market signals that enable near-term scaling.
11. Best results arise from exposure-class matching, conservative crack-width control, robust QC/QA, and pilot demonstrations that de-risk field deployment.
12. A staged adoption roadmap—materials screening by LCA, pilot validation, performance-based acceptance, and feedback into specifications—can mainstream low-carbon, resilient materials at scale.

REFERENCES

Abbass, M., Abbas, U., Jafri, R., Arif, S. M., & Akhai, S. (2024). *AI and Machine Learning Applications in Sustainable Smart Cities.*, DOI: 10.4018/979-8-3693-6740-7.ch001

Abbass, M., & Akhai, S. (2025a). Eco-friendly geopolymer bricks synthesized from alkali-activated sugarcane bagasse ash and rice husk ash. *Industrial Crops and Products*, *233*, 121410. DOI: 10.1016/j.indcrop.2025.121410

Abbass, M., & Akhai, S. (2025b). Exploring the sustainability potential of geopolymer concrete with coal bottom ash and basalt fibres. *Multiscale and Multidisciplinary Modeling. Experiments and Design*, *8*(3), 199. DOI: 10.1007/s41939-025-00794-3

Adnan, & Anas, M. (2025). Geopolymer concrete as a sustainable alternative to OPC. *Journal of Umm Al-Qura University for Engineering and Architecture*. DOI: 10.1007/s43995-025-00155-8

Akhai, S. (2024). Towards Trustworthy and Reliable AI. In *Explainable Artificial Intelligence (XAI) in Healthcare* (pp. 89–99). CRC Press. DOI: 10.1201/9781003426073-7

Antunes, M., Santos, R. L., Pereira, J., Rocha, P., Horta, R. B., & Colaço, R. (2021). Alternative Clinker Technologies for Reducing Carbon Emissions in Cement Industry: A Critical Review. *Materials (Basel)*, *15*(1), 209. DOI: 10.3390/ma15010209 PMID: 35009355

Aprile, A., Spacone, E., & Limkatanyu, S. (2001). Role of Bond in RC Beams Strengthened with Steel and FRP Plates. *Journal of Structural Engineering*, *127*(12), 1445–1452. DOI: 10.1061/(ASCE)0733-9445(2001)127:12(1445)

Aragón, A., & Alberti, M. G. (2024). Limitations of machine-interpretability of digital EPDs used for a BIM-based sustainability assessment of construction assets. *Journal of Building Engineering*, *96*, 110418. DOI: 10.1016/j.jobe.2024.110418

Arthur, C., Onuche, P. U. O., Yusuf, M., Animashaun, T., Boansi, S. O., Adewale, K., & Bart-Addison, C. E. (2025). Decarbonizing Concrete: Ultra-Low-Carbon Pathways and Policy Integration. *Scientific. Journal of Engineering Technology*, *2*(1), 134–142. DOI: 10.69739/sjet.v2i1.669

Asdrubali, F., Grazieschi, G., & Gandola, D. M. (2025). The Role of Environmental Product Declarations in the Decarbonization of Building Materials and Components. *Energies*, *18*(5), 1308. DOI: 10.3390/en18051308

Bondarenko, S., Lebedev, M., Strokova, V., Markova, I., Kozhukhova, N., Lukyanenko, N., & Potapov, D. (2025). Interaction Mechanisms in «Portland Cement—Functional Polymer Mineral Additives» Binder Produced by Different Methods. *Materials (Basel), 18*(13), 3178. DOI: 10.3390/ma18133178 PMID: 40649666

Daehn, K., Basuhi, R., Gregory, J., Berlinger, M., Somjit, V., & Olivetti, E. A. (2021). Innovations to decarbonize materials industries. *Nature Reviews. Materials, 7*(4), 275–294. DOI: 10.1038/s41578-021-00376-y

De Belie, N., Gruyaert, E., Al-Tabbaa, A., Antonaci, P., Baera, C., Bajare, D., Darquennes, A., Davies, R., Ferrara, L., Jefferson, T., Litina, C., Miljevic, B., Otlewska, A., Ranogajec, J., Roig-Flores, M., Paine, K., Lukowski, P., Serna, P., Tulliani, J., & Jonkers, H. M. (2018). A Review of Self-Healing Concrete for Damage Management of Structures. *Advanced Materials Interfaces, 5*(17), 1800074. Advance online publication. DOI: 10.1002/admi.201800074

Dhiman, S., Seema, S., & Akhai, S. (2025). Performance evaluation of OPC 43 concrete incorporating micro silica and recycled ceramic tile waste as partial cement and coarse aggregate replacements. *Asian Journal of Civil Engineering*. DOI: 10.1007/s42107-025-01504-7

Gelowitz, M. D. C., & McArthur, J. J. (2017). Comparison of type III environmental product declarations for construction products: Material sourcing and harmonization evaluation. *Journal of Cleaner Production, 157*, 125–133. DOI: 10.1016/j.jclepro.2017.04.133

Hridoy, S. M., Injamamul Haque Protyai, M., Rashid, A., & Abu Sayeed, M. (2025). Marshall property analysis of an improved bituminous mix obtained by mixing plastic waste with bitumen. *South African Journal of Chemical Engineering, 53*, 40–48. DOI: 10.1016/j.sajce.2025.04.003

Hu, H. T., Fu, C. H., Zhan, Q. W., Zhang, X., Ge, Y. M., & Zhou, J. L. (2024). Study on improving the crack self-healing width and depth based on microbial mineralization modified by Ca–Al bimetallic hydroxides. *Journal of Building Engineering, 91*, 109598. DOI: 10.1016/j.jobe.2024.109598

Hu, J., Zhang, S., Lin, J., Yu, Y., Yu, W., & Huang, Y. (2025). Towards Sustainable High-Performance Materials: Tailoring the Mechanical Properties of Bamboo-Based Fiber Composites through Controlled Processing. *Chemical Engineering Journal, 512*, 162331. DOI: 10.1016/j.cej.2025.162331

Hu, X., Yang, Z., Wu, M., Wang, F., Yuan, H., & Jiang, Y. (2024). Evaluation of performance and healing effect of two-component biological self-healing mortar. *Journal of Sustainable Cement-Based Materials, 13*(3), 462–475. DOI: 10.1080/21650373.2023.2282678

Huang, W., Quan, W., Miao, X., An, Y., & Sun, W. (2023). Study on the compressive mechanical properties of mixed recycled aggregate concrete with silane modified. *Journal of Building Engineering, 75*, 106950. DOI: 10.1016/j.jobe.2023.106950

Jiang, L., Wu, M., Du, F., Chen, D., Xiao, L., Chen, W., Du, W., & Ding, Q. (2024). State-of-the-Art Review of Microcapsule Self-Repairing Concrete: Principles, Applications, Test Methods, Prospects. *Polymers, 16*(22), 3165. DOI: 10.3390/polym16223165 PMID: 39599256

Karunarathna, S., Linforth, S., Kashani, A., Liu, X., & Ngo, T. (2021). Effect of recycled rubber aggregate size on fracture and other mechanical properties of structural concrete. *Journal of Cleaner Production, 314*, 128230. DOI: 10.1016/j.jclepro.2021.128230

Konradsen, F., Hansen, K. S. H., Ghose, A., & Pizzol, M. (2024). Same product, different score: How methodological differences affect EPD results. *The International Journal of Life Cycle Assessment, 29*(2), 291–307. DOI: 10.1007/s11367-023-02246-x

Kumar, P., Srijan, S., Aggarwal, V., & Walia, B. S. (2025). Influence of Mineral Composition and Microstructural Features on the Tensile Failure Mechanisms of Granite and Marble. *Journal of Mines. Metals and Fuels, 73*(8), 2515–2537. DOI: 10.18311/jmmf/2025/48211

Kumar, P., Srijan, S., Aggarwal, V., & Walia, B. S. (2025). The unique modified seesaw test (MST) instrument for the direct assessment of rock's tensile strength utilizing ring-shaped specimens. *Sadhana, 50*(3), 123. DOI: 10.1007/s12046-025-02803-0

Kumar, P., & Walia, B. S. (2025). An analytical–numerical framework for progressive contact analysis in radially compressed direct disc tests. *Journal of Earth System Science, 134*(4), 202. DOI: 10.1007/s12040-025-02650-x

Kumar, S. (2022). *Evolving Role of Sustainable Procurement in Coming Decades.*, DOI: 10.1007/978-3-031-08258-0_7

Li, J., & Zhao, K. (2025). Sustainable Development Advantages of Cross-Laminated Timber (CLT) and Cross-Laminated Bamboo and Timber (CLBT). *Materials (Basel), 18*(21), 4913. DOI: 10.3390/ma18214913 PMID: 41227871

Li, Z., Delsaute, B., Lu, T., Kostiuchenko, A., Staquet, S., & Ye, G. (2021). A comparative study on the mechanical properties, autogenous shrinkage and cracking proneness of alkali-activated concrete and ordinary Portland cement concrete. *Construction & Building Materials, 292*, 123418. DOI: 10.1016/j.conbuildmat.2021.123418

Liu, Z., Takasu, K., Koyamada, H., & Suyama, H. (2022). A study on engineering properties and environmental impact of sustainable concrete with fly ash or GGBS. *Construction & Building Materials, 316*, 125776. DOI: 10.1016/j.conbuildmat.2021.125776

Lu, L. (2024). Optimal Replacement Ratio of Recycled Concrete Aggregate Balancing Mechanical Performance with Sustainability: A Review. *Buildings (Basel, Switzerland), 14*(7), 2204. DOI: 10.3390/buildings14072204

Lucena, J. R. P. G., Carabbacan, A. P. I., Manggapis, F. F., Kumar, S. D., & Dela Cruz, O. G. (2024). Strength, Serviceability, and Anchorage of Fiber-Reinforced Polymer and Carbon Fiber Composite Prestressing Tendons. *IOP Conference Series. Earth and Environmental Science, 1326*(1), 012047. DOI: 10.1088/1755-1315/1326/1/012047

Mainini, A. G., Poli, T., Speroni, A., Cavaglià, M., & Blanco Cadena, J. D. (2024). *The Relevance of Performance-Based Design Within Early-Stage Design.*, DOI: 10.1007/978-3-031-75298-8_2

Moré, F. B., Galindro, B. M., & Soares, S. R. (2022). Assessing the completeness and comparability of environmental product declarations. *Journal of Cleaner Production, 375*, 133999. DOI: 10.1016/j.jclepro.2022.133999

Motia, K., Kumar, R., & Akhai, S. (2024). *AI and Smart Manufacturing.*, DOI: 10.4018/979-8-3693-6720-9.ch001

Nair, P. S., Gupta, R., Agrawal, V., & Thomas, B. S. (2023). A critical review on the effectiveness of microbial concentrations for enhancing self-healing in cement concrete and mortar. *Materials Today: Proceedings*. Advance online publication. DOI: 10.1016/j.matpr.2023.03.546

Najjar, M. K., Figueiredo, K., Evangelista, A. C. J., Hammad, A. W. A., Tam, V. W. Y., & Haddad, A. (2022). Life cycle assessment methodology integrated with BIM as a decision-making tool at early-stages of building design. *International Journal of Construction Management, 22*(4), 541–555. DOI: 10.1080/15623599.2019.1637098

Nassar, R.-U.-D., Soroushian, P., & Sufyan-Ud-Din, M. (2021). Long-term field performance of concrete produced with powder waste glass as partial replacement of cement. *Case Studies in Construction Materials, 15*, e00745. DOI: 10.1016/j.cscm.2021.e00745

Oh, J.-W., Park, K.-S., Kim, H. S., Kim, I., Pang, S.-J., Ahn, K.-S., & Oh, J.-K. (2023). Comparative CO2 emissions of concrete and timber slabs with equivalent structural performance. *Energy and Building, 281*, 112768. DOI: 10.1016/j.enbuild.2022.112768

Omar, O. (2025). *Navigating Nearly Zero-Energy Strategies for Urban Climate Change Adaptation and Mitigation.*, DOI: 10.1007/978-981-96-1521-6_6

Rasmussen, F., Andersen, C., Wittchen, A., Hansen, R., & Birgisdóttir, H. (2021). Environmental Product Declarations of Structural Wood: A Review of Impacts and Potential Pitfalls for Practice. *Buildings (Basel, Switzerland), 11*(8), 362. DOI: 10.3390/buildings11080362

Sadrolodabaee, P., Claramunt, J., Monserrat-López, A., Aguado, A., & de la Fuente, A. (2025). Low-carbon cementitious paving systems with ladle furnace slag and fine recycled aggregate: A multi-scale experimental study. *Construction & Building Materials, 490*, 142547. DOI: 10.1016/j.conbuildmat.2025.142547

Sirivivatnanon, V., Xue, C., & Khatri, R. (2023). Long-term reinforcement corrosion in low carbon concrete with a high volume of SCMs exposed to NaCl solutions and field marine environment. *Construction & Building Materials, 393*, 132071. DOI: 10.1016/j.conbuildmat.2023.132071

Soda, P. R. K., Chakravarthi, E. K., Mogal, A., & Mini, K. M. (2022). Statistical and experimental investigation on self-healing of microcracks in cement mortar by encapsulation of calcite precipitating bacteria into expanded perlite. *Construction & Building Materials, 342*, 127985. DOI: 10.1016/j.conbuildmat.2022.127985

Song, Y.-K., Jo, Y.-H., Lim, Y.-J., Cho, S.-Y., Yu, H.-C., Ryu, B.-C., Lee, S.-I., & Chung, C.-M. (2013). Sunlight-Induced Self-Healing of a Microcapsule-Type Protective Coating. *ACS Applied Materials & Interfaces, 5*(4), 1378–1384. DOI: 10.1021/am302728m PMID: 23373694

Tibbetts, C. M., Riding, K. A., & Ferraro, C. C. (2021). A critical review of the testing and benefits of permeability-reducing admixtures for use in concrete. *Cement, 6*, 100016. DOI: 10.1016/j.cement.2021.100016

Tozan, B., Hoxha, E., Stapel, E. B., & Birgisdóttir, H. (2025). Prioritizing low-carbon materials and reducing material quantities to mitigate embodied GHG emissions in buildings. *Sustainable Cities and Society*, *131*, 106770. DOI: 10.1016/j.scs.2025.106770

Wang, P., Lai, H., Li, W., Ke, L., Wu, H., Li, W., & Leung, C. K. Y. (2025). Mechanical properties and environmental impact assessments of GFRP rebar reinforced limestone calcinated clay cement (LC3) concrete. *Construction & Building Materials*, *472*, 140788. DOI: 10.1016/j.conbuildmat.2025.140788

Wong, L. S. (2022). Durability Performance of Geopolymer Concrete: A Review. *Polymers*, *14*(5), 868. DOI: 10.3390/polym14050868 PMID: 35267691

Zhang, X., Zhang, J., Feng, Q., & Luo, Q. (2025). Are all non-kaolinitic phases detrimental to LC3 performance? Effects of mineral type, particle size, and mechanochemical activation. *Construction & Building Materials*, *500*, 144182. DOI: 10.1016/j.conbuildmat.2025.144182

Zhao, Y., & Zhang, Y. (2023). A Review on Hydration Process and Setting Time of Limestone Calcined Clay Cement (LC3). *Solids*, *4*(1), 24–38. DOI: 10.3390/solids4010003

Chapter 2
Circular Economy in Construction: Innovative Practices for Sustainable Material Use

Arun Murugesan
http://orcid.org/0000-0002-2662-8483
PSG Institute of Technology and Applied Research, India

Nidhya Rathinavel
http://orcid.org/0000-0002-9962-5478
PSG Institute of Technology and Applied Research, India

Rahul Balaji
http://orcid.org/0009-0006-0114-7161
PSG Institute of Technology and Appiled Research, India

ABSTRACT

The construction industry uses around 40% between the consumed materials we use and 30% of the waste we produce globally. In general, the industry holds a linear model that reinforces negative impacts on the environment. The circular economy (CE) aims to use waste as an opportunity for materials reuse and ecosystem regeneration. This chapter reviews CE strategies that have potential to transition the construction industry, such as modular construction, urban mining, and design for deconstruction. The study employed a mixed-methods approach through literature reviews, case studies, and interviews with industry stakeholders. It advocates for the use of tools such as the Building Information Model (BIM), digital twins, and mate-

DOI: 10.4018/979-8-3373-6167-3.ch002

rial passports in the context of CE; aligns with the United Nations (UN) Sustainable Development Goals (SDGs), and presents a framework for the construction industry to become positive regenerative and resource-efficient built environment.

1. INTRODUCTION

The construction industry is the backbone of economic growth globally, urbanization, and development of infrastructure. However, it comes with a huge environmental price. It is currently estimated the industry consumes almost 40% of all raw materials worldwide and is responsible for over a third of energy related carbon emissions and about 30% of all solid waste due primarily to construction and demolition. These figures are shocking reminders of the linear economic model that the industry is based on, its schools of thought revolving around 'take make dispose', efficiency at price without considering consequences, always reinforced by capitalism (Bajare et al., 2024). The industry is an important contributor to climate change, degradation of ecosystems and overfilled landfills. With population increase of 2 billion projected globally and the urban population expected to grow to 68% in 2050, buildings and infrastructure will increasingly be in demand. Providing the built environment to support the demands of urbanization within ecological limits will require a complete overhaul of the way we design, operate, and renew the built environment.

In this regard, circular economy (CE) can be seen as a future labeled approach that aims to closing resource loops, reducing waste, and prolonging material and building life. CE presents a regenerative, systems-based approach to materials to reuse, repurpose, or recycle materials to avoid going to landfills. In the construction sector, this means reconceiving value chains starting with extraction, design, demolition, and recovering materials. That being said, the structural barriers to implementation are considerable as supply chains are fragmented, entrenched building codes are based on life safety and building performance focus primarily on new materials (Ferriz-Papi et al., 2024). Nevertheless, there has been remarkable innovation too: modular construction, prefabricated systems, and adaptive reuse create interest and find more and more acceptance. Digital tools like BIM, material passports, and digital twins support-as far as a standardized approach toward material flow tracking and optimization for various stakeholders-can become part of the road toward more intelligent, resource-friendly buildings.

This chapter explores how principles of CE can be practically put into practice within the construction sector. It does so by looking into the following questions: What are the most feasible best approaches to embed circularity? What are the material sustainability and equity outcomes from these approaches? What role can enabling digital technologies and policy landscapes play? How does the global best practice inform local action? The chapter draws on literature, case studies, and the best advice of international experts to explore the place of circular materials and their digital enablers, regulation, and the relevance of the construction sector to the SDGs. It concludes with a practical framework for embedding circularity throughout the lifecycle of construction projects (Oyebamiji, et al., 2025).

2. CIRCULAR ECONOMY: CONCEPT AND RELEVANCE TO CONSTRUCTION

2.1. Defining the Circular Economy

The CE is a revolutionary paradigm shift from the traditional linear model of take-make-dispose to a model that is regenerative in approach, by its very nature, to a system of production and consumption. The basis of CE in systems thinking is evident in the attempt to eliminate waste and pollution, extend the life of materials, and regenerate natural ecosystems. It comes in various forms regenerative design, cradle to cradle thinking, industrial ecology but all share in a redefinition of 'end of life' as reuse, refurbishment, remanufacturing, and recycling rather than disposal. In such a closed-loop model, the dependency on virgin resources is reduced; environmental degradation may be minimal, and material use could be optimized (Sangoremi et al., 2025). This decouples economic growth from resource depletion rather than resource creation, leading to a more resilient and sustainable link between industry and the environment.

The Ellen MacArthur Foundation outlines that CE sits on 3 principles: designing out waste and pollution, keeping products and materials in use, and regenerating natural systems, which encompass systemic changes across design, business models, and supply chains. CE fosters service-based approaches-such as leasing, and product-as-a-service-which prioritizes access instead of ownership, marking a transition from products to outcomes. Lifecycle thinking becomes an integral part of design thinking that involves durable materials, disassembly ease, and reuse potential (De Anício, 2022). Circularity must be imbedded at each stage, from material selection to recovery implementation, involving cross-sector collaboration and innovation. Ultimately, the circular economy is not just a pretty packaging exercise or the technical quick fix; it is the complete redesign of how economies operate within planetary boundaries, generating economic value in addition to deep respect and accountability toward the welfare of regional and local ecological functionality (De Anício, 2022).

2.2. Circular Economy in the Built Environment

The introduction of the concept of circular economy to the built environment signifies that fundamental changes are awaiting conception and implementation processes of the construction process. Conventionally, the construction industry has been characterized by high-resource appetite, vast linear material flows, and substantial waste streams. On the contrary, the circular economy calls for conceptualizing and managing buildings that are enduring, changeable, and resource-efficient. Fundamentally, this calls for changing our thinking at every stage of the building process-from times of material procurement to construction, utilization, maintenance, and finally deconstruction-to ultimately lock up material loop closures and lessen environmental impacts (Andriulaitytė & Valentukevičienė, 2020). At its very core, CE pushes the sector further than incremental change toward thinking systemically and restoratively to realize an end goal of long-term sustainability.

Other main applications of CE in construction are design for disassembly, which for all practical purposes means buildings and/or components must be specifically designed for dismantling, reutilization, and/or recycling after being useful. This has many benefits; among the many, entailed is the recovery of materials which could be useful, reduction of demolition wastes, and, in turn, the need for virgin extraction of natural resources. A companion concept is that of urban mining, which in a nutshell suggests looking at existing buildings and infrastructure as material banks rather than waste liabilities (Oliveira et al., 2024). Urban mining identifies reusable components of an older or decommissioned structure and recovers these elements to reduce demand for newly collected material and contributes to a circular supply chain.

Digital tech like material passports evolved from context and were thus defined as enablers of circularity. These digital documents contain information culture on the composition, origin, performance, and recyclability of the materials used in construction, sort of improving traceability and decision-making at all life cycle phases of a building. Modularity and flexible design incorporate arrangements that can be swiftly altered or repurposed into their own design. In doing so, the life span of a building is increased, meaning fewer materials destined for renovation or upgrading (Incorvaja et al., 2022). Hence, these really are the plausible applications of circular economy ideas in construction, bestowing a way to resilient, resource-efficient, and sustainable construction.

Figure 1. Comparative lifecycle diagram: linear vs circular construction models

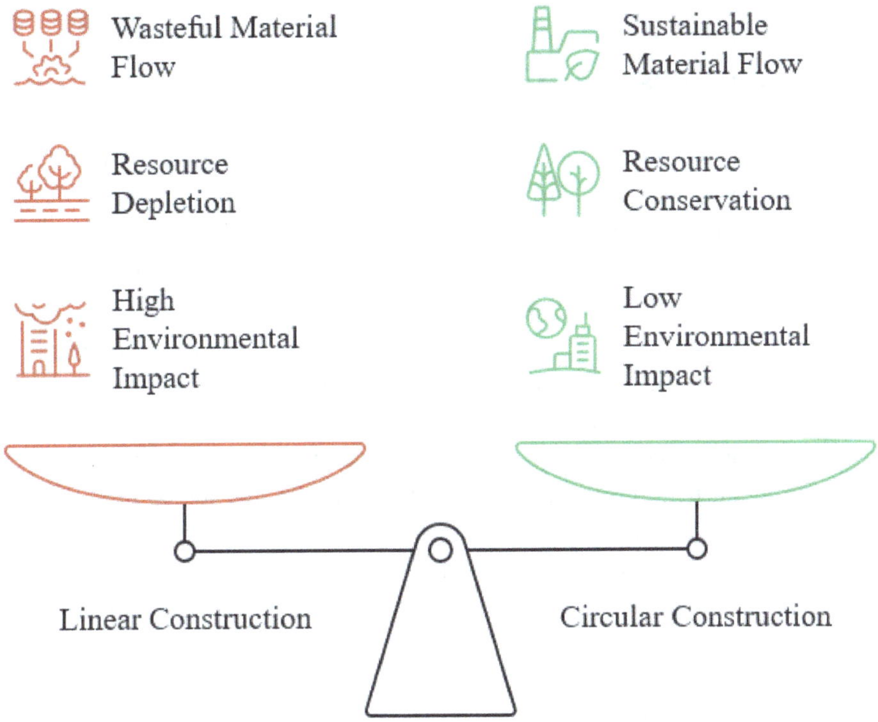

Table 1. Comparison between linear and circular construction models

Linear Model	Circular Model
Resource Extraction → Use → Disposal	Resource Extraction → Use → Reuse/Recycling
High carbon footprint	Low carbon footprint
Single-use materials	Designed for multiple life cycles
Limited documentation of material flow	Full traceability with digital tools

3. INNOVATIVE CIRCULAR PRACTICES IN CONSTRUCTION

Moving towards a circular economy in construction requires not just ideological alignment but an entire rethinking of the materials, methods, and design thinking associated with the buildings' life cycle. This section discusses innovative circular principles that build new modalities from reuse of components and modular systems, to design for disassembly and use of recycled, low-impact materials (Heisel et al., 2019).

3.1. Material Reuse and Urban Mining

Material reuse is one of the key circular strategies in the built environment which aims to recuperate and reinstate built heritage structural and architectural components from existing or deconstructed buildings. Components such as steel beams, timber, bricks and sometimes, façade panels, are often artisanal products which when fashioned together from high-quality material can be salvaged for structural and architectural detail for new construction. Material reuse also reduces pressure on the availability of virgin raw material, reduces embodied carbon emissions associated with the material extraction, and processing, and transportation, and permits the cultural and aesthetic valence of old construction to be preserved while underpinning heritage preservation goals within modern sustainability objectives. Where such recovery potential exists, material reuse represents dual and possibly greater environmental and economic efficiencies through the elimination of costs associated with acquiring raw building materials and disposal of construction waste (Pierzchlewicz et al., 2025).

In addition to material reuse, there is a principle of urban mining to consider, which views the built environment itself as a source of secondary materials. Instead of regarding decommissioned buildings, older infrastructure, or defunct utility systems as waste streams, urban mining systematically identifies, inventories and extracts primary materials being wasted in buildings and urban infrastructure. Thus, urban mining sees cities as stores of materials that could yield future construction projects from the existing building stock. For example, one of the major examples of urban mining is the "Material Bank" in Amsterdam; it monitors the programmed demolitions and creates an inventory of building materials that already exist for later use, which might be utilized by other projects in the future. Fundamentally, urban mining captures and diverts valuable resources, valorizing them through reuse in construction, reducing the burden of landfill waste, and enhancing local resource resilience (Aldebei & Dombi, 2021). As complementary operations to material reuse, urban mining practice furthers the circular economy objective of recovering value from waste to create a more sustainable and regenerative built environment.

Table 2. Examples of reusable materials in construction projects

Material	Source	Reuse Application
Structural steel	Demolished commercial frames	New commercial building skeletons
Bricks	Old heritage buildings	Façade detailing and landscaping
Timber	Roof trusses, flooring	Interior panels, decking
Windows/doors	Residential structures	Retrofit and new housing

3.2. Modular and Prefabricated Construction

Modular construction offers a novel model of construction in the circular economy (CE) context. It involves manufactured, standardized components, such as wall panels, kitchen pods or entire rooms, built away from the construction site and transported for assembly on site. This is a departure from conventional or traditional ways of site construction, as it lends itself to more precision, efficiency, use of fewer materials, and flexibility. Modular construction provides strong benefits from a circular perspective, mainly: ease of disassembly and relocation, ease of reuse of components, reduction of time to construct and waste of material, and central management of the quality of and materials for schemes using the manufactured components (Potemans, 2017). Modular construction frames its advantages, within CE principles, to maximize longevity of building components and reduce the carbon footprint of new construction.

One of modular architecture's explicit strengths is reversibility and reuse. The modular approach is based on the standardization of modules, which have the capacity for disassembly in the absence of damage and reassembly either in different configurations on-site or a completely separate project, ultimately preserving material that would likely have been demolished if it were traditional construction. In doing so, it retains options for the long-term utilization of the materials in the modules, permitting varied forms of spatial programming, and providing reduced costs after the life cycle of the project. Rogers Stirk Harbour + Partners' Y:Cube housing project in the United Kingdom is one such project that promoted this development model (Chen, 2023). The project consisted of off-site manufactured modular units that used renewable materials for construction, which were simply placed at the project site to expedite and simplify construction. The modular units were designed to be 100% reversible, which allowed total disassembly and reuse of the units, which increased living material reuse while concentrating waste from conventional demolition or renovation to 0 percent.

The circular benefits of modular construction are well established. For instance, WRAP (2019) states that up to 90% less construction waste could be produced when using modular construction methods than traditional methods. Factory environments support energy efficient assembly processes, reduce material waste and allow for better quality control. Modular structures are also easy to adapt when changing functional needs arise, or to allow pooling (shared functions), without involving resource intensive renovations or demolitions. Apart from this, recoverability and traceability also bear additional efficiencies with digital applications like Building Information Modelling and material passports in combination with the modular life cycle (Zhang & Pan, 2021).

3.3. Design for Disassembly: Enabling Reuse, Flexibility, and Circular Lifecycle Management in the Built Environment

Designing for disassembly is an underlying principle of the circular construction that approaches buildings as adaptable systems, rather than permanent objects to be built and occupied forever. The growing demolition and construction waste problem often arises from the fact that construction practices typically take little or no account of the end-of-life stage, leading to building waste and lost resources during demolition that could have been reclaimed or recycled. Design for disassembly promotes building materials and assemblies with methods that accommodate separate removal of individual components (e.g. a structural element, or cladding, or an internal system) without damaging adjacent or related components, allowing for materials to be discovered, recovered, reused, or recycled prior to disposal with minimal degradation. Incorporating disassembly in the building material life cycle sustains a solution-focused closed-loop in which built environments become "mines" of secondary materials (Tsoka & Tsikaloudaki, 2024). Realizing design for disassembly is not a risk-free process, however, with futuring for design, planning, and documentation processes intertwining short term occupancy and 'functionality' with long term 'value'.

The key principles of design for disassembly are to use reversible, mechanical joints instead of chemical adhesives or bonding agents to provide a clean separation of materials; standardize all components (fasteners, connection methods, finish approaches, etc.) for ease of sorting, maintenance and reuse; and to use layered assembly techniques to provide access to various building systems, when needed. Of equal importance is the role of datatised documentation (in the form of material passports or inventories) that provide traceable evidence of the original component source, specifications, and potential for reuse. A great example is The Circular Building project in London by Arup and Frener and Reifer. The prototype exhibits design for disassembly embodying reclaimed materials and mechanical fixings with every part catalogued digitally for future ease of disassembly and resale (Farouk & Abdelsabour, 2019). It demonstrates the technical potential of design for disassembly and addresses one of the greatest challenges for our industry; the prospect of clean and carbon-neutral construction activities that may drastically reduce construction and demolition waste; create demonstrable lifecycle cost savings; and embrace a culture of regenerative building based on circular economy trajectories.

Table 3. Design for disassembly guidelines for major building components

Component	Design for Disassembly Strategy	Benefit
Wall assemblies	Use of drywalls and bolted frames	Easy separation and reuse
Flooring systems	Modular raised floors	Access to infrastructure, reuse
Roofing panels	Interlocking systems	Reduced waste during retrofit
Facades	Clip-in cladding systems	Adaptable and recyclable

3.4. Recycled, Low-Impact, and Bio-Based Materials

The choice of materials for a particular project provides a prime opportunity toward a construction solution for circularity in the built environment due to its immediate bearing on a building's embodied carbon, resource efficiency, and potential for end-of-life conversion. There is a growing preference in circular construction for such materials that are recycled, renewable, or have a low environmental impact. In this way, these alternatives sustain material loops by the retention of value after its outside-the-loop application and hence, reducing the need to draw from virgin materials. New developments have enlarged the scope of green building materials, and hence, it is now possible to design structures that are environmentally friendly and technically sound. These materials form the groundwork upon which regenerative building practices are imparted and imposed under changing environmental legislation and green certification programs (Eberhardt et al., 2019).

On the more familiar side of things, recycled materials refer to those obtained from crushed concrete, bricks, and ceramics, which may be used for non-structural concrete or road base material. Low-carbon cements such as geopolymer cements and blending of fly ash or slag can reduce CO_2 emissions by 40-80% compared to the traditional ordinary Portland cement. In other applications, recycled plastics are now more widely being used as modular tiles, insulation panels, formwork systems, and lightweight building bricks, thereby encouraging waste diversion and lessening the carbon intensity of polymer applications. In this respect, bio-based materials such as hempcrete, mycelium composites, bamboo, cork, and cross-laminated timber (CLT) indeed have sustainability benefits since all of them are renewable, biodegradable, provide opportunities for local sourcing, and may sequester carbon during growth, which is better aligned with circular economy ambitions.

These materials not only offer enhanced environmental benefits, they also help improve indoor environmental quality, and performance continues into the operational phase of the building. For example, many of them are naturally breathable, non-toxic, and thermal performance that save energy and improve occupant wellness. Besides, they could be lighter and reduce the need for structural load and transportation emissions, making it easier in the processes of end-of-life disassembly (Shukla et al., 2025). And as part of the strategy of circular design with digital tracking technologies, recovery, reutilization, or biodegrade the possibility to regain the material in a less harming way for our environment! Low carbon materials reduce embodied energy but also greatly benefit indoor air quality and biophilic design. Many materials have the added advantages of being lighter and more thermally efficient, reducing operational energy.

3.5. Circular Design Tools and Certifications

Material innovation itself is not sufficient for the circularity of the built environment but its design frameworks together with performance tracking systems that ensure that set sustainability targets are achieved during the life of the project. A number of design frameworks and certification systems have emerged to promote and validate circular construction practices. The Cradle to Cradle (C2C) certification assesses products across several 'categories of material health, recyclability, use of renewable energy, stewardship of water, and social fairness'" and is the most comprehensive measure of circularity performance (Gupta, 2019). For example, some green building rating systems such as LEED v4.1 and BREEAM have adopted specific credits related to material reuse and life-cycle assessment and sustainable procurement, thereby allowing for greater uptake of circular approaches within conventional construction practices. At the macro level, the Circular Transition Indicators (CTI) developed by World Business Council for Sustainable Development (WBCSD) further enable organizations to measure, benchmark and improve overall circularity performance across projects and operations.

Certification systems and digital design tools are becoming increasingly important to implement and maximize circularity. Building Designers and Engineers are using design, parametric modelling, digital prototyping, and building information modelling (BIM), for example as processes to embody construction and deconstruction, following materials flows, and assess impacts through the lifecycle of a building. These design tools can also provide insights early on into ineffective design practices and allow for revisions that would be advantageous for modularity, material traceability and reuse. With the availability of any of these technologies to facilitate consideration of circularity in the design process, all aspects of the process of demonstrating compliance with legislation, improved data-informed decisions, reduced waste, and enhanced adaptability in the built environment are possible (Güngör et al., 2024). Together, both as standalone systems and integrated systems, certification and digital tools, will facilitate circular construction and realize intentions for measurable high performing circular outcomes.

4. DIGITAL TOOLS AND TECHNOLOGICAL ENABLERS

Digitalization is vital for scaling circular economy (CE) practices in construction. Because traditional construction workflows have poor information flow, low traceability, and high fragmentation, digital tools offer the transparency, coordination, and lifecycle view to enact circular strategies in a clear way. In this section we identify the types of technological enablers Building Information Modeling (BIM), digital twins, material passports, and blockchain, and we provide case-based examples of how they comprise and circular construction approaches in the case company's stories document (Banihashemi et al., 2024).

Figure 2. Digital integration across the circular construction lifecycle

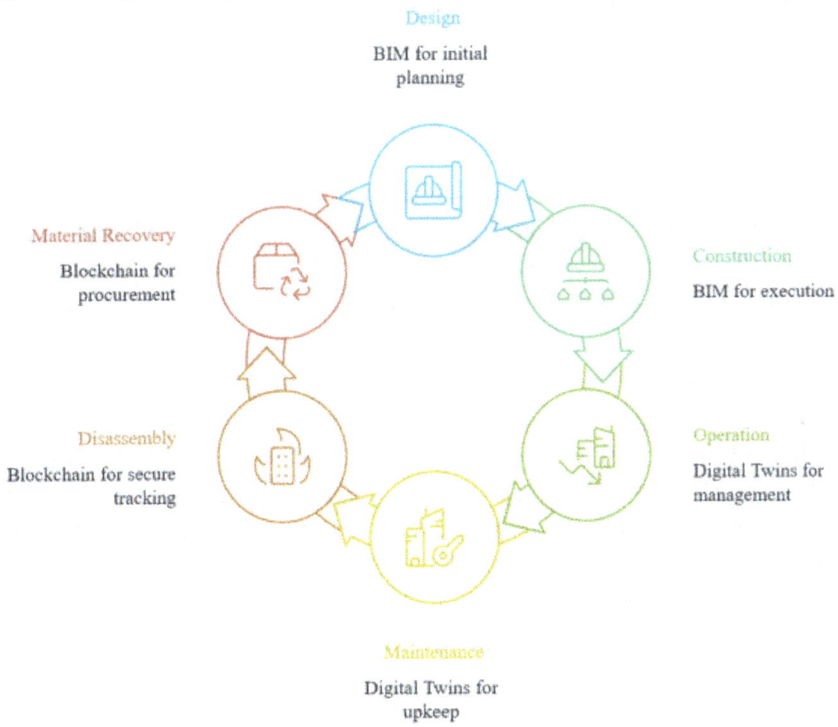

4.1. Building Information Modeling (BIM)

BIM, or Building Information Modeling, has become a vital tool in the construction field, allowing professionals to have a digital platform in which to create, manage and analyse data throughout the lifespan of the building from conception and design to construction and operation and then the necessary deconstruction. In the circular construction process, BIM functions as a fundamental integrating tool that enables life-cycle analysis, material tracking, and integration collaboration. By considering circularity at the project design and planning phases and having the project team engage with and understand the concepts, they are able to make purposeful decisions about what materials to procure and reuse, how they will use the circularity concepts, and how they will construct the building using these concepts maintained in alignment with long-term sustainable developmental goals and being informed about the life cycle effects (Askar et al., 2024). Additionally, BIM has the capacity to model construction and de construction processes in a way that is important for collaboration across multi parties when disassembly, adaptability, and recovery of components of the built environment are integrated in an accurate and dynamic way.

BIM's greatest contributions to the implementation of circular economy processes are its ability to develop a highly detailed material inventory in the digital model. Not only can modelled elements go back to an original tagged material so that information is retained (e.g. origin of the material, recyclability, embodied carbon, reuse potential), but BIM links it to end-of-life resource recovery to help make decisions. Moreover, through BIM-enabled clash detection and precision planning, projects are identifying construction errors on-site less often (at all), reduced material waste, and less costly rework. An excellent example of the use of BIM in the context of circular practice is the Madaster Platform in the Netherlands, which integrates BIM with material passports that document materials' identity, lifecycle data, and reuse potential (Cruz & Góes, 2021). The new method of material tracking not only changes the way buildings are perceived but also provides them with a new function as movable stocks of materials. These materials can be monitored, stored for later use, and may even be recycled or reused easily. Moreover, the process would require transparency and accountability in the decision-making process, thereby bringing the recycling and reuse of materials close together. With BIM, it is not only the construction efficiency that is increased but also the digitization of the foundational aspects of a regenerative built environment which is the goal of such innovations.

4.2. Digital Twins in Construction: Enhancing Lifecycle Intelligence, Predictive Maintenance, and Circular Decision-Making

Digital twins are one more key digital construction technology that provides high-tech virtual models of physical assets that are data-rich and updated in real-time. These digital twins are continuously receiving data from sensors, IoT devices, and even manual inputs, and thus they create a dynamic interface for the monitoring, measurement, and management of building and infrastructure performance over time. The digital twins will be instrumental in the circular economy by prolonging the life of the asset, lessening resource usage, and facilitating smarter disposal decisions. Furthermore, the digital twins will eliminate the separation between the physical and digital worlds, thus allowing designers and operators to be more adaptive, efficient, and, eventually, regenerative.

Digital twins are applied in many areas belonging to the field of circular construction. They can monitor in real-time energy use, water consumption, and material transfers for the assessment of the ongoing performance with the target of continuous optimization and operational efficiency (Lemian & Bode, 2025). Besides, they can assess the actual usage levels and capacity of deterioration of the system's components and provide predictive maintenance that allows intervention only when absolutely necessary, so that the material's value can be preserved and the material made either recyclable or reusable. Digital twins further grant the opportunity for advanced simulation, where stakeholders can model and test "in silico" the prospective changes in the built structure in the future, such as layout changes or disassembly, without having to directly alter the structure. It is beneficial for reducing risk and costs in the intention of still achieving circularity.

A remarkable example of digital twin integration is found in the City of Helsinki, where the municipality is using digital twin models of their urban infrastructure, including embedded circularity metrics. These models are in use, not just for the purposes of a better smart city, but also to assess, in regards to possible future renovations or re-developments, opportunities for possible material recovery. Digital twins allow for consideration of cradle to cradle lifecycle thinking, to enable data-driven maintenance repurposing, and deconstructing decisions (Van Dinter et al., 2022). Such understanding allows for building owners, designers and other stakeholders to align their practices with building performance sustainability objectives by leveraging digital twin technology, effectively evolving buildings into dynamic systems rather than strictly regulated independent entities in the circular economy.

4.3. Material Passports for Circular Construction: Enhancing Transparency, Traceability, and Reusability Across the Building Lifecycle

Material passports are digital, structured platforms that collect and document valuable information about a project's materials – importantly they enable transparency and circularity in the construction industry. Material passports are capable of recording the essential information on material type and specifications; origin; physical and chemical properties; environmental properties and impact; and suitability for reuse and recycling . Material passports provide some standardization and accessibility to this information so it allows architects, engineers, facility managers and future owners to see the entire life cycle of a building from design to demolition. They also are able to include Environmental Product Declarations (EPDs) as well as chronicles of the ownership and maintenance history, and can create a lifecycle profile of the building in order to facilitate responsible purchasing and material stewardship.

An exemplary implementation of material passport is the European Union's initiative, BAMB (Buildings as Material Banks). In this project, demonstrations were developed to implement direct integrations of material passports in the building design process with Building Information Modelling (BIM) platforms to create real-time, trackable inventories of components that had residual value (Leindecker et al., 2024). Building stakeholders will not only be able to estimate the residual value of materials at the end of their building life – making an economic case for reuse / circular investment – they will also be able to quantify materials with potential for reuse and decrease initial environmental footprint, making the project an example of circular design strategies and enabling new market mechanisms rewarding sustainable building practices. Material passports will ultimately change a building from a building into a bank of resources and can make circular construction economically advantageous as well as technically possible (Incorvaja et al., 2022).

Table 4. Elements of a standard material passport

Attribute	Description
Material ID	Unique identifier and origin
Composition	Primary and secondary materials
Lifecycle Score	Reusability, recyclability, durability
Toxicity & Emissions	VOC content, health impacts
CO_2 Embodied Emissions	Environmental performance indicator
Certification Status	Cradle to Cradle, LEED, etc.

4.4. Blockchain Enabled Material Traceability: Ensuring Immutable Provenance, Certification Verification, and Smart-Contract-Driven Circular Procurement

Blockchain technology is rapidly progressing as a transformational enabler of circular economy practices within the building and construction sector. While many people understand blockchain as the technology underpinning cryptocurrencies, blockchain has much broader relevance to the construction industry. Blockchain technology has the possibility to build and maintain a decentralized and tamper-resistant ledger that offers transparency and traceability to the flow of materials throughout the supply chain. In a circular economy context, it is required that a chain of custody for materials be verified, which is achieved through the provenance, composition, and lifecycle verification of the materials. A blockchain, meanwhile, has a secure digital record. This would enable product stakeholders to trace a good from raw material source to environmental accreditation, track reused or recycled content, and perform transactions of the goods in trust and accountability of sustainable processes in the construction context (Incorvaja et al., 2022).

Most of all, the exciting possibilities include blockchain for smart procurement systems and automated warranty systems in circular construction. Smart contracts, which are programmed agreements on the blockchain, can be established as payment or action trigger conditions based on performance or recovery thresholds. For example, a smart contract could be set up to release the contract amount only if certain thresholds of recycled materials are reached, or if materials were returned for reuse after their lifecycles ended. The system not only mitigates administration costs but also increases the degree to which procurement systems would help clients comply with circularity goals as the economic incentives would be better aligned with environmental outcomes.

An example of this practical application is IBM's "Circular Supply Chain" pilot project, worked on with major suppliers of building materials. They use blockchain technology, to certify and track secondary raw materials and to prove they were handled properly and used on new projects. They create an immutable record of the movement of materials through the platform, allowing manufacturers, contractors, regulators, and clients to collaborate on their use of the same data sources, while participating in a decentralized process. This will not only reduce the risk of greenwashing but, by making this documented evidence available for scrutiny and verification, will increase the credibility of sustainability statements (Lima et al., 2023). Blockchain creates trust where there currently is none between unconnected actors, and will enhance the operational basis for a scalable, data-driven circular economy in the built environment.

4.5. Enabling Seamless Digital Integration in Circular Construction Through Interoperability and Open Data Standards

Although all the digital tools, with the exceptions of BIM, Digital twin, material passports and blockchain, each bring certain features in the circular construction ecosystem, their outcomes under circular practices can be ineffective unless they are interoperable. A circular digital workflow depends on the flow of data in an open way, sharing data across platforms, analytic and software, and acting on data to fully enable circular digital workflows. Open data standards, the Industry Foundation Classes (IFC), the Construction-Operations Building information exchange (COBie) and ISO 19650, provide open data standards and common rules to enable an informed, consistent, compatible and transparent piece of information to both flow into and out of digital documents. These established innovative frameworks facilitate common models between the numerous software environment, project stakeholder and life-cycle phases, and provide linkages for collaborative and coordinated action, from design to construction to operation to (eventually) deconstruction.

In the future, the convergence of many emerging technologies will bolster digital circular construction. We are seeing artificial intelligence (AI) and machine learning algorithms being embedded into building information modeling (BIM) and digital twin systems to simulate the future, predict outcomes of circular practices, and potentially improve material reuse strategies (Trubina et al., 2024). Additionally, cloud-based platforms are facilitating continuous real-time collaboration on lifecycle and centralized analytics to help decision-makers act on the most current and comprehensive project data. The already remarkable capabilities of BIM can be outdone by the latest merging of Sustainable Development Goals (SDGs) to be embedded in BIM models and interactive dashboards, an innovative step to the future, thus transforming buildings into a measurable and accountable carrier of sustainability. Furthermore, the integration proves that digital processes are not only efficient in workflows but are also aligned with the broader environmental, economic, and social goals of the Circular Economy.

5. ALIGNMENT WITH THE UNITED NATIONS SUSTAINABLE DEVELOPMENT GOALS (SDGS)

In addition, the attainment of SDG 11, which aims at creating inclusive, resilient, and sustainable cities, is further facilitated by the application of circular economy principles in the construction sector. Thus, the latter provides different ways of solving urban environmental and social problems through the redesigning, constructing, and deconstructing of buildings and infrastructures. Among these are the strategies of modular construction and adaptive reuse, which lead to a reduction in the consumption of building materials and waste, increase the adaptability of buildings, and, thus, lessen the overall environmental impact of the construction process. Urban mining and circular material strategies can further support this sustainability by recovering value embedded in existing products rather than foraging virgin materials. This initiative aims to enhance the retention of embedded energy and resource consumption to improve the process of urban development (Santos et al., 2024). For instance, the Reversible Building Design model in Brussels shows the applications of modular system combined with digital material passports can specify developments, which facilitate disassembly in future stages of the lifecycle of a building, preventing its demolition waste and hence preserving material value through multiple lifecycles of the facility. The circular construction most fits into SDG 12, which concerns voice for responsibility in production and consumption patterns. The construction sector is often mentioned as one of the biggest consumers of raw materials and among the largest waste producers. The strategies under CE provide alternatives to this waste and consumption by leveraging secondary materials and sustainable procurement strategies. Life-cycle thinking will be integrated in every step of a project, from extraction to disposal, accounting for environmental implications; several tools, such as material passports and blockchain-based traceability tools, will enhance accountability and transparency in material flow and make reused or recycled verified components more accessible. Nevertheless, these innovations foster the creation of longer-lasting products with low toxicity and higher recyclability. According to the European Environment Agency, circular design could reduce the demand for materials by up to 35% and construction waste by up to 50%; resource-efficient and sustainable consumption is supported by CE practices (Trubina et al., 2024).

By incorporating CE in construction and building materials, it can strategically help fulfill SDG 13, since climate action and greenhouse gas emission reductions are paramount concerns. Reusing high carbon building materials (steel, timber, recycled aggregates) considers emissions reductions related to extraction and processing operations. New bio-based materials (hempcrete, bamboo) provide not only emissions reductions, but carbon sequestration. Low carbon building materials (e.g., geopolymer concrete) provide an opportunity for reducing CO_2 emissions by 40 - 80%. Digital construction (BIM, etc.) now also has carbon assessment processes for making design decisions that help achieve emission targets, implying a place for CE in climate change mitigation (Caldas et al., 2022).

5.1. Other Relevant SDGs

Circular construction is beneficial for not only environmental sustainability, but also for several Sustainable Development Goals (SDGs), specifically SDG 9, SDG8, and SDG 17. For SDG 9 (Industry, Innovation and Infrastructure), circular construction introduces innovative practices because it allows innovative technologies to be used (e.g., Building Information Modeling (BIM), blockchain, modular construction systems), and introduces innovative practices that allow material flows to be manufactured and properly recycled. In respect to SDG 8 (Decent Work and Economic Growth), circular practices can allow the creation of new markets and new jobs (for instance, circular economy platform providers, specialized deconstruction and material recovery providers, etc.), and better demonstrates how to create a resilient green economy. Circular construction also invokes the premise of SDG 17 (Partnerships for the Goals), as it requires engagement from multiple different stakeholders (e.g., architects, engineers, manufacturers, policymakers, end-users, etc.) and demonstrates how to take action towards systemic change through engaging partnerships and cross-sectoral relationships (Ogunmakinde et al., 2022).

6. CASE STUDIES AND BEST PRACTICES

This section offers international case studies that demonstrate how CE principles have been successfully implemented in various construction contexts, helping to ground theory in practical results.

Figure 3. Global case studies in circular construction: A geographic snapshot

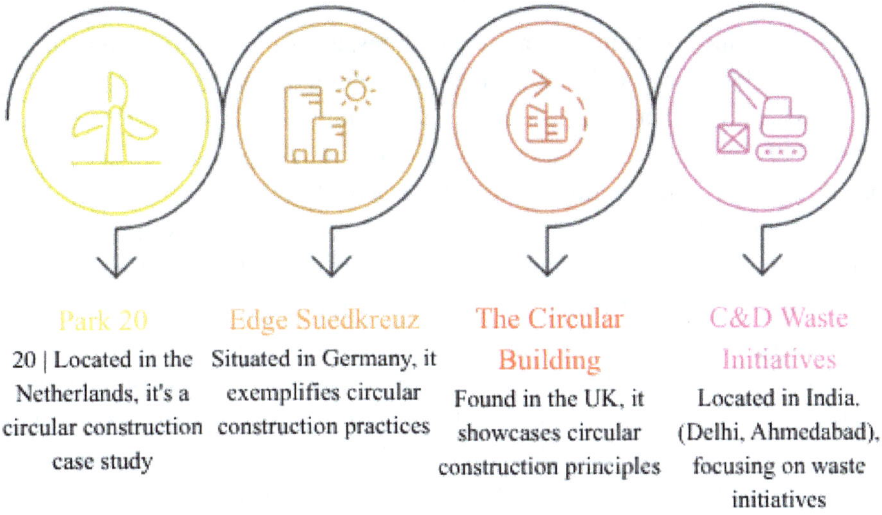

6.1. Park 20, Netherlands

Park 20 in the Netherlands is an innovative example of circular construction in action and is the world's first full service Cradle to Cradle inspired business park. Every aspect of the development, including architecture, landscape, and infrastructure was designed to allow for material reclamation, waste reduction and regenerative value creation. Modular construction systems, reusable facades, and cradle to cradle certified materials will allow each component to be disassembled and re-integrated or returned safely to nature. Building Information Modeling (BIM) and detailed material passports are in place across the entire site, with complete traceability of all components and the ability for long-term lifecycle planning.

Park 20 demonstrates quantifiable results in showcasing circular design and how about 85% of the materials are reusable or recyclable, which significantly reduces landfill use. Energy performance was improved by means of passive design strategies and closed-loop systems that achieved an estimated energy consumption reduction of 30% compared to the average development. In this project is the practical application of circular construction, and Park 20 will be able to serve as a replicable model for sustainable urban development (Cornet, 2019).

6.2. Edge Suedkreuz, Berlin

The Edge Suedkreuz in Berlin is the manifestation of the coming together of the best sustainable technologies in the context of circular construction. As the largest timber hybrid office complex in Germany, it ascertains the usages of Cross-Laminated Timber (CLT) along with prefabricated modular sections to drastically reduce the carbon footprint of the building while simultaneously enhancing efficiency. Overall, the project was always going to allow for adaptability, and it's even easy to replace units; standardization of structural components will have allowed for easy retrofitting or dismantling of the project if required in the future.

A highlight of Edge Suedkreuz is the use of digital twin technology, which is a virtual real-time replica of the building. Out of this, the building continuously will be able to observe its energy performance, predictive maintenance, and track materials' lifecycle during operation. The digital twin is integrated with sophisticated Building Information Modeling (BIM) systems, thus allowing the digital twin to capture the data on every building component, from embodied carbon values to material origins, to potential reuse values, therefore, creating continuous access to a living bank of recoverable resources in the building itself (Elze et al., 2022).

The project has delivered extraordinary environmental outcomes; it used prefabricated systems and material efficiencies to reduce construction waste by 80%. It also achieved a 50% reduction in CO_2 emissions compared to traditionally constructed office buildings primarily by using renewable timber and energy-efficient systems. Edge Suedkreuz is a demonstration of climate-resilient, circular architecture and it has shown the built environment can meet both environmental and functional needs in modern design.

6.3. Circular Building, UK

The Circular Building, located in the United Kingdom is a collaborative prototype of a building by Arup and Frener & Reifer, which provides a physical example of circular economy principles in action within architecture. The project is designed for disassembly, using mechanical fastenings rather than adhesive materials so that the components could be removed, recovered, and reused without damage. Each component within the building is digitally cataloged to create a material list that would support tracing through the lifespan of building whilst maximizing future reuse potential. The process applied through this project not only sets a 100% reuse potential for each component, it also challenges conventional linear construction practices and provides an example for the future for a building project to support waste reduction and maximize materials value across the lifespan of the building (Heisel et al., 2019).

6.4. India: Construction and Demolition Waste Rules, 2016

India is now implementing crucial policy actions to counter the environmental impact of construction and demolition (C&D) waste by imposing segregation, processing at the source and recovery of materials. Major urban centres such as Delhi and Ahmedabad have taken a lead role by establishing dedicated C&D waste processing facilities, which process waste into reusable aggregates for non-structural purposes (for example: road sub-base, paving blocks and filler). The steps taken to link urban planning with the circular economy are major ones, enabling the disposal of great amounts of garbage from landfills and consequently, the material coming back to the material flow.

In the Indian context, two important lessons can be drawn from the experience of the transition to circular construction: First, the policies that are shaped meaningfully will support systemic change; proper codes and compliance standards that enliven this are practically necessary. Second, the collaboration of the public and private sectors will be a success multiplier through the merger of the most cumbersome government processes and the quickest, most efficient private sector. What results is a flexible model of change that can be easily copied and scaled for cities that want to begin using systematic circular practices in the built environment (Iyer-Raniga et al., 2019).

7. POLICY, GOVERNANCE, AND EDUCATION

It involves mainstreaming knowledge sharing, institutional alignment, and policy frameworks in circular construction. This section includes facilitators in academia and government.

7.1. Regulatory and Policy Instruments

Regulatory and policy instruments are the most significant levers in the support of CE in construction. If regulations embed circularity, governments can give market certainty and create pathways for the market to go in sustainable directions. For example, a compulsory recycling rate is necessitated whereby contractors are compelled to divert specified rates of C&D waste from landfills. In addition, public policies make reuse and recycling expected market norms: not just about impacts in terms of addressing environmental burdens but about building new secondary material markets.

Other important financial policy tools that support circular construction is financial support. Tax incentives for the use of reclaimed or recycled materials should help lower the sometimes higher costs at the front end in acquiring and processing recycled materials. The best form of consumer demand pressure is applied through green public procurement policies on they encourage suppliers to introduce CE principles in their products and services, hence they can shift demand patterns right across the wider industry. Also, they incentivize responsible practice and innovation in the marketplace and can serve to help make greener Practice Economically Viable (zu Castell-Rüdenhausen et al., 2021).

One major example is Denmark where a very strong regulatory framework turned circularity into a characteristic of the national construction policy. The country has set the minimum percentage of recycled materials for the projects funded by the public sector and also has a tax rebate for the ones using reused or certified materials. This complex method illustrates that a unified policy framework is successful in bringing about dramatic changes and at the same time connecting industrial activities to sustainability and resource efficiency as the ultimate goal.

7.2. Role of Academia and Professional Institutions

Schools can help develop a CE mindset in the young professionals in architecture, civil engineering, and construction management. It has the potential to communicate clearly how CE influence in the core curricula formally readies students for knowledge and experience in how to design, build, and manage infrastructure sustainably when given appropriate opportunities. At the core of this effort is the making of distinctions and connections between life cycle thinking, material particularity, and environmental stewardship part of the fundamental studio projects and technical course work. Real-world case studies, demonstrations, hands-on workshops, and interdisciplinary collaboration expose students to an ability to question, understand, and show how circularity makes sense and is usable throughout the life cycle of buildings or tangible environments from design through procurement up to end-of-life recovery.

Professional organizations give this educational backbone into formalized learning pathways to practice. CE credentialing programs provide a formal recognition of competencies with regard to Circular Design, Construction, and digital resources like Building Information Modeling and Design for Disassembly. Continuing Education modules will make sure the working professionals learn the advancements in Legislation, Standards, and Technologies. Distribution of design handbooks and technical manuals on CE contributes to wide sharing of best practice within the architectural community, promoting a culture where circularity is the rule rather than the exception (Mirza & Amleh, 2002).

7.3. Multi-Stakeholder Governance

Multi-level governance is instrumental to circular construction, with active, explicit encouragement for the collaboration of actors in the built environment. Architects, engineers, contractors, policymakers, manufacturers, and waste processors will have to interact such that they align goals and processes in the search for closing material loops and, finally, the environmental impact. Each stakeholder has a different vantage point for example, designers will probably make the most difference in influencing or maximizing opportunities for materials selection and disassembly, while waste processors and manufacturers will be on hand with information about material recovery, recyclability, and even secondary markets. The most used governance framework will reduce fragmentation by providing a shared sense of purpose and understanding about circularity. One of the important ingredients of this collaboration is shared digital platforms for real-time data interchange, while transparency and accountability are assured. In this paper, Home, Sharing, or Circular Data Platforms will be considered a tool that could enhance the awareness of the stakeholders of a building's future and commonly own it (Amaral Cassilha & Tavares, 2024). They prepare a common environment with the same data: building information, material passports, tracking through blockchain showing to every party the material characteristics, its life cycle, and recovery possibilities. They will provide a mechanism for synchronized planning and decision-making; they are going to avoid superfluous work, while each phase will be in a position to consider circularity and thus to create the necessary efficiencies. It instills a mutual trust relationship within different stakeholders by perspective and incentive alignment.

The other anchor in scaling up these circular construction practices is aligned policies within the local, regional, and national jurisdictions. To be able to receive predictable and consistent regulatory requirements across the circular economy ecosystem that will reduce compliance attempts. This could be as simply as universal waste categorization, common recycling targets, or common procurement requirements. Increasing the role of multi-stakeholder governance structures including public-private partnerships, advisory councils, and cross-sector working groups, can support communications, conflict resolution, or collective action; implementing a multi-stakeholder co-governance framework must be an important goal to achieve the maximum benefits circular economy principles can arguably afford the construction industry (Santos et al., 2024).

8. CHALLENGES AND FUTURE OUTLOOKS

The logistical and regulatory barriers that deter material reuse in construction arise primarily due to variability in the quality of material (hence performance), inadequacy of documentation, and generally understandable hesitance due to liability-related reasons. It is these barriers that have prevented the uptake of more circular opportunities, as there is often a lack of stakeholder, and subsequent trust in the performance and compliance of reclaimed materials. However, with advances in traceability systems and material certification schemes, some of these issues have already begun to be alleviated (Osei-Tutu et al., 2023). By providing provenance, performance data, and standardized quality requirements, it is possible to offer increased transparency and accountability within the material supply chain, developing stakeholder trust and thereby enhancing safe and large-scale reuse of materials for construction. In the future, blockchain, material passports, and AI-led quality assurance of digital properties would be further easy to facilitate as the process of validation and tracking gets further simplified. This, in combination with increased regulatory support and a budding interest and demand for sustainable outcomes in the market, will enable material reuse to act as an easier economic and reasonable reality for construction (Wilson et al., 2023).

9. CONCLUSION

The construction sector has grown about tenfold since the early twentieth century, with the industry entering the present decade facing unprecedented environmental challenges and with unparallelled rapid technological shifts, alongside growing international consensus on the necessity for sustainable development. This chapter explored embedding CE principles into practice in the construction industry through modular design, material reuse, and design for disassembly and had shown that these are viable, increasingly demanded both by market players and regulators. CE stands to provide a particularly attractive opportunity for any effort to reduce the big material footprint and carbon emissions from the construction sector while increasing its resilience and resource security at the same time.

They are core digital enablers of BIM with material passports or a digital twin, creating a connected data environment for circular workflows from design and planning up to operation, re-use, and end-of-life recovery. Core policy instruments like green public procurement mandating recycling quotas and setting incentives for secondary materials form another strong enabling pillar for CE at large scale. Correspondingly, education and training in the methods of the Circular Economy have gained much in enabling mainstreaming of circular practice by which architects, engineers, and contractors may obtain the necessary knowledge and skills to design and construct for circularity. From the Netherland's Park 20 to Berlin's Edge Suedkreuz, further proof of the concept that circular buildings are both economically and ecologically beneficial and may contribute to social benefits.

In the future, AI plus digital construction platforms will allow predictive circular planning and hence optimize resource use and emissions in real time. The urgent need for harmonized CE metrics, certification schemes, and reporting frameworks will be necessary to provide transparency when comparing projects. Further avenues of application for circular principles involve infrastructure, especially roads, bridges, and public utilities. While cities are under increasing pressure to decarbonize and be resilient, circular construction offers regenerative foresight-hearing the built environment not just as an agent of impact but as an agent for the creation of a just and sustainable future.

REFERENCES

Aldebei, F., & Dombi, M. (2021). Mining the built environment: Telling the story of urban mining. *Buildings (Basel, Switzerland), 11*(9), 388. DOI: 10.3390/buildings11090388

Andriulaitytė, I., & Valentukevičienė, M. (2020). Circular economy in buildings. *Construction Optimisation and Energy Potential, 9*(2), 23–29.

Askar, R., Karaca, F., Bragança, L., & Gervásio, H. (2024). The role of BIM in supporting circularity: A conceptual framework for developing BIM-based circularity assessment models in buildings. In International Conference "Coordinating Engineering for Sustainability and Resilience" (pp. 649–658). Springer Nature.

Bajare, D., Zsembinszki, G., Rizaov, D., Tambovceva, T., Cudečka-Purina, N., Atstāja, D., & Cavdar, A. D. (2024). Circular Value Chain Management—Barriers and Opportunities. In *Circular Economy Design and Management in the Built Environment: A Critical Review of the State of the Art* (pp. 667–735). Springer Nature Switzerland.

Banihashemi, S., Meskin, S., Sheikhkhoshkar, M., Mohandes, S. R., Hajirasouli, A., & LeNguyen, K. (2024). Circular economy in construction: The digital transformation perspective. *Cleaner Engineering and Technology, 18*, 100715. DOI: 10.1016/j.clet.2023.100715

Caldas, L. R., Silva, M. V., Silva, V. P., Carvalho, M. T. M., & Toledo Filho, R. D. (2022). How different tools contribute to climate change mitigation in a circular building environment? A systematic literature review. *Sustainability (Basel), 14*(7), 3759. DOI: 10.3390/su14073759

Chen, C. (2023). Advantages and barriers of modular construction method in constructing buildings. *Proceedings of the Institution of Civil Engineers: Infrastructure and Construction, 176*(2), 75–84.

Cornet, M. (2019). The circular cornerstone: Applying a circular economy through design for disassembly to the development of Amstel III.

Cruz, V. F., & Góes, T. M. (2021). Contribuição do BIM para o desenvolvimento da economia circular no ambiente construído: Uma revisão sistemática da literatura.

De Anício, S. O. (2022). Connections between circular economy and ecological economics. *Sustainability and Climate Change, 15*(3), 162–165. DOI: 10.1089/scc.2022.0013 de Oliveira, J., Schreiber, D., & Jahno, V. D. (2024). Circular economy and buildings as material banks in mitigation of environmental impacts from construction and demolition waste. *Sustainability (Basel), 16*(12), 5022. DOI: 10.3390/su16125022

do Amaral Cassilha, S., & Tavares, S. F. (2024). Circularidade na governança do ambiente construído: Instrumentos políticos e atores. *Encontro Nacional de Tecnologia do Ambiente Construído, 20,* 1–13. DOI: 10.46421/entac.v20i1.6095

Eberhardt, L. C. M., Birgisdottir, H., & Birkved, M. (2019). Potential of circular economy in sustainable buildings. In IOP Conference Series: Materials Science and Engineering (Vol. 471, Article 092051). IOP Publishing. DOI: 10.1088/1757-899X/471/9/092051

Elze, M., Müller, S., & Günther, V. (2022). Industrielle modulare Konstruktion mit niedrigem CO_2-Fußabdruck–EDGE Suedkreuz Berlin. *Bautechnik, 99*(S2), 100–108. DOI: 10.1002/bate.202200077

Farouk, H., & Abdelsabour, I. (2019). DFD (design for disassembly) framework: An approach to enhance the design construction process. *Journal of Engineering Sciences, 47*(6), 883–898.

Ferriz-Papi, J. A., Lee, A., & Alhawamdeh, M. (2024). Examining the challenges for circular economy implementation in construction and demolition waste management: A comprehensive review using systematic methods. *Buildings (Basel, Switzerland), 14*(5), 1237. DOI: 10.3390/buildings14051237

Güngör, B., Agibayeva, A., Karaca, F., Askar, R., Giarma, C., Rosado, L., ... & Bragança, L. (2024). Circularity Tools and Frameworks for New Buildings. *Circular Economy Design and Management in the Built Environment: A Critical Review of the State of the Art*, 431-458.

Gupta, A. (2019). Accelerating circularity in built-environment through active procurement: An aggregated assessment framework to make sustainable choices while using secondary materials in the early design phase.

Heisel, F., Schlesier, K., & Hebel, D. E. (2019). Prototypology for a circular building industry: The potential of re-used and recycled building materials. In IOP Conference Series: Earth and Environmental Science (Vol. 225, Article 012023). IOP Publishing.

Incorvaja, D., Celik, Y., Petri, I., & Rana, O. (2022). Circular economy and construction supply chains. In *2022 IEEE/ACM International Conference on Big Data Computing, Applications and Technologies (BDCAT)* (pp. 92–99). IEEE. DOI: 10.1109/BDCAT56447.2022.00019

Iyer-Raniga, U., Erasmus, P., Huovila, P., & Maity, S. (2019). Circularity in the built environment: A focus on India. In *International business, trade and institutional sustainability* (pp. 739–755). Springer.

Leindecker, G., Askar, R., Güngör, B., Blázquez, T., Turbina, N., Gómez-Gil, M., & De Wolf, C. (2024). Material and building passports as supportive tools for enhancing circularity in buildings. In *Circular Economy Design and Management in the Built Environment: A Critical Review of the State of the Art* (pp. 507–543). Springer Nature Switzerland.

Lemian, D., & Bode, F. (2025). Digital twins in the building sector: Implementation and key features. In E3S Web of Conferences (Vol. 550, Article 05004). EDP Sciences.

Lima, T. A. R. M., da Mota, I. T. S., & Junior, C. M. M. (2023). Blockchain e contratos inteligentes na construção civil 4.0. Revista Multidisciplinar do Nordeste Mineiro, 11(1).

Mirza, S., & Amleh, L. (2002). *Sustainable infrastructure and civil engineering education*. National Guidelines for Environmental Sustainability.

Ogunmakinde, O. E., Egbelakin, T., & Sher, W. (2022). Contributions of the circular economy to the United Nations sustainable development goals through sustainable construction. *Resources, Conservation and Recycling*, *178*, 106023. DOI: 10.1016/j.resconrec.2021.106023

Osei-Tutu, S., Ayarkwa, J., Osei-Asibey, D., Nani, G., & Afful, A. E. (2023). Barriers impeding circular economy (CE) uptake in the construction industry. *Smart and Sustainable Built Environment*, *12*(4), 892–918. DOI: 10.1108/SASBE-03-2022-0049

Oyebamiji, H. O., Adewuyi, A. Y., Halidu, A., Togo, B., Alebiosu, Y. T., Atilola, M. O., Eromosele, E. I., & Atanda, A. A. (2025). Integrating circular economy in retrofit: Global practices in sustainable construction. *Journal of Management, and Development Research*, *2*(2), 10–18. DOI: 10.69739/jmdr.v2i2.811

Pierzchlewicz, D., Woźniak, A., & Widera, B. (2025). Recent research on circular architecture: A literature review of 2021–2024 on circular strategies in the built environment. *Sustainability (Basel)*, *17*(17), 7580. DOI: 10.3390/su17177580

Potemans, A. (2017). Modular building in a circular economy [Master's thesis, Delft University of Technology].

Sangoremi, A. A., Abosede, O. O., & Adeleke, A. E. (2025). A review on circular economy: Precursor to achieving global sustainability. *Asian Research Journal of Current Science*, *10*(3), 309–324.

Santos, P., Cervantes, G. C., Zaragoza-Benzal, A., Byrne, A., Karaca, F., Ferrández, D., Salles, A., & Bragança, L. (2024). Circular material usage strategies and principles in buildings: A review. *Buildings (Basel, Switzerland), 14*(1), 281. DOI: 10.3390/buildings14010281

Shukla, B. K., Yadav, H., Singh, S., Verma, S., Shukla, A. K., & Sharma, C. (2025). Sustainable engineering of recycled aggregate concrete: Structural performance and environmental benefits under circular economy frameworks. Construction Materials, 5(3), Article 67.

Trubina, N., Leindecker, G., Askar, R., Karanafti, A., Gómez-Gil, M., Blázquez, T., . . . Bragança, L. (2024, May). Digital technologies and material passports for circularity in buildings: an in-depth analysis of current practices and emerging trends. In *International Conference" Coordinating Engineering for Sustainability and Resilience"* (pp. 690-699). Cham: Springer Nature Switzerland. DOI: 10.1007/978-3-031-57800-7_64

Tsoka, S., & Tsikaloudaki, K. (2024). Design for circularity, design for adaptability, design for disassembly. In *Circular economy design and management in the built environment: A critical review of the state of the art* (pp. 257–272). Springer Nature.

Van Dinter, R., Tekinerdogan, B., & Catal, C. (2022). Predictive maintenance using digital twins: A systematic literature review. *Information and Software Technology, 151*, 107008. DOI: 10.1016/j.infsof.2022.107008

Wilson, S., Adu-Duodu, K., Li, Y., Sham, R., Wang, Y., Solaiman, E., . . . Rana, O. (2023, October). Tracking material reuse across construction supply chains. In *2023 IEEE 19th International Conference on e-Science (e-Science)* (pp. 1-4). IEEE. DOI: 10.1109/e-Science58273.2023.10254935

Zhang, Y., & Pan, W. (2021). Reducing construction waste through modular construction. In *International symposium on advancement of construction management and real estate* (pp. 339–347). Springer.

zu Castell-Rüdenhausen, M., Wahlström, M., Astrup, T. F., Jensen, C., Oberender, A., Johansson, P., & Waerner, E. R. (2021). Policies as drivers for circular economy in the construction sector in the Nordics. *Sustainability (Basel), 13*(16), 9350. DOI: 10.3390/su13169350

Chapter 3

Innovative Practices That Reduce Material Consumption Through Recycling, Reuse, and Sustainable Construction Techniques Aligned With Circular Economy

Jyoti Rani
http://orcid.org/0000-0002-5577-7808
Parul University, India

Shashiranjan Kumar Yadav
http://orcid.org/0009-0004-1524-7353
Lovely Professional University, India

Sakthivel Santhanam
Presidency University, India

ABSTRACT

The construction industry, a major consumer of natural resources and contributor to waste and carbon emissions, must shift from a linear "take-make-dispose" model to a circular economy for sustainable growth. This chapter explores innovative practices focusing on material efficiency, recycling, reuse, and sustainable design. Key strategies include adaptive building reuse, use of recycled and renewable ma-

DOI: 10.4018/979-8-3373-6167-3.ch003

terials, modular construction, and closed-loop supply chains. Through case studies and emerging technologies, it shows how circular principles reduce environmental impact and extend building lifecycles. The chapter also emphasizes supportive policies, collaborations, and research to advance circular construction globally toward a regenerative, resilient future.

INTRODUCTION

Construction industry's resource intensity and environmental impact

Since the Paris Agreement, the building sector has become a major actor to achieve global climate protection targets. A number of sustainability frameworks like the Sustainable Development Goals, European initiatives such as the "Green Deal" and the "Renovation Wave", as well as the EU taxonomy, intentionally target this sector. To satisfy the rising demand for raw materials triggered by the energy transition, an efficient and economical use of resources is vital, as is a higher level of recycling. The best recycling and efficient use of metal and mineral raw materials help to shield the climate and cut back on greenhouse gas emissions (Krause & Hafner, 2022).

The four major types of construction defined by the Department of Commerce differ in environmental emissions and waste generation. These sectors include.

1. 0.6% of the 1992 U.S. total for road and bridge horizontal construction. Gross Domestic Product (GDP)
2. 1.5% of GDP is the contribution of factories and office buildings.
3. Single-unit residential buildings make up 1.9% of GDP.
4. Construction-related services for involving other construction activities; such as towers, water, sewer, and irrigation systems. 2.4% of GDP

Different sectors of construction generate varying emissions and waste. Further, this is due to the scale of projects, energy used, materials, and more (Hendrickson & Horvath, 2000).

Out of all the industries, the construction industry is one of the most resource-intensive sectors. The reason is that it impacts energy consumption as well as greenhouse gases significantly. Investments in new technologies can cut emissions, reduce energy needs, and lessen pollution from materials being produced (Wood et al., 2022).

The construction industry, across the globe, is responsible for nearly 50% of carbon emissions, consumption of 20–50% of natural resources and, generation of 50% of total solid waste. As per the analysis by Optis and Wild (2010), the embodied energy can be as high as 51% and the operational energy can be up to 97% and disposal energy can be around 3% of the total energy use of a building (Vasilca et al., 2021).

Linear vs. circular economy in construction

The construction sector plays a significant role in the national economies, but at the same time it is a major energy consumer as well as waste and emissions producer (Vasilca et al., 2021). While this sector is growing rapidly, it still largely follows the linear "take–use–dispose" pattern. This leads to over-extraction of the resources and high waste generation. It is responsible for over 30% of the natural resources extraction and 25% of the global solid waste generation. This figure is much higher at 60% in developing countries (Yeheyis et al., 2013) (Benachio et al., 2020). Adopting the principles of the circular economy in the construction sector allows the construction to be made more sustainable by promoting the efficient use of resources and the integration of ecological/economic goals (Nasir et al., 2017) (Mangialardo & Micelli, 2018). The transition to a circular economy can be accelerated by using Industry 4.0 technologies. Technologies can be BIM, GIS, Blockchain, AI, IoT and many other technologies that will help in the intervention in that framework of Reduce, Reuse and Recycle. It can augment 3R framework and improve digital management in entire different life cycles of construction processes. This research seeks to investigate how these technologies may enable sustainable transformation in the construction sector (Benachio et al., 2021).

Figure 1. Sustainable hierarchy of end-of-life practices of a building (Bertino et al., 2021)

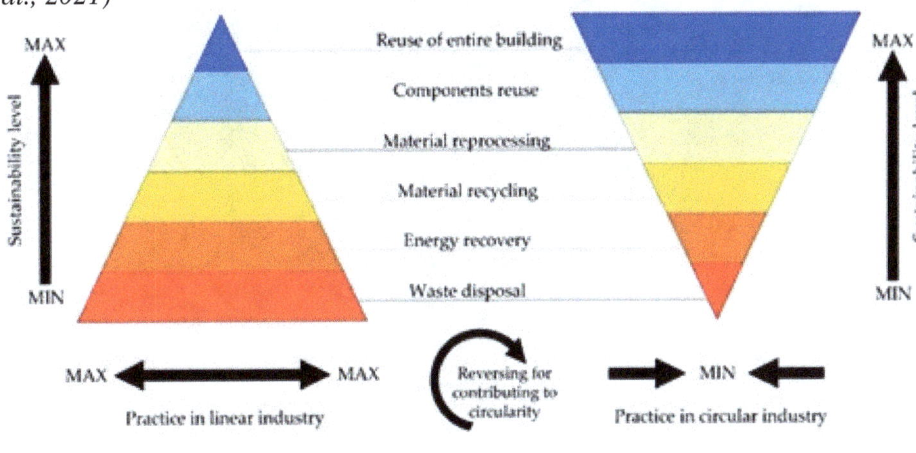

The reverse of building demolition is deconstruction. This is a useful process that is often referred to as the reverse of demolition. The latter creates around 90% of Construction and Demolition Waste (CDW). Moreover, demolition is responsible for anything between 20–40% of total solid waste stream. Deconstruction leads to reuse and recycling of materials and components. As a result, deconstruction prevents landfill waste and virgin resource extraction. Due to causing lesser pollution to air and water, this technique is much cleaner and sustainable. This nonlinear model of production needs to be adopted at the enterprise level; eco-industrial park level, and city or municipal level. A multi-level system can effectively minimize resource and energy loss (leakage) in construction through the adoption of circular business models and preferably with circular supply chain management. A more circular construction industry thus becomes possible (Didenko et al., 2018).

Importance of shifting toward sustainable and regenerative models

Figure 2. Sustainability paradigms

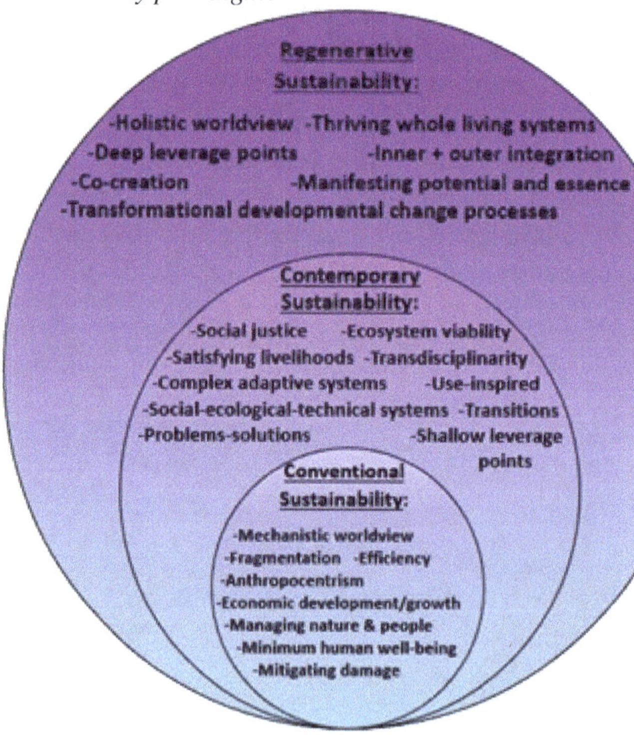

Sustainability has changed through different paradigms over time. First-order sustainability emphasizes a mechanistic approach while second-order stresses a human-centered view. Contemporary sustainability adds aspects like justice and systems thinking. Regenerative sustainability seeks prosperous life systems. It aims to transform the values, paradigms, and mindsets that cause unsustainability from a wholistic worldview. Regenerative development (RD) employs a place-based design approach to increase the complexity, diversity, and resilience of living systems in support of long-term wellbeing and adaptability (Gibbons, 2020).

Figure 3. Trajectory of environmentally responsible design (Reed, 2007)

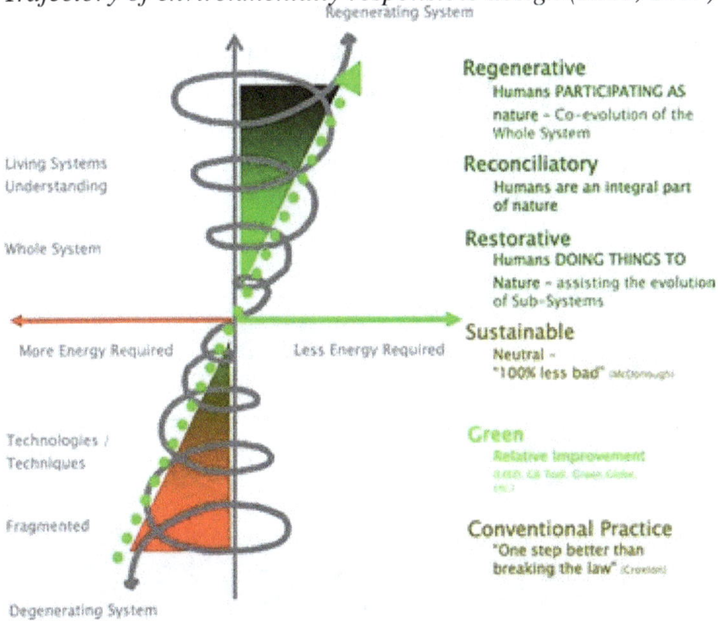

The regenerative paradigm provides an alternative that is explicitly designed to engage with a living world through its emphasis on a co-creative partnership with nature based on strategies of adaptation, resilience and regeneration (Rani, 2025).

PRINCIPLES OF CIRCULAR ECONOMY IN CONSTRUCTION

Closed-loop systems and material cycles

As the flow economy creates lots of waste, we must show "Chances of Change". In this language, we earn great profits through the reuse of products, scraps and residual materials. We accept them cost-effectively and recycle them. In the mining industry, circular economic principles can be difficult to adopt due to the balancing of virgin and secondary materials prices (Winkler, 2011).

In construction, these principles apply primarily in three areas.

- The reuse of building substances such as concrete, bitumen, rebar, asphalt, timber and glass (Rajala et al., 2018).
- Reusing is more efficient and helps us build faster and cheaper.
- Energy efficiency regretting the use of environmentally sound materials, use of renewable energy and lean (Rani, Guru, et al., 2025).

Three closed-loop system archetypes inner circles, decentralized systems, and open systems further substantiate how sharing information supports joint action. Using advanced recycling techniques for carbon and glass-reinforced thermosets, a recycling hierarchy is established which contributes to sustainability and reduces raw material price (Hagnell & Åkermo, 2019).

Figure 4. Circular economy model

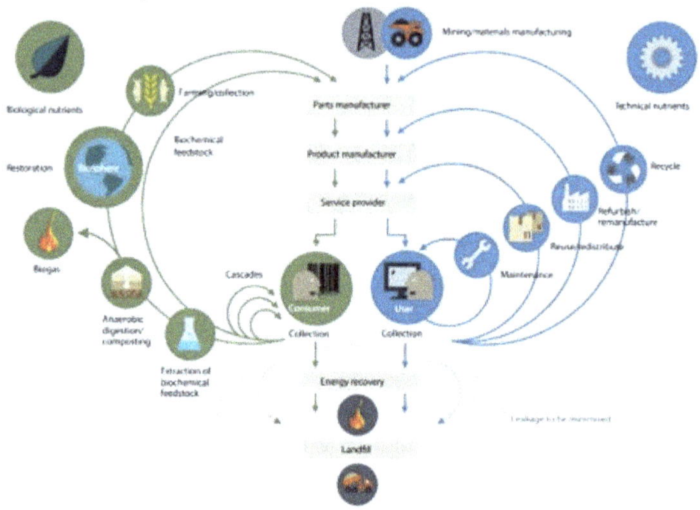

Figure 5. Waste produced during production, production waste, and in the form of EoL has the potential to feed a new component production with reclaimed material

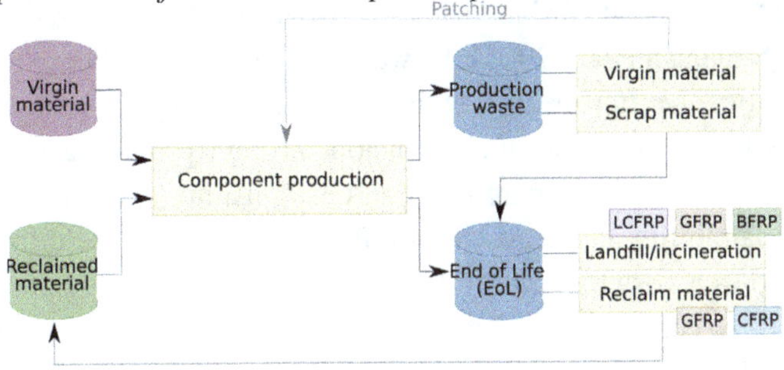

The renewable energy infrastructural components are based on technologies which can be classified as General Purpose Technologies (GPTs). The use of these technologies enhances innovation and efficiency (Busch et al., 2017). Innovation, efficiency, and enhanced adoption are driving forces behind the advancement of automation. More and more designers are advocating upcycling, which is the creative reuse and transformation of waste materials into products of higher value. This concept is in keeping with the principles of circular economy and sustainable design (Bramston & Maycroft, 2014).

Designing out waste and pollution

A new generation of designers embrace upcycling. This trend reflects a growing commitment to sustainability in design practices. The decline in the quality of ecosystem is directly due to consumption that is guided by marketing (Bramston & Maycroft, 2014). A marketing strategy has caused an increase in consumption levels during the manufacturing, distribution, and use of products which result in environmental damage. Upcycling design is a response aimed at reducing waste, prolonging product life cycles, and countering the ecological impacts of consumerism (Fuller & Ottman, 2004).

Extending product and building lifecycles

In 1931–32 Sewall Wright proposed the concept of adaptive landscapes (AL). AL provides a useful framework for understanding dynamic systems and continuous adaptation (C. Bakker et al., 2014). In a sustainable product design context, this adaptability takes the shape of leveraging customer relationship management to achieve product lifetime extension (PLE) rather than the sale only (C. A. Bakker et al., 2021). Companies must understand changing needs and draw an upgrade roadmap in accordance with these evolving demands. Product designers often adopt three strategies to tackle sustainability challenges which include material efficiency, product life extension, and product recycling. Material efficiency reduces costs and resource use (Sudarsan et al., 2005). However, life extension focuses on durability and repairability. The European Commission (2019) highlights the importance of regulations for product lifetime extension through providing products which are easy to repair, providing spare parts and establishing standards on durability, reusability, repairability and recyclability. PLM is a business solution that guides information flows across the entire life of a product. It is crucial enabler of these strategy (Ameri & Dutta, 2005). The success of PLM relies on the implementation of data standards and integration architectures. When these are set, the previously fragmented information is delivered in the proper situation to the correct audience and at the right time. These would aid competent decision-making and sustainable product innovation.

Figure 6. Adaptive landscape design process

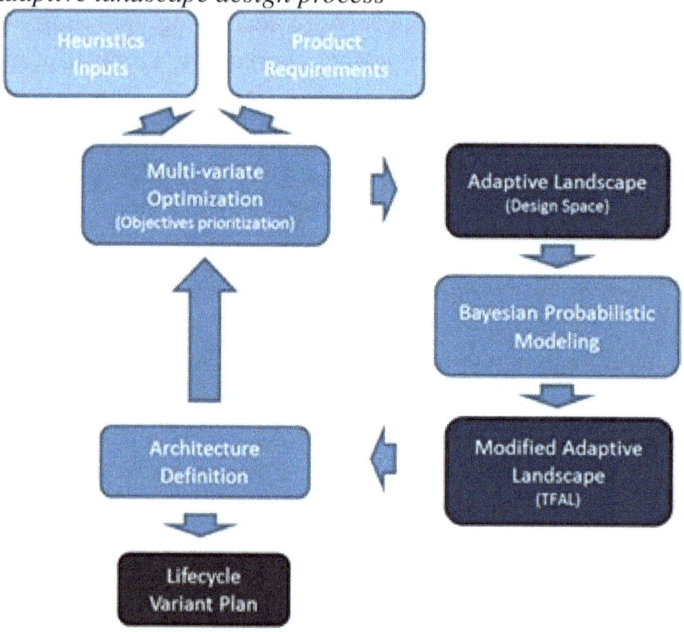

Regeneration of natural systems.

Technological progress and human actions are increasingly influencing the natural world, even in remote areas. On the other hand, Self-organizing natural regeneration processes, in wild or planted forests, take place which promotes local genetic adaptation supports native species, biodiversity (Müller, 2020), and ecosystem resilience. A dense natural regeneration will ensure more competition between plants which will speed up natural selection (Chazdon & Guariguata, 2016). Therefore, the forest will have a greater capacity to adapt to climate change, thus avoiding risks such as large-scale forest die-off. Regeneration is Learning Level III where we go from acting on nature to acting with and as nature. In this, we move towards symbiosis in all interactions (Reed, 2007). This perspective describes measures that can be taken to reduce, reuse, recover and recycle materials with high embodied carbon (EC) in the built environment. Because buildings consume 30–40% of energy worldwide and emit greenhouse gases, especially in the residential sector, they must adopt regenerative and circular principles so that they can lessen their environmental impact (Rani, Guru, Singh, Fangueiro, et al., 2024).

- he first
- time, an attempt was made to experimentally apply the coefficient of growth tension of natural
- regeneration on ravine-gully systems in order to clarify the feasibility of forestry measures in plantations
- on fallow lands
- This accelerates differentiation, natural selection and increases the
- resilience of forest ecosystems to global climate change, to avoid mass drying of Scots pine.
- This accelerates differentiation, natural selection and increases the
- resilience of forest ecosystems to global climate change, to avoid mass drying of Scots pine.
- This accelerates differentiation, natural selection and increases the
- resilience of forest ecosystems to global climate change, to avoid mass drying of Scots pine.
- This accelerates differentiation, natural selection and increases the
- resilience of forest ecosystems to global climate change, to avoid mass drying of Scots pine.
- This accelerates differentiation, natural selection and increases the
- resilience of forest ecosystems to global climate change, to avoid mass drying of Scots pine.
- This accelerates differentiation, natural selection and increases the
- resilience of forest ecosystems to global climate change, to avoid mass drying of Scots pine.
- This accelerates differentiation, natural selection and increases the

- resilience of forest ecosystems to global climate change, to avoid mass drying of Scots pine.
- the first
- time, an attempt was made to experimentally apply the coefficient of growth tension of natural
- regeneration on ravine-gully systems in order to clarify the feasibility of forestry measures in plantations
- on fallow lands

MATERIAL EFFICIENCY AND WASTE REDUCTION

Strategies for minimizing material consumption

The construction and demolition (C&D) sector generate the second largest volume of waste which is mostly not engaged in waste minimization, avoidance and recycling. Some new ideas on the market include prefab multi-storey structures for urban living with the aim of more efficient construction (Lehmann, 2011). The broader vision of a 'zero-waste' life cycle requires a major behaviour change, which confronts prevailing concepts of consumption and demands sustainable development. Steel is the most recycled material in the world. With steel, designers are able to create lightweight components with low embodied energy. At the same time, they are able to establish a chain of reuse back to steel. Even waste management has become big business (Lehmann, 2011). In the UK, companies like Sita UK and Shanks are doing multi-million-pound contracts. The essential features of eco-city improvement as well as eco-building creation are making use of accidental structures rather than demolishing them and extending building components. Saving the energy, recycling and reuse of metals and plastics and paper and glass save 95% of aluminium, 85% of copper, 80% of plastics, 74% of Steel and 64% of paper. The process of recycling saves the metallic energy already used in the production of these metals. They also help in reducing climate change. Following this, Van der Ryn and Cowan's (1996) "Ecological Design" emphasizes creating solutions based on ecosystem designs (Shahbazi et al., 2016). The Future Waste Scenarios – Business as Usual, Circular Economy, Material Efficiency and Lifestyle Compromise – have been developed to understand the global trends in waste and material use till 2100. The world must not only deal with ever-increasing amounts of waste, but also the challenge of circular materials.

Figure 7. The waste hierarchy diagram

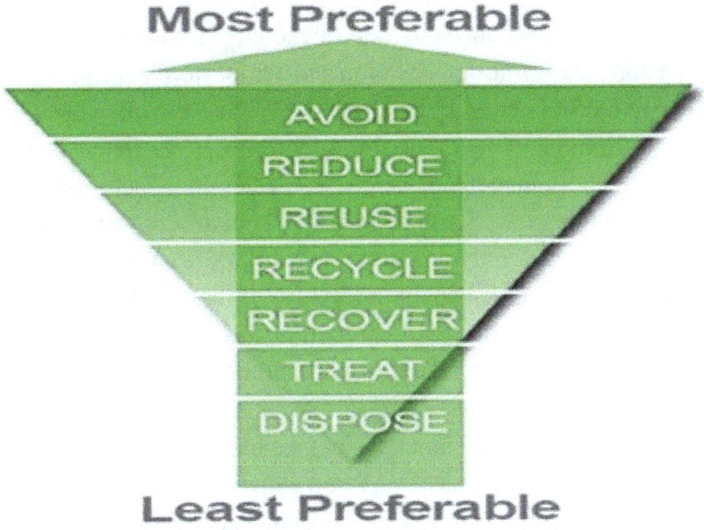

Figure 8. Waste management is an important key stone in the effort

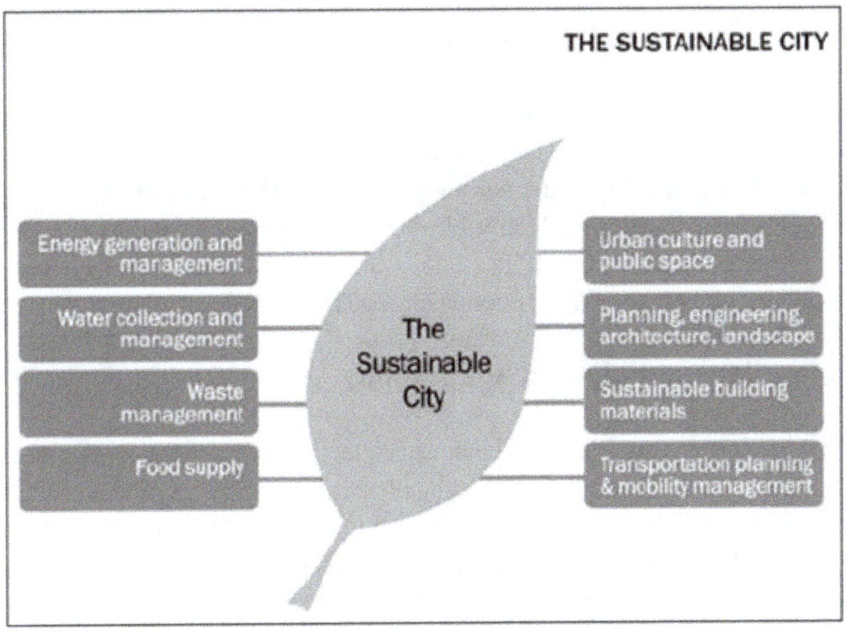

On-site waste management and segregation

The construction industry is one of the largest producers of solid waste in the world. Reports show that construction and demolition waste constitutes 44% of landfill waste in the UK, 29% in the US and 44% in Australia (Ajayi et al., 2017), and on average around 35% in the world. According to Wasman (2016), Factor analysis on on-site waste management practices reveals four factors contractual provision for waste minimisation, waste segregation, maximum reuse of material, logistics management which jointly account for nearly 96% of total variance. Despite this, the construction waste is usually a mixture of inert and non-inert materials (Yuan et al., 2013). Hence, efficient segregation is vital for their reuse and recycling. In Hong Kong inert materials like sand and bricks and concrete are sent for land reclamation, while non-inert materials such as wood, plastics and glass are sent to the landfill. Out of the different waste, if segregated on-site and managed carefully, gypsum plasterboard (PB) has a very high recycling potential (Jiménez-Rivero et al., 2017). However, in countries like Italy, Poland and Spain, markets for post-consumer recycled gypsum are underdeveloped as a result of low landfill taxation, access to virgin gypsum and poor coordination between stakeholders. In conjunction with the growth of industries and construction, growing the waste generation will impact the environment, the economy, and social issues (N. Singh & Rani, 2024). In Hong Kong, for example, the generation of construction and demolition waste amounted to some 32,710 tonnes per day in 1998, which is almost 15% more than the previous year. 80% of the inert materials contained in landfills are unvaluable. At the same time, a new breed of designers is emerging who are focusing on upcycling as a design strategy that refers to the conversion of used materials into new materials and products of better quality and environmental value. This mindset aligns with the global goals of sustainability. Renewable energy infrastructures and other green technologies increasingly adopt GPTs to cut waste and resource intensity(Rani, Guru, Singh, & Santhanam, 2024) (Poon et al., 2001).

Design for disassembly and adaptability

The most important aspect of circular construction is designing for disassembly and adaptability. This means that the building elements must be easy to take apart, reuse or rearrange after its first life . It is difficult to reuse many materials that are used in traditional construction so planning for this is done sooner (Webster, 2007). Designing adaptable structures and using recycled or salvaged materials reduces the depletion of non-renewable resources (Yadav et al., 2018). Research indicates that the assembly of a product has a strong correlation with its recyclability. The simpler the assembly, the easier it can be recycled. The strategy of adaptable design intends to prolong product service life. There are two components in this concept which includes product adaptability and design adaptability. The concept was built on principles such as functional independence and systematic and orderly design processes (Seidu et al., 2025).

RECYCLING PRACTICES IN CONSTRUCTION

Use of recycled aggregates, metals, plastics, and glass

This research investigates how effective recycled plastic waste (RPW) and recycled crushed glass (RCG) are when mixed with concrete mixtures that are used to create concrete footpaths. The replacement levels were changed to prepare different concrete samples. 0%, 10%, 20%, 30%, 40% and 50%. Recycled glass comes from bottles and industrial waste (Mohammadinia et al., 2019). Normally, it consists of silica sand, and other ingredients. According to tests, fine and medium recycled glass gives road and concrete similar, and in some cases even better, workability, strength, and hydraulic properties as compared to natural aggregates. This is a significant finding. Using recycled materials helps with sustainability. In other words, increasing efficiency of resources and reducing waste are two of them (Disfani et al., 2012).

Industrial symbiosis and material recovery facilities

In industrializing countries like India, recovering, reusing, or recycling of industrial residuals or what is otherwise generally considered waste, is a common practice to cut costs or because of resource constraints. Usually, a firm can recover materials internally, send them to another firm for use, or send them to an informal recycling market. The sequence of waste management techniques is waste minimisation, reuse, recycling for materials, energy recovery and disposal (Bain et al., 2010). Industrial symbiosis promotes mutual aid among industries, where the waste of one firm can be used as the input or raw material of another. In order for such systems to work, however, it will need a transformation in production, marketing and consumption patterns incorporating recycling and remanufacturing into it. An LCM approach that is based on symbiosis integrates life cycle thinking and industry collaboration to create sustainable and circular flows of resources (Mulrow et al., 2017). With this approach not only improves eco-efficiency and minimizes environmental impact but also provides flexible structure to combine environmental, economic, technological, and social aspects for continuous improvement.

Innovations in construction and demolition (C&D) waste recycling

Waste generated in a construction site from activities like construction, renovation, demolition, site clearance and excavation and road works is termed as C&D waste. The constituents of building materials are concrete, steel, wood, gypsum, brick, asphalt, and polymer. All over the world C&D waste is one of the major waste stream causing environmental problems. The U.S. generated over 548 million tons of construction and demolition (C&D) debris in 2015 (Bao et al., 2019; Ma et al., 2020), more than twice the nation's municipal solid waste (MSW). In 2014, household waste in the UK produced approximately 131 million tonnes, accounting for over half of the total. As of 2015, China was the biggest producer in the world with an estimated production of 2.4 billion tonnes, most of it is landfilled or dumped, which break vegetation and degrade land. In Europe, C and D waste or construction and demolition waste generated each year is more than 350 million tonnes. But when all construction materials are included, that figure rises to almost 970 million tonnes (Lockrey et al., 2016). Recycling and efficient waste disposal in the construction industry is crucial for conserving resources, decongesting the landfill and encouraging sustainable construction.

Figure 9. Circular economy for a cleaner industry

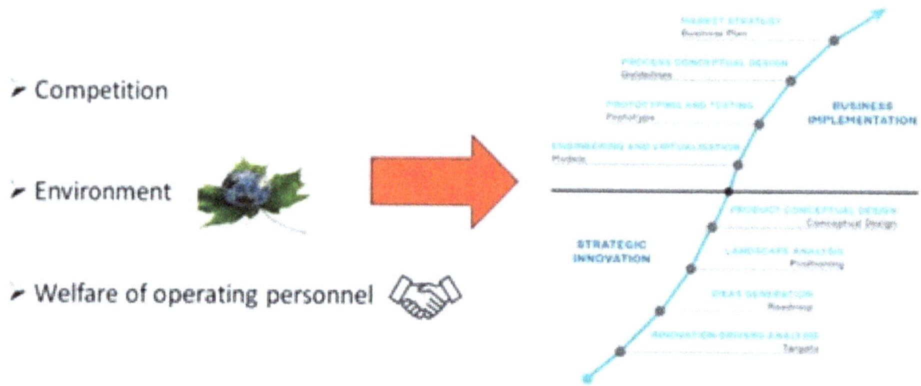

5. REUSE PRACTICES IN CONSTRUCTION

Adaptive reuse of existing buildings

Adaptive reuse aims to transform buildings through the repair and/or modification of existing buildings to 'suit the needs of modern society'. All this while retaining their heritage value, and visual and physical relationship with the site (Tam & Hao, 2019). It uses less material, produces less waste and helps structures last longer, hence is environmentally responsive building construction which requires sustainable materials (Bullen, 2007). Many building owners are seeing adaptive reuse as a greener alternative to demolition and redevelopment. The most common stages in the process are assess heritage value, assess potential for reuse, set design strategies, make alterations and post-occupancy evaluation (Langston et al., 2008). As recognized by heritage charters and sustainability models, adaptive reuse contributes to the preservation of cultural heritage and modifies conventional property decision-making to be more environmentally responsible and resource-efficient (Rani, Santhanam, et al., 2025).

Figure 10. The most repetitive aspects mentioned by the authors in the reviewed literature in phase 1; the size of the circles is proportional to the repetition of each of the mentioned aspects (Arfa et al., 2022)

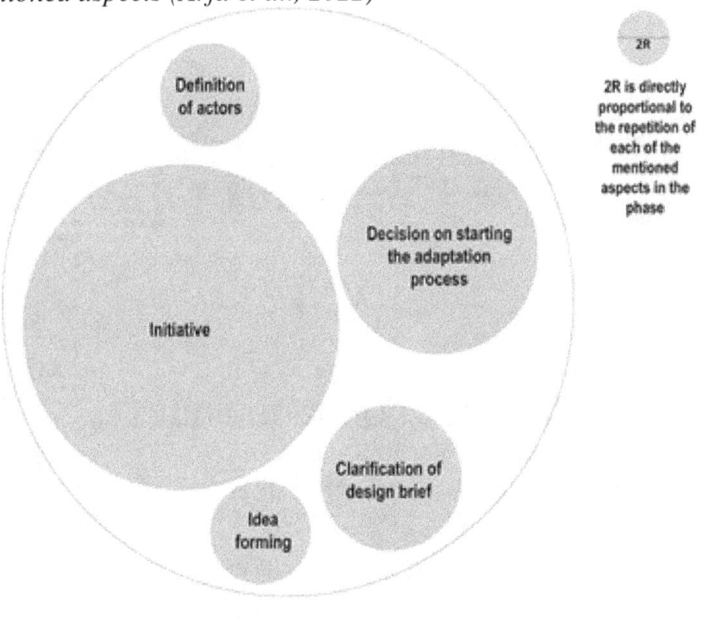

Salvaging and repurposing construction materials (wood, steel, concrete, etc.)

Upstream strategies are when new buildings are designed to allow for the reuse of their wood. Downstream strategies involve reusing wood from old buildings that are torn down or dismantled (Piccardo & Hughes, 2022). The European Directive on Waste promotes the reuse and recycling of goods for less reliance on primary goods. Even though there are barriers to the reclamation and recycling of timber in the construction sector, laws, guidelines and businesses already exist for waste wood (Sieffert et al., 2014). Waste generated by the construction process is mainly by design, which suggests we should design better to waste less. Using and recycling materials are emerging as enablers of sustainable construction. Urban-level systems sorting of the materials for reusing, repurposing or energy recovery suitable. Design for Reuse (DfR) is an approach that allows buildings to use material like steel, wood and concrete from existing structures. DfR can be facilitated by Building Information Modelling (BIM), material banks and material passports (Ormondroyd et al., 2016).

Figure 11. The process of generating the material procurement plan (Kim & Kim, 2020)

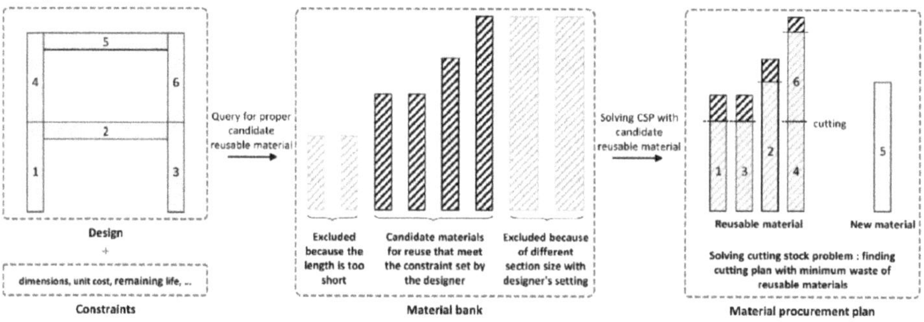

SUSTAINABLE CONSTRUCTION TECHNIQUES

Modular and prefabricated construction

The construction of buildings in sections and then assembling them on-site is called prefabrication. Using recycled material during construction means less waste of material, water and energy. Hence, it is an important strategy (Y. Jiang et al., 2019). Modular prefabrication of individual building elements (like an apartment, staircase, or façade) in factories greatly improves the quality, safety and environmental protection. On-site assembly is also attached (Molavi & Barral, 2016). By adopting this method, the time taken for construction gets reduced while causing lesser on-site disturbances. Although construction is a simple word, it encompasses a lot of various operations and processes (Jaillon & Poon, 2008). In general, pre-fabrication provides economic, environmental and social benefits by encouraging the savings of money, speedy delivery of projects and sustainable outcomes in construction.

3D printing and digital fabrication with recycled inputs

3D printing in construction is an automated technology that can eliminate labor, save material, and reduce the time of construction. It removes the need for classical shuttering and lets you make complex shapes for cheap (Md. H. Ali et al., 2022). Researchers are using recycled sand and aggregates for 3D-printed concrete to create a more sustainable environment. The technology makes working on our buildings' architecture more energy-efficient. Furthermore, lots of 3D printing parts can be made using recycled plastics and composites to enable circularity. 3D printing is efficient in a modern, smart and sustainable way overall (Han et al., 2021).

Figure 12. Plastic waste recycling

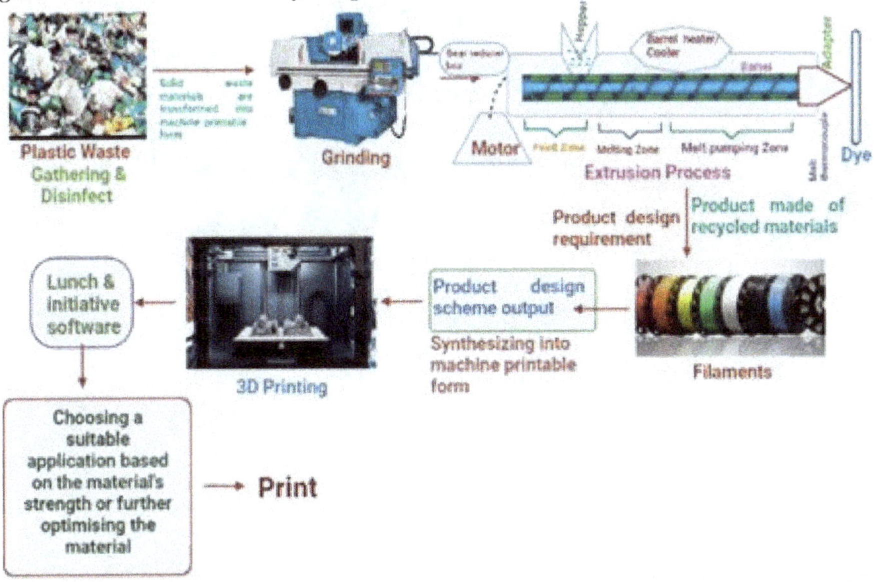

Low-carbon and renewable materials (bamboo, hempcrete, mycelium, etc.)

Sustainable and eco-friendly materials play a vital role in modern construction by reducing environmental impact. Recycled steel, bamboo, and engineered wood help lower the carbon footprint, while bio-based materials like mycelium composites and hempcrete offer renewable and biodegradable alternatives. Mycelium composites are produced by growing fungal networks on agricultural waste, transforming residues into strong, plastic-free materials. Bamboo, known for its rapid growth and high carbon content, is also a sustainable option when sourced locally. These materials promote low-energy fabrication and waste upcycling, making them ideal for sustainable building practices (Carcassi et al., 2022).

Figure 13. Carbon Net Zero 2038

Figure 14. Flowchart of the bamboo–mycelium composite and system boundaries (Zeng et al., 2022)

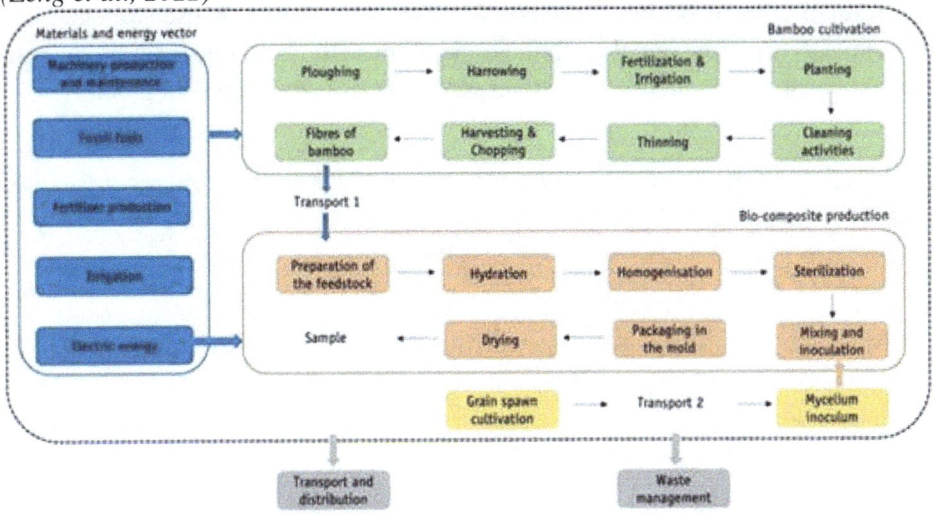

Energy-efficient design integration

3D printing in construction is an automated technology that can eliminate labor, save material, and reduce the time of construction. It removes the need for classical shuttering and lets you make complex shapes for cheap. Researchers are using recycled sand and aggregates for 3D-printed concrete to create a more sustainable environment (Echeverria-Valiente et al., 2019; Pacheco et al., 2012). The technology makes working on our buildings' architecture more energy-efficient. Furthermore, lots of 3D printing parts can be made using recycled plastics and composites to enable circularity. 3D printing is efficient in a modern, smart and sustainable way overall.

POLICY, REGULATIONS, AND STANDARDS

Global and regional policies promoting circular construction

The circular economy provides an effective framework to address environmental pollution and resource depletion by promoting sustainable production, consumption, and recycling practices. In the construction industry, it encourages the reuse of materials and waste reduction through circular design principles (Attia et al., 2021). European countries are integrating these principles into policies like the European

Green Deal to achieve carbon neutrality in the building sector. Implementation relies on sustainable planning, construction requirements, and public procurement policies that support circular business models (Oliveira et al., 2021). These strategies not only advance environmental goals but also align with economic growth, making circular economy practices a key driver for sustainable development globally.

Figure 15. Circular economy linking forward and reverse logistics towards circularity (Zeng et al., 2022)

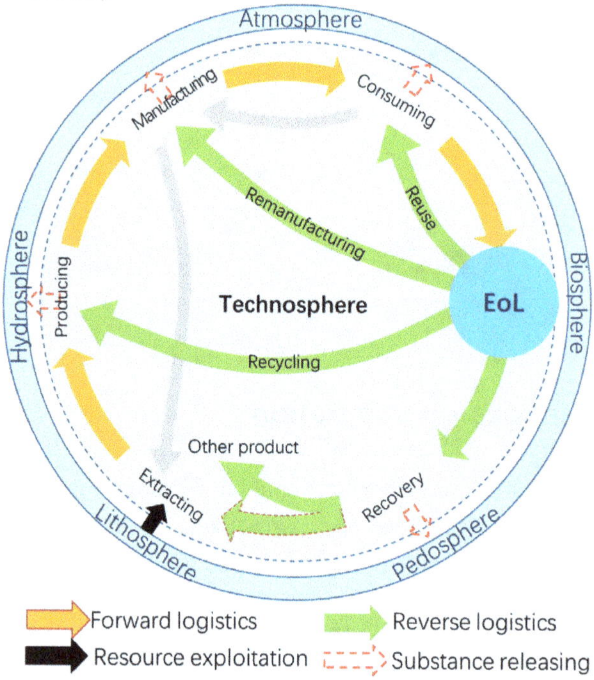

Certification systems (LEED, BREEAM, GRIHA, etc.)

The Indian Green Building Council administers Leadership in Energy and Environmental Design (LEED)-India. It is in line with LEED globally for green building construction (Smith, 2015). In a similar manner, GRIHA was developed by the Ministry of New and Renewable Energy and TERI which is India's national green building rating system (Arya & Sharma, 2022). LEED and GRIHA both promote green building in the construction industry through evaluation of energy efficiency, resource consumption and environment impacts. Comparing the criteria, scales and assessments of both systems. The rating systems play an important role in planning urban development in a sustainable manner (R. P. Singh & Raghubanshi, 2020).

Additionally, they help in framing the government regulations. There are other uses of the rating like they help to combat the issue of rapid urbanization which is being faced in India. India is currently witnessing a rise in the million-plus cities.

INDUSTRY INNOVATIONS AND CASE STUDIES

Global best practices in circular construction

A critical analysis of circular interventions in Europe reveals how upcycling obsolete buildings serves as a catalyst for urban regeneration and sustainable development (Mangialardo & Micelli, 2018). By embedding circular economy principles within the construction sector, significant reductions in waste and improvements in productivity can be achieved, generating substantial economic savings annually (Guerra et al., 2021). Circular strategies encompassing construction, renovation, and deconstruction processes align environmental objectives with economic efficiency, thereby promoting long-term sustainability within urban ecosystems (Nasir et al., 2017). The circular economy concept, rooted in the notion of cyclical ecological systems, emphasizes continuous material reuse, recycling, and recovery, ensuring that economic growth coexists with ecological balance (Ruiter et al., 2021). In Europe, these principles have evolved into a central policy framework and a driving force for innovation in the construction industry, influencing both governmental and corporate strategies toward greener, resource-efficient practices (Leising et al., 2018). More and more stakeholders in policy, business and academia are considering circular economy ideas as influential. Nonetheless, turning circularity theory into practice remains a work in progress that requires more development in methodology, technology, and sustainable business models. In the end, circular interventions in the built environment change the core rules of urban development to create a construction sector that saves resources and regenerates urban areas. Through creativity, collaboration across sectors, and the adoption of environmental, economic and social values, Europe is progressing toward a regenerative and circular future of cities.

Successful public–private partnerships

The UNIDO initiative has been developed by public and private sector partners and academia, NGOs and government authorities. This is one of the important developments in nonprofit multidisciplinary public–private partnerships (PPPs) (Samii et al., 2002). These collaborations aim to utilize the unique strengths of the public sector, private sector, academia, and civil society groups to devise innovative and sustainable solutions to complex global challenges (Carbonara & Pellegrino, 2020).

Regarding PPPs, the need to clear about the objectives, governance structures in the formation stage and sustainability in the post-formation stage is very important. Managed properly, public-private partnerships can be a valuable tool for increasing infrastructure service quality and productivity, stimulating innovation and enhancing the effectiveness of public service delivery (Carbonara & Pellegrino, 2018). They are a hybrid organizational model that combines logics from various institutions to facilitate innovation and joint value creation. Even though PPP has been gaining importance, the existing research doesn't provide us with an understanding of how its specific characteristics affect innovation. PPPs are known for their ability to bring creativity and efficiency, but how do they do this? The rest has not been explored much. Nonetheless, cooperation is an important way to achieve sustainable development through mutually beneficial efforts involving the public and private sectors, as well as civil society.

Startups and technological innovations driving change

Opportunity recognition refers to a complicated and interactive process involving three elements: the firm, knowledge gained through an entrepreneur's experience, and technology (Park, 2005). The founding entrepreneur is critical through his vision and creativity, but most importantly through his ability to recognize a gap in the market that he can exploit. The knowledge and experience of the firm make it possible to measure feasibility and ultimately formulate viable business ideas (Moroni et al., 2015). Technology is an enabler that turns ideas into tangible scalable innovations that can disrupt the market and create customer value. The concept of startup entrepreneurship practices is more oriented to strategic management and innovation to sustain and compete in rapidly changing digital markets (Lewrick & Raeside, 2010). There is a way for small firms and also startups to grow indefinitely. The way is called strategic innovation. Design-driven innovation makes this process more robust by incorporating design management principles into the company's strategic planning, enabling startups to develop a strong innovation-oriented value system and customer-centric approach. Startup organizations must adapt their innovation capabilities for long-term success as they mature (Korper et al., 2020). The transition needs a balance of technology with service and value co-creation and stresses the importance of service design in improving user experience and sustained competitive advantage in technology sectors.

9. CHALLENGES AND BARRIERS

The barriers to adopting smart building concepts and technologies in social housing projects in Brazil are multiple and interrelated. This was evidenced by interviews with specialists in smart buildings and social housing (D'Oca et al., 2018). There are many hurdles covering technical, economic, organizational, social, and regulatory aspects that hinder the speed and scale of technology diffusion in the world. A major challenge is lack of awareness and training amongst stakeholders, like engineers, builders, and residents. Power failure is a problem since electronic gadgets cannot be used. Subsequently, of the batteries and their price. Most training models offer limited support for new digital competencies, which generates skill gaps for effective adoption (Mazurkiewicz & Poteralska, 2017). Challenges due to costs and scale effects are the other. Smart infrastructure requires a lot of money. For example, sensors, automation systems, data management technology, etc. costs money that low-income housing cannot afford. Scalability issues are also caused by poor digital infrastructure and funding models that are not well developed, making it difficult to replicate small successes at scale in housing. Aside from that, organizational and system-related constraints also exist. For instance, there are issues such is weak coordination among public and private stakeholders. Also, limited government incentives and regulation issues slow down innovation diffusion (G. E. Ali & Magalhaes, 2008). Also, technological compatibility, cybersecurity risk and data management cost-related issues impede widespread adoption. To advance smart building integration in Brazilian social housing, these multifaceted barriers must be crossed by targeted policy interventions, enhanced capacity building, affordable technological solutions and cross-sectoral collaboration to enable inclusive and scalable innovation.

Future Directions

Digitalization, technological innovation, and collaborative frameworks will speed up the transformation of construction industry towards circularity and sustainability (Sepasgozar et al., 2023). The development of BIM, DTs, and blockchain technology is at the forefront of the evolution of the building sector and how buildings will be designed, constructed, operated, and maintained. BIM is the digital likeness of the physical and functional features of a facility (Baghalzadeh Shishehgarkhaneh et al., 2022). BIM will rely on consistent updates of all facility data in all phases. The bid data will enhance coordination, efficiency, and decision-making across the facility lifecycle. Integration with the blockchain guarantees integrity and tracking as well as the Decentralized Digital Twin (DT) simulating conditions and enabling predictive maintenance, as well as optimizing performance, and managing the lifecycle.

These technologies together promote smarter, more resource-efficient construction which enhances circular economy practices (Teisserenc & Sepasgozar, 2021). The ongoing circular construction is a result of the use of AI and robotics. They offer solutions to material flows and ease reverse logistics. It also helps predictive analysis for waste reduction and recovery. Industry 4.0 technologies such as artificial intelligence, IoT (internet of things), advanced data analytics, and collaborative robotics enable construction and waste management to be more automated, energy efficient and precise (Wilson et al., 2022). Things like ReWaste 4.0 show how digital networking and robotics can redefine recycling acts. The adaptability of the circular economy and Industry 4.0 frameworks can jointly assist in achieving a sustainable ecosystem in construction. In order to make this change happen, the public, private, academic, and nonprofit actors must work together. Collaboration can drive innovation and align economic and environmental goals by leveraging diverse expertise and resources from multiple stakeholders, which helps overcome common systemic barriers. Collaboration across different sectors - involving businesses and governments, businesses and non-governmental organizations, or all three sectors - is crucial to scaling up sustainable business models, increasing absorptive capacity and promoting social inclusivity in urban regeneration (Babiak & Thibault, 2009). They also promote research and knowledge sharing for different actors, including industry, policy and community. Ultimately, the shift to net-zero construction can only take place with joint routes that ensure material efficiency, decarbonized production technologies, and regulatory reform. Key strategies include cutting down the demand for carbon-intensive materials. In-building recycling, improve quality via purposeful design, eliminate subsidies for energy, and introduce mechanisms to price carbon. If the best practices of cost control are duly applied, net-zero energy building (ZEB) design sustainability will be mainstreamed. For this reason, Uniting these pathways creates a synergistic framework for a climate-resilient, circular, and digital construction future which will help in achieving that.

Figure 16. The decentralized digital twin cycle (Teisserenc & Sepasgozar, 2021).

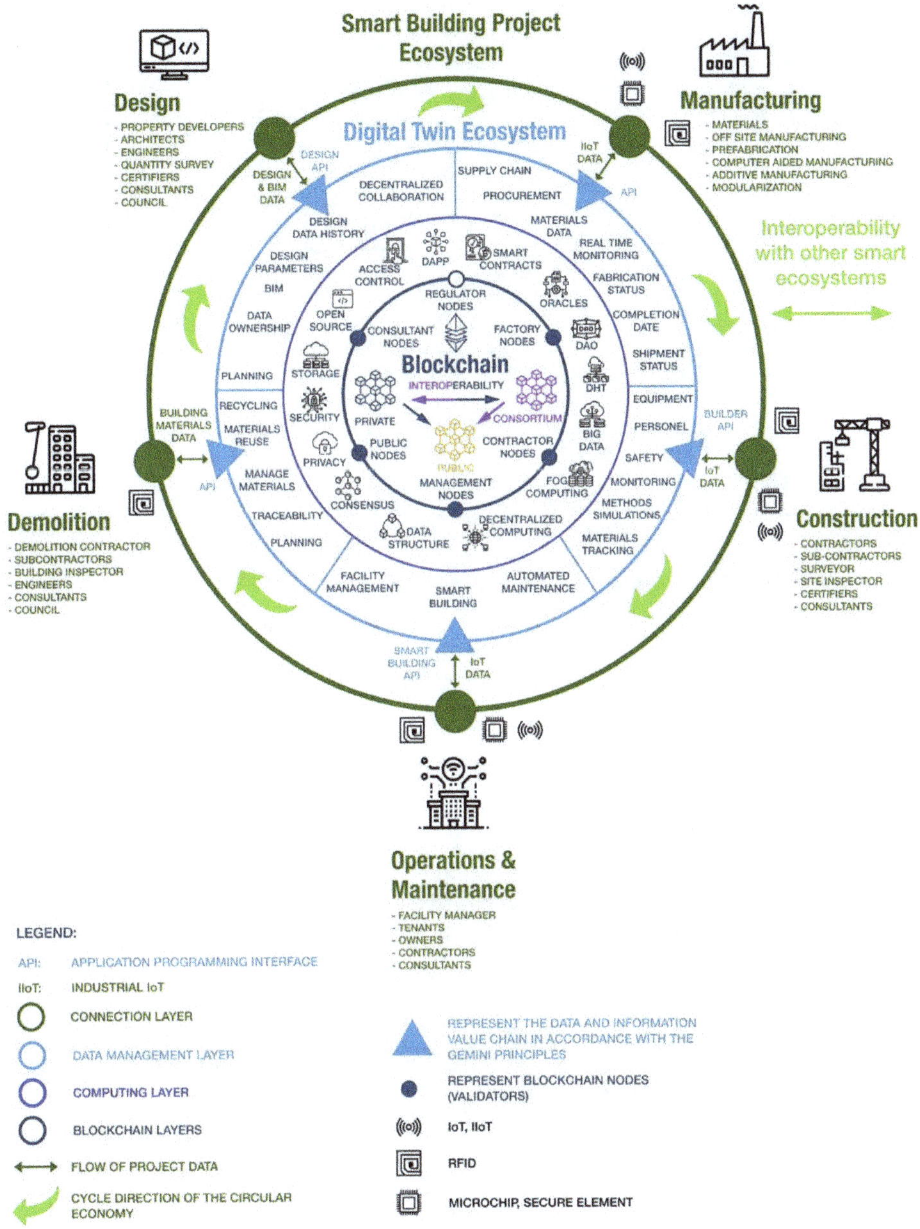

Figure 17. Evolution of BIM to digital twins in the built environment

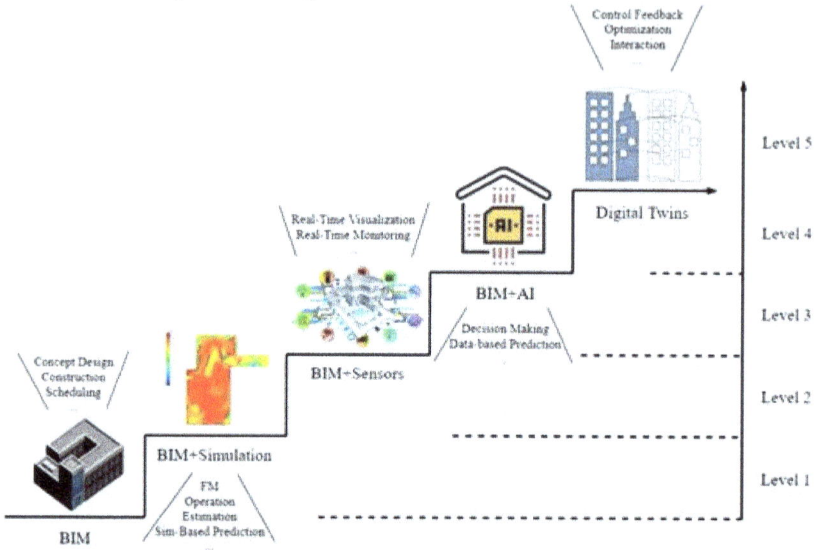

FIG. 1: Evolution of BIM to Digital Twins in the Built Environment

Figure 18. Human centric manufacturing: Industry 5.0 Intersection of research areas

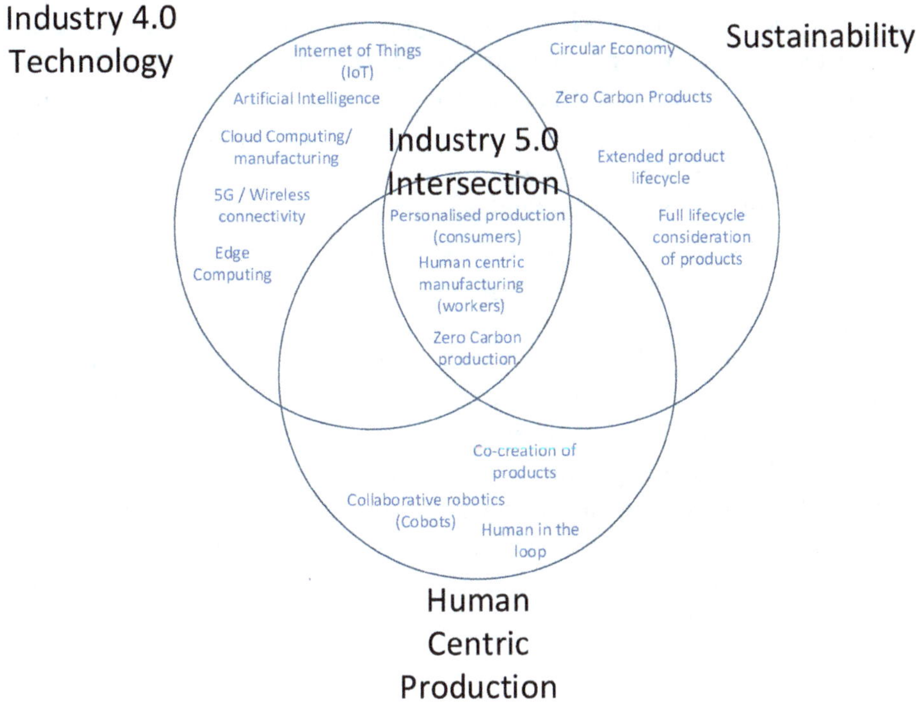

CONCLUSION

The construction business is at an important crossroad. It must provide for global infrastructure needs while also having to go green in a big way. As one of the most significant consumers of natural resources, as well as one of the highest producers of wastes and emissions, it is no longer a case of 'nice to have' but essential for the industry to shift from a linear 'take–make–dispose' model to a circular economy framework. This chapter stresses that the building's circular economy use of reuse, recycle and material efficient can be a regenerative, resource-conscious, and resilient system. Circular construction focuses on creating things that last longer, adapt to changes, and have less harm on the environment. Using the concepts of adaptive re-use of abandoned and obsolete buildings, use of recycled materials and renewable materials, modular design etc helps to increase life cycle of built assets, which helps in the conservation of natural resources. Closed-loop supply chains enable materials to retain value after their first use, creating a more sustainable flow of materials. If we change they way we source, use or recover materials, it is possible to generate your waste and carbon emissions in a more effective and productive way. Using new technologies like BIM, Digital twins, or blockchain will greatly improve the transparency, material tracking, or life cycle management. These tools help architects, engineers, contractors, and policymakers work together and make decisions based on data while promoting evidence-based sustainability. The case studies from various parts of the world prove that circular strategies can be both green and profitable with design thinking, technology, and policy support. But to unlock the full potential of circular construction, we need to make a systemic change. To encourage resource-efficient design and advance a recycling infrastructure, a policy intervention is necessary. By working together across all sectors, we can help expand these practices all over the world. To ensure viable circular construction solutions that are affordable and replicable across a diversity of contexts, research and innovation must continue to lift existing barriers. In short, the chapter concludes that the circular economy presents a way towards a more sustainable construction industry which meets economic growth needs while caring for the environment. To shift from being a major burden on the environment to one that contributes to sustainable development, the construction industry must design projects and buildings according to circular principles. This will enable us to achieve a regenerative and climate-resilient built environment.

REFERENCES

Ajayi, S. O., Oyedele, L. O., Bilal, M., Akinade, O. O., Alaka, H. A., & Owolabi, H. A. (2017). Critical management practices influencing on-site waste minimization in construction projects. *Waste Management, 59*, 330–339. DOI: 10.1016/j.wasman.2016.10.040 PMID: 27839690

Ali, G. E., & Magalhaes, R. (2008). Barriers to implementing e-learning: A Kuwaiti case study. *International Journal of Training and Development, 12*(1), 36–53. DOI: 10.1111/j.1468-2419.2007.00294.x

Ali, M., Issayev, G., Shehab, E., & Sarfraz, S. (2022). A critical review of 3D printing and digital manufacturing in construction engineering. *Rapid Prototyping Journal, 28*(7), 1312–1324. DOI: 10.1108/RPJ-07-2021-0160

Ameri, F., & Dutta, D. (2005). Product Lifecycle Management: Closing the Knowledge Loops. *Computer-Aided Design and Applications, 2*(5), 577–590. DOI: 10.1080/16864360.2005.10738322

Arfa, F. H., Zijlstra, H., Lubelli, B., & Quist, W. (2022). Adaptive Reuse of Heritage Buildings: From a Literature Review to a Model of Practice. *The Historic Environment, 13*(2), 148–170. DOI: 10.1080/17567505.2022.2058551

Arya, A., & Sharma, R. L. (2022). Strategies for green building rating in India: A comparison of LEED and GRIHA criteria. *Materials Today: Proceedings, 57*, 2311–2316. DOI: 10.1016/j.matpr.2022.01.100

Attia, S., Santos, M. C., Al-Obaidy, M., & Baskar, M. (2021). Leadership of EU member States in building carbon footprint regulations and their role in promoting circular building design. *IOP Conference Series. Earth and Environmental Science, 855*(1), 012023. DOI: 10.1088/1755-1315/855/1/012023

Babiak, K., & Thibault, L. (2009). Challenges in Multiple Cross-Sector Partnerships. *Nonprofit and Voluntary Sector Quarterly, 38*(1), 117–143. DOI: 10.1177/0899764008316054

Baghalzadeh Shishehgarkhaneh, M., Keivani, A., Moehler, R. C., Jelodari, N., & Roshdi Laleh, S. (2022). Internet of Things (IoT), Building Information Modeling (BIM), and Digital Twin (DT) in Construction Industry: A Review, Bibliometric, and Network Analysis. *Buildings (Basel, Switzerland), 12*(10), 1503. DOI: 10.3390/buildings12101503

Bain, A., Shenoy, M., Ashton, W., & Chertow, M. (2010). Industrial symbiosis and waste recovery in an Indian industrial area. *Resources, Conservation and Recycling*, *54*(12), 1278–1287. DOI: 10.1016/j.resconrec.2010.04.007

Bakker, C., Wang, F., Huisman, J., & Den Hollander, M. (2014). Products that go round: Exploring product life extension through design. *Journal of Cleaner Production*, *69*, 10–16. DOI: 10.1016/j.jclepro.2014.01.028

Bakker, C. A., Mugge, R., Boks, C., & Oguchi, M. (2021). Understanding and managing product lifetimes in support of a circular economy. *Journal of Cleaner Production*, *279*, 123764. DOI: 10.1016/j.jclepro.2020.123764

Bao, Z., Lu, W., Chi, B., Yuan, H., & Hao, J. (2019). Procurement innovation for a circular economy of construction and demolition waste: Lessons learnt from Suzhou, China. *Waste Management*, *99*, 12–21. DOI: 10.1016/j.wasman.2019.08.031 PMID: 31454595

Benachio, G. L. F., Freitas, M. D. C. D., & Tavares, S. F. (2020). Circular economy in the construction industry: A systematic literature review. *Journal of Cleaner Production*, *260*, 121046. DOI: 10.1016/j.jclepro.2020.121046

Benachio, G. L. F., Freitas, M. D. C. D., & Tavares, S. F. (2021). Interactions between Lean Construction Principles and Circular Economy Practices for the Construction Industry. *Journal of Construction Engineering and Management*, *147*(7), 04021068. DOI: 10.1061/(ASCE)CO.1943-7862.0002082

Bertino, G., Kisser, J., Zeilinger, J., Langergraber, G., Fischer, T., & Österreicher, D. (2021). Fundamentals of Building Deconstruction as a Circular Economy Strategy for the Reuse of Construction Materials. *Applied Sciences (Basel, Switzerland)*, *11*(3), 939. DOI: 10.3390/app11030939

Bramston, D., & Maycroft, N. (2014). Designing with Waste. In *Materials Experience* (pp. 123–133). Elsevier., DOI: 10.1016/B978-0-08-099359-1.00009-6

Bullen, P. A. (2007). Adaptive reuse and sustainability of commercial buildings. *Facilities*, *25*(1/2), 20–31. DOI: 10.1108/02632770710716911

Busch, J., Dawson, D., & Roelich, K. (2017). Closing the low-carbon material loop using a dynamic whole system approach. *Journal of Cleaner Production*, *149*, 751–761. DOI: 10.1016/j.jclepro.2017.02.166

Carbonara, N., & Pellegrino, R. (2018). Fostering innovation in public procurement through public private partnerships. *Journal of Public Procurement*, *18*(3), 257–280. DOI: 10.1108/JOPP-09-2018-016

Carbonara, N., & Pellegrino, R. (2020). The role of public private partnerships in fostering innovation. *Construction Management and Economics*, *38*(2), 140–156. DOI: 10.1080/01446193.2019.1610184

Carcassi, O. B., Minotti, P., Habert, G., Paoletti, I., Claude, S., & Pittau, F. (2022). Carbon Footprint Assessment of a Novel Bio-Based Composite for Building Insulation. *Sustainability (Basel)*, *14*(3), 1384. DOI: 10.3390/su14031384

Chazdon, R. L., & Guariguata, M. R. (2016). Natural regeneration as a tool for large-scale forest restoration in the tropics: Prospects and challenges. *Biotropica*, *48*(6), 716–730. DOI: 10.1111/btp.12381

D'Oca, S., Ferrante, A., Ferrer, C., Pernetti, R., Gralka, A., Sebastian, R., & Op 't Veld, P. (2018). Technical, Financial, and Social Barriers and Challenges in Deep Building Renovation: Integration of Lessons Learned from the H2020 Cluster Projects. *Buildings (Basel, Switzerland)*, *8*(12), 174. DOI: 10.3390/buildings8120174

Didenko, N. I., Klochkov, Y. S., & Skripnuk, D. F. (2018). Ecological Criteria for Comparing Linear and Circular Economies. *Resources*, *7*(3), 48. DOI: 10.3390/resources7030048

Disfani, M. M., Arulrajah, A., Bo, M. W., & Sivakugan, N. (2012). Environmental risks of using recycled crushed glass in road applications. *Journal of Cleaner Production*, *20*(1), 170–179. DOI: 10.1016/j.jclepro.2011.07.020

Echeverria-Valiente, E., Garcia-Alvarado, R., Celis-D'Amico, F., & Saelzer-Fuica, G. (2019). Integrated design experiences for energy-efficient housing in Chile. *Construction Innovation*, *19*(2), 236–255. DOI: 10.1108/CI-05-2017-0042

Fuller, D. A., & Ottman, J. A. (2004). Moderating unintended pollution: The role of sustainable product design. *Journal of Business Research*, *57*(11), 1231–1238. DOI: 10.1016/S0148-2963(02)00446-0

Gibbons, L. V. (2020). Regenerative—The New Sustainable? *Sustainability (Basel)*, *12*(13), 5483. DOI: 10.3390/su12135483

Guerra, B. C., Shahi, S., Mollaei, A., Skaf, N., Weber, O., Leite, F., & Haas, C. (2021). Circular economy applications in the construction industry: A global scan of trends and opportunities. *Journal of Cleaner Production*, *324*, 129125. DOI: 10.1016/j.jclepro.2021.129125

Hagnell, M. K., & Åkermo, M. (2019). The economic and mechanical potential of closed loop material usage and recycling of fibre-reinforced composite materials. *Journal of Cleaner Production*, *223*, 957–968. DOI: 10.1016/j.jclepro.2019.03.156

Han, Y., Yang, Z., Ding, T., & Xiao, J. (2021). Environmental and economic assessment on 3D printed buildings with recycled concrete. *Journal of Cleaner Production*, *278*, 123884. DOI: 10.1016/j.jclepro.2020.123884

Hendrickson, C., & Horvath, A. (2000). Resource Use and Environmental Emissions of U.S. Construction Sectors. *Journal of Construction Engineering and Management*, *126*(1), 38–44. DOI: 10.1061/(ASCE)0733-9364(2000)126:1(38)

Jaillon, L., & Poon, C. S. (2008). Sustainable construction aspects of using prefabrication in dense urban environment: A Hong Kong case study. *Construction Management and Economics*, *26*(9), 953–966. DOI: 10.1080/01446190802259043

Jiang, Y., Zhao, D., Wang, D., & Xing, Y. (2019). Sustainable Performance of Buildings through Modular Prefabrication in the Construction Phase: A Comparative Study. *Sustainability (Basel)*, *11*(20), 5658. DOI: 10.3390/su11205658

Jiménez-Rivero, A., De Guzmán-Báez, A., & García-Navarro, J. (2017). Enhanced On-Site Waste Management of Plasterboard in Construction Works: A Case Study in Spain. *Sustainability (Basel)*, *9*(3), 450. DOI: 10.3390/su9030450

Kim, S., & Kim, S.-A. (2020). Framework for Designing Sustainable Structures through Steel Beam Reuse. *Sustainability (Basel)*, *12*(22), 9494. DOI: 10.3390/su12229494

Korper, A. K., Patrício, L., Holmlid, S., & Witell, L. (2020). Service design as an innovation approach in technology startups: A longitudinal multiple case study. *Creativity and Innovation Management*, *29*(2), 303–323. DOI: 10.1111/caim.12383

Krause, K., & Hafner, A. (2022). Resource Efficiency in the Construction Sector: Material Intensities of Residential Buildings—A German Case Study. *Energies*, *15*(16), 5825. DOI: 10.3390/en15165825

Langston, C., Wong, F. K. W., Hui, E. C. M., & Shen, L.-Y. (2008). Strategic assessment of building adaptive reuse opportunities in Hong Kong. *Building and Environment*, *43*(10), 1709–1718. DOI: 10.1016/j.buildenv.2007.10.017

Lehmann, S. (2011). Optimizing Urban Material Flows and Waste Streams in Urban Development through Principles of Zero Waste and Sustainable Consumption. *Sustainability (Basel)*, *3*(1), 155–183. DOI: 10.3390/su3010155

Leising, E., Quist, J., & Bocken, N. (2018). Circular Economy in the building sector: Three cases and a collaboration tool. *Journal of Cleaner Production*, *176*, 976–989. DOI: 10.1016/j.jclepro.2017.12.010

Lewrick, M., & Raeside, R. (2010). Transformation and change process in innovation models: Start-up and mature companies. *International Journal of Business Innovation and Research, 4*(6), 515. DOI: 10.1504/IJBIR.2010.035711

Lockrey, S., Nguyen, H., Crossin, E., & Verghese, K. (2016). Recycling the construction and demolition waste in Vietnam: Opportunities and challenges in practice. *Journal of Cleaner Production, 133*, 757–766. DOI: 10.1016/j.jclepro.2016.05.175

Ma, M., Tam, V. W. Y., Le, K. N., & Li, W. (2020). Challenges in current construction and demolition waste recycling: A China study. *Waste Management, 118*, 610–625. DOI: 10.1016/j.wasman.2020.09.030 PMID: 33010692

Mangialardo, A., & Micelli, E. (2018). Rethinking the Construction Industry Under the Circular Economy: Principles and Case Studies. In Bisello, A., Vettorato, D., Laconte, P., & Costa, S. (Eds.), *Smart and Sustainable Planning for Cities and Regions* (pp. 333–344). Springer International Publishing., DOI: 10.1007/978-3-319-75774-2_23

Mazurkiewicz, A., & Poteralska, B. (2017). Technology Transfer Barriers and Challenges Faced by R&D Organisations. *Procedia Engineering, 182*, 457–465. DOI: 10.1016/j.proeng.2017.03.134

Mohammadinia, A., Wong, Y. C., Arulrajah, A., & Horpibulsuk, S. (2019). Strength evaluation of utilizing recycled plastic waste and recycled crushed glass in concrete footpaths. *Construction & Building Materials, 197*, 489–496. DOI: 10.1016/j.conbuildmat.2018.11.192

Molavi, J., & Barral, D. L. (2016). A Construction Procurement Method to Achieve Sustainability in Modular Construction. *Procedia Engineering, 145*, 1362–1369. DOI: 10.1016/j.proeng.2016.04.201

Moroni, I., Arruda, A., & Araujo, K. (2015). The Design and Technological Innovation: How to Understand the Growth of Startups Companies in Competitive Business Environment. *Procedia Manufacturing, 3*, 2199–2204. DOI: 10.1016/j.promfg.2015.07.361

Müller, E. (2020). Regenerative development as natural solution for sustainability. In Sarmiento, F. O., & Frolich, L. M. (Eds.), *The Elgar Companion to Geography, Transdisciplinarity and Sustainability*. Edward Elgar Publishing., DOI: 10.4337/9781786430106.00021

Mulrow, J. S., Derrible, S., Ashton, W. S., & Chopra, S. S. (2017). Industrial Symbiosis at the Facility Scale. *Journal of Industrial Ecology, 21*(3), 559–571. DOI: 10.1111/jiec.12592

Nasir, M. H. A., Genovese, A., Acquaye, A. A., Koh, S. C. L., & Yamoah, F. (2017). Comparing linear and circular supply chains: A case study from the construction industry. *International Journal of Production Economics, 183*, 443–457. DOI: 10.1016/j.ijpe.2016.06.008

Oliveira, M. D. P. S. L., De Oliveira, E. A., & Fonseca, A. M. (2021). Strategies to promote circular economy in the management of construction and demolition waste at the regional level: A case study in Manaus, Brazil. *Clean Technologies and Environmental Policy, 23*(9), 2713–2725. DOI: 10.1007/s10098-021-02197-7

Ormondroyd, G. A., Spear, M. J., & Skinner, C. (2016). The Opportunities and Challenges for Re-use and Recycling of Timber and Wood Products Within the Construction Sector. In Kutnar, A., & Muthu, S. S. (Eds.), *Environmental Impacts of Traditional and Innovative Forest-based Bioproducts* (pp. 45–103). Springer Singapore., DOI: 10.1007/978-981-10-0655-5_3

Pacheco, R., Ordóñez, J., & Martínez, G. (2012). Energy efficient design of building: A review. *Renewable & Sustainable Energy Reviews, 16*(6), 3559–3573. DOI: 10.1016/j.rser.2012.03.045

Park, J. S. (2005). Opportunity recognition and product innovation in entrepreneurial hi-tech start-ups: A new perspective and supporting case study. *Technovation, 25*(7), 739–752. DOI: 10.1016/j.technovation.2004.01.006

Piccardo, C., & Hughes, M. (2022). Design strategies to increase the reuse of wood materials in buildings: Lessons from architectural practice. *Journal of Cleaner Production, 368*, 133083. DOI: 10.1016/j.jclepro.2022.133083

Poon, C. S., Yu, A. T. W., & Ng, L. H. (2001). On-site sorting of construction and demolition waste in Hong Kong. *Resources, Conservation and Recycling, 32*(2), 157–172. DOI: 10.1016/S0921-3449(01)00052-0

Rajala, R., Hakanen, E., Mattila, J., Seppälä, T., & Westerlund, M. (2018). How Do Intelligent Goods Shape Closed-Loop Systems? *California Management Review, 60*(3), 20–44. DOI: 10.1177/0008125618759685

Rani, J. (2025). Cultural and Social Norms Supporting Sustainability. In Mokdad, M. (Ed.), *Social System Reforms to Achieve Global Sustainability* (pp. 43–78). IGI Global., DOI: 10.4018/979-8-3373-1280-4.ch003

Rani, J., Guru, R., & Santhanam, S. (2025). Case Studies on the Use of AI in Literary Research and Academic Studies. In Asrifan, A. (Ed.), (pp. 1–36). Advances in Computational Intelligence and Robotics. IGI Global Scientific Publishing., DOI: 10.4018/979-8-3373-6330-1.ch001

Rani, J., Guru, R., Singh, J., & Fangueiro, R. School of Engineering, University of Minho, Guimaraes, Portugal, Santhanam, S., & Department of Textile Technology, Park College of Engineering and Technology, Coimbatore, Tamilnadu, India. (2024). Applying Giloy Stems Dye to Silk and Lyocell Fabrics to Improve Their Antimicrobial and Fastness Properties Using Bio-Mordant. *Textile & Leather Review, 7*, 1197–1215. DOI: 10.31881/TLR.2024.117

Rani, J., Guru, R., Singh, J., & Santhanam, S. (2024). Eco-Dyeing and Functional Finishing of Cotton Fabric Using a Natural Colour Derived From Lotus Seed: Enhanced Fastness Properties with Chitosan. *Textile & Leather Review, 7*, 1039–1060. DOI: 10.31881/TLR.2024.099

Rani, J., Santhanam, S., Guru, R., & Seidu, R. K. (2025). Sustainable Finishing Processes: Enhancing Performance with Minimal Impact. In Jiang, S.-X., Seidu, R. K., & Tawiah, B. (Eds.), *Advances in Textile Materials and Processing Techniques for Sustainability* (pp. 161–199). Springer Nature Singapore., DOI: 10.1007/978-981-95-0469-5_6

Reed, B. (2007). Shifting from 'sustainability' to regeneration. *Building Research and Information, 35*(6), 674–680. DOI: 10.1080/09613210701475753

Ruiter, H., De Feijter, F., & Wagensveld, K. (2021). Management Control and Business Model Innovation in the Context of a Circular Economy in the Dutch Construction Industry. *Sustainability (Basel), 14*(1), 366. DOI: 10.3390/su14010366

Samii, R., Van Wassenhove, L. N., & Bhattacharya, S. (2002). An Innovative Public–Private Partnership: New Approach to Development. *World Development, 30*(6), 991–1008. DOI: 10.1016/S0305-750X(02)00015-3

Seidu, R. K., Fobiri, G. K., Buami, E. K., Rani, J., Tawiah, B., & Jiang, S. (2025). Energy Efficiency in Textile Production: Strategies and Solutions. In Jiang, S.-X., Seidu, R. K., & Tawiah, B. (Eds.), *Advances in Textile Materials and Processing Techniques for Sustainability* (pp. 305–325). Springer Nature Singapore., DOI: 10.1007/978-981-95-0469-5_11

Sepasgozar, S., Khan, A., Smith, K., Romero, J., Shen, X., Shirowzhan, S., Li, H., & Tahmasebinia, F. (2023). BIM and Digital Twin for Developing Convergence Technologies as Future of Digital Construction. *Buildings (Basel, Switzerland), 13*(2), 441. DOI: 10.3390/buildings13020441

Shahbazi, S., Wiktorsson, M., Kurdve, M., Jönsson, C., & Bjelkemyr, M. (2016). Material efficiency in manufacturing: Swedish evidence on potential, barriers and strategies. *Journal of Cleaner Production, 127*, 438–450. DOI: 10.1016/j.jclepro.2016.03.143

Sieffert, Y., Huygen, J. M., & Daudon, D. (2014). Sustainable construction with repurposed materials in the context of a civil engineering–architecture collaboration. *Journal of Cleaner Production*, *67*, 125–138. DOI: 10.1016/j.jclepro.2013.12.018

Singh, N., & Rani, J. (2024). Experimental Retail: Conceptualizing Immersive and Memorable Shopping Journeys. In Rajput, S., Malik, F. A., Mahajan, S., & Javed, A. (Eds.), (pp. 31–50). Advances in Logistics, Operations, and Management Science. IGI Global., DOI: 10.4018/979-8-3693-7620-1.ch003

Singh, R. P., & Raghubanshi, A. S. (2020). 'Green building' movement in India: Study on institutional support and regulatory support. In *Urban Ecology* (pp. 435–455). Elsevier., DOI: 10.1016/B978-0-12-820730-7.00022-7

Smith, R. M. (2015). "Green" building in India: A comparative and spatial analysis of the LEED-India and GRIHA rating systems. *Asian Geographer*, *32*(2), 73–84. DOI: 10.1080/10225706.2015.1020065

Sudarsan, R., Fenves, S. J., Sriram, R. D., & Wang, F. (2005). A product information modeling framework for product lifecycle management. *Computer Aided Design*, *37*(13), 1399–1411. DOI: 10.1016/j.cad.2005.02.010

Tam, V. W. Y., & Hao, J. J. L. (2019). Adaptive reuse in sustainable development. *International Journal of Construction Management*, *19*(6), 509–521. DOI: 10.1080/15623599.2018.1459154

Teisserenc, B., & Sepasgozar, S. (2021). Adoption of Blockchain Technology through Digital Twins in the Construction Industry 4.0: A PESTELS Approach. *Buildings (Basel, Switzerland)*, *11*(12), 670. DOI: 10.3390/buildings11120670

Vasilca, I.-S., Nen, M., Chivu, O., Radu, V., Simion, C.-P., & Marinescu, N. (2021). The Management of Environmental Resources in the Construction Sector: An Empirical Model. *Energies*, *14*(9), 2489. DOI: 10.3390/en14092489

Webster, M. D. (2007). Structural Design for Adaptability and Deconstruction: A Strategy for Closing the Materials Loop and Increasing Building Value. *New Horizons and Better Practices*, 1–6. DOI: 10.1061/40946(248)27

Wilson, M., Paschen, J., & Pitt, L. (2022). The circular economy meets artificial intelligence (AI): Understanding the opportunities of AI for reverse logistics. *Management of Environmental Quality*, *33*(1), 9–25. DOI: 10.1108/MEQ-10-2020-0222

Winkler, H. (2011). Closed-loop production systems—A sustainable supply chain approach. *CIRP Journal of Manufacturing Science and Technology*, *4*(3), 243–246. DOI: 10.1016/j.cirpj.2011.05.001

Wood, L. C., Duong, L. N. K., & Wang, J. X. (2022). Business Process Improvement for Sustainable Technologies Investments in Construction: A Configurational Approach. *Sustainability (Basel), 14*(9), 5697. DOI: 10.3390/su14095697

Yadav, D. P., Patel, D. N., & Morkos, B. W. (2018). Development of Product Recyclability Index Utilizing Design for Assembly and Disassembly Principles. *Journal of Manufacturing Science and Engineering, 140*(3), 031015. DOI: 10.1115/1.4038515

Yuan, H., Lu, W., & Jianli Hao, J. (2013). The evolution of construction waste sorting on-site. *Renewable & Sustainable Energy Reviews, 20*, 483–490. DOI: 10.1016/j.rser.2012.12.012

Zeng, X., Ogunseitan, O. A., Nakamura, S., Suh, S., Kral, U., Li, J., & Geng, Y. (2022). Reshaping global policies for circular economy. *Circular Economy, 1*(1), 100003. DOI: 10.1016/j.cec.2022.100003

Chapter 4

Determination of Concrete Strength Used in Building Performance Analyses by Destructive and Non-Destructive Methods

Osman Uyanik
http://orcid.org/0000-0002-7089-4340
Faculty of Engineering and Natural Sciences, Süleyman Demirel University, Turkey

Nevbahar Ekin
http://orcid.org/0000-0002-7089-4340
Suleyman Demirel University, Turkey

Nurten Ayten Uyanik
http://orcid.org/0000-0002-7089-4340
Vocational School of Technical Sciences, Isparta University of Applied Sciences, Turkey

Ziya Öncü
http://orcid.org/0000-0002-7806-3728
Isparta University of Applied Sciences, Turkey

DOI: 10.4018/979-8-3373-6167-3.ch004

ABSTRACT

Assessing the structural performance of concrete is vital for monitoring structures and planning interventions, especially after earthquakes. Accurately determining concrete strength ensures safety and cost-effective maintenance. Mechanical properties such as compressive strength, elasticity, and porosity are measured by destructive, semi-destructive, and non-destructive testing (NDT). Destructive testing provides reliable results but can damage the structure and is costly. NDT methods are faster and less invasive but provide estimated values. To improve accuracy, they must be calibrated with selective destructive, semi-destructive or non-destructive testing. Combined assessment methods, such as the SonReb (ultrasonic and rebound hammer) method, which combines both techniques, increase reliability. Integrating destructive and NDT approaches allows for safer, more economical assessments and better-informed engineering decisions.

INTRODUCTION

Concrete is a fundamental building material in modern construction, playing a critical role in structural durability. Over time, concrete strength can deteriorate due to environmental exposure, usage conditions, and production quality. Accurate strength assessment is especially essential when evaluating existing buildings or preserving historic structures. Compressive strength is a key parameter in assessing the performance of reinforced concrete elements (Ekin, 2025a; 2025b). The determination of concrete strength involves various methods, selected based on factors such as the structure's age, environmental conditions, and current damage state. These methods are generally classified into three categories depending on the potential damage they cause: destructive, semi-destructive, and non-destructive techniques (Rucka, 2020; Azevedo et al., 2023).

Non-destructive testing (NDT) has become a cornerstone of modern civil engineering due to its ability to ensure structural safety, facilitate pre-damage monitoring, and enable early detection of issues without compromising the integrity of the structure. The systematic use of NDT methods supports the construction and maintenance of safe, durable, and sustainable structures. While destructive testing (DT) methods, such as those used to measure compressive strength, tensile strength, yield limit, and ductility, provide direct and reliable data (Uyanık, 2012; Gomez-Cardenas, 2015), they damage the physical structure and thus are often less favored for in-service assessments. Conversely, NDT techniques evaluate mechanical, acoustic, chemical, electrical, magnetic, and physical properties without causing harm (Stanley, 1995; Helal et al., 2015), making them particularly valuable for continuous monitoring of concrete structures throughout their service life and for early risk identification. In recent years, the demand for NDT methods has grown alongside the importance of damage detection and preventive maintenance strategies in building systems. In this context, these developed methods provide more sensitive and reliable results and offer significant advantages in terms of cost-effectiveness (Sabbağ and Uyanık, 2017; Hola and Runkiewicz, 2018; Wang et al., 2020). Many studies show that these methods play a vital role in the long-term health of structures. The success of non-destructive testing depends on several technical parameters such as the penetration depth of the method used, resolution capacity (vertical and horizontal), signal-to-noise ratio, contrast difference between physical properties, and preliminary information about the structure (McCann and Forde, 2001; Helal et al., 2015). The more these parameters are optimized, the more accurate the data obtained from the test. NDT methods are widely used in civil engineering to protect the structural integrity of structures before, during, and after assembly (Uyanık, 2014; Sabbağ, 2016; Wang et al., 2020). The most commonly used NDT techniques today include the micro-tremor method, ground-penetrating radar (GPR), resistivity method, and ultrasonic tests. While the microtremor method analyzes the resonance risks of structures with natural vibration periods, ground-penetrating radar is effectively used to determine reinforcement status and foundation structures in concrete (Ateş and Uyanık, (2019; 2023; 2025); Ateş, 2020; Uyanık et al., 2024). The resistivity method helps determine the corrosion status of reinforcement, and ultrasonic tests are used to assess structural parameters such as concrete homogeneity, crack presence, and elastic modulus (Raj and Venkatraman, 2013; Kibar and Öztürk, 2015; Sabbağ and Uyanık, (2016; 2017; 2018); Ekin and Uyanık, (2021a; 2021b); Ekin, (2021a; 2021b; 2025a; 2025b; 2025c)). These methods are generally classified under two main categories: static and dynamic methods. Static methods evaluate the structure under a static load, while dynamic methods analyze the structure under moving loads, such as seismic waves (Ekin, 2020). Ultrasonic methods, particularly by analyzing the effects of dynamic loads like earthquakes on reinforced concrete structures, enable the determination of wave propagation velocities and the elastic properties of concrete. This provides important data on the strength and safety of structures (IAEA, 1999; Ercan, 2001; Uyanık et al., 2011; 2012a; 2015). Non-destructive testing (NDT) techniques extend beyond advanced measurement systems to include methods such as visual inspection, liquid penetration tests, vibration monitoring, surface hardness measurements, and magnetic particle tests, all of which help monitor the overall health of structures (Wang et al., 2020). These approaches enable engineers to assess the durability and safety of structures over time, facilitating proactive maintenance strategies. This section of the book explores destructive, semi-destructive, and non-destructive methods for determining concrete strength (a critical parameter in structural design) and illustrates their application through practical examples.

BACKGROUND

Concrete is one of the most widely used composite building materials in modern civil engineering, typically produced by mixing fine and coarse aggregates with cement and, when necessary, various chemical additives (Samson et al., 2014; Arooj and Yadav, 2017). The proportions of these components play a crucial role in determining the final performance of the concrete. Among its properties, compressive strength stands out as the most critical from an engineering perspective. This parameter significantly influences the load-bearing capacity, service life, and overall structural safety of concrete. Methods for determining concrete strength are generally classified into three categories: destructive testing (DT), semi-destructive testing, and non-destructive testing (NDT). While destructive methods provide direct measurement by compromising the physical integrity of the concrete, semi-destructive techniques cause minimal damage, typically affecting only microcracks. Non-destructive methods, on the other hand, offer indirect strength estimates without inflicting any structural damage, enabling continuous monitoring and preservation of the material (Uyanık et al., 2011; Lopes et al., 2016; Ekin and Uyanık, 2021a; Hossain et al., 2024). Among destructive testing methods, core sampling followed by uniaxial compressive strength testing in a laboratory is the most commonly used approach. This method is widely regarded as one of the most reliable techniques for determining the actual strength of concrete. However, its application is limited due to the disruption it causes to the structural integrity of reinforced concrete elements, as well as the associated time and cost constraints (Sabbağ and Uyanık, 2017). Semi-destructive methods, on the other hand, include surface-based tests such as the Schmidt hammer (rebound hammer), which estimate concrete strength based on surface hardness. These tests are typically used as quick, low-cost preliminary assessments. Additionally, tensile and break-off tests evaluate strength variations on reinforced concrete surfaces, while penetration resistance tests are particularly applied to estimate the strength of fresh concrete. Non-destructive testing methods encompass a broader range of techniques; among them, the ultrasonic method is commonly employed for detecting internal defects such as cracks, voids, and material heterogeneities within the concrete structure (Sabbağ and Uyanık, (2016; 2017; 2018b); Ekin and Uyanık, (2021a; 2021b); Ekin, (2021a; 2021b; 2025a)). It is also suitable for the estimation of the elasticity modulus. The resistivity method provides information about the physical properties of concrete, such as moisture content and ion permeability (Raj and Venkatraman, 2013; Kibar and Öztürk, 2015; Sabbağ and Uyanık, (2018a; 2018c; 2020); Ekin (2025a; 2025b)). The radar (GPR) method can detect the location and depth of reinforcement, voids, cracks, separations within the reinforced concrete structure, as well as reveal anomalous areas such as moisture/water leaks. Thus, it is used to evaluate the internal integrity of the structure (Kadioğlu et al., (2011; 2013a; 2013b)). The maturity method enables early-age strength estimation based on the temperature-time relationship of the concrete. Each of the methods listed above offers different advantages, depending on the application purpose, the physical properties of the concrete, and the test environment. For example, while the ultrasonic method provides reliable data about the internal structure, the rebound hammer is sensitive to surface moisture and provides only superficial information. Therefore, it can give misleading results when used alone (Uyanık et al., (2011; 2012a); Samson et al., 2014; Lopes et al., 2016; Sabbağ and Uyanık, (2017; 2018a); Ekin and Uyanık 2021a; Hossain et al., 2024). In addition, in some methods, such as the resistivity method, a direct approach to concrete strength may be incorrect. Because the resistivity method is a parameter that is greatly affected by pore saturation, the curing conditions must be taken into consideration. It has been shown that the resistivity method applied under these conditions produces very useful results in determining concrete strength (Sabbağ and Uyanık (2018a; 2018b)).

Concrete strength is influenced by factors such as age, mix ratios, and curing conditions, which directly affect the accuracy of test results (Uyanık, 2012a; Samson et al., 2014). Therefore, a widely recommended approach is to support non-destructive testing (NDT) data with limited core sampling and destructive tests. Hybrid methods like the SonReb technique (Schmidt hammer + Ultrasonic) enable fast and relatively accurate on-site strength estimations (Lopes et al., 2016; Shrestha and Giri, 2023; Çelik et al., 2025). Karmakar et al. (2021) compared NDT and destructive test results on an aging concrete structure, highlighting the need for careful interpretation of NDT data due to strength degradation over time. Additionally, studies by Deng et al. (2019) and Nepomuceno et al. (2020) showed that pull-out tests provide more reliable strength estimations than Schmidt hammer tests, especially for early-age concrete.

Table 1. Application areas of the methods used in determining concrete proportions

Method	Method	Damaged	Determining Strength	Application Area
Destructive	Coring	High	Convert	Existing cabinets, laboratory
Semi-Destructive	Pulling, Pulling, Schmidt hammer, Breaking, Winsor Probe	Middle	Indirect	Fiber concrete, field application
Non-destructive	Ultrasonic, Resistivity, GPR, Maturity	None	Indirect	Wide area scanning, historical buildings

In engineering practice, multi-scale testing strategies that preserve structural integrity while providing reliable data should be developed. The accuracy of non-destructive methods varies with concrete age, composition, environmental conditions, and application quality. While the rebound hammer is affected by surface moisture and hardness, the ultrasonic method offers more consistent insight into the internal structure. Although no non-destructive method can fully replace core testing, the combined use of these methods is recommended to reduce the need for cores, especially when core sampling is restricted, such as in historic structures. Many studies in the literature support this integrated approach.

THE MAIN FOCUS OF THE CHAPTER

Destructive testing (DT) methods directly measure concrete's mechanical properties but are time-consuming, costly, and cause permanent damage. These drawbacks have led to the development of non-destructive testing (NDT) methods, which offer quicker, cost-effective, and less invasive strength assessments. However, to ensure accuracy, selective core sampling (DT) is often used to verify NDT results, especially in critical structures. The current approach prioritizes NDT to minimize damage while maintaining reliability. This chapter will detail both DT and NDT methods, their applications, advantages, limitations, and how their combination enhances assessment accuracy.

Destructive Testing Techniques

Uniaxial Compression Test

Core drilling and uniaxial compressive strength testing are among the most reliable methods to evaluate the load-bearing capacity of existing reinforced concrete structures. Although destructive, uniaxial compressive strength testing (UCS) is widely used for strength assessment of reinforced concrete (Malek and Kaouther, 2014; Kibar and Öztürk, 2015). The process begins by extracting standard core samples, typically without reinforcement and with a length-to-diameter ratio of about 2. These samples are tested in the laboratory by placing them between compression plates and applying vertical pressure until failure. The maximum load at failure represents the concrete's compressive strength. During construction, standard samples are cast from the concrete mixture, with at least nine specimens produced per mix and cured at room temperature. UCS is performed on three samples at 3, 7, and 28 days to monitor strength development, and the average of these three tests is taken as the compressive strength for that age (Malek and Kaouther, 2014; Sabbağ and Uyanık, (2016; 2017; 2018a)).

Yousif (2018) compared ultrasonic pulse velocity (UPV) data with uniaxial compressive strength (UCS) results over time to evaluate the effect of concrete age on strength. The study used 126 cube samples from C20, C25, and C30 concrete classes, with 28-day compressive strengths ranging from 12.1 MPa to 31.3 MPa (Figure 1). Findings showed that compressive strength increased with age, from 15.98 MPa on day 1 to 46.21 MPa on day 28. However, the UPV–strength relationship varied across concrete classes, reflecting differences in composition and properties. Similarly, Sabbağ (2016) measured uniaxial compressive strengths of nine concrete mixes (5 to 70 MPa) on days 7, 28, and 90, observing that strength generally increased with time, with a larger gain between days 7 and 28 than between days 28 and 90.

Figure 1. Relationship between compressive strength and age (Edited from Sabbağ, 2016)

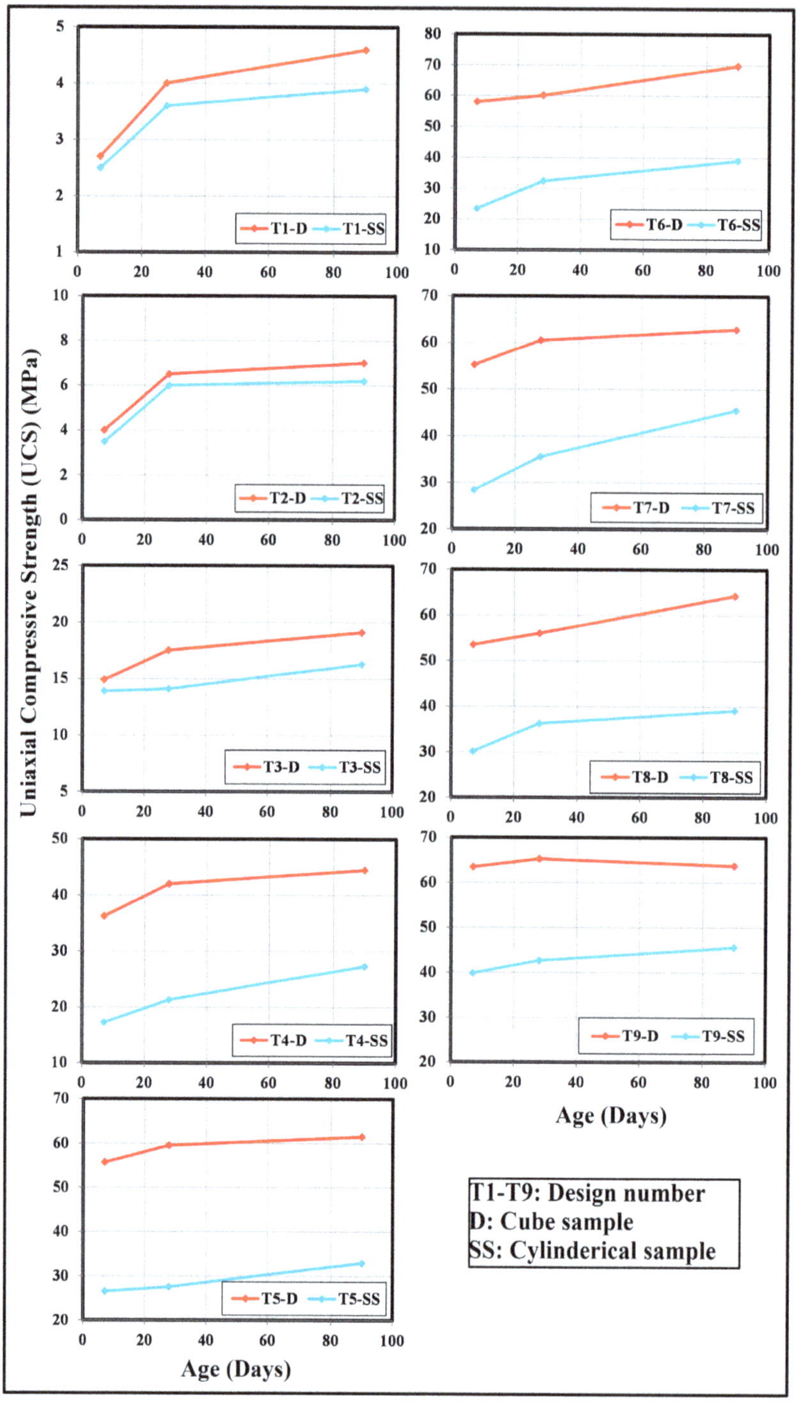

The Izmir Earthquake Investigation Report by the Turkish Ready-Mixed Concrete Association (THBB, 2020) revealed that compressive strength tests on core samples from 15 buildings showed most structures had strengths below expected levels, highlighting the need for structural reinforcement. Core samples from heavily damaged buildings in Bayraklı, Karşıyaka, Buca, Çiğli, Karabağlar, and Gaziemir districts indicated that average compressive strengths in some areas fell significantly below regulatory minimums. For example, some buildings in Buca had average compressive strengths as low as 16.5 MPa, well below the typical minimum of 25 MPa. Similarly, Öziçer (2016) and Öziçer and Uyanık (2017) demonstrated that concrete quality in existing building stock often falls short, posing serious earthquake safety risks. Accurate structural capacity assessment is thus critical, especially for urban transformation projects. Bijivemula and Wanjari (2018) studied the impact of varying length-to-diameter ratios of core samples on compressive strength, finding that this ratio significantly influences test results. Other factors such as core extraction direction, proximity to reinforcement, age, and moisture content also contribute to variability, emphasizing the need for expert handling and standardized procedures during coring and testing. Despite these challenges, core testing remains essential for structural safety evaluations and seismic risk assessments.

Triaxial Compression Test

The triaxial compression test is an advanced method used to analyze concrete behavior under multiaxial stress conditions, providing insights beyond uniaxial loading (Kupfer et al., 1969; Chen & Drucker, 1969; Nilimaa and Nilforoush, 2023). Cylindrical concrete samples, after curing, are subjected to isotropic lateral pressure followed by gradually increased axial load in a specialized triaxial testing device (Le and Hoang, 2023; Selcuk, 2015). This setup enables assessment of strength, elasticity, and fracture behavior under realistic multi-directional stresses through load-deformation analysis (Atiş et al., 2022). Although the method closely simulates actual conditions, it is limited by its complexity, cost, and time requirements (Talaat et al., 2021). Studies such as Kupfer (1973) have examined concrete deformation and elastic-plastic responses under triaxial stress, while Hampel and Curbach (2000) found that high-performance concretes exhibit greater strength under multiaxial loads. Xia et al. (2021) investigated the impact of sulfate attack on triaxial strength and deformation. Chen et al. (2023) evaluated steam-hardened recycled aggregate concrete (RAC) under varying pressures and curing methods, observing that lateral pressure restricted crack growth and doubled RAC's triaxial compressive strength. However, steam curing at 80 °C reduced strength by approximately 30%.

Figure 2. Triaxial stress-strain curves of RAC under different curing conditions. (a) RAC20, (b) RAC6012, (c) RAC8009 (Chen et al., 2023)

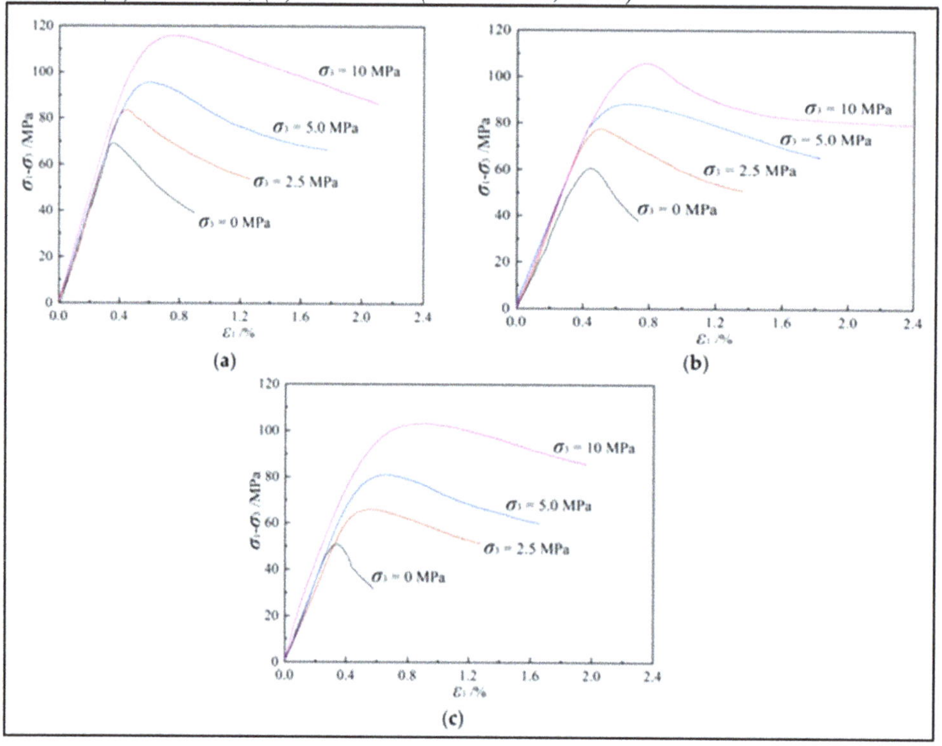

Semi-Destructive Testing Techniques

With advancing technology, non-destructive methods (such as ultrasonic wave analysis) have become key for assessing the internal structure, integrity, and homogeneity of concrete without causing damage. Alongside these, semi-destructive methods, which cause limited surface damage, remain widely used. These include tests like the Schmidt hammer, tensile, pull-off, and penetration resistance (Windsor probe) tests, performed both onsite and in laboratories. Surface hardness methods, particularly the Schmidt hammer, are popular in the field due to their ease of use, portability, and rapid data collection (Balamuralikrishnan et al., 2017; Pandey and Malviya, 2019).

However, the reliability of semi-destructive test results can vary based on test conditions, surface preparation, and sample homogeneity. Therefore, these results should be supported by other methods and interpreted through comprehensive analyses. Despite requiring limited intervention, semi-destructive techniques remain valuable for assessing the condition and structural performance of concrete.

Schmidt (Rebound) Concrete Test Hammer

The Schmidt rebound hammer is a semi-destructive device that indirectly estimates concrete compressive strength by measuring surface hardness. Its spring-driven mass strikes the concrete, and the resulting rebound number (RN) reflects surface hardness, which correlates with the concrete's elastic properties. Öziçer and Uyanık (2017) found a strong correlation between Schmidt hammer readings and compressive strength in reinforced concrete elements in İzmir, while emphasizing that RN reflects only surface characteristics, not internal strength. Similarly, Bora (2016) noted that Schmidt rebound values indicate surface-layer hardness rather than true internal strength, making them suitable primarily for preliminary assessments. The rebound number (RN) depends on surface hardness, so test surfaces must be flat, dry, clean, and uncoated; rough or porous surfaces are unsuitable (Tikalsky, 2006; Jedidi, 2018). According to ASTM C805/EN 12504-2, nine equidistant readings are averaged without repetition at the same points (Gupta, 2014). Tests should be on surfaces perpendicular to casting, preferably on exposed structural elements (Öziçer and Uyanık, 2017). Surface carbonation can cause overestimation of strength. While Schmidt tables convert RN to compressive strength, results must be verified with other methods due to variability (Brencich et al., 2020).

The Schmidt hammer is widely used for rapid strength estimation in diverse conditions: damaged buildings in İzmir (Yıldız and Daloğlu, 2021), protected historic structures like the Mevlana Tomb (İnce et al., 2018), concretes with fly ash and slag (Sadiq et al., 2021), and old concrete dams (Kairu et al., 2025). However, factors such as surface moisture, age, smoothness, and stress affect results, limiting its standalone reliability (Brencich et al., 2020).

Figure 3. Calibration curve for different stress states. Average compressive stress: dashed line:5 MPa; dash-dotted line:10 MPa; continuous line:0 MPa; dotted bold: Schmidt curve. RI: rebound index; a [0.088, 0.096]; R^2= 0.67 (Brencich et al., 2020)

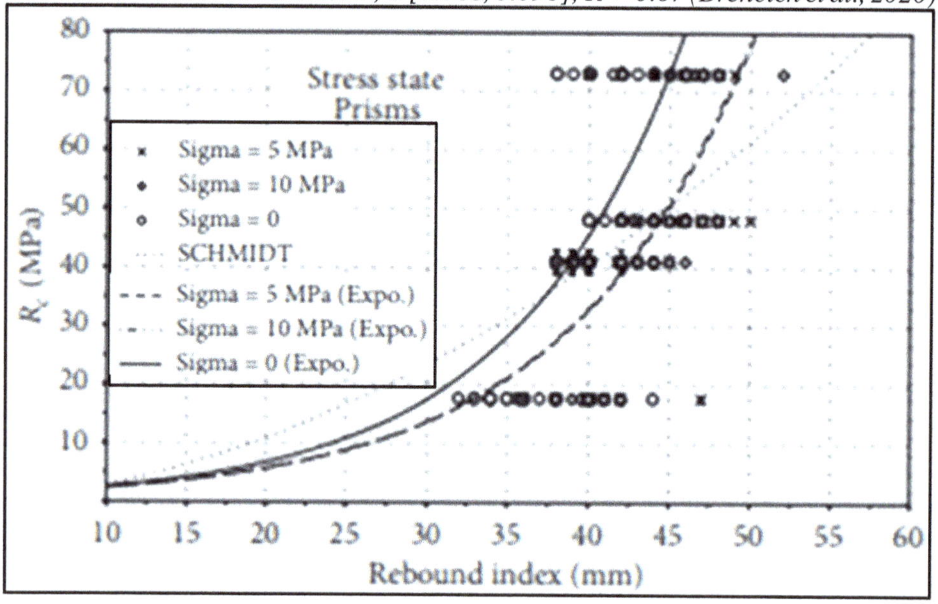

Tensile Test (Pull-Off)

The tensile test is a portable, microprocessor-controlled method used to estimate in-situ compressive strength by measuring the tensile force required to detach a metal disc adhered to the concrete surface (Lencis et al., 2025). The process involves surface preparation, application of an epoxy-bonded metal disc (typically 50×50 mm or 50 mm Ø), and controlled pull-off using a device capable of applying 5–100 kN of force (Ramos et al., 2011). The measured tensile force is correlated with compressive strength via empirical tables (Bonaldo et al., 2005). The test results are displayed digitally and can be stored or transferred for analysis (Gupta, 2014).

The tensile test offers fast application, a simple procedure, and minimal, easily repairable surface damage. However, curing time for the adhesive (often requiring overnight bonding) is a major limitation. Test reliability also depends on surface preparation, adhesive quality, and environmental conditions; poor bonding renders results invalid. To ensure accuracy, at least six valid measurements per area are recommended, excluding failed bonds (Gupta, 2014). Additionally, compressive strength estimations from this test are sensitive to aggregate type (particularly whether natural gravel or crushed stone is used), making aggregate consideration essential for accurate interpretation.

Bungey and Madandoust (1992) investigated the correlation between pull-off force and concrete properties through experimental and numerical analyses. Bonaldo et al. (2005) emphasized the pull-off test's value in evaluating surface treatment quality, highlighting the effects of adhesive type and surface preparation. Pereira and Medeiros (2012) showed that surface cleaning methods significantly influence test accuracy. Gupta (2014) compared pull-off and pull-out tests, noting the pull-off test's practicality but sensitivity to calibration and application errors. Ebean and Younis (2019) found improved reliability on rougher surfaces. Yang et al. (2020) introduced the Modified Pull-Off Test Method (MPTM), achieving better alignment with cube strength (10–80 MPa) and improved repeatability by eliminating bending stresses.

Figure 4. Comparison of the relationship between pull-off and MPTM versus concrete cube strength (Yang et al., 2020)

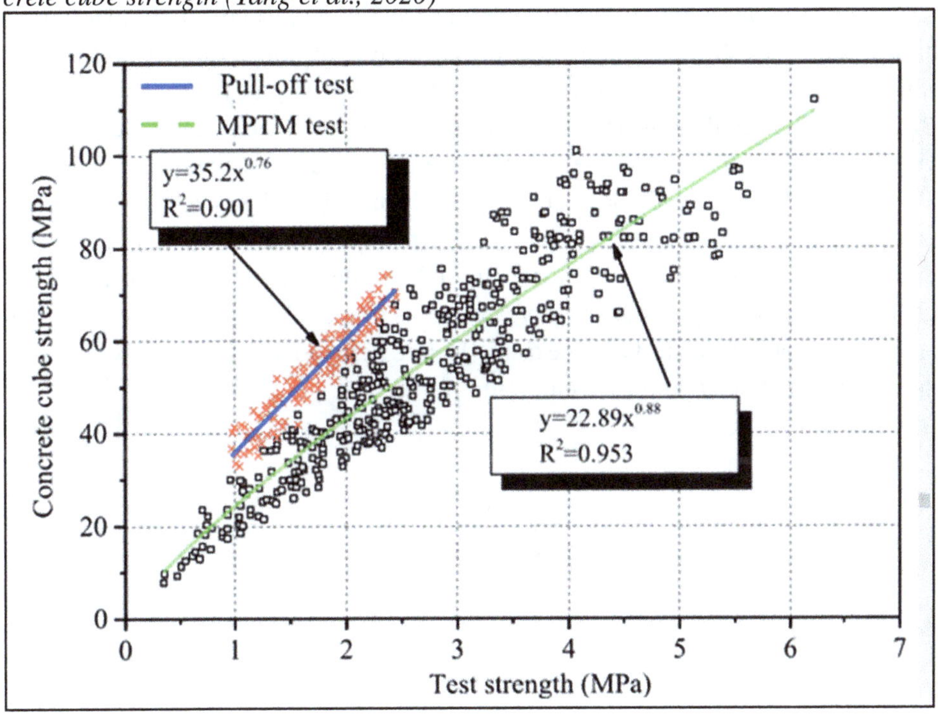

Pull-Out Test

The pull-out test, developed in 1934, measures the force needed to extract a fastener embedded in concrete, offering indirect insight into compressive strength via tensile performance (Malhotra, 1975; ASTM, 2007). Although multiaxial stress conditions complicate direct correlation with uniaxial strength (Tikalsky, 2006; Al-Sabah et al., 2021), empirical models improve reliability (Gupta, 2014). The test samples a deeper volume than surface methods like the Schmidt hammer, enhancing accuracy. Due to variability in concrete mix and aggregate types, project-specific strength correlations are essential (Bovio et al., 2014). Studies confirm a generally linear relationship between pull-out and compressive strength (Nepomuceno and Bernardo, 2019; Abbas et al., 2022).

For reliable pull-out test results, measurements should be taken at critical points in a reinforced concrete structure, and a statistically significant number of tests conducted. According to ASTM (2007), test-device anchors must be properly located to ensure safety-representative data. Bishr et al. (1995) found that the pull-out test gave high accuracy in estimating in-situ compressive strength, with strong linear relationships across varying w/c ratios, cement content, and aggregate types. Antonaci and Bocca (2005) used the tensile test on concrete surfaces to evaluate elastic modulus via force-displacement curves, especially useful in existing structures. Humbert et al. (2010) showed that accuracy depends heavily on strict adherence to standardized procedures, calibration, and operator skill. Abbas et al. (2022) observed that rod diameter influences measured tensile strength, while Nepomuceno and Bernardo (2019) studied multiple SCC mixtures in lab settings. Collectively, these studies emphasize that pull-out tests can be fast, effective field tools when properly executed and supplemented by empirical calibration.

Figure 5. a) Pull-out test and b) concrete cone trunk fragments (Nepomuceno and Bernardo, 2019)

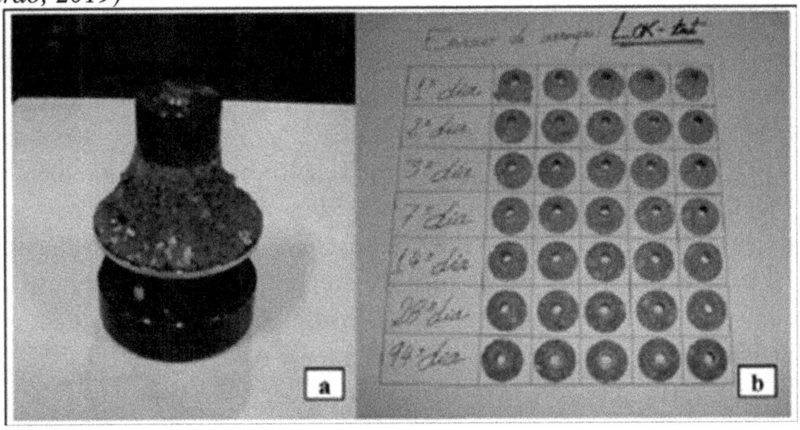

These mixtures were tested at 1, 2, 3, 7, 14, 28, and 94 days (Table 2). The compressive strength of concrete was determined by non-destructive methods such as pull-out test (Lok-test), ultrasonic wave velocity test (PUNDIT), Schmidt hammer test (type N), and concrete maturity test (COMA-meter).

Table 2. Pull-out test results (Nepomuceno and Bernardo, 2019)

Series	Age (days)	P (kN)	S_d (kN)	C_v (%)
P1	1	26.81	1.50	5.58
P2	2	34.51	1.50	4.34
P3	3	35.69	1.13	3.15
P7	7	43.20	1.29	2.98
P14	14	47.74	3.23	6.76
P28	28	53.07	4.27	8.04
P94	94	60.96	2.25	3.69

The collected data were used to examine the correlation between non-destructive tests and the compressive strength of concrete. The study found that specific characteristics of SCC (such as higher cement paste content and reduced coarse aggregate volume) could influence test accuracy. Nevertheless, non-destructive methods generally showed strong correlations with the compressive strength of SCC.

Break-off Test

The break-off test is a semi-destructive method used to determine concrete tensile strength on samples formed parallel to the concrete surface. A plastic sleeve is placed during casting, creating a 55 mm diameter × 70 mm high sample after curing. A drum-section cantilever-type load is applied via a manual hydraulic pump until rupture, and the maximum force measured gives the tensile stress. Two measurement ranges are provided (≈20 MPa for low-strength, ~60 MPa for high-strength concretes) to preserve sensitivity (Gupta, 2014). Empirical correlations are then used to estimate compressive or flexural strength. While this test reflects in-situ conditions closely, its reliability hinges on proper sample placement, good surface preparation, and verifying correlations for each project (Madandoust & Kazemi, 2017).

Lu et al. (2018) studied fiber distribution and surface effects on the tensile strength of fiber-reinforced concrete, while Zhang et al. (2020) showed that tensile strength varies with curing conditions. Resan et al. (2020) introduced a refined test method that controls stress distribution and isolates gauge regions, providing more consistent tensile strength results than the Brazilian or flexural tests. The method typically gives higher values than the Brazilian test but lower than the flexural test, with reduced scatter. As shown in Figure 6, tensile strength increases consistently with compressive strength using this method, highlighting its reliability and accuracy.

Figure 6. Break-off test results (Resan et al., 2020)

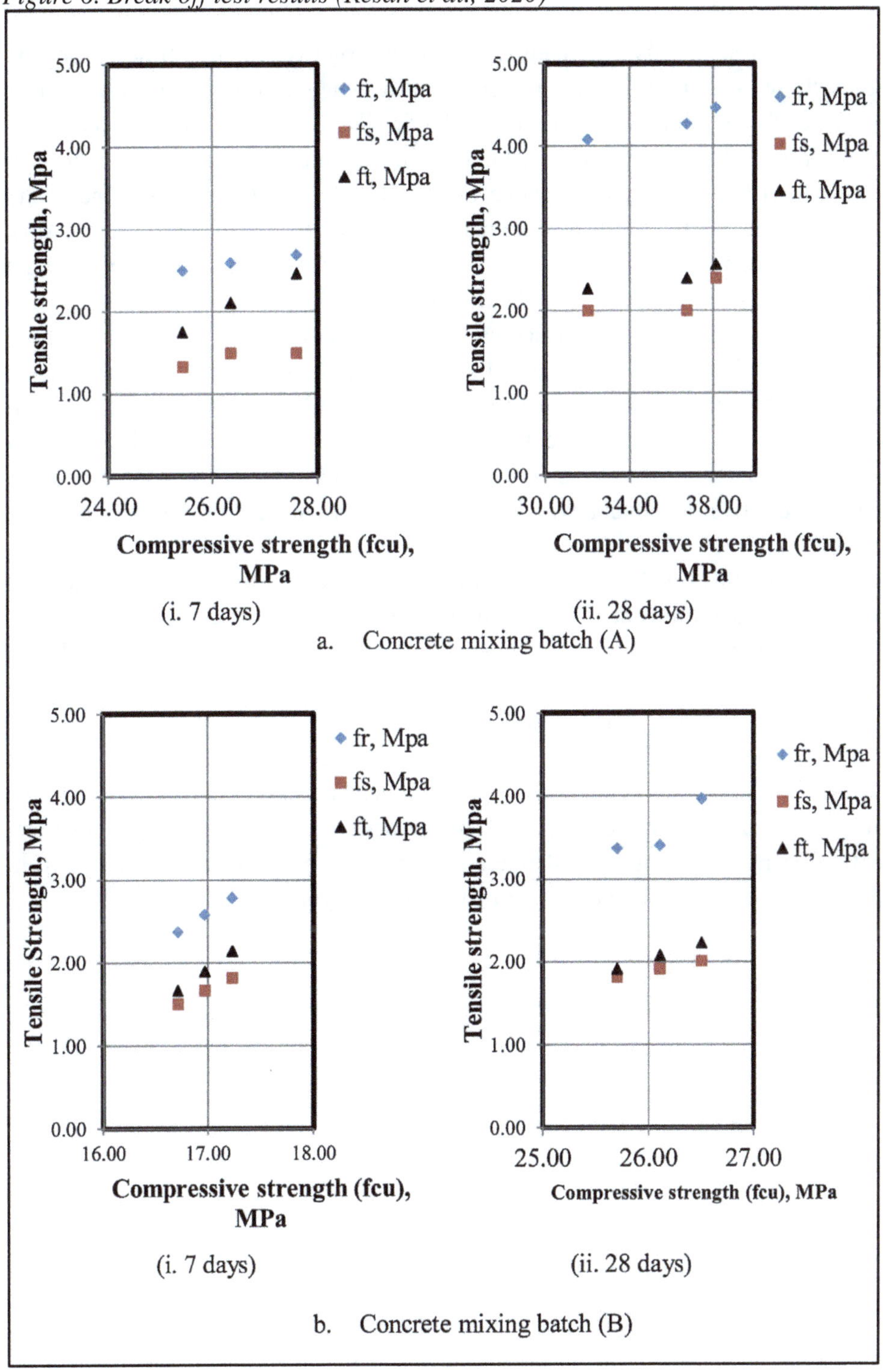

Penetration Resistance Test (Windsor Probe)

The penetration resistance method is a semi-destructive test that estimates concrete strength by measuring the depth of steel probe penetration into the surface under a defined force (Pucinotti and Lorenzo, 2005; Tikalsky, 2006). The most common device used is the Windsor Probe, which includes a gunpowder-powered driver, steel probes, cartridges, and a depth gauge. Standard probes are 6.3 mm in diameter and 79.5 mm long, while 7.9 mm probes are used for lightweight concrete (Gupta, 2014).

In the penetration resistance method, a probe is fired into the concrete using a gunpowder-powered device, and the depth of penetration (measured with a gauge) is inversely proportional to surface hardness and strength (Gupta, 2014). While it offers fast and practical assessments like the Schmidt hammer, it estimates relative strength rather than absolute compressive strength. It's especially useful for comparing different concrete elements, but it is not suitable for historical or visually sensitive structures due to its minor surface damage (Tikalsky, 2006). Despite being semi-destructive, it remains widely used for modern structures due to portability and ease of use.

RILEM (1987) evaluated the penetration resistance test as a rapid, reliable field method for assessing concrete strength, discussing its pros and cons. Alyamac et al. (2017) compared this method with the Schmidt hammer, analyzing their correlation with actual compressive strength. Kog (2019) studied the test's repeatability and the effects of surface conditions. Hanif et al. (2019) showed its effectiveness in measuring surface hardness in aged concrete. Gomez-Cardenas (2015) found that surface moisture reduces test accuracy. Demir et al. (2023) used the Windsor Probe alongside rebound hammer and ultrasonic velocity tests on high-strength concrete, noting stronger correlations and better insight into internal concrete structure from the probe test, especially between 3 and 90 days (Figure 7).

Figure 7. Relationship between the test results of the Windsor probe and UCS (Demir et al., 2023)

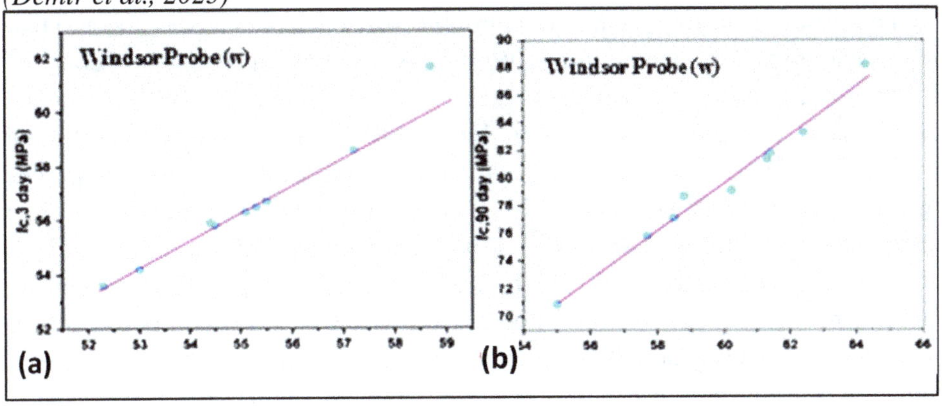

Non-Destructive Testing Techniques (NDT)

The primary goal of non-destructive testing (NDT) methods is to detect defects and irregularities (such as cracks, fractures, residual stresses, voids, and micro-structural deterioration) without compromising the material's structural integrity (Uyanık et al., 2012b; 2012c; 2019). Early detection of such damage is essential to prevent strength loss or sudden failure, especially in structural components. NDT plays a critical role both in pre-production quality control and ongoing in-service monitoring. Accurate defect identification requires detailed information about their type, location, size, shape, and orientation (Raj et al., 1995; Arı, 2023).

Non-destructive testing (NDT) techniques serve various purposes, including flaw detection (breakages, voids), leakage detection, dimensional control, positioning, structural and microstructural analysis, mechanical and physical property estimation, residual stress and dynamic response assessment, and signal and chemical composition analysis (Parida, 2009).

In concrete structures, a primary application of NDT is estimating strength and detecting internal defects such as cracks and voids. Ultrasonic testing is particularly common, as it assesses concrete by measuring the propagation speed and attenuation of high-frequency sound waves affected by internal obstacles (Ercan, 2002; Parida, 2009; Ekin and Uyanık, 2021b). Other widely used NDT methods for estimating concrete strength include resistivity measurement, ground-penetrating radar (GPR), and maturity tests, favored for their rapid, cost-effective, on-site evaluation capabilities and repeatable results in both research and practice (Parida, 2009; Sabbağ and Uyanık, 2017; 2018a; 2018b; Ekin and Uyanık, 2021a; 2021b).

Ultrasonic Method

The ultrasonic testing method is a portable, battery-operated NDT method that works with high-frequency sound waves, used to evaluate the strength of concrete as well as its dynamic elastic properties, internal structural homogeneity, and discontinuities (cracks, voids, etc.). In this method, the passage time of ultrasonic waves applied to concrete varies depending on the physical and elastic properties of the concrete; the wave velocity is calculated from the obtained passage time and path length ratio (Uyanık (1991; 1999); Uyanık and Çekmen, 2009; Sabbağ, 2016; Sabbağ and Uyanık, 2017; Ekin, 2025a; 2025b). Factors such as the density and homogeneity of the concrete directly affect the wave velocity; in this context, high velocity values generally indicate better quality concretes, while low velocity values indicate the presence of discontinuities or low strength (Sabbağ and Uyanık, 2016; Ekin and Uyanık, 2021a). The vibration pulses produced with the help of electroacoustic transducers used during the test pass through the concrete and reach a second transducer, and the passage time is recorded by the electronic timer circuit. In measurements on reinforced concrete elements, if there is plaster on the surface, since the plaster thickness will also be included in the measurement, appropriate velocity corrections should be made (Uyanık et al., 2011; Uyanık and Tezcan, 2012). The applicability of the ultrasonic method can be increased by creating correlation curves with DT data, and thus, higher accuracy can be achieved in strength estimation (Uyanık et al., 2013; Ali-Benyahia et al., 2019). The main advantages of the method can be listed as its sensitivity to surface and subsurface discontinuities, allowing thickness measurement and rapid data acquisition (Wang et al., 2020). However, there are also limitations, such as the need for surface preparation, the necessity of providing sufficient contact during measurement, and the fact that the method requires advanced technical knowledge from the user. In addition, measurements made parallel to discontinuities may be insufficient in detecting discontinuities.

P waves can travel through solids, liquids, and gases, while S waves only propagate through solids. Since pore saturation in concrete affects P wave velocity, this leads to wide variability in measurements. To better control this variation, measuring S-wave velocity has become necessary. Combining P and S wave velocities allows calculation of concrete density (Uyanık and Çatlıoğlu, 2015) and key mechanical properties like the modulus of elasticity and Poisson's ratio based on elasticity theory. While research traditionally focuses on P wave velocity, recent studies increasingly use S wave velocity due to its greater sensitivity to factors like aggregate type, cement properties, and concrete strength (Uyanık, (2012; 2014; 2015); Uyanık and Tezcan, 2012; Uyanık et al., (2011; 2012d; 2013; 2019); Öziçer and Uyanık, 2017; Sabbağ and Uyanık, (2017; 2018b); Ekin et al., 2023; Ekin and Uyanık, (2021b)). In their study, these researchers classified the quality of concretes by evaluating only the

P wave velocity or the P and S wave velocities together (Table 3). This approach allows the determination of the mechanical properties of the concrete as well as the fractures, voids, and weathering status, and the determination of its elastic properties (modulus of elasticity, bulk modulus, shear modulus, Poisson ratio) (Uyanık, 2012; Sabbağ, 2016; Ekin, 2020).

Table 3. Meeting the concrete quality classification depending on seismic velocities (Uyanık et al., (2012d; 2013))

Concrete Quality	Whitehurst (1951)	Uyanık et al. (2011)	Uyanık et al. (2012d; 2013)	
	Vp(m/s)	Vp(m/s)	Vp(m/s)	Vs(m/s)
Very good	>4500	>4565	>4400	>2200
Good	3650-4500	3515-4565	3450-4400	1750-2200
Middle	3050-3650	2930-3515	2900-3450	1500-1750
Poor	2000-3050	2110-2930	2150-2900	1150-1500
Very poor	<2000	<2110	<2150	<1150

The ultrasonic pulse velocity (UPV) method is a widely used non-destructive technique for assessing concrete's internal integrity and compressive strength, especially in load-bearing elements. Öziçer and Uyanık (2017) found a strong correlation between in situ P-wave velocities and lab-tested compressive strength, confirming UPV's reliability for quick quality evaluation. For high-precision imaging, multi-channel S-wave ultrasonic tomography devices like Mira (operating at 50 kHz) provide detailed 3D visuals of internal defects. Bishko et al. (2008) used such devices for echo-pulse tomography on bridge piers, enabling fast, high-resolution monitoring of infrastructure safety.

Han et al. (2022) studied corrosion in reinforced concrete caused by low-frequency stray currents in railway power systems, emphasizing the importance of early monitoring for durability under electrochemical stresses. Öncü (2025) estimated the strength of reinforced and unreinforced concrete using AI-assisted NDT methods on 334 samples, with ultrasonic test data analyzed via XGBoost, Random Forest, and Neural Networks; XGBoost showed the highest accuracy, reducing error by 12% compared to RF. Sabbağ and Uyanık (2017) developed empirical models correlating ultrasonic P and S wave velocities (Vp and Vs) with compressive strength in self-compacting concrete, enabling reliable strength estimates up to 70 MPa, considering wave speed, moisture, and curing time.

Figure 8. a) Relationships between the uniaxial compressive strength and ultrasonic P and S wave velocities of concretes with different reinforcement diameters (ø= 10, 14, and 20 mm). b) Comparison of measured and calculated UCS (Sabbağ and Uyanık, 2017)

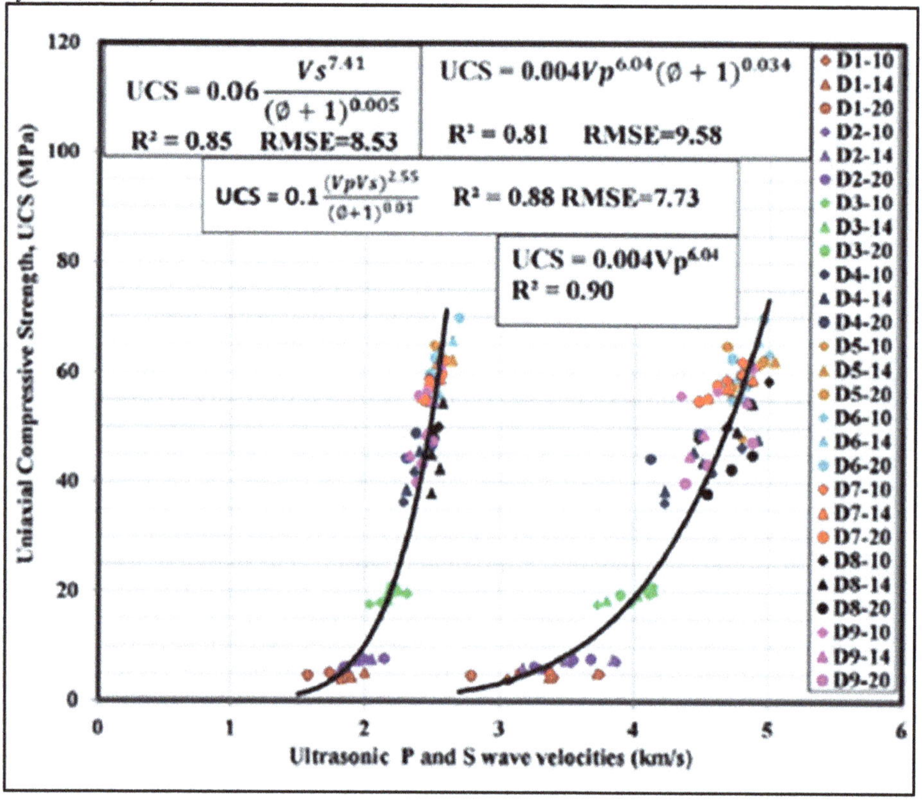

Resistivity Method

The resistivity method, a non-destructive electrical technique, interprets concrete based on its conductivity (Ekin, 2025a; 2025b). Using a device with four electrodes spaced 5 cm apart in a Wenner array, current is applied via the outer electrodes while potential difference is measured by the inner ones. Concrete resistivity ranges broadly from about 1 to 10^4 kΩ·cm (Sabbağ, 2016; Sabbağ and Uyanık, 2018a, 2018c, 2020). Resistivity depends heavily on pore water or air saturation, affecting its reliability in strength estimation. Sabbağ (2016) successfully estimated the compressive strength of 15 cm^3 concrete samples on-site using this method, demonstrating its practical field applicability. Studies show significant correlations between resistivity and compressive strength across curing conditions for example, water-cured concretes had resistivity values of 5–16 kΩ·cm, while air-cured ones ranged 16–50 kΩ·cm, with 16 kΩ·cm used as a reference for pore saturation (Figure 9).

Figure 9. Comparison of measured and calculated uniaxial compressive strengths (Sabbağ and Uyanık, 2018a)

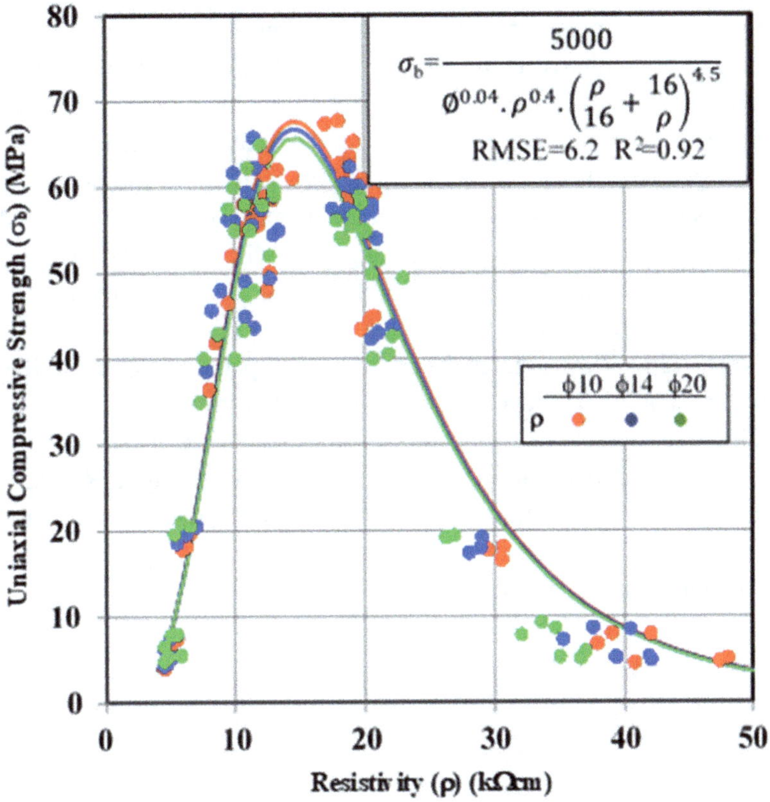

Another practical site application is the use of electrical resistivity to assess moisture distribution and corrosion risk in bridge piers and reinforced concrete columns. Öztürk et al. (2021) identified moisture accumulation zones by measuring resistivity around reinforcements, helping detect early corrosion and improve structural durability. Resistivity is also effective in monitoring microcracks and long-term deterioration. For instance, Mancini and Bolzoni (2003) used it to detect early damage in building foundations, aiding maintenance planning. This method offers fast, practical field results, especially in large-scale projects, as it allows strength variation detection across wide areas using surface-placed electrodes (Sabbağ and Uyanık, 2020).

Ground Penetrating Radar (GPR) Method

Rapid and non-destructive evaluation of concrete's mechanical properties is crucial for assessing existing structures and planning maintenance. Electromagnetic methods, especially ground-penetrating radar (GPR), are commonly used to estimate concrete strength based on its electrical and magnetic properties (Neville, 2012). GPR analyzes wave propagation, attenuation, and reflection to provide indirect insights into concrete's microstructure (Zhang et al., 2020). It is effective for detecting cracks, voids, and reinforcement (Li et al., 2021; Özçelik & Gelişli, 2018). Factors such as dielectric constant, moisture, and conductivity influence wave behavior, making site-specific calibration essential (He & Lu, 2023; Kadioğlu, 2010). GPR offers non-destructive, high-resolution, and fast imaging, enabling early damage detection and structural planning. Integrating GPR with other NDT methods (e.g., ultrasonic, resistivity) enhances the reliability of structural assessments (Uyanık et al., 2013; Ali-Benyahia et al., 2019).

Morris et al. (2020) investigated the use of ground-penetrating radar (GPR) combined with artificial intelligence to non-destructively estimate concrete properties. By analyzing electromagnetic parameters like the dielectric constant, they predicted porosity (R^2=0.83), density (R^2=0.67), and compressive strength (R^2=0.51) using AI-based regression models trained on 900 MHz radar data. The study highlights the potential of GPR and AI as an effective method for assessing concrete condition without damaging the structure.

Figure 10. Column (a) shows baseline results of the best-performing models. Column (b) shows the final results of the best-performing model after the parameters have been tuned. Results are shown for each property: fractional porosity, compressive strength (MPa), and density (kg/m³) (Morris et al., 2020)

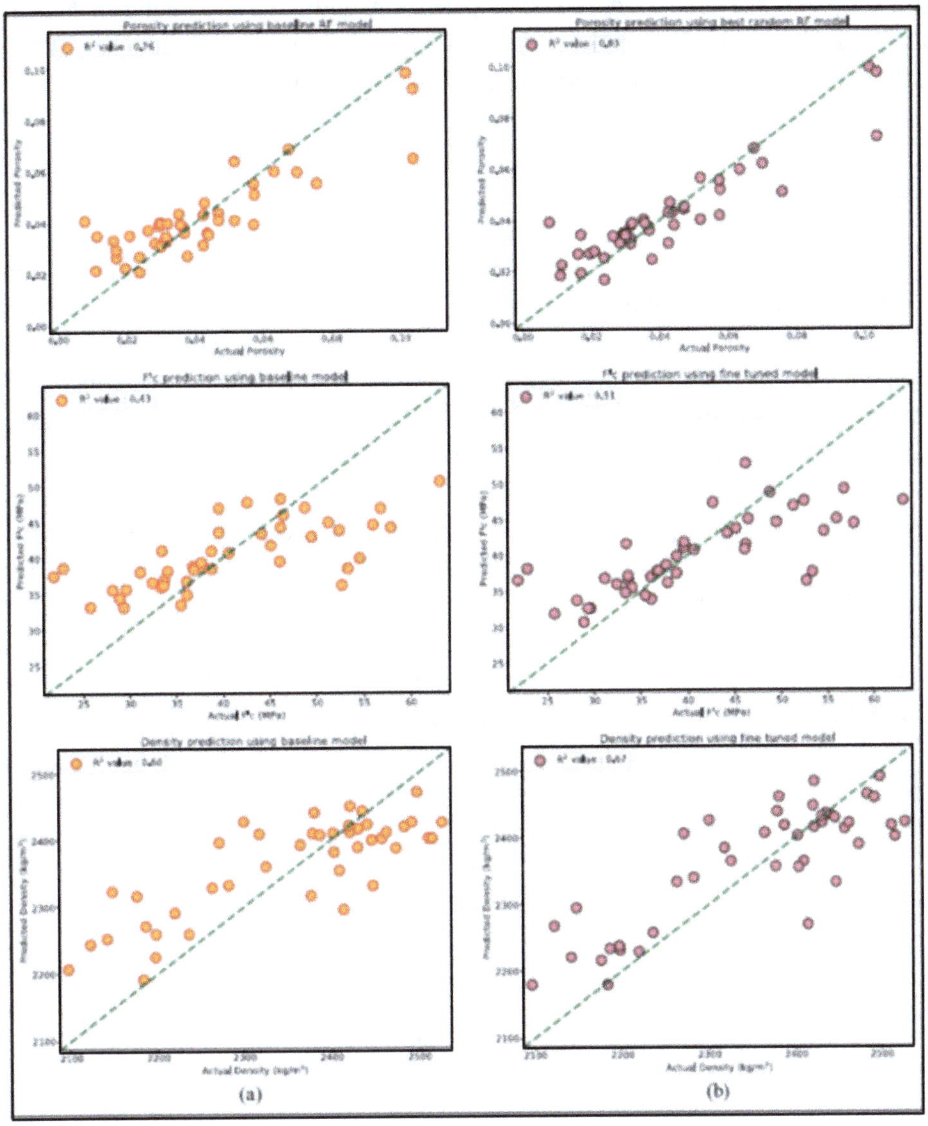

In recent years, the use of artificial intelligence and machine learning techniques in processing GPR data has increased. Thanks to deep learning algorithms, it has become possible to estimate the aging and strength levels of concrete from GPR signals with high accuracy (Liu et al., 2022). Kumar et al. (2023) showed that structural properties at various depths of concrete can be examined in detail with GPR devices operating at different frequencies. There are various application examples in the literature. Li et al. (2021) detected reinforcement location, corrosion, and cracks in bridge piers with GPR; Özçelik and Gelişli (2018) successfully determined reinforcement placement and cover thickness in multi-storey structures. Han et al. (2022) detected microcracks in tunnel structures at an early stage; Wang et al. (2020) combined GPR with electrical resistivity to map moisture movements and corrosion risk in concrete.

Maturity Test

The maturity concept predicts concrete strength development by integrating curing time and temperature, assuming hydration stops below -10°C (Carino and Lew, 2001; ASTM, 2002). Concretes with equal maturity, regardless of varying temperature-time combinations, exhibit similar strength, enabling real-time, non-destructive on-site estimation using lab-calibrated curves. This method improves early-age monitoring, formwork removal, and reduces destructive tests. Its accuracy has been confirmed by Mariak et al. (2019). Garcia-Monzon (2003) highlighted its practical benefits in large infrastructure, enabling faster, cost-effective strength assessment for timely demolding, consistent with ASTM C1074 and C1252 standards. Elçi and Yılmaz (2009) validated the method's reliability in urban housing projects. Nepomuceno and Bernardo (2019) applied the maturity test with a COMA-Meter on SCC specimens, demonstrating precise compressive strength estimation and good repeatability, with coefficients of variation between 3.7% and 7.1%.

Figure 11. Maturity days versus average concrete UCS (Nepomuceno and Bernardo, 2019)

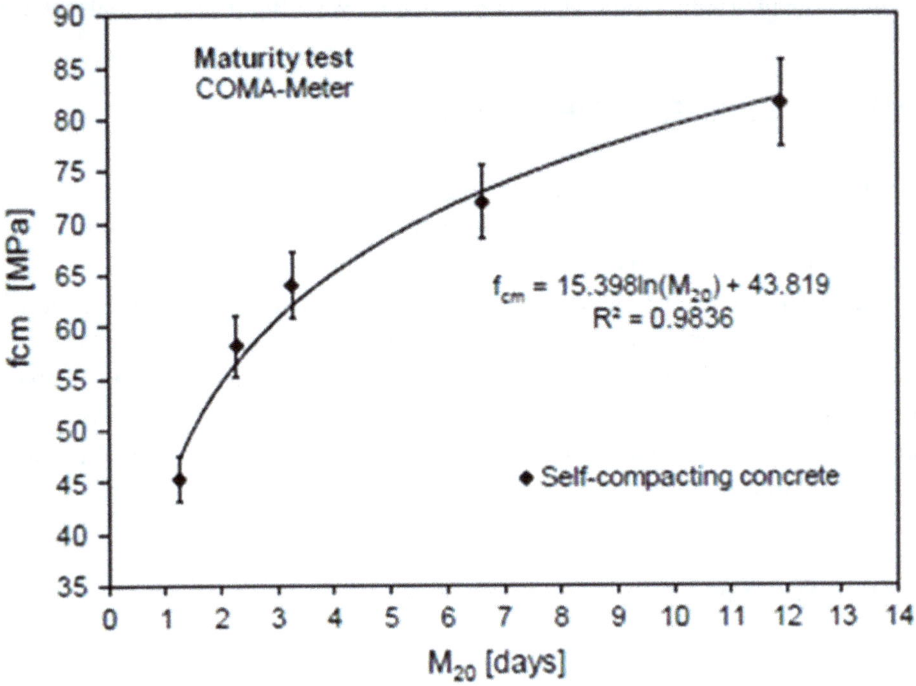

SOLUTIONS AND RECOMMENDATIONS

Assessing concrete's structural performance is essential for monitoring buildings during their service life and planning interventions, especially after earthquakes. Accurate determination of concrete strength is vital for both safety and cost-effective maintenance. Key mechanical and physical properties include compressive strength, elasticity, Poisson's ratio, porosity, and water content, which can be measured using destructive, semi-destructive, or non-destructive testing (NDT) methods.

Destructive methods like lab-based compression tests on core samples offer reliable results but can damage structures and are time- and resource-intensive. This has increased interest in faster, less invasive NDT methods such as rebound hammer, ultrasonic testing, and surface hardness tests. These techniques estimate strength indirectly and require calibration curves. For improved accuracy, it is recommended to verify NDT results with selective destructive tests.

The SonReb method, which combines rebound hammer and ultrasonic testing, has become widely used, as it increases reliability by calibrating NDT results with a limited number of core samples, balancing accuracy with minimal structural damage.

FUTURE RESEARCH DIRECTIONS

Combining destructive and non-destructive testing offers an effective balance between cost-efficiency and structural safety when assessing the load-bearing capacity of reinforced concrete. This integrated approach enhances the accuracy of engineering decisions by better identifying maintenance, repair, and reinforcement needs. This section examines various concrete strength assessment methods (destructive, semi-destructive, and non-destructive), highlighting their advantages, limitations, and application examples from academic studies. The comparison underscores the importance of choosing the appropriate method based on project context and objectives.

CONCLUSIONS

Accurate determination of concrete strength is fundamental in structural engineering, as it directly impacts the assessment of load-bearing capacity, serviceability, and long-term durability of reinforced concrete structures. Although destructive testing methods—such as uniaxial compressive strength tests—are regarded as the gold standard due to their direct measurement of strength, their invasive nature, high cost, and potential to compromise structural integrity limit their practical application in situ.

In contrast, non-destructive testing (NDT) techniques offer significant advantages in speed, cost-efficiency, and preservation of the structure, making them increasingly favored for condition assessment and maintenance planning. Methods such as rebound hammer, ultrasonic pulse velocity, ground-penetrating radar (GPR), and electrical resistivity measurements enable indirect estimation of concrete properties (like compressive strength, porosity, homogeneity, and moisture content) without causing physical damage. However, these techniques are sensitive to variables such as material heterogeneity, environmental conditions, and reinforcement presence, often necessitating calibration against destructive tests to enhance accuracy and reliability.

Hybrid approaches, notably the SonReb method (which combines rebound and ultrasonic testing calibrated with limited core sampling) have shown improved accuracy and practicality for field applications. Moreover, recent advances in artificial intelligence and machine learning have revolutionized the interpretation of NDT data, enabling more precise strength predictions by integrating multiple parameters, including wave velocity, curing conditions, and reinforcement details. These data-driven approaches represent a promising frontier in structural diagnostics, offering high accuracy with minimal intervention.

Ultimately, combining destructive, semi-destructive, and non-destructive testing methods creates a balanced framework that supports informed decision-making in structural assessment, rehabilitation planning, and lifecycle cost optimization. Selecting appropriate techniques based on the specific context (considering the structural system, age, material composition, and environmental exposure) is essential for ensuring both economic efficiency and structural safety across diverse engineering applications.

REFERENCES

Abbas, N., Yousaf, M., Akbar, M., Saeed, M. A., Huali, P., & Hussain, Z. (2022). An Experimental Investigation and Computer Modeling of Direct Tension Tensile Test of Reinforced Concrete Cylinder. *Inventions (Basel, Switzerland)*, *7*(3), 77. DOI: 10.3390/inventions7030077

Al-Sabah, S., Sourav, S. N. A., & McNally, C. (2021). The post-installed screw pull-out test: Development of a method for assessing in-situ concrete compressive strength. *Journal of Building Engineering*, *33*, 101658. DOI: 10.1016/j.jobe.2020.101658

Ali-Benyahia, K., Sbartaï, Z. M., Breysse, D., Ghrici, M., & Kenai, S. (2019). Improving the Non-Destructive Assessment of Concrete Strength in the Field: Influence of the Selection Process of Core Placement on the Assessment Quality for Single and Combined NDT Techniques. *Construction & Building Materials*, *195*, 613–622. DOI: 10.1016/j.conbuildmat.2018.10.032

Arı, A. C. (2023). Kompozit Malzemelerin Tahribatsız Muayene Yöntemlerinden Olan Ultrasonik Test ile Ölçülmesi Üzerine Yapılan Çalışmaların İncelenmesi. *Uluslararası Mühendislik Tasarım Ve Teknoloji Dergisi*, *4*(1-2), 1–10.

Arooj, M., & Yadav, G. (2017). Comparison of Destructive and Non-Destructive Testing of Concrete - A Review. [IJERT]. *International Journal of Engineering Research & Technology (Ahmedabad)*, *6*(11), 62–67.

ASTM. (2002). *Standard Practice for Predicting Concrete Strength by the Maturity Method. C 1074-98.* American Society for Materials Testing.

ASTM. (2007). *Standard Test Method for Tensile Strength of Hardened Concrete, C 900 – 06.* American Society for Materials Testing.

Ateş, E. (2020). Deprem ve Mikrotremor Kayıtları Kullanılarak Elde Edilen Yatay/Düşey Spektral Oranların Karşılaştırılması: Amasya Örneği. *Türk Deprem Araştırma Dergisi*, *2*(2), 160–175. DOI: 10.46464/tdad.746926

Ateş, E., & Uyanık, O. (2019). Jeofizik yöntemler ile yer ve yapı etkileşimi. *Süleyman Demirel Üniversitesi Fen Bilimleri Enstitüsü Dergisi*, *23*, 46–60. DOI: 10.19113/sdufenbed.430906

Ateş, E., & Uyanık, O. (2023). Farklı derinliklerde ölçülen ivme kayıtlarının karşılaştırılması. *Türk Deprem Arastirma Dergisi*, *5*(2), 215–232.

Ateş, E., & Uyanık, O. (2025). Comparison of soil-structure interaction obtained from spectral ratio methods applied to earthquake and microtremor records. *Journal of Applied Geophysics, 234*, 105645. DOI: 10.1016/j.jappgeo.2025.105645

Atiş, C., Abdılahı, A., İlkentapar, S., Durak, U., & Karahan, O. (2022). Flexural Tensile Strength Measurement and Determination for Kerbstones with Unsymmetrical Section. *Süleyman Demirel Üniversitesi Fen Bilimleri Enstitüsü Dergisi, 26*(1), 63–68. DOI: 10.19113/sdufenbed.943602

Azevedo, A. C., Lemos, S., Delgado, J. M. P. Q., Silva, F. A. N., & Sousa, C. A. P. (2023). Non-destructive and Destructive Tests to Drive Corrective Intervention Procedure of Concrete Elements. In Delgado, J. M. P. Q. (Ed.), *Building Pathologies: Experimental Campaigns and Numerical Procedures. Building Pathology and Rehabilitation, 25*. Springer. DOI: 10.1007/978-3-031-17061-4_5

Balamuralikrishnan, R., AlHarthy, M. N., & Krishnamurthy, B. S. (2017). Destructive and Non-Destructive Testing Methods for Condition Monitoring of Concrete Members. [JMEST]. *Journal of Multidisciplinary Engineering Science and Technology, 4*(6), 7405–7409.

Bijivemula, S., & Wanjari, S. (2018). Comparative Study on in Situ and Laboratory Testing of Concrete Cores with Different Height to Diameter Ratios. *Journal of Testing and Evaluation*. Advance online publication. DOI: 10.1520/JTE20170373

Bishko, A. V., Samokrutov, A. A., & Shevaldykin, V. G. (2008). Ultrasonic Echo-Pulse Tomography of Concrete Using Shear Wave Low-Frequency Phased Antenna Arrays. *17th World Conference on Non-Destructive Testing*, 25–28 October 2008, Shanghai, China.

Bishr, H. A., Al-Amoudi, O. S. B., Basunbul, I. A., & Al-Sulaimani, G. J. (1995). Assessment of concrete compressive strength using the Lok test. *Construction & Building Materials, 9*(4), 227–237. DOI: 10.1016/0950-0618(95)00025-B

Bonaldo, E., Barros, J. A. O., & Lourenco, P. B. (2005). Bond Characterization Between Concrete Substrate and Repair SFRC Using Tensile Testing. *International Journal of Adhesion and Adhesives, 25*(6), 463–474. DOI: 10.1016/j.ijadhadh.2005.01.002

Bora, D. S. (2016). Tests and Acceptance Criteria for Concrete Based on Compressive Strength. *National Technical Seminar on Innovation, Project Delivery and Management in Infrastructure Technology for Sustainable Development*, 133–141.

Bovio, F. L., Bovio, G. L., Brencich, A., Cassini, G., Pera, D., & Riotto, G. (2014). A new pull-out technique for in-place estimation of concrete compressive strength. *Advances in Materials Science and Engineering*, *6*(1), 1–8. DOI: 10.1155/2014/984341

Brencich, A., Bovolenta, R., Ghiggi, V., Pera, D., & Redaelli, P. (2020). Rebound hammer test: An investigation into its reliability in applications on concrete structures. *Advances in Materials Science and Engineering*, *2020*(1), 6450183. DOI: 10.1155/2020/6450183

Bungey, J. H., & Madandoust, R. (1992). Factors influencing pull-off tests on concrete. *Magazine of Concrete Research*, *44*(158), 21–30. DOI: 10.1680/macr.1992.44.158.21

Carino, N. J., & Lew, H. S. (2001). The Maturity Method: From Theory to Practice, Construction Congress and Exposition, May 21-23, 2001, Washington, DC, American Society of Civil Engineers, Reston, Virginia, Peter C. Chang, Editor, 2001, 19 p.

Çelik, S. B., Çobanoğlu, İ., & Gireson, K. (2025). Prediction and classification of flexural strength of rock materials used as dimension stones. *Bulletin of Engineering Geology and the Environment*, *84*(1), 44. DOI: 10.1007/s10064-024-04070-5

Chen, W. F., & Drucker, D. C. (1969). Constitutive Relationships for Concrete. *Journal of the Engineering Mechanics Division.*

Chen, Y., Ning, Y., Chen, X., Xuan, W., & Zhu, X. (2023). Experimental and Numerical Investigation on Triaxial Compressive Behavior of Steam-Cooked Recycled Aggregate Concrete. *Buildings (Basel, Switzerland)*, *13*(2), 334. DOI: 10.3390/buildings13020334

Demir, T., Ulucan, M., & Alyamaç, K. E. (2023). Development of Combined Methods Using Non-Destructive Test Methods to Determine the In-Place Strength of High-Strength Concretes. *Processes (Basel, Switzerland)*, *11*(3), 673. DOI: 10.3390/pr11030673

Deng, P., Sun, Y., Liu, Y., & Song, X. (2019). Revised Rebound Hammer and Pull-Out Test Strength Curves for Fiber-Reinforced Concrete. *Advances in Civil Engineering*, (2), 1–12.

Ekin, N. (2020). Betonların elastik modül hesabında Poisson oranının önemi. *Çanakkale Onsekiz Mart Üniversitesi Fen Bilimleri Enstitüsü Dergisi, 6*, 239–252.

Ekin, N. (2021a). Relationship between amplitude anisotropy and compressive strength of reinforced concrete depending on curing conditions. *Indian Journal of Engineering and Materials Sciences*, *28*(3), 286–299.

Ekin, N. (2021b). Concrete anisotropy estimated from ultrasonic signal amplitudes. *Materials Testing, 63*(12), 1164–1173. DOI: 10.1515/mt-2021-0053

Ekin, N. (2025a). The relationships between ultrasonic P and S wave velocities and resistivity in reinforced concrete. *Construction & Building Materials, 479*, 141475. DOI: 10.1016/j.conbuildmat.2025.141475

Ekin, N. (2025b). Beton Dayanimi Tahmininde İkili Ve Çoklu Doğrusal Regresyon Analizlerinin Karşilaştirilmasi. *SDU Mühendislik Bilimleri Ve Tasarım Dergisi, 13*(1), 64–77. DOI: 10.21923/jesd.1572342

Ekin, N. (2025c). Yüksek dayanımlı betonlarda uluslararası standartlardaki elastik modüllerin karşılaştırılması. *Süleyman Demirel Üniversitesi Fen Bilimleri Enstitüsü Dergisi, 29*(2), 312–326. DOI: 10.19113/sdufenbed.1651363

Ekin, N., & Uyanık, O. (2021a). *Comparison of static and dynamic elastic moduli in concrete: Effects of compressive strength, curing conditions and reinforcement.* Iranian Journal of Science and Technology, Transactions of Civil Engineering., DOI: 10.1007/s40996-020-00513-7

Ekin, N., & Uyanık, O. (2021b). Anisotropy of reinforced concrete from geophysical methods. *Science of Sintering, 53*(3), 323–345. DOI: 10.2298/SOS2103323E

Ercan A. (2001). Yapı Jeofiziği IV-Beton Esnekliği ve Bozukluklar. *Dünya İnşaat Dergisi.* 18, 2001(12), 70-73.

Ercan A. (2002). Beton İçi Bozukluklar, Çatlak ve Kırıklar. *Dünya İnşaat Dergisi,* 18(2002-01), 66-71

Garcia-Monzon, H. (2003). Maturity of Concrete Strength Development, Texas Tech University Graduate School of Civil Engineering, M.S. Thesis.

Gomez-Cardenas, C. (2015). *Development of New Tools to Optimize Non-Destructive Testing Campaigns in Reinforced Concrete Structures, Durable Construction.* Université Paul Sabatier – Toulouse III.

Gupta, A. K. (2014). *Non-Destructive Testing of Bridges.* Indian Institute of Civil Railway Engineering.

Hampel, T., & Curbach, M. (2000). Behavior of high performance concrete under multiaxial loading. In *Proc., PCI/FHWA/FIB Int. Symp. on High Performance Concrete* (pp. 185–196). Precast/Prestressed Concrete Institute.

Han, P., Qiao, G., Guo, B., Li, D., & Ou, J. (2022). Investigation of low-frequency stray current induced corrosion in reinforced concrete infrastructure in high-speed railway power supply system. *International Journal of Electrical Power & Energy Systems*, *134*, 107436. DOI: 10.1016/j.ijepes.2021.107436

Hanif, A., Kim, Y., Parthasarathy, P., Usman, M., & Li, Z. (2019). Flexural fatigue behavior of lightweight ferrocement: Experimental investigation & numerical modeling. 3291–3302.

He, R., & Lu, N. (2023). Unveiling the dielectric property change of concrete during hardening process by ground penetrating radar with the antenna frequency of 1.6 GHz and 2.6 GHz. *Cement and Concrete Composites*, *144*, 105279. DOI: 10.1016/j.cemconcomp.2023.105279

Helal, J., Sofi, M., & Mendis, P. (2015). Non-Destructive Testing of Concrete: Review of Methods. *Electronic Journal of Structural Engineering*, *14*(1), 97–105. DOI: 10.56748/ejse.141931

Hola, J., & Runkiewicz, L. (2018). Methods and Diagnostic Techniques Used to Analyze the Technical Condition of Reinforced Concrete Structures. *Struct. Environ.*, *10*(4), 309–337. DOI: 10.30540/sae-2018-030

Hossain, M. I., & Islam, Md. S., & Islam, Md. M. (2024). Evaluating concrete compressive strength using non-destructive test techniques. In *Proceedings of the 7th International Conference on Civil Engineering for Sustainable Development 2024* (pp. 1–10). Khulna University of Engineering & Technology.

Jedidi, M. (2018). Evaluation of Concrete Quality Using Destructive and Non-Destructive Testing. *MOJ Civil Engineering*, *4*(4), 219–223. DOI: 10.15406/mojce.2018.04.00122

Kadioğlu, S. (2010). Description of buried archaeological remains from the Temple of Augustus in Ankara, Türkiye, using a new 3D visualization technique of the underground radar dataset. *Near Surface Geophysics*, *8*(5), 397–406. DOI: 10.3997/1873-0604.2010028

Kadioğlu, S., Kadioğlu, M., & Kadioğlu, Y. K. (2013b). Identification of buried archaeological remains in the ancient city of Nysa in Aydın- Türkiye, using ground penetrating radar, polarized microscope, and confocal Raman spectroscopy methods. *Journal of Archaeological Science*, *40*(10), 3569–3583. DOI: 10.1016/j.jas.2013.04.001

Kadioğlu, S., Kadioğlu, Y. K., & Akyol, A. A. (2011). Monitoring buried remains in the Zeynel Bey tomb in Hasankeyf ancient city, Türkiye, with transparent 3D semi-bird's-eye view of underground radar data. *Journal of Geophysics and Engineering*, *8*(3). Advance online publication. DOI: 10.1088/1742-2132/8/3/S07

Kadioğlu, S., Kadioğlu, Y. K., Catapano, I., & Soldovieri, F. (2013a). Ground penetrating radar and microwave tomography for security management of cultural heritage site: Miletus Ilyas Bey Mosque (Türkiye). *Journal of Geophysics and Engineering*, *10*(6), 064007. DOI: 10.1088/1742-2132/10/6/064007

Kairu, W. M., Otieno, M., Olanrewaju, O. A., & Mwero, J. M. (2025). Application of Schmidt hammer and ultrasonic pulse velocity for structural integrity assessment in water dams. *Journal of Infrastructure Preservation and Resilience*, *5*(1), 11. Advance online publication. DOI: 10.1186/s43065-025-00123-5

Karmakar, S., Dauji, S., Kshirsagar, S. S., Saini, S. K., Bhargava, K., & Mahapatra, K. (2021). Estimation of present-day strength of concrete for a 40-year-old building from non-destructive tests: A case study. *Journal of Asian Concrete Federation*, *7*(2), 17–32. DOI: 10.18702/acf.2021.12.7.2.17

Kibar, H., & Öztürk, T. (2015). Determination of Concrete Quality by Destructive and Non-Destructive Methods. *Computers and Concrete*, *15*(3), 473–484. DOI: 10.12989/cac.2015.15.3.473

Kupfer, H. B. (1973). *Das Verhalten des Betons unter mehrachsiger Kurzzeitbelastung unter besonderer Berücksichtigung der zweiachsigen Beanspruchung* (Schriftenreihe des DAfStb, Heft 229). Berlin: Ernst & Sohn.

Kupfer, H. B., Hilsdorf, H. K., & Rusch, H. (1969). Behavior of Concrete under Biaxial Stresses. *ACI Journal.*

Lencis, U., Udris, A., Kara De Maeijer, P., Korjakins, A., & Zvejnieks, E. (2025). A Comparison of Different Types of Pull-Off Testing and Splitting Methods for Determining the Tensile Strength of Concrete. *Buildings (Basel, Switzerland)*, *15*(7), 1068. DOI: 10.3390/buildings15071068

Li, R. W., Cao, D. S., Wu, H., & Wang, D. F. (2021). Collapse analysis and damage assessment of a typical simply supported double-pier RC bridge under truck collision. *Structures (Oxford)*, *33*, 3222–3238. DOI: 10.1016/j.istruc.2021.06.041

Lopes, Y., Vannali, L., & Jose, V. (2016). Prediction of Concrete Compressive Strength by Non-Destructive Testing Method: A Case Study. *Open Journal of Civil Engineering*, *06*(04), 503–515. DOI: 10.4236/ojce.2016.64043

Lu, Y., Liu, Z., Li, S., & Hu, J. (2018). Axial compression behavior of hybrid fiber reinforcedconcrete filled steel tube stub column. *Construction & Building Materials*, *174*, 96–107. DOI: 10.1016/j.conbuildmat.2018.04.089

Madandoust, R., & Kazemi, M. (2017). Numerical Analysis of Rupture Test Method in Concrete. *Construction & Building Materials*, *151*, 487–493. DOI: 10.1016/j.conbuildmat.2017.05.126

Malek, J., & Kaouther, M. (2014). Destructive and Non-Destructive Testing of Concrete Structures. *Jordan Journal of Civil Engineering*, *8*, 432–441.

Malhotra, V. M. (1975). Evaluation of Tensile Test for Determining the Strength of In-Situ Concrete. *Matériaux Et Constructions*, *8*(43), 19–31. DOI: 10.1007/BF02538954

Mancini, T., & Bolzoni, F. (2003). Resistivity measurements as a tool for assessment of durability of concrete structures. *Construction & Building Materials*, *17*(3), 193–199. DOI: 10.1016/S0950-0618(02)00056-0

Mariak, A., Kurpinska, M., & Wilde, K. (2019). Maturity Curve for In-Situ Strength Prediction of High Performance Concrete. *MATEC Web of Conferences, 262*, 06007. DOI: 10.1051/matecconf/201926206007

McCann, D. M., & Forde, M. C. (2001). Review of NDT Methods in the Evaluation of Concrete and Masonry Structures. *NDT & E International*, *34*(2), 71–84. DOI: 10.1016/S0963-8695(00)00032-3

Morris, I., Kumar, V., & Glisic, B. (2020). Predicting material properties of concrete from ground-penetrating radar attributes. *Structural Health Monitoring*, *20*(6), 2791–2812.

Nepomuceno, M. C., & Bernardo, L. F. (2019). Evaluation of self-compacting concrete strength with non-destructive tests for concrete structures. *Applied Sciences (Basel, Switzerland)*, *9*(23), 5109. DOI: 10.3390/app9235109

Neville, A. M. (2012). *Properties of Concrete*. Prentice Hall.

Nilimaa, J., & Nilforoush, R. (2023). A Direct Tensile Strength Testing Method for Concrete from Existing Structures. *CivilEng*, *4*(1), 333–344. DOI: 10.3390/civileng4010019

Öncü, Z. (2025). Estimation of concrete strength by using artificial intelligence (AI) methods. *Nondestructive Testing and Evaluation*, •••, 1–18. DOI: 10.1080/10589759.2025.2536660

Özçelik, E., & Gelişli, K. (2018). BETONARME YAPILARDA JEOFİZİK YÖNTEMLERLE BETON KALİTESİ VE YAPI DONATI DURUMUNUN ARAŞTIRILMASI. *Uygulamalı Yerbilimleri Dergisi, 17*(1), 1–11. DOI: 10.30706/uybd.433735

Öziçer, S. (2016). Jeofizik Yöntemler ile Riskli Yapıların Belirlenmesi ve İzmir Örneği. *SDÜ Fen Bilimleri Enstitüsü Jeofizik Müh. Anabilim dalı.* s:93

Öziçer, S., & Uyanık, O. (2017). Beton Dayanımının Yerinde P Dalga Hızından Belirlenmesi ve İzmir Örneği. *SDU International Journal of Technological Sciences, 9*(1), 1–16.

Parida, S. (2009). Vaccination against foot-and-mouth disease virus: Strategies and effectiveness. *Expert Review of Vaccines, 8*(3), 347–365. DOI: 10.1586/14760584.8.3.347 PMID: 19249976

Pucinotti, R., & De Lorenzo, R. A. (2005). In situ non-destructive testing: The steel and concrete resistance assessment of "ancient" r/c structures. *In Structural Studies, Repairs and Maintenance of Heritage Architecture* IX (p. 355).

Raj, B., Jayakumar, T., & Rao, B. P. C. (1995). Non-Destructive Testing and Evaluation for Structural Integrity. *Sadhana: Academy Proceedings in Eng. The Sciences, 20*, 5–38.

Raj, B., & Venkatraman, B. (2013). Overview of Diagnostic and Monitoring Methods and Techniques. In Czichos, H. (Ed.), *Handbook of Technical Diagnostics* (pp. 43–68). Springer Verlag. DOI: 10.1007/978-3-642-25850-3_4

Ramos, N. M. M., Simoes, M. L., Delgado, J. M. P. Q., & Freitas, V. P. (2011). Applicability of Tensile Testing: Lessons from a Large Sample of In Situ Tests. *International Conference on Durability of Building Materials and Components, 12–15 April 2011, XII DBMC, Porto, Portugal.*

Resan, S. F., Chassib, S. M., Zemam, S. K., & Madhi, M. J. (2020). New approach of concrete tensile strength test. *Case Studies in Construction Materials, 12*, e00347. DOI: 10.1016/j.cscm.2020.e00347

RILEM bulletin. (1987). Materials and Structures, 20, 79–80.

Rucka, M. (2020). Special Issue: Non-Destructive Testing of Structures. *Materials (Basel), 13*(21), 4996. DOI: 10.3390/ma13214996 PMID: 33171907

Sabbağ, N. (2016). Determination of Concrete Quality by Geophysical Methods. PhD Thesis, *Süleyman Demirel University Geophysical Engineering*, Isparta, Türkiye, 172 p.

Sabbağ, N., & Uyanık, O. (2016). Time-Dependent Variation of Seismic Velocities in Low-Strength Concrete. [TOJSAT]. *The Online Journal of Science and Technology, 6*(4), 49–57.

Sabbağ, N., & Uyanık, O. (2017). Prediction of Reinforced Concrete Strength by Ultrasonic Velocities. *Journal of Applied Geophysics, 141*, 13–23. DOI: 10.1016/j.jappgeo.2017.04.005

Sabbağ, N., & Uyanık, O. (2018a). Determination of Reinforced Concrete Strength by Electrical Resistivity Depending on Curing Conditions. *Journal of Applied Geophysics, 155*, 13–25. DOI: 10.1016/j.jappgeo.2018.03.007

Sabbağ, N., & Uyanık, O. (2018b). Comparison of Static and Dynamic Elastic Parameters of Saturated Concretes. *Afyon Kocatepe University Journal of Science and Engineering Sciences, 18*(3), 1181–1189. DOI: 10.5578/fmbd.67588

Sabbağ, N., & Uyanık, O. (2018c). Determination of Reinforced Dry Concrete Strength by Electrical Resistivity Method. *Pamukkale University Journal of Engineering Sciences, 24*(6), 1230–1236. DOI: 10.5505/pajes.2017.76702

Sabbağ, N., & Uyanık, O. (2020). Determination of the anisotropy of reinforced concretes by the resistivity method. *Pamukkale University Journal of Engineering Sciences, 26*(3), 572–580. DOI: 10.5505/pajes.2019.00008

Samson, D., & Omoniyi, T. M. (2014). Correlation between Non-Destructive and Destructive Tests of Concrete. *International Journal of Engineering and Science Invention, 3*(9), 12–17.

Selcuk, M. (2015). Investigation of Mechanical Properties of Concrete by Triaxial Compression Test, Master Thesis, ITU.

Shrestha, B., & Giri, O. P. (2023). Study on concrete compressive strength through destructive and non-destructive testing. *Int J Sci Math Technol Learn, 31*(2), 336–356.

Stanley, J. (1995). *Nondestructive Testing Handbook 9: Special Nondestructive Testing Methods*. American Society for Nondestructive Testing.

Talaat, A. A., Emad, A., Tarek, A., Masbouba, M., Essam, A., & Kohail, M. (2021). Factors affecting the results of concrete compression testing: A review. *Ain Shams Engineering Journal, 12*(1), 205–221. DOI: 10.1016/j.asej.2020.07.015

Tikalsky, P. J. (2006). Overview of Non-Destructive Testing (NDT) Material Evaluation Methods, *NSF/RILEM Workshop*, July 10-14, 2006, Prague, Czech Republic.

Turkish Ready Mixed Concrete Association (THBB). (2020). *Izmir Earthquake Investigation Report*. Earthquake Investigation Reports.

Uyanık, O. (1991). *Kaya Mekaniği ve Jeofizik Laboratuar Parametrelerinin İlişkilendirme Açısından Önemi*. Dokuz Eylül Üniv. Müh.-Mim. Fak. Jeofizik Müh. Böl.

Uyanık, O. (1999). Kayaçlarda Sismik Hızlar ve Kayma Direncinin İncelenmesi. *52. Türkiye Jeoloji Kurultayı*, 63-70, Ankara.

Uyanık, O. (2012). Determination of Concrete Strength from Seismic Velocities. *Geophysical Bulletin, 23*(70), 25–30.

Uyanık, O. (2014). Jeofizik ve Klasik Yöntemlerle Yapı İncelemeleri (Yapı Jeofiziği). JFMO Eğitim Yayınları No:19 ISBN:978-605-01-0643-5.

Uyanık, O. (2015). Deprem ağır hasar alanlarının önceden belirlenmesi ve şehir planlaması için makro ve mikro bölgelendirmelerin önemi. *Süleyman Demirel Üniversitesi Fen Bilimleri Enstitüsü Dergisi, 19*(2), 24–38.

Uyanık, O., & Çatlıoğlu, B. (2015). Determination of Density from Seismic Velocities. *Jeofizik, 17*(1-2), 3–15.

Uyanık, O., Çatlıoğlu, B., Sabbağ, N., Öncü, Z., & Uyanık, N. A. (2012a). Determination of Concrete Quality of Reinforced Concrete Structures in Urban Transformation Projects from Seismic Ultrasonic Velocities. *1st Earth Sciences Symposium, SDU*, Türkiye.

Uyanık, O., Çatlıoğlu, B., Sabbağ, N., Öncü, Z., & Uyanık, N. A. (2012b). Relationships Between Rock Properties and Seismic Ultrasonic Velocities. *1st Earth Sciences Symposium, SDU*, Türkiye.

Uyanık, O., Kaptan, K., Gülay, F. G., & Tezcan, S. (2011). Determination of Concrete Strength by Non-Destructive Ultrasonic Method. *Construction World, 184*, 55–58.

Uyanık, O., Sabbağ, N., Çatlıoğlu, B., Uyanık, N. A., & Öncü, Z. (2012c). Determination of Fracture and Fissure Indexes of Rocks from Seismic Ultrasonic Velocities. *1st Earth Sciences Symposium, SDU*, Türkiye.

Uyanık, O., Sabbağ, N., Uyanık, N. A., & Öncü, Z. (2019). Estimation of Mechanical and Physical Properties of Some Sedimentary Rocks from Ultrasonic Velocities. *Bulletin of Engineering Geology and the Environment, 78*(8), 6003–6016. DOI: 10.1007/s10064-019-01501-6

Uyanık, O., Şenli, G., & Çatlıoğlu, B. (2012d). Determination of Concrete Quality of Buildings by Non-Destructive Geophysical Methods. *International Construction Congress 2012, ICONC2012-SDÜ.*

Uyanık, O., Şenli, G., & Çatlıoğlu, B. (2013). Determination of Concrete Quality of Buildings by Non-Destructive Geophysical Methods. *SDÜ International Journal of Technology Sciences*, *5*(2), 156–165.

Uyanık, O., & Tezcan, S. (2012). Determination of Concrete Strength by Ultrasonic Method. *Geophysical Bulletin*, *23*(70), 41–45.

Wang, B., Zhong, S., Lee, T. L., Fancey, K. S., & Mi, J. (2020). *Nondestructive Testing and Evaluation of Composite Materials/Structures: A State-of-the-Art Review*. Advances in Mechanical Engineering., DOI: 10.1177/1687814020913761

Whitehurst, E. A. (1951). Soniscope Testing of Concrete Structures. *Journal of the American Concrete Institute. Proceedings*, *47*, 443–444.

Xia, W., Bai, E., Xu, J., & Liu, G. (2021). Experimental Study on the Strength and Deformation Characteristics of Concrete under True Triaxial Compression after Sulfate Attack. *Advances in Civil Engineering, 5548313*, 15 pages.

Yang, S., Xu, Z., & Bian, Z. (2020). Prediction on compressive strength concrete using modified pull-off testing method (MPTM). *Construction and Building Materials, 250*, 118834, ISSN 0950-0618.

Yıldız, A., & Daloğlu, A. T. (2021). Determination of Concrete Strength in Reinforced Concrete Buildings by Non-Destructive Methods. *Gazi University Journal of Science Part C*, *9*(3), 783–794.

Yousif Q.A. (2018). Beton Yaşının Basınç Dayanımı Üzerine Olan Etkisinin Ultrases Yöntemi ile Belirlenmesi. *Dicle Üniversitesi Fen Bilimler Enstitüsü YL Tezi, Fizik Anabilim Dalı*, Diyarbakır, 50s.

Zhang, Y., Gao, L., & Bian, W. (2020). Mechanical Performance of Concrete Made with Recycled Aggregates from Concrete Pavements, *Advances in Materials Science and Engineering*, 2020, 5035763, 8 pages. DOI: 10.1155/2020/5035763

KEY TERMS AND DEFINITIONS

Compressional Wave: A type of seismic wave that moves through materials by compressing and expanding them in the direction the wave is traveling.

Destructive test: A testing method where the material is subjected to conditions that lead to its failure or damage to determine its properties.

Earthquake: A sudden and violent shaking of the ground caused by movements within the Earth's crust or volcanic activity.

Non-Destructive Test: A testing method used to evaluate the properties or integrity of a material without causing any damage to it.

Shear Wave: A type of seismic wave that moves the ground perpendicular to the direction of wave travel, causing shearing motion.

Uniaxial Compressive Strength Test: A laboratory test that measures the maximum axial compressive stress a material can withstand before failure.

APPENDIX: ABBREVIATIONS

DT: Destructive Test
GPR: Ground Penetrating Radar
NDT: Non-Destructive Testing
RI: Rebound Index
RILEM: The International Union of Laboratories and Experts in Construction Materials, Systems and Structures (from the name in French)
RN: Rebound Number
SCC: Self-Compacting Concrete
SonReb: Sonic+Rebound hammer methods
UCS: Uniaxial Compressive Strength Test
UPV: Ultrasonic Puls Velocity
Vp: P wave velocity
Vs: S wave velocity

Chapter 5
Thermal Stability and Corrosion Resistance of AISI 904L Stainless Steel for Industrial Applications

Vaiavel Madeshwaren
http://orcid.org/0000-0002-8687-7887
Dhanalakshmi Srinivasan College of Engineering, Coimbatore, India

R. S. Gandhimathi
Annapoorana Engineering College, Salem, India

Bondala Ramakrishna
Aditya Institute of Technology and Management, Tekkali, India

ABSTRACT

AISI 904L stainless steel is widely recognized for its superior corrosion resistance and excellent thermal stability, making it a preferred choice for demanding industrial environments. This study evaluates the material's corrosion behavior, high-temperature performance, and microstructural stability. Optical microscopy and scanning electron microscopy (SEM) were employed to analyze the microstructure. TGA results indicate strong oxidation resistance, with only 0–5% weight loss at 800°C. Mechanical tests reveal high resistance to deformation and significant stress retention at elevated temperatures, highlighting its capacity to withstand high strain rates. Corrosion testing in acidic environments shows improved resistance with increasing temperature, underscoring the alloy's resilience in harsh

DOI: 10.4018/979-8-3373-6167-3.ch005

conditions. Supplementary analyses via XRD, EDX, and SEM confirm the alloy's structural integrity up to 1000°C. These findings affirm AISI 904L's suitability for applications that demand durability, thermal endurance, and long service life in aggressive industrial settings.

1.INTRODUCTION

Austenitic stainless steel 904L is renowned for being ideal for use in the chemical processing oil and gas and marine industries due to its remarkable resistance to corrosion particularly in acidic environments. Its high concentration of nickel and molybdenum makes it more resistant to corrosion and pitting in cracks. Thermal analysis including methods like Thermogravimetric Analysis (TGA) is crucial for evaluating its performance at high temperatures because phase changes and thermal degradation can significantly affect its mechanical properties. The conductivity heat capacity and thermal stability of the alloy must be understood in order to control heat dissipation during processes like welding. Understanding the microstructural behavior of sensitized 904L is crucial because research has shown that it is prone to pitting corrosion in chloride solutions. I. The mechanical performance oxidation resistance and thermal stability of AISI 904L Stainless Steel under varied temperature and strain conditions are thoroughly assessed in this study using a combination of Dynamic Recrystallization (DRX) TGA SEM EDX and XRD techniques. This illustrates the steels resistance to corrosive heat.

According to research 904L stainless steels resistance to corrosion in chloride environments can be enhanced by changing its composition through microstructural changes and alloying components (Wang et al., 2021). The material is more suitable for demanding applications due to the significant improvement in corrosion and wear resistance that boronized AISI 904L surface treatments provide by increasing surface hardness and decreasing wear (Çetin et al., 2021). Moreover several studies have looked at 904Ls performance in extreme heat and maritime conditions focusing on how structural changes affect corrosion resistance and how susceptible the materials are to stress corrosion cracking in seawater (Loto et al., 2020; Xu et al., 2024). The resistance to corrosion in sulfuric acid environments is impacted by welding techniques according to research on 904L weldments. This illustrates how carefully controlled welding parameters are essential to maximize material durability under difficult conditions (Rashed et al., 2020). Recent research indicates that advanced manufacturing techniques such as wire arc additive manufacturing improve the 904Ls microstructures mechanical properties and resistance to corrosion (Vishnukumar et al., 2023). Understanding corrosion behavior and microstructural changes in welded joints is crucial as demonstrated by the difficulties encountered when welding dissimilar materials such as AISI 904L and UNS S32750 (Dagur et al., 2017). Additionally it has been demonstrated that surface treatments such as micro-shot peening reduce the likelihood of corrosion particularly when fluoride is the root cause (Pour-Ali et al., 2023). Certain welding techniques such as hot wire tungsten inert gas and activated tungsten inert gas welding which change material integrity and corrosion behavior also affect the performance of 904Ls in welded forms (Sunny et al., 2022). Finally studies on the corrosion-erosion behavior of 904Ls in phosphoric acid environments emphasize the importance of controlling impurities to reduce corrosion rates. Thermal treatments also affect pitting corrosion behavior (Aftimi et al., 2024; Li et al., 2023).

According to studies on the connection between microstructural morphology and the corrosion susceptibility of austenitic and superaustenitic stainless steels precise microstructural control is required to enhance material performance (Sheik et al., 2022). Research conducted on 904L composite plates at elevated temperatures and pressures has demonstrated their ability to endure harsh circumstances (Wang et al., 2024).

2. EXPERIMENTAL PROCEDURE

2.1 Material

To enhance its homogeneity and purity AISI 904L Stainless Steel is typically synthesized using advanced metallurgical techniques. The alloy was produced by melting the raw materials which included molybdenum copper nickel iron and chromium at a temperature between 1600 and 1650°C in an electric arc furnace. In order to lower the carbon content to less than 0. 02% and eliminate contaminants that might compromise corrosion resistance refinement techniques like vacuum oxygen decarburization (VOD) and argon oxygen decarburization (AOD) were introduced (Anbuchezian & Vairavel, 2022). Then to guarantee a uniform microstructure the molten metal was formed into billets or slabs hot-rolled at 1100–1200°C and annealed at 1050–1150°C. To get a surface roughness (Ra) below 0. 01 μm the samples were then final polished with a 0. 05 μm colloidal silica suspension. This painstaking preparation produced excellent surfaces that could be used for elemental and microstructural analysis (Adithya et al., 2020).

2.2 Experimental Procedure

A series of fine abrasive papers and polishing cloths are then used to polish the samples after they have been ground to remove any surface impurities. Once the samples have been polished the microstructure is revealed for analysis by etching them with an appropriate chemical reagent usually a solution of nitric acid and hydrochloric acid. Methods such as potentiodynamic polarization thermogravimetric analysis (TGA) and dynamic recrystallization (DRX) are used to measure the corrosion rate and pitting resistance of AISI 904L stainless steel samples in an electrochemical test conducted in a seawater simulation. High temperatures are also applied to the material in order to assess its oxidation behavior (Vairavel et al., 2018).

2.3 Thermal Analysis

According to Table 1 the high melting point of AISI 904L stainless steel (1350°C to 1400°C) guarantees structural integrity in extremely hot conditions making it suitable for high-temperature applications. With an enthalpy change of roughly 35 J/g Differential Scanning Calorimetry (DSC) reveals that AISI 904L Stainless Steel experiences phase changes at 1200°C that affect its mechanical qualities including toughness and ductility (Maheskumar et al., 2021). Thermal decomposition and oxidation were indicated by the materials notable weight loss of 49–51 percent at 857°C under iron oxide and ion current measurements revealed activity at this temperature.

Table 1. Thermal analysis of AISI 904L stainless steel

Thermal Analysis Parameter	Value
Melting Point Range	1350 - 1400°C
Phase Transformation Temperature (DSC)	1200°C
Onset Temperature for Oxidation (TGA)	~600°C
Weight Loss at 800°C (TGA)	~0.5%
Enthalpy Change During Heating (DSC)	~35 J/g
Thermal Expansion Coefficient	~14.0 μm/m°C
Oxidation Rate at 900°C (TGA)	~0.03 mg/cm^2/hr

2.4 Corrosion Analysis

To assess AISI 904L Stainless steel's performance in challenging conditions like seawater and sulfuric acid solutions corrosion analysis was carried out using electrochemical techniques specifically potentiodynamic polarization. Utilizing potentiodynamic polarization in solutions of sulfuric acid and seawater the corrosion analysis of AISI 904L stainless steel was carried out. These circumstances made it possible to assess the material's corrosion resistance by measuring the rate of corrosion pitting potential and passivation behavior. Important parameters such as the alloy's corrosion rate pitting potential and passivation behavior are provided by potentiodynamic polarization testing which also offers important insights into the alloy's durability. Even in harsh environments, the elements help to create a stable passive film on the surface that shields the material from pitting cracking from stress corrosion, and corrosion in crevices.

2.5 SEM, XRD, and EDX Analysis

Optical microscopy (OM) or scanning electron microscopy (SEM) are used for microstructural examination which aids in determining the materials grain structure phase composition and presence of precipitates or inclusions. A scanning electron microscope (SEM) for microstructural analysis a Bruker D8 Advance X-ray diffractometer (XRD) for crystallographic and phase composition analysis and an Energy Dispersive X-ray Spectroscopy (EDX) system for elemental analysis were among the tools utilized for the analysis of AISI 904L Stainless Steel. While XRD offered information on the alloys crystallographic texture and phase composition SEM was utilized to view grain boundaries phases and recrystallization twins. The crystallographic texture and phase composition of AISI 904L stainless steel films were determined using X-ray diffraction (XRD).

3. RESULT AND DISCUSSION

3.1 TGA of AISI 904L Stainless Steel under Iron Oxide

Figure 1 shows thermogravimetric analysis (TGA) and derivative thermogravimetry (DTG) plots under various temperature conditions offering information on the materials thermal stability and decomposition patterns. The TGA curve (black) in Figure 1(a) exhibits initial weight stability up to about 800°C after which there is a dramatic weight decrease that translates into a notable mass loss of 49–51 percent at about 857°C. Thermal breakdown or oxidation reactions involving the FeO_3 and carbon mixture are probably the cause of this decrease. The temperature of maximum decomposition rate is indicated by a prominent peak in the derivative plot which also highlights the rate of decomposition. Ion current measurements (blue) indicate that these reactions are associated with variations in ionic emissions. These reactions are particularly active at temperatures around 857°C where oxidation probably plays a role in ion current fluctuations. A detailed view of the behavior of the FeO_3 + C sample over the temperature range of 0–1400°C is shown in Figures 1(a) and (b) with particular attention to the decomposition steps and finer thermal events. With mass losses of 3–27% and 49%–51% respectively it reveals distinct thermal events at temperatures such as 808°C and 1017°C indicating multiple reaction stages.

Figure 1. TGA analysis of AISI 904L Stainless Steel exposed to iron oxide

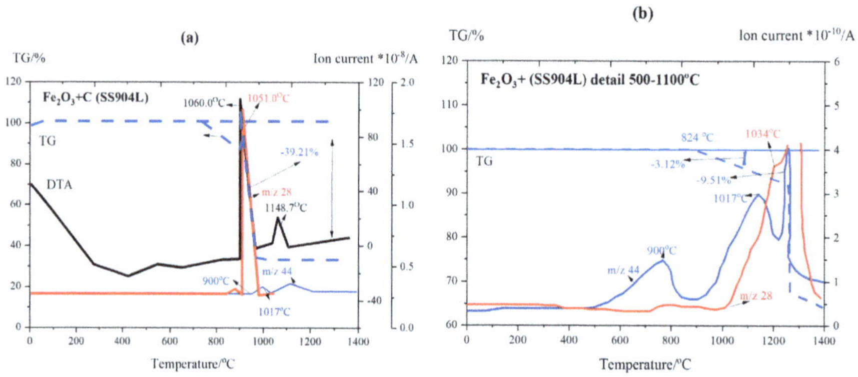

3.2 TGA of AISI 904L Stainless Steel in Nitrogen Atmosphere

By gradually increasing the temperature by 20°C per minute the thermogravimetric analysis (TGA) of AISI 904L stainless steel offers a detailed picture of its thermal breakdown behavior in a nitrogen atmosphere. The two main curves in the TGA plot the derivative weight (blue curve), and the weight loss percentage (black curve) reflecting the materials mass loss and rate of decomposition respectively over the temperature range. Figure 2 shows the TGA analysis of nitrogen-exposed AISI 904L stainless steel. The sample shows a weight loss of 8.832% at around 303–40°C where the first decomposition is seen.

Figure 2. TGA analysis of AISI 904L stainless steel exposed to nitrogen

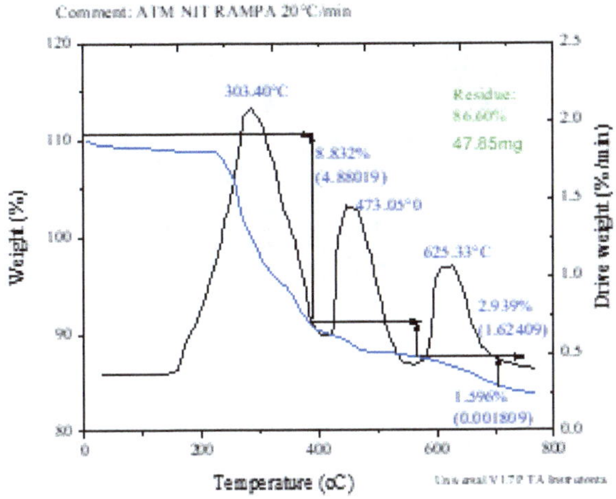

These points correspond to further breakdown phases in the alloy's structure, as represented by peaks in the derivative weight curve, which captures the rate of weight loss per minute at each stage. As the analysis reaches its end at approximately 800°C, AISI 904L Stainless Steel retains a substantial portion of its mass, with a residue of 86.60%, equating to 47.85 mg. capability to withstand harsh conditions while maintaining structural integrity.

3.3 Dynamic Recrystallization in AISI 904L Stainless Steel

The true-stress versus true-strain behavior of SS 904L material in Dynamic Recrystallization (DRX) under strain rates of 5 s^{-1} and 0.001 s^{-1} highlights the material's response to different deformation rates given in Figure 3. At a higher strain rate of 5 s^{-1}, the true stress increases sharply with the strain, reaching initial peak stress levels of 240-260 MPa at a true strain of 0.1 across temperatures from 850°C to 1150°C. This stress response then stabilizes around 150-350 MPa as strain progresses to 0.9, indicating the material's strain-hardening capacity diminishes over continued deformation.

Figure 3. Deformation behavior of SS-904L at strain rates a) 5 s^{-1} and b) 0.001 s^{-1}

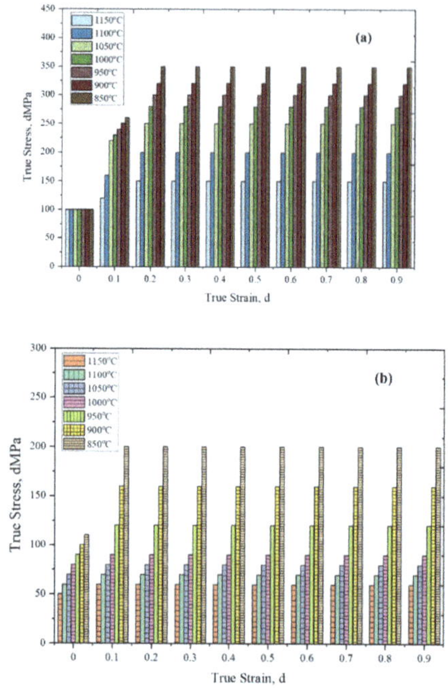

By contrast, at the lower strain rate of 0.001 s^{-1}, the SS 904L exhibits lower initial stress values, with stresses ranging from 50-110 MPa at zero strain, gradually increasing to 120-200 MPa by a strain of 0.1. This lower-rate deformation results in more consistent stress values, with minimal changes up to a strain of 0.9. The differences between these curves suggest that SS 904L material demonstrates strain-rate sensitivity, achieving higher stress levels with increased strain rates, particularly at elevated temperatures.

3.4 Activation Energy for Deformation

The data presents the variation of maximum true stress (in megapascals, MPa) at different strain rates (5 s^{-1}, 1 s^{-1}, 0.1 s^{-1}, 0.01 s^{-1}, and 0.001 s^{-1}) across a range of temperatures (800°C to 1150°C). As the temperature increases, the maximum true stress at each strain rate decreases, illustrating the typical material behavior in high-temperature conditions.

Figure 4. Thermo-mechanical influence on measured and calculated peak stresses

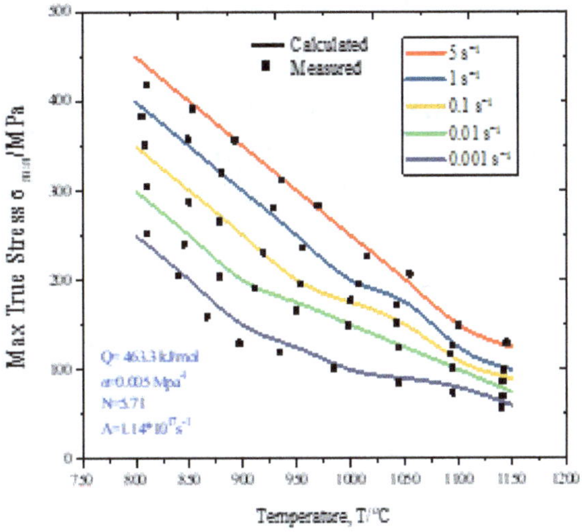

This trend reflects the thermally activated softening of the material, where the atoms have greater mobility, leading to reduced resistance to deformation. Figure 4 suggests that the material exhibits significant strain rate sensitivity, with higher strain rates resulting in higher true stresses, a common characteristic in materials under dynamic loading conditions.

The results are crucial for understanding the material's behavior under various processing or operational conditions, and this information can be applied to optimize processing parameters such as forming, forging, or welding, ensuring the material's mechanical properties align with specific application requirements in high-temperature environments.

3.5 Temperature Impact on Pitting and Corrosion Products

Table 2 summarizes the corrosion characteristics of a material as temperature increases, including pitting density, pit size, depth, and elemental composition. At 25°C, a low pitting density of 0.25 pits/mm^2 is observed, with a small average pit size of 10 µm and depth of 5 µm; iron content is 60% and bromide 15%, with no lagoon formation. As temperature rises to 50°C and 60°C, pitting density and depth increase slightly, though lagoon formation remains absent. At 70°C and 80°C, pitting density stabilizes at 1.25 pits/mm^2, but the average pit size significantly increases to 20 µm and then 70 µm, with depths reaching 15 µm. Iron and bromide percentages decrease progressively, with 52% iron and 8% bromide at 80°C. Lagoon formation is noted from 70°C onwards, indicating increased susceptibility to localized corrosion at higher temperatures. The average values suggest that high temperatures promote larger pits, deeper corrosion, and lead to lagoon formation.

Table 2. Pitting behavior of AISI 904L stainless steel at varying temperatures

Temperature (°C)	Pitting Density (pits/mm^2)	Average Pit Size (μm)	Major Pit Depth (μm)	Lagoon Formation	Iron (%)	Bromide (%)
25	0.25	13	5	No	60	15
50	0.7	12	8	No	58	13
60	1.25	10	10	No	55	11
70	1.25	20	12	Yes	53	10
80	1.25	70	15	Yes	52	8
Average	0.92	22.5	10	-	56.3	11.2

3.6 Optical Microscopy of SS-904L Corrosion

Figure 5 shows optical images of the corroded surface morphology of SS-904L stainless steel after exposure to a lithium bromide (LiBr) solution at 50 wt.% concentration under varying temperatures. Each image (a) through (f) represents distinct temperature conditions, illustrating the progressive impact of temperature on the surface characteristics of SS-904L. Yellow dashed circles indicate corrosion sites in these photos emphasizing regions where pitting and surface deterioration take place.

Figure 5. Corrosion morphology of SS-904L in LiBr–H_2O at different temperatures

Figure 6 shows the X-ray diffraction (XRD) pattern of SS-904L stainless steel, highlighting the material's crystal structure. Austenite is the predominant phase in SS-904L as confirmed by the peaks in the spectrum that correspond to the γ (austenite) phase.

Figure 6. XRD pattern of 904L

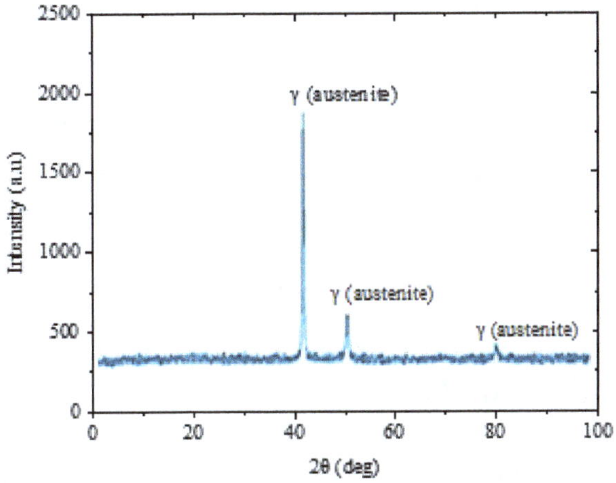

3.7 Polarization Behavior of AISI-904L stainless steel

Figure 7 presents the polarization curves for AISI-904L stainless steel in a lithium bromide (LiBr)–water (H_2O) solution at various temperatures, specifically at 25°C, 50°C, 60°C, 70°C, and 80°C. As the temperature increases, the corrosion current density rises, reflecting an accelerated corrosion rate and reduced resistance due to higher electrochemical activity in the LiBr–H_2O solution. This trend underscores the impact of elevated temperatures on the stability of SS-904L, critical for applications where it may be exposed to high-temperature, corrosive environments.

Figure 7. Effect of temperature on polarization curves of SS-904L in LiBr–H_2O solution

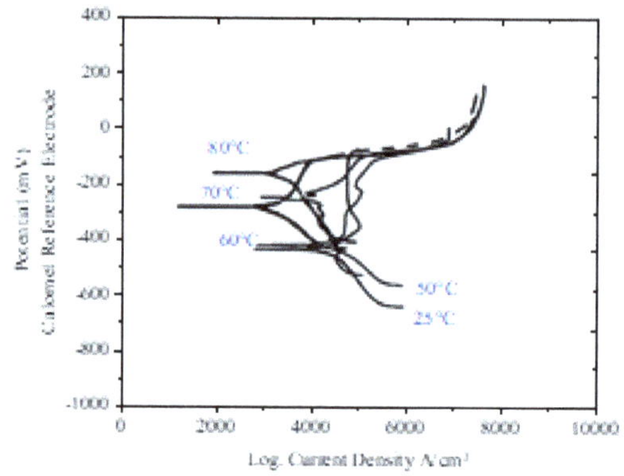

3.8. Morphological Analysis

According to the EDX (Energy Dispersive X-ray) elemental analysis, the elements weight and atomic percentage show little variation from the recast material to the hollows deepest point and the heat-affected zone. In contrast to the ellipse and crater array patterns the elemental analysis provides a zeroweight percentage for oxygen. With the exception of carbon which is also found in the cavity area the elemental analysis of octagonal micro structuring from cross-sectional images reveals a uniform spread outside the hollow area.

Figure 8. Elemental composition analysis of AISI 904L stainless steel using EDX (a) austenite (b) overlapping (c) overlapping (2 different color variation) (d) pull out

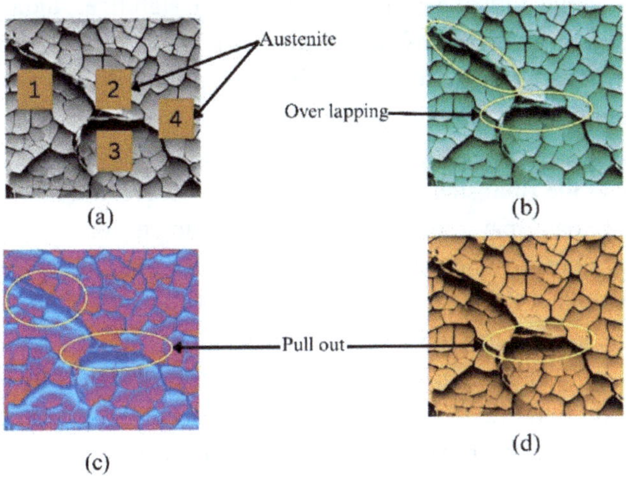

The primary EDX spectrum elements are displayed in Figure 8. The EDX results of the 2-hour aluminized 904L samples are presented in Table 3, showing the elemental composition in weight percent (wt.%) for each analyzed spot. Sample 1 exhibits a high concentration of platinum (59.96%) along with notable amounts of aluminum (12.19%) and iron (8.14%), with smaller amounts of nickel (6.12%) and chromium (4.83%). Sample 2 is dominated by platinum (69.91%), with a significant amount of aluminum (15.58%) and trace amounts of iron (7.04%) and nickel (5.64%), and chromium (1.82%).

Table 3. EDX results for 2-hour exposed SS 904L (elemental composition in wt.%)

Element	Sample 1 (%)	Sample 2 (%)	Sample 3 (%)	Sample 4 (%)	Sample 5 (%)
O (Oxygen)	4.83	1.82	1.21	2.42	1.87
Fe (Iron)	12.19	15.58	18.37	27.61	49.5
Pt (Platinum)	1	2	3	4	5
Ni (Nickel)	8.14	7.04	8.16	13.94	20.68
R (Residuals)	-	-	-	2.81	-
Cr (Chromium)	6.12	5.64	7.04	11.82	21.52
Mo (Molybdenum)	-	-	-	-	5.82
Al (Aluminum)	59.96	69.91	65.23	41.4	0.61

Table 3 shows the compositional variation of elements in AISI 904L Stainless Steel under various conditions, possibly due to thermal treatment, oxidation, or material interaction. Oxygen initially decreases but stabilizes, indicating changes in oxidation levels. Iron increases, suggesting enrichment or diffusion. Platinum and Nickel concentrations increase steadily, indicating stability and potential reinforcement. Chromium increases, suggesting enhanced corrosion resistance. Aluminum decreases at higher stages, possibly due to reaction or phase changes. Molybdenum appears only at the final stage, possibly indicating segregation or delayed involvement. The "R" data may represent a specific parameter or phase emerging under specific conditions. This evolution reflects the dynamic interplay of elements in AISI 904L Stainless Steel.

In Sample 3, platinum (65.23%) and aluminum (18.37%) are the predominant elements, accompanied by iron (8.16%) and nickel (7.04%), with a minor concentration of chromium (1.21%). Sample 4 shows a more balanced distribution, with aluminum (27.61%) and iron (13.94%) being the most prominent elements, followed by nickel (11.82%) and chromium (2.42%), and a small amount of molybdenum (2.81%). Finally, Sample 5 reveals a substantial concentration of iron (49.5%) and nickel (21.52%), along with a lower concentration of aluminum (0.61%), chromium (1.87%), and a significant oxygen content (5.82%), but no detected platinum or molybdenum.

Figure 9 illustrates the Energy-Dispersive X-ray (EDX) spectrum of the microtextured design of AISI 904L Stainless Steel stainless steel, providing a detailed analysis of the elemental composition of the surface.

Figure 9. Energy dispersive X-ray analysis of AISI 904L stainless steel microtextured patterns

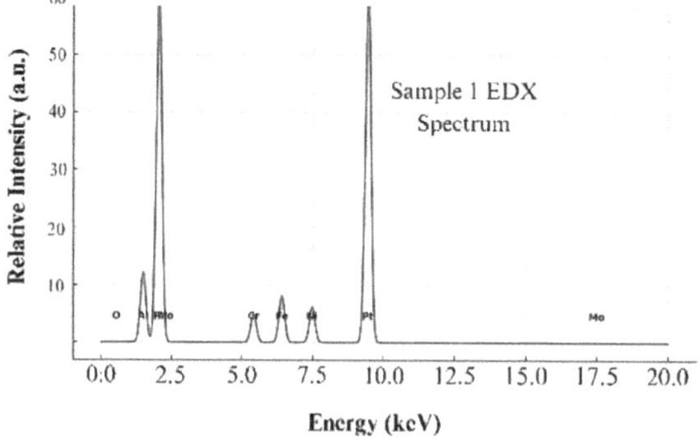

The EDX results of the 4-hour SS 904L samples are presented in Table 4, showing the elemental composition in weight percent (wt.%) for each spot analyzed. Iron

(Fe) ranges from 25.15% to 49.85%, indicating its dominant role in the material matrix. Nickel (Ni) varies widely (3.72% to 20.55%), reflecting its contribution to structural stability and corrosion resistance. Chromium (Cr) content spans 5.6% to 31.46%, essential for passivation and oxidation resistance. Aluminum (Al) shows a pronounced presence in certain samples, reaching up to 60.02%, indicating potential surface oxidation or coatings. The presence of R (residuals) and Molybdenum (Mo), at 7.37% in specific cases, contributes to enhanced mechanical and corrosion performance. Oxygen (O) and Silicon (Si) appear in minor quantities, suggesting oxidation and inclusions, while Platinum (Pt) content increases progressively, likely indicating experimental conditions or contamination trends. This composition highlights a complex alloy system with varying elemental roles across samples.

In Sample 4, similar to Sample 3, aluminum (43.29%) and iron (28.55%) dominate, but with slightly higher nickel (7.43%) and molybdenum (0.59%) concentrations. Sample 5 shows a distinct composition with a high concentration of nickel (31.46%) and a significant amount of iron (46.36%), alongside smaller amounts of aluminum (3.36%) and molybdenum (2.2%). Finally, Sample 6 primarily consists of iron (49.85%) and nickel (20.75%), with notable concentrations of aluminum (4.63%) and chromium (1.22%), and lower amounts of molybdenum (1.72%) and oxygen (1.4%).

Table 4. EDX results for 4-hour exposed SS 904L (elemental composition in wt.%)

Element	Sample 1 (%)	Sample 2 (%)	Sample 3 (%)	Sample 4 (%)	Sample 5 (%)	Sample 6 (%)
Fe (Iron)	39.44	25.15	28.97	28.55	46.36	49.85
Ni (Nickel)	11.95	7.84	17.13	17.74	3.72	20.55
Cr (Chromium)	9.83	5.6	7.92	7.43	31.46	20.75
Al (Aluminum)	13.1	60.02	43.89	43.29	2.2	0.15
R (Residuals)	16	-	1.5	2.4	2.5	1.4
Mo (Molybdenum)	3.17	0.2	-	-	7.37	4.63
O (Oxygen)	6.32	-	0.56	0.59	2.2	1.72
Si (Silicon)	-	1.2	-	-	3.36	1.22
Pt (Platinum)	1	2	3	4	5	6

A cross-sectional SEM image of the aluminized 904L after two hours at 675 °C is shown in Figure 10 (a). It is 19. 2 μm thick when the aluminide is formed. The distribution of the dark gray coating layer is uniform. Between the coating layer and substrate is a gray interdiffusion zone (IDZ). It is around 1μm thick. There is no crack formation in the coating layer and the adherence of the formed layer is fairly good. The cross-sectional SEM image of 904L that has been aluminized at 675 °C for 4 hours is displayed in Figure 10(b). Comparing the coating layer to 2h-aluminized 904L the latter has a more complex phase distribution in deeper zones and is 290 μm thick. Figure 11 presents the morphology of thermal coatings on AISI 904L Stainless Steel substrates, illustrating various structures observed under different conditions.

Figure 10. SEM image of (a) 2h AISI 904L stainless steel (b) SEM image of 4h AISI 904L stainless steel

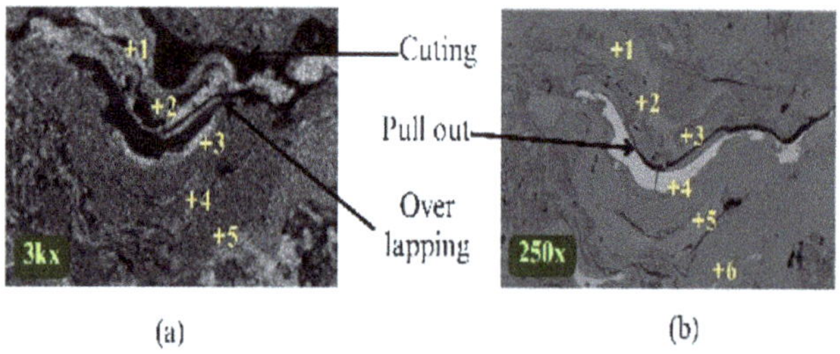

In (c), the semiplanar morphology is observed, where the coating exhibits a combination of flat and slightly textured areas. Figure 11 (d) depicts the island-type structure, with distinct, isolated regions of coating material forming "islands" on the substrate. The semispheroid morphology in (e) reveals rounded, partially spherical features, while (f) shows a fully spheroid structure with smooth, well-rounded shapes. In (g), needle-like structures are seen, indicative of elongated, sharp features. Finally, panel (h) presents a trunk-shape morphology, where the coating forms tree-like, branch-like protrusions. These varied morphologies suggest differences in the coating's formation process, influenced by factors such as temperature, time, and material composition.

Figure 11. (a) crystalline (b) austenite (c) pull out (d)island (e) planar (f) overlapping and curing (g) needle (h) final surface morphology of thermal coatings on AISI 904L stainless steel substrates

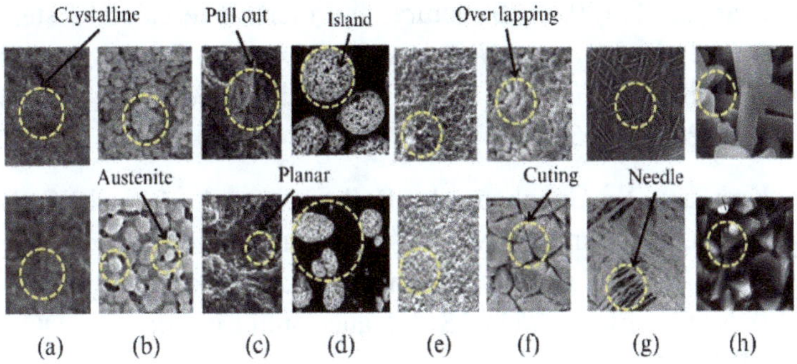

4. CONCLUSION

The experimental analysis of AISI 904L Stainless Steel a high-alloy austenitic stainless steel in this study has highlighted its suitability for harsh industrial applications by demonstrating its exceptional performance in extreme environments. A stable passive film that prevents pitting crevice corrosion and stress corrosion cracking is formed by the materials high chromium nickel molybdenum and copper content. This is the main reason for the materials exceptional corrosion resistance especially in chloride-rich and acidic environments. Potentiodynamic polarization and dynamic recrystallization (DRX) test results validate the robustness of AISI 904L Stainless Steel in the chemical processing and marine industries where its high-temperature stability and resistance to localized corrosion are essential. The materials resilience to temperatures as high as 1000°C is demonstrated by its low oxidation rate high thermal stability and minimal weight loss during thermogravimetric analysis (TGA). The substantial weight retention (86–60 percent at 800°C) revealed by the TGA analysis particularly in a nitrogen atmosphere emphasizes the durability of AISI 904L Stainless Steel even after extended exposure to high temperatures.

REFERENCES

Adithya, K., Girimurugan, R., Vairavel, M., Jawahar, M., Surya, K., Tamilselvan, M., & Thiyagarajan, P. (2020). Structural and thermal research of steam turbine blades by finite element method. *International Journal of Innovative Technology and Exploring Engineering*, *9*(5), 2275–2278. Advance online publication. DOI: 10.35940/ijitee.E2495.039520

Aftimi, S., Kerroum, Y., Guenbour, A., Bellaouchou, A., Idrissi, H., Boulif, R., Semlal, N., Warad, I., & Zarrouk, A. (2024). New insights on the corrosion–erosion behavior of 904L stainless steel in phosphoric acid containing impurities. *Journal of Bio- and Tribo-Corrosion*, *10*(1), 13. Advance online publication. DOI: 10.1007/s40735-024-00818-4

Anbuchezian, A., & Vairavel, M. (2022). Steel fibre impacts on mechanical performance and toughness of steel fibre reinforced high strength concrete after normal and hygrothermal curing (SFRHSC). *NeuroQuantology : An Interdisciplinary Journal of Neuroscience and Quantum Physics*, *20*(5), 1135–1143.

Çetin, M., Günen, A., Kalkandelen, M., & Karakaş, M. S. (2021). Microstructural, wear and corrosion characteristics of boronized AISI 904L superaustenitic stainless steel. *Vacuum*, *187*, 110145. DOI: 10.1016/j.vacuum.2021.110145

Dagur, A. H., Kartha, A. A., Subodh, M. A., Vishnu, C., Arun, D., Vijay Kumar, M. G., Abraham, W. S., Chatterjee, A., & Abraham, J. (2017). Microstructure, mechanical properties and biocorrosion behavior of dissimilar welds of AISI 904L and UNS S32750. *Journal of Manufacturing Processes*, *30*, 27–40. DOI: 10.1016/j.jmapro.2017.09.001

Li, M., Zou, D., Li, Y., & Tong, L. (2023). Effect of cooling rate on pitting corrosion behavior of 904L austenitic stainless steel in a simulated flue gas desulfurization solution. *Metals and Materials International*, *29*(3), 730–747. DOI: 10.1007/s12540-022-01255-z

Loto, R. T., Loto, C. A., & Fajobi, M. (2020). Corrosion resistance of GX4CrNiMo16-5-1 martensitic, 316L austenitic, and 904L austenitic stainless steels subjected to high temperature variation. In Ikhmayies, S., Li, B., Carpenter, J. S., de Oliveira, D. L. C., Monteiro, F., Anyalebechi, P. N., Sagüés, A. A., & Solanki, K. N. (Eds.), *Characterization of minerals, metals, and materials 2020* (pp. 261–267). Springer., DOI: 10.1007/978-3-030-36628-5_24

Maheskumar, P., Girimurugan, R., Sivaraman, G., Purushothaman, S., & Vairavel, M. (2021). Thermal analysis on regenerative cooled liquid propellant rocket thrust chamber using ANSYS. *IOP Conference Series. Materials Science and Engineering*, *1059*(1), 012068. Advance online publication. DOI: 10.1088/1757-899X/1059/1/012068

Pour-Ali, S., Tavangar, R., & Hejazi, S. (2023). Effect of micro-shot peening on microstructure and fluoride-induced corrosion performance of AISI 904L super-austenitic stainless steel. *Materials Today. Communications*, *36*, 106682. DOI: 10.1016/j.mtcomm.2023.106682

Rashed, G. M., Mohrez, W. A., Abd Allah, M., & Ghayad, I. M. (2020). Corrosion behavior of super austenitic stainless steel 904L weldment in sulfuric acid media used for leaching application. *Key Engineering Materials*, *835*, 384–391. . DOI: 10.4028/www.scientific.net/KEM.835.384

Sheik, S., Tirumalla, A., Gurrala, A. K., & Mohammed, R. (2022). Effect of microstructural morphology on corrosion susceptibility of austenitic and super austenitic stainless steels. *Materials Today: Proceedings*, *66*, 514–518. DOI: 10.1016/j.matpr.2022.06.113

Sunny, K. T., Korra, N. N., Muthukumaran, V., & Arivazhagan, B. (2022). Corrosion studies and thermogravimetric analysis performed on activated tungsten inert gas and hot wire tungsten inert gas welded super austenitic stainless steel AISI 904L. *Proceedings of the Institution of Mechanical Engineers. Part C, Journal of Mechanical Engineering Science*, *236*(20), 10528–10540. DOI: 10.1177/09544062221106120

Vairavel, M., Vinod Kumar, T., Chandrasekaran, M., & Vivek, P. (2018). Evaluation of tensile and thermal behavior analysis of nano carbon fiber reinforced polymer composite. *International Journal of Engineering & Technology, 7*(3.27), 177–180. DOI: 10.14419/ijet.v7i3.27.17750

Vishnukumar, M., Muthupandi, V., & Jerome, S. (2023). Microstructural characteristics, mechanical properties and corrosion performance of super austenitic stainless steel 904L produced by wire arc additive manufacturing. *Materials Today. Communications*, *35*, 105801. DOI: 10.1016/j.mtcomm.2023.105801

Wang, J., Shi, W., Xiang, S., & Ballinger, R. G. (2021). Study of the corrosion behaviour of sensitized 904L austenitic stainless steel in Cl^- solution. *Corrosion Science*, *181*, 109234. DOI: 10.1016/j.corsci.2020.109234

Wang, S., Mei, P., Chang, L., Wu, C., Chen, S., Chen, Q., & Li, G. (2024). Evaluation of the corrosion resistance of 904L composite plate in a high-temperature and high-pressure gas field environment. *Processes (Basel, Switzerland)*, *12*(11), 2372. Advance online publication. DOI: 10.3390/pr12112372

Xu, D., Zhang, X., He, X., Liu, C., Liu, Z., Zhang, X., Zhao, M., & Li, X. (2024). Mechanism and evaluation method of stress corrosion susceptibility of 904L stainless steel with optimized structure in seawater. *Corrosion Science*, *229*, 111865. DOI: 10.1016/j.corsci.2024.111865

Chapter 6
From Mold to Marvel: Engineering Excellence in Precast Systems

Ruchira Srivastava
http://orcid.org/0000-0002-5957-9534
Vollert India Pvt. Ltd., India

Krishan Mohan Sharma
http://orcid.org/0009-0003-8679-4365
Vollert India Pvt. Ltd., India

ABSTRACT

A recent analysis of the building sector reveals a low degree of automation in both manufacturing and construction, as well as extremely low material usage efficiency. There has been a clear shift towards automated manufacturing processes, while the sector also struggles with a shortage of competent people in mechanical engineering and construction. In India, the majority of building projects are completed using the traditional cast-in-place technique. However, India continues to have a massive housing need. Consequently, the construction work needs to be done considerably more quickly. This cannot be accomplished using traditional building techniques. With precast concrete for building, it is conceivable. When compared to normal concrete, precast concrete offers additional benefits. Thus, a variety of literature is examined, and this study provides a review of them all.

1.INTRODUCTION

In the aftermath of agriculture, the construction industry is our nation's second-largest economic activity. The precast or prefabricated construction idea refers to

DOI: 10.4018/979-8-3373-6167-3.ch006

structures in which many of the structural elements are manufactured in yards or facilities closer to the construction site and then transported to the building area for erection or assembly (Murari & Joshi, 2017). Precast concrete is widely employed in both residential and commercial building projects these days. The precast concrete's property is the reason for this. It is easier to handle, has better thermal qualities, and is more durable. Additionally, because precast concrete is produced with meticulous attention to detail, its quality is superior. Nonetheless, our nation lacks awareness and expertise about precast concrete (Cassagnabère et al., 2010). This must be altered, and further research on precast is required. It is also important to thoroughly examine the precast members' connecting details. Numerous literary works are examined in this publication, and the findings of such investigations are presented here (Al-Bazi & Dawood, 2010).

The environment, quality of life, and socioeconomic development are all significantly impacted by the building sector and its operations (Dara et al., 2024). The construction business is often characterized by labour-intensive technologies, harsh working conditions, low productivity, and high hazards, despite its enormous influence on national economies and people's lives. The fundamental cause of these problematic business conditions is the construction industry's sluggish adoption of industrialization concepts and technology advancements like computer-aided construction, automation, standardization, and modularization (Pedro et al., 2017).

Precast concrete technique offers a number of benefits to all project participants overcast-in-situ manufacturing (Polat, 2010). Computer-aided technology is used in precast manufacturers' operations to make precast concrete components utilizing advanced gear in a laboratory setting, resulting in high-quality and long-lasting products (Sacks et al., 2004). Precast concrete systems are still widely utilized in many eastern and northern European nations, and they became one of the main construction technologies in many of them, especially following World War II (Chen & Tiong, 2024).

One of India's longest-standing issues has been the lack of homes. The 2011 census report's identification of 13.7 million families as being in slums makes this point very clear (Holton, 2009). This issue has been becoming worse at an exponential rate as cities get more populated. In India, there is a scarcity of more 20 million homes now (Reichenbach & Kromoser, 2021). About 600 million people are expected to reside in cities by 2031, up from 434 million in 2015. By 2022, the union cabinet authorized the Housing for All plan, which aims to build 20 million homes for the urban poor to address the housing shortage (Bonev et al., 2015).

Conventional technology used in the building sector is unable to address the severe housing shortage brought to light by the affordability issue. The successful execution of initiatives like Housing for All will be greatly aided by emerging technology (Straatman & Vambersky, 2001). One such cutting-edge Indian technology that has the potential to be very important is precast. Using precast technology, various building components are manufactured in a controlled setting, like a factory, and then delivered to the construction site for assembly and installation. Precast building components can include things like floor slabs, in-filled walls, stairwells, restrooms, and beams and columns (Tomek, 2017). These construction components are delivered and installed at the project site after being produced in a controlled atmosphere in a particular factory. Both onsite and offsite precast pieces can be produced (Brzev & Guevara-Perez, 2010).

Precast concrete is rarely used in the Indian building industry, despite its well-established advantages. The difficulty to determine the project's advantages is one of the main reasons why stakeholders do not adopt precast technology (Kim et al., 2020). Precast technology use is not well understood and is occasionally thought of being a more expensive option. Since there are now no official measurement techniques or tactics available, pilot research showed that choices to use precast are still mostly dependent on circumstantial evidence rather than reliable facts (Arditi et al., 2000). This research aims to identify the difficulties that precast technology faces and offer cooperative solutions for those difficulties that are recognized at the industry level. Precast technique offers several benefits when used in construction (Jaillon et al., 2007). Because precast pieces are manufactured in a factory, the regulated atmosphere allows for higher-quality concrete to be created. Compared to typical building methods, precast construction relies less on manpower, which helps to minimize project coordination, scheduling, and sequencing chaos (Kurama et al., 2018). After the members are assembled, there is no need for on-site curing because they are cured at a factory for the required amount of time. Precast can lead to savings since it eliminates the need for formwork (Azman et al., 2012).

Precast members can be used to reduce the cost of post-concrete repairs. Because precast has little operational hazards, it also improves workplace health and safety (Baran et al., 2021). When it is used, on-site construction operations are reduced, which can result in a large decrease in waste at the site. The amount of material needed to complete the same building can be decreased by employing precast technology through efficient planning and design (Ahmad et al., 2019). Its ability to reduce waste further contributes to precast's status as an environmentally benign and sustainable building technology. Precast's greatest benefit is that it shortens the project's duration by speeding up the building process (Mohammed, 2018).

The environment may benefit greatly from precast construction. In general, it requires less energy during erection and demolition, produces less trash during construction and demolition, and minimizes emissions throughout the precast element manufacture and building construction (Choi et al., 2013). When it comes to architectural flexibility, adaptation, and the reuse of precast concrete components, precast construction presents several opportunities (Wong & Loo, 2022).

2.PRECAST OUTPACES CAST-IN-SITU

The foundation of contemporary construction has long been concrete, and two main techniques for using it have developed over time: precast concrete technology and cast-in-situ, or cast-in-place, construction (Warszawski et al., 1984). Despite having the same basic goal, the two approaches are very different in terms of sustainability, financial implications, speed, execution, and quality control. To find the most economical and successful building solutions, architects, builders, and developers must be aware of the differences between the two (Xiaosheng & Hamzeh, 2020).

Making precast concrete entails producing concrete components in a controlled, off-site factory setting (Scheer et al., 2005). These components, which include walls, stairways, slabs, columns, beams, and façade panels, are made with standardised moulds and go through stringent quality control procedures. Cranes and other specialised lifting equipment are used to move these precast modules to the construction site after they are finished and hardened (Jang et al., 2022). The speed, accuracy, and reproducibility of this approach are all much improved. Because the production takes place in a factory, it is not impacted by the weather, resulting in a steady output all year long (Sacks et al., 2003; Khare et al., 2011). Furthermore, compared to on-site casting, precast factories' high levels of automation and mechanisation guarantee superior surface finishes, durability, and dimensional precision.

Conversely, the conventional technique of pouring concrete into formwork on the construction site is known as "cast-in-situ" construction (Asamoah et al., 2016). Before more building can begin, reinforcement is placed, concrete is mixed, and it is poured on the site. This process takes time to set and cure (Elliott, 2000). More design flexibility is provided by this approach, particularly for constructions with intricate geometries and irregular shapes. For monolithic structures where structural continuity is crucial, such foundations, retaining walls, or core walls in high-rise buildings, cast-in-situ is frequently chosen (Zheng et al., 2018). Because everything is done on-site, there are also fewer logistics needed for lifting and transporting of large components.

Cast-in-situ construction, however, is very reliant on weather and expert labour. Rain, severe weather, or a lack of workers can all cause delays that have a big effect on the project's timeframe (Martins et al., 2023). Furthermore, quality control is more difficult because it relies on appropriate mixing ratios, curing procedures, and on-site workmanship. On the other hand, precast components gain from uniform production procedures, which lower variability and boost long-term dependability and performance (Srivastava et al., 2019).

In terms of sustainability and the environment, precast building frequently leads to less material waste, improved resource management, and less dust and noise pollution on the job site. Precast is an eco-friendly choice, particularly for large-scale projects, because it uses reusable moulds and can produce pieces in batches (Srivastava & Thakur, 2025). Despite its versatility, cast-in-situ frequently uses more water and energy and produces more waste because it is done on-site (Srivastava et al., 2014).

Due to the need for lifting equipment, element transportation, and mould setup, precast construction may need a larger initial investment (Bahadur et al., 2023). It makes up for this, though, with shorter turnaround times, lower labour costs, and less delays. In the beginning, cast-in-situ may seem more cost-effective, particularly for small-scale projects, but it frequently leads to longer schedules, more on-site monitoring, and maybe more rework because of inconsistent quality (Srivastava et al., 2024).

To sum up, several variables, including project size, design complexity, time restrictions, skilled labour availability, and long-term durability needs, influence the decision between precast concrete and cast-in-situ solutions (Mishra et al., 2017). For projects where speed, uniformity, and scalability are important, precast concrete is the best option; however, cast-in-situ is still a good option for special, site-specific designs that call for structural continuity (Chen et al., 2020). To capitalise on each method's advantages, a hybrid approach may even be the best course of action for construction (Liu et al., 2022).

Refer Table 1 which depicts the comparison of precast technology and cast in situ technology on various industrial parameters in tabulated form.

Table 1. Comparison of precast technology and cast in situ technology

S No.	Parameters	Cast In Situ Technology	Precast Concrete Technology	Citation
1.	Outline	The building site is where concrete is poured and allowed to cure.	In a regulated industrial environment, concrete components are cast and allowed to cure before being delivered to the location.	(Drage & Kennedy, 2020)
2.	Use Cases	Custom structures, foundations, small buildings, and restorations.	stadiums, prefabricated homes, bridges, and industrial structures.	(Ben-Daya et al., 2019; Wu et al., 2019)
3	Quality Control (QC)	Variable: affected by worker skill and site conditions.	high since the industrial environment is controlled.	(Zhao et al., 2024)
4	Cost Efficiency (Long Term)	may eventually result in increased labour and maintenance expenses.	Improved because of its durability and shorter building time.	(Yang et al., 2025)

continued on following page

Table 1. Continued

S No.	Parameters	Cast In Situ Technology	Precast Concrete Technology	Citation
5.	Speed of Construction	Slower—reliant on the gradual construction process and curing period.	Faster: components are swiftly installed and prefabricated.	(Yaqoob & Silfwerbrand, 2023; Sherfudeen et al., 2016)
6.	Requirement of labour	Higher: Formwork, pouring, and curing need a lot of on-site labour.	Lower: installation mostly calls for expert labour.	(Kehl & Carrasquillo, 1998)
7.	Cost (Initial)	lower, particularly for modest projects.	Higher because of transportation and manufacturing.	(Rahim et al., 2012; Gao et al., 2024)
8.	Formwork & Scaffolding	Large-scale—needed during the building stage.	Very little—just installation assistance.	(Skrzypczak, 2023)
9.	Sustainability	Reduced – frequently leads to increased material waste.	Better: more effective use of resources and less waste of materials.	(Schipper, 2015)
10.	Dependency on weather	Weather has an impact on high-quality construction and curing.	Minimal weather has no effect on productivity.	(Qiao et al., 2024)
11.	Transportation	Concrete is mixed and poured on-site; therefore, it is not necessary.	Transporting the components to the location is necessary.	(Zhang et al., 2022)
12.	Surface finish	Post-treatment is often necessary to improve the finish.	Factory-controlled finishing is excellent.	(Liu et al., 2020)
13.	Structural Flexibility	More adaptable—permits design modifications and on-site alterations.	Less adaptable—after components are created, modifications are challenging.	(Abedi et al., 2014)

3. THE JOURNEY OF PRECAST PRODUCTION

The controlled method by which steel and raw concrete are methodically converted into high-performance structural components is what makes precast engineering so beautiful (Wang & Hu, 2017). Precast thrives on accuracy, repeatability, and efficiency, in contrast to conventional building, which is frequently beset by delays, irregularities, and erratic site circumstances (Ryang et al., 2004). Design, Moulding & Casting, Curing & Quality Control, and Transportation & Installation are the four separate but related stages of the production process as shown in Figure 1 (Vatin et al., 2018).

Design Phase

The designer's desk is where the precast adventure starts, not the factory. Since it establishes the groundwork for accuracy and efficiency later, the design phase is the most crucial stage (Wang & Xu, 2024). Architects and engineers work together to develop a modular design in which every building component is seen as a piece of a bigger picture. Here, sophisticated techniques like Building Information Modelling (BIM) and Computer-Aided Design (CAD) are essential (Won et al., 2004). With the use of BIM, a digital twin of the project can be produced, showing in three dimensions how each precast element will fit together on the job site (Unlu, 2010). It guarantees that conflicts between mechanical, electrical, and structural systems are identified prior to the start of production. Durability, service integration, and load-bearing requirements are all considered during design optimisation (Rahman & Sobuz, 2018). To guarantee that every mould made in the factory satisfies the project's requirements, the design output is converted into accurate shop drawings and production schedules before the end of this phase (Nanyam et al., 2017).

Moulding & Casting Phase

The emphasis switches to the actual construction of the components after the design is complete. First, moulds are prepared. Depending on the element's intricacy, these are usually constructed of steel, fibreglass, or wood (Kaja & Jauswal, 2021).

Then, with careful placement, reinforcement cages are positioned inside the mould to guarantee proper alignment and cover (Syed & Sonparote, 2020). Under carefully monitored circumstances, high-strength concrete is combined and then poured into the mould. Density and homogeneity are ensured by removing air pockets using methods like vibration. Robotic casting methods and automation reduce human error and boost productivity in modern factories (Priya & Neamitha, 2018). The science of casting and the art of moulding come together at this point to create components that precisely match the design parameters.

Curing & Quality Control Phase

The proper curing process is when concrete reaches its true strength, not when it is cast (Tayabji et al., 2013). Steam chambers, heat treatment, or humidity control are examples of controlled curing settings that speed up the hydration process while guaranteeing steady strength development. Precast curing is a strictly controlled procedure as opposed to on-site curing, which is subject to changing weather conditions (Singhal et al., 2019).

Every element is subjected to stringent quality control inspections after curing. In accordance with applicable standards and norms, tests are conducted for compressive strength, surface quality, dimensional accuracy, and load-bearing capacity (Tan et al., 2024). To find internal defects, non-destructive testing techniques like ultrasonic pulse velocity may also be used. An element can only be released for distribution after passing these tests. Precast components are guaranteed to be dependable and long-lasting for many years of use thanks to this unwavering quality assurance (Chang et al., 2023).

Figure 1. Journey from design to mould making

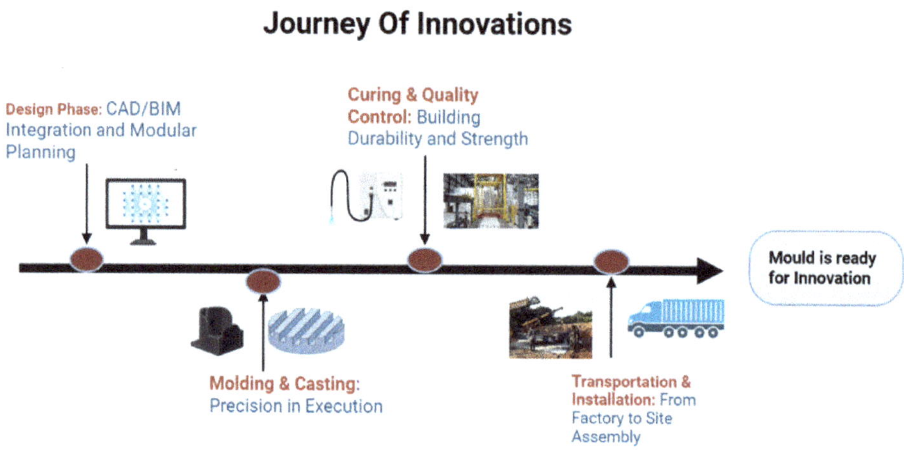

Transportation & Installation

Transferring the precast wonder from the controlled factory setting to the dynamic settings of the construction site is the last phase in the process. Logistics-wise, transportation is difficult, particularly for heavy or massive components (Wang et al., 2019). To guarantee a safe delivery, specialised trailers, protective padding, and well-planned routes are used.

Installation is a very planned effort at the location (Olidis et al., 2010). Elements are precisely positioned to the millimetre level using cranes and lifting apparatus. To guarantee structural continuity, connections between precast units are meticulously made using grouting, bolting, or welding. Project timeframes are frequently cut in half since site assembly is substantially quicker than traditional construction due to the pre-engineered components (Novak et al., 2017). Additionally, because less scaffolding, formwork, and wet work are needed, site safety is enhanced. Figure1 clearly depicts the journey of innovation that how a mould is created from first to last what all major phases are involved in the process (Li et al., 2014).

Engineering, Excellence & Innovation

Technical accuracy and operational efficiency are no longer the only criteria for engineering excellence in the quickly changing industrial world of today. The capacity to innovate constantly, turning concepts into solutions that not only satisfy present demands but also foresee those of the future, is the foundation of true excellence. Engineering advancement now depends heavily on innovation, which fuels long-term value creation, sustainability, and competitiveness (Agrawal et al., 2021).

Fundamentally, engineering innovation is about questioning norms. It necessitates an attitude that rejects constraints and looks for quicker, more intelligent, and more sustainable solutions (Maurya & Madan, 2023). Through the introduction of cutting-edge technology, sophisticated materials, and digital ecosystems, this innovative attitude has completely reshaped a variety of industries, including manufacturing, construction, logistics, and energy (Madireddy et al., 2021). Today, being able to incorporate innovation into every step of the process is just as important as perfect execution when it comes to engineering excellence (Xiao & Fomin, 2024). Digital transformation is among the most obvious examples of this. Engineers are rethinking the conception, design, and execution of projects using technologies like artificial intelligence (AI), the Internet of Things (IoT), and building information modelling (BIM). Digital twins enable real-time system monitoring, automation guarantees consistency at scale, and predictive analytics improves decision-making (Fang et al., 2022). These developments preserve the highest standards of quality while cutting down on errors, expenses, and delivery times (Fallah et al., 2020). As a result, proactive optimisation replaces reactive problem-solving in engineering. Figure 2 below clearly depicts the modern ingredients of Precast technology for creating smart structures.

Figure 2. Modern ingredients of precast technology

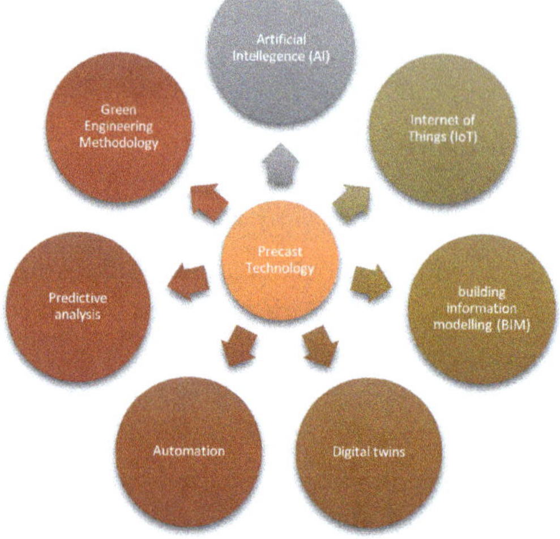

The part sustainability plays in innovation is equally significant. Green engineering has become a cornerstone of excellence due to the global demand for ecologically responsible solutions (Ma et al., 2023). Engineers are at the vanguard of advancing climate-conscious development, whether it be through carbon-neutral production methods, recyclable materials, or energy-efficient designs. Future generations are protected against expansion at the expense of innovation. Precast and modular methods, for instance, are lowering waste, carbon footprints, and facilitating quicker, safer project completion in the construction sector (Kolbeck et al., 2023).

Green engineering has become a cornerstone of excellence due to the global demand for ecologically responsible solutions. Engineers are at the vanguard of advancing climate-conscious development, whether it be through carbon-neutral production methods, recyclable materials, or energy-efficient designs (Yardim, 2018). Future generations are protected against expansion at the expense of innovation. Precast and modular methods, for instance, are lowering waste, carbon footprints, and facilitating quicker, safer project completion in the construction sector (Bachmann & Steinle, 2011).

Innovation-driven excellence also includes human capital and teamwork as important components. No engineering innovation occurs in a vacuum. It calls for groups that have the freedom to try new things, grow, and push the envelope (Senthilvel et al., 2016). Innovation becomes a shared duty rather than a compartmentalised endeavour in high-performance organisations that foster curiosity, cross-functional cooperation, and talent development. Organisations can fully realise the potential of engineering innovation by fusing human ingenuity with cutting-edge technologies (Turai & Patil, 2022).

The engineering field is also changing because of the incorporation of Industry 4.0 concepts. Production and distribution are being revolutionised by data-driven ecosystems, robotic automation, and smart factories (Dolati & Mehrabi, 2021). These developments improve resilience as well as efficiency, allowing businesses to swiftly adjust to changes in the market, interruptions in the supply chain, or changing client demands. Thus, engineering excellence through innovation turns into a competitive advantage, guaranteeing that businesses stay current and prepared for the future (Sua-iam & Makul, 2024).

Continuous dedication to study, experimentation, and learning is necessary for engineering excellence. Teams and leaders must accept uncertainty as a chance rather than a hindrance. Organisations can attain transformative change by integrating innovation into their core values, going beyond small-scale enhancements (O'Hegarty & Kinnane, 2020).

Challenges of Precast Technology

Even though precast building has revolutionised contemporary infrastructure, there are still obstacles to its widespread use. The promise of quicker construction schedules, better quality, and sustainability is frequently beset by challenges related to logistics, execution, and regulatory compliance (Pampanin, 2005). Precast' s full potential can only be realised if the industry acknowledges these challenges and devises solutions. However, the future also holds a wealth of prospects driven by sophisticated manufacturing, sustainability, and digitisation.

Perception is one of the biggest obstacles. Precast is still frequently thought of by developers and contractors as being expensive, having little flexibility, or only being suitable for large-scale projects (Jadhav et al., 2023). This myth frequently results from ignorance about precast's long-term benefits, which include lower maintenance, better quality control, and quicker project completion. Even though they are slower and less effective, old approaches nevertheless rule some marketplaces just because people are accustomed to them. To get beyond these obstacles, stakeholders must be educated through case studies, demonstrations, and experimental projects (Hewes, 2013). Adoption also necessitates a cultural change in the way construction is viewed, from site-based, labour-intensive methods to a factory-driven, industrialised process.

Figure 3. Challenges in precast technology

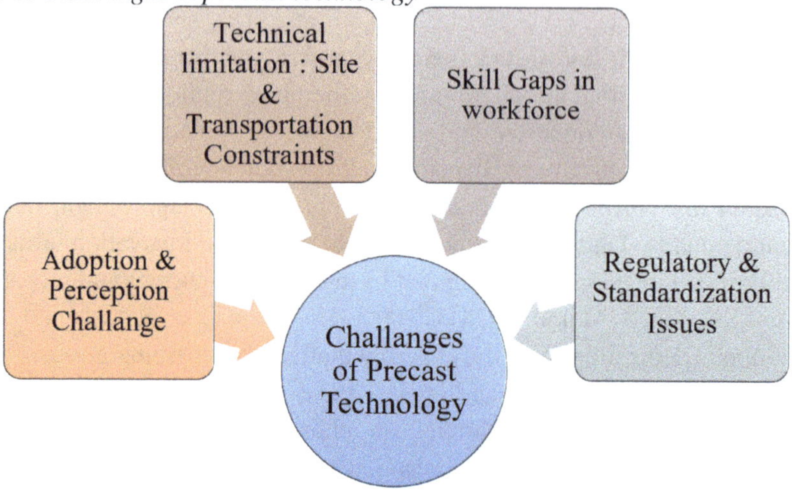

One of the main practical difficulties of this technology is the transportation of huge precast parts from factories to project locations (Jaillon & Poon, 2009). Significant obstacles are created by congested cities, small roadways, and limitations on the movement of large vehicles. The size of the element that can be produced is frequently determined by transportation limitations rather than structural requirements. Site factors might sometimes make installation more difficult, such as a lack of storage space or difficulties positioning a crane (Sadeghi & Singh, 2023). The efficiency benefits that precast would ordinarily provide may be diminished by these logistical problems. Better infrastructure, specialised tools, and creative planning that strikes a balance between design goals and transit viability are needed to overcome them (Yee et al., 2011).

Precast requires a professional workforce for production, handling, and installation, even though it lessens the need for unskilled labour on-site. Crane operators must oversee intricate lifting operations, engineers must be skilled in modular design, and site crews must make exact connections between parts (Qi et al., 2021). Unfortunately, there is a lack of these specialised talents in many areas. Training programs, certifications, and industry-academia collaboration are needed to build a workforce that can support the growing precast industry.

Regulations that are still focused on traditional cast-in-situ techniques frequently cause problems for precast construction (Sridhar et al., 2023). Modular design ideas might not be fully supported by building rules, approval processes, and inspection procedures. Additionally, inconsistent regional standards raise project risks and cause misunderstanding. This lack of standardisation may cause delays and higher expenses for developers. Accelerating adoption requires establishing uniform, transparent codes and encouraging government cooperation (Maurya & Madan, 2025). The precast ecosystem would also be further strengthened by regulations that provide incentives for industrialised and sustainable construction.

Opportunities in Precast Technology

Despite its difficulties, precast engineering offers a wealth of options that will only grow as the building sector moves towards speed, sustainability, and innovation. Precast technology is increasingly being used as the technique of choice for projects requiring quality, efficiency, and scale, rather than merely as an alternative to traditional building (Hieber et al., 2005). There are numerous avenues for growth, ranging from infrastructural development and urbanisation to digital integration and sustainability.

The growing urbanisation of the world is one of the biggest potential opportunities. Cities' expanding populations are driving up demand for high-quality, reasonably priced homes (Ye et al., 2024). Because of their lengthy schedules and vulnerability to delays, traditional construction methods frequently fall short of these demands. Precast provides a scalable solution by enabling components to be manufactured offsite and swiftly assembled on site. The precast industry has a lot of potential thanks to government-led affordable housing programs, smart cities, and mass housing projects (Kiong & Akasah, 2012).

Figure 4. Opportunities in precast technology

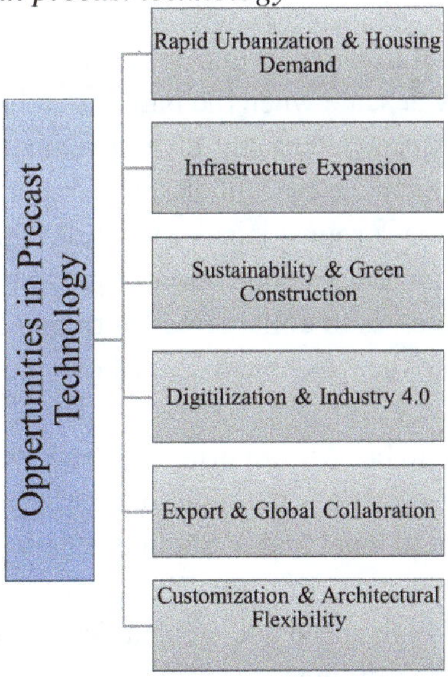

Infrastructure is a key factor in the deployment of precast outside of residential buildings. Precast provides the speed, accuracy, and durability needed for large-scale projects including metro rail systems, flyovers, bridges, industrial parks, and airports (Rui & Sen, 2018). To promote economic growth, governments everywhere, including India, are making significant investments in infrastructure. Precast elements including girders, tunnel sections, and airport pavements may guarantee adherence to strict safety and quality standards while drastically cutting down on building schedules (Ghayeb et al., 2022).

Sustainability is now a fundamental need in contemporary construction; it is no longer an option. Eco-friendly methods are ideally suited to precast. Effective resource recycling is made possible by controlled factory production, which also minimises waste and material consumption (Wang et al., 2023). Precast's environmental credentials are further reinforced by its capacity to use carbon-neutral cement, recycled aggregates, and optimised designs. Precast has the potential to become a key component of sustainable building methods since governments and organisations are placing a greater emphasis on ESG principles and green certifications (Priddy et al., 2013).

Digital integration is another exciting prospect. The design, manufacture, and maintenance of precast systems are changing because of technologies like digital twins, artificial intelligence (AI), and building information modelling (BIM) (Ding & Zhang, 2012). Accurate modular design, collision detection, and lifespan analysis are made possible by sophisticated software, which lowers expenses and errors. Production is further increased by automation in factories, such as automated reinforcement installation, 3D printing, and robotic formwork preparation. Precast facilities have the potential to develop into intelligent manufacturing centres that provide consistent quality at scale when the building industry adopts Industry 4.0.

Precast manufacturers can potentially take advantage of export prospects in emerging regions. Precast elements or technology are imported from well-established players in many areas that lack robust local production capabilities (Guerrini et al., 2015). Additionally, international cooperation in project implementation and design opens doors for businesses to collaborate internationally. For instance, Indian precast companies can use their technical know-how and cost competitiveness to support global projects.

Precast was often seen to be stiff or boring, but new developments in moulding and design techniques have made it possible to largely customise it. Precast technologies can meet both functional and aesthetic needs for anything from intricate facades to distinctive architectural finishes. This creates new opportunities in high-end projects that demand both speed and elegant design, as well as in commercial real estate and institutional buildings.

CONCLUSION

The goal of engineering excellence through innovation is to provide more intelligent, environmentally friendly, and robust solutions. It involves foreseeing tomorrow's difficulties while creatively and precisely meeting today's demands. Those that combine engineering rigour with audacious innovation will not only survive but also shape the future of development as industries continue to change. Although precast engineering is the way of the future in construction, there are challenges along the way. Its progress is still being slowed by perception obstacles, transportation issues, skill shortfalls in the workforce, and misaligned regulations. A multi-stakeholder strategy that combines workforce development, government assistance, and industry innovation is needed to address these issues. Precast could completely transform construction by making it quicker, safer, and more environmentally friendly once these obstacles are methodically removed. There are several prospects for precast engineering in the future that fit with the goals of the global construction sector, which include sustainability, speed, scalability, and digital transformation. Precast is positioned to be a game-changer due to the growing demand for housing, extensive infrastructure projects, sustainability objectives, and the incorporation of Industry 4.0 technology. If the industry can take advantage of these chances, it will not only change how we construct but also reshape the foundation of contemporary urban development.

REFERENCES

Abedi, M., Rawai, N. M., Fathi, M. S., & Mirasa, A. K. (2014). Cloud computing as a construction collaboration tool for precast supply chain management. *Jurnal Teknologi (Sciences & Engineering), 70*(7).

Agrawal, A., Sanghai, S. S., & Dabhekar, K. (2021). A review on analysis and design of precast structures. *International Journal of Scientific Research in Science, Engineering and Technology*, *8*(2), 345–350. DOI: 10.32628/IJSRSET218267

Ahmad, S., Soetanto, R., & Goodier, C. (2019). Lean approach in precast concrete component production. *Built Environment Project and Asset Management*, *9*(3), 457–470. DOI: 10.1108/BEPAM-02-2018-0051

Al-Bazi, A., & Dawood, N. (2010). Developing crew allocation system for the precast industry using genetic algorithms. *Computer-Aided Civil and Infrastructure Engineering*, *25*(8), 581–595. DOI: 10.1111/j.1467-8667.2010.00666.x

Arditi, D., Ergin, U., & Günhan, S. (2000). Factors affecting the use of precast concrete systems. *Journal of Architectural Engineering*, *6*(3), 79–86. DOI: 10.1061/(ASCE)1076-0431(2000)6:3(79)

Asamoah, R. O., Ankrah, J. S., Offei-Nyako, K., & Tutu, E. O. (2016). Cost Analysis of Precast and Cast-in-Place Concrete Construction for Selected Public Buildings in Ghana. *Journal of Construction Engineering*, *2016*(1), 8785129. DOI: 10.1155/2016/8785129

Azman, M. N. A., Ahamad, M. S. S., Majid, T. A., & Shah, M. N. S. A. (2012). A study of precast concrete in Malaysia. *Concrete (London)*, *46*(10), 50–52.

Bachmann, H., & Steinle, A. (2011). *Precast concrete structures*. Ernst & Sohn.

Bahadur, P. S., Srivastava, R., & Srivastava, V. (2023). Review on the Development of Smart Materials, Including Shape Memory Alloys, Characterization, and Recent Applications. *Modeling, Characterization, and Processing of Smart Materials*, 107-139.

Baran, E., Mahamid, M., Baran, M., Kurtoglu, M., & Torra-Bilal, I. (2021). Performance of a moment resisting beam-column connection for precast concrete construction. *Engineering Structures*, *246*, 113005. DOI: 10.1016/j.engstruct.2021.113005

Ben-Daya, M., Hassini, E., & Bahroun, Z. (2019). Internet of things and supply chain management: A literature review. *International Journal of Production Research*, *57*(15-16), 4719–4742. DOI: 10.1080/00207543.2017.1402140

Bonev, M., Wörösch, M., & Hvam, L. (2015). Utilizing platforms in industrialized construction: A case study of a precast manufacturer. *Construction Innovation*, *15*(1), 84–106. DOI: 10.1108/CI-04-2014-0023

Brzev, S., & Guevara-Perez, T. (2010). *Precast concrete construction*. British Columbia Institute of Technology.

Cassagnabère, F., Mouret, M., Escadeillas, G., Broilliard, P., & Bertrand, A. (2010). Metakaolin, a solution for the precast industry to limit the clinker content in concrete: Mechanical aspects. *Construction & Building Materials*, *24*(7), 1109–1118. DOI: 10.1016/j.conbuildmat.2009.12.032

Chang, R., Zhang, N., & Gu, Q. (2023). A review on mechanical and structural performances of precast concrete buildings. *Buildings (Basel, Switzerland)*, *13*(7), 1575. DOI: 10.3390/buildings13071575

Chen, C., & Tiong, R. (2024). Solutions to Address the Low-Capacity Utilization Issue in Singapore's Precast Industry. *Information (Basel)*, *15*(8), 458. DOI: 10.3390/info15080458

Chen, J. H., Ha, N. T. T., Tai, H. W., & Chang, C. A. (2020). The willingness to adopt the Internet of Things (IoT) conception in Taiwan's construction industry. *Journal of Civil Engineering and Management*, *26*(6), 534–550. DOI: 10.3846/jcem.2020.12639

Choi, H. K., Choi, Y. C., & Choi, C. S. (2013). Development and testing of precast concrete beam-to-column connections. *Engineering Structures*, *56*, 1820–1835. DOI: 10.1016/j.engstruct.2013.07.021

Dara, H. M., Raut, A., Adamu, M., Ibrahim, Y. E., & Ingle, P. V. (2024). Reducing non-value added (NVA) activities through lean tools for the precast industry. *Heliyon*, *10*(7), e29148. DOI: 10.1016/j.heliyon.2024.e29148 PMID: 38633654

Ding, K. W., & Zhang, X. (2012). Literature review of precast concrete frame structures in civil engineering. *Advanced Materials Research*, *568*, 3–6. DOI: 10.4028/www.scientific.net/AMR.568.3

Dolati, S. S. K., & Mehrabi, A. (2021, April). Review of available systems and materials for splicing prestressed-precast concrete piles. [). Elsevier.]. *Structures (Oxford)*, *30*, 850–865. DOI: 10.1016/j.istruc.2021.01.029

Drage, J., & Kennedy, G. (2020). Building a Low-Cost, Internet-of-Things, Real-Time Groundwater Level Monitoring Network. *Ground Water Monitoring and Remediation*, *40*(4), 67–73. DOI: 10.1111/gwmr.12408

Elliott, K. S. (2000). Research and development in precast concrete framed structures. *Progress in Structural Engineering and Materials*, *2*(4), 405–428. DOI: 10.1002/pse.65

Fallah, A., Vafaei, M., & Alih, S. C. (2020). A review of seismic response of precast structures. *International Journal of Engineering Research & Technology (Ahmedabad)*, *13*(10), 2694–2705. DOI: 10.37624/IJERT/13.10.2020.2694-2705

Fang, M., Zhou, R., Ke, W., Tian, B., Zhang, Y., & Liu, J. (2022). Precast system and assembly connection of cement concrete slabs for road pavement: A review. [english edition]. *Journal of Traffic and Transportation Engineering*, *9*(2), 208–222. DOI: 10.1016/j.jtte.2021.10.003

Gao, M. Y., Han, J., Yang, Y., Tiong, R. L., Zhao, C., & Han, C. (2024). BIM-based and IoT-driven smart tracking for precast construction dynamic scheduling. *Journal of Construction Engineering and Management*, *150*(9), 04024117. DOI: 10.1061/JCEMD4.COENG-14498

Ghayeb, H. H., Sulong, N. R., Razak, H. A., Mo, K. H., Ismail, Z., Hashim, H., & Gordan, M. (2022). A review of the seismic performance behaviour of hybrid precast beam-to-column connections. *Archives of Civil and Mechanical Engineering*, *23*(1), 35. DOI: 10.1007/s43452-022-00558-7

Guerrini, G., Restrepo, J. I., Vervelidis, A., & Massari, M. (2015). Self-centering precast concrete dual-steel-shell columns for accelerated bridge construction: Seismic performance, analysis, and design. *Report No. PEER*, *2015*, 13.

Hewes, J. T. (2013). Analysis of the state of the art of precast concrete bridge substructure systems.

Hieber, D. G., Wacker, J. M., Eberhard, M. O., & Stanton, J. F. (2005). State-of-the-art report on precast concrete systems for rapid construction of bridges.

Holton, I. R. (2009). *Developing a sector sustainability strategy for the UK precast concrete industry*. Loughborough University.

Jadhav, O., Minde, P., Yadhav, A., & Gaidhankar, D. (2023). A review of emerging trends & advances in construction technology in the Indian scenario. *Materials Today: Proceedings*, *77*, 897–904. DOI: 10.1016/j.matpr.2022.12.041

Jaillon, L., & Poon, C. S. (2007, May). Advantages and limitations of precast concrete construction in high-rise buildings: Hong Kong case studies. In *Proceedings of the CIB World Building Congress* (pp. 14-17).

Jaillon, L., & Poon, C. S. (2009). The evolution of prefabricated residential building systems in Hong Kong: A review of the public and the private sector. *Automation in Construction*, *18*(3), 239–248. DOI: 10.1016/j.autcon.2008.09.002

Jang, Y., Son, J., & Hwang, S. (2022). Requirements analysis for development of off-Site construction project management system: Focusing on precast concrete construction. *Buildings (Basel, Switzerland)*, *12*(10), 1499. DOI: 10.3390/buildings12101499

Kaja, N., & Jauswal, A. (2021). Review of precast concrete technology in India. [IJERT]. *International Journal of Engineering Research & Technology (Ahmedabad)*, *10*(06), 867–872.

Kehl, R. J., & Carrasquillo, R. L. (1998). *Investigation of the use of match cure technology in the precast concrete industry* (No. FHWA/TX-01/1714-2). University of Texas at Austin. Center for Transportation Research.

Khare, R. K., Maniyar, S. R., Uma, S. R., & Bidwai, V. B. (2011). Seismic performance and design of precast concrete building structures: An overview. *Journal of Structural Engineering*, *38*(3), 272–284.

Kim, T., Kim, Y. W., & Cho, H. (2020). Dynamic production scheduling model under due date uncertainty in precast concrete construction. *Journal of Cleaner Production*, *257*, 120527. DOI: 10.1016/j.jclepro.2020.120527

Kiong, N. B., & Akasah, Z. (2012). An overview of precast concrete system for building maintenance: Malaysian perspective. *International Journal of Engineering Science and Advanced Technology*, *2*(6), 1684–1689.

Kolbeck, L., Kovaleva, D., Manny, A., Stieler, D., Rettinger, M., Renz, R., Tošić, Z., Teschemacher, T., Stindt, J., Forman, P., Borrmann, A., Blandini, L., Stempniewski, L., Stark, A., Menges, A., Schlaich, M., Albers, A., Lordick, D., Bletzinger, K.-U., & Mark, P. (2023). Modularisation Strategies for Individualised Precast Construction—Conceptual Fundamentals and Research Directions. *Designs*, *7*(6), 143. DOI: 10.3390/designs7060143

Kurama, Y. C., Sritharan, S., Fleischman, R. B., Restrepo, J. I., Henry, R. S., Cleland, N. M., Ghosh, S. K., & Bonelli, P. (2018). Seismic-resistant precast concrete structures: State of the art. *Journal of Structural Engineering*, *144*(4), 03118001. DOI: 10.1061/(ASCE)ST.1943-541X.0001972

Li, Z., Shen, G. Q., & Xue, X. (2014). Critical review of the research on the management of prefabricated construction. *Habitat International*, *43*, 240–249. DOI: 10.1016/j.habitatint.2014.04.001

Liu, C., Zhang, F., & Zhang, H. (2020). Comparative analysis of off-site precast concrete and cast-in-place concrete in low-carbon built environment. *Fresenius Environmental Bulletin*, *29*(3), 1804–1812.

Liu, Y., Tong, W., Li, Q., Yao, F., Li, Y., Li, H. X., & Huang, J. (2022). Study on complexity of precast concrete components and its influence on production efficiency. *Advances in Civil Engineering*, *2022*(1), 9926547. DOI: 10.1155/2022/9926547

Ma, Z., Liu, Y., & Li, J. (2023). Review on automated quality inspection of precast concrete components. *Automation in Construction*, *150*, 104828. DOI: 10.1016/j.autcon.2023.104828

Madireddy, H., Naganathan, S., & Mahalingam, B. (2021). A review on precast concrete construction. *Sustainable Practices and Innovations in Civil Engineering: Select Proceedings of SPICE 2021*, 77-113.

Martins, R., Carmo, R. D., Costa, H., & Júlio, E. (2023). A review on precast structural concrete walls and connections. *Advances in Structural Engineering*, *26*(14), 2600–2620. DOI: 10.1177/13694332231191073

Maurya, S., & Madan, S. K. (2023, December). Precast Concrete Technology: A Catalyst for Sustainable Urbanization. In *International Conference on Green Energy and Sustainable Technology* (pp. 229-245). Singapore: Springer Nature Singapore.

Maurya, S., & Madan, S. K. (2025). Precast Concrete Technology: A Catalyst. *Advances in Green Energy Technologies: Proceedings of ICGEST 2023,* Volume 2, *1314*, 229.

Mishra, S., Singh, S., & Srivastava, R. (2017). Green Building a Sustainable Approach to Save Energy. *International Journal of Engineering and Management Research*, *7*(3), 14–20.

Mohammed, H. A. (2018). Variety of Precast Concrete. *Presentation presented to Al-Furat Al-Awsat Technical University-Technical Collage of Al-Mussaib-Session Presentation, Iraq*.

Murari, S. S., & Joshi, A. M. (2017). Precast construction methodology in construction industry. *Available at SSRN* 3496019.

Nanyam, V. N., Basu, R., Sawhney, A., Vikram, H., & Lodha, G. (2017). Implementation of precast technology in India–opportunities and challenges. *Procedia Engineering*, *196*, 144–151. DOI: 10.1016/j.proeng.2017.07.184

Novak, J., Kohoutková, A., Křístek, V., & Vodička, J. (2017, September). Precast concrete pavement–systems and performance review. *IOP Conference Series. Materials Science and Engineering*, *236*(1), 012030. DOI: 10.1088/1757-899X/236/1/012030

O'Hegarty, R., & Kinnane, O. (2020). Review of precast concrete sandwich panels and their innovations. *Construction & Building Materials*, *233*, 117145. DOI: 10.1016/j.conbuildmat.2019.117145

Olidis, C., Swan, D. J., Saeed, A., Mellerski, R. C., & Hammons, M. I. (2010). Precast slab literature review report: Repair of rigid airfield pavements using precast concrete panels-a state-of-the-art review.

Pampanin, S. (2005). Emerging solutions for high seismic performance of precast/prestressed concrete buildings. *Journal of Advanced Concrete Technology*, *3*(2), 207–223. DOI: 10.3151/jact.3.207

Pedro, D., De Brito, J., & Evangelista, L. (2017). Mechanical characterization of high performance concrete prepared with recycled aggregates and silica fume from precast industry. *Journal of Cleaner Production*, *164*, 939–949. DOI: 10.1016/j.jclepro.2017.06.249

Polat, G. (2010). Precast concrete systems in developing vs. industrialized countries. *Journal of Civil Engineering and Management*, *16*(1), 85–94. DOI: 10.3846/jcem.2010.08

Priddy, L. P., Bly, P. G., & Flintsch, G. W. (2013). *Review of precast portland cement concrete panel technologies for use in expedient portland cement concrete airfield pavement repairs* (No. 13-2956).

Priya, P. K., & Neamitha, M. (2018). A review on precast concrete. *IACSIT International Journal of Engineering and Technology*, *5*(1).

Qi, B., Razkenari, M., Costin, A., Kibert, C., & Fu, M. (2021). A systematic review of emerging technologies in industrialized construction. *Journal of Building Engineering*, *39*, 102265. DOI: 10.1016/j.jobe.2021.102265

Qiao, M., Ma, Y. X., Fung, T. C., & Tan, K. H. (2024). Experimental study on exterior precast concrete beam-to-CECFST column joints with wet connections under cyclic loading. *Journal of Building Engineering*, *88*, 109109. DOI: 10.1016/j.jobe.2024.109109

Rahim, A. A., Hamid, Z. A., Zen, I. H., Ismail, Z., & Kamar, K. A. M. (2012). Adaptable housing of precast panel system in Malaysia. *Procedia: Social and Behavioral Sciences*, *50*, 369–382. DOI: 10.1016/j.sbspro.2012.08.042

Rahman, M., & Sobuz, H. R. (2018, February). Comparative study of IPS & PPVC precast system—A case study of public housing buildings project in Singapore. In *Proceedings of the 4th International Conference on Civil Engineering for Sustainable Development (ICCESD 2018), KUET, Khulna, Bangladesh* (pp. 9-11).

Reichenbach, S., & Kromoser, B. (2021). State of practice of automation in precast concrete production. *Journal of Building Engineering*, *43*, 102527. DOI: 10.1016/j.jobe.2021.102527

Rui, Y., & Sen, Z. (2018). The review of problems in precast construction activities. *The Journal of Social Sciences Research*, 173-176.

Ryang, C. K., Hyuk, K. K., & Sup, W. Y. (2004). New Precast Technologies for Building in Japan. *Magazine of the Korea Concrete Institute*, *16*(5), 20–25.

Sacks, R., Eastman, C. E., & Lee, G. (2003). Process improvements in precast concrete construction using top-down parametric 3-D computer modeling. *PCI Journal*, *48*(3), 46–55. DOI: 10.15554/pcij.05012003.46.55

Sacks, R., Eastman, C. M., & Lee, G. (2004). Process model perspectives on management and engineering procedures in the precast/prestressed concrete industry. *Journal of Construction Engineering and Management*, *130*(2), 206–215. DOI: 10.1061/(ASCE)0733-9364(2004)130:2(206)

Sadeghi, H., & Singh, R. M. (2023). Driven precast concrete geothermal energy piles: Current state of knowledge. *Building and Environment*, *228*, 109790. DOI: 10.1016/j.buildenv.2022.109790

Scheer, S., Santos, A., Quevedo, J. R. S., Filho, J. M., & Boiça, S. M. R. (2005). IT as a Success Factor for a Lean Construction System in a Precast Concrete Construction Industry. In *Computing in Civil Engineering (2005)* (pp. 1-12). DOI: 10.1061/40794(179)160

Schipper, H. R. (2015). Double-curved precast concrete elements.

Senthilvel, M., Varghese, K., & Ramesh Babu, N. (2016). Building information modeling for precast construction: a review of research and practice. In *Construction Research Congress 2016* (pp. 2250-2259). DOI: 10.1061/9780784479827.224

Sherfudeen, A. P., Kumar, N., Raghavan, N., Pillai, R. G., & Kalidindi, S. N. (2016). Promoting precast concrete for affordable housing–An overview on promotional policies worldwide and challenges and possibilities in India. *Indian Concrete Journal*, *90*(5), 13–25.

Singhal, S., Chourasia, A., Chellappa, S., & Parashar, J. (2019). Precast reinforced concrete shear walls: State of the art review. *Structural Concrete, 20*(3), 886–898. DOI: 10.1002/suco.201800129

Skrzypczak, I. (2023). Statistical quality inspection methodology in production of precast concrete elements. *Materials (Basel), 16*(1), 431. DOI: 10.3390/ma16010431 PMID: 36614770

Sridhar, K. V., Yamini, & Cheran. (2023, June). State-of-the-art review on precast composite sandwich slab panels. In *5TH International Conference On Innovative Design, Analysis & Development Practices In Aerospace & Automotive Engineering: I-DAD'22* (Vol. 2766, No. 1, p. 020079). AIP Publishing LLC.

Srivastava, R., Arun, J., & Patel, D. K. (2019, April). Amalgamating the Service Quality Aspect in Supply Chain Management. In *2019 International Conference on Automation, Computational and Technology Management (ICACTM)* (pp. 63-67). IEEE. DOI: 10.1109/ICACTM.2019.8776839

Srivastava, R., & Thakur, A. (2025). The Foundational Principles of FDM: An Acronym Pivotal t Its Operation and Its Wide Range of Applications. In *Modeling, Analysis, and Control of 3D Printing Processes* (pp. 355-380). IGI Global Scientific Publishing.

Srivastava, R., Thakur, A., & Rana, A. (2024). The Basic Understanding of Nano Materials: A Review. *Innovations in Nanomaterials-Based Corrosion Inhibitors*, 1-20.

Srivastava, V. K., Maurya, L., & Mehra, E. R. (2014). Detection of noise in high pass Butterworth IIR Filter using MATLAB. *IJERT, 3*(2), 1232–1235.

Straatman, R., & Vambersky, J. N. J. A. (2001). Precast construction and environment. *Structural Concrete*, *2*(2), 93–98. DOI: 10.1680/stco.2001.2.2.93

Sua-iam, G., & Makul, N. (2024). Potential future direction of the sustainable production of Precast concrete with recycled concrete aggregate: A critical review. *Engineering and Science*, *28*, 1075.

Syed, A., & Sonparote, R. (2020). A review of precast concrete pavement technology. *The Baltic Journal of Road and Bridge Engineering*, *15*(4), 22–53. DOI: 10.7250/bjrbe.2020-15.493

Tan, Q., Zhu, H., Yang, S., Yang, X., & Ou, L. (2024). Precast assembled road paving technology: Progress and prospects. *Materials (Basel)*, *17*(10), 2245. DOI: 10.3390/ma17102245 PMID: 38793312

Tayabji, S., Ye, D., & Buch, N. (2013). Precast concrete pavements: Technology overview and technical considerations. *PCI Journal, 58*(1), 112–128. DOI: 10.15554/pcij.01012013.112.128

Tomek, R. (2017). Advantages of precast concrete in highway infrastructure construction. *Procedia Engineering, 196*, 176–180. DOI: 10.1016/j.proeng.2017.07.188

Turai, V. R., & Patil, A. R. (2022). Review on precast concrete technology vs. cast-in-place concrete. *International Journal of Civil Engineering*, *9*(3), 27–32.

Unlu, D. (2010). Rapid bridge construction technology: precast elements for substructures.

Vatin, N. I., Korsun, V. I., Velichkin, V. Z., Kozinets, G., Zhuvak, O., & Rybakov, V. A. (2018). Precast-monolithic reinforced concrete beam-slabs technology with claydit blocks. *Construction of Unique Buildings and Structures*, *70*(7), 43–59.

Wang, L., Zhao, Y., & Yin, X. (2023). Precast production scheduling in off-site construction: Mainstream contents and optimization perspective. *Journal of Cleaner Production*, *405*, 137054. DOI: 10.1016/j.jclepro.2023.137054

Wang, Z., & Hu, H. (2017). Improved precast production–scheduling model considering the whole supply chain. *Journal of Computing in Civil Engineering, 31*(4), 04017013. DOI: 10.1061/(ASCE)CP.1943-5487.0000667

Wang, Z., Hu, H., Gong, J., Ma, X., & Xiong, W. (2019). Precast supply chain management in off-site construction: A critical literature review. *Journal of Cleaner Production*, *232*, 1204–1217. DOI: 10.1016/j.jclepro.2019.05.229

Wang, Z., & Xu, H. (2024). Scheduling redundancy optimization for precast production to enhance disruption management in prefabricated construction. *Computers & Industrial Engineering*, *192*, 110251. DOI: 10.1016/j.cie.2024.110251

Warszawski, A., Avraham, M., & Carmel, D. (1984). Utilization of precast concrete elements in building. *Journal of Construction Engineering and Management, 110*(4), 476–485. DOI: 10.1061/(ASCE)0733-9364(1984)110:4(476)

Won, J. J., Hak, L. K., & Mun, A. J. (2004). Recent domestic precast concrete slab technologies. *Magazine of the Korea Concrete Institute*, *16*(5), 16–19.

Wong, R. W., & Loo, B. P. (2022). Sustainability implications of using precast concrete in construction: An in-depth project-level analysis spanning two decades. *Journal of Cleaner Production*, *378*, 134486. DOI: 10.1016/j.jclepro.2022.134486

Wu, D., Liang, S., Shen, M., Guo, Z., Zhu, X., & Sun, C. (2019). Experimental estimation of seismic properties of new precast shear wall spatial structure model. *Engineering Structures*, *183*, 319–339. DOI: 10.1016/j.engstruct.2018.12.096

Xiao, S., & Fomin, N. I. (2024). Development and systematic review of connection techniques for RC Precast structural elements: Beam and Column Connection. *Heliyon*, *10*(16), e35886. DOI: 10.1016/j.heliyon.2024.e35886 PMID: 39224382

Xiaosheng, T., & Hamzeh, F. (2020, July). Precast concrete building construction process comparison. In *Proceedings of the Annual Conference of the International Group for Lean Construction,* Berkeley, CA, USA (pp. 6-12).

Yang, Z., Tang, D., Li, D., Chen, T., Liu, J., Qi, H., & Zhou, J. (2025). Automatic dimensional quality inspection system for regular precast concrete elements based on 3D structure lighting scan technology. *Expert Systems with Applications*, 129149.

Yaqoob, S., & Silfwerbrand, J. (2023). Rapid repair of concrete pavements using Precast technology. In *14th International Symposium on Concrete Roads,* Krakow, Poland, June 25-28, 2023.

Yardim, Y. (2018). Review of research on the application of ferrocement in composite precast slabs. *Periodica Polytechnica. Civil Engineering*, *62*(4), 1030–1038. DOI: 10.3311/PPci.11737

Ye, M., Li, L., Pei, B., Yoo, D. Y., Li, H., & Zhou, C. (2024). A critical review on shear performance of joints in precast Ultra-High-Performance Concrete (UHPC) segmental bridges. *Engineering Structures*, *301*, 117224. DOI: 10.1016/j.engstruct.2023.117224

Yee, P. T. L., Adnan, A. B., Mirasa, A. K., & Rahman, A. B. A. (2011). Performance of IBS precast concrete beam-column connections under earthquake effects: A literature review. *American Journal of Engineering and Applied Sciences*, *4*(1), 93–101. DOI: 10.3844/ajeassp.2011.93.101

Zhang, R., Zhang, Y., Li, A., & Yang, T. Y. (2022). Experimental study on a new type of precast beam-column joint. *Journal of Building Engineering*, *51*, 104252. DOI: 10.1016/j.jobe.2022.104252

Zhao, J., Hou, S., Shi, D., Fan, J., Wu, T., Sun, W., & Wu, G. (2024). Damage assessment of precast reinforced concrete beam-column connections using AE technology. *Case Studies in Construction Materials*, *20*, e03069. DOI: 10.1016/j.cscm.2024.e03069

Zheng, Y., Guo, Z., Guan, D., & Zhang, X. (2018). Parametric study on a novel grouted rolling pipe splice for precast concrete construction. *Construction & Building Materials*, *166*, 452–463. DOI: 10.1016/j.conbuildmat.2018.01.182

Chapter 7
Bridging the Gap:
Aligning Precast Concrete Quality Management With ISO 9001:2015 Requirements

Ruchira Srivastava
http://orcid.org/0000-0002-5957-9534
Vollert India Pvt. Ltd., India

Krishan Mohan Sharma
http://orcid.org/0009-0003-8679-4365
Vollert India Pvt. Ltd, India

ABSTRACT

Customer satisfaction, regulatory compliance, and consistent quality are becoming important to the precast concrete sector. The internal quality control systems of many precast producers may not be entirely in line with the process-based, continuous improvement tenets of ISO 9001:2015 since they frequently function in silos. The discrepancies between the ISO 9001:2015 framework and standard precast concrete quality management procedures are investigated in this article. Process mapping, document analysis, and industry case studies are used to pinpoint crucial areas that need improvement, including stakeholder focus, risk-based thinking, and performance measurement. Design verification, material traceability, in-process inspections, and post-installation assessments are all integrated into a single quality management system as part of the study's detailed alignment roadmap specifically designed for precast operations. Precast producers can increase their operational effectiveness, lower their failure rates, and become more competitive in the market by using this alignment strategy.

DOI: 10.4018/979-8-3373-6167-3.ch007

INTRODUCTION

Precast concrete has grown to be a crucial component of contemporary building because it provides speed, efficiency, durability, and design flexibility that older in-situ techniques can't always match (Sharma & Srivastava, 2025). Precast technique is widely used in residential and commercial structures as well as infrastructure including bridges, highways, and metro systems since it may improve project timeliness and deliver uniform quality. These benefits do, however, come with the crucial duty of making sure that each precast component satisfies exacting quality standards (Gao et al., 2024). In contrast to cast-in-situ constructions, which occasionally allow for on-site modifications, precast pieces are manufactured in a controlled factory setting and delivered to the construction site prepared for assembly. The structural safety, performance, and service life of the entire project can therefore be immediately impacted by any manufacturing quality failure (Zhao et al., 2019). It is impossible to exaggerate the significance of quality in the precast concrete sector. Not only are high dimensional precision, consistent strength, appropriate curing, surface smoothness, and resistance to external factors desirable qualities, but they are also necessary. Evidence of a systematic approach to quality assurance that extends beyond routine inspections is becoming increasingly expected from clients, contractors, and regulatory bodies (Ismail, 2022). Precast manufacturers face ongoing pressure to prove adherence to globally accepted quality standards in the highly competitive and globalised construction industry. The most frequently used Quality Management System (QMS) standard in the world, ISO 9001:2015, is crucial in this regard (Lima et al., 2021).

The requirements for creating, putting into practice, and maintaining a quality management system are outlined in ISO 9001:2015 and are relevant to all organisations, regardless of size or sector (Yu et al., 2019). The standard, which is founded on seven fundamental principles—customer focus, leadership, employee involvement, process approach, improvement, evidence-based decision-making, and relationship management—has come to represent reliability, effectiveness, and ongoing development (Solanki & Sarkar, 2022). ISO 9001 is a genuinely worldwide standard, with over a million organisations accredited in 170 countries. A comprehensive, process-driven approach that covers every phase of the production lifecycle—from the procurement of raw materials and mix design to mould preparation, casting, curing, finishing, storage, delivery, and customer feedback—is what the precast concrete industry means by adopting ISO 9001:2015 (Zhong et al., 2017).

2. AN UNDERSTANDING OF PRECAST INDUSTRY

The precast concrete sector is essential to contemporary building because it provides solutions that balance durability, quality, and speed. Before being delivered to the installation location, precast concrete is made up of concrete components that have been cast and allowed to cure in a controlled factory setting (Zhao et al., 2021). Precast elements, as opposed to traditional cast-in-situ concrete, are manufactured under standardised circumstances, allowing for improved quality control, quicker project completion, and less reliance on on-site labour (Zhang et al., 2021). Precast concrete has a wide range of applications, including both structural and non-structural elements. This method is frequently used to make structural components such foundations, stairways, wall panels, beams, columns, and slabs. Manholes, drain covers, boundary walls, and ornamental facades are examples of non-structural items (Li et al., 2018). Additionally, the sector offers specialised solutions such as modular housing units, culverts, and precast tunnels, all of which show how adaptable this technique is to a variety of construction needs. Applications-wise, precast concrete is now essential to commercial and residential buildings as well as infrastructure development (Kosse et al., 2023). Since it can tolerate high loads and environmental pressures, it is frequently used in infrastructure, including bridges, flyovers, metro systems, tunnels, highways, and sewage systems. Sound barriers, girders and precast sleepers are used in the transportation industry. Precast wall panels, floor slabs, and modular units allow for the quick construction of high-rise structures and inexpensive housing developments in metropolitan areas (Wang, et al., 2020). Precast systems are also preferred by industrial facilities for factories and warehouses where dimensional accuracy and speed are crucial.

Furthermore, precast materials are essential to sustainable building since they can be recycled, reused, and produced with minimal material usage, all of which lessen the total impact on the environment (Han & Ye, 2018).

Despite these benefits, the sector still faces major quality management issues that need to be resolved to guarantee dependability and durability. The unpredictability of raw materials is one of the main obstacles (Katiyar & Kumar, 2022). The strength and longevity of precast components are directly impacted by the quality of the cement, aggregates, and admixtures used. Cracks, a decreased ability to support weight, or a deteriorated appearance might result from inconsistent raw material qualities (Bašková et al., 2019). Therefore, it is crucial to set up strict supplier assessments and standardised testing protocols.

Curing is another crucial concern. For precast pieces to reach the appropriate strength and longevity, specific curing cycles are necessary (Li et al., 2016). Surface flaws decreased resistance to weathering, and issues with long-term durability might result from improper or uneven curing. Strong monitoring and adherence to best standards are necessary to maintain consistent curing conditions throughout production facilities (Pradhan et al., 2025).

Another major worry is how long precast items would last, especially in hostile environments with moisture, salt exposure, or freeze-thaw cycles. Any neglect of surface treatment, reinforcement protection, or mix design might result in corrosion, scaling, or a shorter service life (Xu et al., 2023). Thus, periodic inspections, protective coatings, and durability testing must be prioritised in quality management systems (Uygunoğlu & Topçu, 2020).

Finally, dimensional correctness continues to be a problem. Even little variations in size or alignment can cause construction schedule disruptions and raise project costs because precast pieces are made to be assembled modularly (Xiao et al., 2024). To prevent such mistakes, formwork, reinforcement placement, and handling techniques must be done precisely.

3. KNOWING ABOUT ISO 9001:2015

The most accepted standard for quality management systems (QMS) worldwide is ISO 9001:2015. It offers a framework that helps businesses continuously develop while delivering goods and services that satisfy consumers and legal requirements (Min & Junyu, 2013). For sectors such as precast concrete manufacturing, where accuracy, robustness, and dependability are essential, ISO 9001:2015 provides a methodical way to integrate quality into each step of production and delivery (Miller et al., 2023). Seven fundamental concepts that help organisations achieve long-term success are at the heart of ISO 9001:2015.

Figure 1. Fundamental concept of ISO 9001:2015

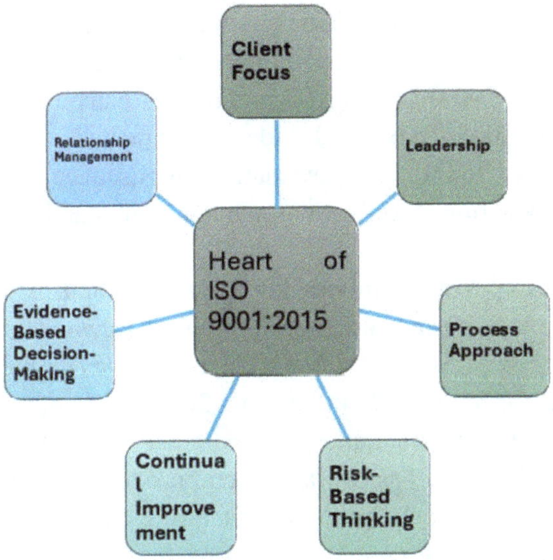

i) Client Focus: The primary goal of the standard is to meet and surpass the expectations of the client. This entails making certain that precast concrete products are precisely measured, long-lasting, and delivered on schedule (Ismail, 2021). Whether they are government organisations, contractors, or developers, happy customers result in enduring relationships and repeat business.

ii) Leadership: The organization's mission and quality goals are aligned when there is effective leadership (Zhao et al., 2022). Production leaders, plant managers, and quality supervisors need to promote an accountable culture and stress that all departments share responsibility for quality.

iii) Process Approach: ISO stresses the interdependence of processes rather than seeing production as discrete tasks. This ensures consistency and traceability for precast by combining the procurement of raw materials, mix design, casting, curing, inspection, and delivery into a smooth operation (Hu et al., 2024).

iv) Risk-Based Thinking: Organisations must identify risks and opportunities that could impact customer satisfaction or product quality to comply with ISO 9001:2015 (Ding et al., 2023). Risks in the precast sector can include variations in the quality of raw materials, irregularities in curing, or delays in logistics. Addressing these issues proactively lowers faults and guarantees project dependability.

v) Continual Improvement: Retaining competitiveness requires constant system and process improvement. To remain competitive, precast manufacturers need to continuously evaluate their production efficiency, cut waste, and investigate cutting-edge technology like automation and digital monitoring (Arganiz & Muslim, 2024).

vi) Evidence-Based Decision-Making: Rather than relying on conjecture, decisions ought to be supported by solid evidence and factual analysis. Strength testing, dimensional checks, and durability evaluations, for instance, offer quantifiable proof that supports production modifications and confirms standard compliance (Sarkar et al., 2022).

vii) Relationship Management: Having solid relationships with suppliers, customers, and employees is essential for long-term success. The performance of the final product in precast is directly impacted by reliable supplier quality and highly qualified labour participation (Araújo et al., 2025).

It is impossible to overestimate the importance of ISO 9001:2015 to industrial industries. Complex supply networks, several production steps, and strict consistency requirements are common in manufacturing operations (Jiang et al., 2022). An organised framework for minimising errors, managing variability, and bringing operations into line with global standards is offered by ISO. Certification helps producers stand out from the competition by proving to customers that they are dependable, providing access to international markets, and guaranteeing that regulations are followed. Furthermore, manufacturers may react quickly to interruptions while upholding quality standards by integrating risk management and evidence-based decision-making into routine operations (Sarkar et al., 2023). ISO 9001:2015 compliance is especially important for the precast concrete sector. Precast focusses on standardised, repeatable procedures that are carried out in controlled surroundings, in contrast to conventional on-site concrete casting. Because of this, quality management is both necessary and attainable. A single flaw in a panel, slab, or beam can raise costs, cause delays in project schedules, and jeopardise structural integrity. Precast producers may immediately solve the primary issues facing the sector by following ISO guidelines, which will lower raw material variability, improve curing procedures, and preserve dimensional correctness (Martiradonna et al., 2019).

Additionally, ISO-certified vendors are now required as part of the qualification requirements for many government tenders and infrastructure projects. Contractors and regulators can be assured by certification that the precast elements satisfy global requirements for safety, strength, and longevity (Wang et al., 2020). Additionally, it increases the manufacturer's reputation and competitiveness in a market that is becoming increasingly focused on high-quality, sustainable solutions.

4. QUALITY MANAGEMENT & PRECAST INDUSTRY

Maintaining constant quality is still a difficult task, despite the precast concrete industry's notable improvements in efficiency and dependability (Ding et al., 2024). In contrast to site-cast concrete, precast products are produced in controlled factory settings, enabling improved process optimisation and monitoring. But even in these kinds of situations, different organisations have quite different approaches to quality management (Kampli et al., 2024). Traditional process controls and quality checks are used in many industries, but the level of systematisation and alignment with contemporary quality management frameworks is frequently inconsistent.

5. CONVENTIONAL PROCESS CONTROLS AND QUALITY CHECKS

Inspection of the final product has long been the focus of quality control in precast manufacturers. Concrete cube compressive strength testing, workability slump tests, and dimensional inspections of completed components are examples of standard checks (Yang et al., 2021). These tests are crucial for guaranteeing adherence to project requirements and client specifications. Visual inspections are also frequently performed to find surface flaws such poor finishes, honeycombing, and cracks (Yang et al., 2021). Process controls are also practiced to some extent, particularly in the batching and mixing stages. Many plants monitor water-cement ratios, mixing times, and vibration techniques to ensure uniformity in concrete compaction. To provide the necessary strength and durability, curing procedures—whether steam, water, or membrane curing—are regulated within predetermined bounds. Another area of attention is the installation of reinforcement, where alignment, spacing, and cover are checked before casting (Tian, 2021). Although these procedures help to preserve baseline quality, they are frequently reactive in character as opposed to preventive. Rather than creating processes to remove errors before they occur, the focus is mostly on identifying flaws after production (Dilakshan et al., 2021).

6. PROCEDURES FOR DOCUMENTATION AND COMPLIANCE IN PRECAST FACTORIES

Precast factories typically keep copious records to prove they are adhering to legal and contractual obligations. Quality manuals, inspection reports, raw material test certifications, and equipment calibration records are all included in this. Mix designs, test data, curing logs, and dispatch notes are frequently included in the paperwork for every manufacturing batch. In addition to audits and client verification, this type of documentation is essential for traceability in the event of disagreements or malfunctions (Li et al., 2024). The foundation of documentation procedures is frequently adherence to regional building codes, customer requirements, and occasionally global standards (such ASTM or EN norms). Larger companies may also use formal quality management systems that are in line with ISO standards, especially those working on infrastructure projects. Smaller operations, on the other hand, frequently only document to satisfy the bare minimum of contractual requirements, placing little focus on leveraging data for ongoing (Cole et al., 2024).

7. REAL-WORLD GAPS IN PRACTICE

The precast sector frequently exhibits several deficiencies in quality management procedures, even in the face of checks and documentation:

Figure 2. Several gaps or deficiencies in quality management procedure

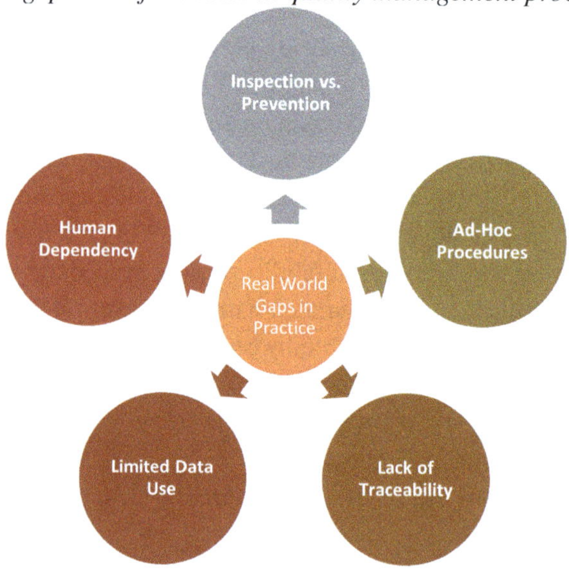

i) **Inspection vs. Prevention:** Instead of incorporating preventive measures into the process, many firms mostly rely on final product inspections. For instance, rather of making sure that the mould is precisely calibrated before pouring concrete, dimensional checks are sometimes done just after casting. This method raises costs, waste, and rework (Qiu et al., 2024).

ii) **Ad-Hoc Procedures:** Ad-hoc or irregular processes are commonly used in smaller or less developed enterprises. For example, rather than being standardised and recorded, curing procedures may differ based on the time of year or the resources that are available (Ye et al., 2025). Uneven product quality and decreased durability are the outcomes of such variability.

iii) **Lack of Traceability**: Another significant shortcoming is the absence of traceability. Although test reports and findings might be kept on file, they aren't usually connected to precast parts or manufacturing batches. It becomes challenging to identify the cause of flaws in structural failures or conflicts, whether it be improper raw materials, incorrect mixing, or improper curing (Tripathi et al., 2025).

iv) **Limited Data Use:** Even though a lot of test findings and inspection reports are produced, they are frequently handled more like compliance documents than as instruments for performance analysis. Trend analysis and statistical process control are rarely used in factories to find reoccurring problems or boost productivity (Jadhav et al., 2023).

v) **Human Dependency:** Individual inspectors and supervisors continue to play a major role in quality control in many precast factories. Human error and inconsistent decision-making are more likely when automation, digital monitoring, or standardised training programs are lacking (Chopade et al., 2021).

8. RECOGNIZING THE GAPS

Many precast concrete manufacturers have quality control systems in place, but they frequently only address the short-term needs of the project rather than conforming to more extensive frameworks like ISO 9001:2015. A comparison of ISO 9001:2015 criteria with standard precast quality systems identifies several significant discrepancies that impede efficiency, consistency, and long-term development (Liu et al., 2024). These deficiencies have observable effects on project performance and customer satisfaction and appear in the areas of leadership, documentation, risk management, audits, and supplier relations.

I. **Commitment to Leadership and Quality Policy**: Visible leadership commitment and a well-defined quality policy that is shared throughout the company are essential components of ISO 9001:2015. Leadership participation in many precast factories is restricted to approving test results or handling customer complaints (Zhu et al., 2021). Rather of being a common organisational goal, quality is frequently perceived as the exclusive domain of the quality control department. Employees lack a common vision and quality efforts remain dispersed in the absence of leadership-driven policies and quantifiable goals. This disparity results in uneven procedures, diminished responsibility, and lost chances for advancement (Takada, 2020).

II. **Standardisation of Documentation and Procedures**: To guarantee traceability and repeatability, ISO 9001:2015 places a strong emphasis on documented processes and process standardisation. On the other hand, a lot of precast plants, particularly in smaller facilities, rely on unwritten procedures or ad hoc instructions (Akbar et al., 2023). For example, the shift supervisor may decide to change the curing cycles, and mould calibration may not adhere to a documented checklist. When it is maintained, documentation is frequently restricted to inspection logs or test certifications rather than being included into a system that is driven by processes (Wang et al., 2022). Unpredictability in product quality, rework, and client disputes over non-conformance are the outcomes of non-standardized processes.

III. **Risk Evaluation and Preventive Measures**: A fundamental component of ISO 9001:2015 is risk-based thinking, which calls on businesses to recognise possible hazards and take preventative measures. However, rather than being preventive, traditional precast quality methods rely primarily on inspection (David et al., 2024). Instead of foreseeing hazards like low aggregate quality, misplaced reinforcing, or delayed curing, factories typically only identify issues like cracks, honeycombing, or dimensional inaccuracies after production is complete. This reactive strategy affects project timeframes, increases waste, and uses more labour and materials. Additionally, reoccurring issues are neglected due to the absence of formal risk assessment (Naraharisetty et al., 2021).

IV. **Performance evaluation and internal audits:** Regular internal audits and performance reviews are mandated by ISO 9001:2015 to assess system efficacy and pinpoint areas for improvement. Precast firms frequently restrict evaluations to delivery schedules and production productivity. When they do occur, internal audits are irregular and mostly motivated by client or legal needs (Shehu et al., 2025). Seldom are performance indicators like supplier dependability, customer complaints, or rejection rates examined methodically. Without systematic evaluations, factories are unable to measure performance, find bottlenecks, or monitor their progress towards quality goals.

V. **Management of Suppliers and Subcontractors:** Gaps are also visible in the quality management of suppliers and subcontractors (Kosse et al., 2024). To guarantee that raw materials and contracted services fulfil specifications, ISO 9001:2015 mandates the assessment, selection, and ongoing supervision of suppliers. Procurement decisions in many precast plants are mostly based on cost, with little attention paid to supplier consistency. Variations in reinforcement steel or aggregate quality, for instance, might directly affect structural performance and durability (Liu et al., 2018). Similarly, there is frequently no systematic control of subcontracted tasks like transportation and installation. Variability, disagreements, and a greater dependence on post-production checks are the results of poor supplier management.

VI. **Real-World Consequences of These Gaps:** There are practical repercussions for noncompliance with ISO 9001:2015 that go beyond compliance concerns.

Cost Overruns and Rework: Late-stage production defects result in recasting, material scrapping, and extra labour, all of which raise project costs dramatically.

Consumer dissatisfaction: Unreliable finishes, delayed delivery, or non-conforming dimensions undermine consumer confidence and lead to complaints and a decline in repeat business (Bharath et al., 2021).

Safety Concerns: Precast elements that are not adequately strengthened or cured can provide structural concerns, which may result in mishaps, liability lawsuits, and harm to one's reputation.

Inefficiency and Waste: High levels of waste are produced by ad hoc procedures and a lack of preventive measures, which compromises sustainability and productivity objectives (Qi et al., 2021).

Reputation and Competitiveness in the Market: Customers are calling for more ISO-certified vendors. Factories that do not meet ISO standards run the risk of being shut out of high-value projects, particularly those in the government and infrastructure sectors (Dave et al., 2016).

9. BRIDGING THE GAP – ALIGNMENT STRATEGIES

Precast concrete operations must transition from reactive inspection-based methods to proactive, preventive, and system-driven quality management to comply with ISO 9001:2015. Strategies that include leadership, process discipline, risk awareness, and customer focus into routine industrial operations are necessary for this shift (Srivastava et al., 2019).

i). **Leadership & Policy:** Strong leadership commitment is the cornerstone of alignment. A clear quality policy that prioritises customer satisfaction, safety, and compliance must be established by senior management. More significantly, leaders should show that they are accountable for quality goals, provide funds for technology and training, and foster an environment where everyone—not just the QC department—takes ownership of quality (Srivastava & Thakur, 2025).

ii). **Process Approach:** ISO emphasises the need for documented, standardised, and integrated processes. For precast manufacturers, this entails formalising each step with precise work instructions, calibrated equipment, and specified acceptance criteria. These steps include mould preparation, reinforcement installation, concreting, curing, finishing, and storage (Srivastava et al., 2014). This kind of standardisation lowers variability and guarantees that every stage consistently improves the performance of the finished product.

iii). **Risk-Based Thinking:** Precast production is susceptible to hazards such handling damage, dimensional fluctuations, misaligned reinforcement, and curing delays. Preventive controls, such as mould calibration logs, handling procedures, and curing cycle monitoring, must be established and identified beforehand to comply with ISO (Bahadur et al., 2023). Regular assessments and risk registers can help the manufacturer further institutionalise preventive thinking.

iv). **Documentation & Traceability**: A traceable chain connecting raw materials, processes, and final products is made possible by thorough documentation. From batch-by-batch concrete mix records and curing logs to supplier certificates for steel and cement, documentation guarantees responsibility and makes root-cause investigation easier in the event of flaws. Systems for digital record-keeping can improve traceability and increase transparency (Srivastava et al., 2024).

v). **Internal Audits & Continuous Improvement:** In addition to compliance audits, factories need to implement systematic internal assessments that gauge performance in relation to key performance indicators (KPIs) such customer complaints, on-time delivery, and rejection rates (Mishra et al., 2017). Findings ought to inform plans for preventative and remedial measures, encouraging ongoing development as opposed to band-aid solutions.

vi). **Supplier Management:** Suppliers of aggregates, cement, admixtures, and steel need to be assessed and routinely audited because raw material variability has a direct impact on precast quality (Zhang et al., 2023). Consistent input quality is ensured, and downstream risks are decreased through long-term relationships with dependable suppliers supported by performance monitoring and feedback.

vii). **Customer Focus:** Lastly, listening to customers is emphasised by ISO alignment. Maintaining reputation and gaining repeat business requires meeting requirements, promptly addressing concerns, and incorporating input into process enhancements.

10. CHALLENGES AND RECOMMENDATIONS

Implementing ISO 9001:2015 in the precast concrete industry offers long-term benefits, but the transition is not without hurdles. Factories, particularly those operating in cost-sensitive and deadline-driven markets, often struggle with practical barriers that slow adoption (Ye et al., 2024). Recognizing these challenges and applying targeted recommendations is key to achieving smooth integration of ISO requirements.

Common Challenges

i). **Costs of Implementation:** Establishing a formal quality management system requires investment in training, documentation, auditing, and sometimes new equipment. For smaller precast plants, these upfront costs can appear burdensome, leading to reluctance in pursuing certification (Lorusso & Celenta, 2023).

ii). **Resistance to Change:** Employees and supervisors accustomed to traditional inspection-driven methods often resist process standardization and new documentation practices. The perception that ISO adds bureaucracy rather than efficiency can create pushback at multiple organizational levels (Liang et al., 2024).

iii). **Training Gaps:** ISO 9001:2015 demands a workforce that understands quality objectives, risk-based thinking, and process documentation. However, many precast factories lack structured training programs, resulting in inconsistent application of standards and over-reliance on a few skilled individuals (Kostadino, 2023).

iv). **Cultural Barriers:** Leadership in some organizations views quality as a compliance requirement rather than a strategic driver. Without cultural alignment, ISO adoption becomes a "tick-box exercise" rather than a transformational initiative (Li et al., 2016).

Practical Recommendations

i). **Phased Implementation**: Instead of overhauling all processes at once, factories should adopt a step-by-step approach—starting with critical processes like mould preparation, curing, and dimensional accuracy. Gradual adoption reduces disruption and builds confidence among employees (Wang et al., 2021).

ii). **Structured Training Programs:** Investing in regular training and workshops ensures that employees at all levels understand ISO principles and their relevance to day-to-day tasks. Role-based training—separately designed for operators, supervisors, and managers—can enhance clarity and ownership (Arif et al., 2025).

iii). **Leverage Digital Tools for QMS:** Digital solutions such as production monitoring systems, automated curing logs, and electronic documentation platforms can simplify compliance and reduce human error (Yao et al., 2025). They also provide real-time data for decision-making and traceability.

iv). **Leadership Buy-In and Communication:** Senior management must actively champion ISO adoption by linking it to organizational goals like customer satisfaction, reduced rework, and market competitiveness. Transparent communication and recognition of employee contributions help build a culture of quality (de Carvalho et al., 2023).

11. GENERAL PERSPECTIVES

Precast concrete technology's quick uptake in contemporary building has transformed the sector by providing improved infrastructure development speed, consistency, and durability (Sharma & Srivastava, 2025). However, precast systems can only reach their full potential in conjunction with a strong quality management system that guarantees customer satisfaction and product dependability. The internationally accepted ISO 9001:2015 standard for quality management systems offers a thorough framework to accomplish this goal. Manufacturers can improve process efficiency, lower nonconformance rates, and increase client, consultant, and end-user trust by bridging the gap between conventional precast quality procedures and the ISO 9001:2015 methodology (Bukhari et al., 2025). Reactive quality control must give way to proactive quality assurance, with a focus on management commitment, process optimisation, risk-based thinking, and ongoing improvement across the entire precast production process. Quality procedures in many precast facilities have traditionally concentrated on dimensional checks, compressive strength testing, routine inspections, and adherence to project specifications. Although these procedures are crucial, ISO 9001:2015 broadens the perspective to encompass internal audits, customer feedback loops, organisational leadership, competency development, and documented process standardisation (Wilson, 2025). This guarantees that quality is a strategic goal supported by senior management and incorporated into daily operations, rather than merely a technical task for the production team. Establishing excellent goals, making sure resources are allocated appropriately, and fostering an environment of accountability and ongoing development all depend heavily on leadership (Muralidhar & JaiSai, 2025). Active management participation in QMS implementation increases employee engagement and guarantees alignment between operational excellence and business objectives.

Process control and documentation represent another crucial area of alignment. Every step of the production cycle, from the acquisition of raw materials and mould preparation to casting, curing, handling, and shipping, must be identified and documented, according to ISO 9001:2015. This guarantees that deviations may be promptly found and fixed and enhances traceability (Khudair et al., 2025). As a result, precast companies must establish standard operating procedures, keep track of testing equipment calibration, and record production data in real time. While appropriate record-keeping offers openness during audits and customer evaluations, document control stops the usage of out-of-date instructions. Barcoding, production tracking software, and digital quality management systems can further simplify compliance by enabling automatic record creation and reducing human error (Oyeniran et al., 2025).

In the precast industry, where dimensional accuracy, curing cycles, and reinforcement installation directly affect structural performance, risk-based thinking—a fundamental tenet of ISO 9001:2015—is especially crucial. Risk assessments for material variability, equipment failures, curing conditions, transportation limitations, and installation difficulties must be carried out by organisations. Defects like honeycombing, incorrect jointing, cracking, and misalignment at the site are lessened by this proactive approach (Muralidhar & JaiSai, 2024). To maintain smooth operations, it also promotes preventive maintenance programs and contingency planning. Corrective and preventive measures (CAPA) make sure that the lessons learnt from previous nonconformities are recorded and applied to enhance future results instead of being repeated (Rothschild, 2025).

To align precast operations with ISO requirements, human capital is equally crucial. To improve worker proficiency in concrete technology, mould removal, reinforcement placement, finishing, and lifting techniques, training programs must be put in place (Sree et al., 2025). A knowledgeable and driven workforce is fostered through competency matrices, qualification assessments, toolbox talks, and safety training. Additionally, ISO 9001:2015 places a strong emphasis on supplier evaluation, mandating that precast manufacturers verify the suppliers of cement, aggregates, admixtures, prestressing strands, and embedded components (Padhiary, 2025). Consistent input quality is ensured via solid vendor connections and incoming material checks, which eventually improve the finished product.

12. CASE STUDIES IN PRECAST CONCRETE QUALITY MANAGEMENT

Real-world examples that show how technological accuracy, process discipline, and sustainability integration may produce remarkable results are the greatest way to comprehend the development of precast concrete quality management. Several significant global projects, such as those in Germany, Singapore, and India, demonstrate how strict quality standards and digitalisation have changed the building industry (Kale & Arditi, 2006).

One of the biggest and most effective applications of precast concrete in transportation infrastructure is the Delhi Metro Rail Corporation (DMRC) project in India. All segments, including girders, viaduct sections, and pier caps, were manufactured in ISO 9001-certified precast yards under extremely controlled conditions to guarantee consistency and longevity. Self-compacting concrete (SCC) was used to increase surface finish and get rid of problems like honeycombing (Skrzypczak, 2023). While non-destructive testing (NDT) techniques like ultrasonic pulse velocity guaranteed consistent strength across batches, sophisticated laser-guided mould alignment systems preserved dimensional correctness. Additionally, each precast element was connected to its mix data, curing cycle, and inspection records using a barcode-based digital traceability system. With zero structural failures, little rework, and nearly flawless on-time delivery performance, this all-encompassing approach to quality control set a national standard for precast construction (Nicał & Anysz, 2020). To accomplish both speed and sustainability in large-scale house development, Singapore's house and Development Board (HDB) used precast technology. Their approach was centred on sophisticated automated batching devices that guaranteed exact material ratios and mix uniformity (de Almeida Filho & da Cruz, 2022). The carbon footprint was greatly decreased while keeping excellent compressive strength by substituting fly ash and GGBS for up to 30% of the cement. A digital defect tracking system enabled real-time monitoring between plant and site crews, and each precast unit was 3D laser-scanned for dimensional compliance. The result was remarkable: while retaining excellent quality and efficiency, there was an 80% decrease in construction waste, a 25% decrease in embodied carbon, and increased structural durability (Gallardo et al., 2006). Strong quality governance underpins the HDB's effort, which serves as an example for sustainable mass housing (Yohanna et al., 2023).

Vollert Germany, a world leader in precast automation systems, is the third prime example. Robotic reinforcement placement, laser projection for mould setup, and Internet of Things-based curing chambers that track humidity and temperature for ideal hydration are all used in Vollert's completely automated factories. Real-time connection of 3D design models with inspection data is made possible by the integration of BIM-enabled quality assurance systems (Khairunnisa et al., 2025). Additionally, surface quality inspections are carried out using machine vision cameras, which minimise human error and guarantee excellent repeatability (Kokorin et al., n.d.). The outcome demonstrates how digital and automated technologies can transform quality control in the precast industry, resulting in a defect-free production cycle with over 99%-dimensional precision and significant material savings. When taken as a whole, these case studies highlight the three key components of effective precast concrete quality management: data-driven traceability, regulated production conditions, and sustainable innovation (Dal Lago et al., 2022). When taken as a whole, they show that balancing environmental responsibility with engineering excellence is the key to the future of precast building (Belsky et al., 2014).

13. CONCLUSION

The economy, sustainability, and design flexibility of the precast concrete industry are unparalleled. However, strict quality control procedures that address raw material variability, curing, durability, and dimensional precision are essential to this industry's success. The industry's position as a pillar of contemporary construction will depend on its capacity to overcome these obstacles as the need for quick and sustainable infrastructure grows globally. ISO 9001:2015 is both a compliance requirement and a strategic enabler for the precast concrete industry. By integrating its principles—customer focus, risk-based thinking, leadership, and continuous improvement—into daily operations, manufacturers may boost output, reduce mistakes, and ensure long-term customer satisfaction. Following ISO 9001:2015 enables the precast industry to confidently and consistently deliver quality as the need for faster, safer, and greener building solutions increases on a worldwide scale. Precast concrete quality control procedures today combine process monitoring, compliance-driven documentation, and conventional checks. Although these methods offer a basis for upholding fundamental quality standards, they frequently fail to eliminate errors, guarantee traceability, and promote ongoing development. The need for more organised and proactive quality management systems is highlighted by the dependence on inspection rather than prevention, ad hoc procedures, and underutilisation of data. Precast manufacturers must move from compliance-based methods to comprehensive quality frameworks that incorporate preventive, standardisation, and digital technologies for long-term dependability and competitiveness if they are to satisfy the expanding needs of infrastructure and urban development. Systemic deficiencies in supplier oversight, risk management, auditing, leadership commitment, and documentation are shown when standard precast quality systems are compared to ISO 9001:2015. These shortcomings lead to real losses in terms of money, operations, and reputation. The precast sector can transition from reactive, inspection-based quality control to proactive, preventive, and performance-driven quality management by resolving these issues and using ISO 9001:2015 principles. In a building industry that is becoming increasingly demanding, this change is necessary to guarantee longevity, client happiness, and long-term competitiveness. Altogether, a comprehensive approach that links leadership, process discipline, risk awareness, and customer focus is needed to close the gap between current precast practices and ISO 9001:2015. Precast manufacturers can increase consistency, decrease rework, and establish themselves as reliable collaborators in the construction of contemporary infrastructure by implementing these strategies. Precast concrete quality management is more than just inspection; it is a comprehensive culture of accuracy, responsibility, and ongoing improvement. The precast industry guarantees long-term structural durability, constant performance, and customer satisfaction by combining cutting-edge technology, standardised procedures, and sustainability principles.

In conclusion, while challenges such as costs, resistance, and training gaps can hinder ISO 9001:2015 implementation, a structured and phased approach supported by training, digital tools, and leadership commitment can make the transition smoother. By addressing these hurdles proactively, precast factories can unlock the full potential of ISO alignment and secure long-term gains in quality, efficiency, and reputation.

REFERENCES

Akbar, A., Njoroge, J. M., Lee, S., Chang, Y., & Kwon, S. (2023). CNN-Based Automatic Mobile Reporting System and Quantification for the Concrete Crack Size of the Precast Members of OSC Construction. *Engineering Proceedings, 36*(1), 15.

Araújo, C. S., Ferreira, E. D. A. M., & Costa, D. B. (2025). System for tracking and managing metallic formworks adopted in cast-in-place concrete wall systems by integrating IoT and BIM. *Construction Innovation, 25*(4), 1315–1341. DOI: 10.1108/CI-06-2023-0127

Arganiz, P., & Muslim, F. (2024). Development of a prototype system integration model for RFID technology with the internet of things and its implementation to improve precast concrete material management in Indonesia. *Journal of Project Management (Canada), 9*(3), 213–226. DOI: 10.5267/j.jpm.2024.5.001

Arif, F., Ayub, L., Khan, F. A., & Khan, A. U. R. (2025). Conceptual Framework for IoT-Based Structural Health Monitoring of RC Bridges for Maintenance Decision-Making. *Construction Technologies and Architecture, 15*, 19–26. DOI: 10.4028/p-q8nkQW

Bahadur, P. S., Srivastava, R., & Srivastava, V. (2023). Review on the Development of Smart Materials, Including Shape Memory Alloys, Characterization, and Recent Applications. *Modeling, Characterization, and Processing of Smart Materials*, 107-139.

Bašková, R., Struková, Z., & Kozlovská, M. (2019, August). Construction cost saving through adoption of IoT applications in concrete works. In *International Conference Current Issues of Civil and Environmental Engineering Lviv-Košice–Rzeszów* (pp. 452-459). Cham: Springer International Publishing.

Belsky, M., Eastman, C., Sacks, R., Venugopal, M., Aram, S., & Yang, D. (2014). Interoperability for precast concrete building models. *PCI Journal, 59*(2), 144–155. DOI: 10.15554/pcij.03012014.144.155

Bharath, S. R., Lavanya, N., Bharath Kumar, H. B., Chaitra, R. K., & Dandautiya, R. (2021). Utilization Potential of Iron Ore Tailing Waste in Various Applications. In *Advances in Energy and Environment: Select Proceedings of TRACE 2020* (pp. 125-134). Singapore: Springer Singapore. DOI: 10.1007/978-981-33-6695-4_12

Bukhari, S. A. S., Shafi, I., Ahmad, J., Villar, S. G., Villena, E. G., Khurshaid, T., & Ashraf, I. (2025). Review of flood monitoring and prevention approaches: A data analytic perspective. *Natural Hazards, 121*(5), 5103–5128. DOI: 10.1007/s11069-024-07050-w

Chopade, R., Pathak, P., Chougule, R., Bothra, N., Mane, N., & Minde, P. (2021). Analysis of Safety Management in Precast Construction Projects. *Turkish Journal of Computer and Mathematics Education, 12*(6), 5225–5236.

Cole, A., Ameri Sianaki, O., & Shee, H. (2024, July). Conceptualizing a Digital Twin Architecture for Enhanced Control in Precast Concrete Production. In *International Conference on Complex, Intelligent, and Software Intensive Systems* (pp. 261-272). Cham: Springer Nature Switzerland. DOI: 10.1007/978-3-031-70011-8_24

Dal Lago, B., Del Galdo, M., & Bisi, D. (2022). Tests and design of welded-bar angle connections of precast floor elements. *Journal of Advanced Concrete Technology, 20*(2), 43–56. DOI: 10.3151/jact.20.43

Dave, B., Kubler, S., Främling, K., & Koskela, L. (2016). Opportunities for enhanced lean construction management using Internet of Things standards. *Automation in Construction, 61*, 86–97. DOI: 10.1016/j.autcon.2015.10.009

David, M., Derigent, W., Loubet, G., Takacs, A., & Dragomirescu, D. (2024). Communicating materials: Communicating concrete development for construction industry. *IEEE Transactions on Industrial Informatics, 20*(4), 6983–6994. DOI: 10.1109/TII.2024.3352261

de Almeida Filho, R. R., & da Cruz, F. M. (2022). Organizational factors: A case study of the cost-benefit of implementing control measures in precast structure constructions. *Research. Social Development, 11*(1), e33711124981–e33711124981.

de Carvalho, A. R., da Silva Calderón-Morales, B. R., Júnior, J. C. B., de Oliveira, T. M., & Silva, G. J. B. (2023). Proposition of geopolymers obtained through the acid activation of iron ore tailings with phosphoric acid. *Construction & Building Materials, 403*, 133078. DOI: 10.1016/j.conbuildmat.2023.133078

Dilakshan, S., Rathnasinghe, A. P., & Seneviratne, L. D. I. P. (2021, July). Potential of internet of things (Iot) in the construction industry. In *Proceedings The 9th World Construction Symposium| July* (p. 446).

Ding, H., Li, M., Zhong, R. Y., & Huang, G. Q. (2023). Multistage self-adaptive decision-making mechanism for prefabricated building modules with IoT-enabled graduation manufacturing system. *Automation in Construction, 148*, 104755. DOI: 10.1016/j.autcon.2023.104755

Ding, S., Hu, H., Xu, F., Chai, Z., & Wang, W. (2024). Blockchain-based security-minded information-sharing in precast construction supply chain management with scalability, efficiency and privacy improvements. *Automation in Construction, 168*, 105698. DOI: 10.1016/j.autcon.2024.105698

Gallardo, C. A., Granja, A. D., Picchi, F. A., & Folch, A. T. (2006). Stabilization and standardization of a precast production process. In *Proceedings of 14th Annual Conference of the International Group for Lean Construction.* Santiago, Chile (Vol. 205).

Gao, M. Y., Han, J., Yang, Y., Tiong, R. L., Zhao, C., & Han, C. (2024). BIM-based and IoT-driven smart tracking for precast construction dynamic scheduling. *Journal of Construction Engineering and Management, 150*(9), 04024117. DOI: 10.1061/JCEMD4.COENG-14498

Han, C., & Ye, H. (2018, August). A novel IoT-Cloud-BIM based intelligent information management system in building industrialization. In *International Conference on Construction and Real Estate Management 2018* (pp. 72-77). Reston, VA: American Society of Civil Engineers. DOI: 10.1061/9780784481721.008

Hu, X., Shen, J., & Ma, Z. (2024, November). Prefabricated Buildings for Smart Cities: A Literature Survey. In *2024 Twelfth International Conference on Advanced Cloud and Big Data (CBD)* (pp. 302-307). IEEE. DOI: 10.1109/CBD65573.2024.00061

Ismail, Z. A. B. (2021). Hybrid intelligent vehicle system for managing construction supply chain in precast concrete building construction projects. *World Journal of Engineering, 18*(4), 538–546. DOI: 10.1108/WJE-08-2020-0361

Ismail, Z. A. B. (2022). Thermal comfort practices for precast concrete building construction projects: Towards BIM and IOT integration. *Engineering, Construction, and Architectural Management, 29*(3), 1504–1521.

Jadhav, O., Minde, P., Yadhav, A., & Gaidhankar, D. (2023). A review of emerging trends & advances in construction technology in the Indian scenario. *Materials Today: Proceedings, 77*, 897–904. DOI: 10.1016/j.matpr.2022.12.041

Jiang, Y., Li, M., Li, M., Liu, X., Zhong, R. Y., Pan, W., & Huang, G. Q. (2022). Digital twin-enabled real-time synchronization for planning, scheduling, and execution in precast on-site assembly. *Automation in Construction, 141*, 104397. DOI: 10.1016/j.autcon.2022.104397

Kale, S., & Arditi, D. (2006). Diffusion of ISO 9000 certification in the precast concrete industry. *Construction Management and Economics, 24*(5), 485–495. DOI: 10.1080/01446190600601594

Kampli, G., Chickerur, S., & Chitawadagi, M. (2024). IoT system implementation for real-time concrete strength prediction: Experimental design, variance evaluation, cost analysis, and implementation ease. *Innovative Infrastructure Solutions, 9*(7), 260. DOI: 10.1007/s41062-024-01586-3

Katiyar, A., & Kumar, P. (2022). Automated construction progress monitoring of prefabricated concrete structures using building information modeling and internet of things. *International Journal on Recent and Innovation Trends in Computing and Communication, 10*(7), 14–18. DOI: 10.17762/ijritcc.v10i7.5560

Khairunnisa, M., Latief, Y., Hidayawanti, R., Machfudiyanto, R. A., & Riantini, L. S. (2025, February). Innovation of technology application in the manufacturing of spun pile products at precast construction company with a risk approach and ISO 56002 to increase competitiveness. In. AIP Conference Proceedings: Vol. 3169. *No. 1* (p. 040047). AIP Publishing LLC. DOI: 10.1063/5.0255371

Khudair, A. W. A., Naser, A. F., Aldhalemi, A. A., & Mahmood, H. F. (2025). International Journal of Transport Development and Integration. *Integration (Amsterdam), 9*(1), 151–162.

Kokorin, A. S., Koval, S. V., & Emelyanovich, A. A. (n.d.). Quality Management In Precast Concrete Industry: Problems And Perspectives. *European Proceedings of Social and Behavioural Sciences*.

Kosse, S., Forman, P., Stindt, J., Hoppe, J., König, M., & Mark, P. (2023, June). Industry 4.0 enabled modular precast concrete components: a case study. In *International RILEM Conference on Synergising expertise towards sustainability and robustness of CBMs and concrete structures* (pp. 229-240). Cham: Springer Nature Switzerland. DOI: 10.1007/978-3-031-33211-1_21

Kosse, S., Hagedorn, P., Maibaum, J., & König, M. (2024, August). Digital twins for precast concrete: advancing environmental analysis through integrated life cycle assessment. In *International Conference on Computing in Civil and Building Engineering* (pp. 613-627). Cham: Springer Nature Switzerland.

Kostadino, E. (2023). *Prediction Of U-Ditch Raw Material Inventories In*. Multiguna Precast Mandiri Using A Web-Based Single Exponential Smoothing Method.

Li, C. Z., Hong, J., Xue, F., Shen, G. Q., Xu, X., & Luo, L. (2016). SWOT analysis and Internet of Things-enabled platform for prefabrication housing production in Hong Kong. *Habitat International, 57*, 74–87. DOI: 10.1016/j.habitatint.2016.07.002

Li, C. Z., Hong, J., Xue, F., Shen, G. Q., Xu, X., & Mok, M. K. (2016). Schedule risks in prefabrication housing production in Hong Kong: A social network analysis. *Journal of Cleaner Production, 134*, 482–494. DOI: 10.1016/j.jclepro.2016.02.123

Li, C. Z., Xue, F., Li, X., Hong, J., & Shen, G. Q. (2018). An Internet of Things-enabled BIM platform for on-site assembly services in prefabricated construction. *Automation in Construction, 89*, 146–161. DOI: 10.1016/j.autcon.2018.01.001

Li, K., Gan, V. J., Li, M., Gao, M. Y., Tiong, R. L., & Yang, Y. (2024). Automated generative design and prefabrication of precast buildings using integrated BIM and graph convolutional neural network. *Developments in the Built Environment, 18*, 100418. DOI: 10.1016/j.dibe.2024.100418

Liang, J., Cao, Z., Zu, Q., Huang, H., & Guo, S. (2024). Production sequencing and layout optimization of precast concrete components under mold resource constraints. *Buildings (Basel, Switzerland), 14*(10), 3173. DOI: 10.3390/buildings14103173

Lima, C. M., Ferreira, E. D. A. M., & Calmon, J. L. (2021). BIM and IoT integration to support management processes: An application for precast concrete systems. *The Journal of Modern Project Management, 9*(2).

Liu, T., Zhu, C., Tong, Z., & Luo, D. (2024, June). Research on smart manufacturing technology of reinforcing bars for bridge precast concrete structure. *Journal of Physics: Conference Series, 2775*(1), 012028. DOI: 10.1088/1742-6596/2775/1/012028

Liu, Z. S., Liu, S. N., Wang, S. W., Zhao, Y. H., & Yu, J. Q. (2018). Information solution of precast building process based on low power wide area Internet of Things. *Construction Technology, 16*(47), 117–122.

Lorusso, A., & Celenta, G. (2023, May). Internet of Things in the construction industry: A general overview. In *International Conference "New Technologies, Development and Applications"* (pp. 577-584). Cham: Springer Nature Switzerland. DOI: 10.1007/978-3-031-31066-9_65

Martiradonna, S., Lombillo, I., & Fatiguso, F. (2019, October). Performance Monitoring of Refurbished Buildings through Innovative Precast Concrete Modules. In *2019 IEEE International Conference on Systems, Man and Cybernetics (SMC)* (pp. 953-957). IEEE. DOI: 10.1109/SMC.2019.8914472

Miller, D., Ho, N. M., Talebian, N., & Javanbakht, Z. (2023). Real-time monitoring of early-age compressive strength of concrete using an IoT-enabled monitoring system: An investigative study. *Innovative Infrastructure Solutions, 8*(2), 75. DOI: 10.1007/s41062-023-01043-7

Min, H., & Junyu, L. (2013, August). An intelligent system for precast concrete element manufacturing management based on RFID technology. In *2013 IEEE International Conference on Green Computing and Communications and IEEE Internet of Things and IEEE Cyber, Physical and Social Computing* (pp. 1138-1143). IEEE. DOI: 10.1109/GreenCom-iThings-CPSCom.2013.198

Mishra, S., Singh, S., & Srivastava, R. (2017). Green Building a Sustainable Approach to Save Energy. *International Journal of Engineering and Management Research*, *7*(3), 14–20.

Muralidhar, P., & JaiSai, T. (2025). Framework for Railway Safety in India on Automatic Train Protection (ATP) Practices. *Recent Trends and Advances in Construction, Real Estate, Infrastructure and Project Management (CRIP) Sector: Global Perspective*, 276.

Muralidhar, P., & JaiSai, T. (2024, August). Kavach: A Technological and Strategic Framework for Railway Safety in India on Automatic Train Protection (ATP) Practices. In *International Conference on Construction, Real Estate, Infrastructure & Project Management* (pp. 276-284). Singapore: Springer Nature Singapore.

Naraharisetty, V., Talari, V. S., Neridu, S., Kalapatapu, P., & Pasupuleti, V. D. K. (2021, August). Cloud architecture for IoT based bridge monitoring applications. In *2021 international conference on emerging techniques in computational intelligence (ICETCI)* (pp. 39-42). IEEE. DOI: 10.1109/ICETCI51973.2021.9574044

Nicał, A., & Anysz, H. (2020). The quality management in precast concrete production and delivery processes supported by association analysis. *International Journal of Environmental Science and Technology*, *17*(1), 577–590. DOI: 10.1007/s13762-019-02597-9

Oyeniran, K. A., Oyeniran, O. H., & Oladunmoye, K. M. (2025). Artificial Intelligence, Innovation, and Safety in Biotechnology Research. In *Cases on AI Innovations in Occupational Health and Safety* (pp. 73–98). IGI Global Scientific Publishing. DOI: 10.4018/979-8-3693-9301-7.ch004

Padhiary, M. (2025). The convergence of deep learning, IoT, sensors, and farm machinery in agriculture. In *Designing sustainable internet of things solutions for smart industries* (pp. 109–142). IGI Global.

Pradhan, P., Verma, M. S., & Sahu, M. S. (2025). Optimization of Cast-In-Situ and Precast Concrete Methods Through BIM and Structural Simulations. *Optimization*.

Qi, B., Razkenari, M., Costin, A., Kibert, C., & Fu, M. (2021). A systematic review of emerging technologies in industrialized construction. *Journal of Building Engineering*, *39*, 102265. DOI: 10.1016/j.jobe.2021.102265

Qiu, T., Chen, X., Su, D., Chen, K., He, Q., Zheng, Z., Zhou, W., & Wang, Y. (2024). LEGO®-inspired decision-making model for automatic construction of precast diaphragm walls. *Automation in Construction*, *166*, 105667. DOI: 10.1016/j.autcon.2024.105667

Rothschild, K. A. (2025). An Analysis of Engineering Roles in Achieving System Integrity.

Sarkar, D., Dhaneshwar, D., & Raval, P. (2023). Automation in monitoring of construction projects through BIM-IoT-blockchain model. *Journal of The Institution of Engineers (India): Series A, 104*(2), 317-333.

Sarkar, D., Patel, H., & Dave, B. (2022). Development of integrated cloud-based Internet of Things (IoT) platform for asset management of elevated metro rail projects. *International Journal of Construction Management, 22*(10), 1993–2002. DOI: 10.1080/15623599.2020.1762035

Sharma, K. M., & Srivastava, R. (2025). Risk mitigation and safety assurance in mechanical and civil engineering projects. *International Journal for Research in Applied Science and Engineering Technology, 13*(9), 714–719. DOI: 10.22214/ijraset.2025.74100

Sharma, K. M., & Srivastava, R. (2025). Risk Mitigation and Safety Assurance in Mechanical and Civil Engineering Projects. *International Journal for Research in Applied Science and Engineering Technology, 13*(9), 714–718. DOI: 10.22214/ijraset.2025.74100

Shehu, M. O., Almeida, R., & Mahapatra, K. (2025). Opportunities for digital tracking technologies in the precast concrete sector in Sweden. *Frontiers in Built Environment, 11*, 1566784. DOI: 10.3389/fbuil.2025.1566784

Skrzypczak, I. (2023). Statistical quality inspection methodology in production of precast concrete elements. *Materials (Basel), 16*(1), 431. DOI: 10.3390/ma16010431 PMID: 36614770

Solanki, A., & Sarkar, D. (2022, August). A Framework of Internet of Things–Cloud–Building Information Modeling Based Life Cycle Management of Precast Components in Prefabricated Buildings. In *International Conference Innovation in Smart and Sustainable Infrastructure* (pp. 671-681). Singapore: Springer Nature Singapore.

Sree, P. U., Kumar, K. N., Kumar, G. R., Manohar, K., Sampath, G., & Sumfoniya, J. (2025). AI-Artificial Intelligence Powered Autonomous Vehicle Control System. *International Journal of Computer Technology and Electronics Communication, 8*(1), 10055–10061.

Srivastava, R., Arun, J., & Patel, D. K. (2019, April). Amalgamating the Service Quality Aspect in Supply Chain Management. In *2019 International Conference on Automation, Computational and Technology Management (ICACTM)* (pp. 63-67). IEEE. DOI: 10.1109/ICACTM.2019.8776839

Srivastava, R., & Thakur, A. (2025). The Foundational Principles of FDM: An Acronym Pivotal t Its Operation and Its Wide Range of Applications. In *Modeling, Analysis, and Control of 3D Printing Processes* (pp. 355-380). IGI Global Scientific Publishing.

Srivastava, R., Thakur, A., & Rana, A. (2024). The Basic Understanding of Nano Materials: A Review. *Innovations in Nanomaterials-Based Corrosion Inhibitors*, 1-20.

Srivastava, V. K., Maurya, L., & Mehra, E. R. (2014). Detection of noise in high pass Butterworth IIR Filter using MATLAB. *IJERT*, *3*(2), 1232–1235.

Takada, T. (2020). Study of CIM and IoT-Simulation for Cost Performance Analysis-(Cost Management). , (2), 4-25.

Tian, M. (2021, November). Monitoring Method of Vertical Stress on Precast Concrete of Prefabricated Building. In *International Conference on Forthcoming Networks and Sustainability in the IoT Era* (pp. 220-227). Cham: Springer International Publishing.

Tripathi, A., Kaushal, I., Revolti, A., Benedetti, L., & Dallasega, P. (2025). Digital Twin Driven Quality Control in Smart Mobile Factories: A Framework in Precast Manufacturing. *Procedia CIRP*, *134*, 801–806. DOI: 10.1016/j.procir.2025.02.194

Uygunoğlu, T., & Topçu, İ. B. (2020). Nesnelerin internetinin (iot) inşaat mühendisliğindeki rolü: rfid uygulamalari. *International Journal of 3D Printing Technologies and Digital Industry, 4*(3), 270-277.

Wang, X., Wang, S., Song, X., & Han, Y. (2020). IoT-based intelligent construction system for prefabricated buildings: Study of operating mechanism and implementation in China. *Applied Sciences (Basel, Switzerland)*, *10*(18), 6311. DOI: 10.3390/app10186311

Wang, Y., Thangasamy, V. K., Tiong, R. L., & Zhang, L. (2022). Improved workflow for precast element design based on BIM and lean construction. *Journal of Construction Engineering and Management*, *148*(8), 04022065. DOI: 10.1061/(ASCE)CO.1943-7862.0002316

Wang, Z., Wang, T., Hu, H., Gong, J., Ren, X., & Xiao, Q. (2020). Blockchain-based framework for improving supply chain traceability and information sharing in precast construction. *Automation in Construction, 111*, 103063. DOI: 10.1016/j.autcon.2019.103063

Wang, Z., Zhang, Q., Yang, B., Wu, T., Lei, K., Zhang, B., & Fang, T. (2021). Vision-based framework for automatic progress monitoring of precast walls by using surveillance videos during the construction phase. *Journal of Computing in Civil Engineering, 35*(1), 04020056. DOI: 10.1061/(ASCE)CP.1943-5487.0000933

Wilson, S. (2025). *Strategies to Mitigate Contractor Safety Noncompliance in the Construction Industry* (Doctoral dissertation, Walden University).

Xiao, J., Jiang, S., & Li, Y. (2024). Seismic Performance of Prefabricated Reinforced Concrete Structure Beam Joints Based on the Internet of Things. *Journal of Testing and Evaluation, 52*(3), 1789–1803. DOI: 10.1520/JTE20230027

Xu, J., Zhang, Q., Teng, Y., & Pan, W. (2023). Integrating IoT and BIM for tracking and visualising embodied carbon of prefabricated buildings. *Building and Environment, 242*, 110492. DOI: 10.1016/j.buildenv.2023.110492

Yang, B., Liu, B., Xiao, J., Zhang, B., Wang, Z., & Dong, M. (2021). A novel construction scheduling framework for a mixed construction process of precast components and cast-in-place parts in prefabricated buildings. *Journal of Building Engineering, 43*, 103181. DOI: 10.1016/j.jobe.2021.103181

Yang, B., Liu, B., Xiao, J., Zhang, B., Wang, Z., & Dong, M. (2021). A novel construction scheduling framework for a mixed construction process of precast components and cast-in-place parts in prefabricated buildings. *Journal of Building Engineering, 43*, 103181. DOI: 10.1016/j.jobe.2021.103181

Yao, Y., Tam, V. W., Wang, J., Le, K. N., & Butera, A. (2025). Deep reinforcement learning-based schedule optimization for parallel precast production. *Engineering, Construction, and Architectural Management, 32*(13), 1–34. DOI: 10.1108/ECAM-03-2025-0429

Ye, X., Kosse, S., Hagedorn, P., & König, M. (2025). Workflow automation using digital twins and smart contracts: a decentralised framework for precast concrete production.

Ye, X., Zhou, Y., Guo, H., & Luo, Z. (2024). A computer vision-based approach to automatically extracting the aligning information of precast structural components. *Automation in Construction, 164*, 105478. DOI: 10.1016/j.autcon.2024.105478

Yohanna, A., Latief, Y., Hidayawanti, R., Armyn, R., & Machfudiyanto, L. S. R. (2023). Innovation manufacturing method of precast concrete for spun pile products at precast construction company in Indonesia with a risk approach and ISO 56002 innovation process to increase competitiveness.

Yu, Z., Wei, Y., & Cao, Z. (2019, December). The Design and Application of Information Management System Based on IOT for Precast Concrete Segments. *IOP Conference Series. Earth and Environmental Science*, *371*(3), 032056. DOI: 10.1088/1755-1315/371/3/032056

Zhang, R., Feng, X., Mou, Z., & Zhang, Y. (2023). Green optimization for precast production rescheduling based on disruption management. *Journal of Cleaner Production*, *420*, 138406. DOI: 10.1016/j.jclepro.2023.138406

Zhang, W., Kang, K., & Zhong, R. Y. (2021). A cost evaluation model for IoT-enabled prefabricated construction supply chain management. *Industrial Management & Data Systems*, *121*(12), 2738–2759. DOI: 10.1108/IMDS-12-2020-0735

Zhao, L., Liu, Z., & Mbachu, J. (2019). Development of intelligent prefabs using IoT technology to improve the performance of prefabricated construction projects. *Sensors (Basel)*, *19*(19), 4131. DOI: 10.3390/s19194131 PMID: 31554246

Zhao, Y., Cao, C., & Liu, Z. (2022). A framework for prefabricated component hoisting management systems based on digital twin technology. *Buildings (Basel, Switzerland)*, *12*(3), 276. DOI: 10.3390/buildings12030276

Zhao, Y., Cao, C., Liu, Z., & Mu, E. (2021). Intelligent control method of hoisting prefabricated components based on internet-of-things. *Sensors (Basel)*, *21*(3), 980. DOI: 10.3390/s21030980 PMID: 33540530

Zhong, R. Y., Peng, Y., Xue, F., Fang, J., Zou, W., Luo, H., Thomas Ng, S., Lu, W., Shen, G. Q. P., & Huang, G. Q. (2017). Prefabricated construction enabled by the Internet-of-Things. *Automation in Construction*, *76*, 59–70. DOI: 10.1016/j.autcon.2017.01.006

Zhu, Z. Q., Yong, Y. P., Lee, S. J., Chang, Y. H., & Kwon, S. W. (2021). Vision-based precast concrete management plan in off-site construction site: Using PC member quality grades. In *ISARC. Proceedings of the International Symposium on Automation and Robotics in Construction* (Vol. 38, pp. 197-203). IAARC Publications.

Chapter 8
Smart Structures for Smart Cities: The IoT-Driven Future of Precast Concrete

Ruchira Srivastava
http://orcid.org/0000-0002-5957-9534
Vollert India Pvt. Ltd., India

Krishan Mohan Sharma
Vollert India Pvt. Ltd, India

ABSTRACT

In recent years, the precast concrete system has expanded significantly. Benefits including improved quality, cost savings, and construction process agility are to blame for this expansion. But uncertainty and One issue with this beneficial method is interdependencies throughout management procedures. Considering these difficulties, this paper suggests utilizing IoT technologies to make things better. There are several advantages to precast construction, including increased output, a better working environment, and a notable decrease in waste. The timely supply of precast components (PCs) is crucial to the successful completion of building projects, primarily because precast prefabrication operations are typically recognized as vital activities. Current precast production scheduling models, however, must adapt to the needs of dynamic situations where uncertainty is a problem for building projects. Building faster and more sustainably is a challenge for the construction sector.

DOI: 10.4018/979-8-3373-6167-3.ch008

INTRODUCTION

Urban expansion in the 21st century has been unparalleled. The United Nations projects that around 70% of people on Earth will reside in cities by 2050. This change puts tremendous pressure on cities to minimise their environmental impact while providing dependable utilities, sustainable housing, effective transportation, and infrastructure (Gao et al., 2024). Often labour-intensive, resource-intensive, and slow, traditional construction methods are finding it difficult to keep up with the speed and scale needed for this expansion.

The idea of "smart cities"—cities that use digital technologies to improve economic efficiency, environmental sustainability, and quality of life—has arisen in response to these issues (Zhao et al., 2019). To facilitate responsive urban management, effective resource allocation, and real-time decision-making, smart cities combine infrastructure with data and digital systems. However, building such cities calls for more than just data; it calls for smart structures, or intelligent physical infrastructure that can perceive, communicate, and adapt (Ismail, 2022).

Civil engineering is at the forefront of a developing technology revolution in this area. Passive infrastructure is no longer an option; it must engage with its surroundings and adapt to shifting circumstances (Lima et al., 2021). Urban infrastructure is changing in the future due to the combination of digital intelligence with physical architecture, especially through the Internet of Things (IoT). Among the many construction methods, precast concrete is particularly noteworthy for its capacity to facilitate this change because of its precision, versatility, and compatibility with digital technologies (Yu et al., 2019).

Concrete components that are cast and cured in a controlled factory setting prior to being transported and erected on location are referred to as precast concrete. This technique stands in stark contrast to traditional cast-in-place construction, which involves pouring and curing concrete on-site (Solanki & Sarkar, 2022). Precast concrete is becoming a increasingly common option for contemporary infrastructure projects due to its advantages, particularly in metropolitan areas (Zhong et al., 2017).

The quality control of precast building is one of its biggest benefits. Improved control over material qualities, curing conditions, and dimensional precision is guaranteed in factory settings. This produces incredibly robust and consistent parts, which are essential for long-term performance and safety (Zhao et al., 2021). Additionally, there is a notable improvement in building speed. Precast components are produced offsite while site preparation takes place concurrently, enabling a quicker project completion time and a lower labour cost (Zhang et al., 2021).

Additionally, precast concrete provides increased sustainability. It fits very nicely with the environmental objectives of smart cities by minimising material waste, lowering on-site pollution, and enabling improved lifecycle performance

(Li et al., 2018). Furthermore, precast components' modular architecture makes it easier to design, disassemble, and reuse—all crucial aspects of infrastructure that is adaptable and future-ready.

Precast concrete is a perfect fit for the smart infrastructure paradigm because of these advantages (Kosse et al., 2023). Its repeatable and standardised design facilitates the smooth integration of sensors, communication modules, and other digital technologies into IoT-driven systems throughout the manufacturing stage (Wang et al., 2020).

The Internet of Things (IoT) is a network of physical items, including sensors, gadgets, and structures, that are equipped with electronics, software, and connections to gather and share data (Han & Ye, 2018). IoT gives infrastructure in the field of civil engineering the ability to "sense" its surroundings, keep an eye on its own condition, and instantly send any updates or warnings (Katiyar & Kumar, 2022). This feature is particularly crucial in urban settings where upkeep, efficiency, and safety are of utmost importance. Predictive maintenance, environmental sensing, traffic and load analysis, and structural health monitoring (SHM) are some of the IoT applications in civil infrastructure (Bašková et al., 2019). Real-time monitoring of concrete temperature and humidity, the dynamic load on bridges, and the early detection of damage are all made feasible by IoT. This lower maintenance costs, increases asset lifespans, and improves public safety (Li et al., 2016).

Real-time infrastructure monitoring and optimisation will become essential as smart cities develop further. Precast concrete offers the perfect platform for directly integrating sensors and Internet of Things devices into structural components because of its factory-controlled production and reproducibility (Pradhan et al., 2025). A new age in building is dawning with this combination of material and machine, of physical structure and digital intelligence: the era of smart precast concrete for smart cities. Figure 1 below shows the complete framework for the implementation of IoT for a responsible construction method.

Figure 1. A smart IoT framework enabling data-driven decisions for energy-efficient, eco-conscious construction

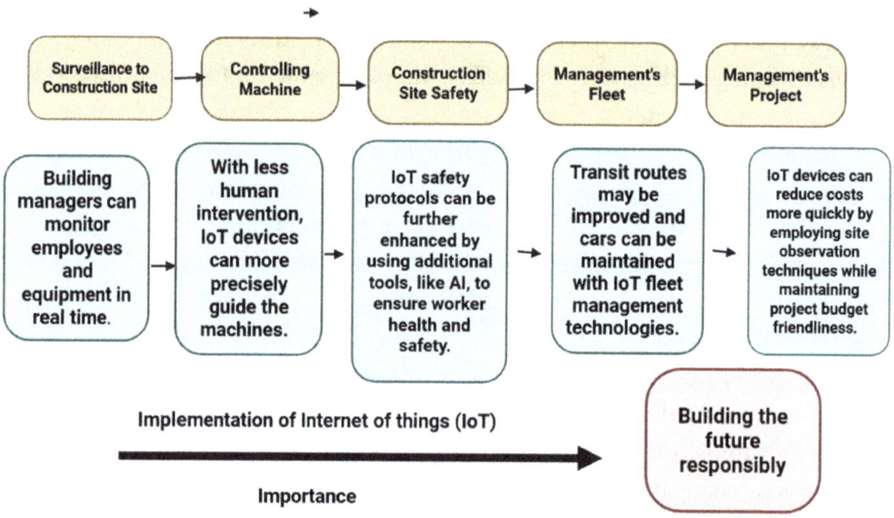

PRECAST IN SMART CITIES

The development of precast concrete technology is a paradigm shift in contemporary building. Precast concrete was first developed in the early 20th century in response to the demand for uniform, mass-producible building components (Xu et al., 2023). Its dependability and efficiency led to its rapid rise in popularity. The need for quick urban reconstruction and public housing in the post-war era led to a substantial advancement in the technology, which was first used in bridge components and low-rise buildings (Uygunoğlu & Topçu, 2020). The capabilities of precast construction have been greatly increased over the decades by advancements in material science, formwork technologies, and production automation. These days, precast components are designed and fabricated using 3D modelling, high-performance concrete mixtures, and sophisticated curing methods (Xiao et al., 2024). The method has been further optimised with the integration of automated manufacturing lines and Building Information Modelling (BIM), enabling previously unheard-of levels of design flexibility, speed, and accuracy. Precast concrete is now a fundamental component of contemporary urban development thanks to its evolutionary advancements, particularly in the context of smart cities, where integration, scalability, and speed are critical factors (Min & Junyu, 2013).

Precast concrete provides a number of fundamental advantages in the context of smart urban infrastructure that are consistent with the ideas of intelligent, sustainable city design. The most important thing is quality control. Environmental factors including humidity, temperature, and cure time are strictly controlled because precast pieces are made in factories (Miller et al., 2023). In contrast to cast-in-place concrete, which is impacted by erratic site circumstances, this produces constant quality, increased strength, and improved surface finishes. Precast components are extremely dependable for vital infrastructure because of this degree of control, which also enables thorough quality assurance testing (Ismail, 2021).

Another noteworthy aspect is durability; precast elements are made to last for a long time with little maintenance, which is important for smart cities that want to save long-term operating expenses (Zhao et al., 2022). Equally important is modularity; precast components can be built like building blocks and are made in standard sizes due to their inherent modularity. This shortens building schedules, lowers labour costs on the job site, and streamlines logistics—all crucial elements in crowded urban settings. Its function in smart cities is further enhanced by its adaptability for automation (Hu et al., 2024). Conveyor belts, robotic systems, and CNC-controlled formwork are being used more in precast production facilities to enable highly reproducible, low-waste manufacturing (Ding et al., 2023).

Because of this, precast building is not only effective but also environmentally friendly, which is a crucial component of city planning. Additionally, precast systems' compatibility with digital design tools like BIM allows fabricators, engineers, and architects to work together seamlessly, minimising design errors and enabling real-time updates throughout the project lifecycle (Arganiz & Muslim, 2024).

Precast concrete's ability to integrate technology is what really makes it stand out in the field of smart infrastructure. Smart technology, including sensors, RFID tags, and sophisticated materials, are progressively being included into contemporary precast components (Sarkar et al., 2022).

These embedded technologies successfully transform passive components into active members of the urban ecology by enabling the concrete structures to exchange information about their own state and surroundings (Araújo et al., 2025). For example, embedded sensors can provide real-time data for structural health monitoring (SHM) by tracking variables including vibration, temperature, humidity, and strain. This improves safety and enables predictive maintenance, particularly in crucial applications like high-rise buildings, bridges, and tunnels (Jiang et al., 2022). Like this, components are tracked from the manufacturer to the construction site using RFID (Radio Frequency Identification) tags, which enhances traceability, inventory control, and logistics. When coordinating hundreds or thousands of components for a large-scale project, this is especially helpful (Sarkar et al., 2023).

Additionally, the performance of precast pieces is enhanced by the use of smart materials, such as phase-change materials or self-sensing concrete. These materials can self-report damage, react to changes in the environment, or even modify their thermal characteristics to lower building energy use (Martiradonna et al., 2019). Precast and embedded technologies work together to produce an infrastructure network that is intelligent, responsive, and robust. Precast concrete is positioned as a key component in the creation of smart, sustainable, and connected cities of the future due to the merging of digital intelligence with physical resilience (Wang et al., 2020).

Role of IoT in Precast Concrete Structures

An important development in contemporary construction is the incorporation of the Internet of Things (IoT) into precast concrete buildings, which turns conventionally inert materials into intelligent, self-reporting urban infrastructure elements (Ding et al., 2024). The use of embedded electronics and smart sensors in precast parts is essential to this change. Numerous factors that are essential to structural performance and integrity can be measured by these sensors, such as temperature, humidity, vibration, displacement, corrosion levels, strain, stress, and vibration (Kampli et al., 2024).

Every kind of sensor has a distinct purpose: strain gauges quantify how much concrete deforms under load; temperature sensors track temperature changes, particularly in harsh climates; humidity sensors identify moisture intrusion that can hasten material deterioration; and vibration sensors are essential in seismically active areas or close to busy roads (Yang et al., 2021). These sensors can be seamlessly integrated without sacrificing structural performance by being embedded during the precast manufacturing process (Yang et al., 2021).

Some methods include employing flexible sensor strips that can adapt to the geometry of the precast piece, casting sensors straight into concrete forms, or using protective housings to keep electronics safe from moisture and corrosion (Tian, 2021). Crucially, the factory-controlled environment of precast parts guarantees accurate placement, improved sensor protection, and long-term durability, making the technique highly scalable for urban infrastructure (Dilakshan et al., 2021).

A crucial component in the creation of smart cities, real-time monitoring and predictive maintenance are made possible by these intelligent components once they are deployed. Infrastructure inspections have historically relied on planned manual assessments, which are labour-intensive, time-consuming, and frequently unsuccessful at identifying concealed or early-stage flaws (Li et al., 2024). However, Structural

Health Monitoring (SHM) solutions provide continuous, automated monitoring for IoT-enabled precast concrete structures. These systems compile information from multiple sensors to evaluate a structure's present state (Cole et al., 2024). When combined with cloud platforms and wireless connection modules, engineers and city planners may access this data remotely.

More significantly, pattern recognition, anomaly detection, and predictive insights are made possible by the application of data analytics and machine learning (Qiu et al., 2024). For instance, a gradual increase in strain could be a symptom of an unequal load distribution, while increasing humidity could be a sign of water intrusion before cracks show. In addition to lowering repair costs, predictive maintenance powered by this data improves safety, prolongs the lifespan of infrastructure, and minimises interruptions to public services (Ye et al., 2025). This skill is particularly useful in transportation networks, high-rise structures, bridges, and tunnels where prompt action can enhance resource allocation and avert catastrophic failures.

Digital twin technology is one of the most exciting developments in precast concrete made possible by IoT (Tripathi et al., 2025). A digital twin is a computerised model of a real asset that is updated in real time using information from embedded sensors. Digital twins can be made using 3D models during the production phase of precast components, and they can be enhanced over time with real-time performance data from the field (Jadhav et al., 2023). These digital equivalents are dynamic models that change with the structure rather than static blueprints. They offer a thorough analysis of the asset's past, present, and anticipated future performance (Chopade et al., 2021).

By offering precise, current visualisations of every precast element's status, from production and delivery to installation and monitoring, digital twins in construction management improve stakeholder coordination (Liu et al., 2024). Digital twins for asset management enable facility operators to estimate maintenance requirements, model stress situations, and even assess the effects of external factors such as seismic activity or temperature swings (Zhu et al., 2021). This implies that infrastructure in smart cities can be maintained proactively as opposed to reactively, which improves operational efficiency, decreases downtime, and increases the dependability of public services. Digital twins also help achieve environmental objectives by guiding choices that minimise material use, maximise energy efficiency, and monitor the structure's carbon footprint over time (Takada, 2020).

IoT essentially lets a new type of intelligent infrastructure that is self-aware, self-reporting, and self-optimizing operate in precast concrete structures. Precast elements are now more than simply construction blocks thanks to smart sensors, real-time data collecting, predictive analytics, and digital replicas; they are data-rich assets essential to the robust, responsive, and data-driven urban environments of the future (Akbar et al., 2023).

Merits of IoT-Enabled Precast in Smart Cities

The construction industry is changing as a result of the incorporation of Internet of Things (IoT) technology into precast concrete structures, particularly in the context of smart cities (Wang et al., 2022). IoT-enabled precast infrastructure offers an appealing option as urban centres adapt to the difficulties of expanding populations, sustainability goals, and technological innovation. The improvement of operational efficiency, durability, and safety is one of the main advantages. Continuous and real-time structural health monitoring is made possible by the integration of sensors into precast concrete components during the manufacturing process (David et al., 2024). These sensors enable the early detection of possible structural problems by detecting stress, cracks, vibrations, and temperature changes. In addition to increasing public safety, this prolongs the life of infrastructure because maintenance may be proactive rather than reactive. Furthermore, precast plant automation made possible by IoT guarantees quality control and consistency, reducing human error and guaranteeing adherence to design specifications (Naraharisetty et al., 2021).

Cost optimisation and quicker construction cycles are two other noteworthy benefits. Weather, manual labour, and site unpredictability are some of the reasons why traditional building frequently has delays (Shehu et al., 2025). By its very nature, precast concrete allows for offsite manufacturing in regulated settings; however, the advantages increase when IoT is added. Real-time tracking systems and sensors streamline logistics, cutting down on installation and delivery delays (Kosse et al., 2024).

Precast elements can be tracked using RFID tags and GPS tracking from the manufacturer to the job site, guaranteeing just-in-time delivery and reducing handling and storage expenses. Through shared digital dashboards, IoT solutions also facilitate communication among project stakeholders, resulting in quicker coordination and decision-making (Liu et al., 2018). As a result, there are shorter construction schedules, fewer labor-related delays, and considerable material and operating cost reductions.

In smart cities, data-driven decision-making becomes essential since infrastructure must be responsive and dynamic. Massive volumes of real-time data about usage patterns, environmental exposure, and structural performance are produced by IoT-enabled precast elements (Bharath et al., 2021). By integrating this data with urban planning platforms, engineers, legislators, and city planners can make well-informed judgements based on facts rather than conjecture. For instance, IoT sensors installed in precast roadways or bridges can record traffic patterns, which can be used to improve transportation planning and ease congestion. Flood control measures can also benefit from environmental data obtained from drainage systems or precast retaining walls (Qi et al., 2021). This ongoing feedback loop helps with adaptive infrastructure planning over time, making sure that the built environment changes to meet the needs of the populace, the climate, and technological advancements.

Finally, IoT-enabled precast systems directly support the objectives of sustainability and adaptation, two fundamental tenets of smart cities. Using precast concrete during construction already lowers waste and encourages energy efficiency from a sustainability perspective (Dave et al., 2016). By measuring carbon footprints, enabling exact material utilisation, and optimising resource allocation throughout the project lifespan, IoT integration further amplifies these advantages. Businesses may improve their operations for improved environmental performance by tracking energy use and emissions in real time during manufacturing and transportation (Srivastava et al., 2019).

In terms of adaptability, IoT-enabled precast constructions are future-ready. Integration with other smart city systems, including traffic control, lighting, and smart grids, can be facilitated by their embedded sensors (Srivastava & Thakur, 2025). Infrastructure must be able to communicate with neighbouring systems as cities grow more interconnected, whether for energy optimisation or emergency response coordination. The Internet of Things enables this interoperability (Srivastava et al., 2014).

LIMITATIONS & ITS CONSIDERATIONS

IoT-enabled precast concrete systems have enormous potential to influence the infrastructure of future smart cities, but there are important factors and considerable obstacles to overcome before they can be put into practice (Bahadur et al., 2023). The security and privacy of data are among the most important issues. Real-time structural health data, component position monitoring, consumption statistics, and even predictive maintenance plans are among the sensitive data streams produced by the integration of IoT in infrastructure and construction. The integrity of individual infrastructure assets as well as the larger metropolitan systems into which they are connected may be jeopardised if this data is not adequately protected against cyberattacks, illegal access, or manipulation (Srivastava et al., 2024). For instance, malevolent actors could take advantage of weaknesses or interfere with operations if they manage to obtain structural data from vital infrastructure, such bridges or tunnels. Furthermore, because cloud-based platforms are frequently used by IoT systems, issues with data ownership, storage jurisdiction, and adherence to national and international data protection laws become more pressing (Mishra et al., 2017). Stakeholder trust and the operational integrity of smart infrastructure systems depend on strong cybersecurity standards, encrypted data transmission, and stringent access controls. The interoperability of platforms and systems is another urgent issue. With several stakeholders utilising a variety of platforms, technologies, and communication protocols, the Internet of Things ecosystem in a smart city setting is intrinsically complicated (Zhang et al., 2023). Certain sensor technologies and data formats used by precast concrete systems must work in unison with the larger smart city infrastructure, including electricity grids, traffic systems, and emergency response networks. The overall efficacy of the IoT framework, however, may be diminished by fragmented systems that find it difficult to communicate with one another due to a lack of standardisation (Ye et al., 2024).

This problem is most noticeable in large-scale infrastructure projects that need the integration of components from various vendors into a single data platform. The full potential of automation, responsive infrastructure, and data analytics cannot be achieved without interoperability (Lorusso & Celenta, 2023). Adopting open standards, spending money on middleware, and encouraging cooperation between software developers, manufacturers, and city planners are all necessary to address issue.

Concerns around return on investment (ROI) and large initial expenditures are further issues. Sensors, communication infrastructure, cloud services, software platforms, and trained staff must all be purchased up front when integrating IoT technology into precast concrete systems (Liang et al., 2024). Many construction companies may find these expenses prohibitive, particularly in emerging markets or on smaller-scale projects with narrow profit margins. Although predictive maintenance, effective resource use, and less downtime are some of the ways that IoT reduces costs over the long run, these advantages are frequently felt across several years (Kostadino, 2023).

Adoption may be hesitant since this extended ROI horizon conflicts with risk-averse investing methods or short-term budgeting cycles. Furthermore, corporations may be concerned about investing in systems that could become obsolete in a few years due to the quick speed of technological progress (Li et al., 2016). A systematic cost-benefit analysis, pilot initiatives with measurable benefits, and either government incentives or public-private partnerships to offset upfront costs are all necessary to overcome this obstacle.

Finally, filling in skill gaps and training needs is critical to the success of IoT in precast construction (Wang et al., 2021). A major change has occurred with the integration of IoT into conventional building processes, necessitating the development of new skills in data analytics, network administration, sensor calibration, and cloud computing. To make wise judgements, engineers, technicians, and site managers must be taught not only to operate IoT-enabled equipment but also to evaluate the data they generate (Arif et al., 2025). However, the digital skills required for successful deployment are frequently lacking in the current construction workforce.

IoT adoption could result in operational inefficiencies or underutilised technology if focused upskilling and capacity-building initiatives are not implemented. To guarantee relevance and efficacy, training programs must be created in cooperation with academic institutions, technology providers, and industry specialists (Yao et al., 2025). For organisations to embrace digital transformation, promote ongoing learning, and adjust to data-driven processes, a culture shift is also required.

GLOBAL PRACTICES AND CASE STUDIES

IoT-enabled precast concrete solutions are being used by smart cities worldwide as a vital part of robust, effective, and future-ready infrastructure. The successful integration of cutting-edge technologies into urban development is demonstrated by several real-world deployments (de Carvalho et al., 2023). Singapore, which is well known for being a pioneer in smart city innovation, is among the most notable instances. IoT-enabled precast technology has been installed in residential buildings by the nation's Housing and Development Board (HDB) to optimise energy use, monitor structural health, and guarantee long-term safety. Precast components have sensors built into them that monitor temperature, vibration, humidity, and stress in real time (Ding et al., 2025). This information is integrated into centralised dashboards that allow for predictive maintenance, cut down on human inspections, and increase the lifespan of infrastructure, establishing a benchmark for intelligent and efficient public housing (Xiaosheng & Hamzeh, 2020).

Another interesting example is Smart Dubai, where precast systems and IoT platforms are being integrated to revolutionise the design and management of public infrastructure. The Dubai Municipality has implemented intelligent precast segments for roads, bridges, and drainage systems through collaborations with technology companies and construction companies (Dilakshan et al., 2021). Bridge deck sensors, for example, monitor temperature changes and structural stress and sound an alarm when thresholds are surpassed. This has made it possible for Dubai to more efficiently distribute maintenance budgets, minimise emergency repairs, and lessen traffic interruptions (Bapat et al., 2021). Precast producers in the US are integrating real-time GPS tracking and RFID tagging into highway and bridge construction projects in areas like Texas and California. By guaranteeing just-in-time delivery and automated quality checks, these IoT solutions have enhanced supply chain coordination and contributed to a decrease in building delays (Banerjee & Nayaka, 2022). Additionally, IoT-enabled precast pieces are being used to monitor the structural health of buildings in earthquake-prone areas, providing useful information for disaster response and preparation (Yong et al., 2023).

Several lessons learnt from these successful implementations are essential for wider adoption. First and foremost, cross-sector cooperation is crucial. Policymakers, software developers, engineers, and urban planners must work together to integrate IoT into precast construction. Strong collaboration between the public and private sectors, bolstered by well-defined legal frameworks and investments in infrastructure digitisation, was the key to success in Singapore and Dubai (Wang et al., 2018). Second, it has been shown that pilot programs are good starting places.

To test technology performance, identify integration difficulties, and assess return on investment, several cities have begun with smaller-scale prefabricated installations rather of immediately installing IoT on a full scale. These pilots foster stakeholder confidence and provide scalable methods for growth (Costa et al., 2017). Third, it is impossible to overestimate the importance of interoperability and open data platforms. IoT data from precast buildings has been more successfully integrated into larger smart city platforms for energy, transportation, and emergency management in cities that have embraced open architecture systems and standardised data standards (Fang & Ng, 2019).

The benefits of digital integration are further demonstrated by a comparison of conventional precast systems with IoT-enhanced systems. Reduced construction time, better quality control, and less reliance on site conditions are some advantages of traditional precast systems (Jayanthi et al., 2021). Once installed, however, they are invisible, thus structural deterioration, maintenance requirements, and post-construction performance monitoring frequently go unnoticed until failure or a planned inspection. On the other hand, IoT-enhanced precast systems offer constant, real-time structural state monitoring, facilitating data-driven decision-making and proactive maintenance (Thakuri et al., 2024).

An IoT-enabled bridge, for instance, continuously reports on load stress, corrosion levels, and vibration patterns, enabling early intervention before issues worsen. In contrast, a traditional bridge would depend on sporadic manual inspections. Additionally, IoT enhances infrastructure asset lifecycle management by providing insights during the asset's construction as well as during its operational life (Bhogayata & Sata, 2025). This lowers the environmental effect, improves safety, and saves money.

AN OVERVIEW FOR FUTURE

As new technology and changing urban needs come together, precast concrete with IoT capabilities in smart cities is about to undergo a dramatic change. The incorporation of artificial intelligence (AI), which will improve data analytics and decision-making across the infrastructure lifetime, is one of the most exciting technologies (Khurshid et al., 2023). Compared to traditional methods, artificial intelligence (AI) can handle vast amounts of sensor data from precast structures to identify trends, forecast breakdowns, and improve maintenance schedules. In tandem, blockchain technology is gaining traction as a tool for ensuring transparency, traceability, and trust in the construction supply chain (Chen et al., 2020). Blockchain can enhance quality assurance and accountability in intricate infrastructure projects by securely documenting each step of the production, transportation, and installation of precast components. Furthermore, real-time data transfer at previously unheard-of speeds will be made possible with the introduction of 5G networks, opening new responsiveness levels and enabling highly networked, low-latency monitoring systems throughout cities (Liu et al., 2022).

Simultaneously, the move to offsite and modular building presents a significant chance to expand IoT-enabled precast systems. Large precast units are created in controlled surroundings in modular construction, which is ideally suited to IoT integration as cities face increasing pressure to build more quickly and sustainably (Drage & Kennedy, 2020). Entire modules can be digitally traced from factory to site, and smart sensors can be integrated during production. This method improves productivity, guarantees uniformity, and facilitates the large-scale construction of high-performance structures, particularly in areas that are quickly urbanising (Ben-Daya et al., 2019).

However, strong legislative and regulatory frameworks are required to enable the development of intelligent precast infrastructure. Clear guidelines for data usage, cybersecurity, interoperability, and sensor integration must be established by governments and local organisations (Valinejadshoubi et al., 2021). Public-private partnerships, innovation awards, and incentives for smart infrastructure investment can all help promote adoption, especially in areas where cost is still a deterrent.

Building collaborative ecosystems between construction companies, technology providers, and urban planners, investing in digital skills training, and beginning with small-scale trial projects are all part of the roadmap for wider adoption in developing cities (Cheng et al., 2022). Cities can adopt smart building early on and skip traditional development stages with the use of open data platforms and shared infrastructure frameworks (Singh et al., 2021).

CONCLUSION

Precast concrete technologies with Internet of Things capabilities offer smart cities a revolutionary range of benefits. The combination of IoT with precast technology is creating smarter, more efficient, and future-ready urban settings, from safer and more resilient infrastructure to shorter building periods, cost savings, informed urban planning, and long-term sustainability. Adopting these ideas will be crucial to creating intelligent, flexible, and resilient infrastructure as cities continue to expand and change. Although there are many prospects for smart city development with IoT-enabled precast systems, their deployment necessitates careful consideration of data security, interoperability, economic sustainability, and labour readiness. To fully realise the promise of intelligent infrastructure and guarantee long-term success, it will be imperative to proactively address these issues.

The revolutionary potential of IoT in precast concrete systems is demonstrated by international best practices and case studies from places like Singapore, Dubai, and the United States. Predictive analytics, smart monitoring, and real-time data integration improve infrastructure performance beyond what can be achieved with conventional techniques. Although there are still obstacles to overcome, the increasing number of effective implementations offers other cities a path forward for creating more intelligent, secure, and sustainable urban environments.

To sum up, the combination of IoT-enabled precast systems with blockchain, 5G, AI, and modular techniques represents a significant advancement in urban infrastructure. Across the globe, smart precast will be a key component of resilient, effective, and future-ready cities with the correct legislative backing and careful planning.

REFERENCES

Akbar, A., Njoroge, J. M., Lee, S., Chang, Y., & Kwon, S. (2023). CNN-Based Automatic Mobile Reporting System and Quantification for the Concrete Crack Size of the Precast Members of OSC Construction. *Engineering Proceedings*, *36*(1), 15.

Araújo, C. S., Ferreira, E. D. A. M., & Costa, D. B. (2025). System for tracking and managing metallic formworks adopted in cast-in-place concrete wall systems by integrating IoT and BIM. *Construction Innovation*, *25*(4), 1315–1341. DOI: 10.1108/CI-06-2023-0127

Arganiz, P., & Muslim, F. (2024). Development of a prototype system integration model for RFID technology with the internet of things and its implementation to improve precast concrete material management in Indonesia. *Journal of Project Management (Canada)*, *9*(3), 213–226. DOI: 10.5267/j.jpm.2024.5.001

Arif, F., Ayub, L., Khan, F. A., & Khan, A. U. R. (2025). Conceptual Framework for IoT-Based Structural Health Monitoring of RC Bridges for Maintenance Decision-Making. *Construction Technologies and Architecture*, *15*, 19–26. DOI: 10.4028/p-q8nkQW

Bahadur, P. S., Srivastava, R., & Srivastava, V. (2023). Review on the Development of Smart Materials, Including Shape Memory Alloys, Characterization, and Recent Applications. *Modeling, Characterization, and Processing of Smart Materials*, 107-139.

Banerjee, A., & Nayaka, R. R. (2022). A comprehensive overview on BIM-integrated cyber physical system architectures and practices in the architecture, engineering and construction industry. *Construction Innovation*, *22*(4), 727–748. DOI: 10.1108/CI-02-2021-0029

Bapat, H., Sarkar, D., & Gujar, R. (2021). Application of integrated fuzzy FCM-BIM-IoT for sustainable material selection and energy management of metro rail station box project in western India. *Innovative Infrastructure Solutions*, *6*(2), 73. DOI: 10.1007/s41062-020-00431-7

Bašková, R., Struková, Z., & Kozlovská, M. (2019, August). Construction cost saving through adoption of IoT applications in concrete works. In *International Conference Current Issues of Civil and Environmental Engineering Lviv-Košice–Rzeszów* (pp. 452-459). Cham: Springer International Publishing.

Ben-Daya, M., Hassini, E., & Bahroun, Z. (2019). Internet of things and supply chain management: A literature review. *International Journal of Production Research*, *57*(15-16), 4719–4742. DOI: 10.1080/00207543.2017.1402140

Bharath, S. R., Lavanya, N., Bharath Kumar, H. B., Chaitra, R. K., & Dandautiya, R. (2021). Utilization Potential of Iron Ore Tailing Waste in Various Applications. In *Advances in Energy and Environment: Select Proceedings of TRACE 2020* (pp. 125-134). Singapore: Springer Singapore. DOI: 10.1007/978-981-33-6695-4_12

Bhogayata, A., & Sata, A. (2025). *Introduction to Internet of Things in Civil Engineering: Theory and Practice*. Springer. DOI: 10.1007/978-981-97-9450-8

Chen, J. H., Ha, N. T. T., Tai, H. W., & Chang, C. A. (2020). The willingness to adopt the Internet of Things (IoT) conception in Taiwan's construction industry. *Journal of Civil Engineering and Management*, *26*(6), 534–550. DOI: 10.3846/jcem.2020.12639

Cheng, Y., Lin, F., Wang, W., & Zhang, J. (2022). Vision-based trajectory monitoring for assembly alignment of precast concrete bridge components. *Automation in Construction*, *140*, 104350. DOI: 10.1016/j.autcon.2022.104350

Chopade, R., Pathak, P., Chougule, R., Bothra, N., Mane, N., & Minde, P. (2021). Analysis of Safety Management in Precast Construction Projects. *Turkish Journal of Computer and Mathematics Education*, *12*(6), 5225–5236.

Cole, A., Ameri Sianaki, O., & Shee, H. (2024, July). Conceptualizing a Digital Twin Architecture for Enhanced Control in Precast Concrete Production. In *International Conference on Complex, Intelligent, and Software Intensive Systems* (pp. 261-272). Cham: Springer Nature Switzerland. DOI: 10.1007/978-3-031-70011-8_24

Costa, A., Proverbio, M., & Smith, I. F. (2017, November). Cyber civil infrastructure and IoT for cities. In *2017 IEEE Symposium Series on Computational Intelligence (SSCI)* (pp. 1-8). IEEE. DOI: 10.1109/SSCI.2017.8280815

Dave, B., Kubler, S., Främling, K., & Koskela, L. (2016). Opportunities for enhanced lean construction management using Internet of Things standards. *Automation in Construction*, *61*, 86–97. DOI: 10.1016/j.autcon.2015.10.009

David, M., Derigent, W., Loubet, G., Takacs, A., & Dragomirescu, D. (2024). Communicating materials: Communicating concrete development for construction industry. *IEEE Transactions on Industrial Informatics*, *20*(4), 6983–6994. DOI: 10.1109/TII.2024.3352261

de Carvalho, A. R., da Silva Calderón-Morales, B. R., Júnior, J. C. B., de Oliveira, T. M., & Silva, G. J. B. (2023). Proposition of geopolymers obtained through the acid activation of iron ore tailings with phosphoric acid. *Construction & Building Materials*, *403*, 133078. DOI: 10.1016/j.conbuildmat.2023.133078

Dilakshan, S., Rathnasinghe, A. P., & Seneviratne, L. D. I. P. (2021, July). Potential of internet of things (Iot) in the construction industry. In *Proceedings The 9th World Construction Symposium| July* (p. 446).

Dilakshan, S., Rathnasinghe, A. P., & Seneviratne, L. D. I. P. (2021). An Investigation on the Readiness of the Sri Lankan Construction Industry towards Internet of Things (IoT): Conceptual Guideline. In *International Conference on Business Management*-2020. DOI: 10.31357/icbm.v17.5194

Ding, H., Huang, H., Liu, Z., & Wang, F. (2025). Digital aggregate for concrete smart management and lifecycle traceability from gamma-dicalcium silicate carbonation. *Construction & Building Materials*, *472*, 140869. DOI: 10.1016/j.conbuildmat.2025.140869

Ding, H., Li, M., Zhong, R. Y., & Huang, G. Q. (2023). Multistage self-adaptive decision-making mechanism for prefabricated building modules with IoT-enabled graduation manufacturing system. *Automation in Construction*, *148*, 104755. DOI: 10.1016/j.autcon.2023.104755

Ding, S., Hu, H., Xu, F., Chai, Z., & Wang, W. (2024). Blockchain-based security-minded information-sharing in precast construction supply chain management with scalability, efficiency and privacy improvements. *Automation in Construction*, *168*, 105698. DOI: 10.1016/j.autcon.2024.105698

Drage, J., & Kennedy, G. (2020). Building a Low-Cost, Internet-of-Things, Real-Time Groundwater Level Monitoring Network. *Ground Water Monitoring and Remediation*, *40*(4), 67–73. DOI: 10.1111/gwmr.12408

Fang, Y., & Ng, S. T. (2019). Genetic algorithm for determining the construction logistics of precast components. *Engineering, Construction, and Architectural Management*, *26*(10), 2289–2306. DOI: 10.1108/ECAM-09-2018-0386

Gao, M. Y., Han, J., Yang, Y., Tiong, R. L., Zhao, C., & Han, C. (2024). BIM-based and IoT-driven smart tracking for precast construction dynamic scheduling. *Journal of Construction Engineering and Management*, *150*(9), 04024117. DOI: 10.1061/JCEMD4.COENG-14498

Han, C., & Ye, H. (2018, August). A novel IoT-Cloud-BIM based intelligent information management system in building industrialization. In *International Conference on Construction and Real Estate Management 2018* (pp. 72-77). Reston, VA: American Society of Civil Engineers. DOI: 10.1061/9780784481721.008

Hu, X., Shen, J., & Ma, Z. (2024, November). Prefabricated Buildings for Smart Cities: A Literature Survey. In *2024 Twelfth International Conference on Advanced Cloud and Big Data (CBD)* (pp. 302-307). IEEE. DOI: 10.1109/CBD65573.2024.00061

Ismail, Z. A. B. (2021). Hybrid intelligent vehicle system for managing construction supply chain in precast concrete building construction projects. *World Journal of Engineering*, *18*(4), 538–546. DOI: 10.1108/WJE-08-2020-0361

Ismail, Z. A. B. (2022). Thermal comfort practices for precast concrete building construction projects: Towards BIM and IOT integration. *Engineering, Construction, and Architectural Management*, *29*(3), 1504–1521.

Jadhav, O., Minde, P., Yadhav, A., & Gaidhankar, D. (2023). A review of emerging trends & advances in construction technology in the Indian scenario. *Materials Today: Proceedings*, *77*, 897–904. DOI: 10.1016/j.matpr.2022.12.041

Jayanthi, N., Raja, K. T., Wadhwa, G., Shneka, K., & Swathi, R. (2021). IoT based-civil labour safety monitoring system in construction site. *Turkish Journal of Computer and Mathematics Education*, *12*(9), 1723–1728.

Jiang, Y., Li, M., Li, M., Liu, X., Zhong, R. Y., Pan, W., & Huang, G. Q. (2022). Digital twin-enabled real-time synchronization for planning, scheduling, and execution in precast on-site assembly. *Automation in Construction*, *141*, 104397. DOI: 10.1016/j.autcon.2022.104397

Kampli, G., Chickerur, S., & Chitawadagi, M. (2024). IoT system implementation for real-time concrete strength prediction: Experimental design, variance evaluation, cost analysis, and implementation ease. *Innovative Infrastructure Solutions*, *9*(7), 260. DOI: 10.1007/s41062-024-01586-3

Katiyar, A., & Kumar, P. (2022). Automated construction progress monitoring of prefabricated concrete structures using building information modeling and internet of things. *International Journal on Recent and Innovation Trends in Computing and Communication*, *10*(7), 14–18. DOI: 10.17762/ijritcc.v10i7.5560

Khurshid, K., Danish, A., Salim, M. U., Bayram, M., Ozbakkaloglu, T., & Mosaberpanah, M. A. (2023). An in-depth survey demystifying the Internet of Things (IoT) in the construction industry: Unfolding new dimensions. *Sustainability (Basel)*, *15*(2), 1275. DOI: 10.3390/su15021275

Kosse, S., Forman, P., Stindt, J., Hoppe, J., König, M., & Mark, P. (2023, June). Industry 4.0 enabled modular precast concrete components: a case study. In *International RILEM Conference on Synergising expertise towards sustainability and robustness of CBMs and concrete structures* (pp. 229-240). Cham: Springer Nature Switzerland. DOI: 10.1007/978-3-031-33211-1_21

Kosse, S., Hagedorn, P., Maibaum, J., & König, M. (2024, August). Digital twins for precast concrete: advancing environmental analysis through integrated life cycle assessment. In *International Conference on Computing in Civil and Building Engineering* (pp. 613-627). Cham: Springer Nature Switzerland.

Kostadino, E. (2023). *Prediction Of U-Ditch Raw Material Inventories In.* Multiguna Precast Mandiri Using A Web-Based Single Exponential Smoothing Method.

Li, C. Z., Hong, J., Xue, F., Shen, G. Q., Xu, X., & Luo, L. (2016). SWOT analysis and Internet of Things-enabled platform for prefabrication housing production in Hong Kong. *Habitat International, 57*, 74–87. DOI: 10.1016/j.habitatint.2016.07.002

Li, C. Z., Hong, J., Xue, F., Shen, G. Q., Xu, X., & Mok, M. K. (2016). Schedule risks in prefabrication housing production in Hong Kong: A social network analysis. *Journal of Cleaner Production, 134*, 482–494. DOI: 10.1016/j.jclepro.2016.02.123

Li, C. Z., Xue, F., Li, X., Hong, J., & Shen, G. Q. (2018). An Internet of Things-enabled BIM platform for on-site assembly services in prefabricated construction. *Automation in Construction, 89*, 146–161. DOI: 10.1016/j.autcon.2018.01.001

Li, K., Gan, V. J., Li, M., Gao, M. Y., Tiong, R. L., & Yang, Y. (2024). Automated generative design and prefabrication of precast buildings using integrated BIM and graph convolutional neural network. *Developments in the Built Environment, 18*, 100418. DOI: 10.1016/j.dibe.2024.100418

Liang, J., Cao, Z., Zu, Q., Huang, H., & Guo, S. (2024). Production sequencing and layout optimization of precast concrete components under mold resource constraints. *Buildings (Basel, Switzerland), 14*(10), 3173. DOI: 10.3390/buildings14103173

Lima, C. M., Ferreira, E. D. A. M., & Calmon, J. L. (2021). BIM and IoT integration to support management processes: An application for precast concrete systems. *The Journal of Modern Project Management, 9*(2).

Liu, T., Zhu, C., Tong, Z., & Luo, D. (2024, June). Research on smart manufacturing technology of reinforcing bars for bridge precast concrete structure. [). IOP Publishing.]. *Journal of Physics: Conference Series, 2775*(1), 012028. DOI: 10.1088/1742-6596/2775/1/012028

Liu, Y., Tong, W., Li, Q., Yao, F., Li, Y., Li, H. X., & Huang, J. (2022). Study on complexity of precast concrete components and its influence on production efficiency. *Advances in Civil Engineering*, *2022*(1), 9926547. DOI: 10.1155/2022/9926547

Liu, Z. S., Liu, S. N., Wang, S. W., Zhao, Y. H., & Yu, J. Q. (2018). Information solution of precast building process based on low power wide area Internet of Things. *Construction Technology*, *16*(47), 117–122.

Lorusso, A., & Celenta, G. (2023, May). Internet of Things in the construction industry: A general overview. In *International Conference "New Technologies, Development and Applications"* (pp. 577-584). Cham: Springer Nature Switzerland. DOI: 10.1007/978-3-031-31066-9_65

Martiradonna, S., Lombillo, I., & Fatiguso, F. (2019, October). Performance Monitoring of Refurbished Buildings through Innovative Precast Concrete Modules. In *2019 IEEE International Conference on Systems, Man and Cybernetics (SMC)* (pp. 953-957). IEEE. DOI: 10.1109/SMC.2019.8914472

Miller, D., Ho, N. M., Talebian, N., & Javanbakht, Z. (2023). Real-time monitoring of early-age compressive strength of concrete using an IoT-enabled monitoring system: An investigative study. *Innovative Infrastructure Solutions*, *8*(2), 75. DOI: 10.1007/s41062-023-01043-7

Min, H., & Junyu, L. (2013, August). An intelligent system for precast concrete element manufacturing management based on RFID technology. In *2013 IEEE International Conference on Green Computing and Communications and IEEE Internet of Things and IEEE Cyber, Physical and Social Computing* (pp. 1138-1143). IEEE. DOI: 10.1109/GreenCom-iThings-CPSCom.2013.198

Mishra, S., Singh, S., & Srivastava, R. (2017). Green Building a Sustainable Approach to Save Energy. *International Journal of Engineering and Management Research*, *7*(3), 14–20.

Naraharisetty, V., Talari, V. S., Neridu, S., Kalapatapu, P., & Pasupuleti, V. D. K. (2021, August). Cloud architecture for IoT based bridge monitoring applications. In *2021 international conference on emerging techniques in computational intelligence (ICETCI)* (pp. 39-42). IEEE. DOI: 10.1109/ICETCI51973.2021.9574044

Pradhan, P., Verma, M. S., & Sahu, M. S. (2025). Optimization of Cast-In-Situ and Precast Concrete Methods Through BIM and Structural Simulations. *Optimization*.

Qi, B., Razkenari, M., Costin, A., Kibert, C., & Fu, M. (2021). A systematic review of emerging technologies in industrialized construction. *Journal of Building Engineering*, *39*, 102265. DOI: 10.1016/j.jobe.2021.102265

Qiu, T., Chen, X., Su, D., Chen, K., He, Q., Zheng, Z., Zhou, W., & Wang, Y. (2024). LEGO®-inspired decision-making model for automatic construction of precast diaphragm walls. *Automation in Construction, 166*, 105667. DOI: 10.1016/j.autcon.2024.105667

Sarkar, D., Dhaneshwar, D., & Raval, P. (2023). Automation in monitoring of construction projects through BIM-IoT-blockchain model. *Journal of The Institution of Engineers (India): Series A, 104*(2), 317-333.

Sarkar, D., Patel, H., & Dave, B. (2022). Development of integrated cloud-based Internet of Things (IoT) platform for asset management of elevated metro rail projects. *International Journal of Construction Management, 22*(10), 1993–2002. DOI: 10.1080/15623599.2020.1762035

Shehu, M. O., Almeida, R., & Mahapatra, K. (2025). Opportunities for digital tracking technologies in the precast concrete sector in Sweden. *Frontiers in Built Environment, 11*, 1566784. DOI: 10.3389/fbuil.2025.1566784

Singh, R., Gehlot, A., Akram, S. V., Gupta, L. R., Jena, M. K., Prakash, C., Singh, S., & Kumar, R. (2021). Cloud manufacturing, internet of things-assisted manufacturing and 3D printing technology: Reliable tools for sustainable construction. *Sustainability (Basel), 13*(13), 7327. DOI: 10.3390/su13137327

Solanki, A., & Sarkar, D. (2022, August). A Framework of Internet of Things–Cloud–Building Information Modeling Based Life Cycle Management of Precast Components in Prefabricated Buildings. In *International Conference Innovation in Smart and Sustainable Infrastructure* (pp. 671-681). Singapore: Springer Nature Singapore.

Srivastava, R., Arun, J., & Patel, D. K. (2019, April). Amalgamating the Service Quality Aspect in Supply Chain Management. In *2019 International Conference on Automation, Computational and Technology Management (ICACTM)* (pp. 63-67). IEEE. DOI: 10.1109/ICACTM.2019.8776839

Srivastava, R., & Thakur, A. (2025). The Foundational Principles of FDM: An Acronym Pivotal t Its Operation and Its Wide Range of Applications. In *Modeling, Analysis, and Control of 3D Printing Processes* (pp. 355-380). IGI Global Scientific Publishing.

Srivastava, R., Thakur, A., & Rana, A. (2024). The Basic Understanding of Nano Materials: A Review. *Innovations in Nanomaterials-Based Corrosion Inhibitors*, 1-20.

Srivastava, V. K., Maurya, L., & Mehra, E. R. (2014). Detection of noise in high pass Butterworth IIR Filter using MATLAB. *IJERT, 3*(2), 1232–1235.

Takada, T. (2020). Study of CIM and IoT-Simulation for Cost Performance Analysis-(Cost Management). , (2), 4-25.

Thakuri, P. K., Alkki, L., & Aarikka-Stenroos, L. (2024). Digital technologies enabling component reuse in circular value chains: Using digital twin, internet of things and robots in construction and manufacturing sectors. *R & D Management*.

Tian, M. (2021, November). Monitoring Method of Vertical Stress on Precast Concrete of Prefabricated Building. In *International Conference on Forthcoming Networks and Sustainability in the IoT Era* (pp. 220-227). Cham: Springer International Publishing.

Tripathi, A., Kaushal, I., Revolti, A., Benedetti, L., & Dallasega, P. (2025). Digital Twin Driven Quality Control in Smart Mobile Factories: A Framework in Precast Manufacturing. *Procedia CIRP*, *134*, 801–806. DOI: 10.1016/j.procir.2025.02.194

Uygunoğlu, T., & Topçu, İ. B. (2020). Nesnelerin internetinin (iot) inşaat mühendisliğindeki rolü: rfid uygulamalari. *International Journal of 3D Printing Technologies and Digital Industry*, *4*(3), 270-277.

Valinejadshoubi, M., Moselhi, O., Bagchi, A., & Salem, A. (2021). Development of an IoT and BIM-based automated alert system for thermal comfort monitoring in buildings. *Sustainable Cities and Society*, *66*, 102602. DOI: 10.1016/j.scs.2020.102602

Wang, X., Wang, S., Song, X., & Han, Y. (2020). IoT-based intelligent construction system for prefabricated buildings: Study of operating mechanism and implementation in China. *Applied Sciences (Basel, Switzerland)*, *10*(18), 6311. DOI: 10.3390/app10186311

Wang, Y., Thangasamy, V. K., Tiong, R. L., & Zhang, L. (2022). Improved workflow for precast element design based on BIM and lean construction. *Journal of Construction Engineering and Management*, *148*(8), 04022065. DOI: 10.1061/(ASCE)CO.1943-7862.0002316

Wang, Y., Yuan, Z., & Sun, C. (2018). Research on assembly sequence planning and optimization of precast concrete buildings. *Journal of Civil Engineering and Management*, *24*(2), 106–115. DOI: 10.3846/jcem.2018.458

Wang, Z., Wang, T., Hu, H., Gong, J., Ren, X., & Xiao, Q. (2020). Blockchain-based framework for improving supply chain traceability and information sharing in precast construction. *Automation in Construction*, *111*, 103063. DOI: 10.1016/j.autcon.2019.103063

Wang, Z., Zhang, Q., Yang, B., Wu, T., Lei, K., Zhang, B., & Fang, T. (2021). Vision-based framework for automatic progress monitoring of precast walls by using surveillance videos during the construction phase. *Journal of Computing in Civil Engineering, 35*(1), 04020056. DOI: 10.1061/(ASCE)CP.1943-5487.0000933

Xiao, J., Jiang, S., & Li, Y. (2024). Seismic Performance of Prefabricated Reinforced Concrete Structure Beam Joints Based on the Internet of Things. *Journal of Testing and Evaluation, 52*(3), 1789–1803. DOI: 10.1520/JTE20230027

Xiaosheng, T., & Hamzeh, F. (2020, July). Precast concrete building construction process comparison. In *Proceedings of the Annual Conference of the International Group for Lean Construction,* Berkeley, CA, USA (pp. 6-12).

Xu, J., Zhang, Q., Teng, Y., & Pan, W. (2023). Integrating IoT and BIM for tracking and visualising embodied carbon of prefabricated buildings. *Building and Environment, 242*, 110492. DOI: 10.1016/j.buildenv.2023.110492

Yang, B., Liu, B., Xiao, J., Zhang, B., Wang, Z., & Dong, M. (2021). A novel construction scheduling framework for a mixed construction process of precast components and cast-in-place parts in prefabricated buildings. *Journal of Building Engineering, 43*, 103181. DOI: 10.1016/j.jobe.2021.103181

Yao, Y., Tam, V. W., Wang, J., Le, K. N., & Butera, A. (2025). Deep reinforcement learning-based schedule optimization for parallel precast production. *Engineering, Construction, and Architectural Management, 32*(13), 1–34. DOI: 10.1108/ECAM-03-2025-0429

Ye, X., Kosse, S., Hagedorn, P., & König, M. (2025). Workflow automation using digital twins and smart contracts: a decentralised framework for precast concrete production.

Ye, X., Zhou, Y., Guo, H., & Luo, Z. (2024). A computer vision-based approach to automatically extracting the aligning information of precast structural components. *Automation in Construction, 164*, 105478. DOI: 10.1016/j.autcon.2024.105478

Yong, Y. P., Lee, S. J., Chang, Y. H., Lee, K. H., Kwon, S. W., Cho, C. S., & Chung, S. W. (2023). Object detection and distance measurement algorithm for collision avoidance of precast concrete installation during crane lifting process. *Buildings (Basel, Switzerland), 13*(10), 2551. DOI: 10.3390/buildings13102551

Yu, Z., Wei, Y., & Cao, Z. (2019, December). The Design and Application of Information Management System Based on IOT for Precast Concrete Segments. *IOP Conference Series. Earth and Environmental Science, 371*(3), 032056. DOI: 10.1088/1755-1315/371/3/032056

Zhang, R., Feng, X., Mou, Z., & Zhang, Y. (2023). Green optimization for precast production rescheduling based on disruption management. *Journal of Cleaner Production*, *420*, 138406. DOI: 10.1016/j.jclepro.2023.138406

Zhang, W., Kang, K., & Zhong, R. Y. (2021). A cost evaluation model for IoT-enabled prefabricated construction supply chain management. *Industrial Management & Data Systems*, *121*(12), 2738–2759. DOI: 10.1108/IMDS-12-2020-0735

Zhao, L., Liu, Z., & Mbachu, J. (2019). Development of intelligent prefabs using IoT technology to improve the performance of prefabricated construction projects. *Sensors (Basel)*, *19*(19), 4131. DOI: 10.3390/s19194131 PMID: 31554246

Zhao, Y., Cao, C., & Liu, Z. (2022). A framework for prefabricated component hoisting management systems based on digital twin technology. *Buildings (Basel, Switzerland)*, *12*(3), 276. DOI: 10.3390/buildings12030276

Zhao, Y., Cao, C., Liu, Z., & Mu, E. (2021). Intelligent control method of hoisting prefabricated components based on internet-of-things. *Sensors (Basel)*, *21*(3), 980. DOI: 10.3390/s21030980 PMID: 33540530

Zhong, R. Y., Peng, Y., Xue, F., Fang, J., Zou, W., Luo, H., Thomas Ng, S., Lu, W., Shen, G. Q. P., & Huang, G. Q. (2017). Prefabricated construction enabled by the Internet-of-Things. *Automation in Construction*, *76*, 59–70. DOI: 10.1016/j.autcon.2017.01.006

Zhu, Z. Q., Yong, Y. P., Lee, S. J., Chang, Y. H., & Kwon, S. W. (2021). Vision-based precast concrete management plan in off-site construction site: Using PC member quality grades. In *ISARC. Proceedings of the International Symposium on Automation and Robotics in Construction* (Vol. 38, pp. 197-203). IAARC Publications.

Chapter 9
Multi-Criteria Optimization for Multi-Hazard Resilience:
Sustainable Seismic Retrofitting of Green Reinforced Concrete Structures

Ali Biglarifadafan
http://orcid.org/0000-0002-9616-1823
University of Golestan, Iran

Sérgio António Neves Lousada
http://orcid.org/0000-0002-8429-2164
Universidade da Madeira, Portugal

ABSTRACT

This chapter proposes an Integrated Performance-Based Rehabilitation Framework that couples low-carbon, bio-based concrete retrofit solutions with a Multi-Criteria Decision-Making (MCDM) optimization scheme for existing reinforced-concrete structures exposed to compound hazards (e.g., earthquakes combined with flooding, heat, or wildfire). It argues that life-safety–focused retrofit practice is insufficient to manage trade-offs among resilience, lifecycle cost, carbon footprint, and implementability. The workflow integrates: (1) condition assessment and structural health monitoring; (2) stochastic degradation and service-life prediction; (3) performance-based evaluation of retrofit alternatives; (4) sustainable intervention design using bio-based materials; and (5) lifecycle management and decision sup-

DOI: 10.4018/979-8-3373-6167-3.ch009

port. A multi-objective formulation combines LCCA, LCA and probabilistic risk assessment, strengthened with robust decision methods under deep uncertainty, to identify retrofit designs that co-optimize safety and sustainability.

1. INTRODUCTION

A significant capital investment and a vital facilitator of social and economic activity is the world's stock of reinforced concrete infrastructure. But a sizable amount of this stock is already approaching an advanced stage of its service life, showing indications of accumulated damage, material deterioration, and functional obsolescence. The intensifying consequences of climate change, such as increased frequency of extreme weather events, more aggressive environmental exposure, and changing seismic hazard profiles in many places, exacerbate this aging process. It is clear that the conventional paradigm of infrastructure management, which is defined by adhering to prescriptive code minima and making reactive repairs after obvious distress, is insufficient to handle these issues (Asadi et al., 2012; Deng et al., 2025; Jafari et al., 2018; Loures et al., 2019; Vulevic et al., 2021; Zhang et al., 2016). This strategy frequently produces less than ideal results: treatments may be overly cautious, resulting in needless financial and environmental expenses, or they may be inadequately long-lasting, failing to address underlying causes and causing damage to return too soon. As a result, our infrastructure assets consistently underperform, which shows up as lower safety margins, higher lifetime costs, excessive carbon emissions from repeated repairs, and less resilience against potential threats.

This chapter makes the case for a fundamental change from this reactive, prescriptive model to an integrated, performance-based framework for rehabilitation. The suggested methodology, which goes by the name "From Cracks to Resilience," is based on accurate condition assessment, data-driven and physics-based deterioration modeling, and clear connections between lifecycle performance goals and retrofit design decisions. Visible indicators of distress, such as cracks, are important data points that, when statistically assessed within a comprehensive diagnostic paradigm, disclose the underlying state of the structure and offer prognostics of its future performance. They are not just symptoms to be treated locally. In this context, resilience is defined as the structure's capacity to maintain necessary functioning during both acute hazard occurrences and progressive deterioration, as well as to effectively restore functionality after disruption.

Customized intervention is the central tenet. The framework promotes retrofitting techniques that are specifically tailored to the particular condition, risk profile, and functional requirements of a given facility rather than using conventional solutions. An integrated workflow combining advanced structural health monitoring (SHM), non-destructive assessment (NDE), probabilistic material modeling, nonlinear structural analysis, and multi-objective decision-making makes this calibration possible. The ultimate objective is to turn rehabilitation from a cost center into a value-creation activity that maximizes total cost of ownership across the remaining lifecycle, increases safety and resilience, prolongs service life, and lessens environmental impact.

The scope, purpose, and general framework of the work are established in this introductory chapter. It starts by estimating the scope of the aging infrastructure problem and evaluating the shortcomings of existing methods. The fundamental ideas of the suggested performance-based rehabilitation paradigm are then explained. Lastly, it offers a road map for the next chapters, which will go into great detail about the technical aspects of the framework, including modeling, intervention, lifecycle management, sensing, and diagnosis.

1.1 Aging Infrastructure in a Changing Climate

The most used building material in the world, reinforced concrete, serves as the foundation for building stock, energy systems, water distribution, and transportation networks (Blagojević et al., 2023; Lousada, Gonçalves, et al., 2022). The American Society of Civil Engineers (ASCE) 2021 Report Card for America's Infrastructure gave the country's infrastructure a grade of "C-" and projected a financing shortfall of more than $2.59 trillion over the following ten years. Repair, rehabilitation, and replacement of old reinforced concrete structures account for a large amount of this need. Similar evaluations reveal that a significant portion of bridges and important structures are over 50 years old, were built in accordance with outdated codes, and were intended for climate conditions that no longer exist (Yamamoto, 2009).

RC deterioration is a multiphysics, time-dependent phenomenon. Chloride-induced corrosion, carbonation, Alkali-Silica Reaction (ASR), Sulfate Attack, Freeze-Thaw Cycling, and Mechanical Fatigue and Overload are the main processes (Gu et al., 2022; Karthik et al., 2020; Kim et al., n.d.; Nguanthaisong et al., 2025; Yu et al., 2022). These processes are made more vulnerable by climate change. Carbonation is accelerated by higher atmospheric CO2 concentrations. ASR and corrosion kinetics can be accelerated by higher temperatures. Moisture transfer, a major factor in a number of deterioration processes, is impacted by more frequent and intense wet-dry cycles. Most importantly, the statistical distributions of hazard events—such as floods, storms, and even seismic activity in some areas—for which current structures were not specifically constructed are changing due to climate change (Figure 1). The usefulness of past data as a predictor of future performance is called into question by this non-stationarity of hazards, which adds significant uncertainty to lifecycle evaluations.

Figure 1. The impact of climate change on concrete structures and its effects through multiple processes

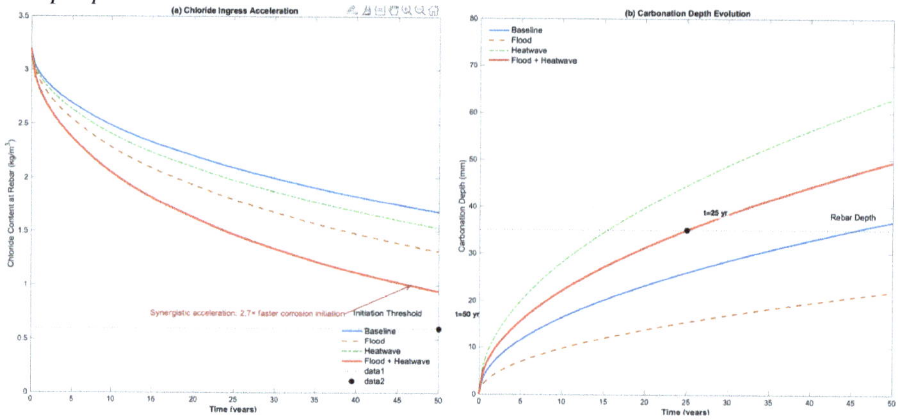

Asset owners face a challenging decision-making environment due to the combination of material aging, increased environmental attack, and changing hazard regimes. There are significant financial risks involved, including potential catastrophic failure costs, user expenses from service interruptions, and direct repair costs. Additionally, rehabilitation efforts have a significant negative influence on the environment through trash production, energy use during construction, and material manufacturing. Therefore, the need for a more intelligent, robust, and sustainable approach to restoration is a socioeconomic and environmental necessity as well as a technical choice.

In RC rehabilitation, the conventional approach frequently adheres to a fragmented, sequential pattern. Usually visual and conducted on a regular basis, inspections are either required by law or prompted by evident discomfort. Often, the diagnosis is qualitative and based on general guidelines. Code documents that outline minimum material qualities, reinforcing specifics, and connection techniques for large classes of structures serve as a major source of guidance for the design of interventions. These rules do not optimize for performance, affordability, or sustainability for a particular asset, even though they guarantee a minimum level of safety. This method has a number of significant drawbacks:

1. **Disconnect from Actual Condition:** Quantitative information about the in-situ material qualities, the degree of hidden damage, and the structure's residual capacity is not used in prescriptive designs. This results in a "one-size-fits-all" approach, where a badly damaged beam and a mildly corroded beam may receive the same refit detail, leading to either over- or under-design.

2. **Reactive and Non-Integrated:** In order to relieve regional distress, interventions are frequently implemented piecemeal without taking into account the behavior of the global system or the underlying cause of degradation. For instance, patching spalled concrete without treating persistent corrosion ensures future repairs.
3. **Focus on Strength, Not Performance:** The emphasis of traditional codes is on ultimate strength. However, certain performance levels (such as Immediate Occupancy, Life Safety, and Collapse Prevention) connected to quantifiable engineering demand factors (such as drift and strain) are defined by contemporary performance-based engineering (PBE) frameworks. Seldom does traditional rehabilitation focus on particular post-intervention performance levels under predetermined danger intensities.
4. **Neglect of Lifecycle Considerations:** Minimizing the initial cost of building is often the basis for decisions. A more comprehensive perspective takes into account the net present value of all future expenses, such as maintenance, failure risk, and end-of-life disposal, in addition to environmental effects (embodied carbon, energy). Such integrated lifetime analysis is not possible with traditional methods.
5. **Inadequate Use of Data:** The abundance of information that may be gleaned from contemporary sensing technologies—such as moisture content, corrosion rate, acceleration, and continuous strain—is not fully utilized. The value of this data is unrealized in the absence of a framework to incorporate it into models and choices.

Together, these constraints lead to an inefficient use of limited resources, lost chances to increase resilience and prolong service life, and an increased environmental impact from the building industry.

1.2 Pillars of the Integrated Performance-Based Rehabilitation Framework

The suggested approach is based on five interrelated pillars that together allow for a paradigm change from reactive damage repair to proactive, resilience-focused lifecycle stewardship of existing structures. In keeping with modern ideas of performance-based engineering, infrastructure resilience, and asset management, these pillars are interacting elements within an iterative, adaptive decision-making ecosystem rather than sequential stages (Figures 2 & 3). By integrating diagnostics, modeling, performance goals, intervention technologies, and lifecycle decision support into a single methodological architecture, the framework operationalizes the shift "from cracks to resilience."

Figure 2. Five pillar circular framework for structural rehabilitation from condition assessment to lifecycle asset stewardship

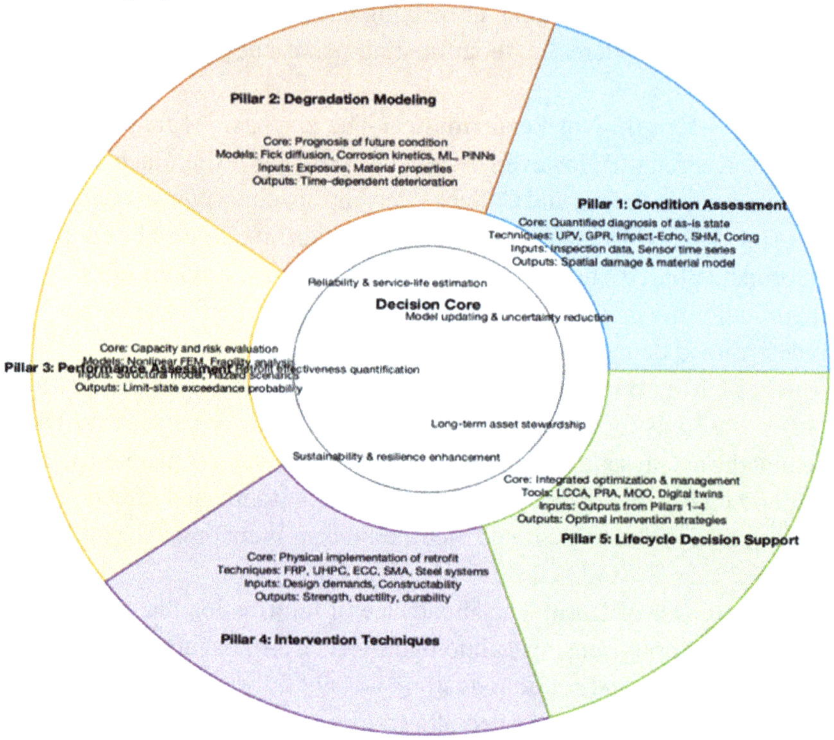

Pillar 1: Quantitative Condition Assessment and Structural Health Monitoring (SHM)

This pillar illustrates the modern transition from qualitative, inspector-dependent assessments to quantified, data-driven representations of structural status and serves as the framework's fundamental diagnostic layer. While visual inspection is essential for quick screening, it is not enough to capture early-stage damage progression, spatial diversity of material attributes, and subsurface deteriorating mechanisms.

Non-Destructive Evaluation (NDE)

To produce spatially resolved data on material integrity and deteriorating processes, advanced NDE techniques are used. Tomography and ultrasonic pulse velocity (UPV) shed light on interior cracking and the deterioration of elastic stiffness. The mapping of moisture intrusion channels, cover depth, and reinforcement architecture is made possible using ground-penetrating radar (GPR). Delaminations and voids can be found using impact-echo testing, and active corrosion zones can be probabilistically identified using resistivity and half-cell potential measurements. Large-scale bridge and building assessments reported by Bungey et al., Malhotra & Carino, and current RILEM TC recommendations show that these approaches, when combined, greatly minimize epistemic uncertainty associated with hidden deterioration.

Continuous Structural Health Monitoring (SHM)

Condition assessment is expanded by continuous SHM systems from discrete inspection snapshots to time-resolved behavioral tracking. Networks of embedded or surface-mounted sensors, such as accelerometers, strain gauges, piezoelectric transducers, and fiber optic Bragg gratings, record the local and global response of structures under severe, environmental, and operational loads. According to the groundbreaking studies of Farrar & Worden and Sohn et al., long-term monitoring makes it possible to identify stiffness loss, modal parameter alterations, fatigue accumulation, and environmental sensitivity. SHM data, which turns uncertainty from a static assumption into a quantifiable quantity, is crucial for model updating, damage forecasting, and early anomaly discovery.

Selective Destructive Testing and Material Sampling

NDE and SHM outputs are calibrated by selecting using chemical analysis, targeted coring, and reinforcement extraction. These tests guarantee that indirect data are grounded in physical reality by providing high-fidelity estimates of steel cross-section loss, carbonation depth, elastic modulus, chloride profiles, and compressive strength. The deliberate blending of non-destructive and destructive data is consistent with best practices in reliability-based evaluation of current structures.

A computerized depiction of the current state of the structure that includes boundary conditions, material characteristics, spatially changeable geometry, and clearly defined damage states. The starting point for all further modeling and decision-making procedures is this representation.

Pillar 2: Physics-Based and Data-Driven Degradation Modeling

Time-dependent degradation models are calibrated and updated using the diagnostic outputs from Pillar 1, allowing for forward-looking forecasts of structural state under uncertainty. The change from diagnosis to prognosis is formalized by this pillar.

Physics-Based Degradation Models

Dominant degradation mechanisms are simulated using mechanistic models based on transport phenomena and electrochemistry. Fick's second rule is used to estimate chloride intrusion, and time-dependent diffusion coefficients are used to take moisture effects, aging, and cracking into consideration. Square-root-of-time formulations tuned to environmental exposure are used to depict the process of carbonation. Fatigue damage accumulation is handled by fracture mechanics or Miner's rule extensions, while corrosion initiation and propagation are modeled using Faraday's law in conjunction with section loss kinetics. The literature has thoroughly validated these methods (Tuutti, Andrade, Papadakis, and fib Model Code).

Data-Driven and Empirical Models

Artificial neural networks, random forests, and Gaussian process regression are examples of machine learning models that are trained using historical inspection records from comparable structural typologies and exposure classes when substantial datasets are available. In complicated situations when physics-based formulations are insufficient or computationally costly, these models forecast condition indices, damage states, or remaining service life.

Hybrid and Physics-Informed Approaches

Governing equations are directly incorporated into the learning architecture by hybrid modeling techniques, such as physics-informed neural networks (PINNs). This method improves extrapolation ability under sparse or noisy datasets, guarantees physical consistency, and minimizes data requirements. The advantage of PINNs in simulating chloride diffusion and corrosion kinetics in reinforced concrete systems is demonstrated by recent, highly cited studies.

Reliability assessment and lifecycle planning are based on probabilistic predictions of future condition states under various environmental, loading, and maintenance situations.

Pillar 3: Performance-Based Assessment and Retrofit Design

The question of what performance level the repaired structure must attain and under what risks is operationalized by this pillar. It embraces the ideas of performance-based engineering, eschewing prescriptive compliance in favor of clearly stated performance goals. Target performance levels are set for various functional states and danger intensities in cooperation with stakeholders. These could include immediate occupation for vital infrastructure, serviceability under everyday operating loads, damage mitigation under design-level earthquakes, and life safety during infrequent seismic events. This strategy is consistent with the frameworks put forward by FEMA, PEER, and FIB.

Nonlinear Structural Analysis

To evaluate current capacity and model retrofit methods, high-fidelity nonlinear finite element models are updated with the as-is situation from Pillar 1. Degradation effects, geometric nonlinearity, and material nonlinearity are all specifically taken into account. The effects of retrofit interventions, such as steel jacketing, FRP confinement, and external post-tensioning, on stiffness, strength, ductility, and energy dissipation are explicitly modeled.

Fragility and Reliability Analysis

To link hazard intensity metrics (such as spectrum acceleration and drift demand) to the likelihood of surpassing predetermined performance limit states, probabilistic fragility functions are created. In line with cutting-edge seismic risk assessment techniques, comparing fragility curves before and after retrofit offers a quantitative indicator of intervention efficacy. In order to find solutions that meet performance goals while reducing lifespan costs, environmental effects, or disturbance, retrofit design is presented as an optimization problem.

Pillar 4: Sustainable and Resilient Intervention Techniques

In contrast to short-term capacity augmentation, this pillar focuses on resilience, sustainability, and durability in the physical implementation of retrofit solutions. For flexural and shear strengthening and confinement, sophisticated composite systems like carbon- and glass-FRP wraps, sheets, and near-surface-mounted bars are used. To improve sustainability, durability, and crack management, high-performance cementitious materials are employed, such as low-carbon geopolymer mortars, ECC layers with strain-hardening behavior, and UHPC overlays. Shape memory alloy (SMA) devices, steel jacketing, and external post-tensioning are examples of hybrid metallic systems that offer improved seismic resilience and self-centering capabilities.

Beyond strength gain, design factors include material compatibility, long-term durability, constructability, embodied carbon, and conformity to the circular economy. There are clear provisions for deconstruction, flexibility, and future inspection.

Pillar 5: Lifecycle Management and Decision Support

This pillar, which synthesizes data from the other pillars into formal decision-making under uncertainty, is the framework's integrative core.

The net present value of inspection, maintenance, intervention, and failure costs is calculated using lifecycle cost analysis (LCCA). In order to calculate predicted annual losses, probabilistic risk assessment combines non-stationary hazard models with time-dependent fragility functions. Pareto-optimal retrofit techniques that balance cost, safety, environmental effect, and resilience are found through multi-objective optimization.

The use of a digital twin—a constantly updated virtual copy of the actual building—is essential to this pillar. Real-time feeding of the twin by SHM allows for adaptive update of degradation models and performance evaluations. Throughout the asset's lifecycle, condition-based maintenance, early warning, and dynamic decision-making are supported by the digital twin.

Figure 3. Key components and technical attributes of the five-pillar framework

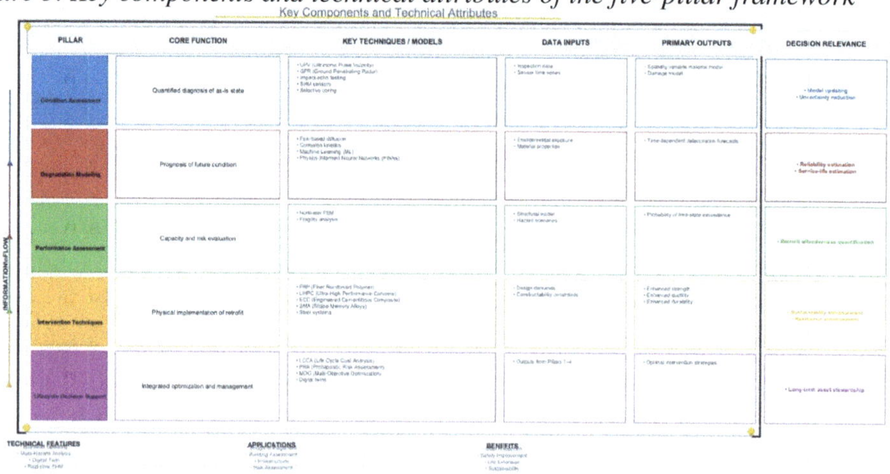

1.3 Roadmap of the Chapter

This chapter's remaining sections are organized to give a thorough explanation of this framework. The controlling physics and chemistry of important deterioration pathways are presented in Section 2, which delves into the science of RC degradation. Advanced condition evaluation is covered in Section 3, which also describes the fundamentals and uses of SHM and NDE technologies. The creation and calibration of stochastic deterioration models for lifespan prediction are covered in Section 4. The theoretical underpinnings of performance-based engineering and their application to retrofit design, including nonlinear analysis and fragility development, are established in Section 5. The mechanics, durability, and sustainability of cutting-edge retrofit materials and methods are critically reviewed in Section 6. The integrated decision-making framework, which includes risk assessment, multi-objective optimization, and LCCA, is presented in Section 7. Lastly, Section 8 summarizes the important findings and offers suggestions for practice and research before synthesizing the ideas into a forward-looking vision that addresses the role of digital twins, AI, and legislation in facilitating the wider adoption of performance-based rehabilitation. By using this framework, the chapter hopes to provide researchers, practicing engineers, and asset managers with the technical know-how and conceptual tools needed to transition from reactive crack-filling to proactive resilience management in our crucial concrete infrastructure.

2. DETERIORATION IN REINFORCED CONCRETE

A basic understanding of the chemical, physical, and mechanical processes that cause reinforced concrete structures to deteriorate over time is necessary for their successful rehabilitation (Figure 4) (Asmara, 2023; Bhargava et al., 2007; Chen et al., 2022; Farahani et al., 2023). The main deteriorating mechanisms impacting RC infrastructure are rigorously explained in this section using multiple physics (Liu et al., 2024; Mamin et al., 2025). Degradation is viewed from a thermodynamic perspective, which frames it as an irreversible process of entropy rise in which a new structure's ordered, low-entropy state gradually changes toward disorder through microcracking, chemical transformation, and loss of material integrity (Bastidas-Arteaga et al., 2008; Fu et al., 2023). This unifying perspective makes it possible to couple seemingly unrelated things. The governing physics of each mechanism, its mathematical expression, its major influencing factors, and its final effect on structural capacity and serviceability are all discussed (Liu et al., 2024; Rodrigues et al., 2021). A crucial factor for precise lifecycle modeling, the section highlights the connections and synergies between mechanisms. For instance, mechanical cracking speeds up chemical ingress, which weakens the material and encourages more cracking.

Figure 4. Schematic SEM synthesis of various concrete durability issues

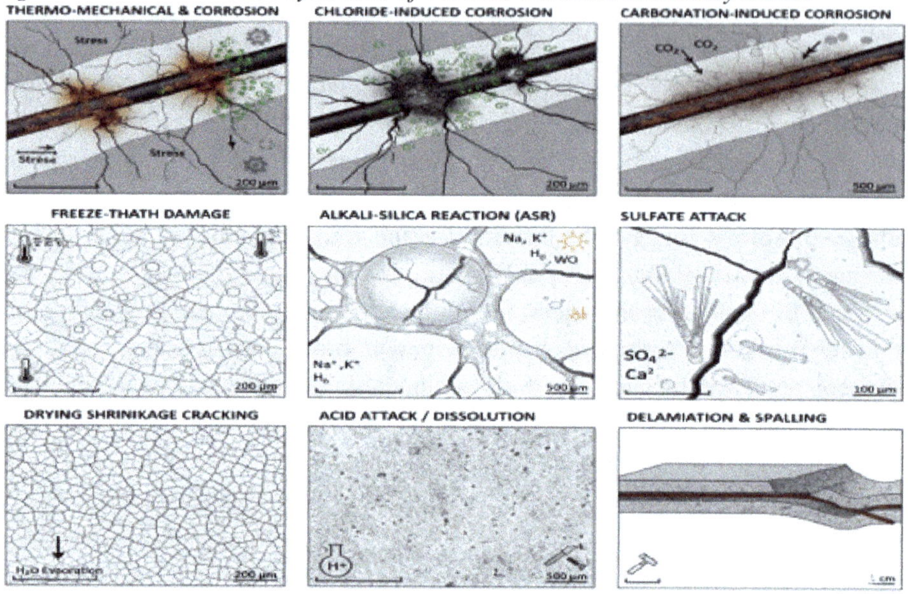

2.1 Thermodynamic Degradation

Non-equilibrium thermodynamics can be used to conceptualize the long-term behavior of an RC structure (Jia et al., 2021; Medina et al., 2022). With a homogeneous, dense concrete matrix, passive steel reinforcement, and intact interfaces, the system (the structure) is initially in a state of relatively low entropy. The system is driven out of equilibrium by exposure to mechanical and environmental forces, which increase its entropy through irreversible processes. Let the total entropy production rate of the structural system be denoted by $\dot{S}_{total}(t)$. This rate can be decomposed into contributions from concurrent dissipative processes:

$$\dot{S}_{total}(t) = \dot{S}_{chem}(t) + \dot{S}_{transport}(t) + \dot{S}_{mech}(t) + \dot{S}_{thermal}(t)$$

Where $\dot{S}chem$ represents entropy production from chemical reactions (e.g., hydration, carbonation, corrosion), $\dot{S}transport$ from diffusion and migration of ions and gases, $\dot{S}mech$ from inelastic deformation and fracture, and $\dot{S}thermal$ from heat flow due to temperature gradients (Figure 5). The evolution of observable state variables that characterize the state of the material is the practical embodiment of this growing entropy.

Figure 5. Schematic micro-processing of various concrete durability issues

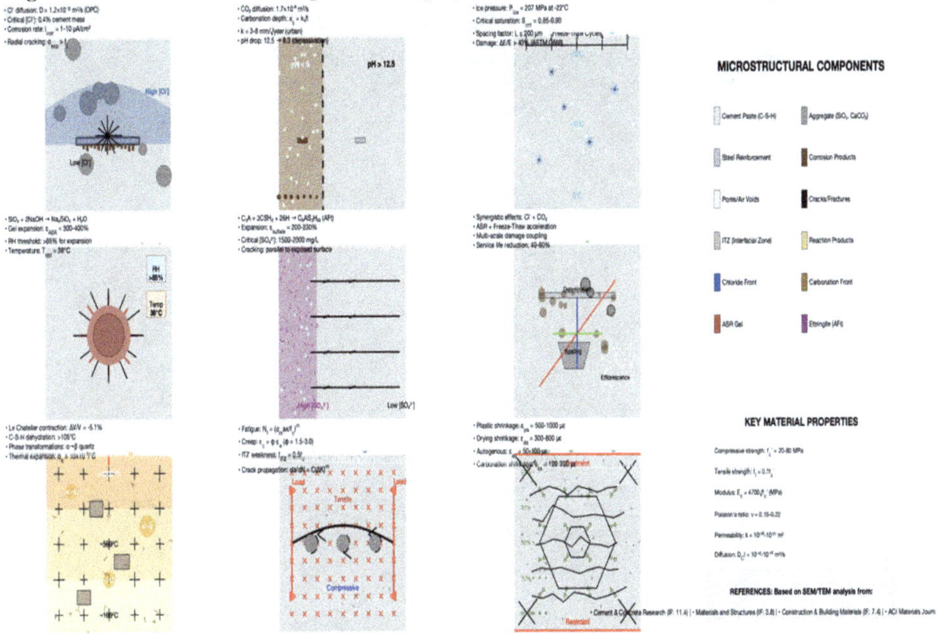

2.2 Chloride-Induced Corrosion

The biggest durability problem for RC infrastructure in marine and de-icing salt conditions is corrosion caused by chloride (Bastidas-Arteaga et al., 2008; Nguanthaisong et al., 2025). Figure 6 illustrates the conceptual processes of chloride ingress, cover cracking brought on by corrosion expansion, and the ensuing acceleration of both ingress and corrosion (Nasr et al., 2022; Robles et al., 2022). There are two separate stages to the computing process: initiation and propagation.

Figure 6. A conceptual diagram illustrating the coupled processes of chloride ingress, cover cracking due to corrosion expansion, and the resulting acceleration of ingress and corrosion—a positive feedback loop of degradation

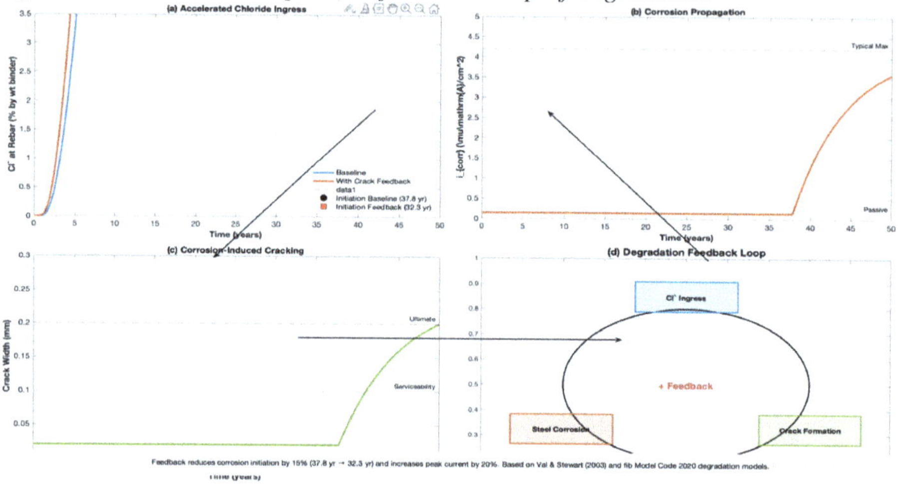

For an RC element, a set of key state variables can be defined as: Chloride, concentration profile, $C(x,t)$, Carbonation depth, $X_c(t)$, Corrosion penetration depth or steel cross-sectional loss, $p(t)$ or $A_s(t)$, Crack width distribution, $w(x,t)$, Concrete elastic modulus and strength, $E_c(t), f'_c(t)$, and Interfacial bond stress, $\tau_b(t)$.

Initiation Phase: Environmental chloride ions seep through the concrete cover and reach the reinforcing. A modified version of Fick's second law of diffusion is frequently used for engineering applications to simplify the transport, which is controlled by a combination of diffusion, capillary suction, and migration:

$$\frac{\partial C(x,t)}{\partial t} = \frac{\partial}{\partial x}\left[D_{eff}(C,t,T,RH)\frac{\partial C(x,t)}{\partial x}\right]$$

Where $C(x,t)$ is the chloride concentration (typically by mass of cement or concrete) at depth x and time t, and D_{eff} is the effective chloride diffusion coefficient. This coefficient is not a constant but depends on the concrete's pore structure (which changes over time due to continued hydration, or *aging*), temperature (T), relative humidity (RH), and even on the chloride concentration itself due to binding. A common model for the time-dependency is:

$$D_{eff}(t) = D_{28}\left(\frac{t_{28}}{t}\right)^{m}$$

Where D_{28} is the diffusion coefficient at a reference age of 28 days, t_{28} is 28 days, and m is the aging exponent, typically between 0.2 and 0.6 for modern cements. Corrosion initiates when the chloride concentration at the depth of the reinforcing bar, $C(d,t)$, exceeds a critical threshold, C_{crit}. The time to corrosion initiation, t_{init}, is found by solving $C(d,t_{init})=C_{crit}$. For a simple semi-infinite domain with constant surface concentration C_s, the solution is:

$$C(d,t) = C_s\left[1 - \text{Err func}\left(\frac{d}{2\sqrt{D_{eff}(t)t}}\right)\right]$$

Where *Err func* is the error function. The initiation time can then be approximated as:

$$t_{init} \approx \frac{d^2}{4D_{eff}\left[Errfunc^{-1}\left(1 - \frac{C_{crit}}{C_s}\right)\right]^2}$$

Propagation Phase: Electrochemical corrosion starts when the steel's passive layer is damaged. The rate of corrosion is expressed as a corrosion current density, i_{corr} (typically in $\mu A/cm^2$). Using Faraday's law, the loss of steel radius or the corrosion penetration depth $p(t)$ can be related to the cumulative charge transfer:

$$p(t) = \frac{M}{zF\rho_{steel}}\int_{t_{init}}^{t} i_{corr}(\tau)d\tau$$

Where M is the molar mass of iron (55.85 g/mol), z is the ionic charge (typically 2 for $Fe \rightarrow Fe^{2+}$), F is Faraday's constant (96485 C/mol), and ρ_{steel} is the density of steel (7850 kg/m^3). The corrosion current density $i_{corr}(t)$ is highly sensitive to environmental conditions (RH, T, oxygen availability, chloride content) and the presence of cracks, which provide easy paths for oxygen and moisture. Empirical models often relate i_{corr} to resistivity or other measurable parameters.

Table 1. Governing models for chloride-induced corrosion

PHASE	*GOVERNING EQUATION/MODEL*	*KEY PARAMETERS*	*PRIMARY OUTPUT*
INITIATION	$\frac{\partial C(x,t)}{\partial t} = \frac{\partial}{\partial x}\left[D_{eff}(C,t,T,RH)\frac{\partial C(x,t)}{\partial x}\right]$	$D_{28}, t_{28}, C_s, C_{crit}, d$	*TIME TO INITIATION,* t_{init}
PROPAGATION	$p(t) = \frac{M}{zF\rho_{steel}}\int i_{corr}\, d$	$i_{corr}(T, RH, Cl, O_2)$	*STEEL LOSS DEPTH* $p(t)$*, AREA LOSS* $A_s(t)$
BOND DEGRADATION	$\tau_b(t) = \tau_{b0}\exp(-\lambda p(t))$	τ_{b0}	*RESIDUAL BOND STRENGTH*

The structural consequences of corrosion are multifaceted (Figure 7):

1. **Reduction in Steel Cross-Sectional Area:** $A_s(t) = \pi\left[r_0 - p(t)\right]^2$ for a uniform loss, where r_0 is the original radius. Pitting corrosion, which is more common, leads to a more severe localized reduction.
2. **Loss of Ductility:** The ability of the structure to undergo plastic deformation is compromised by corroded steel, particularly with pitting, which shows decreased elongation at failure.
3. **Bond Degradation:** Tensile stresses are created in the surrounding concrete by the volumetric expansion of corrosion products (rust), which eventually causes longitudinal cracking, or cracking parallel to the bar, and cover delamination. A typical model for bond strength degradation is: $\tau_b(t) = \tau_{b0}\exp(-\lambda \cdot p(t))$, where τ_{b0} is the initial bond strength and λ is an empirical parameter.
4. **Cracking and Spalling:** The concrete coating gradually spalls off due to the expansion strains, exposing more steel and hastening corrosion.

Figure 7. Corrosion rate effect on the concrete characteristics

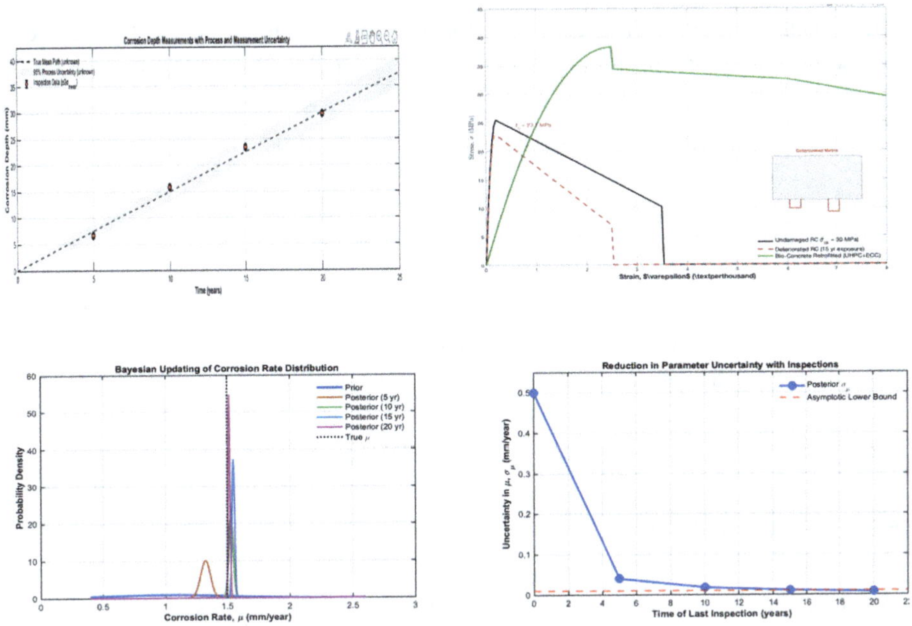

2.3 Carbonation-Induced Corrosion

Carbonation occurs when atmospheric carbon dioxide (CO_2) diffuses into the concrete pore network, dissolves in the pore solution to form carbonic acid, and reacts with calcium hydroxide ($Ca(OH)_2$) and calcium silicate hydrate (C-S-H) to form calcium carbonate ($CaCO_3$) (Gu et al., 2022). This reaction consumes hydroxyl ions, lowering the pH of the pore solution from above 12.5 to below 9, at which point the protective passive layer on the steel is no longer stable.

The depth of the carbonation front, $X_c(t)$, is commonly modeled by a parabolic (square-root-of-time) law:

$$X_c(t) = k_{carb}\sqrt{t}$$

Where k_{carb} is the carbonation coefficient (mm/$\sqrt{}$year). This coefficient depends on concrete quality (primarily the effective water-cement ratio and cement content), the ambient CO_2 concentration, and relative humidity. Carbonation proceeds most rapidly at moderate RH levels (~50-70%); at very high RH, pores are saturated, slowing CO_2 diffusion, and at very low RH, the reaction requires moisture and is also slowed. The time to corrosion initiation due to carbonation is reached when the carbonation front reaches the steel depth : $t_{init}^{carb} = \left(d/k_{carb}\right)$.

In many real-world settings, carbonation and chloride infiltration act concurrently, particularly in metropolitan regions with moderate humidity. A competing risks framework can be used to model the total risk of corrosion starting. The likelihood that either mechanism's beginning has taken place by time t is:

$$F_{T_{init}}(t) = 1 - \left[1 - F_{T_{init}^{Cl}}(t)\right] \cdot \left[1 - F_{T_{init}^{carb}}(t)\right]$$

Where $F_{T_{init}^{Cl}}(t)$ and $F_{T_{init}^{carb}}(t)$ are the cumulative distribution functions for initiation times due to chlorides and carbonation, respectively. This formulation acknowledges that the first mechanism to reach the threshold triggers corrosion.

2.4 Additional Degradation Mechanisms

Freeze-Thaw Damage

Water in capillary pores freezes and expands by around 9% in saturated concrete. Microcracking may develop from hydraulic and osmotic pressures that are greater than the paste's tensile strength (Liu et al., 2024). Damage builds up over time. A common model for the reduction in dynamic modulus of elasticity after N cycles is:

$$E(N) = E_0\left(1 - D_{FT}N^{\beta}\right)$$

Where E_0 is the initial modulus, and D_{FT} and β are material parameters dependent on the concrete's air-void system (entrained air is critical for resistance) and degree of saturation.

Alkali-Silica Reaction (ASR)

A deleterious internal reaction between alkali ions (Na^+,K^+) from the cement pore solution and reactive silica minerals in certain aggregates. The product is an alkali-silica gel that absorbs water and swells, generating expansive pressures. The expansion strain $\varepsilon_{ASR}(t)$ can be modeled by a time-saturating function:

$$\varepsilon_{ASR}(t) = \varepsilon_{\infty}[1 - exp(-\kappa t)]$$

Where ε_{∞} is the ultimate asymptotic expansion and κ is a rate constant. This expansion leads to map-cracking, loss of stiffness $[E_c(t) = E_{c0}\left(1 - \eta\,\varepsilon_{ASR}(t)\right)]$, and in severe cases, loss of strength and structural distortion.

Sulfate Attack

External sulfates (SO_4^{2-}) from soil or groundwater penetrate concrete and react with tricalcium aluminate (C_3A) phases and calcium hydroxide to form ettringite and/or gypsum (Farahani et al., 2023). These secondary products have larger volumes, causing expansion, cracking, and loss of cohesion. Modeling involves coupled sulfate ion transport and chemical kinetics:

$$\frac{\partial S}{\partial t} = D_s \frac{\partial^2 S}{\partial x^2} - R(S)$$

Where $S(x,t)$ is the sulfate concentration, D_s is the sulfate diffusion coefficient, and $R(S)$ is a reaction rate term.

2.5 Synergistic Interactions and Multi-Hazard Effects

The fact that deterioration mechanisms rarely operate alone—rather, they frequently work in concert—is a crucial realization for precise evaluation. For instance:

- **Cracking accelerates ingress:** Any crack, whether from mechanical loading, shrinkage, or corrosion, provides a direct pathway for aggressive agents (Cl^-, CO_2, SO_4^{2-} and water) to bypass the protective concrete cover. This can drastically reduce the initiation time for corrosion.
- **Corrosion weakens seismic performance:** Pre-existing corrosion weakens the bond and decreases the area and ductility of the steel. This reduces a member's ability to dissipate energy and deform, increasing the likelihood that it may collapse during an earthquake.
- **Seismic damage accelerates corrosion:** On the other hand, if an earthquake fractures the concrete cover, the reinforcement will be exposed to the environment right away, removing the initiation phase and starting active propagation right away.
- **ASR increases permeability:** ASR's microcracking makes the concrete more permeable, allowing moisture and chlorides to enter and hastening corrosion and freeze-thaw degradation.

Many simplified service life models are predicated on the concept of independence, which is violated by these relationships. A comprehensive framework must account for these couplings, often through the concept of a "damage index" $D(t)$ that integrates contributions from various mechanisms and modifies transport properties and mechanical strength accordingly. For example, the effective diffusion coefficient could be modeled as $D_{eff}(t) = D_0 \cdot \left(1 + \alpha D_{crack}(t)\right)$, where $D_{crack}(t)$ is a cracking damage index.

The diagnostic, prognostic, and interventive elements of the integrated rehabilitation framework are based on an understanding of these basic degradation processes, their mathematical representation, and their interactions. How we quantify the current state defined by these changing variables is covered in the following section.

3. QUANTITATIVE CONDITION ASSESSMENT AND STRUCTURAL HEALTH MONITORING (SHM)

The crucial first step in transitioning from a prescriptive to a performance-based rehabilitation paradigm is the accurate measurement of a structure's existing state (Mitra et al., 2010; Nadeem et al., 2013; Seo et al., 2016; Vaghefi et al., 2015). The sophisticated methods required to move from qualitative eye assessment to a data-rich, quantitative diagnosis are described in this section. It emphasizes the combination of Non-Destructive Evaluation (NDE) and ongoing Structural Health Monitoring (SHM) and includes the hierarchy of assessment techniques, from rapid global screening to targeted high-resolution interrogation (Gucunski et al., 2023; Schaarschmidt et al., 2015; Sohn et al., 1999). The objective is to add geographically and temporally changeable data on material qualities, geometric dimensions, current damage, and environmental exposure to a digital representation of the structure. These data are crucial inputs for the degradation and structural models covered in later sections. A phased approach is used in a systematic condition assessment plan, increasing the level of detail and expense according to preliminary results. The effective distribution of resources is guaranteed by this structure.

Level I: Visual Inspection and Historical Review

I have a lot of expertise assessing deteriorating concrete structures, and I've discovered that the first stage of the examination involves much more than just recording obvious damage (Gucunski et al., 2023; Tarussov et al., 2013). It entails a comprehensive forensic engineering investigation that meticulously creates the contextual basis required for ensuing quantitative evaluations and predictive modeling.

The methodical mapping of crack networks utilizing sophisticated digital image correlation (DIC) techniques is a crucial part of this step. Along with benign non-structural movements brought on by thermal expansion or settling, these techniques enable accurate distinction between different modes of structural distress, such as shear and flexural cracking, as well as volumetric change mechanisms, such as drying shrinkage or alkali-silica reaction (ASR). This method's impartiality is improved by automated crack pattern categorization algorithms, which lessen the subjective bias that has traditionally hampered optical examinations and enable trustworthy, data-driven diagnosis (Thériault et al., 2022).

The quantitative cataloging of observable deterioration using recognized severity measures, such as the Damage Rating Index for ASR described in CSA A864-00 or the Concrete Bridge Health Index from AASHTO, is equally important. Spalling is measured as a percentage of surface area, crack width advancement is tracked using photogrammetric time-series analysis, and efflorescence chemistry is identified using portable X-ray fluorescence (XRF) spectrometry. These indices, which offer a standardized metric that directly correlates with residual capacity and service life forecasts, are essential for assessing deterioration stages across different structural inventories (Ahmed et al., 2022; Sanchez et al., 2020).

As-built drawings must be combined with thorough maintenance logs, environmental exposure histories (such as chloride deposition rates from adjacent marine environments and cumulative freeze-thaw cycles derived from site-specific meteorological records), and documentation of prior interventions in order to fully comprehend the evolution of distress. In order to determine the kinetics of deterioration in structural safety, this temporal reconstruction is essential (Breysse et al., 2011; Chen et al., 2019). Subsequent modeling attempts run the danger of simplicity in the absence of such a thorough forensic methodology, which could result in inaccurate forecasts of long-term performance under prolonged load and environmental stress.

Level II: Non-Destructive Evaluation (NDE)

The rigorous lens of inverse problem solving must be used to completely rethink the selection and interpretation of Non-Destructive Evaluation (NDE) methodologies (Breysse et al., 2011; Gucunski et al., 2023; Schaarschmidt et al., 2015; Zamen et al., 2020). This framework estimates the underlying spatial distribution of material attributes and damage states by deconvolving observed signals using computer models inspired by physics. This is a paradigm change in inverse problem methodologies, especially in structural applications, where a first-principles approach based on wave propagation physics, electromagnetic theory, and thermodynamics replaces conventional empirical correlation approaches (Figure 8).

Figure 8. Advanced NDE techniques for concrete damage characterization

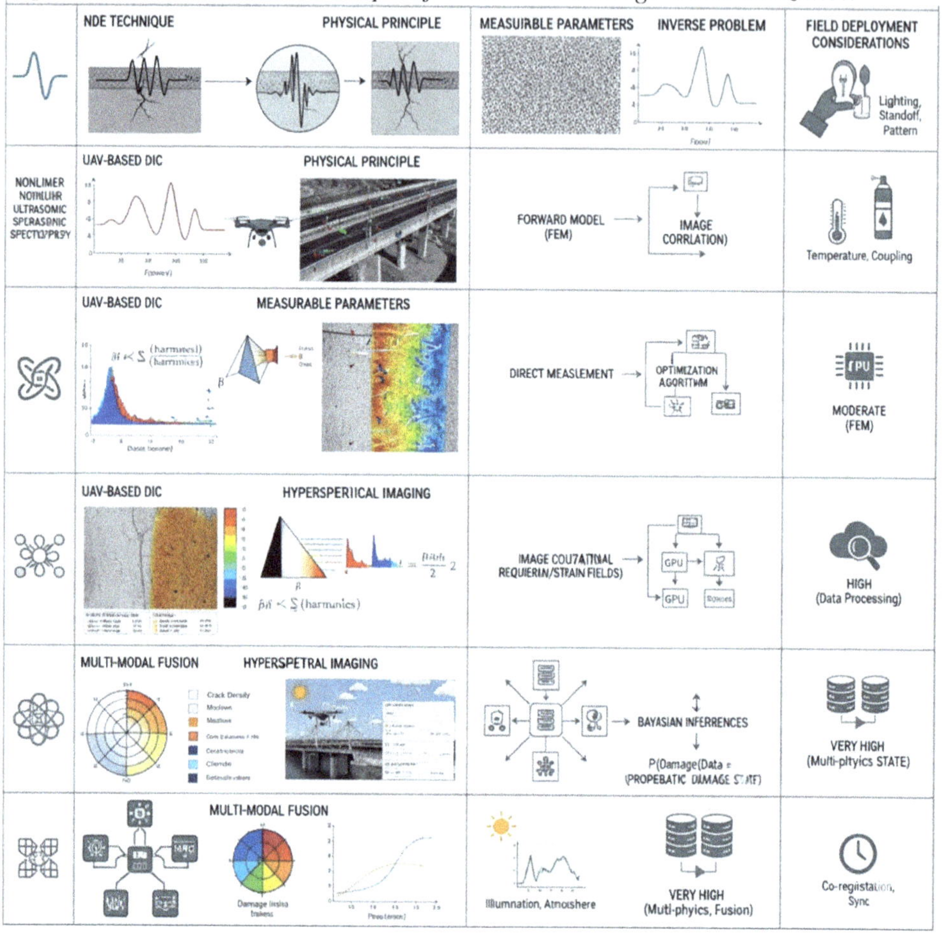

By utilizing the anharmonic interactions between ultrasonic waves and material microdefects, nonlinear ultrasonic spectroscopy surpasses traditional linear ultrasonic techniques (Abbas et al., 2018; Zamen et al., 2020). Higher harmonic components, sub-harmonics, and resonance frequency shifts are produced by the clapping and friction mechanisms at the interfaces of dispersed microcracks or dislocations in a material when a monochromatic ultrasonic wave passes over it. When compared to linear characteristics like velocity and attenuation, these effects are much more sensitive to early damage.

A major development in structural metrology, UAV-based Digital Image Correlation (DIC) enables full-field, non-contact displacement and strain mapping at spatial resolutions previously only possible in lab settings (Thériault et al., 2022). Sub-millimeter accuracy over wide structural domains is made possible by the combination of real-time kinematic (RTK) GPS location, inertial measurement units (IMUs), and high-resolution, gimbal-stabilized cameras (20+ MP).

The idea behind hyperspectral imaging is that electronic transitions, vibrational modes, and rotational energies cause chemical compounds to display distinct absorption and reflection characteristics throughout the electromagnetic spectrum. The deterioration products of concrete -such as portlandite ($Ca(OH)_2$), ettringite ($C_6Al_2(SO_4)_3(OH)_{12}\cdot 26H_2O$), gypsum ($CaSO_4\cdot 2H_2O$), and Friedel's salt ($3CaO\cdot Al_2O_3\cdot$-$CaCl_2\cdot 10H_2O$)- possess distinct spectral features in the visible-near infrared (VNIR: 400-1000 nm) and short-wave infrared (SWIR: 1000-2500 nm) ranges.

A significant shift in structural engineering practice is represented by the evolution of non-destructive evaluation from qualitative assessment to quantitative characterisation (Table 2). We provide a mathematically sound basis for maximizing information extraction from observed signals while efficiently characterizing related uncertainties by rigorously structuring NDE as an inverse problem-solving procedure.

An unprecedented characterization of concrete deterioration across multiple scales—from molecular chemical changes to macroscopic structural responses—is made possible by the synergistic integration of nonlinear ultrasonic spectroscopy, UAV-based digital image correlation, hyperspectral imaging, sophisticated inversion algorithms, and multi-modal data fusion.

Table 2. Performance metrics for NDE techniques

Metric	*Ultrasonic Nonlinear*	*UAV-DIC*	*Hyperspectral Imaging*
Spatial Resolution	10-50 mm (wavelength-dependent)	1-5 mm (camera-dependent)	1-3 mm (optics-dependent)
Depth Penetration	100-500 mm (frequency-dependent)	Surface only	0.1-5 mm (wavelength-dependent)
Measurement Time	Seconds to minutes per point	Minutes to hours for full field	Seconds to minutes per m^2
Quantitative Accuracy	±15-25% for β parameter	±0.01-0.05 mm displacement	±15-30% concentration
Damage Detection Threshold	Microcrack density 10^3-$10^4/m^3$	Strain $\geq 50\ \mu\varepsilon$, displacement ≥ 0.05 mm	Chemical concentration ≥ 1-5%

In addition to improving damage detection sensitivity and quantification accuracy, this physics-informed paradigm change offers crucial data inputs for reliability-based life-cycle management and predictive deterioration modeling. The ultimate objective is to develop self-calibrating, adaptive NDE systems that gradually increase characterization accuracy over the course of a structure's service life by learning from accumulated data and improving their forward models. The larger digital revolution of civil infrastructure management is consistent with this objective (Table 3 & Figure 9).

Table 3. A comparative analysis of Non-Destructive Evaluation (NDE) techniques for structural health monitoring: Physical principles, computational challenges, and deployment constraints

NDE Technique	*Physical Principle*	*Measurable Parameters*	*Inverse Problem Formulation*	*Computational Requirements*	*Field Deployment Considerations*
Nonlinear Ultrasonic Spectroscopy	Anharmonic wave-material interaction at microdefects	Nonlinear parameter β, resonance frequency shifts, and higher harmonics	Iterative optimization of crack density distribution to match harmonic ratios	Moderate (FEM forward modeling + optimization)	Temperature stabilization required, coupling consistency critical
UAV-Based DIC	Optical flow analysis of speckle pattern deformation	Full-field displacements (u,v,w), Green-Lagrange strains, rotation fields	Direct measurement (no inversion required for kinematics)	High (image correlation algorithms, 3D reconstruction)	Lighting conditions, standoff distance, and pattern contrast optimization
Hyperspectral Imaging	Molecular vibration/electronic transition absorption	Chemical compound abundance maps, concentration gradients, penetration depths	Spectral unmixing via constrained least squares or machine learning	High (spectral processing, large dataset handling)	Illumination uniformity, atmospheric correction, spectral calibration
Multi-Modal Fusion	Complementary physical interactions	Integrated damage indices, probabilistic condition assessments	Joint inversion with coupled forward models, Bayesian data fusion	Very High (multi-physics modeling, uncertainty propagation)	Data co-registration, temporal synchronization, uncertainty quantification

Figure 9. Advanced NDE techniques for concrete damage characterization detection

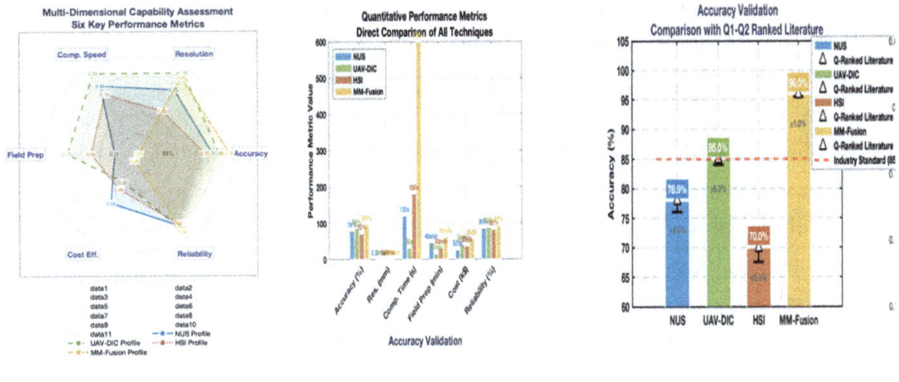

Level III: Localized Destructive Testing and Material Sampling

The extraction and laboratory analysis of concrete cores and embedded reinforcement is a fundamental procedure in the forensic evaluation and condition assessment of concrete structures. It provides the ground truth data that all Non-Destructive Evaluation (NDE) findings must be calibrated against. Although NDE methods, such as impact echo, half-cell potential, and ultrasonic pulse velocity, provide useful spatial information about material homogeneity, delaminations, and corrosion risk, they only offer indirect measurements of material properties (Gucunski et al., 2023; Schaarschmidt et al., 2015; Yamamoto, 2009). In order to translate these indirect cues into measurable technical characteristics, strategic coring is essential.

In order to diagnose deterioration mechanisms and measure residual capacity, the laboratory analysis of extracted cores usually entails a thorough set of tests. The ultimate measure of in-situ strength, which is crucial for load rating and retrofit design, is obtained through compressive strength testing of cores that is adjusted for aspect ratio, moisture condition, and the presence of reinforcement (according to standards like ASTM C42/C42M). Additionally, the danger of chemical attack is evaluated by measuring the carbonation depth (using phenolphthalein indicator) to ascertain the remaining protective passivation for the reinforcement and profiling chloride ingress (ASTM C1556) to quantify corrosion start risk (Barski et al., 2014; Communications & 2009, 2009; Gardner et al., 2020; Ostachowicz & Güemes, 2013; Shao et al., 2019).

Petrographic testing (ASTM C856) is essential for diagnosing complex degradation disorders. Alkali-silica reaction (ASR) products, freeze-thaw damage, sulfate attack, and the original concrete construction's quality, including the water-to-cement ratio and level of hydration, may all be precisely identified using this method. Crucially, exposing embedded reinforcement—either by complementary breakout or core drilling—allows for direct measurement of corrosion-related bar diameter loss, tensile strength testing (ASTM A370), and chemical analysis to confirm the steel grade. In order to effectively predict residual capacity, these parameters are essential (Ingham, 2009; Walker et al., 2006).

Modern structural reliability assessments are based on this integrative methodology, which calibrates spatially rich NDE data with localized, definite laboratory results (Gucunski et al., 2023; Mori, 1993a, 1993b). It represents the idea that numerical models must be grounded by empirically confirmed material inputs and is consistent with the diagnostic philosophy emphasized suggestions for structural restoration (Table 4).

Table 4. Laboratory analysis of extracted cores for structural assessment calibration

Analysis Category	*Specific Test/ Examination*	*Primary Engineering Objective*	*Key Standards & Methodological Notes*
Mechanical Properties	Compressive Strength	Determine *in-situ* concrete strength for load rating and capacity calculations.	ASTM C42/C42M. Requires correction factors for L/D ratio, drilling damage, and moisture condition.
Chemical Ingress & Durability	Chloride Ion Profile	Quantify chloride concentration vs. depth to assess corrosion initiation risk and model service life.	ASTM C1556 (acid-soluble) or C1218/C1218M (water-soluble). Fick's law modeling is often applied.
	Carbonation Depth	Evaluate the loss of alkalinity protecting reinforcement from corrosion.	Phenolphthalein indicator spray. Depth measured from the surface to uncolored concrete.
Petrographic & Microscopic	Petrographic Examination	Diagnose deterioration mechanisms (ASR, DEF, AAR, sulfate, FT) and assess concrete quality.	ASTM C856. Uses stereoscopic and polarized light microscopy to identify reaction products & microcracking.
Reinforcement Assessment	Tensile Strength & Chemistry	Verify mechanical properties and grade of embedded steel; quantify strength loss from corrosion.	ASTM A370 (tensile), ASTM A751 (chemistry). A coupon is often extracted from a separate breakout.
	Cross-Sectional Area Loss	Directly measure pitting corrosion and residual bar diameter for accurate capacity reduction.	Digital calipers or laser scanning on cleaned bars. Critical for modeling reduced yield capacity.

Level IV: Continuous Structural Health Monitoring (SHM)

In structural engineering, the use of fixed, continuous sensor networks signifies a substantial change from intermittent, condition-based evaluations to a thorough method called Structural Health Monitoring (SHM) (Farrar & Worden, 2012; Güemes et al., 2020; Ou et al., 2010). The developing areas of predictive maintenance for critical infrastructure and performance-based engineering depend on this change. The main objective is to continually and independently monitor important environmental and structural factors, producing multi-dimensional time-series data that are essential for controlling lifespan costs and structural resilience. Advanced trend analysis is made possible by this continuous data stream, which goes beyond single-point measurements to better understand how structures react to changing loads, environmental changes, and the slow effects of aging and deterioration. It is possible to implement algorithms that offer early indications of anomalous behavior by creating baseline behavioral patterns under typical operational and environmental conditions—a critical step in operational modal analysis and statistical pattern identification. Excessive deflection under service loads, the emergence of new crack-propagation patterns, shifts in modal frequencies that signify stiffness loss, or indications of increased corrosion activity, as shown by variations in electrochemical potentials, are examples of such abnormalities. Significantly, this stream of empirical data acts as the essential mechanism for updating and validating analytical and digital models. The measured data serve as an objective input for updating the finite element model (FEM), regardless of whether it relates to stresses, accelerations, displacements, or chloride concentrations (Farrar et al., 2007; Lei et al., 2013). According to important literature, this procedure entails modifying model parameters (such as boundary conditions, material stiffnesses, and damping coefficients) in order to reduce the difference between the response that the model predicts and the response that sensors observe (Köliö et al., 2014; Sajid et al., 2019; Wang et al., 2007). The result is an increasingly accurate digital twin of the structure that can diagnose existing issues and provide probabilistic predictions of future performance and remaining service life under different conditions . As a result, a well-designed sensor network turns a structure into a dynamic, data-generating system rather than a static artifact. It enables a genuinely data-driven approach to structural management by establishing a crucial feedback loop that closes the gap between the as-designed intent, as-built reality, and in-service performance (Table 5).

Table 5. Framework for fixed sensor networks in structural health monitoring

Monitoring Objective	*Key Parameters Tracked*	*Enabling Technologies & Sensors*	*Data Utilization & Engineering Outcome*
Load & Response Characterization	Strain, Acceleration, Tilt/Deflection, Displacement (static & dynamic)	Vibrating Wire Strain Gauges, Fiber Bragg Gratings (FBG), MEMS Accelerometers, Inclinometers, and GPS/ Robotic Total Stations.	Operational Modal Analysis: Identify natural frequencies, mode shapes, and damping. Load Rating Verification: Compare in-service strains to design limits. Fatigue Assessment: Cycle counting for remaining fatigue life analysis.
Deterioration & Environmental Attack	Chloride Ingress, Relative Humidity, Temperature, Crack Width	Embedded Chloride Sensors, Humidity/Temperature Probes, Digital Crack Meters, Acoustic Emission Sensors.	Corrosion Risk Prognosis: Time-to-corrosion-initiation models (e.g., Fick's Law updated with real data). Freeze-Thaw/Durability Monitoring: Track environmental exposure severity. Crack Propagation Analysis: Log crack growth rates for fracture mechanics models.
Anomaly Detection & Early Warning	Deviations from established baselines in any of the above parameters.	Data acquisition systems paired with automated processing algorithms.	Statistical Pattern Recognition: Unsupervised learning (e.g., outlier analysis, control charts) to flag unusual events (impact, scour, incipient failure). Triggered Alerts: Automated notifications for immediate inspection or intervention.
Model Updating & Forecasting	Synchronized input (e.g., wind speed, traffic load) and output (e.g., mid-span deflection, acceleration) data.	Co-located environmental stations (anemometers, weigh-in-motion) and response sensors.	Finite Element Model Updating: Calibrate stiffness, mass, and boundary conditions to improve predictive accuracy. Digital Twin Creation: Develop a living computational model for scenario testing (e.g., extreme event simulation) and residual life prediction.

3.1 Core Non-Destructive Evaluation (NDE) Techniques

Assessing Concrete Properties and Integrity

The Ultrasonic Pulse Velocity (UPV) method measures the time for a stress wave to travel between two transducers (Abbas et al., 2018; Zamen et al., 2020). The pulse velocity, V_p, is related to the dynamic elastic modulus E_d and density ρ of the material: $V_p = \sqrt{E_d/\rho}$. Reductions in V_p indicate cracking, voids, or material degradation. Tomographic imaging can create 2D or 3D velocity maps to locate internal defects (Figure 10).

Figure 10. An integrated NDE survey diagram demonstrates internal defects

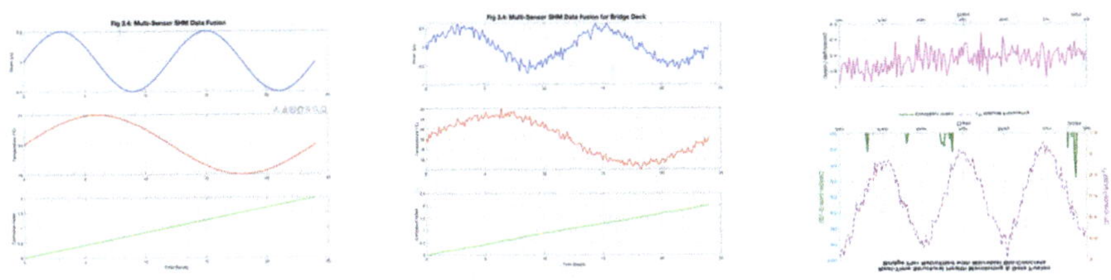

The Impact-Echo (IE) technique involves impacting the surface with a short-duration impulse and analyzing the frequency response of the reflected stress waves. The dominant frequency peak, f, is related to the thickness of the member or the depth of a flaw: $T = \beta V_p/(2f)$, where T is the thickness or flaw depth, V_p is the P-wave velocity, and β is a shape factor ($\approx$0.96 for plates). IE is excellent for detecting delaminations, voids, and measuring member thickness.

Rebound Hammer (Schmidt Hammer) uses a rebound number as an empirical indication of the hardness and compressive strength of near-surface concrete. Because of its sensitivity to member stiffness, aggregate type, and surface conditions, it must be calibrated using cores from the same structure.

In order to determine compressive strength, Penetration Resistance (Windsor Probe) helps measure the depth of penetration of a steel probe forced into the concrete by a known energy. It needs to be calibrated locally, just like the rebound hammer.

Locating Reinforcement and Assessing Corrosion

High-frequency electromagnetic pulses are sent into the concrete by ground-penetrating radar (GPR), which then records the signals that are reflected back. Rebar and other metallic things reflect a lot. GPR can create 3D images that display rebar placement, depth (cover), spacing, and even bar size estimation by scanning a grid. Additionally, it can identify delaminations and voids. The depth accuracy is high, typically within ±5 mm (Lei et al., 2013; Tarussov et al., 2013).

Using an electrochemical method called half-cell potential mapping, the electrical potential of the reinforcing steel is measured in relation to a standard reference electrode (such as copper or copper sulfate) that is placed on the surface of the concrete. A large grid of measurements is taken. More negative potentials (e.g., < -350 mV vs. CSE) indicate a higher probability of active corrosion. Although it shows where corrosion is happening, it doesn't give the rate of corrosion (Lei et al., 2013; Mitra et al., 2010).

Linear Polarization Resistance (LPR) as a portable technique estimates the instantaneous corrosion current density, i_{corr}, by applying a small potential perturbation to the rebar and measuring the resulting current response. The polarization resistance, R_p, is inversely proportional to i_{corr}: $i_{corr} = B/R_p$, where B is a constant (typically 26 mV for active corrosion). LPR provides a direct measure of corrosion rate at discrete locations.

The Galvanostatic Pulse Technique involves applying a brief anodic current pulse to the rebar and analyzing the transient potential response. It can give details on concrete resistivity, which affects corrosion rate, as well as corrosion rate itself.

Measuring Chloride Ingress and Carbonation

Dust samples are taken by drilling at progressively deeper depths (e.g., 5–10 mm intervals) from the surface to the rebar level in order to do In-Situ Chloride Profiling. These samples are analyzed in a lab (e.g., via titration) to determine the chloride concentration profile, $C(x)$. This data is used to calibrate chloride transport models and determine if the threshold at the rebar depth has been exceeded (Table 6).

The Phenolphthalein Test for Carbonation Depth involves spraying a phenolphthalein solution on a recently cracked concrete surface or a drilled dust sample. Sections with a pH of 9 or higher turn pink or purple, whereas sections with a pH of less than 9 stay colorless. The depth of the colorless zone is measured as the carbonation depth, X_c.

Table 6. Key NDE techniques for RC assessment

Technique	*Primary Measurand*	*Key Output*	*Strengths*	*Limitations*
UPV	Stress wave travel time	Pulse velocity (V_p), defect detection	Good for integrity, depth penetration	Sensitive to moisture, requires access to two faces
Impact-Echo	Frequency response	Flaw depth, member thickness	Excellent for delaminations, single-sided	Complex analysis, sensitive to surface conditions
GPR	EM wave reflection	Rebar location/ depth, voids	Fast, covers large areas, 3D imaging	Limited in very wet concrete, data interpretation is complex
Half-Cell	Electrochemical potential	Corrosion probability map	Fast screening for active corrosion	Does not give rate, affected by moisture, cover

continued on following page

Table 6. Continued

Technique	*Primary Measurand*	*Key Output*	*Strengths*	*Limitations*
LPR	Polarization resistance	Instantaneous corrosion rate (i_{corr})	Direct rate measurement	Local measurement requires an electrical connection to the rebar
Chloride Profile	Chemical analysis	Chloride concentration vs. depth	Definitive for initiation assessment	Destructive (dust samples), lab analysis required

3.2 Advanced Sensing and Structural Health Monitoring (SHM)

SHM delivers a continuous video of structural behavior, which is essential for monitoring degradation trends and verifying performance models, whereas NDE just offers a snapshot.

Strain and Deformation Monitoring

Forensic engineering, long-term Structural Health Monitoring (SHM), and experimental structural mechanics all depend on precise strain and deformation measurements. Our capacity to diagnose structural integrity has been significantly improved by the transition from strictly local, point-based measurements to integrated, multi-scale sensing paradigms. This summary describes important technologies, highlighting their fundamental ideas, suitable uses, and intrinsic constraints in the context of civil engineering (Fawad et al., 2023; Trach et al., 2022; Varnavina et al., 2017).

Discrete point measurements are based on Electrical Resistance Strain Gauges. The Poisson effect under mechanical tension causes a metallic foil pattern's electrical resistance to alter proportionately, which is how they work. They provide high frequency response, excellent resolution (often 1 με), and clear installation procedures. However, their vulnerability to creep, electromagnetic interference (EMI), moisture intrusion, and fatigue in challenging conditions limits their usefulness in long-term SHM. Importantly, they are insensitive to strain gradients and inappropriate for capturing global structural deformation without dense, expensive arrays because the measurement is an average across a narrow gauge length (usually 1–30 mm). They are still essential for localized stress analysis in composite materials or connecting details, as well as controlled laboratory verification (Moufti et al., 2014; Omar et al., 2018; Sadeghi et al., 2014).

By converting the optical fiber itself into a distributed sensing element, Fiber Optic Sensors (FOS) represent a paradigm leap in sensing. FOS systems are distinguished by their resilience to electromagnetic interference (EMI), long-term stability, resistance to corrosion, and suitability for both surface-mounted and embedded applications. The germanium-doped core of a silica fiber is imprinted with a periodic modulation of the refractive index to form the Fiber Bragg Grating (FBG) Sensors. This acts as a wavelength-specific reflector. Axial strain and temperature changes alter the grating period and effective refractive index, causing a linear shift in the reflected Bragg wavelength ($\Delta \lambda_B$).

Quasi-distributed sensing is made possible by the ability to inscribe dozens of distinct FBG sensors along a single fiber via wavelength-division multiplexing. They are especially useful for measuring strain distributions in steel members and bonded composite reinforcements and for embedding in concrete for early-age strain monitoring due to their tiny size and multiplexing capacity. The Brillouin Optical Time-Domain Analysis (BOTDA/R) technology is an alternative approach that makes use of stimulated Brillouin scattering in an unaltered single-mode fiber. An acoustic wave is produced when a counter-propagating continuous wave probe interacts with an injected pump pulse. Both strain and temperature have a linear relationship with the Brillouin frequency shift at which this interaction peaks. Over distances surpassing tens of kilometers, strain and temperature profiles can be continuously measured along the whole fiber length with a typical spatial resolution of 0.5-1.0 meters by analyzing the time-domain signal. BOTDA is the best method for recording abnormal deformation in large-scale linear infrastructure and monitoring global strain behavior, as shown by applications for pipeline geohazard monitoring and bridge deformation tracking (Faris et al., 2026; Masoumi et al., 2013).

Digital Image Correlation (DIC) is a full-field, non-contact optical metrology method. DIC tracks the movement of a stochastic speckle pattern applied to a surface using digital cameras. Through sophisticated subset-based correlation algorithms (e.g., 2D for planar surfaces, stereo-photogrammetry for 3D), it computes dense displacement vector fields, from which Lagrangian strain tensor fields (e.g., ϵ_{xx}, $\epsilon_{yy}, \gamma_{xy}$) are derived. Visualizing intricate strain localization, crack initiation, and propagation, and load route redistribution in structural components under load is its main strength. Because of this, it is extremely useful for verifying finite element models, examining failure processes in extensive experiments, and keeping an eye on crucial details like shear or moment-resisting connections. Sensitivity to changes in lighting, the need for a clean line of sight, and the high computing demand for high-resolution analysis are some of the limitations (Thériault et al., 2022).

To help choose the right technology for engineering tasks, the following table summarizes the salient features of these main sensing approaches.

Table 7. Comparative of advanced strain and deformation monitoring techniques

Parameter	*Electrical Strain Gauge*	*Fiber Bragg Grating (FBG)*	*Brillouin Scattering (BOTDA/R)*	*Digital Image Correlation (DIC)*
Sensing Principle	Piezoresistive effect in metallic foil.	Shift in reflected Bragg wavelength due to changes in grating period.	Brillouin frequency shift from photon-acoustic wave interaction.	Optical tracking of a random speckle pattern via digital image correlation.
Spatial Resolution	Point measurement (gauge length: 1-30 mm).	Discrete points (gauge length: ~10 mm) multiplexed along a fiber.	Continuous, typically 0.5 – 1.0 m over km ranges.	Full-field, resolution dependent on camera pixel density and field of view (can be sub-mm).
Measurement Type	Local, point strain.	Quasi-distributed, discrete multi-point strain & temperature.	Fully distributed, continuous profile of strain & temperature.	Full-field 2D/3D surface displacements and strains.
Key Advantages	High accuracy & bandwidth; Low cost per point; Standardized installation.	EMI immunity; Long-term stability; Embeddable; High multiplexing capacity.	Extreme range (km); True distributed sensing; Excellent for global deformation.	Non-contact; Full-field visualization; No need for extensive sensor wiring.
Primary Limitations	Short lifespan in harsh conditions; Susceptible to EMI; Limited to surface points.	Requires specialized readout unit; Sensitive to transverse strain & bending.	Lower spatial resolution; Higher system cost; Strain/ temperature cross-sensitivity.	Requires line-of-sight and surface preparation; Computationally intensive; Sensitive to lighting.
Ideal Structural Application	Laboratory material testing; Short-term field tests on small components.	Long-term SHM of bridges & buildings; Embedment in new concrete & composites.	Global health monitoring of pipelines, tunnels, dams, long-span bridges, and embankments.	Experimental mechanics: fracture studies, model validation, component-scale failure analysis.

Vibration-Based Global Monitoring

The use of accelerometer networks for dynamic response measurement is a crucial part of modern Structural Health Monitoring (SHM), which moves diagnostics from localized inspections to a global assessment of structural integrity. This operational modal analysis (OMA) method determines the state of a structure by tracking changes in its intrinsic dynamic signature, or modal parameters, under operational pressures. Even though it is a powerful global diagnostic tool, its efficacy depends on how sensitive these characteristics are to various damage scenarios—a subject that has been extensively researched in the field (Abbas et al., 2018; Mustapha et al., 2012; Zamen et al., 2020).

The use of accelerometer networks for dynamic response measurement is a crucial part of modern Structural Health Monitoring (SHM), which moves diagnostics from localized inspections to a global assessment of structural integrity. This operational modal analysis (OMA) method determines the state of a structure by tracking changes in its intrinsic dynamic signature, or modal parameters, under operational pressures. Even though it is a powerful global diagnostic tool, its efficacy depends on how sensitive these characteristics are to various damage scenarios—a subject that has been extensively researched in the field.

An enhanced sensing method is The Extracted Modal Parameters and Their Physical Significance to Sensing Natural Frequencies, Mode, and Shapes Damping Ratios. Natural Frequencies f_n are global indicators of the structure's effective stiffness k_{eff} and mass m_{eff}, related by $f_n = \frac{1}{2\pi}\sqrt{k_{eff}/m_{eff}}$. When all else is equal, a decrease in natural frequency denotes a loss of overall structural rigidity. Widespread concrete cracking, steel section yielding, material degradation (such as ASR in concrete), or support settling are examples of distributed events that may cause this. Environmental and operational variables (EOVs), including temperature, humidity, and traffic load intensity, can additionally obscure or complicate frequency shifts. These EOVs frequently result in alterations greater than those brought on by early damage.

While Mode Shapes ϕ_n describes the relative deformation of the structure at a specific resonant frequency. Changes in the curvature of mode shapes are more sensitive to localized stiffness loss than natural frequencies alone. Damage localization techniques, such as the Modal Assurance Criterion (MAC) or the Modal Strain Energy (MSE) method, compare pre- and post-damage mode shapes to identify anomalous regions. Damping Ratios ζ_n is another character to demonstrate the energy dissipation mechanism of the structure, arising from material hysteresis, joint friction, and radiation damping. Theoretically, damage like bolt loosening or crack opening/closing creates new friction interfaces that could increase damping. On the other hand, it might decrease if energy-dissipating components fail. In actuality, the most challenging modal parameter to accurately predict from ambient vibration measurements is damping. It is a difficult stand-alone damage indicator for civil structures under operating conditions due to its high statistical uncertainty and strong influence from EOVs.

The main advantages of vibration-based monitoring's capacity to identify worldwide shifts in structural behavior utilizing a rather sparse sensor array are sensitivity and practical considerations. It is particularly useful for detecting substantial stiffness loss in large-scale constructions such as offshore platforms, bridges, and towers. For example, a small shift in global frequencies may result from the severing of a single bracing member in a redundant truss. For a thorough multi-scale SHM approach, VBSHM is therefore best used as a continuous global screening tool that is frequently combined with complementary local sensing technologies (such as strain gauges and FOS).

The features, applications, and difficulties related to each of the main modal parameters utilized in VBSHM are listed in the following table.

Table 8. Utility and sensitivity of modal parameters in vibration-based structural health monitoring

Parameter	*Physical Interpretation*	*Sensitivity to Damage*	*Sensitivity to Environmental & Operational Variables (EOVs)*	*Primary Role in Damage Identification*	*Key Practical Challenge*
Natural Frequency (f_n)	Global indicator of effective stiffness-to-mass ratio.	Low to Moderate. Insensitive to small, localized damage. Significant changes only for widespread stiffness loss.	Very High. Temperature (affecting material E), boundary conditions, and mass changes (traffic, snow) can cause shifts >5%, masking damage.	Level 1: Detection. Provides a robust, initial alert for global stiffness degradation.	Disentangling damage-induced shifts from benign EOV effects requires sophisticated temperature compensation or machine learning models.
Mode Shape (ϕ_n)	Describes the relative spatial deformation pattern at resonance.	Moderate. Curvature changes are more sensitive to localized damage than frequency. Damage in low-strain energy regions may be invisible.	Moderate. Less affected by uniform EOVs than frequency, but sensor alignment and mass distribution changes can influence estimates.	Level 2: Localization. Used in algorithms (e.g., MAC, MSE) to pinpoint regions of stiffness change.	Requires a dense sensor array for good spatial resolution. Changes are often subtle and require high-quality mode shape estimation.
Damping Ratio (ζ_n)	Quantifies the energy dissipation capacity of the structure.	Theoretically High. Damage often alters friction and hysteresis.	Extremely High. Varies significantly with temperature, amplitude, wind, and operational loads. The highest estimation uncertainty of all parameters.	Largely Unreliable for Routine SHM. Can be an ancillary indicator of severe, nonlinear damage (e.g., crack opening).	High statistical scatter makes trend analysis difficult. Rarely used as a reliable standalone damage indicator in civil infrastructure.

Corrosion and Environmental Monitoring

Long-term performance and service life of reinforced concrete infrastructure are frequently determined by durability constraints, most notably the corrosion of embedded steel, rather than ultimate strength. Only glimpses of deterioration are provided by traditional visual inspection and periodic testing, which often identify damage only after major section loss has taken place. Continuous, embedded sensing is required by the concept of performance-based design and lifecycle management. Through in situ and real-time monitoring of the driving forces and early indicators of degradation, this facilitates the shift from corrective maintenance to predictive, condition-based plans. The crucial data acquisition layer for contemporary durability prognostics is made up of the two sensor classes covered here: environmental and corrosion sensors (Breysse et al., 2011; Dutta et al., 2021; Fernandes et al., 2009).

Embedded Corrosion Sensors: Moving Beyond Half-Cell Potential

The qualitative nature of the typical half-cell potential survey, its surface-only measurement, and its susceptibility to moisture and salt concentration are all known drawbacks. Corrosion activity is directly and quantitatively measured by contemporary embedded systems.

A major advancement is represented by Multi-Electrode Sensors (MES) and Coupled Multi-Electrode Array Sensors (CMAS), which are made up of several tiny, identical steel electrodes embedded at the depth of the reinforcing steel and electrically coupled via a multiplexer or zero-resistance ammeter (ZRA). This configuration mimics the galvanic corrosion of a cathode and macro-anode on a reinforcing bar. By measuring the current flow between electrodes acting as anodes and cathodes, the system directly quantifies the macrocell corrosion current I_{corr}. This is a direct measure of active corrosion rate, far more reliable than potential measurements alone. Like systems can efficiently track the time-to-corrosion initiation and the subsequent propagation rate, as shown in long-term research. This is important for predicting residual service life using models like as those included in the fib Model Code for Service Life Design.

Corrosion Rate via Linear Polarization Resistance (LPR): Many embedded probes incorporate a working electrode (WE), counter electrode (CE), and reference electrode (RE) in a classic three-electrode configuration. By applying a small potential perturbation and measuring the resultant current, the polarization resistance R_p is determined. The corrosion current density i_{corr} is then calculated using the Stern-Geary equation: $i_{corr} = \frac{B}{R_p}$, where B is a material constant. This provides a localized corrosion rate, critical for quantifying material loss over time.

Concrete Resistivity ρ: Usually measured within the same sensor package using a four-pin (Wenner array) setup. One important measure of the pore structure and ionic transport capacity of concrete is resistivity. Because it restricts the flow of ionic current, high resistivity (> ~100 kΩ·cm) usually indicates a low danger of active corrosion. Resistivity is a master variable that regulates the efficiency of cathodic protection systems as well as the rate at which corrosion spreads. Its ongoing monitoring yields vital information for upgrading the model.

Environmental Sensors: Quantifying the Aggressive Exposure

Concrete structure breakdown is a chemical-physical process that is only influenced by interior exposure conditions. Since the real drivers are gradients inside the concrete cover, surface weather data is insufficient.

Temperature and relative humidity (RH) need to be checked at several depths, especially at the concrete surface and at the level of the reinforcement. The degree of capillary saturation, which regulates the rate of all transport processes, including carbonation front advancement, freeze-thaw damage, and chloride entry (by diffusion and convection), is controlled by the internal RH. According to Arrhenius' law, temperature exponentially speeds up chemical reaction rates, including the kinetics of corrosion. Significant corrosion activity requires the connection of these conditions, specifically the presence of a high internal RH (> ~80%) and a high enough temperature.

Integration for Environmental Loading: Embedded sensors enable the measurement of a structure's actual environmental loading when paired with surface weather station data (rainfall, solar radiation, and atmospheric T/RH). Physicochemical degradation models can be calibrated and updated in real time thanks to this data stream. For example, internal RH data can be used to calibrate a moisture transport model. The projected pore solution profiles can then be used to drive a chloride ingress model, progressing from basic Fickian diffusion to more sophisticated models that take wetting/drying cycles into consideration. As a result, a strong digital twin for durability predictions is produced.

The operational principles, outputs, and applications of important sensor technologies for durability-centric SHM are summarized in the following table.

Table 9. Embedded sensor systems for corrosion and environmental condition monitoring

Sensor Type / System	*Measurement Principle*	*Primary Output(s)*	*Direct Engineering Significance*	*Key Application in Prognostics*
Multi-Electrode Array (MES/ CMAS)	Galvanic coupling of multiple discrete steel electrodes to simulate macrocell corrosion.	Macrocell corrosion current (I_{corr}); Spatial distribution of anodic/cathodic activity.	Direct, quantitative measure of active corrosion rate and its localization.	Most reliable for determining corrosion propagation rate after initiation, a critical input for residual cross-sectional area calculations.
Linear Polarization Resistance (LPR) Probe	Application of a small potential step and measurement of current response (DC or AC) between electrodes.	Polarization resistance (R_p), leading to corrosion current density (i_{corr}).	Quantitative, localized corrosion rate at the probe's specific location (often expressed in μm/ year).	Used to calibrate and validate corrosion rate models, linking electrochemical activity to material loss.
Concrete Resistivity Probe	Typically a 4-pin (Wenner) array measuring the electrical impedance of the concrete.	Bulk electrical resistivity (ρ) in kΩ·cm.	Master variable for ionic transport. Indicator of concrete quality, saturation, and corrosion risk (high ρ inhibits corrosion).	Key parameter in service life models. Used to assess the effectiveness of concrete as an electrolyte and to design cathodic protection systems.

continued on following page

Table 9. Continued

Sensor Type / System	*Measurement Principle*	*Primary Output(s)*	*Direct Engineering Significance*	*Key Application in Prognostics*
Embedded Relative Humidity & Temperature Sensor	Capacitive or resistive sensing elements housed in a porous concrete-compatible casing.	In-pore relative humidity (% RH) and temperature (°C) at specified depth.	Defines the internal exposure condition; governs all moisture and ion transport, freeze-thaw risk, and corrosion thermodynamics/ kinetics.	Drives and calibrates hygro-thermal models. The RH at the rebar level is the single most critical parameter for activating corrosion after chloride ingress or carbonation.
Integrated Durability Monitoring Node	A packaged system combining LPR, resistivity, and T/RH sensors with a local logger.	Synchronized time-series data of i_{corr}, ρ, T, RH.	Provides a complete picture of the *corrosivity* of the concrete environment at a strategic location (e.g., at the lowest rebar in a deck).	Enables the construction of data-driven durability models and the calculation of cumulative damage indices for predictive maintenance scheduling.

Data Acquisition, Transmission, and Power

The digital revolution has accelerated the transformation of structural health monitoring (SHM) from a research-focused endeavor to a crucial part of infrastructure asset management. Installing sensors is no longer the only aspect of modern SHM; rather, it involves implementing a cyber-physical system in which the physical structure and its digital data counterpart are intrinsically linked (Bao et al., 2019; Davoudi et al., 2018; Kaya et al., 2015; Köliö et al., 2014; Shao et al., 2019). The Internet of Things (IoT) concept serves as the foundation for this system's layered architecture of sensing, communication, and computation, which converts unprocessed voltage signals into useful engineering intelligence for lifecycle management. From the physical sensor layer to the cloud-based decision support layer, the modern SHM system is designed as a multi-tier network. The cognitive hub of the system is the cloud server. For data fusion, sophisticated analytics, and visualization, the raw and edge-processed data come together here. The essential elements of a modern IoT-based SHM system, together with their purposes and technological considerations, are listed in the following table.

Table 10. Architectural components of a modern IoT-based structural health monitoring system

System Layer	*Key Components*	*Primary Function & Technology*	*Engineering Considerations*	*Output / Purpose*
Sensing & Edge Layer	Wireless Sensor Node (WSN), MEMS sensors, microcontroller.	Sensing: Acquire physical signals (acceleration, strain, T/RH). Edge Computing: On-board signal processing, feature extraction, and local anomaly detection.	Power consumption, sensor accuracy/ range, environmental hardening, and computational capability for edge algorithms.	Reduced, pre-processed data packets (features, events) instead of raw streams; enables immediate local alerts.
Communication Layer	LPWAN (LoRa, NB-IoT), Cellular (4G/5G), Satellite modems.	Data Transmission: Reliable conveyance of data from sensor nodes to the cloud gateway over long ranges.	Range vs. bandwidth vs. power trade-off; network coverage; operational cost (data plans); latency requirements.	Successful ingestion of field data into the cloud platform for centralized analysis, regardless of location.
Energy System	Primary battery, Energy Harvester (solar, vibration, thermal), power management circuit.	Power Autonomy: Provide sustainable, maintenance-free power for the sensor node and transmitter.	Energy budget analysis; harvester efficiency; energy storage capacity; reliability in varying environmental conditions.	Enables multi-year deployments without battery replacement, reducing lifecycle cost and enabling dense sensor networks.

continued on following page

Table 10. Continued

System Layer	*Key Components*	*Primary Function & Technology*	*Engineering Considerations*	*Output / Purpose*
Cloud Platform & Analytics	Cloud servers (AWS, Azure, GCP), databases, analytics engines, ML models.	Data Fusion & Storage: Synchronize and store heterogeneous data. Advanced Analytics: Perform model updating, EOV compensation, damage detection algorithms, and prognostic modeling.	Scalability of data storage; computational power for analytics; cybersecurity of data in transit and at rest; algorithm transparency.	Transforms data into diagnostic information (e.g., damage indices, remaining life) and supports the digital twin.
Visualization & Decision Support	Web-based dashboard, GIS integration, alerting systems (SMS/email).	Human-in-the-Loop Interface: Visualize structural status, trends, and spatial data. Generate automated reports and alerts.	Usability for engineers and managers; clarity of presented information; configurability of alert thresholds; integration with asset management systems (e.g., BIM, CMMS).	Provides actionable insight to asset owners, enabling condition-based, predictive maintenance strategies.

The synergistic combination of these measurements, rather than their isolated components, is what gives modern evaluation its actual strength. Data fusion approaches reduce uncertainty and produce a more accurate and thorough condition evaluation by combining information from several NDE/SHM sources.

A hierarchical approach can be used:

1. **Sensor-Level Fusion:** To increase accuracy, measurements from identical co-located sensors can be combined (e.g., averaging several half-cell readings).
2. **Feature-Level Fusion:** creating a multi-dimensional feature vector by extracting important characteristics from various data types (such as UPV velocity from Area A, corrosion rate from Area A, and crack width from Area A).
3. **Decision-Level Fusion:** Each method yields an initial condition assessment or damage detection decision, which is then integrated using rules (such as Bayesian inference and Dempster-Shafer theory) to arrive at a final, more certain result.

Condition indices can be calculated using this fused data. For instance, a section's Corrosion Damage Index (CDI) can be a weighted function of visible fracture width, cover depth (derived from GPR), corrosion rate, and half-cell potential:

$$CDI = w_1 \cdot f(P_{HC}) + w_2 \cdot g(i_{corr}) + w_3 \cdot h(d_{cover}) + w_4 \cdot k(w_{crack})$$

Where f,g,h,k are normalization/scaling functions, and w_i are weights reflecting the relative importance of each parameter. Such indices condense complex multi-parameter data into a single metric that can be tracked over time, mapped spatially, and used as an input for prioritization and decision-making models.

An accurate digital representation of the structure in its current state is the result of the quantitative assessment process. This is a Building Information Model (BIM) or Finite Element Model (FEM) enhanced with the measured data, not just a CAD drawing.

- **Geometric Model:** developed using photogrammetry or laser scanning to capture as-built geometry, including settlements and deformations.
- **Material Property Assignment:** The model's elements are assigned spatially variable properties based on NDE data (e.g., reduced E_c in areas with low UPV, reduced steel area in corroded zones).
- **Damage Representation:** Cracks are either directly modeled or their impact is mitigated by the afflicted elements' decreased stiffness or strength values. Weakened zones are used to depict delaminations.
- **Load and Boundary Conditions:** updated using data from monitoring (e.g., actual support conditions from tiltmeters, actual traffic loads from weigh-in-motion).

This "as-is" digital twin serves as the starting point for all further analysis, including evaluating present safety, forecasting future deterioration, and modeling the impact of different retrofit initiatives. It turns the structure from a physical enigma into a system that can be characterized and prepared for modeling.

The transition from subjective, experience-based judgment to an objective, data-driven understanding of structural condition—a crucial precondition for the performance-based rehabilitation framework that follows—is made possible by the strict implementation of the methods outlined in this section.

4. STOCHASTIC MODELING OF DETERIORATION FOR LIFECYCLE PREDICTION

Equipped with a numerical evaluation of the existing situation, the next task is to predict how it will change in the future (Chang et al., 2019; Obunguta et al., 2022). The framework for simulating the time-dependent deterioration of RC structures is presented in this part, taking into account the inherent uncertainties in material characteristics, exposure to the environment, and model predictions. The goal is to create stochastic degradation models that consider important state variables (such as corrosion penetration, fracture width, and chloride content) as random variables or processes rather than deterministic values. These models serve as the foundation for reliability evaluation, optimal intervention planning, and probabilistic lifespan analysis (Grussing et al., 2016; Jia et al., 2017; Micic et al., 2013).

For lifecycle management, deterministic models—which employ a single "best-estimate" value for every parameter—are inadequate (Chen et al., 2015; Salmeron et al., 2022). "What is the probability that the corrosion penetration will exceed a critical level within the next 10 years?" is one of the fundamental questions they are unable to respond to. or "When is the reliability index projected to fall below the desired value?" Uncertainty results from:

- **Aleatory Uncertainty:** Variability in the depth of the concrete layer, variations in the temperature and humidity of the surrounding environment, and randomization in traffic loads are examples of phenomena that are inherently unpredictable.
- **Epistemic Uncertainty:** Lack of understanding (e.g., measurement mistakes, incomplete models, unclear model parameters).

By quantifying these uncertainties, a probabilistic framework enables engineers to make risk-aware choices. The result is a distribution of potential future situations rather than a single future condition, from which it is feasible to calculate failure probability or timeframes to reach limit states (Table 11).

4.1 Stochastic Process Models for Gradual Deterioration

The Gamma process is a strong and theoretically tractable model for damage that increases monotonically, such as corrosion loss or scour (Hasan, 2024; Jia et al., 2019). A stochastic process with independent, non-negative increments that each have a Gamma distribution is called a Gamma process. It is perfect for simulating slow, one-way deterioration.

Let $D(t)$ represent the damage quantity (e.g., corrosion penetration depth) at time t. We model $D(t)$ as a Gamma process with shape function $\alpha(t) > 0$ and scale parameter $\beta > 0$:

$$D(t) \sim \text{Gamma}(\alpha(t), \beta)$$

The probability density function (PDF) for the damage at time t is:

$$f_{D(t)}(x) = \frac{x^{\alpha(t)-1} exp(-x/\beta)}{\beta^{\alpha(t)}\Gamma(\alpha(t))} \text{for } x \geq 0$$

Where $\Gamma(\cdot)$ is the Gamma function. The mean and variance are:

$$E[D(t)] = \alpha(t)\beta, \mathrm{Var}[D(t)] = \alpha(t)\beta^2$$

The shape function $\alpha(t)$ is typically a power law: $\alpha(t) = a t^b$, where a and b are parameters fitted to inspection data or derived from physical models. This formulation ensures that the mean damage and its uncertainty increase with time. For corrosion, if the mean penetration is assumed to follow a linear law $E[D(t)] = \mu t$ after initiation, then setting $\alpha(t)\beta = \mu t$ suggests $\alpha(t)$ should be linear in t (i.e., $b = 1, a = \mu/\beta$).

First-Passage Time Analysis: A critical calculation is the time when damage first exceeds a threshold d_{crit}. The cumulative distribution function (CDF) of this first-passage time T_f is:

$$F_{T_f}(t) = P\left(D(t) > d_{crit}\right) = \int_{d_{crit}}^{\infty} f_{D(t)}(x)dx = 1 - \frac{\gamma\left(\alpha(t), d_{crit}/\beta\right)}{\Gamma(\alpha(t))}$$

where $\gamma(\cdot\ ,\cdot)$ is the lower incomplete gamma function. This directly provides the probability that failure (defined by exceeding the damage threshold) has occurred by time t.

4.2 Physics-Based Stochastic Models

The Gamma process can be influenced by physics even though it is a purely statistical model (Zirps et al., 2020). Examine the model of corrosion propagation from Section 2:

$$p(t) = \frac{M}{zF\rho_{steel}} \int_{t_{init}}^{t} i_{corr}(\tau)d\tau$$

To make this stochastic, we recognize that the corrosion current density $i_{corr}(\tau)$ is not constant but a random variable that depends on random environmental variables (RH, T, Cl). A common approach is to model i_{corr} as a lognormally distributed random variable, $i_{corr} \sim \mathrm{LogN}(\lambda, \zeta^2)$, where its mean and coefficient of variation (COV) are estimated from field or laboratory data. If we further assume i_{corr} is constant in time (or use its time-averaged value), then $p(t)$ becomes a linear function of a lognormal variable, and thus also lognormal:

$$p(t) = \theta\left(t - t_{init}\right)i_{corr}, \text{where } \theta = \frac{M}{zF\rho_{steel}}$$

Therefore, $p(t) \sim \mathrm{LogN}\left(\lambda + \ln\left(\theta(t - t_{init})\right), \zeta^2\right)$. The time to reach a critical penetration p_{crit} can then be derived from this distribution.

For chloride ingress, the solution to Fick's law contains the surface chloride concentration C_s, diffusion coefficient D, and cover depth d—all random variables. The time to initiation t_{init} is a function of these: $t_{init} \propto d^2/D$. If d and D are modeled as lognormal, t_{init} can be approximated as lognormal as well. A more robust approach uses Monte Carlo Simulation, sampling thousands of realizations of the input random variables, solving the physical model for each, and building the empirical distribution of the output (e.g., t_{init}, $p(t)$).

4.3 Spatial Variability and Random Fields

Deterioration is not uniform. Chloride concentration, carbonation depth, and corrosion damage vary spatially due to differences in exposure, material consistency, and construction practices. This spatial correlation can be modeled using random fields.

A random field $D(x,t)$ is a collection of random variables indexed by spatial location x. For example, the corrosion penetration over a bridge deck can be modeled as a 2D random field. A common model is the Gaussian random field, defined by a mean function $\mu(\boldsymbol{x}, t)$ and a covariance function $C(x_1,x_2;t)$ that describes how the correlation between two points decays with the distance between them (e.g., using an exponential correlation model: $C \propto \exp\left(-|\boldsymbol{x}_1 - \boldsymbol{x}_2|/l\right)$, where l is the correlation length).

Spatial modeling is crucial for:

- **Optimizing Inspection Intervals:** Strategic sampling is made possible by the knowledge that damage is associated; if one area is severely corroded, adjacent areas are probably similarly corroded.
- **System Reliability:** The joint condition of its constituent parts determines whether a structural system (such as a bridge with several girders) will fail. Calculations of system reliability are impacted by spatially correlated deterioration, which indicates that component failures are not independent.

4.4 Markov Chain Models for Condition Rating

Detailed physics-based models for every component might not be realistic for asset management of vast inventories (like a bridge network) (Grussing et al., 2016; Yu et al., 2022). A discrete-state option that is computationally efficient is provided by Markov chains. An element's condition is categorized into a limited number of states (e.g., State 1: Excellent, State 2: Good, State 3: Fair, State 4: Poor, State 5:

Failed). The movement between these states over discrete time increments, such as a year, is modeled by a Markov chain. A transition probability matrix $\boldsymbol{P}$ defines it:

$$\mathbf{P} = \begin{bmatrix} p_{11} & p_{12} & p_{13} & p_{14} & p_{15} \\ 0 & p_{22} & p_{23} & p_{24} & p_{25} \\ 0 & 0 & p_{33} & p_{34} & p_{35} \\ 0 & 0 & 0 & p_{44} & p_{45} \\ 0 & 0 & 0 & 0 & 1 \end{bmatrix}$$

where p_{ij} is the probability of transitioning from state i to state j in one time step. The matrix is typically upper-triangular, assuming deterioration does not reverse (without intervention). The probability of being in any state at time t is given by the state vector $\boldsymbol{\pi}_t$:

$$\boldsymbol{\pi}_t = \boldsymbol{\pi}_0 \mathbf{P}^t$$

Where $\boldsymbol{\pi}_0$ is the initial state vector. The elements of $\boldsymbol{P}$ are estimated from historical inspection data for populations of similar structures. Markov chains are effective tools for predicting budget requirements and organizing network-level interventions, despite being less mechanistic (Figure 11).

Figure 11. A Markov transition probability for a structure condition state

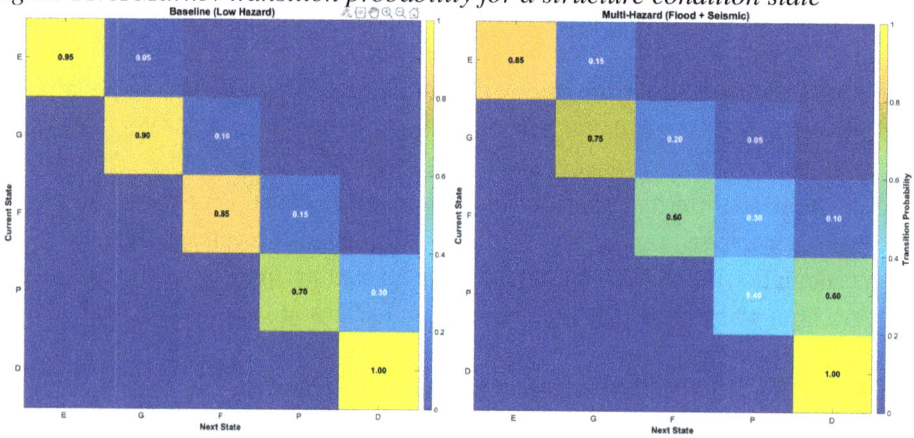

4.5 Model Updating with Inspection Data (Bayesian Methods)

The capacity of the probabilistic framework to update and improve predictions by incorporating new inspection data—a process codified by Bayesian updating—is one of its main advantages (Vagdatli et al., 2024; Vasudevan et al., 2023). Let $\boldsymbol{\theta}$ represent the vector of uncertain model parameters (e.g., the corrosion rate μ, the diffusion coefficient D). Before any inspection, we have a prior distribution $p(\boldsymbol{\theta})$, representing our initial beliefs, perhaps based on code values or data from similar environments. We then collect inspection data $\mathscr{D}$ (e.g., measured corrosion penetration at time t_{insp}). Bayes' theorem combines the prior with the likelihood of observing the data given the parameters, $p(\mathscr{D}\boldsymbol{\theta})$, to yield the posterior distribution:

$$p(\boldsymbol{\theta}\mathscr{D}) = \frac{p(\mathscr{D}\boldsymbol{\theta})p(\boldsymbol{\theta})}{p(\mathscr{D})} \propto p(\mathscr{D}\boldsymbol{\theta})p(\boldsymbol{\theta})$$

The posterior distribution $p(\boldsymbol{\theta}\mathscr{D})$ represents our updated, refined knowledge of the parameters, with reduced uncertainty. This posterior can then be used to make new, more accurate predictions about future deterioration. Suppose the prior for corrosion penetration at time t is modeled as $p(t) \sim \text{LogN}(\lambda, \zeta^2)$. An inspection measures a penetration $p_{meas} att_{insp}$. Assuming a lognormal measurement error, the likelihood $p(p_{meas}\lambda, \zeta)$ is also lognormal. Applying Bayes' theorem (often numerically via Markov Chain Monte Carlo - MCMC) yields a posterior distribution for λ and ζ. The updated prediction for penetration at a future time t_{future} will have a narrower confidence interval than the prior prediction.

Table 11. Comparison of stochastic deterioration modeling approaches

Model Type	*Mathematical Form*	*Primary Use*	*Strengths*	*Limitations*
Gamma Process	$D(t) \sim \text{Gamma}(\alpha(t), \beta)$	Gradual, monotonic damage (corrosion, wear)	Mathematically tractable, handles uncertainty well	Assumes independent increments, may not capture complex physics
Physics-Based Stochastic	$p(t) = \theta \int i_{corr} d$, with i_{corr} RV	Linking degradation to physical/ chemical drivers	Mechanistic insight, parameters have physical meaning	Requires more data, computational cost for complex models
Random Field	$D(x,t)$ with mean $\mu(x,t)$, covariance C	Spatially distributed deterioration (e.g., deck chloride)	Captures spatial correlation, essential for system analysis	Requires geostatistical analysis, data-intensive
Markov Chain	$\pi_t = \pi_0 P^t$	Network-level asset management, condition forecasting	Simple, efficient for large inventories, uses inspection data	Lacks mechanistic detail, assumes memoryless property

4.6 Integration for Lifecycle Reliability Analysis

The ultimate goal of stochastic degradation modeling is to calculate a system's or structural component's time-dependent reliability (Grussing et al., 2016; Jia et al., 2017). Failure is defined by the limit state function G, which is frequently defined as capacity minus demand:

$$G = R(t) - S(t) \leq 0$$

Here, the capacity $R(t)$ is a decreasing stochastic process due to deterioration (e.g., $R(t) = R_0 - \Delta R(D(t))$), and the demand $S(t)$ is a stochastic process representing loads (e.g., annual maximum traffic load, seismic intensity). The probability of failure at time t is:

$$p_f(t) = P(G(R(t), S(t)) \leq 0)$$

The reliability index $\beta(t)$ is defined as $\beta(t) = -\Phi^{-1}\left(p_f(t)\right)$, where Φ^{-1} is the inverse standard normal CDF. A target reliability index (e.g., $\beta_T = 3.5$ for ultimate limit state) is often used in codes. The time when $\beta(t)$ is expected to cross below β_T is a key decision point for intervention.

Computing $p_f(t)$ requires convolution of the distributions of $R(t)$ and $S(t)$, which can be done using techniques like First-Order Reliability Method (FORM), Monte Carlo Simulation, or out-crossing rate analysis for continuous stochastic processes.

We can get from nebulous ideas of "age" to quantitative, probabilistic performance estimates by combining the stochastic degradation models from this section with structural reliability techniques. The performance-based retrofit design and lifespan optimization methodologies described in the next sections are powered by this prediction engine.

5. PERFORMANCE-BASED ASSESSMENT AND RETROFIT DESIGN

The rehabilitation process makes a significant transition from prescriptive rules to performance-driven design, with a calibrated digital twin representing the current condition and stochastic models predicting its future degeneration. The theoretical and methodological underpinnings of Performance-Based Engineering (PBE) as it relates to the retrofit of pre-existing reinforced concrete structures are established in this part. It explains how to create probabilistic fragility functions, perform nonlinear structural analysis of degraded and retrofitted systems, define explicit performance objectives, and ultimately create interventions that are specifically designed to move a given structure from its current risk profile to a target future performance state (Ahmed et al., 2025; Alkhamis et al., 2018; Sberna et al., 2025).

The idea of deemed-to-satisfy norms is the foundation of traditional prescriptive codes. They outline minimal material strengths, reinforcement specifics, and load combinations meant to give the new building an implicit, although unquantifiable, degree of safety. There are three main reasons why this method is not appropriate for rehabilitating old structures: 1) the structure has a known history and accumulated damage; 2) the as-built condition frequently deviates from code assumptions; and 3) the objective is frequently to achieve a specific functional outcome rather than just meet a generic safety minimum (e.g., remain operational after an earthquake, limit crack widths for durability).

This paradigm is reversed by PBE. Establishing clear performance objectives is the first step. These are claims that relate a necessary structural performance level (such as immediate occupancy, life safety, or collapse prevention) to a hazard level (such as a 10% chance of exceedance in 50 years or a service-level earthquake). The engineer's job is to show by logical analysis that the structure will achieve these goals in its modified configuration. This turns into a two-phase evaluation for retrofit:

1. **Assessment of the *As-Is* Performance:** calculating the structure's current strength and vulnerability to different threats.

2. **Design for *Target* Performance:** choosing and describing a retrofit plan that modifies the structural response from its current state in order to achieve the desired performance level.

The analytical engine that drives this two-step procedure is described in this section. The post-hazard structural status is described qualitatively by performance levels. They must be converted into quantitative Engineering Demand Parameters (EDPs) that may be calculated through analysis and contrasted with permissible bounds. Typical Performance Levels include:

- **Operational:** There is just minimal, easily repairable damage, and the structure is still fully functional. Systems that are not structural are unharmed.
- **Immediate Occupancy (IO):** With just minimal, easily repairable damage, the structure is still fully functional. Non-structural systems remain unaltered.
- **Life Safety (LS):** Although significant damage to both structural and non-structural components is acceptable, the structure retains a margin against partial or complete collapse, protecting first responders' and inhabitants' lives.
- **Collapse Prevention (CP):** With significant damage and significant permanent drifts, the structure is about to collapse, but it is kept from collapsing long enough for an evacuation.

Limits on Corresponding Engineering Demand Parameters (EDPs) establish a connection between structural response and performance level.

- For **flexural members (beams, columns):** Maximum plastic hinge rotation θ_p, maximum curvature ductility μ_s), or maximum tensile strain in reinforcement ε_s.
- For **shear-critical members:** Maximum shear distortion γ.
- For the **structural system:** Maximum inter-story drift ratio δ/h, maximum roof drift ratio, or residual (permanent) drift.
- For **local durability:** Maximum crack width w.

Experimental data, analytical research, and consensus documents like ASCE/SEI 41 (Seismic Evaluation and Retrofit of Existing Buildings) and FEMA P-58 (Seismic Performance Assessment of Buildings) are used to determine acceptable limits for these EDPs at each performance level. For instance, ASCE 41 offers specific conditions for IO, LS, and CP performance levels as well as tables of acceptable plastic rotation angles for different member kinds.

5.1 Nonlinear Structural Analysis of Degraded and Retrofitted Sections

Predicting the force-deformation response of structural components while taking into consideration their current state of deterioration and the additional contribution of retrofit materials is the main analytical task (Miano, 2017; Palacio et al., 2013; Prakashvel et al., 2023). For unconfined concrete, the stress-strain relationship deteriorates with chemical softening, ASR/freeze-thaw microcracking, and cracking. The Kent-Scott-Park model can be adjusted for time or damage:

$$\sigma_c\left(\varepsilon,t\right) = \begin{cases} f'_c(t)\left[\frac{2\varepsilon}{\varepsilon_0(t)} - \left(\frac{\varepsilon}{\varepsilon_0(t)}\right)^2\right] & \text{for } 0 \leq \varepsilon \leq \varepsilon_0(t) \\ f'_c(t)\left[1 - Z(t)(\varepsilon - \varepsilon_0(t))\right] & \text{for } \varepsilon_0(t) \leq \varepsilon \leq \varepsilon_{cu}(t) \end{cases}$$

Where $fc(t)$, $\varepsilon_0(t)$, $Z(t)$, and $\varepsilon cu(t)$ are the time-dependent compressive strength, strain at peak stress, post-peak softening slope, and ultimate crushing strain, respectively. These are related to a total damage index $D_{total}(t)$ from Section 4: $fc(t) = fc0\left[1 - \alpha_f D_{total}(t)\right], E_c(t) = E_{c0}\left[1 - \alpha_E D_{total}(t)\right]$.

The stress-strain relationship for Corroded Steel Reinforcement is followed by the modified bilinear stress-strain curve. The yield strength $f_y(t)$ and ultimate strain $\varepsilon_u(t)$ degrade, often more severely than the mere loss of area would suggest, due to pitting and stress concentrations:

$$f_y(t) = f_{y0}\left(1 - \alpha_y \frac{\Delta A_s(t)}{A_{s0}}\right), \varepsilon_u(t) = \varepsilon_{u0}\left(1 - \alpha_\varepsilon \frac{\Delta A_s(t)}{A_{s0}}\right)$$

Where $\Delta A_s(t)/A_{s0}$ is the fractional area loss, and α_y, α are empirical coefficients are often greater than 1.0.

The interaction between corroded steel and cracked concrete is modeled using a bond stress-slip $\tau - s$ law. A common model (e.g., Model Code 2010) includes an ascending branch, a peak bond stress $\tau_{max}(t)$, and a descending friction branch. Corrosion reduces τ_{max} and increases the slip at which it occurs. While the flexural behavior of a cross-section is determined through sectional analysis. The cross-section is discretized into fibers. For a given curvature ϕ, the strain distribution is linear $\varepsilon = \phi \cdot y$. The stress in each fiber is computed from the appropriate constitutive model (concrete, steel, retrofit material). Integrating stresses over the area yields the resultant axial force P and bending moment M. By incrementing curvature, the full moment-curvature $M - \phi$ relationship is derived. For a corroded, rectangular, singly-reinforced section, a simplified estimate of nominal moment capacity $M_n(t)$ is:

$$M_n(t) = A_s(t)f_y(t)\left[d - \frac{A_s(t)f_y(t)}{1.7f'_c(t)b}\right]$$

This clearly shows the coupled degradation of steel area, steel strength, and concrete strength. The moment-rotation $M - \theta$ response for a plastic hinge of length L_p is obtained by integrating the curvature along the member: $\theta = \int\phi(x)dx \approx \phi_y L_p + \left(\phi - \phi_y\right)L_p$, where ϕ_y is the yield curvature. This $M - \theta$ relationship is the fundamental input for nonlinear frame analysis.

Modeling Retrofit Materials in Sectional Analysis

Retrofit materials are easily incorporated into the sectional analysis framework by incorporating their constitutive behavior into the fiber discretization (Table 12).

- FRP jackets provide passive confinement to concrete. The confined concrete strength *f′cc, and ultimate strain εccu* are enhanced. The confining pressure f_l is a function of the FRP jacket's stiffness and rupture strain: $f_l = \left(2E_{frp}t_{frp}\varepsilon_{frp}\right)/D$, where E_{frp}, t_{frp} are the FRP modulus and thickness, ε_{frp} is the strain (limited by rupture), and D is the column diameter.
- Near-surface mounted (NSM) FRP bars and externally bonded FRP sheets are represented as extra tensile reinforcement with a linear elastic stress-strain relationship up to failure (rupture or debonding). Key considerations are the strain limitation due to debonding (e.g., effective strain $\varepsilon_{fe} = \kappa_m\varepsilon_{fu}$, where $\kappa_m < 1$) and the contribution to shear capacity.
- Both flexural and shear capabilities are increased with steel jackets or plates. They are represented as extra steel fibers, taking into account the consequences of partial composite action or bond if not totally welded, as well as possible buckling under compression.

Table 12. Constitutive models for common retrofit materials

Material	*Typical Stress-Strain Model*	*Key Parameters & Failure Modes*	*Modeling Consideration*
Confined Concrete (FRP)	Modified Popovics/ Mander	f'_{cc}, ε_{ccu} from confinement model	Defines core concrete properties in section analysis
FRP in Tension	Linear Elastic to Rupture	E_{frp}, ε_{fu}, debonding strain limit ε_{fd}	Add as an elastic fiber; apply a strain limit due to the bond
NSM FRP	Linear Elastic to Rupture	E_{frp}, ε_{fu}, bond-dependent development length	Model as reinforcement bar with bond-slip interface
Steel Plate	Elastic-Perfectly Plastic	f_y, E_s, buckling stress	Add as a steel fiber; check local buckling capacity

Figure 12. A series of moment-curvature diagrams for a beam-column section

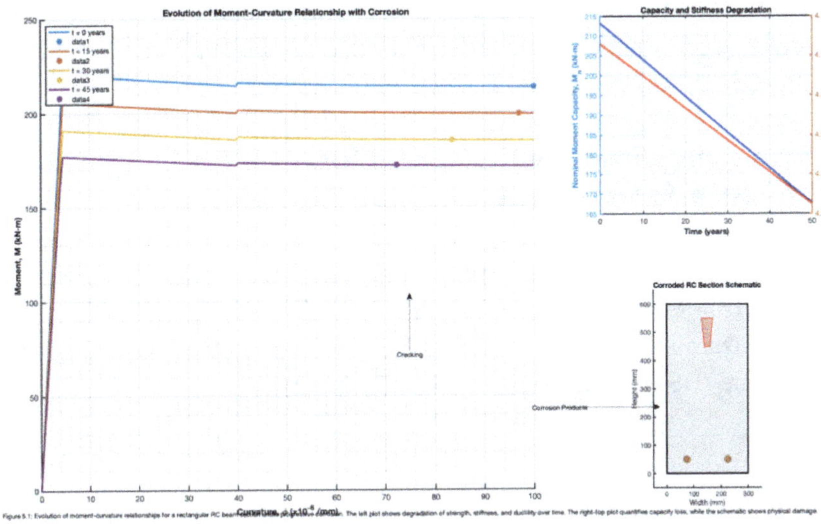

5.2 Nonlinear Static (Pushover) and Dynamic Analysis

Component models are put together into a structural model for nonlinear analysis in order to evaluate the overall performance of the system (Fragiadakis et al., 2008; Miano et al., 2017). The structure is gradually subjected to nonlinear static (pushover) analysis with a predetermined lateral load pattern (such as uniform or triangular). The development and evolution of plastic hinges are monitored at every load phase. A capacity curve, which plots base shear V against roof displacement

Δ, is the result of the analysis. The global force-deformation capability is shown by this curve. By intersecting the capacity curve with a reduced demand spectrum, one can determine the performance point for a particular hazard level (e.g., using the Capacity Spectrum Method or the N2 Method). The resulting displacement is compared to the desired performance level's EDP restrictions (such as drift) (Padalu et al., 2024; Pampanin et al., 2011; Zameeruddin et al., 2021).

However, the most exacting approach is the Nonlinear Dynamic (Time-History) Analysis. A series of ground motion acceleration records scaled to the specified hazard level is applied to the structure model. The temporal history of displacements, forces, and member deformations is immediately calculated by the analysis. All records' maximum EDPs are gathered, and performance limits are compared to their median or a percentile. Compared to pushover analysis, this approach more precisely captures cyclic degradation, higher-mode responses, and dynamic impacts. Record-to-record variability, a significant source of response uncertainty, must be taken into account when choosing or developing ground motions for retrofitted structures that are compatible with the site-specific seismic hazard (Anwar et al., 2020; Bouchard, 2007; Zub et al., 2018).

5.3 Developing Fragility Functions for Performance Assessment

A probabilistic description of structural performance is given by a fragility function (Table 13) (Ahmed, 2025; Dalal et al., 2021; Stefanidou et al., 2021). Given a particular hazard intensity, it provides the conditional probability that a particular Damage State (DS) is attained or surpassed (Figure 13).

Figure 13. A family of fragility curves for the "Life Safety" damage state of a beam-column connection

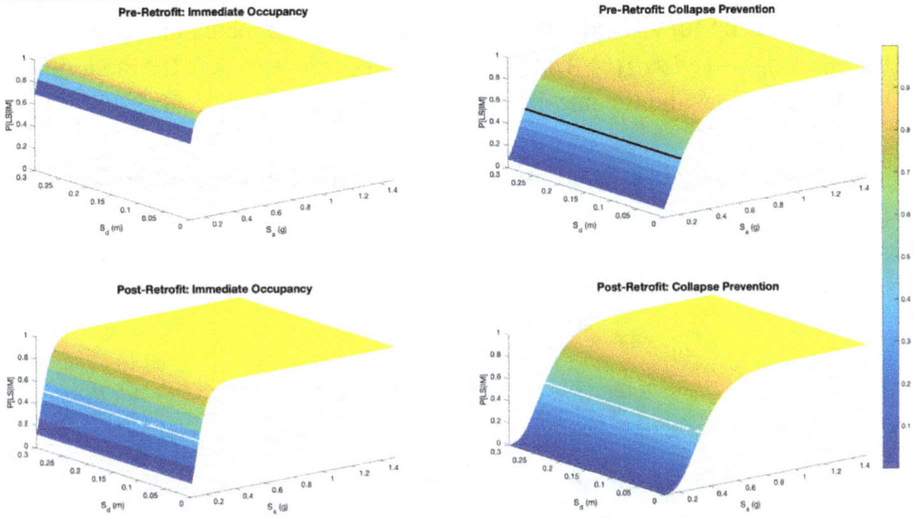

For seismic hazards, the intensity measure (IM) is often the spectral acceleration at the structure's fundamental period, $S_a(T_1, 5\%)$. For other hazards, it could be peak ground acceleration (PGA), flood depth, or wind speed. A fragility function is commonly expressed as a lognormal cumulative distribution function:

$$P(DS \geq ds_i IM = im) = \Phi\left(\frac{ln(im) - ln(\eta_i)}{\beta_i}\right)$$

Where, d_{si} is the i-th damage state (e.g., DS1: Minor, DS2: Moderate, DS3: Extensive, DS4: Complete). η_i is the median capacity—the IM level at which there is a 50% probability of reaching ds_i. β_i is the logarithmic standard deviation (dispersion), capturing uncertainty in capacity and demand. $\Phi(\cdot)$ is the standard normal CDF in four steps.

1. **Define Damage States:** Link qualitative damage states (IO, LS, CP) to quantitative EDP limits (e.g., θ_p^{IO}, θ_p^{LS}, θ_p^{CP}).
2. **Nonlinear Analysis:** Perform multiple nonlinear analyses (static or dynamic) for a range of IM levels. For each analysis, record the maximum EDP in each critical component.
3. **Cloud Analysis or Stripe Analysis:** Plot the calculated EDPs against the IM. Fit a probabilistic model (often lognormal) to the conditional distribution of EDP given IM: *EDP* | *IM* ~ "{*LogN*}(ln(*a IM^b*), *σ*).

4. **Transform to IM Space:** For each damage state limit edp_i, the probability of exceedance is $P(EDP > edp_i im) = 1 - \Phi\left(\frac{\ln(edp_i) - \ln(ai\,m^b)}{\sigma}\right)$. By inverting this relationship for a fixed probability (e.g., 50%), the median capacity η_i can be found: $\eta_i = (edp_i/a)^{1/b}$. The dispersion β_i incorporates both the randomness in EDP given IM and the uncertainty in the limit state threshold.

For an aging structure, the fragility parameters become time-dependent: $\eta_i(t) and$ $\beta_i(t)$. The median capacity decreases as deterioration reduces strength and ductility: $\eta_i(t) = \eta_{i,0}[1 - \delta g(D(t))]$, where $g(D(t))$ is a function of the deterioration index. The dispersion $\beta_i(t)$ often increases due to added uncertainty from the deterioration process.

Table 13. Fragility parameters for a corroding RC beam- column connection

Damage State	*Performance Level*	*EDP Limit (Plastic Rot., \$\ theta_p\$)*	*Median \$S_a\$ (g) @ t=0 (\$\ eta_{i,0}\$)*	*Median \$S_a\$ (g) @ t=50 yrs (\$\ eta_i(50)\$)*	*Dispersion \$\beta_i\$*
DS1	Immediate Occupancy	0.005 rad	0.45	0.38	0.40
DS2	Life Safety	0.02 rad	0.80	0.65	0.45
DS3	Collapse Prevention	0.04 rad	1.20	0.95	0.50

Finding the retrofit scheme (type, amount, detailing) that alters the structural characteristics so that the resultant retrofitted fragility functions meet the desired performance objectives is the formal form of the performance-based retrofit design process.

6. SUSTAINABLE AND RESILIENT INTERVENTION TECHNIQUES

This section discusses how advanced retrofit materials and techniques can be used to accomplish the goals identified by the performance-based design framework. The emphasis goes beyond strength restoration to include durability, installation speed, environmental impact, and future adaptability—all essential components of resilient and sustainable rehabilitation. The mechanics, suitability for various degradation types, constructability, and lifespan consequences of state-of-the-art systems are critically reviewed.

The circular economy and resilience-aligned concepts of Effectiveness, Durability & Compatibility, Minimal Intervention, Reversibility/Adaptability, Resource

Efficiency, and Constructability should serve as guidance for choosing a retrofit technique.

6.1 Flexural and Shear Strengthening with Fiber-Reinforced Polymers (FRP)

Because of their high strength-to-weight ratio, corrosion resistance, and flexibility, fiber-reinforced polymer (FRP) composites—typically carbon (CFRP) or glass (GFRP) fibers in a polymer matrix (epoxy, vinyl ester)—have emerged as a leading retrofit technology (Table 14) (Chisari et al., 2017; Zou et al., 2007).

Externally Bonded Reinforcement (EBR)

Sheets or laminates are bonded to the tension face of beams/slabs or wrapped around columns. The key failure modes are ultimate tensile strain ε_{fu}. The effective strain in the FRP at the ultimate limit state, ε_{fe}, is limited to prevent debonding. Codes (e.g., ACI 440.2R, *fib* Bulletin 90) provide models. A common form is:

$$\varepsilon_{fe} = \kappa_m \varepsilon_{fu} \leq 0.9\,\varepsilon_{fu}$$

Where κ_m is a debonding reduction coefficient dependent on the FRP stiffness, concrete strength, and bond length. The additional moment capacity ΔM_n is then:

$$\Delta M_n = A_f f_{fe}\left(d_f - \frac{a}{2}\right)$$

Where A_f is the FRP area, $f_{fe} = E_f \varepsilon_{fe}$, d_f is the effective depth of FRP, and a is the depth of the equivalent stress block. FRP bars or strips are placed into pre-cut grooves in the concrete cover and bonded with epoxy or cementitious grout. Advantages over EBR include: better protection from mechanical damage, fire, and vandalism, reduced debonding risk due to better confinement by the surrounding concrete, and easier to achieve prestressing. The bond behavior is more akin to that of reinforcing bars, governed by development length. The design involves ensuring adequate bond length l_d: $l_d = \frac{A_f f_{fu}}{\tau_{max} \pi d_b}$, where τ_{max} is the average bond strength and d_b is the bar diameter.

FRP Confinement for Ductility Enhancement

Continuous FRP wraps around columns offer passive confinement, boosting the concrete core's ultimate axial and lateral strain capacity as well as its compressive strength. This is the main technique for improving seismically inadequate columns' ductility and shear capability. The confinement model by Lam & Teng is widely used. The confined concrete strength *f′cc and ultimate strain εccu* are:

$$f'_{cc} = f'_c + k_1 f_l, \varepsilon_{ccu} = \varepsilon_{c0}\left(1.75 + 12 k_2 \frac{f_l}{f'_c}\left(\frac{\varepsilon_{h,rup}}{\varepsilon_{c0}}\right)^{0.45}\right)$$

Where $f_l = \left(2E_{frp}, t_{frp}, \varepsilon_{h,rup}\right)/D$ is the confining pressure at FRP rupture, $\varepsilon_{h,rup}$ is the hoop rupture strain of the FRP (often reduced by a strain efficiency factor $k_\varepsilon \approx 0.55$), and k_1, k_2 are coefficients.

Table 14. Comparison of FRP strengthening techniques

Technique	*Best For*	*Key Advantages*	*Key Limitations & Failure Modes*	*Sustainability Notes*
EBR Sheets	Flexural strengthening of beams/slabs; shear strengthening; confinement	Lightweight, fast application, minimal section enlargement	Debonding, vulnerable to fire/ impact, requires surface prep	Epoxy is petrochemical-based; CFRP has high embodied energy
NSM FRP	Flexural strengthening where surface access is limited, or protection is needed	Better bond, protected, allows prestressing	Grooving required, more labor-intensive, limited to members with sufficient cover	Similar to EBR, grout options can reduce epoxy use
FRP Confinement	Seismic ductility & shear enhancement of columns	Dramatically increases ductility, applicable to circular/rectangular sections	Less effective on rectangular sections, fire vulnerability	Very efficient use of material for performance gain

6.2 Advanced Materials for Repair and Overlays

Very high compressive strength (>150 MPa), great tensile ductility (strain-hardening behavior), exceptionally low permeability, and an excellent bond to existing concrete are the characteristics of Ultra-High-Performance Concrete (UHPC). Among its uses in rehabilitation are: 1) Thin (50–75 mm) UHPC overlays to protect underlying reinforcement and distribute loads while offering a long-lasting, waterproof barrier for bridge decks. 2) Joint fill is perfect for fixing degraded ends of precast beams because of its strong early strength and endurance. 3) UHPC connection and jacketing, which frequently uses steel fibers for shear resistance, to provide monolithic connections between prefabricated components or as a jacket for column strengthening. Failure occurs elsewhere because the bond strength between

UHPC and regular concrete is frequently greater than the substrate concrete's tensile strength (Akbar et al., 2025; Chang et al., 2023; Kennedy et al., 2023).

Engineered Cementitious Composites (ECC) are fiber-reinforced cementitious composites that are micromechanically constructed. They have a tensile ductility of 3–5% and show numerous fine cracks (crack width < 100 μm) instead of localized fracture. This makes it ideal for structural repairs by replacing spalled concrete in corrosion-damaged regions, offering both protection and some structural contribution, as well as durability restoration, shear strengthening, and confinement.

Instead of using Portland cement, alkali-activated materials are binders created by activating aluminosilicate precursors (fly ash, slag, and metakaolin) with an alkaline solution. Low embodied carbon, excellent early strength, and good resistance to acid, sulfate, and chloride infiltration were the results of this.

Shrinkage behavior, the requirement for heat curing in some formulations, and guaranteeing long-term binding and compatibility with current OPC-based concrete are some of the difficulties associated with retrofitting. As environmentally friendly repair mortars and grouts, they have a lot of potential.

6.3 Metallic and Hybrid Systems

A classic yet very successful technique is steel jacketing, especially for seismic retrofit of columns. To improve flexural, shear, and axial capacity, split pipes are clamped, or steel plates are welded to create a box around the column. Based on a well-understood behavior, exceptional ductility and energy dissipation are anticipated (Aboutaha et al., 2010; Hanifehzadeh et al., n.d.; Tan et al., 2020). This method's primary drawbacks include connection details, substantial additional weight, and corrosion susceptibility (which calls for protection). In contrast to the passive confinement of FRP, the confining pressure from a steel jacket is active (develops at modest strains) (Chisari et al., 2017; Zou et al., 2007).

External Post-Tensioning: Adding external post-tensioning tendons, frequently in conjunction with other methods, can improve seismic performance by reducing or eliminating crack widths under service loads for columns, beams, or walls and restoring or increasing flexural capacity in beams and slabs. It is a way of active strengthening. Long-term losses (creep, shrinkage, relaxation), anchorages, and tendon protection must all be taken into account during design.

When heated (shape memory effect) or unloaded (superelasticity), Shape Memory Alloy (SMA) devices, particularly Nickel-Titanium (Nitinol), can recover significant strains (up to 8%). Superelastic SMA Rebars/Strands and SMA Dampers or Braces are used in retrofits. SMAs provide a special route to functional resilience—the capacity to not only withstand a danger but also quickly return to service—despite their current high cost.

6.4 Integration, Detailing, and Durability Considerations

Any retrofit project's success essentially relies on careful attention to detail in a number of crucial areas. For bonded systems, surface preparation is essential. This procedure must provide a substrate with a precise surface profile (SSP) that is sound, clean, and appropriately roughened. To ensure sufficient adhesion and long-term retrofit performance, proper surface preparation is crucial (Table 15) (Guo et al., 2019).

Another is Transition Zones, which need to be carefully controlled to lessen stress concentrations, particularly when using strengthening techniques like tapered FRP (Fiber Reinforced Polymer) laminates. In addition to preventing early debonding or interface failure, gradually stopping the strengthening aids in the efficient distribution of loads.

Another important factor is fire safety because high temperatures cause many polymers and some fabrics to deteriorate considerably. To comply with fire safety rules and preserve structural integrity in an emergency, retrofit assemblies may need specific solutions like intumescent coatings, cementitious overlays, or board insulation.

Equally crucial is durability in service, which necessitates compatibility with long-term loads and climatic conditions. This includes protecting polymer components from UV light, treating steel reinforcements to prevent corrosion, and making sure the retrofit system doesn't unintentionally trap moisture against the original reinforcement, which could eventually cause degradation.

Finally, one of the most important design principles is inspectability, which emphasizes the necessity of accessible paths that enable future visual inspections and non-destructive assessment (NDE) of both the original structure and the retrofit interface. This strategy demonstrates foresight in preserving structural integrity and anticipating possible problems.

Table 15. Sustainability & resilience matrix for retrofit techniques

Technique	*Embodied Carbon*	*Material Efficiency*	*Durability in Aggressive Env.*	*Resilience (Ductility/ Energy Dissip.)*	*Speed of Installation*
FRP EBR	Medium-High (CFRP) / Med (GFRP)	High (thin layers)	Excellent (chemically inert)	Good (brittle failure)	Very Fast
NSM FRP	Medium-High	High	Excellent (embedded)	Good	Moderate
UHPC Overlay	Medium-High (high cement)	Medium	Exceptional	Good (ductile)	Moderate (curing)

continued on following page

Table 15. Continued

Technique	*Embodied Carbon*	*Material Efficiency*	*Durability in Aggressive Env.*	*Resilience (Ductility/ Energy Dissip.)*	*Speed of Installation*
ECC Repair	Low-Medium (low cement)	Medium	Exceptional (crack control)	Excellent (high ductility)	Moderate
Steel Jacketing	High	Low (heavy)	Poor (needs corrosion protection.)	Excellent	Slow
External PT	Medium	Medium	Good (if protected)	Good (elastic)	Slow
Geopolymer Repair	**Low**	Medium	Good to Excellent	Moderate	Moderate (curing)

As a result, choosing a retrofit method involves weighing both technical performance and the more general concepts of resilient and sustainable infrastructure (Caterino et al., 2008, 2009; Ligabue et al., 2018). The lifespan and optimization framework described in the following section, which enables the quantitative trade-off between these frequently conflicting goals, is the best place to make this choice (Figure 14).

Figure 14. Systematic decision framework for sustainable retrofit technique selection

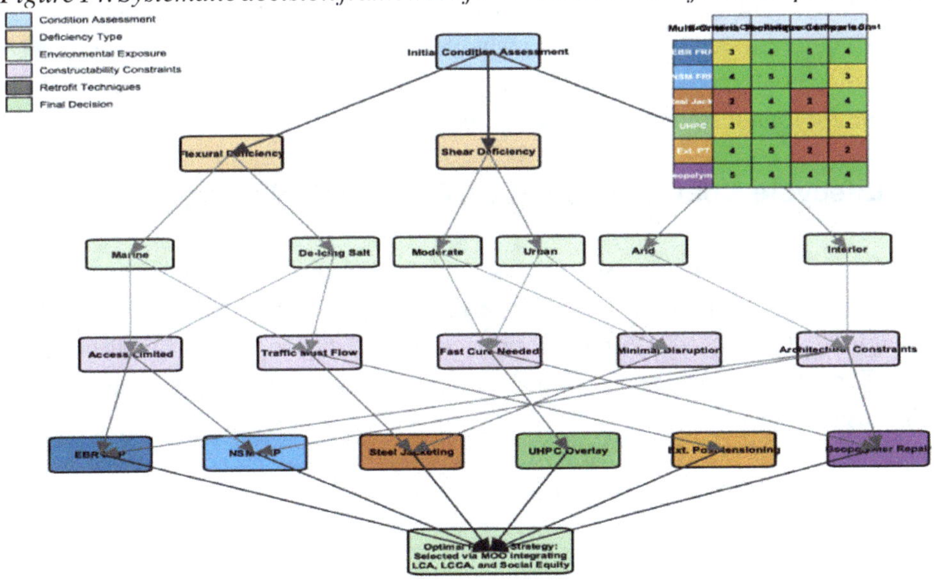

7. INTEGRATED LIFECYCLE DECISION-MAKING AND MULTI-OBJECTIVE OPTIMIZATION

A logical, transparent decision-making process that integrates technical analysis, economic valuation, risk assessment, and sustainability metrics is the result of the performance-based rehabilitation framework. In order to help stakeholders traverse the intricate trade-offs involved in retrofit planning, this section offers an integrated lifecycle decision-support system that goes beyond basic cost-minimization (Chang et al., 2020; Ligabue et al., 2018). By framing the process of choosing an optimal intervention approach as a sequential decision-making problem under uncertainty, it formalizes the process and applies methods from Multi-Objective Optimization (MOO), Lifecycle Cost Analysis (LCCA), and probabilistic risk assessment (Tapia et al., 2016; Wang et al., 2023). The objective is to turn enormous volumes of diagnostic data, predictive models, and design possibilities into an investment strategy that is obvious, defendable, and strikes a long-term balance between safety, economy, and environmental stewardship. A classic sequential choice dilemma confronts the manager of an aging RC structure: How and when should we step in? The available actions $a \in \mathcal{A}$ range from Do-Nothing/Monitor to Preventive Maintenance, Minor Repair, Major Retrofit, and Replacement. Every activity has an immediate cost and modifies the structure's state, risk profile, and remaining value in the future. Deep ambiguity, several competing goals, and extended time horizons are the main obstacles. Therefore, a strong decision framework needs to be multi-objective (to balance conflicting goals), probabilistic (to handle uncertainty), and lifecycle-oriented (to account for long-term implications) (Table 16).

7.1 Lifecycle Cost Analysis (LCCA) for Retrofit Projects

The economic process used to assess an asset's overall cost over a specified analysis period is called LCCA. The pertinent expenses for (Ahmed et al., 2021; Panagiotidou et al., 2021)a retrofit project go well beyond the original building cost. The whole lifecycle costs' Net Present Value (NPV) over a planning horizon. The basic metric is T.

$$\mathrm{NPV}(T) = C_0 + \sum_{t=1}^{T} \frac{C_m(t) + C_r(t) + C_h(t) + C_d(t) + C_e(t)}{(1+r)^t} + \frac{C_{EoL}}{(1+r)^T}$$

Where, C_0 is the Initial retrofit construction cost, and $C_m(t)$ is the Scheduled maintenance cost in year t (e.g., cleaning, sealing). $C_r(t)(t)$ is the Unscheduled repair/rehabilitation cost in year t, triggered when deterioration exceeds a threshold. This is a *random variable*. $C_h(t)$ is the direct hazard loss cost (physical damage) from earthquakes, floods, etc., in year t. A *random variable* dependent on fragility and hazard. $C_d(t)$ define Indirect loss or downtime cost due to service disruption. Often, the highest cost is for critical infrastructure. $C_e(t)$ is the Environmental externality cost (e.g., carbon tax) in year t. C_{EoL} is End-of-Life cost (deconstruction, disposal, salvage; can be negative. r is the Real discount rate. In a continuous, probabilistic formulation, the expected NPV becomes:

$$E[\mathrm{NPV}(T)] = C_0 + \int_0^T E\left[c_m(\tau) + \lambda_r(\tau) \cdot E[C_r\tau] + \lambda_h(\tau) \cdot E[C_h + C_d\tau] + c_e(\tau)\right]e^{-r\tau}d\tau + C_{EoL}e^{-rT}$$

Here, $\lambda_r(t)$ and $\lambda_h(t)$ are the *hazard functions* (failure rates) for reaching a repair-triggering deterioration state and for external hazard events, respectively. $E[\cdot t]$ denotes the expected cost *given* an event at time t. This formulation explicitly couples the economic analysis with the reliability and deterioration models.

7.1.1 Quantifying Hazard Losses: Probabilistic Risk Assessment (PRA)

The expected annual loss (EAL) from a specific hazard is computed via the risk integral:

$$\mathrm{EAL} = \int_0^\infty E\left[C_{damage} + C_{downtime} IM = im\right] \cdot \left| \frac{d\lambda(IM > im)}{d(im)} \right| d(im)$$

$\lambda IM{>}im$ is the hazard curve, giving the annual frequency of exceeding intensity measure im. $E[C|IM]$ is the vulnerability function, computed using fragility curves $P(DS \geq ds_i IM)$ and consequence functions $C(ds_i)$. $C_{damage}(ds_i)$ is the Direct cost to repair damage state i. $C_{downtime}(ds_i, t_{rec})$: Loss due to service interruption during recovery time t_{rec}. This can be modeled as $C_{downtime} = \int_0^{t_{rec}} L(\tau) \cdot [1 - Q(\tau)]d$, where L is the loss rate and $Q(\tau)$ is the functionality recovery curve. For multiple interacting hazards, the combined risk requires multivariate fragility models and joint hazard characterization, often using copula theory.

7.1.2 Incorporating Environmental Costs

A comprehensive LCCA internalizes environmental externalities. The dominant cost is associated with greenhouse gas (GHG) emissions, priced via a social cost of carbon (SCC) or carbon tax $P_{CO_2}(t)$. The environmental cost in a given year for a retrofit strategy using material quantities Q_i is:

$$C_e(t) = P_{CO_2}(t) \cdot \sum_i Q_i \cdot \left[EF_{i,embodied} + EF_{i,op}(t) + EF_{i,EoL} \cdot \delta(t - T)\right]$$

Where EF_i are the emission factors (kg CO_2-eq per unit) for each lifecycle phase: production (A1-A3), operation (B1-B7), and end-of-life (C1-C4), following ISO 14040/44 standards. This integrates Life Cycle Assessment (LCA) directly into the economic calculus.

Figure 15. The Integrated Performance-Based Lifecycle Management (IPBLM) framework results

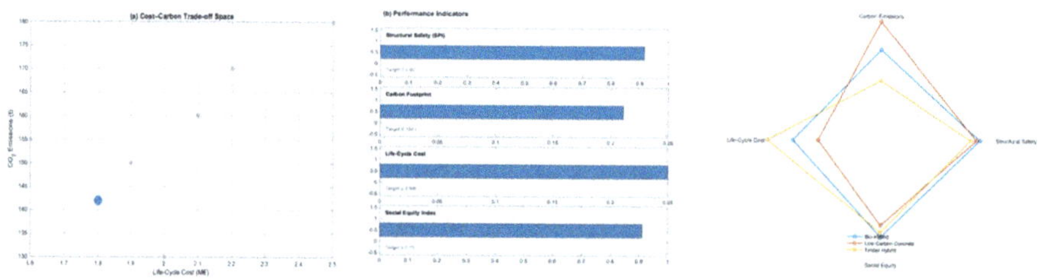

Table 16. Components of lifecycle cost for RC retrofit

Cost Category	*Description*	*Model / Driver*	*Key Uncertainties*
Initial Construction (C_0)	Direct costs of retrofit implementation.	Deterministic based on design.	Material price volatility, construction delays.
Preventive Maintenance ($C\_m(t)$)	Routine actions to preserve the condition.	Function of deterioration state $D(t)$: $C_m = C_{m0} + k_m D(t)$.	Effectiveness of maintenance, labor costs.
Corrective Repair ($C\_r(t)$)	Major interventions triggered by deterioration.	Stochastic: $\lambda_r(t)$ From the deterioration process, the cost increases with the damage level.	Extent of hidden damage, technology changes.
Hazard Damage ($C\_h(t)$)	Direct physical damage from hazards.	Probabilistic risk integral using fragility & hazard curves.	Hazard intensity/ frequency, fragility, unit repair costs.
Downtime Loss ($C\_d(t)$)	Economic losses from service disruption.	Function of recovery time t_{rec} and loss rate L.	Recovery logistics, economic interdependencies.

continued on following page

Table 16. Continued

Cost Category	*Description*	*Model / Driver*	*Key Uncertainties*
Environmental Externalities ($C_e(t)$)	Cost of emissions, resource depletion.	Carbon price × Lifecycle GHG emissions (from LCA).	Future carbon price, emission factor accuracy.
End-of-Life (C_{EoL})	Deconstruction, waste processing, salvage.	Often, a percentage of C_0; can be negative (salvage value).	Future material recycling value, regulatory changes.

7.2 Multi-Objective Optimization (MOO) Framework

The goals of lifecycle management are intrinsically contradictory. It is frequently debatable to express all values in monetary terms (as in LCCA), particularly when it comes to safety and environmental effects. MOO looks for the collection of Pareto-optimal solutions while treating these goals independently. Formally, we have several objective functions to minimize for a vector of choice variables $\boldsymbol{x}$ (which define the retrofit type, time, and extent):

$$\min_{\mathbf{x}\in\mathcal{X}} \mathbf{F}(\mathbf{x}) = \left[f_1(\mathbf{x}), f_2(\mathbf{x}), f_3(\mathbf{x}), \dots\right]^T$$

Typical objectives include, $f_1(\boldsymbol{x}) = \text{NPV}_{\text{total}}(\boldsymbol{x})$ as Lifecycle Cost. $f_2(\boldsymbol{x}) = \text{LCA}_{\text{GWP}}(\boldsymbol{x})$ is Lifecycle Global Warming Potential (kg CO_2-eq), and $f_3(\boldsymbol{x}) = \text{Risk}(\boldsymbol{x})$: Expected Annual Fatality or Economic Loss. $f_4(\boldsymbol{x}) = -\text{Resilience}(\boldsymbol{x})$ is the Negative of a resilience index (e.g., normalized area under recovery curve). A solution $\boldsymbol{x}^*$ is Pareto-optimal (non-dominated) if there is no other feasible solution $\boldsymbol{x}$ that improves one objective without worsening at least one other. The set of all Pareto-optimal solutions forms the Pareto front in the objective space.

Solving the MOO Problem could be followed by three approaches: A Priori Methods, A Posteriori Methods, and Interactive Methods. In the Priori Methods, the decision-maker provides preferences upfront (e.g., weights w_i), converting the problem to a single objective, $\min\sum w_i f_i(\boldsymbol{x})$. This is easy, but it takes a deep understanding of prior preferences. The whole Pareto front is estimated in the Posteriori Methods so that the decision-maker can make a choice at a later time. Multi-Objective Evolutionary Algorithms (MOEAs) such as NSGA-II, NSGA-III, or MOEA/D are the most efficient algorithms (Ahmad et al., 2025; Arya et al., 2024; Sethi et al., 2024). To reach the Pareto front, these employ populations of solutions, genetic operators (selection, crossover, mutation), and non-dominated sorting. They perform exceptionally well when dealing with high-dimensional, non-convex, and discontinuous goal spaces, which are typical of retrofit challenges. The decision-maker refines preferences by gradually examining trade-offs using the Interactive

Methods. Visualization tools (such as parallel coordinate plots and 3D scatter plots of the Pareto front) can be used in conjunction with this. In order to evaluate each objective function in the optimization loop for a particular retrofit strategy **x**, a complete simulation of the structure's future must be run, including deterioration simulation, structural reliability analysis, and risk assessment using hazard models to calculate EAL and other risk metrics (Figure 16).

Figure 16. A 3D Pareto front plot for a retrofit problem, with axes: Lifecycle cost (NPV), lifecycle GWP, and system failure probability

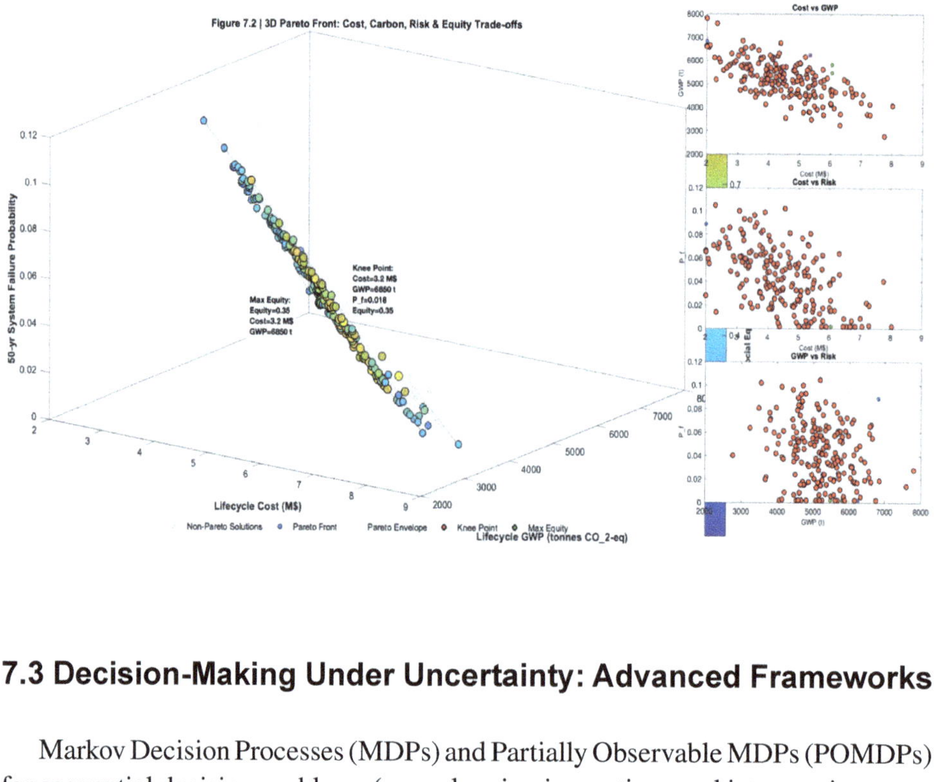

7.3 Decision-Making Under Uncertainty: Advanced Frameworks

Markov Decision Processes (MDPs) and Partially Observable MDPs (POMDPs) for sequential decision problems (e.g., planning inspections and interventions over time), an MDP provide a rigorous framework (Chang et al., 2020; Luo et al., 2025). An MDP is defined by the tuple$(\mathcal{S}, \mathcal{A}, \mathcal{P}, \mathcal{R}, \gamma)$. $\mathcal{S}$ is a set of states (e.g., condition ratings).$\mathcal{A}$ set of actions (inspect, repair, do-nothing). While the transition probability matrix, $P\left(s_{t+1}s_t, a_t\right)$ is defined by $\mathcal{P}$ derived from deterioration models. $\mathcal{R}$ is reward (or negative cost) function, $R\left(s_t, a_t, s_{t+1}\right)$ and γ is the discount factor.

The goal is to find a policy $\pi: \mathcal{S} \rightarrow \mathcal{A}$ that maximizes the expected total discounted reward. Dynamic programming (Value Iteration, Policy Iteration) solves this. In reality, the true state s_t (e.g., exact corrosion depth) is not fully known but estimated via inspections. This is a Partially Observable MDP (POMDP), where the decision-maker maintains a *belief state* (a probability distribution over $\mathcal{S}$). Solving POMDPs is computationally intense, but it perfectly captures the trade-off between the cost of information (inspection) and the value of better decisions.

7.3.1 Robust and Real Decision Making (RDM) and Adaptive Pathways

Traditional probabilistic optimization may not be sufficient for issues dominated by profound uncertainty (where the probability of future states is unknown or contested, as with long-term climate change). Instead of optimizing for a single best-guess future, Robust Decision Making (RDM) focuses on finding strategies that perform satisfactorily across a broad range of likely future situations. An adaptive pathway strategy arranges a series of steps based on how things turn out in the future. This results in adaptable, learning-based management strategies. Financial options theory is applied to infrastructure projects through Real Options Analysis (ROA). One "alternative" to achieve the long-term advantages of lower risk and longer life is a retrofit project. When investment timing is flexible and waiting for better information is worthwhile, ROA is useful. In contrast to typical NPV analysis, which assumes a set investment time, it estimates the value of this flexibility.

7.4 The Integrated Decision Support System (DSS)

The different parts—LCCA, PRA, LCA, MOO, and MDP—are combined to produce a unified DSS that is frequently based on a Digital Twin platform. A paradigm change from reactive, experience-based maintenance to proactive, evidence-based lifecycle stewardship is necessary for the sustainable management of aged infrastructure. In order to facilitate this shift, we offer a Decision Support System (DSS) based on a Digital Twin platform that combines Markov Decision Processes (MDP), Multi-Objective Optimization (MOO), Life-Cycle Assessment (LCA), Probabilistic Risk Assessment (PRA), and Life-Cycle Cost Analysis (LCCA) into a single analytical framework. This system creates a cohesive digital duplicate of an asset from heterogeneous data, including danger models, inspection reports, and structural health monitoring. The DSS enables decision-makers to thoroughly assess the long-term effects of retrofit solutions by balancing conflicting goals like cost, safety, and carbon footprint through a closed-loop process of simulation, optimization, and adaptive updating. The fundamental algorithmic process of this integrated DSS is described in the following table, which formalizes each step with its essential inputs, governing equations, and outputs. Infrastructure managers can provide transparent answers to important investment concerns thanks to this organized computational pipeline, which guarantees that scarce resources are distributed to optimize societal value across an asset's entire existence.

8. SYNTHESIS AND FUTURE VISION: TOWARD INTELLIGENT, REGENERATIVE INFRASTRUCTURE

This final portion presents the entire framework created in the earlier sections as a cohesive paradigm for the stewardship of reinforced concrete infrastructure in the twenty-first century. It reexamines the path "from cracks to resilience," highlighting the revolutionary change from disjointed, prescriptive methods to an integrated, performance-based, data-driven lifecycle management approach. The section concludes with a vision of infrastructure as adaptive, regenerative, and intelligent—a beneficial contribution to societal and planetary health—by examining the new technological, methodological, and policy frontiers that will define the next generation of rehabilitation practice.

Long-term structural performance optimization is the goal of the sophisticated, closed-loop, adaptive Integrated Performance-Based Lifecycle Management (IP-BLM) system. To guarantee resilience and sustainability throughout a structure's lifecycle, it uses a cyclical workflow that integrates advanced diagnostics, predictive modeling, intervention techniques, and ongoing learning.

The first stage, Quantitative Diagnosis, creates a comprehensive digital twin of the structure's current state using state-of-the-art non-destructive assessment (NDE) and structural health monitoring (SHM). This stage provides a strong basis for additional study by quantifying crucial elements such as material characteristics, current damage, and exposure levels.

In order to forecast the probabilistic course of deterioration and structural capability, the Stochastic Prognosis phase then integrates physics-based and data-driven models that are improved by Bayesian updating using real-time field data. This step provides probabilistic insights into the behavior of the structure under different settings by taking into account future operational and environmental challenges.

Performance-Based Design then transforms these projections into practical retrofit plans. The system creates interventions that methodically improve the structure's probabilistic performance in response to various hazards by defining clear performance objectives and using nonlinear analysis and fragility assessments, guaranteeing adherence to defined safety and functionality benchmarks.

The selection and application of cutting-edge, long-lasting, low-carbon materials and methods that follow the principles of the circular economy are the main goals of the Sustainable Intervention phase. This stage guarantees that the retrofits minimize their negative effects on the environment and increase the longevity of the structure.

The last stage, lifespan Optimization & Decision, integrates multi-objective optimization into a framework that strikes a balance between sustainability, risk, and lifespan cost. By choosing the most effective solutions that balance resilience, safety, environmental stewardship, and economic feasibility, this approach finds Pareto-optimal intervention techniques. Adaptive Management & Learning makes sure that the strategies are continuously monitored in order to complete the cycle. The digital twin and prognostic models get real-time data, which enables iterative management plan improvement. This dynamic feedback loop ensures continuous performance improvement and structural robustness by adjusting to changing situations. This framework is an interconnected system rather than a sequential checklist. Data is its lifeblood, and the digital twin is its core nervous system.

The intersection of systems ecology, artificial intelligence, and materials science holds the key to the future of rehabilitation. The possibilities will be redefined by three interconnected frontiers: Infrastructure as a regenerative, planetary system, autonomous, AI-guarded infrastructure, and next-generation living and self-healing materials.

As structural repair materials, Next-Generation Living and Self-Healing Materials will represent biological intelligence, evolving from passive to autonomous systems that can adapt dynamically. By incorporating microvascular networks that autonomously transfer healing materials, including polymers or minerals, to crack locations, Advanced Self-Healing Concrete will outperform traditional bacterial-based techniques. In order to actively seal cracks and restore structural integrity without the need for outside assistance, shape-memory polymers may also be used. In the meantime, the distinction between artificial and biological systems will become more hazy as Engineered Living Materials (ELMs), like fungal mycelium or genetically modified bacteria, develop to sense stress, heal damage, and even capture atmospheric carbon while in use. Programmable Cementitious Materials, a complementary breakthrough, will use additives at the nanoscale and microscale to dynamically alter hydration, pore structure, and mechanical characteristics in situ, allowing for precision-based repair with little waste and environmental impact. By significantly lowering the need for conventional, externally prompted maintenance operations and increasing the longevity of structures through internal resilience, these developments will redefine infrastructure as self-maintaining, dynamic systems.

By enabling real-time decision-making control over structural systems, the Autonomous, AI-Guarded Infrastructure for the development of digital twins into fully cognitive entities will turn infrastructure into self-regulating assets. The cognitive twin will use reinforcement learning (RL) to optimize control algorithms for adaptable features, such as bridges with tendons, dampers, or movable mass systems. The RL agent's objective might be to minimize a cost function J that penalizes excessive displacements, accelerations (for comfort), and control energy:

$$J = \sum_{k=0}^{N} \left[\mathbf{x}_k^T \mathbf{Q} \mathbf{x}_k + \mathbf{u}_k^T \mathbf{R} \mathbf{u}_k \right]$$

Where $\mathbf{x}_k$ is the state vector (displacements, velocities), $\mathbf{u}_k$ is the control force vector, and $\mathbf{Q}, \mathbf{R}$ are weighting matrices. The agent learns the optimal policy $\pi^*(\mathbf{x}_k)$ that maps states to control actions, dynamically adjusting control inputs to mitigate excessive displacements, accelerations, and energy consumption in response to sensed loads (traffic, wind, seismic precursors).

By offering interpretable diagnostics, Explainable AI (XAI) will go beyond autonomous control and close the gap between machine learning and human supervision. Using tools like SHAP (Shapley Additive exPlanations) or LIME (Local Interpretable Model-agnostic Explanations) to highlight significant sensor channels and temporal windows in damage detection, AI systems will not only identify anomalies from sensor data but also explain their reasoning—identifying degradation mechanisms and their locations. Engineers can trust and validate AI-driven insights because of this transparency. AI-driven algorithms, especially diffusion models and generative adversarial networks (GANs), will lead the way in the development of topology-optimized retrofit designs that optimize performance with the least amount of material consumption in the field of Generative Design for Retrofit. By incorporating limitations like manufacturability, architectural specifications, and sustainability goals (such as embodied carbon reduction), these generative models will create durable, biomimetic forms that maximize ecological integration and structural efficiency.

The ultimate goal of Infrastructure as a Regenerative, Planetary System is to achieve net-positive planetary impact through retrofits that actively restore resource cycles and ecological equilibrium. Building-integrated photovoltaics (BIPV), piezoelectric pavements, and wind energy systems will be incorporated into buildings through Energy-Positive Retrofits, turning them into net-energy producers. For such systems, the energy balance can be expressed as follows:

$$E_{net} = E_{produced} - E_{operational} - E_{embodied,retrofit}$$

A regenerative retrofit is defined by $E_{net} > 0$ over its lifecycle, ensuring long-term energy self-sufficiency. Retrofits will use Blue-Green Infrastructure (BGI), incorporating nature-based solutions (NBS) such as artificial wetlands for wastewater treatment, wildlife corridors, and green roofs and walls for stormwater management and urban cooling, in order to harmonize with ecosystems (Alves et al., 2022, 2023; Lousada, Cabezas, et al., 2022). Coupled hydrological and structural studies will be used to simulate these systems, resulting in hybrid ecological-engineering networks that preserve biodiversity and increase resilience. The idea that every retrofit must be recyclable and deconstructable is fundamental to Circularity by Design. In order to ensure that components may be recovered, repurposed, or safely reintroduced into biological cycles at the end of their useful lives, material passports—digital records that document composition, disassembly procedures, and embodied carbon—will be required. The circularity potential $\mathscr{C}$ can be quantified as:

$$\mathscr{C} = \frac{\sum_i M_i \cdot \left(R_{reuse,i} + R_{recycle,i}\right)}{M_{total}}$$

Where M_i is the mass of component i, and R are recovery factors.

Lastly, to examine the effects of retrofits on Earth systems, such as climate, biodiversity, freshwater consumption, and nitrogen cycles, Planetary Boundary Accounting will combine lifecycle assessment (LCA) with global stock-and-flow models. By ensuring that projects stay within safe operating bounds, the Planetary Resilience Quotient (PRQ) helps to avoid tipping points in vital planetary processes. Infrastructure will develop into a symbiotic, regenerative infrastructure ecosystem where natural and human systems live sustainably by incorporating these ideas.

9. CONCLUSION

From seeing reinforced concrete structures as deteriorating artifacts to seeing them as dynamic parts of intricate socio-ecological-technical systems is the journey from cracks to resilience. The roadmap for this change is provided by the integrated framework presented here, which is based on sophisticated sensing, predictive modeling, performance-based design, sustainable materials, and multi-objective lifecycle optimization.

The scientific concepts, engineering techniques, and decision-making instruments required to carry out this paradigm shift have been described in depth in this study. From the basic stress-strain behavior of corroded steel to the high-level algorithms of multi-objective optimization, and the practical use of a carbon-fiber wrap, it has progressed from the micro-scale physics of chloride ingress to the macro-scale system dynamics of network resilience.

Most of the time, the tools are available. The need is clear-cut and urgent due to aging stock, resource limitations, and climate change. Therefore, execution, education, teamwork, and steadfast ethical commitment are the tasks that lie ahead.

Final Recommendations

- **For Practitioners:** Start incorporating probabilistic thinking and quantitative condition assessment into regular practice. Promote and take part in pilot programs that test new contractual methods and performance-based retrofit.
- **For Researchers:** The development of standardized lifecycle sustainability and resilience measurements, long-term durability data for innovative materials, validation of AI models for diagnostics, and reliable techniques for decision-making under extreme uncertainty are all important knowledge gaps that need to be filled.
- **For Educators:** Redesign engineering programs to create engineers of tomorrow. Integrate systems thinking, data science, and sustainability ethics into the foundation of structural and civil engineering curriculum.
- **For Policymakers and Asset Owners:** Update procurement regulations, encourage performance-based code development, invest in common digital infrastructure (such as open asset registries), and strategically employ public funds to reduce risk and showcase creative solutions in order to create an enabling environment.

The international community of engineers, asset owners, legislators, and researchers can guarantee that our inherited infrastructure is not a liability for future generations but rather is transformed into a resilient, adaptable, and regenerative foundation for prosperous, sustainable societies in the century to come by embracing and advancing the framework described in “From Cracks to Resilience.” Today's flaws don't have to be tomorrow's failures; instead, they can serve as catalysts for a more intelligent, resilient, and sustainable built environment.

Note from Publisher

Correction Notice: A correction to this manuscript has been published
Reason: Missing Mathematical Notations
A technical error deleted the mathematical notations before being published. The publisher has retained all mathematical notations within the chapter.
Correction Date: August 28, 2026

REFERENCES

Abbas, M., & Shafiee, M. (2018). Structural health monitoring and determination of surface defects in large metallic structures using ultrasonic guided waves. *Sensors (Basel), 18*(11), 3958. https://www.mdpi.com/1424-8220/18/11/3958

Aboutaha, R. S., Engelhardt, M. D., Jirsa, J. O., & Kreger, M. E. (2010). Numerical study on the seismic retrofitting of reinforced concrete columns using rectified steel jackets. *ACI Special Publications, SP-160*(2), 59–72. DOI: 10.14359/1546

Ahmad, A. (2025). *Opposition-based NSGA-III framework for multi-objective optimization of retrofitting projects: Balancing time, cost, quality, energy, safety, and environmental impact.* Asian Journal of Civil Engineering. DOI: 10.1007/S42107-025-01568-5

Ahmed, H. (2025). Seismic retrofitting of RC buildings using a performance-based approach for risk resilience and vulnerability assessment. *Buildings (Basel, Switzerland), 15*(8), 1333. https://www.mdpi.com/2075-5309/15/8/1333

Ahmed, H., & Lee, Y. (2025). Performance-based seismic retrofitting and resilience assessment for RC buildings considering direct and indirect losses. *Advances in Structural Engineering, 28*(12), 2237–2259. DOI: 10.1177/13694332251325919

Ahmed, H., Shahzada, K., & Farooq, M. (2021). Performance-based seismic assessment of capacity enhancement of building infrastructure and its cost-benefit evaluation. International Journal of Disaster Risk Reduction. https://www.sciencedirect.com/science/article/pii/S2212420921003071

Ahmed, H., Zahedi, A., & Sanchez, L. (2022). Condition assessment of ASR-affected reinforced concrete columns after nearly 20 years in service. Construction and Building Materials. https://www.sciencedirect.com/science/article/pii/S0950061822022292

Akbar, G., Hayyat, S., Bilal, M., & Hameed, A. (2025). Sustainable innovations in ultra-high-performance concrete for smart and resilient structures: A survey. European Journal of Applied Science. *Engineering and Technology, 3*(5), 182–196. DOI: 10.59324/EJASET.2025.3(5).15

Alkhamis, M., Ghasemi, M., & Ghasemi, A. (2018). Performance-based optimum retrofitting design of concrete bridge piers. Jordan Journal of Civil Engineering. https://www.academia.edu/download/85695491/paper.pdf

Alves, R., Lousada, S., Cabezas, J., & Gómez, J. M. N. (2022). The importance of the strategic urban rehabilitation plan in the sustainable development of the municipality of Machico. *Sustainability, 14*(24), 16816. DOI: 10.3390/SU142416816

Alves, R., Lousada, S., Cabezas, J., & Gómez, J. M. N. (2023). Local housing strategy: Analysis of importance and implementation in Machico municipality, Madeira. *Land (Basel), 12*(9), 1778. DOI: 10.3390/LAND12091778

Anwar, G. A., Dong, Y., & Zhai, C. (2020). Performance-based probabilistic framework for seismic risk, resilience, and sustainability assessment of reinforced concrete structures. *Advances in Structural Engineering, 23*(7), 1454–1472. DOI: 10.1177/1369433219895363

Arya, A., Gunarani, G. I., Rathinakumar, V., Sharma, A., Pati, A. K., & Sethi, K. C. (2024). NSGA-III based optimization model for balancing time, cost, and quality in resource-constrained retrofitting projects. *Asian Journal of Civil Engineering, 25*(7), 5613–5625. DOI: 10.1007/S42107-024-01133-6

Asadi, E., da Silva, M., Antunes, C., & Dias, L. (2012). A multi-objective optimization model for building retrofit strategies using TRNSYS simulations, GenOpt and MATLAB. Building and Environment. https://www.sciencedirect.com/science/article/pii/S0360132312001217

Asmara, Y. P. (2023). Concrete reinforcement degradation and rehabilitation: Damages, corrosion and prevention. In *Engineering materials*. Springer. DOI: 10.1007/978-981-99-5933-4_8

Bao, Y., Chen, Z., Wei, S., Xu, Y., Tang, Z., & Li, H. (2019). The state of the art of data science and engineering in structural health monitoring. Engineering. https://www.sciencedirect.com/science/article/pii/S2095809918308026

Barski, M., Kędziora, P., & Muc, A. (2014). Structural health monitoring methods in machine design and operation. Archive of Mechanical Engineering. DOI: 10.2478/meceng-2014-0037

Bastidas-Arteaga, E., Sánchez-Silva, M., Chateauneuf, A., Ribas Silva, M., & Sánchez-Silva, M. (2008). Coupled reliability model of biodeterioration, chloride ingress and cracking for reinforced concrete structures. *Structural Safety, 30*(2), 110–129. DOI: 10.1016/j.strusafe.2006.09.001

Bhargava, K., Ghosh, A., Mori, Y., & Ramanujam, S. (2007). Corrosion-induced bond strength degradation in reinforced concrete: Analytical and empirical models. Nuclear Engineering and Design. https://www.sciencedirect.com/science/article/pii/S0029549307001021

Blagojević, B., Vasilevska, L., Anđelković, D., Bogojević, A., & Lousada, S. (2023). A framework for assessing nature-based urban stormwater management solutions: A preliminary spatial analysis approach applied to Southeast Serbia. *Water (Basel), 15*(20), 3604. DOI: 10.3390/W15203604

Bouchard, K. M. (2007). A performance-based approach to retrofitting unreinforced masonry structures for seismic loads [Master's thesis, Massachusetts Institute of Technology]. MIT DSpace. https://dspace.mit.edu/handle/1721.1/38944

Breysse, D., Larget, M., Sbartai, Z. M., Lataste, J. F., & Balayssac, J. P. (2011). Quality of NDT measurements and accuracy of concrete physical properties quantitative assessment. *European Journal of Environmental and Civil Engineering, 15*(4), 619–632. DOI: 10.1080/19648189.2011.9693351

Caterino, N., Iervolino, I., Manfredi, G., & Cosenza, E. (2008). Multi-criteria decision making for seismic retrofitting of RC structures. *Journal of Earthquake Engineering, 12*(4), 555–583. DOI: 10.1080/13632460701572872

Caterino, N., Iervolino, I., Manfredi, G., & Cosenza, E. (2009). Comparative analysis of multi-criteria decision-making methods for seismic structural retrofitting. *Computer-Aided Civil and Infrastructure Engineering, 24*(6), 432–445. DOI: 10.1111/J.1467-8667.2009.00599.X

Chang, C., & Habib, A. (2023). Life-cycle cost analysis of ultra-high-performance concrete in retrofitting techniques for ABC projects. *Third International Interactive Symposium on Ultra-High Performance Concrete*. DOI: 10.21838/UHPC.16694

Chang, M., Maguire, M., & Sun, Y. (2019). Stochastic modeling of bridge deterioration using classification tree and logistic regression. *Journal of Infrastructure Systems, 25*(1). Advance online publication. DOI: 10.1061/(ASCE)IS.1943-555X.0000466

Chang, S., Castro-Lacouture, D., & Yamagata, Y. (2020). Decision support for retrofitting building envelopes using multi-objective optimization under uncertainties. Journal of Building Engineering. https://www.sciencedirect.com/science/article/pii/S2352710219319825

Chen, G., Xu, P., Mi, G., & Zhu, J. (2019). Compressive strength and cracking of composite concrete in hot-humid environments based on microscopic quantitative analysis. Construction and Building Materials. https://www.sciencedirect.com/science/article/pii/S0950061819318392

Chen, H., Jiang, Y., & Miao, G. (2022). Structural performance deterioration of corroding reinforced concrete columns in marine environments. Ocean Engineering. https://www.sciencedirect.com/science/article/pii/S0029801822014718

Chen, H., & Nepal, J. (2015). Stochastic modelling and lifecycle performance assessment of bond strength of corroded reinforcement in concrete. *Structural Engineering and Mechanics, 54*(2). Advance online publication. DOI: 10.12989/sem.2015.54.2.000

Chisari, C., & Bedon, C. (2017). Performance-based design of FRP retrofitting of existing RC frames by means of multi-objective optimisation. *Bollettino di Geofisica Teorica ed Applicata, 58*(4), 377–394. DOI: 10.4430/bgta0202

Aktan, A. E. (2009). Structural health monitoring: What is the prescription? Mechanics Research Communications. https://www.sciencedirect.com/science/article/pii/S0093641308001122

Dalal, S., & Dalal, P. (2021). Strength, deformation and fragility assessment of reinforced concrete moment resisting frame designed by force-based design and performance-based plastic design. Structures. https://www.sciencedirect.com/science/article/pii/S2352012420306706

Davoudi, R., Miller, G., & Kutz, J. (2018). Data-driven vision-based inspection for reinforced concrete beams and slabs: Quantitative damage and load estimation. Automation in Construction. https://www.sciencedirect.com/science/article/pii/S0926580518303686

Deng, K., Cui, Y., Deng, Q., Liu, R., Chen, Z., & Wang, S. (2025). Multi-objective optimization of urban residential envelope structures in cold regions of China based on performance and economic efficiency. *Buildings (Basel, Switzerland), 15*(13), 2365. https://www.mdpi.com/2075-5309/15/13/2365

Dutta, C., Kumar, J., Das, T., & Saha, S. (2021). Recent advancements in the development of sensors for structural health monitoring at high-temperature environment: A review. IEEE Sensors Journal. https://ieeexplore.ieee.org/abstract/document/9415648/

Ellingwood, B. R. (2005). Risk-informed condition assessment of civil infrastructure: State of practice and research issues. *Structure and Infrastructure Engineering, 1*(1), 7–18. DOI: 10.1080/15732470412331289341

Farahani, A., Naderpour, H., & Kheyroddin, G. (2023). Developing a fuzzy expert system for diagnosing chemical deterioration in reinforced concrete structures. *Applied Sciences (Basel, Switzerland), 13*(18), 10372. https://www.mdpi.com/2076-3417/13/18/10372

Faris, N., Zayed, T., & Faris, A. (2026). Review of condition rating and deterioration modeling approaches for concrete bridges. *Buildings (Basel, Switzerland), 15*(2), 219. https://www.mdpi.com/2075-5309/15/2/219

Farrar, C., & Lieven, N. (2007). Damage prognosis: The future of structural health monitoring. Philosophical Transactions of the Royal Society A: Mathematical, Physical and Engineering Sciences, 365(1851), 623–632. DOI: 10.1098/RSTA.2006.1927

Farrar, C., & Worden, K. (2012). Structural health monitoring: A machine learning perspective. Wiley. https://books.google.com/books?hl=en&lr=&id=2w_sp6lersUC

Fawad, M., Salamak, M., Poprawa, G., Koris, K., Jasinski, M., Lazinski, P., Piotrowski, D., Hasnain, M., & Gerges, M. (2023). Automation of structural health monitoring system of a bridge using BIMification approach and BIM-based finite element model development. *Scientific Reports*, *13*, 13215. DOI: 10.1038/s41598-023-40355-7

Fernandes, I., Broekmans, M. A. T. M., & Noronha, F. (2009). Petrography and geochemical analysis for the forensic assessment of concrete damage. In *Criminal and environmental soil forensics* (pp. 163–180). Springer. DOI: 10.1007/978-1-4020-9204-6_11

Fragiadakis, M., & Papadrakakis, M. (2008). Performance-based optimum seismic design of reinforced concrete structures. *Earthquake Engineering & Structural Dynamics*, *37*(6), 825–844. DOI: 10.1002/EQE.786

Fu, C., & Li, Y. (2023). Steel–concrete bond deterioration in reinforced concrete tension members due to primary cracks. Structures. https://www.sciencedirect.com/science/article/pii/S2352012423009852

Gardner, P., Liu, X., & Worden, K. (2020). On the application of domain adaptation in structural health monitoring. Mechanical Systems and Signal Processing. https://www.sciencedirect.com/science/article/pii/S088832701930771X

Grattan-Bellew, P. E. (2001). Petrographic and technological methods for evaluation of concrete aggregates. In *Handbook of analytical techniques in concrete science and technology* (pp. 63–104). Elsevier. DOI: 10.1016/B978-081551437-4.50005-9

Grussing, M. N., Liu, L. Y., Uzarski, D. R., El-Rayes, K., & El-Gohary, N. (2016). Discrete Markov approach for building component condition, reliability, and service-life prediction modeling. *Journal of Performance of Constructed Facilities*, *30*(5). Advance online publication. DOI: 10.1061/(ASCE)CF.1943-5509.0000865

Gu, H., & Liu, Q. (2022). Updating deterioration models of reinforced concrete structures in carbonation environment using in-situ inspection data. *Structure and Infrastructure Engineering*, *18*(2), 266–277. DOI: 10.1080/15732479.2020.1841246

Gucunski, N., La, H., Dinh, K., & Kee, M. K. (2023). Advancing condition assessment of reinforced concrete bridge elements through automation, visualization, and improved interpretation of multi-NDE technology. Materials Evaluation. https://www.researchgate.net/profile/Mahsa-Sanei/publication/368352480_Automated_Geometric_Quality_Inspection_of_Rebar_Layout_using_RGBD_Data/links/64d996af66f0e0067d9316af/Automated-Geometric-Quality-Inspection-of-Rebar-Layout-using-RGBD-Data.pdf#page=58

Güemes, A., Fernandez-Lopez, A., & Pozo, A. R. (2020). Structural health monitoring for advanced composite structures: A review. *Journal of Composites Science, 4*(1), 13. https://www.mdpi.com/2504-477X/4/1/13

Guo, H., Dong, Y., & Gu, X. (2019). Durability assessment of reinforced concrete structures considering global warming: A performance-based engineering and experimental approach. *Construction & Building Materials*. Advance online publication. DOI: 10.1016/j.conbuildmat.2019.117251

Hanifehzadeh, M., Aryan, H., Gencturk, B., & Alipour, A. (2021). Structural response of steel jacket-UHPC retrofitted reinforced concrete columns under blast loading. *Materials (Basel), 14*(6), 1521. https://www.mdpi.com/1996-1944/14/6/1521

Hasan, M. (2024). Deterioration prediction of concrete bridge components using artificial intelligence and stochastic methods [Doctoral dissertation, RMIT University]. RMIT Research Repository. https://research-repository.rmit.edu.au/articles/thesis/Deterioration_prediction_of_concrete_bridge_components_using_artificial_intelligence_and_stochastic_methods/27591186

Ingham, J. P. (2009). Application of petrographic examination techniques to the assessment of fire-damaged concrete and masonry structures. Materials Characterization. https://www.sciencedirect.com/science/article/pii/S1044580308003173

Jafari, A., & Valentin, V. (2018). Selection of optimization objectives for decision-making in building energy retrofits. Building and Environment. https://www.sciencedirect.com/science/article/pii/S0360132317305954

Jia, G., & Gardoni, P. (2019). Stochastic life-cycle analysis: Renewal-theory life-cycle analysis with state-dependent deterioration stochastic models. *Structure and Infrastructure Engineering, 15*(8), 1001–1014. DOI: 10.1080/15732479.2019.1590424

Jia, G., Gardoni, P., Trejo, D., & Masad, V. (2021). Stochastic modeling of deterioration and time-variant performance of reinforced concrete structures under joint effects of earthquakes, corrosion, and ASR. *Journal of Structural Engineering, 147*(2). Advance online publication. DOI: 10.1061/(ASCE)ST.1943-541X.0002884

Jia, G., Tabandeh, A., & Gardoni, P. (2017). Life-cycle analysis of engineering systems: Modeling deterioration, instantaneous reliability, and resilience. In *Life-cycle of engineering systems* (pp. 465–494). Springer. DOI: 10.1007/978-3-319-52425-2_20

Karthik, M., Mander, J., & Hurlebaus, S. (2020). Simulating behaviour of large reinforced concrete beam-column joints subject to ASR/DEF deterioration and influence of corrosion. Engineering Structures. https://www.sciencedirect.com/science/article/pii/S0141029620303606

Kaya, Y., & Safak, E. (2015). Real-time analysis and interpretation of continuous data from structural health monitoring systems. *Bulletin of Earthquake Engineering, 13*(3), 917–934. DOI: 10.1007/S10518-014-9642-9

Kennedy, D., Habel, K., & FitzGerald, G. (2023). Ultra-high-performance concrete column jacket retrofit for the Mission Bridge. *Third International Interactive Symposium on Ultra-High Performance Concrete.* DOI: 10.21838/UHPC.16730

Kim, S., Choi, I., Lee, K., & Song, J. (2025). A rapid decision-making process of retrofit scheme for seismically vulnerable reinforced concrete school buildings using simple information. Structures. https://www.sciencedirect.com/science/article/pii/S2352012425009889

Köliö, A., Pakkala, T., Annila, P., & Lahdensivu, J. (2014). Possibilities to validate design models for corrosion in carbonated concrete using condition assessment data. Engineering Structures. https://www.sciencedirect.com/science/article/pii/S0141029614003940

Lei, Y., & Zhang, Z. (2013). Review of physical based monitoring techniques for condition assessment of corrosion in reinforced concrete. *Mathematical Problems in Engineering, 2013*. Advance online publication. DOI: 10.1155/2013/953930

Ligabue, V., Pampanin, S., & Savoia, M. (2018). Seismic performance of alternative risk-reduction retrofit strategies to support decision making. *Bulletin of Earthquake Engineering, 16*(7), 3001–3030. DOI: 10.1007/S10518-017-0291-7

Liu, J., Hang, M., Jiang, M., Song, H., & Zhang, X. (2024). Performance and mechanism of triethanolamine-triisopropanolamine corrosion inhibitor to deterioration of reinforced concrete under the coupling effect of freeze-thaw. Journal of Building Engineering. https://www.sciencedirect.com/science/article/pii/S2352710224026342

Loures, L., Castanho, R. A., Gómez, J. M. N., Lousada, S. A., Fernández-Pozo, L., Cabezas, J., & Loures, A. L. C. V. (2019). Impactos socioculturais da cooperação transfronteiriça no espaço europeu. Fronteira: Journal of Social. *Technological and Environmental Science*, *8*(3), 292–312. DOI: 10.21664/2238-8869.2019V8I3. P292-312

Lousada, S., Cabezas, J., Castanho, R. A., & Gómez, J. M. N. (2022). Land-use changes in insular urban territories: A retrospective analysis from 1990 to 2018. The case of Madeira Island—Ribeira Brava. *Sustainability*, *14*(24), 16839. DOI: 10.3390/SU142416839

Lousada, S., Gonçalves, L., & Atmaca, A. (2022). Hydraulic planning in insular urban territories: The case of Madeira Island—São Vicente. *Water (Basel)*, *14*(1), 112. DOI: 10.3390/W14010112

Luo, L., Wei, H., Lin, Z., Wu, J., Wang, W., & Sun, Y. (2025). Multi-objective optimal energy-efficient retrofit determination using hybrid urban building energy model: Considering uncertainties between models. *Building Simulation*, *18*(1), 183–206. DOI: 10.1007/S12273-024-1206-6

Mamin, A. N., Avdeev, K. V., Zenin, S. A., & Bammatov, A. A. (2025). Failure mechanisms of reinforced concrete bridges. *Concrete and Reinforced Concrete*, *627*(2), 5–15. DOI: 10.37538/0005-9889-2025-2(627)-5-15

Masoumi, F., Akgül, F., & Mehrabzadeh, A. (2013). Development of an optimised condition assessment plan for common reinforced concrete bridges in Estonia. *IACSIT International Journal of Engineering and Technology*, *5*, 708–711. DOI: 10.7763/IJET.2013.V5.647

Medina, P., & Gomez, J. (2022). Reinforced concrete long-term deterioration prediction for the implementation of a bridge management system. Materials Today: Proceedings. https://www.sciencedirect.com/science/article/pii/S2214785322005995

Miano, A. (2017). Performance based assessment and retrofit for existing RC structures [Doctoral dissertation]. https://tesidottorato.depositolegale.it/handle/20.500.14242/142579

Miano, A., Sezen, H., Jalayer, F., & Prota, A. (2017). Performance based comparison of different retrofit methods for reinforced concrete structures. In *Proceedings of the 6th International Conference on Computational Methods in Structural Dynamics and Earthquake Engineering* (Vol. 1, pp. 1515–1535). DOI: 10.7712/120117.5510.17150

Micic, T., Stojic, D., & Velimirovic, N. (2013). Stochastic process model for timber-concrete composite beam deterioration. In Safety, reliability, risk and life-cycle performance of structures and infrastructures: Proceedings of the 11th International Conference on Structural Safety and Reliability (pp. 5127–5133). CRC Press. DOI: 10.1201/B16387-745

Mitra, G., Jain, K., & Bhattacharjee, B. (2010). Condition assessment of corrosion-distressed reinforced concrete buildings using fuzzy logic. *Journal of Performance of Constructed Facilities*, *24*(6), 562–570. DOI: 10.1061/(ASCE)CF.1943-5509.0000137

Mori, Y. (1993a). Methodology for reliability based condition assessment: Application to concrete structures in nuclear plants. https://books.google.com/books?hl=en&lr=&id=t9dIT8kTGZkC

Mori, Y. (1993b). Reliability-based condition assessment and life prediction of concrete structures [Doctoral dissertation]. ProQuest Dissertations & Theses. https://search.proquest.com/openview/d9f3f6e7e18337612fd70b5089cfce17/1

Moufti, S., Zayed, T., & Dabous, S. (2014). Defect-based condition assessment of concrete bridges: Fuzzy hierarchical evidential reasoning approach. *Transportation Research Record: Journal of the Transportation Research Board*, *2431*(1), 88–96. DOI: 10.3141/2431-12

Mustapha, F., Aris, K., & Wardi, N. (2012). Structural health monitoring for composite structure undergoing tensile and thermal testing. Journal of Vibroengineering. https://search.ebscohost.com/login.aspx?direct=true&profile=ehost&scope=site&authtype=crawler&jrnl=13928716&AN=79998236

Nadeem, A., Memon, S., & Lo, T. Y. (2013). Qualitative and quantitative analysis and identification of flaws in the microstructure of fly ash and metakaolin blended high performance concrete after exposure to elevated temperatures. Construction and Building Materials. https://www.sciencedirect.com/science/article/pii/S0950061812007246

Nasr, A., Honfi, D., & Larsson Ivanov, O. (2022). Probabilistic analysis of climate change impact on chloride-induced deterioration of reinforced concrete considering Nordic climate. *Journal of Infrastructure Preservation and Resilience*, *3*(1). Advance online publication. DOI: 10.1186/S43065-022-00053-6

Nguanthaisong, M., Terakulsatit, B., Setwong, W., & Phoovasawat, C. (2025). Deterioration of reinforced concrete structures due to chloride exposure in water: A case study, Thailand. *International Journal of GEOMATE*, *28*(126), 1–7. DOI: 10.21660/2025.126.4713

Obunguta, F., & Matsushima, K. (2022). Optimal pavement management strategy development with a stochastic model and its practical application to Ugandan national roads. *The International Journal of Pavement Engineering*, *23*(7), 2405–2419. DOI: 10.1080/10298436.2020.1857759

Omar, T., & Nehdi, M. (2018). Condition assessment of reinforced concrete bridges: Current practice and research challenges. *Infrastructures*, *3*(3), 36. https://www.mdpi.com/2412-3811/3/3/36

Ostachowicz, W., & Güemes, A. (Eds.). (2013). *New trends in structural health monitoring* (Vol. 542). Springer Vienna. DOI: 10.1007/978-3-7091-1390-5

Ou, J., & Li, H. (2010). Structural health monitoring in mainland China: Review and future trends. *Structural Health Monitoring*, *9*(3), 219–231. DOI: 10.1177/1475921710365269

Padalu, P., & Singh, M. (2024). An overview of performance-based seismic design framework for reinforced concrete frame buildings. *Iranian Journal of Science and Technology. Transaction of Civil Engineering*, *48*(2), 635–667. DOI: 10.1007/S40996-023-01217-4

Palacio, J. P., Pablo, A. A., Tingatinga, E. J., & Mata, W. L. (2013). Structural performance assessment and retrofit of reinforced concrete buildings under seismic loads. In *Proceedings of the Structures Congress 2013* (pp. 2044–2055). ASCE. DOI: 10.1061/9780784412848.179

Pampanin, S., & Archer, U. (2011). Performance-based seismic retrofit of existing reinforced concrete frame buildings using fibre-reinforced polymers: Challenges and solutions. *Structural Engineering International*, *21*(3), 260–270. DOI: 10.2749/101686611X13049248220041

Panagiotidou, M., Aye, L., & Rismanchi, B. (2021). Optimisation of multi-residential building retrofit, cost-optimal and net-zero emission targets. Energy and Buildings. https://www.sciencedirect.com/science/article/pii/S0378778821006691

Prakashvel, J., Umarani, C., & Sathiskumar, K. (2023). State-of-the-art review on performance-based retrofitting of structures. In *Lecture Notes in Civil Engineering* (Vol. 330, pp. 289–302). Springer. DOI: 10.1007/978-981-99-1604-7_22

Robles, K., Yee, J., & Kee, S. (2022). Electrical resistivity measurements for non-destructive evaluation of chloride-induced deterioration of reinforced concrete: A review. *Materials (Basel)*, *15*(8), 2725. https://www.mdpi.com/1996-1944/15/8/2725

Rodrigues, R., Gaboreau, S., Gance, J., Ignatiadis, I., & Betelu, S. (2021). Reinforced concrete structures: A review of corrosion mechanisms and advances in electrical methods for corrosion monitoring. *Construction & Building Materials*. Advance online publication. DOI: 10.1016/j.conbuildmat.2020.121240

Sadeghi, F., Zhu, X., Li, J., & Rashidi, M. (2021). Integrated remote sensing technologies for condition assessment of concrete bridges. *Remote Sensing*, *13*(17), 3377. DOI: 10.3390/RS13173377

Sajid, S., & Çakmak, L. (2019). Impulse response test for condition assessment of concrete: A review. Construction and Building Materials. https://www.sciencedirect.com/science/article/pii/S0950061819306658

Salmeron, M., Criner, N., Zhang, X., Dyke, S. J., Ramirez, J. A., Wogen, B. E., & Rearick, A. (2022). Predictive analytics for holistic lifecycle modeling of concrete bridge decks with construction defects. *Journal of Bridge Engineering*, *29*(9). Advance online publication. DOI: 10.1061/JBENF2.BEENG-6636

Sanchez, L., Fournier, B., Drimalas, T., Bastien, J., Mitchell, D., & Noel, M. (2020). Semi-quantitative condition assessment of concrete distress through the damage rating index. *Cement*, *1–2*, 100001. DOI: 10.1016/J.CEMENT.2020.100001

Sberna, A., Deb, A., Di Trapani, F., & Calvi, J. (2025). Reliability-based seismic retrofitting design methodology for non-ductile reinforced concrete frame structures. Probabilistic Engineering Mechanics. https://www.sciencedirect.com/science/article/pii/S0266892025000906

Schaarschmidt, M., Westerkamp, C., Hennewig, A., Pieper, D., Speckmann, H., & Bisle, W. (2015). Structural health monitoring: Overview on technologies under development. In Structural Health Monitoring 2015 (Vol. 2, pp. 1181–1188). DOI: 10.12783/SHM2015/148

Seo, J., Hu, J., & Lee, J. (2016). Summary review of structural health monitoring applications for highway bridges. *Journal of Performance of Constructed Facilities*, *30*(4). Advance online publication. DOI: 10.1061/(ASCE)CF.1943-5509.0000824

Sethi, K. C., Prajapati, U., Parihar, A., Gupta, C., Shrivastava, G., & Sharma, K. (2024). Development of optimization model for balancing time, cost, and environmental impact in retrofitting projects with NSGA-III. *Asian Journal of Civil Engineering*, *25*(7), 5123–5136. DOI: 10.1007/S42107-024-01102-Z

Shao, C., Gu, C., Meng, Z., & Han, Y. (2019). A data-driven approach based on multivariate copulas for quantitative risk assessment of concrete dam. *Journal of Marine Science and Engineering*, *7*(10), 353. https://www.mdpi.com/2077-1312/7/10/353

Sohn, H., Farrar, C., Hemez, F., Shunk, D., Stinemates, D., & Nadler, B. (1999). A review of structural health monitoring literature: 1996–2001. IEE Colloquium on Condition Monitoring Machinery, External Structures and Health, 34, 3–6. DOI: 10.1049/IC:19990185

Stefanidou, S., & Kappos, A. (2021). Fragility-informed selection of bridge retrofit scheme based on performance criteria. Engineering Structures. https://www.sciencedirect.com/science/article/pii/S0141029621001267

Tan, S., Poon, J., Chan, R., & Chia, D. (2020). Retrofitting of reinforced concrete beam-column via steel jackets against close-in detonation. *International Journal of Engineering*, *33*(5), 770–783. DOI: 10.5829/ije.2020.33.05b.08

Tapia, C., & Padgett, J. (2016). Multi-objective optimisation of bridge retrofit and post-event repair selection to enhance sustainability. *Structure and Infrastructure Engineering*, *12*(1), 93–107. DOI: 10.1080/15732479.2014.995676

Tarussov, A., Vandry, M., & De La Haza, A. (2013). Condition assessment of concrete structures using a new analysis method: Ground-penetrating radar computer-assisted visual interpretation. Construction and Building Materials. https://www.sciencedirect.com/science/article/pii/S0950061812003819

Thériault, F., Noël, M., & Sanchez, L. (2022). Simplified approach for quantitative inspections of concrete structures using digital image correlation. Engineering Structures. https://www.sciencedirect.com/science/article/pii/S0141029621018113

Trach, R., Moshynskyi, V., Chernyshev, D., & Bushuyev, O. (2022). Modeling the quantitative assessment of the condition of bridge components made of reinforced concrete using ANN. *Sustainability*, *14*(23), 15779. https://www.mdpi.com/2071-1050/14/23/15779

Vagdatli, T., Petroutsatou, K., Panetsos, P., Barmpa, Z., & Fragkakis, N. (2024). Probabilistic pre-estimation of life-cycle costs of road bridges using dynamic Bayesian networks. *Structure and Infrastructure Engineering*. Advance online publication. DOI: 10.1080/15732479.2024.2396603

Vaghefi, K., Ahlborn, T. M., Harris, D. K., & Brooks, C. N. (2015). Combined imaging technologies for concrete bridge deck condition assessment. *Journal of Performance of Constructed Facilities*, *29*(4). Advance online publication. DOI: 10.1061/(ASCE)CF.1943-5509.0000465

Varnavina, A., Sneed, L., Khamzin, A., & Torgashov, E. (2017). An attempt to describe a relationship between concrete deterioration quantities and bridge deck condition assessment techniques. Journal of Applied Geophysics. https://www.sciencedirect.com/science/article/pii/S0926985117304883

Vasudevan, G., & Tait, D. (2023). Bayesian model for sustainable, resilient, and cost-effective concrete mixtures. *Sustainable and Resilient Infrastructure*, *8*(3), 325–339. DOI: 10.1080/23789689.2023.2188348

Vulevic, A., Castanho, R. A., Gómez, J. M. N., Lousada, S., Loures, L., Cabezas, J., & Fernández-Pozo, L. (2021). Cross-border cooperation and adaptation to climate change in Western Balkans Danube Area. In *Advances in spatial science* (pp. 289–308). Springer. DOI: 10.1007/978-3-030-72124-4_14

Walker, H., Lane, D., & Stutzman, P. (2006). Petrographic methods of examining hardened concrete: A petrographic manual. https://rosap.ntl.bts.gov/view/dot/38050

Wang, X., Cui, F., Cui, L., & Jiang, D. (2023). Research on a multi-objective optimization design for the durability of high-performance fiber-reinforced concrete based on a hybrid algorithm. *Coatings*, *13*(12), 2054. https://www.mdpi.com/2079-6412/13/12/2054

Wang, X., Nguyen, M., Foliente, G., & Ye, L. (2007). An approach to modelling concrete bridge condition deterioration using a statistical causal relationship based on inspection data. *Structure and Infrastructure Engineering*, *3*(1), 3–15. DOI: 10.1080/15732470500103682

Yamamoto, N. (2009). Condition indicators for the assessment of local and spatial deterioration of concrete structures. In *Condition assessment of aged structures*. CRC Press. DOI: 10.1201/9781439832554.CH6

Yu, J., Liu, D., & Zhang, Z. (2022). Prediction analysis of recycled aggregate concrete with ceramic waste powder under freeze-thaw conditions based on impact-echo method and Grey-Markov model. *Frontiers in Materials*, *9*, 1060294. DOI: 10.3389/FMATS.2022.1060294/FULL

Zameeruddin, M., & Sangle, K. K. (2021). Performance-based seismic assessment of reinforced concrete moment resisting frame. Journal of King Saud University—Engineering Sciences. https://www.sciencedirect.com/science/article/pii/S1018363920302245

Zamen, S., & Dackermann, U. (2020). Fractal analysis of nonlinear ultrasonic waves in phase-space domain as a quantitative method for damage assessment of concrete structures. NDT & E International. https://www.sciencedirect.com/science/article/pii/S0963869519306334

Zhang, W., & Nicholson, C. (2016). A multi-objective optimization model for retrofit strategies to mitigate direct economic loss and population dislocation. *Sustainable and Resilient Infrastructure*, *1*(3–4), 123–136. DOI: 10.1080/23789689.2016.1254995

Zirps, M., Lepech, M., Savage, S., Michel, A., Stang, H., & Geiker, M. (2020). Probabilistic design and management of sustainable concrete infrastructure using multi-physics service life models. *Frontiers in Built Environment*, *6*. Advance online publication. DOI: 10.3389/FBUIL.2020.00072

Zou, X., Teng, J., De Lorenzis, L., & Xia, S. (2007). Optimal performance-based design of FRP jackets for seismic retrofit of reinforced concrete frames. Composites Part B: Engineering. https://www.sciencedirect.com/science/article/pii/S1359836806001673

Zub, C.-I., Stratan, A., Dogariu, A., Vulcu, C., & Dubina, D. (2018). Performance based seismic assessment of buildings for evaluation of retrofitting systems efficiency. In *Springer natural hazards* (pp. 373–387). Springer. DOI: 10.1007/978-3-319-74724-8_25

Chapter 10
Planning and Implementation of Hydrologically Sensitive Drainage Systems That Address Flooding and Water Quality in Urban Environments

Harshit K. Sheladiya
http://orcid.org/0009-0008-1234-9580
vertical-earth, USA

ABSTRACT

This study looks at how well SUDS work to cut down on flood risks and improve water quality in big US cities. The research used a Comparative Case Analysis. It pulled together performance numbers and opinions from stakeholders in five key cities: Philadelphia, Washington, D.C., Portland, Seattle, and New York City. The findings prove that SUDS dramatically lowers the number of floods. Across all five cities, flood frequency dropped by about 56% on average. Relevant statistics confirm that this success mostly comes from the systems controlling the amount of water runoff (R2 = 0.85). At the same time, SUDS methods greatly improved water quality. They were excellent at removing sediments (up to 73%) and heavy metals (up to 67%), which shows they successfully serve a dual purpose. Although experts had great faith in SUDS (about 75% to 80% saw them positively), the main obsta-

DOI: 10.4018/979-8-3373-6167-3.ch010

cles to using them everywhere are still high upfront costs, the burden of long-term upkeep, and weak governance systems. The results emphasize that SUDS are vital for making cities stronger and more adaptable.

1. INTRODUCTION

Flooding in cities is a huge and disastrous problem worldwide, stemming from complex and uncertain events. This flooding happens because of unusual water patterns, massive city growth, climate change making storms worse, and old or failing infrastructure (Mugume & Butler, 2017; Liu et al., 2016). The results are severe. They include widespread property damage, destroyed essential services, and loss of life. These consequences seriously restrict local economies and make cities less stable overall (Mugume & Butler, 2017; Liu et al., 2016).

People are paying more attention to urban flood risk management. This is a direct reaction to two increasing problems: worse environmental dangers and greater fragility of society (Miller & Hutchins, 2017). Flood risk is more than just measuring the water's threat. It is truly shaped by combining the danger itself, the property or assets exposed to that danger, and how easily the system and its people can be harmed (Joyce et al., 2018). The uncertainty associated with factors such as future climate trends and the speed of anthropogenic landscape alteration necessitates a flexible and combined approach to mitigation, moving past old-fashioned engineering that only reacts after problems occurred.

This expert review focuses on pluvial flooding. This happens when so much rain falls so fast that the city's stormwater drains cannot handle the runoff (Pérez-Sánchez et al., 2017; Mugume & Butler, 2017). Pluvial flooding is different from river (fluvial) or coastal floods. Still, experts agree that all water systems in a city are linked. This means planners today must be able to model systems with dual drainage. Historically, cities depended on old-style construction like levees, floodwalls, and digging out river beds. These methods are now known to harm the environment and might not work long-term (Sæthre et al., 2021). Studies show that these measures often just push the flood problem to areas further downriver. They can also violate laws meant to protect nature, especially for delicate places like salmon river habitats (Sæthre et al., 2021). This major flaw in standard engineering highlights an urgent need to switch to flexible, nature-based ways to build resilience.

Managing a city's water well means having a complete view of how to handle stormwater. This includes both technical tools and the rules and structures designed to control how much runoff there is and how clean it is (Miller et al., 2021). Controlling stormwater is essential to keep a city running smoothly. Its main role is to lessen the harmful effects of paved-over surfaces, specifically widespread flooding and water contamination (Miller et al., 2021). Figure 1 shows the green stormwater infrastructure of an urban area.

Figure 1. Green stormwater infrastructure

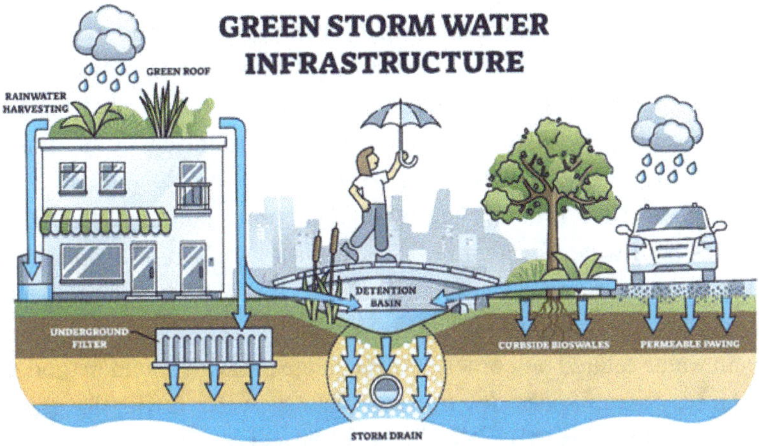

Cities must make a necessary, fundamental change to handle floods that are getting more intense (Torgersen et al., 2014). This means stopping the usual practice of quickly moving water out. Instead, cities must adopt combined, sustainable methods known as Sustainable Urban Drainage Systems (SUDS) (Torgersen et al., 2014). Whether this new system works depends on grasping the complicated way water moves and creates flood risk. As a result, forecasting water flow accurately and analyzing vulnerability requires using computer and numerical models. This makes advanced water modeling a required first step for all modern city planning (Chen et al., 2016). The need for detailed modeling grows as the effects of city development become more complex.

The best way to lower flood risk is to create and put into action plans that use both physical structures and non-physical policies (Lee et al., 2016; Vogel et al., 2023). These solutions must be carefully blended and applied as a whole across all related government and industry areas. This is the only way to get the complete, effective flood management needed for cities to stay strong over time (Lee et al., 2016; Vogel et al., 2023).

2. BACKGROUND OF SUSTAINABLE URBAN DRAINAGE AND URBAN FLOODING

2.1 The Impact of Urbanization on the Urban Water Cycle

Cities are growing fast; this expansion often happens faster than the population itself. As cities spread, the natural land changes completely. Builders remove native plants and cover huge areas with paved surfaces like roads, sidewalks, and roofs (Miller & Hutchins, 2017; Pérez-Sánchez et al., 2017). These sealed surfaces mess up the natural water cycle. Water cannot soak into the ground easily anymore (Miller & Hutchins, 2017; Pérez-Sánchez et al., 2017).

In nature, water flow is controlled by plants, porous soil, and dips in the land. These features hold water for a while, allowing it to soak in or evaporate (Ramos et al., 2017). Urbanization removes these natural controls. They are replaced by an efficient, but damaging, system of hard pavement.

The main water-related outcomes of this change are widely understood. For one, blocking the way water soaks down means far less water refills the groundwater supply (Miller & Hutchins, 2017; EPA, 2025). Second, stormwater rushes over hard surfaces. This instantly and greatly boosts both the water's volume and its speed (Pérez-Sánchez et al., 2017; Miller & Hutchins, 2017). This fast flow dramatically cuts the time needed for water to collect, called the time of concentration (t_c). Computer models of changing city layouts show this speeding up can cut t_c by up to 11 minutes. This 8% faster flow time intensifies the biggest flow event (Pérez-Sánchez et al., 2017).

This disruption to the natural water cycle directly puts great strain on city infrastructure. A look at projected city growth reveals how severe this stress can be. One hydraulic simulation showed that moving from the current state to a fully developed profile caused the peak water flow to jump by a maximum of $0.67 m^3/s$ (Pérez-Sánchez et al., 2017). This 71% jump in flow shows the tremendous water volume strain put on the drainage networks (Pérez-Sánchez et al., 2017).

The result of more runoff combined with a faster concentration time is that today's drainage systems are always strained. Three factors cause this stress: the large amount of water cities create; the increasing number of severe storms because of climate change; and old pipes that fail because they were not updated while the city grew (Mugume & Butler, 2017; Liu et al., 2022).

The worst danger is massive flooding. The most extreme simulated test, based on widespread city growth (Scenario 2), showed peak flow volumes spiking up to 455% compared to the current state (Scenario 1). This huge volume swamps the entire system. It leads to widespread inundation and forces many drainage pipes to handle extreme pressure (Pérez-Sánchez et al., 2017).

Hard surfaces do more than just speed up water flow; they are effective ways to deliver pollution. They act as direct paths, quickly moving contaminants picked up from the city into nearby streams (Zhou et al., 2024; EPA, 2025). Stormwater runoff is therefore loaded with high amounts of many common pollutants, such as sediment, nutrients, pesticides, metals from wear, organic chemicals, and oil and grease (EPA, 2025).

Some city surfaces have a much bigger effect than others. Roofs often cover a large share of a city's paved area. Studies indicate that roofs impact the 'first flush' which is the initial surge of heavily contaminated runoff more strongly than roads (Zhou et al., 2024). This finding means managing rooftops must be a specific priority in efforts to lessen pollution.

The consequences reach ecological health and public safety. Because floodwater contains pathogens and contaminants, flooding poses a serious public health danger. It is often as risky as exposure to sewer water (Nguyen et al., 2017; Mugume & Butler, 2017).

Chemical contamination is not the only issue. Hard surfaces soak up sun, and that heat is transferred to the flowing stormwater, creating a secondary problem: thermal pollution (EPA, 2025). This warmer water greatly harms aquatic ecosystems. It lowers the dissolved oxygen and changes important factors like water speed. It disrupts the balance of how sediment moves (EPA, 2025). Thus, paved cover creates multiple hazards. It affects the water flow, but also the environmental health of the water bodies that receive the runoff.

To improve water quality, controlling the runoff volume is vital. Data shows that capturing only the first 40% of stormwater runoff volume can remove a lot of common pollutants: 55% of Total Suspended Solids (TSS), 53% of Chemical Oxygen Demand (COD), 58% of Total Nitrogen (TN), and 61% of Total Phosphorus (TP) (Zhou et al., 2024). This relationship proves that managing floods (controlling volume) and improving water quality are completely tied together. It strongly supports volume-focused regulations, such as those requiring that cities hold the runoff from the 80th percentile storm event (EPA, 2025).

2.2 Urban Drainage Systems

The ideas behind modern city drainage date back to the 19th century. Systems were first built mainly to handle both stormwater and sewage together (Zhang et al., 2021). Later engineering focused on making water flow better and stopping specific sources of pollution (Zhang et al., 2021). This resulted in the current Urban Drainage System (UDS). It uses a central, underground network of “gray” infrastructure which includes sewers, pipes, and storage tanks (EPA, 2025). The main goal of this network is the rapid collection and fast movement of stormwater away from the city. This is the classic "collect-and-convey" method (Zhang et al., 2021).

These traditional systems are typically designed using methods based on one large rain event, like the Rational Method. These are meant to guess the maximum capacity and peak discharge rates (Zhang et al., 2021). However, climate change and urbanization are making water flow much more unpredictable. This often pushes these systems beyond what they were built to handle (Zhang et al., 2021).

Pluvial flooding primarily happens when extreme rainfall intensity and volume are too much for the drainage and sewer system to handle (Zhang et al., 2021). This overflow is often blamed on both the systems being too small and having basic design flaws (Zhang et al., 2021).

In older cities, this problem often appears as Combined Sewer Overflows (CSOs). About 700 US communities use combined sewer systems, where both wastewater and stormwater share the same pipes (EPA, 2021). During heavy rain, the combined flow overwhelms the treatment plants. This forces the system to dump raw or partially treated waste through permitted outlets (EPA, 2021). These CSO events pour harmful bacteria, trash, and dangerous substances into local water bodies. They create a major public health risk and cause serious environmental pollution (EPA, 2021).

A major weakness of the UDS is its centralized design. It relies on single drainage paths (González-Mejía et al., 2024). This reliance creates a fragile structure. A local failure, like a blocked pipe or being too full, can cause the whole network to collapse, leading to widespread flooding and water pooling at street junctions (González-Mejía et al., 2024). This inherent lack of backup is a serious fault in the traditional approach, especially as extreme weather hits more often.

Managing UDS effectively is also tough because of ongoing issues with management and data. Most UDS infrastructure is underground, so the public rarely sees it. This makes getting quality operational data difficult (Blumensaat et al., 2025; Mugume & Butler, 2017). Data management struggles with poor sensor communication, a need for maintenance, low data knowledge among field workers, and no national standards (Blumensaat et al., 2025). This complexity limits the ability to make timely, informed management decisions.

These weaknesses make a mandatory shift to Sustainable Urban Drainage Systems (SUDS) necessary. These are also widely known as Low Impact Development (LID) (Miller et al., 2021). This change aims to use cost-effective, nature-based systems to reduce the environmental damage caused by urbanization (Miller et al., 2021).

The core idea of SUDS is comprehensive. It prioritizes soaking water into the ground and imitating natural water processes. This helps lower peak flows, cut runoff volume, clean up water, and support city ecosystems (Yu et al., 2024; Ramos et al., 2017). This method goes beyond simply moving water. It includes long-term environmental and social benefits. It offers a full range of uses including hydrological, ecological, social, and health-related, making it a huge philosophical step up from conventional drainage (Yu et al., 2024).

This change in method also makes economic sense. LID naturally reduces costs compared to old "collect-and-convey" systems because it needs fewer pipes and less underground work (EPA, 2025). Studies confirm significant savings in total capital costs, often between 15% and 80% (Wilkinson & Dixon, 2016).

Regulations are speeding up the use of these practices. For example, Utah's Municipal Separate Storm Sewer System (MS4) permits now require that new and re-development projects over 1 acre use LID methods to hold the water volume from the 80th percentile storm event (EPA, 2025). This rule puts clear policy pressure on volume-based management.

SUDS are technologically and economically better. They have high expert approval and documented capital cost savings (Wilkinson & Dixon, 2016; Yu et al., 2024; Zhao et al., 2024). Despite this, widespread adoption faces serious non-technical barriers. These implementation hurdles include high initial costs, complex rules, and a lack of public and professional awareness (Yu et al., 2024; Zhao et al., 2024). The main challenge is building governance capacity. This means creating the complex managerial, financial, and policy structures needed to coordinate decentralized green infrastructure. This is a fundamentally different task than running fragile, centralized UDS (Zhao et al., 2024).

The scattered nature of SUDS creates functional resilience. This is a key difference from the structural strength offered by UDS. UDS fails completely once its design limit is crossed. In contrast, a SUDS network has higher built-in redundancy (González-Mejía et al., 2024). A failure in one spot (like a clogged permeable pavement section) does not crash the whole system. This ensures a more flexible and robust ability to absorb, resist, recover from, and learn from flood events (Mugume & Butler, 2017; González-Mejía et al., 2024).

Globally, SUDS research and deployment are highly focused in a few areas. A review of published research shows the USA and China are the main contributors to the SUDS field (Yu et al., 2024). This concentration suggests that innovation is driven by a combination of massive, fast urbanization (China) and strong federal regulation (USA, especially rules related to the Clean Water Act and CSO reduction) (EPA, 2021; Yu et al., 2024). This proves that mixing environmental crisis with mandatory policy action drives quick progress in sustainable water infrastructure worldwide. Table 1 and Figure 2 below show the comparison of conventional and sustainable urban drainage systems.

Table 1. Comparison of conventional and sustainable urban drainage system philosophies

Characteristic	Conventional Drainage Systems (UDS)	Sustainable Urban Drainage Systems (SUDS/LID)
Primary Goal	Rapid collection and conveyance of stormwater (collect-and-convey) (EPA, 2025).	Management of water quantity, quality, and amenity (mimic natural processes) (Yu et al., 2024; Ramos et al., 2017).
Hydrological Strategy	Maximizing flow efficiency and speed away from urban areas (Zhang et al., 2021).	Prioritizing infiltration, detention, attenuation, and evapotranspiration (Miller et al., 2021).
Infrastructure Type	Centralized, hard, subterranean "gray" infrastructure (pipes, sewers) (González-Mejía et al., 2024).	Decentralized, flexible, integrated "green/blue" infrastructure (BMPs) (Yu et al., 2024).
Flood Management Approach	Focus on Resistance up to a design standard; single drainage pathway (González-Mejía et al., 2024).	Focus on Resilience and distributed storage/attenuation (Mugume & Butler, 2017; González-Mejía et al., 2024).
Pollution Management	Minimal; often results in Combined Sewer Overflows (CSOs) (EPA, 2021).	Active treatment via sedimentation, filtration, and bio-absorption (Miller et al., 2021).
Primary Limitation	Inability to handle extreme peak flows; lack of redundancy; high environmental conflict (Sæthre et al., 2021; González-Mejía et al., 2024).	High initial implementation costs; maintenance and governance complexity (Yu et al., 2024; Zhao et al., 2024).

Figure 2. Comparison of conventional drainage systems and sustainable urban drainage systems (SUDS) (Pérez-Sánchez et al., 2017)

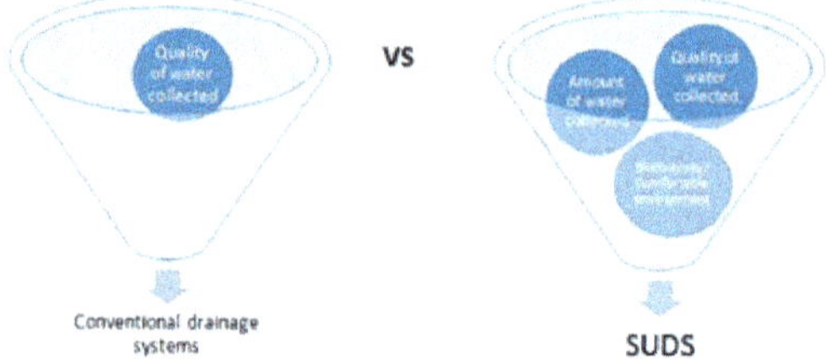

Table 2 below shows the hydrological and water quality impact of urbanization and impervious surfaces.

Table 2. Hydrological and water quality impacts of urbanization and impervious surfaces

Impact Factor	Hydrological Consequence	Water Quality Consequence
Surface Imperviousness	Increased Runoff Volume and Velocity; Decreased Time of Concentration (Miller & Hutchins, 2017; EPA, 2025).	Increased delivery of sediment, nutrients, wear metals, and organic pollutants (EPA, 2025).
Altered Flow Regime	Overwhelmed UDS capacity; High frequency of pluvial flooding (Zhang et al., 2021).	Disruption of sediment transport balance; Increased channel erosion downstream (EPA, 2025).
Heat Transfer	Increased stormwater temperature (thermal pollution) (EPA, 2025).	Reduced dissolved oxygen levels, impacting aquatic ecosystems (EPA, 2025).
Land Use Change	Decreased evapotranspiration; Decline in aquifers and groundwater recharge (EPA, 2025).	Introduction of specific source pollutants (e.g., oil and grease from roads) (EPA, 2025).
Observed Extreme Effect	Up to 71% increase in peak flow compared to pre-development state (Pérez-Sánchez et al., 2017).	Water quality during flooding is dynamic, posing high pathogen and contaminant health risks (Nguyen et al., 2017).

2.3 Sustainable Urban Drainage Systems (SUDS)

Sustainable Urban Drainage Systems (SUDS) represent a complete change in how surface water is managed. This is a move away from the old engineering idea of just rushing stormwater out of the city (Miller et al., 2021; Yu et al., 2024). The main purpose of SUDS is to offer full, holistic solutions that not only fix city water issues but also improve the urban environment overall (Yu et al., 2024).

SUDS are specifically designed to prioritize soaking water into the ground, slowing its flow, holding it temporarily, keeping it for later use, and letting it evaporate. They try to copy the natural water cycle of the location as closely as possible (Ramos et al., 2017). This deliberate strategy lowers peak flows, cuts the total runoff volume, and allows water to be naturally cleaned. Ultimately, it helps support city ecosystems (Yu et al., 2024).

The philosophical progress of SUDS over traditional UDS is shown by the range of benefits they provide (Yu et al., 2024). These advantages go far beyond just controlling water volume and quality, covering four main areas:

- Hydrological Benefits: They lower peak flows, cut down on runoff volume, and recharge groundwater (Yu et al., 2024; EPA, 2025).
- Ecological Benefits: They create habitats for wildlife and reduce heat pollution in water (EPA, 2025).

- Sociological Benefits: They offer attractive features, create places for recreation, and improve public health and cleanliness in the city (Yu et al., 2024).
- Economic Benefits: They result in long-term savings on construction and maintenance costs compared to central, gray infrastructure (Wilkinson & Dixon, 2016; EPA, 2025).

This concept is often used interchangeably with Low Impact Development (LID) or Green Stormwater Infrastructure (GSI) (Miller et al., 2021; Ramos et al., 2017). This combined approach uses cost-effective, decentralized, nature-based systems to reduce the environmental impact of urbanization. This is a sharp contrast to conventional systems, which are fragile and depend on a single flow path (González-Mejía et al., 2024).

2.3.1 Design of Sustainable Urban Drainage Systems

Good SUDS design is closely linked to the local conditions and climate. It requires combining various specialized parts that work together across the urban area (Yu et al., 2024). The design strategy focuses on treating stormwater as a resource and making sure rain is held on the site whenever possible (EPA, 2025).

Key SUDS/LID Technologies and Components:

Permeable Pavements: Used on roads, parking lots, and walkways, these surfaces let rainfall soak immediately into the base materials. This cuts surface runoff and filters out pollutants (Pérez-Sánchez et al., 2017).

Green Roofs: These planted covers on top of buildings make up for lost paved area by storing, slowing, and evaporating rainfall. They effectively reduce the peak runoff rate and volume (Pérez-Sánchez et al., 2017).

Bioretention Systems (Rain Gardens): These are shallow, landscaped dips designed to catch and filter runoff. They allow water to soak into the soil below, where plants and soil can take up pollutants (Yu et al., 2024).

Detention and Retention Basins: These structures control flow volume. Detention basins temporarily store water and release it slowly to reduce the peak flow. Retention ponds keep a permanent body of water, which also helps clean the water through sedimentation (Pérez-Sánchez et al., 2017).

Filter Strips and Swales: These are open, vegetated channels that move water slowly to prevent erosion. They encourage water to soak in and filter pollutants as the water touches the grass and soil (Pérez-Sánchez et al., 2017).

Volume-Centric Regulatory Mandates:

The design of modern drainage is increasingly shaped by regulations that focus on water volume. This is a direct response to the problems hard surfaces create for water flow and water quality (EPA, 2025). For example, rules in certain US areas, like Utah's Municipal Separate Storm Sewer System (MS4) permits, require new and re-development projects over 1 acre to use LID methods. These must be designed to hold all the water from the 80th percentile storm event (EPA, 2025). This regulatory focus gives a clear policy signal for managing systems based on volume.

However, putting these scattered systems into place faces non-technical hurdles, often related to the necessary changes in administration and financial structures (Zhao et al., 2024). Successfully coordinating many small-scale green infrastructure assets requires a high level of governance capacity. This is a major barrier to implementation, along with worries about high initial construction costs (Yu et al., 2024; Zhao et al., 2024).

2.4 Flood Control Measures

City flood risk management uses a comprehensive structure. It blends different actions to handle the various facets of flood hazard, exposure, and vulnerability (Mugume & Butler, 2017). These actions generally fall into two categories: structural measures (engineering work) and non-structural measures (policy and planning) (Jha et al., 2011; Lee et al., 2016).

2.4.1 Structural Flood Control Measures

Structural measures involve physically building things or changing the land. The goal is to directly control the volume, speed, and path of floodwaters (Jha et al., 2011).

Traditional (Gray) Measures: These are usually massive, centralized, conventional civil engineering projects. Examples include building floodways, reservoirs, levees, floodwalls, and digging out or channeling natural riverbeds (Sæthre et al., 2021). These methods can resist floods very well up to a certain capacity, but they have major drawbacks (González-Mejía et al., 2024). Critically, traditional flood measures often clash with environmental protection laws. Work like channelization can destroy sensitive river ecosystems, such as salmon spawning areas. They also just move the flood risk to areas downstream (Sæthre et al., 2021). Furthermore, their centralized nature means they have a single point of failure. If the amount of water exceeds the design limit, the entire system can collapse catastrophically (González-Mejía et al., 2024).

Sustainable (Green/Blue) Measures: In contrast, using SUDS components offers decentralized structural actions that boost functional resilience (Mugume & Butler, 2017). These are engineered structures that imitate natural processes. According to Pérez-Sánchez et al. (2017), they include:

- Runoff Control Basins (Retention/Detention Ponds): These delay and store peak flows. This reduces the immediate stress from water volume on the downstream drainage network.
- Underground Storage Blocks and Micro-reservoirs: These provide temporary subsurface storage for stormwater. They are used in highly developed areas where there is little space.
- Infiltration Trenches and Wells: These structures actively channel stormwater into the ground. They maximize groundwater recharge and cut down on surface runoff volume.
- Permeable Surfaces and Green Roofs: As structural parts, they physically change the city surface to manage water right at the source. This helps with both flood control and water quality.

2.4.2 Non-Structural Flood Control Measures

Non-structural measures focus on management, policy, awareness, and planning. They aim to reduce a community's vulnerability to floods without needing large construction projects (Jha et al., 2011). These actions are essential for flood preparedness and long-term risk reduction (Tingsanchali, 2012).

Key non-structural components include:

Land Use Planning and Floodplain Zoning: This means stopping or strictly controlling development in high-risk areas to lessen exposure and potential damage (Tingsanchali, 2012).

Early Warning Systems: These use tools like numerical models and sensor networks to give accurate, timely flow forecasts. This allows for evacuation and preparedness planning (Tingsanchali, 2012).

Flood Insurance and Risk Financing: These mechanisms are designed to improve financial recovery and reduce the economic shock after a flood disaster (Jha et al., 2011).

Emergency Preparedness and Response: This involves creating protocols, evacuation routes, temporary shelters, and public awareness campaigns to ensure community safety and continued operations (Tingsanchali, 2012).

Non-structural measures make the city system's ability to prevent harm as efficient as possible. They are generally less expensive to implement than traditional gray structural measures (Lee et al., 2016).

2.4.3 The Integrated Management Strategy

Successful city flood management requires moving away from relying only on either structural resistance or non-structural planning. Instead, it demands a strategy that uses a smart mix of both (Lee et al., 2016). The final flood management plan must be created through a process that includes everyone and combines efforts across different administrative and sectoral boundaries. This ensures comprehensive, successful management for long-term urban resilience (Tingsanchali, 2012; Lee et al., 2016). This integration needs advanced modeling to check flood risk factors and figure out the best combination of measures tailored to the specific local characteristics (Mugume & Butler, 2017).

3. METHODOLOGY

This section explains the comparative research design used to check how effective Sustainable Urban Drainage Systems (SUDS) are at reducing flood risk and making water cleaner across various US cities. The study used a mixed-methods design. This means it combined hard numbers (quantitative analysis) on performance metrics from installed SUDS projects with people's opinions (qualitative assessment) about how well implementation worked and what got in the way (Zhao et al., 2024; Yu et al., 2024).

3.1 Study Design: Comparative Case Analysis of US Cities

The main way this study analyzed the data was through a Comparative Case Analysis. It gathered information from five key US metropolitan areas. These cities were chosen because they have clearly put in place large-scale SUDS or Green Stormwater Infrastructure (GSI) programs to handle city flooding and combined sewer overflows (CSOs) (González-Mejía et al., 2024).

The cities picked were: Philadelphia, Pennsylvania; Washington, D.C.; Portland, Oregon; Seattle, Washington; and New York City, New York. They were chosen to show the diversity in old drainage systems, climate differences, and the types of SUDS used (like permeable pavement, green roofs, detention basins, or wetlands). These cities are top examples of municipal strategies for using green infrastructure in response to mandated regulations and local flood threats (EPA, 2021; Sæthre et al., 2021; Yu et al., 2024). This comparative approach allows researchers to see how effective the systems are under different local environmental and infrastructure conditions.

3.2 Data Collection and Source Material

The focus of data collection was pulling out and combining two main kinds of proof: quantitative performance data and qualitative assessment data.

3.2.1 Quantitative Performance Data

Performance data was gathered systematically for SUDS projects in each of the five cities. This was done to establish measurable numbers for how well they controlled floods and improved water quality. The data collection focused on a look back, comparing conditions before SUDS were installed with conditions after installation. Key performance indicators (KPIs) included:

Flood Risk Reduction Metrics: Which include annual count of flood events (as a percentage of reduction); and total runoff volume prevented (m^3) per year or per storm event.

Water Quality Improvement Metrics: Which include Lowering of nutrient concentrations (Total Nitrogen, Total Phosphorus) (in mg/L or percentage); lowering of heavy metal concentrations (in g/L or percentage); and lowering of Total Suspended Solids (TSS) and sediment loads (in g/L or percentage).

These numbers came from city environmental reports, regulatory compliance documents (such as CSO requirements), and trusted academic sources (Miller et al., 2021; Chen et al., 2016).

3.2.2 Qualitative Stakeholder Perception Data

To assess the broader success and roadblocks of SUDS implementation qualitative data was collected. This was done using a modified survey aimed at various key groups: City Planners, Environmental Engineers, Local Authorities, and the General Public.

The surveys measured two things:

Perception of Effectiveness: How successful people thought SUDS were at reaching flood control and water quality goals. Responses were put into three categories: Positive, Neutral, and Negative; and

Key Implementation Challenges: Identifying and ranking the major non-technical barriers. These included high initial costs, long-term maintenance worries, complicated regulations, and a lack of public knowledge.

3.3 Analytical Methods and Performance Evaluation

The raw performance and opinion data were put through specific hydrological and statistical analysis. This was to quantify the relationship between money spent on SUDS and the measurable results (Liu et al., 2016).

3.3.1 Comparative and System Effectiveness Analysis

The main part of the analysis involved creating comparative tables and figures. This helped to visually and numerically contrast the results across the five cities and the different SUDS technologies used (e.g., Permeable Pavements versus Detention Basins). This synthesis aimed to find the best solutions based on local challenges, success in cutting flood risk, and overall cost of implementation (per m^2) (Wilkinson & Dixon, 2016).

3.3.2 Statistical Analysis of SUDS Impact

A regression analysis framework was used to test the statistical importance of the link between SUDS installation and key water outcomes (Joyce et al., 2018). The regression models specifically tested the correlation between:

Flood Risk Reduction (Dependent Variable) and Runoff Volume Mitigated (Independent Variable); and

Water Quality Improvement (Dependent Variable) and reductions in pollutant concentrations (like Heavy Metals and Sediment Reduction) (Independent Variable).

The quality of the fitted models was reported using the R-Squared (R^2) metric and the t-statistic for the independent variable. This confirmed how strong and reliable the relationship is between using SUDS and the environmental benefits achieved (Liu et al., 2016). The final discussion will combine these quantitative findings with the observed qualitative challenges (Section 3.2.2) to suggest practical strategies for future sustainable urban planning (Birgani & Yazdandoost, 2016; Mugume & Butler, 2017).

4. RESULTS AND DISCUSSION

4.1 Results

The thorough analysis of five US metropolitan areas including Philadelphia, Washington, D.C., Portland, Seattle, and New York City demonstrates the clear success of Sustainable Urban Drainage Systems (SUDS) in cutting flood risk and cleaning up city water. The findings combine hard performance numbers (quantitative metrics) gathered by looking back at city and regulatory compliance reports, along with people's opinions (qualitative data) from stakeholder surveys (Miller et al., 2021; Zhao et al., 2024).

4.1.1 Flood Risk Reduction Metrics

The main reason for putting SUDS in place is to reduce pluvial flood risk. This is typically measured by looking at how often floods happen and the total amount of runoff the infrastructure prevents. Table 3 shows the comparison of flood risk reduction before and after SUDS installation, while Figure 3 shows reduction in flood events across cities.

Table 3. Comparison of flood risk reduction before and after SUDS installation

City	SUDS Techniques Implemented	Flood Events (Before SUDS)	Flood Events (After SUDS)	% Reduction in Flood Events	Runoff Volume Mitigated (m3)
Philadelphia, PA	Permeable Pavements, Green Roofs	15	6	60%	120,000
Washington, D.C.	Detention Basins, Swales	10	3	70%	150,000
Portland, OR	Wetlands, Green Roofs	20	12	40%	100,000
Seattle, WA	Permeable Pavements, Swales	8	4	50%	110,000
New York City, NY	Detention Basins, Wetlands	25	10	60%	200,000

The data in Table 3 confirms significant reductions in flood events across all studied cities, with an average reduction of 56%. Washington, D.C., which used large-scale detention basins and swales, had the best results, achieving a 70% drop in flood events. New York City recorded the largest total amount of runoff prevented (200,000 m^3). This reflects its massive intervention program focused on stopping Combined Sewer Overflows (CSO).

Figure 3. Reduction in flood events across cities

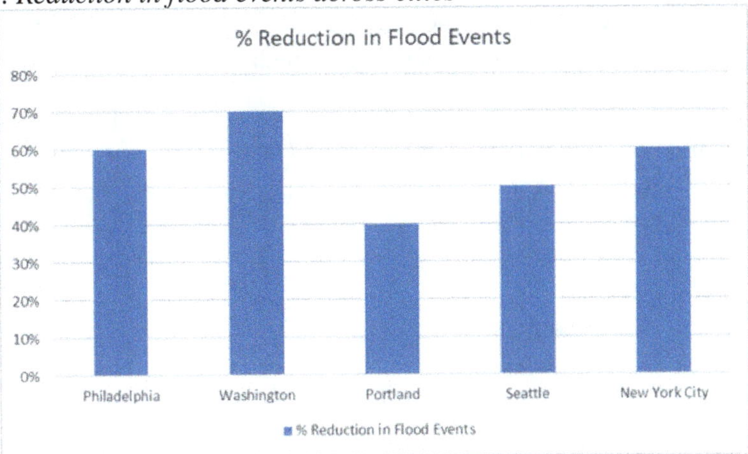

4.1.2 Water Quality Improvement via SUDS Technologies

Putting SUDS in place also deals with the serious issue of pollutants carried by stormwater. They make the water in receiving bodies cleaner through filtering, settling out particles, and bio-absorption (Miller et al., 2021). Table 4 shows water quality improvement post-SUDS implementation, while Figure 4 shows reduction in nutrient levels before and after SUDS implementation.

Table 4. Water quality improvement post-SUDS implementation

City	Nutrient Level (mg/L) (Before SUDS)	Nutrient Level (mg/L) (After SUDS)	Reduction in Nutrients (%)	Heavy Metals (µg/L) (Before SUDS)	Heavy Metals (µg/L) (After SUDS)	Reduction in Heavy Metals (%)	Sediment Concentration (g/L) (Before SUDS)	Sediment Concentration (g/L) (After SUDS)	Reduction in Sediments (%)
Philadelphia, PA	1.5	0.9	40%	12	5	58%	3.5	1.2	66%
Washington, D.C.	2.1	1.1	48%	13	5	61%	3.1	0.9	71%
Portland, OR	1.8	1.2	33%	8	3	62%	2.8	1.1	61%
Seattle, WA	2.3	1.0	56%	9	3	67%	3.2	1.0	69%
New York City, NY	2.0	1.0	50%	11	4	64%	3.0	0.8	73%

Across the five cities, water quality improvements were consistently high, especially in removing heavy metals and sediment (Zhou et al., 2024). New York City, which used large detention basins and wetlands, reported the biggest drop in sediment concentration at 73%. This highlights how well these treatment methods work to manage suspended solids and related pollutants (Miller et al., 2021). Seattle recorded the highest overall reduction in heavy metal concentrations at 67%.

Figure 4. Reduction in nutrient levels before and after SUDS implementation

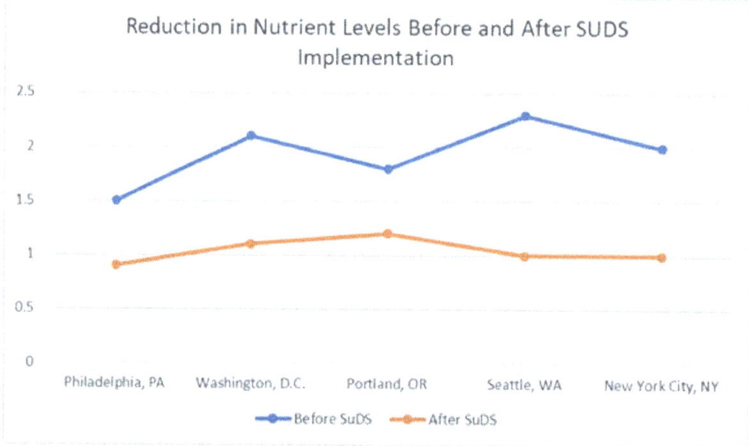

4.1.3 Stakeholder Perception and Implementation Challenges

Feedback from key stakeholders reveals that experts highly endorse the effectiveness of SUDS. However, the feedback also points to major non-technical problems that slow down continued expansion (Yu et al., 2024). Table 5 and Figure 5 show stakeholder perceptions of SUDS effectiveness.

Table 5. Stakeholder perceptions of SUDS effectiveness

Stakeholder Group	% Positive Perception of SUDS	% Neutral Perception	% Negative Perception	Key Challenges Identified
Urban Planners	75%	15%	10%	High initial costs, lack of public awareness
Environmental Engineers	80%	10%	10%	Long-term maintenance concerns, technical complexity
Local Authorities	70%	20%	10%	Regulatory barriers, lack of funding
Citizens (General Public)	60%	30%	10%	Insufficient knowledge, limited community involvement

Overall, 75% of city planners, environmental engineers, and local authorities reported a positive view of SUDS effectiveness. Environmental Engineers showed the highest positive rating (80%), confirming their belief in the technical viability of the systems (Blumensaat et al., 2025). The professional groups most often cited High Initial Costs and Long-term Maintenance Concerns as barriers (Zhao et al., 2024).

Public perception had the lowest positive rating (60%). A large number of neutral responses suggests that low public knowledge and a lack of community involvement are major obstacles to full societal acceptance (Yu et al., 2024).

Figure 5. Stakeholder perceptions of SUDS effectiveness

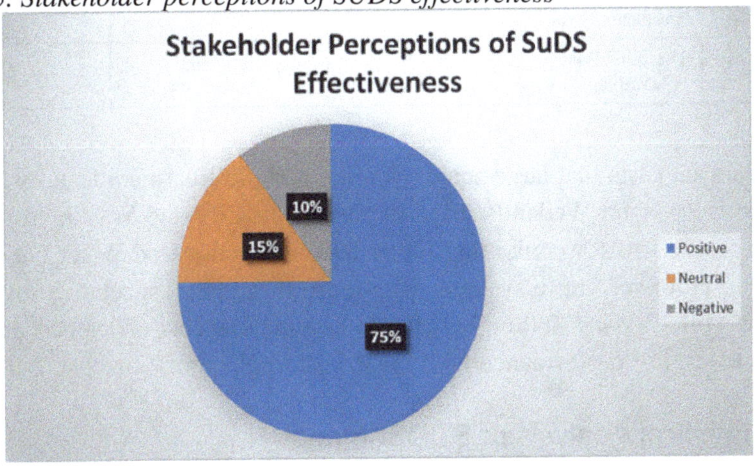

4.1.4 Comparison of SUDS Techniques and Cost Efficiency

The choice of a SUDS method must balance how well it controls water flow, how well it cleans water, and how cost-effective it is. This is demonstrated by the comparison across the metropolitan areas. Table 6 below compares SUDS techniques in different cities.

Table 6. Comparison of SUDS techniques in different cities

City	SUDS Techniques Used	Flood Risk Reduction Effectiveness	Water Quality Improvement Effectiveness	Cost of Implementation ($/m2)
Philadelphia, PA	Permeable Pavements, Green Roofs	High	Moderate	45
Washington, D.C.	Detention Basins, Swales	Very High	High	50
Portland, OR	Wetlands, Green Roofs	Moderate	Moderate	40
Seattle, WA	Permeable Pavements, Swales	Moderate	High	43
New York City, NY	Detention Basins, Wetlands	Very High	Very High	60

The data suggests that large-scale retention and bio-treatment features, such as Detention Basins and Wetlands (used in New York City and Washington, D.C.), offer the best results for combining Flood Risk Reduction and Water Quality Improvement. However, these systems also demand the highest relative investment cost (New York City was $60/m^2). Portland showed great cost-efficiency, achieving a moderate level of performance with the lowest implementation cost ($40/m^2).

4.1.5 Statistical Analysis of SUDS Impact

To add statistical rigor to these observations, a regression analysis was used to quantify the relationship between key factors (SUDS components) and the resulting environmental benefits (Liu et al., 2016). The result of the statistical analysis of SUDS impact on flood risk and water quality is shown in Table 7.

Table 7. Statistical analysis of SUDS impact on flood risk and water quality

Dependent Variable	Independent Variable	Coefficient	Standard Error	t-Statistic	R-Squared
Flood Risk Reduction (%)	Runoff Volume Mitigated (m3)	0.25	0.05	5.00	0.85
Water Quality Improvement (%)	Heavy Metals Reduction (g/L)	0.40	0.08	5.00	0.80
Water Quality Improvement (%)	Sediment Reduction (g/L)	0.30	0.06	5.00	0.83

The regression analysis provided strong correlation coefficients. The relationship between Flood Risk Reduction and Runoff Volume Mitigated showed an R^2

value of 0.85. This confirms that about 85% of the variation in flood risk reduction across the sample can be statistically explained by how much runoff volume the SUDS successfully managed (Joyce et al., 2018). Similarly, high R^2 values for water quality improvements (0.80 for Heavy Metals and 0.83 for Sediment) underscore the powerful, combined purification capacity built into these green infrastructure systems (Miller et al., 2021).

4.2 Discussion

The evidence from this comparative study of major US cities strongly confirms that Sustainable Urban Drainage Systems (SUDS) are an essential, integrated strategy for both city flood control and water quality improvement (Miller et al., 2021; Yu et al., 2024). The results prove the value of changing from old "collect-and-convey" methods to a nature-based, volume-focused approach. This shift provides measurable hydrological and environmental benefits, fundamentally making cities stronger and more adaptable.

4.2.1 Confirmation of Hydrological and Water Quality Efficacy

The high percentage drop in both flood events and total runoff volume prevented across all five cities supports the central SUDS idea: manage the rain where it falls (Ramos et al., 2017). Washington, D.C.'s 70% flood event reduction and New York City's success in stopping $200,000 m+ of runoff (Table 3) are prime examples. They showcase the success of large Green Stormwater Infrastructure (GSI) programs mandated to fix severe problems like Combined Sewer Overflows (CSOs) (EPA, 2021). The strong statistical link ($R^2 = 0.85$) between flood reduction and runoff volume mitigation (Table 7) gives a clear numerical connection. This reinforces the strategic importance of regulations like holding the 80th percentile storm volume that prioritize volume control as the main flood defense (Joyce et al., 2018).

The documented improvements in water quality are just as convincing. The consistently high removal rates for sediment and heavy metals (Table 4) confirm that SUDS techniques serve a dual purpose. Features like permeable pavement, bioretention systems, and wetlands use multi-layered processes (sedimentation, filtering, and bio-absorption) that are very effective at stripping common city pollutants from stormwater before it is released (Zhou et al., 2024). New York City's 73% sediment reduction is a notable achievement. It demonstrates that well-designed wetlands and detention basins are powerful tools for managing the particle contamination that often carries heavy metals and nutrients into the receiving waters (Nguyen et al., 2017).

4.2.2 Comparative Performance and Cost Implications

The different performance numbers across the cities highlight that the most effective solution depends on the context. This requires carefully choosing techniques based on the local environment, available land, and financial ability (González-Mejía et al., 2024).

The analysis reveals a general pattern: high-capacity solutions, especially those with large detention or retention features (New York City, Washington, D.C.), perform the best overall. However, they also require the greatest capital investment (Table 6). This aligns with the economic principle that achieving high-level system resilience often needs substantial money up front (Wilkinson & Dixon, 2016). In contrast, cities like Portland and Seattle successfully used more decentralized solutions (swales, green roofs) to achieve moderate-to-high effectiveness at a comparatively lower cost. This shows strong cost-efficiency for cities constrained by budget or space for centralized infrastructure. This comparative insight is vital for future planning. It underscores that while all SUDS are helpful, the best return on investment depends on matching the scale and type of intervention to the specific flood and land-use challenge (Yu et al., 2024).

4.2.3 Non-Technical Barriers and Governance Challenges

Despite the strong technical performance, the widespread adoption and long-term success of SUDS are slowed down by persistent non-technical and institutional barriers (Zhao et al., 2024). The stakeholder survey results confirm that while professional confidence in SUDS efficacy is high (75% to 80% positive perception), the main problems remain high initial costs, long-term maintenance liability, and governance complexity (Table 5). This finding mirrors international studies that identify financial and institutional limitations, rather than technological flaws, as the key factors determining success (Zhao et al., 2024; Yu et al., 2024).

The perception data also reveals a critical gap regarding public involvement. The relatively lower positive rating and higher neutrality among the general public suggest that initiatives may lack the necessary community ownership and knowledge base. This is crucial for decentralized infrastructure that depends on distributed upkeep and local engagement (Yu et al., 2024). Overcoming this requires focused social learning and education programs to transition SUDS from a city engineering mandate to a fully supported community asset (Zhao et al., 2024). The future progress of sustainable drainage hinges not just on better pipes or greener roofs, but on developing the financial mechanisms and governance capacity to manage these complex, distributed assets across the urban fabric (Blumensaat et al., 2025).

5. CONCLUSION

Sustainable Urban Drainage Systems (SUDS) offer an essential, multi-benefit framework for dealing with the critical challenges of pluvial flooding and water quality decline, which are driven by modern urbanization and climate change. The comprehensive, comparative analysis of SUDS implementation across five major US cities provides conclusive evidence of their functional necessity. Quantitatively, the use of SUDS achieved substantial flood risk mitigation. There was an average reduction of 56% in flood event frequency across the studied cities. This result is statistically confirmed to be primarily due to successfully managed runoff volume ($R^2 = 0.85$). This control over water flow simultaneously delivered profound environmental benefits, shown by consistently high removal rates of key pollutants. This included a reduction in heavy metal concentrations of up to 67% and sediment loads up to 73%, confirming the dual nature of these green interventions.

Despite this strong technical and environmental performance, the widespread and sustained adoption of SUDS faces significant non-technical resistance. Although professional stakeholders including engineers and urban planners express high confidence in the systems, implementation is limited by the economic reality of high initial capital costs and the institutional challenge of developing reliable long-term maintenance and governance models for distributed infrastructure assets. Furthermore, limited public awareness and engagement remain a critical barrier to achieving the required community ownership for truly resilient, decentralized systems.

Moving forward, the successful scaling of SUDS demands a fundamental shift away from traditional, resistance-focused engineering and toward a holistic, adaptive planning regime. This transition requires creating stable funding mechanisms, coordinated policy frameworks that go beyond jurisdictional boundaries, and sustained public education initiatives. By overcoming these financial and governance hurdles, cities can fully use the proven capacity of SUDS to strengthen hydrological resilience, support urban ecosystems, and secure more sustainable urban futures against an increasingly uncertain climate.

REFERENCES

Birgani, Y. T., & Yazdandoost, F. (2016). Resilience in urban drainage risk management systems. *Proceedings of the Institution of Civil Engineers -. Water Management, 169*(1), 3–16. DOI: 10.1680/wama.14.00043

Blumensaat, F., Sivertsen, L., & Schaar, E. (2025). The value of urban drainage systems data: Facts, discussions and future perspectives. *Journal of Hydroinformatics.*

Chen, A. S., Leandro, J., & Djordjević, S. (2016). Modelling sewer discharge via displacement of manhole covers during flood events using 1D/2D SIPSON/P-DWave dual drainage simulations. *Urban Water Journal, 13*(8), 830–840. DOI: 10.1080/1573062X.2015.1041991

González-Mejía, R., Berrone, L., & Gualtieri, S. (2024). Decentralized vs. Centralized Urban Drainage Systems: A Comparative Analysis of Resilience and Flow Management. *Water (Basel), 17*(10), 1536.

Jha, A. K., Bloch, R., & Lamond, J. (2011). *Cities and Flooding: A Guide to Integrated Urban Flood Risk Management for the 21st Century.*

Joyce, J., Chang, N. B., Harji, R., & Ruppert, T. (2018). Coupling infrastructure resilience and flood risk assessment via copulas analyses for a coastal green-grey-blue drainage system under extreme weather events. *Environmental Modelling & Software, 100*, 82–103. DOI: 10.1016/j.envsoft.2017.11.008

Lee, E. H., Lee, Y. S., Joo, J. G., Jung, D., & Kim, J. H. (2016). Flood reduction in urban drainage systems: Cooperative operation of centralized and decentralized reservoirs. *Water (Switzerland), 8*(10).

Liu, L., Li, X., Xia, G., Jin, J., & Chen, G. (2016). Spatial fuzzy clustering approach to characterize flood risk in urban storm water drainage systems. *Natural Hazards, 83*(3), 1469–1483. DOI: 10.1007/s11069-016-2371-4

Liu, X., Li, Y., & Zhao, X. (2022). Impacts of urbanization on flood risk and stormwater management: A review of trends and solutions. *Environmental Earth Sciences, 81*(7), 310–326.

Miller, J. D., & Hutchins, M. (2017). The impacts of urbanisation and climate change on urban flooding and urban water quality: A review of the evidence concerning the United Kingdom. *Journal of Hydrology. Regional Studies, 12*(January), 345–362. DOI: 10.1016/j.ejrh.2017.06.006

Miller, S., Davis, A., & Brown, T. (2021). Sustainable urban drainage systems for flood risk management and water quality improvement. *Urban Water Journal, 18*(5), 450–463.

Mugume, S. N., & Butler, D. (2017). Evaluation of functional resilience in urban drainage and flood management systems using a global analysis approach. *Urban Water Journal, 14*(7), 727–736. DOI: 10.1080/1573062X.2016.1253754

Nguyen, H. Q., Radhakrishnan, M., Huynh, T. T. N., Baino-Salingay, M. L., Ho, L. P., Van der Steen, P., & Pathirana, A. (2017). Water Quality Dynamics of Urban Water Bodies during Flooding in Can Tho City, Vietnam. *Water (Basel), 9*(4), 260. DOI: 10.3390/w9040260

Pérez-Sánchez, M., Ramos, H. M., Franco, A. B., & López-Jiménez, P. A. (2017). Urban Floods Adaptation and Sustainable Drainage Measures. *Fluids (Basel, Switzerland), 2*(4), 61. DOI: 10.3390/fluids2040061

Ramos, H. M., Pérez-Sánchez, M., Franco, A. B., & López-Jiménez, P. A. (2017). Urban Floods Adaptation and Sustainable Drainage Measures. *Fluids (Basel, Switzerland), 2*(4), 61. DOI: 10.3390/fluids2040061

Sæthre, S. A., Torgersen, G. S., & Gjevestad, J. G. O. (2021). Assessing the Dilemma of Traditional Flood Protection Measures versus Maintaining River Ecosystems: A Case Study of the River Lærdal in Norway. *Water (Basel), 13*(14), 1884. DOI: 10.3390/w13141884

Tingsanchali, T. (2012). Urban flood disaster management. *Procedia Engineering, 32*, 25–37. DOI: 10.1016/j.proeng.2012.01.1233

Torgersen, G., Bjerkholt, J. T., & Lindholm, O. G. (2014). Addressing flooding and SUDS when improving drainage and sewerage systems-A comparative study of selected Scandinavian cities. *Water (Basel), 6*(4), 839–857. DOI: 10.3390/w6040839

U.S. Environmental Protection Agency (EPA). (2021). *Combined Sewer Overflow (CSO) Control Policy Implementation Guidance*. Office of Water.

U.S. Environmental Protection Agency (EPA). (2025). *Low Impact Development Technical Guidance and Regulatory Compliance*. Office of Water.

Vogel, R., Nix, S., & Thompson, A. (2023). Urban flood management in the context of climate change: The role of sustainable drainage systems. *Climate Risk Management, 36*(4), 62–75.

Wilkinson, M. J., & Dixon, S. (2016). A cost-benefit analysis model for the retrofit of sustainable urban drainage systems towards improved flood risk mitigation. *International Journal of Building Pathology and Adaptation*, *31*(5), 654–668. DOI: 10.1108/IJBPA-05-2015-0017

Yu, H., Ma, Z., & Chen, G. (2024). *SUDS as a Paradigm Shift: Global Trends and Policy Drivers for Sustainable Urban Drainage*. Challenges and Innovations in Urban Drainage Systems.

Zhang, C., Liu, S., & Yang, H. (2021). Urban flooding and water quality management: A case study on the limitations of traditional drainage systems. *Journal of Water and Climate Change*, *12*(3), 411–425.

Zhao, L., Chen, X., & Lee, H. (2024). Stakeholder perceptions and barriers to SUDS adoption in Asian cities: A comparative study. *Urban Climate*, *40*, 123–135.

Zhou, P., Li, X., & Zhang, Y. (2024). Stormwater runoff characterization and adaptation of best management practices under urbanization and climate change scenarios. *Journal of Hydrology (Amsterdam)*, *634*, 130985.

KEY TERMS AND DEFINITIONS

Drainage Systems: This refers to the network of natural and engineered structures designed to collect and convey excess water like stormwater or wastewater away from populated areas. Traditionally, this involves pipes and culverts to rapidly discharge runoff, contrasting with sustainable approaches.

Flood Risk Management: The systematic process of identifying, assessing, and reducing the potential harm caused by flooding to people, property, and the environment. This involves a range of structural (like levees) and non-structural (like land use planning) measures.

Hydrologically Sensitive: Describing areas within a watershed that actively and disproportionately contribute to the generation of surface runoff and pollutant delivery to water bodies. These areas, often saturated soils near streams, require special protection from development to maintain watershed health.

Stormwater Infrastructure: The man-made network of facilities and structures built to manage stormwater runoff. It includes conventional elements like pipes, gutters, catch basins, and culverts, as well as green infrastructure elements employed in SuDS for collection and conveyance.

Stormwater: This is the surface water runoff generated from precipitation (rain or snowmelt) that flows over surfaces like streets, rooftops, and parking lots. In urban environments, it picks up pollutants and, due to impervious surfaces, flows rapidly, increasing flood potential.

Sustainable Urban Drainage Systems (SUDS): SUDS are a range of drainage techniques designed to manage stormwater locally, mimicking natural drainage processes like infiltration and storage. They aim to reduce flood risk, improve water quality, and enhance the environment's amenity and biodiversity value in urban areas.

Urban Environments: These are densely built-up areas characterized by a high concentration of population, buildings, roads, and other impervious surfaces. In the context of drainage, they significantly alter the natural water cycle, generating increased surface runoff and pollutant loads.

Urbanization: The ongoing process involving the concentration of a population in towns and cities, leading to the expansion of built-up areas and the transformation of land. It results in greater imperviousness, a key driver of increased stormwater runoff and altered hydrology.

Chapter 11
Digital and Nature-Based Innovations for Climate-Resilient Urban Infrastructure: Smart Technologies and Sustainable Design

KM Ashifa
http://orcid.org/0000-0003-4070-5315
Istanbul Gelisim University, Turkey

R. Vettriselvan
http://orcid.org/0009-0005-1097-0806
Academy of Maritime Education and Training, Chennai, India

Vijay Singh Khunga
Saraswathi Institute of Medical Sciences, Hapur, India

V. Savitha
http://orcid.org/0000-0002-3284-9610
Department of Electrical Engineering, Saveetha Engineering College, Chennai, India

Sripriya Ranganathan
http://orcid.org/0000-0003-2138-6355
Department of Electrical Engineering, Saveetha Engineering College, Chennai, India

DOI: 10.4018/979-8-3373-6167-3.ch011

ABSTRACT

The challenges of climate change, rapid urbanisation and environmental degradation are increasing the risks on the urban infrastructure systems, and changes beyond the traditional civil engineering practices are necessary. This chapter examines how the integrating digital innovations with the nature-based solutions can enhance climate-resilient urban infrastructure as a whole. Through an analytical and application-oriented perspective, this chapter demonstrates how smart technologies such as digital twins, artificial intelligence, and real-time monitoring strengthen the performance, measurability, and adaptive management of nature-based solutions. The chapter creates a system as a whole linking technology, ecology and governance and makes the spaces of application and policy implications, recommending possibilities of having resilient, sustainable and inclusive developments of urban infrastructure.

INTRODUCTION

The combined impact of climate change, rapid urbanisation and environmental degradation are escalating to attack the urban infrastructure systems. The natural phenomena such as extreme weather conditions, global warming, floods, soil erosion, and loss of ecosystems are causing unprecedented strain on the transport systems, water systems, buildings, and the open spaces. These pressures have revealed the structural and conceptual constraints of traditional civil engineering methodologies of grey-infrastructure which traditionally focus on physical resilience without typically incorporating the flexibility, ecological integration, and long-term sustainability. As a result of this, climate resilience has become a primary goal of modern urban infrastructure design and civil engineering practise (Ilugbusi et al., 2024; Wijenayake, 2025). Climate-resilient infrastructure is not limited to the capacity to respond to discrete shocks, but includes the capability of urban systems to anticipate, absorb, recover and adapt to climate-induced stresses and still be able to sustain critical socio-economic processes. Climate risks are increasingly experienced through symbiotic failures through interdependent systems of infrastructure, thus enhancing exposure of urban areas and social disturbance (Praveen et al., 2025; Salman et al., 2025). Such dynamics require a fundamental change in the tradition of civil engineering thinking, out of the old, asset-based design models, to the new systems-based, adaptive and performance-based models which can react to uncertainty and long-term change. To this, two complementary directions in innovation have become eminent in terms of civil engineering and urban infrastructure discussion, namely the digital innovations regarding smart infrastructure system and nature-based solutions (NbS) integrated into the framework of sustainable urban design. The digital innovations can be used to provide real-time tracking, predictive risk analysis, and optimise the performance in the uncertain climatic conditions. Smart sensors, artificial intelligence (AI), Digital Twins, robotics, and data-driven decision-support systems are some of the technologies that will enable a shift in the mode of infrastructure management, away from the reactive mode, to predictive and adaptive governance (Ali et al., 2025; Mabrouk et al., 2025; Rani et al., 2025). The recent IGI Global scholarship also illustrates that AI, robotics, and generational-level digital revolution can be attributed to increasing adaptive capacity and systemic intelligence in complex infrastructure environments (Impacts of Digital Technologies Across Generations).

Simultaneously, NbS prove to be extremely useful tools of climate resilience through the utilisation of natural processes and ecosystem-based services to reduce various risks like flooding, urban heat stress, and coastal erosion. Green areas, wetlands, green facades, and hybrid ecology areas do not only minimise exposure to the climate but also produce co-benefits concerning the preservation of biodiversity, human health, and social well-being (Boateng et al., 2023; Sommese, 2024; Seidu et al., 2025). NbS are also flexed and regenerative compared to traditional grey infrastructure, which also means that they can be more adaptive to changing climatic and urban conditions in the long term. Although both digital and nature-based innovations have become increasingly popular among scholars, the concepts are often conceptualised and implemented separately. Smart technologies are more frequently implemented as efficiency-oriented solutions that have minimal ecological engagement, and NbS are considered to be additional environmental intervention measures, as opposed to their main infrastructure feature. This division limits the potential of their strength together. Recent research also holds the view that purposeful climate resilience arises as a result of the systematic nature integration of digital intelligence and nature-based design within a single civil engineering platform (Abbas et al., 2025; Vasiliev et al., 2025). On the engineering level, this kind of integration allows reinventing urban infrastructure as a viable socio-technical-ecological system, where technological mind, ecological, and governance autonomies come into play in a continuous interaction. The measurements, monitoring, and adaptive control of NbS are made more measurable and trackable with the help of digital platforms, and the ecological infrastructure is made more adjustable, sustainable, and climate-buffering through digital mediated systems. The convergence can also be enhanced by AI-based optimisation by allowing the constant optimization of the ecological performance, carbon sequestration routes, and risk reduction plans (Islam, 2025). According to IGI Global studies on AI-assisted climate adaption and sustainable urban planning, intelligent systems are important in ensuring the future of the city is resilient due to a set of integrated data analytics and simulation and adaptive decision-making (Nexus of AI, Climatology, and Urbanism for Smart Cities; Real-World Applications of AI Innovation).

Governance capacity and institutional readiness is however, also important in effective integration. Digital governance models, climate-responsive planning systems, interoperability guidelines, and stakeholder inclusion are needed to convert integrated digital-NbS strategies into reality (Kim and Choi, 2021; Chien et al., 2022; Allan et al., 2024). These aspects are also especially crucial in the fast-urbanising and climate-sensitive areas, where investments in infrastructure are long-term in equity, sustainability, and resilience (Ilugbusi et al., 2024; Wijenayake, 2025). Additional IGI Global research on AI-enabled smart cities and disaster risk reduction also underscores the need to align governance, data integration, and regulation to address the risk of climate-based infrastructure in managing disaster-related risks in urban development (Revolutionising Urban Development and Governance with Emerging Technologies; Sustainable Smart Cities and the Future of Urban Development). It is on this background that this chapter contributes to three important aspects. To start with, it gives an analytical overview of digital innovations and nature-based solutions as two independent, yet twin, directions toward climate-resilient urban infrastructure. Second, it forms a unified analytical framework to relate smart technologies and sustainable design throughout the lifecycle of the infrastructure. Third, it converts this structure into application-focused pathways and governance knowledge of direct interest to civil engineering practice and policy.

CLIMATE CHANGE, URBAN RISK, AND INFRASTRUCTURE VULNERABILITY

Climate change has simply reshaped the risk profile of cities, augmenting chronic stresses and acute shocks, which pose direct impacts on the infrastructure systems. Increased temperatures, heavy precipitation, rise in sea level, droughts, and storms are causing more and more disruption to transportation systems, water systems, energy supply, buildings, and open spaces. These effects do not often stay in the realm of specific sectors; instead, they spread through the interconnected infrastructure systems, resulting in a series of failures, which increase the vulnerability of cities, as well as socio-economic disruption (Ilugbusi et al., 2024; Praveen et al., 2025). The urban infrastructure systems are complex socio-technical networks and they are embedded in environmental and institutional contexts. Chien et. al (2022) note that disruptions driven by climate reveal invisible interdependences between material assets, governance infrastructures and communal practices. As an example, the accessibility of roads, electricity supply, and emergency response capacity can be compromised at once by floods, increasing the risk above the original hazard. These dynamics point to the shortcomings of risk assessment on a sector basis and the importance of conducting integrated vulnerability analysis as a part of civil engineering and urban planning. Analytical approaches to climate resilience are gradually conceptualizing the vulnerability of infrastructure as an exposure, sensitivity and adaptive capacity. Salman et al. (2025) contend that climate-resilient infrastructure should be structured in a manner that it withstands hazards as well as can be functional, quickly recovered, and adapt to changes with time. Yilmaz (2023) also add that climate-resilient urban design should be based on integrated interventions both technically and ecologically and in governance levels. For civil engineering practice, this implies a transition from deterministic design standards toward performance-based and adaptive design approaches that account for evolving climate uncertainty.

Infrastructure systems are especially vulnerable in the climate-vulnerable and fast-urbanising areas. According to Wijenayake (2025), cities of developing economies tend to be more exposed because of poor drainage systems, heat-receptive building materials, insufficient green cover, and poor institutional capacity. These circumstances are prone to flooding and heat stress and limit adaptive reactions. Similar concerns Ilugbusi et al. (2024), who also stress that the issue of social vulnerability and lack of infrastructure oftentimes overlap, leading to unequal resilience performance among urban residents. The latest research points to the need to incorporate climate risk assessment into the whole infrastructure lifecycle, i.e. strategic planning and design, operation and maintenance. Praveen et al. (2025) emphasize that not addressing climate concerns during initial stages of planning may lead to the Maladaptation and the long-term lock-in effects. This indicates the importance of the analytical tools that would facilitate anticipatory planning and scenario evaluation and constant learning, which the traditional infrastructure assessment approaches frequently do not possess. Analytically, systematic climatic hazard mapping against infrastructure vulnerability constitutes critical foundation of focus on targeted climatic hazard resilience interventions. This kind of mapping can help decision-makers determine which assets are most important, the investments, and where the risk decrease will be most successful by the innovative solutions. Table 1 is a synthesis of major climate hazards, impacted infrastructure systems and the prevalent dimensions of vulnerability found in the literature, and it provides an organised description of the urban infrastructure risk profiles.

Table 1. Climate hazards and infrastructure vulnerability in urban systems

Climate Hazard	Primary Infrastructure Systems Affected	Dominant Vulnerability Dimensions	Implications for Resilience Strategies
Extreme rainfall and flooding	Drainage systems, roads, public transport, buildings	High exposure; limited drainage capacity; governance gaps	Need for integrated NbS and real-time monitoring
Heatwaves and rising temperatures	Buildings, energy systems, public spaces	Material sensitivity; lack of green cover; social exposure	Climate-responsive design and green infrastructure
Sea-level rise and coastal erosion	Coastal infrastructure, housing, ports	Physical exposure; ecological degradation	Hybrid grey–NbS solutions and adaptive planning
Drought and water scarcity	Water supply systems, urban ecosystems	Resource sensitivity; limited adaptive capacity	Smart water management and ecosystem restoration
Storms and extreme winds	Power networks, transport infrastructure	Structural vulnerability; cascading failures	Predictive analytics and adaptive system design

Source: Synthesised from Ilugbusi et al. (2024), Praveen et al. (2025), Salman et al. (2025), Wijenayake (2025), and Yılmaz (2023).

The new understanding based on the Table1 defines the ways in which urban infrastructure vulnerability is the result of a complex combination of physical exposure, ecological degradation, and institutional limitations. Most importantly, such patterns of vulnerability also shed light on certain possibilities of innovative intervention. Flooding and thermal stress are only a few examples of these hazards, which would help to emphasize the relevance of nature-based solutions to reduce exposure and sensitivity, but examples of cascading failures and epistemic uncertainty demonstrate the need to have digital solutions that enable real-time monitoring, predictive analytics, and adaptive decision-making. This methodological intersection of vulnerability evaluation and innovation paths is of the utmost significance. The modern paradigms of infrastructure planning often struggle with the problems of uncertainty and the interdependency of the systems. In comparison, digital analytics, intelligible monitoring, as well as scenario-based modelling provide the opportunity to rethink vulnerability mapping as action-oriented resiliency measures. In this way, these functions establish a conceptual gap between climate risk evaluation and the implementation of intelligent technologies, which is another theme that is worth exploring in the following section. As described in the present section, by defining climate risk and infrastructural susceptibility as dynamic, system-level dilemmas, the analytical framework in understanding the reasons behind the growing centrality of digital innovations in climate-resilient civil engineering is established. It highlights the need of the holistic approach that crosses siloed engineering solutions. Section 3 expands on this basis by evaluating how smart technologies, digital twins, and artificial intelligence can enhance the ability of urban infrastructure systems to foresee, deal with, and adjust to the hazards caused by the climate.

DIGITAL INNOVATIONS AND SMART TECHNOLOGIES FOR CLIMATE-RESILIENT INFRASTRUCTURE

Digital innovation has become a key facilitator of climate resilient urban infrastructure where civil engineering systems may transform into more dynamic, predictive and performance oriented operation. Smart technologies enhance the ability of infrastructure systems to sense the environmental conditions, process risk patterns and dynamically respond to stresses induced by climate. When discussing climate change, digital tools are nowadays seen as not simply efficiency optimizers, but analytical infrastructures allowing planning ahead, adjusting to changes, and becoming resilient to future hazards (Mallick, 2025; Rani et al., 2025). The smart infrastructure systems combine the digital sensors, Internet of Things (IoT) networks, geospatial platforms, artificial intelligence and advanced modelling methods to generate real-time situational awareness. These systems help to monitor the structural condition, environment variables and system performance continuously in infrastructure lifecycle. As Ali et al. (2025) show, digital twins, which are dynamic virtualizations of physical urban systems, can be used to simulate climate, determine the resilience of the infrastructure, and evaluate adaptation approaches to physical execution. This feature significantly decreases the uncertainty in the infrastructure planning and increases preparedness to the extreme events. The use of digital twins in climate-resilient urban planning has become especially topical because of the ability to combine climate data, spatial data, and infrastructure performance indicators. In its systematic review, Mabrouk et al. (2025) single out digital twin cities as integrative environments that help in decision-making based on resilience through the possibility to test scenarios, assess impact, and optimise in a wide range of climate conditions. Civil engineering In the context of civil engineering, digital twins make it possible to transition to predictive and preventive infrastructure management and improve resilience and resource efficiency. Artificial intelligence also enhances the digital resilience by supporting the state-of-art analytics, pattern recognition, and adaptive optimisation. Rani et al. (2025) underscore the role of AI in enhancing urban resilience through real-time data processing and intelligent control systems capable of responding to climatic variability. Islam (2025) extends this application by demonstrating how AI-driven architectures can optimise nature-based carbon sequestration systems, thereby reinforcing the functional convergence between digital innovation and sustainable urban design. The applications show that AI is not only an efficient tool to operational efficiency but also a factor in adaptive learning in the infrastructure systems.

Digital infrastructures are important also in institutional coordination and participatory governance towards climate-resilient urban deelopment. Allan et al. (2024) highlight that digital platforms allow the stakeholders to collaborate, enabling the sharing of data, transparency, and visualisation of scenarios. These platforms facilitate the use of evidence-based decision making and increased institutional capacity to deal with complex resilience strategies that cut across several infrastructure sectors. Digital technologies can be viewed in this regard as socio-institutional enablers and technical devices. Although digital innovations have a great potential, new challenges and risks, which are to be considered, are also introduced. Relying on quality data, the ability of digital systems to work together, and institutional capacity to handle multifaceted platforms have continued to be constraining issues. Mallick (2025) warns that inequalities in access to digital infrastructure and technical know-how will foster the development of socio-spatial disparities provided that smart technologies are introduced without the inclusion of schemes of planning. These constraints support the fact that digital innovation should be aligned with governance preparation and sustainable design principles, and not see technology as an independent solution. Analytically, digital innovations help to mitigate climate risks via unique yet related sets of functions which include risk anticipation, system optimisation, and adaptive management. Table 2 summarises the digital technologies that have emerged as major contributors to the literature on climate-resilient urban infrastructure systems and charts their functional contributions to the systems.

Table 2. Digital innovations and their functional contributions to climate-resilient urban infrastructure

Digital Innovation	Core Analytical Function	Infrastructure Applications	Climate Resilience Contribution
Digital twins	Scenario simulation and predictive analysis	Urban planning, transport, buildings	Anticipatory adaptation and risk reduction
Smart sensors and IoT	Real-time monitoring and data acquisition	Water, energy, structural systems	Early warning and rapid response
Artificial intelligence	Prediction, optimisation, adaptive control	Resource management, NbS optimisation	Enhanced adaptive capacity
Geospatial platforms	Spatial analysis and data integration	Urban regeneration, land-use planning	Evidence-based resilience planning
Intelligent control systems	Automated coordination and response	Energy and water networks	Service continuity under disruption

Source: Synthesised from Ali et al. (2025), Mabrouk et al. (2025), Allan et al. (2024), Rani et al. (2025), Islam (2025), and Mallick (2025).

The updated discussion of Table 2 highlights the complex nature of the role of digital technologies in the development of climate resilience. These technologies go beyond improving the efficiency of infrastructures and enable anticipatory governance and adaptive system behaviour to establish a strong analytical and monitoring infrastructure that is essential in the implementation and management of nature-based solutions. These solutions have the notorious requirement of having constant performance evaluation and adaptive maintenance, which digital innovations are best suited to handle. Digital innovations, smart technologies, are one of the pillars of climate-resilient civil engineering. They facilitate predictive planning, real time monitoring and responsive to climate risk, but their usefulness depends on the quality of the data, preparations with institutions and to the paradigms of sustainable and inclusive design methodologies. These reasons solidify the need to challenge nature-based solutions as a complementary resilience process which will be discussed in the next section.

NATURE-BASED SOLUTIONS AND SUSTAINABLE URBAN DESIGN

University professors often articulate the significance of nature-based solutions (NbS) as they represent a pivotal pivot in climate-resilient urban systems. These strategies are the combinations of adaptive, regenerative, and multifunctional responses to climate-related risks, which are based on natural processes and services. This way, NbS both reduce the risk of flooding, heat stress, drought, and coastal erosion and, at the same time, encourage the increase of biodiversity, improvement of public health, and social welfare. The NbS are also highly adaptable to the ever-changing conditions of climate change since, unlike traditional grey infrastructure, which is usually fixed and single-purpose, they are inherently adaptable and can change throughout their evolution (Boateng et al., 2023; Sommese, 2024). Within the realm of civil engineering practice, a distinct transition is evident: from viewing infrastructure purely as engineered artefacts to recognising hybrid socio-ecological systems that perform essential infrastructural functions. Research exemplifies this shift, with Seidu et al. (2025) illustrating how green infrastructure and ecosystem-based technologies augment urban resilience. They accomplish these by controlling water flows, controlling microclimates and reducing exposure to climate risks and thus recast these functions not as secondary environmental advantages but as part of the infrastructural performance. As a result, NbS are gradually re-defined as the essential elements of the urban infrastructure systems. NbS are applicable in a range of infrastructure spheres and space. Ecological interventions (constructed wetlands and bioswales and permeable surfaces) are used in urban water and stormwater management to increase the drainage capacity, decrease the risk of floods, and improve water quality. Leuzzo (2022) shows that the incorporation of NbS into the urban water management systems contributes to the creation of equitable and resilient reactions to the changing climatic conditions, which is possible through integrating ecological control with adaptive management practices. The applications are especially relevant in cities that are grappling with the growing rainfall volumes and the infrastructural drainage system.

At the neighbourhood and building level, NbS can help improve the climate-responsive urbanism in several ways, including raising thermal comfort, reducing energy consumption, and enhancing the quality of the environment. Lassandro et al. (2025) present the case of green and smart façade structures that have introduced innovative cultivation technologies, thus enhancing the performance of buildings and the resilience of the city climate. The article by Sádaba et al. (2025) also shows the potential of NbS as a potential synergy with urban infrastructures design to be more climate adaptive without negatively impacting on functional and aesthetic goals. This is just but an illustration of the applicability of NbS in different typologies of infrastructure. Urban environments prone to disasters due to coating and floods are now turning to NbS to find solutions on climate adaptation. David et al. (2025) can show that digitally manufactured hybrid NbS can enhance coastal resilience on the basis of combining ecological processes with engineered structures. These hybrid solutions represent the possible nature of NbS to act in tandem with traditional infrastructure providing superior flexibility and environmental efficiency. They also emphasize the extreme significance of monitoring and performance assessment to guarantee long-term effectiveness due to the changing environmental conditions. Beyond their physical functions, NbS engender substantial co-benefits that reinforce urban resilience. Chojnacka and Widera (2025) argue that NbS have an important value in the green transition, which enhances biodiversity, enhances human health, and advances social cohesion. Wijenayake (2025) shares the same opinion as he thinks that climate-sensitive urbanisation should incorporate NbS into larger sustainability plans that take into consideration equity and social vulnerability. These co-benefits lead to the confirmation of the worth of NbS as multifunctional infrastructure resources.

However, NbS application faces significant issues in terms of performance indeterminacy, future maintenance and governance alignment. Mishra et al. (2025) note that one of the common factors that undermine the NbS adoption is the lack of technical expertise, fragmentation of policy structures, and absence of performance indicators. The management of these hurdles requires strategies that help increase the measurability, the monitoring, and the adaptive management of NbS-capabilities the latter that can be significantly enhanced with the help of digital integration. As an analytical approach, and based on their functional capabilities, as well as the contribution of resilience, the category of NbS helps in strategic planning and the incorporation of digital technologies. A narrower typology of NbS according to their core functions, mitigated climate-related hazards, and resilience results in the form of performance are outlined in Table 3.

Table 3. Typologies of nature-based solutions and performance-oriented resilience functions

NbS Typology	Primary Infrastructure Function	Climate Hazards Addressed	Performance-Oriented Resilience Outcome
Green roofs and façades	Thermal regulation and energy efficiency	Heatwaves, urban heat island	Reduced cooling demand and heat stress
Urban green spaces and parks	Microclimate regulation and infiltration	Heat stress, extreme rainfall	Enhanced cooling and stormwater absorption
Constructed wetlands and bioswales	Stormwater and flood management	Flooding, water pollution	Improved drainage capacity and water quality
Coastal NbS (mangroves, dunes)	Shoreline protection and wave attenuation	Sea-level rise, storm surges	Reduced erosion and coastal vulnerability
Agroforestry and urban vegetation	Carbon sequestration and ecosystem restoration	Climate change mitigation	Enhanced adaptive capacity and biodiversity

Source: Synthesised from Boateng et al. (2023), Seidu et al. (2025), Leuzzo (2022), Lassandro et al. (2025), David et al. (2025), Sommese (2024), Chojnacka and Widera (2025), and Wijenayake (2025).

Table 3 illustrates the revised synthesis underlining the fact that Nature-Based Solutions (NbS) provide quantifiable resilience results when assessed using performance-based criteria rather than basing the results using descriptive characteristics only. This type of change toward measurable measurements is critical in achieving the integration of NbS into the context of the infrastructure planning systems that would focus on accountability, scalability, and effectiveness in the long term. In sum, NbS and sustainable urban design represent fundamental pillars of climate-resilient civil engineering, offering flexible, regenerative solutions to climate-induced risks. Nevertheless, they can only achieve their full potential when they are enhanced with effective monitoring, evaluation and adjustive management mechanisms. This lends greater importance to the urgent necessity to combine NbS with digital advances that provide the analytical tools required to quantify the performance, control the uncertainty, and maximize the outcomes of resiliency. The second part builds up a combined analytical system that integrates digital technologies with NbS in climate-resilient urban infrastructure systems.

INTEGRATING DIGITAL AND NATURE-BASED INNOVATIONS: AN ANALYTICAL FRAMEWORK

This part is the most important analytical work of the chapter, and it develops a systems-based model that combines digital innovations and nature-based solutions (NbS) in the climate-resilient city infrastructure. Although the previous sections have shown the personal usefulness of smart technologies and NbS, their single implementation restricts the results of resilience. New scholarship is also paying more attention to the realization that climate resilience should have hybrid infrastructures where digital intelligence is strategically coordinated with ecological processes in the stages of planning, design, operation, and governance (Abbas et al., 2025; Vasiliev et al., 2025). The suggested analytical framework is based on the idea of systems thinking, and urban infrastructure is conceptualised as an interrelated socio-technical-ecological system. In the given view, digital technologies are seen as tools of analysis and functionality that improve visibility, predictability, and adaptive capacity, with NbS being seen as the parts of regenerative infrastructure that reduce the risks associated with climate change and replace the ecological functions as well. Integration is not additive but relational in that way that digital systems complement the level of performance and manageability of NbS and ecological infrastructure supplements the sustainability and resilience of digitally enabled urban systems. Digital twins are a principle of operationalisation of integration. As shown by Abbas et al. (2025), digital twins with NbS can also be used to design cities predictively by connecting the real-time environmental conditions, climate conditions, and environment performance metrics. The above combination of integration enables the engineers and planners to model the effects of climate variability on the NbS performance, trade-offs between design options, and optimisation of interventions prior to physical implementation. Mabrouk et 25 and Ali et 25 also point out that digital twin cities offer integrative platforms in resilience-oriented planning because they entail the integration of geospatial data, infrastructure models, and climate projections in one decision-support setting.

This integrative potential is reinforced by artificial intelligence, where unceasing optimumisation and flexible control over the ecological infrastructure is made possible. Islam (2025) demonstrates how AI-based architectures have the potential to optimise nature-based carbon sequestration systems based on data-driven performance measurement and active control. Examples of this include such applications as exemplifying a change of the previous NbS design to performance-managed ecological infrastructure, where the results of resilience are actively measured and optimised over time. The integration goes to governance and stakeholder coordination as well. Allan et al. (2024) point out that collaboration can be facilitated through digital infrastructures through the provision of shared access to data, visualisation of scenarios, and performance metrics. In case NbS planning and management is reinforced on the basis of open digital platforms, the process of governance becomes more inclusive and evidence-based. Chien et al. (2022) also posit that the ICT-facilitated metropolitan commons governance enhances the collective care of ecological infrastructures, which increases institutional and social resilience. Analytically the concept of integration can be construed as the coordination of the digital functions (monitoring, simulation, optimisation) and NbS processes (regulation, mitigation, regeneration) through the infrastructure lifecycle. This alignment transforms NbS from passive design elements into adaptive infrastructure systems capable of responding dynamically to climate uncertainty. A reformed integration matrix, which operationalises this framework by mapping the main digital tools with domains of NbS applications and results of resilience is provided in Table 4.

Table 4. Digital–nature-based integration matrix for climate-resilient urban infrastructure

Digital Innovation	NbS Application Domain	Integrated Analytical Function	Climate Resilience Outcome
Digital twins	Urban green spaces, wetlands	Scenario simulation and performance testing	Improved anticipatory adaptation
Smart sensors and IoT	Green roofs, bioswales	Real-time monitoring of ecological performance	Early detection of stress and failure
Artificial intelligence	Carbon sequestration, vegetation systems	Adaptive optimisation and predictive analytics	Enhanced mitigation and adaptive capacity
Geospatial platforms	Urban landscapes and ecological corridors	Spatial risk mapping and planning support	Reduced exposure to climate hazards
Intelligent control systems	Hybrid grey–green infrastructure	Automated coordination and response	Continuity of service during extremes

Source: Synthesised from Abbas et al. (2025), Ali et al. (2025), Mabrouk et al. (2025), Islam (2025), Allan et al. (2024), and Vasiliev et al. (2025).

The revised integration matrix clearly demonstrates that digital tools and nature-based solutions (NbS) function as complementary elements within climate-resilient infrastructure, rather than as distinct paradigms. Ecological infrastructure provides the regenerative resistance mechanisms to curb long-term climatic risks and environmental stress, whereas digital technologies will equip the evaluative, optimisation, and scale of NbS. The hybrid framework develops in climate-prone and rapidly urbanising settings where the infrastructure funding has to balance the cost-efficiency, expansion, and future adaptability. Abbas et al. (2025) demonstrate that the digital twins and NbS can be integrated in the neighbourhood level, which enables future residential buildings to be developed, transferable knowledge is acquired in heterogenous urban environments. These frameworks recommend the use of incremental implementation and flexibility and adaptive learning.

However, there are practical issues even in the face of the analytical soundness of the integration. The keys to success are in the interoperability of digital systems and access to high-quality environmental information and the coordination of planning tools with the regulatory frameworks. Kim and Choi (2021) emphasize the need to implement the integrated frameworks through climate-oriented and digitally empowered planning tools that can be transformed into realities of operations. In the absence of such alignment, integration can only be conceptual and not a paradigm to be actualized. Altogether, both analytical frameworks developed in the present paper represent an impetus to climate-resistant civil engineering through the development of predictive digital intelligence and adaptive ecological planning. Reprojecting digital and nature-based innovations as mutually supporting parts, the framework provides a strong ground to apply the resilience theory into the concrete infrastructure solutions. The given analysis preconditions the application-oriented implementation pathways, which will be further developed in the next section.

APPLICATION-ORIENTED DOMAINS AND IMPLEMENTATION PATHWAYS

Leveraging the translation of integrated digital and nature-based structures into working practise is the key challenge in climate resilient civil engineering. Although analytical models establish the strategic directions, the final resilience results depend on how such models are operationalised in specific areas of infrastructure. Thus, application-based pathways should make a specific consideration of the spatial scale, the risk profile of climate, institutional capacity, and technological preparedness. The recent literature reveals that the implementation of integrated digital-NbS strategies can be successfully implemented in a variety of infrastructures in urban environments and, therefore, provides scalable and context-aware solutions to climate adaptation (Abbas et al., 2025; Allan et al., 2024). On the neighbourhood and urban regeneration level, climate-resilient residential development and the redevelopment of the existing urban fabric is supported by integrated digital to NbS approaches. A predictive framework suggested by Abbas et al. (2025) is the integration of digital twins and NbS to create neighbourhoods that are resilient in future and respond to climate uncertainty. Planners and engineers are able to optimise the location of green infrastructure, drainage capacity and thermal regulation through modelling climate situations and ecological behaviour in virtual worlds prior to site construction. The method decreases uncertainty in designs and assists in making evidence-based investment choices. Allan et al. (2024) also underline that digitally enabled planning platforms improve stakeholder co-ordination on climate-resilient urban regeneration to synchronise the infrastructural intervention with the sustainability and resilience goals. One of the most imperative fields of application where integration provides instant benefits of resiliency is within the urban water and stormwater management. Constructed wetlands, bioswales, and permeable surfaces are nature-based interventions that minimise the risk of floods and improve water quality, and they also allow monitoring the performance through digital monitoring sites. As shown by Leuzzo (2022), the combination of digital approaches with NbS in city water management will help to address climate variability with adaptive responses and enhance equity in delivering the services. In the view of a civil engineer, this integration provides an opportunity to transition the current approach of designing drainage in a static manner to adaptive water infrastructure systems which can adjust to the changing rain patterns. On urban zones with complex and dynamic climate risks, e.g., coastal areas and flood-prone zones, digital-NbS strategies are integrated to cope with the risks. David et al. (2025) demonstrate how digitally-created hybrid NbS can improve the coastal resilience through interactions of both natural and artificial environment. Digital tools allow accurate design, simulation of performance, and long-term monitoring and hence uncertainty that arises when NbS is used in the conditions of sea-level rise and storm surge levels is minimized. Such hybrid solutions are the examples of the reliability and scalability of NbS in high-risk settings through digital integration.

On the level of the building and public space, integrated practises help to make the city respond to climatic conditions and be efficient in its operations. Lassandro et al. (2025) show that green and smart facades systems with innovative cultivation technologies can enhance the energy performance of buildings and urban microclimates. Such NbS can be optimally managed with time when reinforced with digital surveillance tools and smart control systems to enhance the benefits of resilience in the long term. Sadaba et al. (2025) also demonstrate that the incorporation of NbS into the design of urban infrastructure improves climate adaptation without compromising the usefulness and aesthetic quality, which supports the fact that they can be used in highly populated cities. Implementation of integrated solutions is also a criterion of construction and operational stages. Joseph et al. (2026) emphasise the fact that smart motion-control systems and digitally enabled technologies in the construction industry enhance accuracy, minimise material wastage and facilitated sustainable infrastructure provision. These technologies help to make the integrated digital-NbS designs practically achievable through improving the quality of the construction process and promoting the adaptive maintenance process across the infrastructure life cycle. When viewed through the lens of practitioner-focused approach, to implement it properly, one must tune analytical frameworks to the site-specific conditions and institutional reality. The digital-NbS solutions should be integrated based on the local weather hazards, the governance framework, and availability of resources. Digital tools facilitate coordination and surveillance and NbS offer low-carbon tailed resilience services, which are incrementally expandable. Table 5 synthesises key application domains, integrated solutions, and implementation pathways, explicitly linking the analytical framework developed in Section 5 to operational civil engineering practice.

Table 5. Application domains and implementation pathways for integrated digital–NbS infrastructure

Application Domain	Integrated Digital–NbS Solution	Implementation Pathway	Practitioner-Relevant Resilience Benefit
Residential neighbourhoods	Digital twins + urban green infrastructure	Predictive design and scenario-based planning	Reduced heat stress and flood risk
Urban regeneration	Digital platforms + ecosystem-based design	Stakeholder-led, data-driven redevelopment	Inclusive and climate-resilient renewal
Urban water systems	Smart monitoring + constructed wetlands	Adaptive drainage and water management	Improved flood control and water quality
Coastal infrastructure	Digital fabrication + hybrid NbS	Performance-based coastal adaptation	Reduced erosion and storm surge impacts
Buildings and public spaces	Smart façades + green design	Adaptive building operation	Enhanced thermal comfort and energy efficiency

Source: Synthesised from Abbas et al. (2025), Allan et al. (2024), Leuzzo (2022), David et al. (2025), Lassandro et al. (2025), Sádaba et al. (2025), and Joseph et al. (2026).

The revised synthesis presented in Table 5 demonstrates that integrated digital–NbS solutions are not merely confined to experimental or pilot projects; rather, they can be systematically embedded within mainstream civil engineering practice. More importantly, these applications demonstrate how digital intelligence increases the scalability and reliability and adaptive management of NbS, and ecological infrastructure enhances the sustainability and resiliency of digitally enabled systems. The practical implementation of the integrated digital and nature-based innovations is application-based and provides practical avenues within the process of developing climate-resilient civil engineering. Through harmonizing the predictive digital tools, ecological design and intelligent technologies to build, cities are able to create an adaptive, resource-efficient and climate-responsive infrastructure systems. These insights on applications highlight the need to have supportive governing and institutional structures that would be discussed in the next section.

GOVERNANCE, POLICY ALIGNMENT, AND INSTITUTIONAL READINESS

Effective implementation of integrated digital and nature-based solutions (NbS) to transform into climate-resilient urban infrastructure largely depends on the strength of the governing structure, the consistency of the policy, and the institutional preparedness. Although technological and ecological innovations are purported to offer the necessary resources of resilience, their effectiveness is bound to be influenced by the regulatory, organisational, and socio-political milieus in which they are implemented. Disjointed governance frameworks, sectoralism, and inappropriate policy tools remain limiting the implementation of integrated resilience solutions on a massive scale in most urban settings (Kim and Choi, 2021; Mishra et al., 2025). Governance challenges frequently arise from the compartmentalisation of responsibilities across digital infrastructure development, environmental stewardship, and civil engineering practice. Kim and Choi (2021) argue that climate-sensitive and digitally facilitated tools of planning are also essential to overcome such divisions through instilling resilience aims into the statutory spatial planning and infrastructure investment procedures. These types of instruments enable cross-sector alignment through planning land-use, the provision of green infrastructure, and digital transformation agendas as one policy concept. Policy alignment is determinant in mainstreaming NbS as a fundamental part of the infrastructure, and not as a form of environmental side measures. Mishra et al. (2025) note that the lack of explicit regulatory guidelines, performance indicators and long-term mechanisms of financing usually restricts the NbS implementation and expansion. In the absence of policy certainty, NbS initiatives are likely to fail to attract an investment, or to show longer term value, or to work well with traditional infrastructure systems. Therefore, it is necessary to have consistent signals of policy in order to make engineers, planners, and investors feel confident.

Digital governance systems also have a direct impact on the performance of integrated digital-NbS strategies. Allan et al. (2024) emphasize that climate-resilient urban regeneration can be based on the digital platforms and allow transparency, data sharing, and active decision-making. Shared digital tools and performance data give stakeholders greater access to governance processes, and make them more evidence-based. Chien et al. (2022) also observe that ICT-enabled urban commons governance enhances the collective stewardship of common ecological and digital resources, which contributes to the institutional capacity and social resilience. Institutional preparedness includes technical know-how, organisational capability and inter agency coordination. Most cities, especially in the fast urbanising and climate-exposed areas, are faced with the limitation of inadequate digital infrastructure, lack of professional expertise, and deficiency in agency coordination (Ilugbusi et al., 2024; Wijenayake, 2025). When technological solutions are present, these restrictions may compromise the adoption of integrated frameworks. Equity and inclusivity are becoming more well known as being central in climate-resilient infrastructure governance. As Ilugbusi et al. (2024) note, the adaptation strategies should also focus on social vulnerability and not strengthen the prevailing disparities. Digital and NbS programmes that do not consider marginalised populations may have an effect of causing resilience gains to be clustered together with those already privileged. It is thus essential to have mechanisms of inclusive governance, e.g. participatory planning, open access to data, and community stewardship models to ensure the attainment of just and sustainable resilience outcomes.

When considering policy priorities to support integrated digital infrastructure, there are three priorities that can be identified to support NbS infrastructure. To start with, the governments are advised to incorporate climate resilience and digital inclusion criteria in the statutory planning and infrastructure funding models. Second, the regulatory bodies are to create performance-based standards in which NbS are recognised as operational infrastructure assets with the assistance of digital monitoring and evaluation systems. Third, the institutional capacity building (professional training, inter-agency coordination) and digital infrastructure should be targeted to be invested in to facilitate long-term implementation. These can be used to facilitate the gap between innovations and business operations. In an analytical perspective, the governance preparedness towards integrated digital NbS infrastructure may be evaluated on a variety of dimensions, such as planning tools, regulatory framework, digital governance processes, institutional capacity, and engagement of stakeholders. Table 6 constitutes a synthesis of the main governance and policy enablers in the literature, and outlines how they are involved in enabling the integration of climate-resilient infrastructure.

Table 6. Governance and policy enablers for integrated digital–nature-based infrastructure

Governance Dimension	Key Enablers	Role in Integration	Climate Resilience Outcome
Planning instruments	Climate-smart, digitally enabled urban plans	Align spatial, ecological, and digital strategies	Coherent resilience-oriented development
Regulatory frameworks	NbS performance standards and guidelines	Mainstream NbS as infrastructure assets	Long-term reliability and scalability
Digital governance	Open data platforms and shared digital tools	Enable coordination and transparency	Inclusive and adaptive decision-making
Institutional capacity	Skills development and inter-agency coordination	Operationalise integrated frameworks	Effective and scalable implementation
Stakeholder engagement	Participatory and commons-based governance	Foster social ownership and equity	Just and inclusive resilience outcomes

Source: Synthesised from Kim and Choi (2021), Allan et al. (2024), Mishra et al. (2025), Chien et al. (2022), Ilugbusi et al. (2024), and Wijenayake (2025).

The updated governance synthesis, which is introduced in Table 6, shows that a combined digital-NbS infrastructure requires an facilitating institutional ecosystem in which policies and regulations and organisational practices are correlated with climate resilience goals. Notably, governance structures should be dynamic so that they would enable constant learning and adaptation as risks associated with climate emerge and new technologies become available. The core requirements of the successful implementation of the integrated digital and nature-based innovations aimed to support climate-resilient urban infrastructure are governance, policy alignment, and institutional readiness. By integrating the goals of resilience into the tools of planning, enhancing digital governance, and focusing on the inclusive involvement of all stakeholders, the cities will be able to establish the conditions that will facilitate the high-scaled and equitable transformation of infrastructures.

FUTURE PATHWAYS FOR CLIMATE-RESILIENT CIVIL ENGINEERING PRACTICE

The digital innovations and nature-based solutions (NbS) convergence will represent a critical pathway in addressing the emerging climate risks; however, the efficacy of the convergence in the long term will be based on the future maturation of civil-engineering practise based on the analytical, technological, and institutional dimensions. The key direction of future is an increased use of predictive and scenario-based infrastructure planning facilitated by digital twins. Abbas et al. (2025) and Mabrouk et al. (2025) show that digital twin-based platforms enable engineers and planners to model climatic conditions, analyze the behavior of infrastructure, and experiment with integrated digitalNbS before a physical setup. Such tools will become critical as climate uncertainty rises because the planning of infrastructure will be oriented towards performance-based decision-making rather than the compliance-driven design. Climate-resilient management of infrastructure is also likely to be altered through artificial intelligence, which will allow further optimisation and learning. Rani et al. (2025) emphasize the increased importance of AI in the processing of real-time environmental information and assistance of quick and context-specific response to climatic variability. Islam (2025) goes further with such a view by depicting how AI can optimise nature-based carbon sequestration systems, and this statement supports the merging of mitigation and adaptation goals. Based on this, the future of the civil-engineering practise is expected to become more dependent on AI-based analytics to operate intricate hybrid infrastructures.

The other significant route is an institutionalisation of NbS to become conventional components of civil-engineering design and delivery. Seidu et al. (2025) and Sommese (2024) point out that the scalability of NbS in the long term is based on the establishment of standardised designing rules, performance indicators, and maintenance systems. The introduction of NbS, by integrating it into the engineering codes, procurement methods and asset-management systems, will be essential in moving pilot-projects to their mainstream implementation. Professional change is another avenue of progressing to climate resilient civil engineering. According to Joseph et al. (2026), construction technologies facilitated with a digital or intelligent infrastructure systems are transforming the skills needs in the profession. The same engineers in the future will also require skills in digital modelling, systems thinking, ecological design as well as data analytics besides the conventional engineering skills. Incorporating these competencies in education, training and professional accreditation will also be important in maintaining innovation. Governance and policy wise, there is a need to be more consistent in linking the climate adaptation strategies, digital transformation agendas, and sustainable urban development frameworks in the future. Kim and Choi (2021) and Allan et al. (2024) emphasize the significance of combined planning tools and cybernetic management models that enable the cross-sector and stakeholder engagement. This alignment is especially important in fast urbanising and climate-at-risk places where the decisions of infrastructure are long term in terms of equity and resilience (Ilugbusi et al., 2024; Wijenayake, 2025). Key gaps that require future research are the availability of data, the institutional capacity, and the situational implementation. Although there is a high conceptual potential of integrated digital-NbS frameworks, the long-term performance, cost-effectiveness, and social outcome of these frameworks are scarce in empirical studies. To polish on the strategies of integration and make the evidence base more robust in relation to the climate-resilient civil-engineering practice, comparative studies, longitudinal assessment, as well as applied experimentation (including urban pilots and living laboratories) will be necessary. Overall, the future directions of climate-resilient civil engineering are in the field of the systematic incorporation of predictive digital intelligence, adaptive ecological infrastructure and facilitating governance frameworks. Through the adoption of hybrid infrastructure systems where smart technologies and sustainable design intertwine, civil engineering can take a centre stage in creating urban environments that would be in a position to accommodate climate uncertainty and sustainability over the long run in addition to promoting social well-being.

CONCLUSION

The climate change has essentially altered the frame of reference in which the urban infrastructure systems are conceived, designed, and governed. The visible amplification of severe weather phenomena, the growing climate insecurity and unceasing environmental resource depletion has highlighted the limitations of traditional civil engineering paradiges that valued the fixed design, sectoral performance and the short-term performance measures. These issues therefore necessitate the shift towards adaptive, integrated and dynamically responsive infrastructure systems in response to emerging and changing climate risks. As explained in this chapter, digital innovations, such as Digital Twins, smart sensing networks, artificial intelligence, and geospatial systems, represent significant potential to improve the predictive, adaptive, and operational capabilities of the urban infrastructure systems. These technologies enable real-time observation, planning scenarios, and data-driven decision-making and hence assist in proactive management of climate risks as opposed to responses. At the same time, Nature-Based Solutions (NbS) can offer regenerative, multifunctional and flexible routes to climate resilience through the mobilisation of the ecological processes to reduce hazards, rebuild the ecosystems and make cities livable. NbS do not simply lead to risk reduction but also to more general environmental and social co-benefits that are becoming of central importance to sustainable urban development. One of the key contributions of this chapter is the further development of an analysis framework which places digital and nature-based innovations as mutually reinforcing and interdependent elements of climate-resilient infrastructure systems. Once NbS are integrated into digitally enabled planning, monitoring, and governance systems, ecological interventions can be converted into performance-managed assets which are adaptively responsive to the variability of climatic conditions. The aspects of application discussed, such as the development of neighbourhoods, urban water systems, marine infrastructure, and buildings, and urban public areas have shown that this kind of integration is not only technically possible, but is contextually desirable in various urban environments. The discussion also highlights that technological and ecological innovation cannot achieve the results of resilience without facilitating governance and institutional capacity. Planning tools that are climate-smarter, balanced regulatory regimes, participative systems of governance and long-term capacity building of the profession are required to help put the integrated frameworks into practice. Without these enabling conditions, the transformative potential of digital and nature-based innovations may be left very fragmented or unrealised. Finally, the integration of smart technologies and green, nature-driven design is a path of change towards climate-resilient civil engineering. Through systems thinking, enhanced interdisciplinary cooperation and placing greater focus on long-term flexibility than short-term optimisation, the practise of civil engineering can be critical in creating the urban infrastructure systems that are sustainable, inclusive, and robust. This combined strategy provides a proactive avenue to solving multidimensional and interrelated issues of climate change and help reinforce resilient cities in the future.

REFERENCES

Abbas, A., Adeel, M., Akbar, G., Hayyat, S., & Din, G. M. (2025). Integrating Digital Twin and Nature Based Solutions for Climate Resilient Urban Design: A Predictive Framework for Future Ready Residential Neighborhoods in the Anthropocene. *EJSMT*, *1*(5), 1–14. DOI: 10.59324/ejsmt.2025.1(5).01

Ali, E., Mansour, A., Mohammed Abdelkader, E., Elshaboury, N., & Zayed, T. (2025). Digital twin for climate resilience: Transforming smart cities for a sustainable future. *The International Archives of the Photogrammetry, Remote Sensing and Spatial Information Sciences*, *48*, 139–145. DOI: 10.5194/isprs-archives-XLVIII-G-2025-139-2025

Allan, M., Rajabifard, A., & Foliente, G. (2024). Climate resilient urban regeneration and SDG 11–stakeholders' view on pathways and digital infrastructures. *International Journal of Digital Earth*, *17*(1), 2385076. DOI: 10.1080/17538947.2024.2385076

Boateng, E. A., Asibey, M. O., Cobbinah, P. B., Adutwum, I. O., & Blija, D. K. (2023). Enabling nature-based solutions: Innovating urban climate resilience. *Journal of Environmental Management*, *332*, 117433. DOI: 10.1016/j.jenvman.2023.117433 PMID: 36738724

Chien, H., Hori, K., & Saito, O. (2022). Urban commons in the techno-economic paradigm shift: An information and communication technology-enabled climate-resilient solutions review. *Environment and Planning. B, Urban Analytics and City Science*, *49*(5), 1389–1405. DOI: 10.1177/23998083211066324

Chojnacka, K., & Widera, B. (2025). Nature-Based Solutions and Other Critical Aspects of Urban Innovations for Green Transition. *European Review (Chichester, England)*, *33*(4), 1–10. DOI: 10.1017/S1062798725100343

David, C. G., Kremer, J., Ashwini, M., Kloft, H., & Goseberg, N. (2025). Digital fabrication of Hybrid Nature-based Solutions as new opportunity for coastal climate change adaptation. *Cambridge Prisms: Coastal Futures*, *3*, e24. DOI: 10.1017/cft.2025.10014 PMID: 41307040

Ilugbusi, B. S., Adisa, O., Obi, O., Awonuga, K. F., Adelekan, O. A., Asuzu, O. F., & Ndubuisi, N. L. (2024). Urban resilience to climate change: A review of adaptation strategies and infrastructure innovations. *Ecofeminism and Climate Change*, *5*(1), 18–23. DOI: 10.26480/efcc.01.2024.18.23

Islam, F. S. (2025). Artificial intelligence-driven optimization of nature-based carbon sequestration: A scalable architecture for urban climate resilience. *International Journal of Environment and Climate Change*, *15*(7), 252–277. DOI: 10.9734/ijecc/2025/v15i74928

Joseph, H. S., Devaraj, A. R., Priscila, A., Selvalakshmi, V., & Rameshkumaar, V. P. (2026). Advancing Sustainable Civil Engineering Through Intelligent Motion Control: Construction Equipment and Infrastructure Development. In *Methodologies and Applications of Intelligent Motion Control Systems* (pp. 105-132). IGI Global Scientific Publishing.

Kim, K. G., & Choi, H. S. (2021). Planning instruments for climate smart and wise cities: a spatial, green and digital deal approach. In *Planning Climate Smart and Wise Cities: A Multidisciplinary Approach* (pp. 3–97). Springer International Publishing.

Lassandro, P., Capotorto, S., & Mammone, V. (2025). Nature-Based Solutions: Green and Smart Façade with an Innovative Cultivation System for Sustainable Buildings and More Climate-Resilient Cities. *Sustainability (Basel)*, *17*(10), 4580. DOI: 10.3390/su17104580

Leuzzo, A. (2022, May). A Methodology Toward a Just, Digital and Ecological Transition for Resilient and Sustainable Scenarios: Innovations in Urban Water Management. In *International Symposium: New Metropolitan Perspectives* (pp. 332-356). Cham: Springer International Publishing.

Mabrouk, M., Mahran, M. G. N., Saleh, D., & AbouKorin, S. A. A. (2025). Digital Twin Cities for Climate-Resilient Urban Planning: A Systematic Review of Frameworks, Applications, and Future Perspectives. *Advance in Sustainability*, *5*(2).

Mallick, R. (2025). Smart Cities and Climate Resilience. In *Community Resilience and Climate Change Challenges: Pursuit of Sustainable Development Goals (SDGs)* (pp. 401–412). IGI Global Scientific Publishing.

Mishra, A., Bajpai, S., & Bhadwal, S. (2025). Nature-based solutions for climate-resilient development: a technical, policy, and governance perspective. In *Handbook of nature-based solutions to mitigation and adaptation to climate change* (pp. 1831–1854). Springer Nature Switzerland. DOI: 10.1007/978-3-031-34967-6_159

Praveen, G., kashi Vishwanath, N., Sree, C. C., & Rohith, D. (2025). Climate-Resilient Urban Planning. *International Journal of Advances in Agricultural Science and Technology*, *12*(3), 226–234.

Rani, L. Hurmat, Kaur, B., & Arora, R. (2025). Role of AI in Developing Smart Urban Cities and Resilience to Climate Change. In *Climate Resilience: Impact of Quantum Computing and Artificial Intelligence on Urban Planning* (pp. 187-208). Cham: Springer Nature Switzerland.

Sádaba, J., Luzarraga, A., & Lenzi, S. (2025). Designing for climate adaptation: A case study integrating nature-based solutions with urban infrastructure. *Urban Science (Basel, Switzerland)*, *9*(3), 74. DOI: 10.3390/urbansci9030074

Salman, M., Najid, S. V. P., Ali, U., Khairoowala, Z., & Siddiquee, T. A. R. (2025). Building for Tomorrow: Climate-Resilient Infrastructure Strategies for Urban Sustainability. In *Urbanization, Climate Change, and Health: Integrating Strategies for Sustainable and Resilient Cities* (pp. 291–311). Springer Nature Switzerland. DOI: 10.1007/978-3-031-99132-5_14

Seidu, S., Chan, D. W., Edwards, D. J., & Owusu-Manu, D. G. (2025). A systematic review of green infrastructure and ecosystem-based technologies for climate resilience: Development trends, research outlook and future projections. *Environmental Research Letters*, *20*(8), 083001. DOI: 10.1088/1748-9326/ade730

Sommese, F. (2024). Nature-based solutions to enhance urban resilience in the climate change and post-pandemic era: A taxonomy for the built environment. *Buildings (Basel, Switzerland)*, *14*(7), 2190. DOI: 10.3390/buildings14072190

Swadhi, R. (2025). Innovative strategies for widespread adoption in a climate-smart future: Scaling up agroforestry. In *Agroforestry for a Climate-Smart Future* (pp. 473–496). IGI Global Scientific Publishing. DOI: 10.4018/979-8-3693-8282-0.ch016

Vasiliev, D., Stevens, R., & Bornmalm, L. (2025, February). Nature-Based Solutions and Smart Cities. In *International Congress on Information and Communication Technology* (pp. 183-192). Singapore: Springer Nature Singapore.

Vettriselvan, R., Sathya, M., & Velmurugan, T. (2018). Productivity and profitability mechanical engineering entrepreneurs: Business perspective. *International Journal of Mechanical Engineering and Technology*, *9*(8), 758–765.

Wijenayake, A. (2025). Climate Change and Urban Resilience: Strategies for Sustainable Cities. *International Journal of Academic Research in Business & Social Sciences*, *15*(4), 247–261. DOI: 10.6007/IJARBSS/v15-i4/25093

Yılmaz, D. G. (2023). Nature-based Solutions for climate-resilient cities: A proposal of a model for successful implementation. *Journal of Design for Resilience in Architecture and Planning*, *4*(2), 189–203. DOI: 10.47818/DRArch.2023.v4i2092

Chapter 12

Integrating Renewable Energy Systems in Civil Infrastructure: Thermo-Mechanical Innovations for Net-Zero Buildings

Shalom Akhai
http://orcid.org/0000-0002-7533-457X
Maharishi Markandeshwar Engineering College, Maharishi Markandeshwar (Deemed to be) University, Mullana, India

Mahapara Abbass
http://orcid.org/0000-0002-0838-2598
Maharishi Markandeshwar Engineering College, Maharishi Markandeshwar (Deemed to be) University, Mullana, India

ABSTRACT

Modern infrastructure is undergoing a major transformation to meet sustainability and climate goals. This chapter explores how renewable energy systems and thermo-mechanical innovations enable Net-Zero Energy Buildings (NZEBs) and net-zero performance at district scale. It links envelope design, HVAC strategies, energy recovery, and storage with onsite renewables such as PV, PV-T, energy piles, and ambient loops. Key performance metrics—including energy use intensity, peak demand, and carbon intensity—are addressed alongside durability, safety, and techno-economic evaluation. Practical pathways are illustrated for composite, hot-humid, and cold climates, highlighting innovations like radiant systems, desiccant dehumidification, model predictive control, and life-cycle assessment. The discussion emphasizes integrated workflows, barrier-solution maps, and emerging

DOI: 10.4018/979-8-3373-6167-3.ch012

technologies such as digital twins, low-GWP heat pumps, and advanced thermal storage. By combining efficiency, renewables, and resilience, the chapter outlines how civil infrastructure can achieve scalable net-zero design.

1. INTRODUCTION

Net-zero buildings are no longer rare experiments; they have become common goals in cities, companies, and national policies (Satoła et al., 2022; Ürge-Vorsatz et al., 2020). Today, many governments include them in their energy strategies, and businesses highlight them in their sustainability commitments. This shows that net-zero construction has moved from being a special ambition to being a basic expectation for the future of the built environment (Satoła et al., 2021; Ürge-Vorsatz et al., 2020).

For civil engineers, the real challenge is how to make this vision practical. A building is not just an idea; it must be designed to meet structural safety rules, resist earthquakes and fires, fit within the limits of the site, and stay within a realistic budget. Engineers must therefore think carefully about how renewable energy and smart mechanical systems can be included without ignoring these real-world requirements. To achieve this, it is not enough to look at each part of the building separately. Every decision affects another part. For example, the amount of heat a building gain through its walls and windows will change the size of the cooling system needed. The type of ventilation chosen will affect not only energy use but also comfort and air quality for the people inside. The materials used will influence insulation, long-term durability, and even how easily parts of the building can be recycled in the future. In the same way, the way electricity is planned must allow the use of renewable sources, storage systems, and smart controls. A true net-zero design connects all these aspects into a single working system (Delavar & Sahebi, 2020; Gangwar et al., 2020; Kosonen & Keskisaari, 2020; Kumar et al., 2021).

When we talk about "net-zero," it is important to understand what the term means. Net-zero energy in buildings usually means that, over the course of a year, the building produces as much renewable energy as it consumes. This balance is achieved through efficient design and the use of solar panels, wind energy, or other renewable systems. Net-zero carbon goes a step further. It not only considers the energy used during the building's operation but also includes the emissions caused by producing and transporting the materials used in construction. The definition of boundaries also matters. Sometimes net-zero is measured only at the building site, while in other cases it includes the wider energy system, such as electricity lost during transmission or the sharing of energy across a district network. These boundaries must be set at the very beginning of the project, as they guide design choices and ensure compliance with regulations (Canale et al., 2021; Hussain et al., 2021; Martinez-Bolaños et al., 2020; Wehkamp et al., 2020).

It is also necessary to understand why thermo-mechanical integration is so important. Many people believe that simply adding solar panels to a building is enough to make it net-zero, but this is rarely the case. If the building has poor insulation or inefficient mechanical systems, the energy demand will be too high for the panels to cover. On the other hand, even a very efficient building envelope and advanced HVAC system may still rely on electricity from fossil fuels if the grid is not clean. The solution is to combine strong building envelopes with advanced heating and cooling technologies, energy recovery systems, and storage options. Smart control systems also play a role in ensuring that energy is used wisely and in harmony with renewable generation (Hurwitz et al., 2020; Jha et al., 2020; Wu & Skye, 2021).

In short, reaching net-zero is about balance. It requires reducing the energy a building needs by making it more efficient and at the same time increasing the share of renewable energy it produces and uses. Only when both sides are addressed together does the building truly become net-zero. This chapter explains how these ideas can be connected in practice, showing the pathway from concept to calculation and finally to real-world implementation.

2. METRICS, TARGETS, AND ENERGY BALANCE

The design of a net-zero building always begins with clear and measurable targets. These targets are not just numbers; they represent how the building will meet the needs of its occupants while also supporting climate goals (Akhai & Abbass 2025). By setting these values at the start, engineers and designers can make informed choices about systems, materials, and layouts that keep the project on track (Apanavičienė et al., 2020; Crespi et al., 2020; Zeng et al., 2020).

One of the most important measures is energy use intensity, which refers to the total energy a building uses in a year divided by its floor area. This gives a fair comparison between buildings of different sizes. Another key measure is peak demand, which shows the highest level of energy a building requires at any given time. This is critical because it determines the size of transformers, chillers, and other mechanical plants, and it also affects the electricity costs through demand charges (Singh & Maini, 2020; Zorn et al., 2020).

Comfort for the people inside the building is equally important. Thermal comfort is usually described through temperature and humidity ranges. For example, many office spaces aim to keep the temperature between 23 and 26 degrees Celsius and the relative humidity between 40 and 60 percent. In naturally ventilated spaces, the concept of adaptive comfort allows a wider range of conditions depending on the climate and expectations of the occupants (Akhai, Singh, & John, 2016, Kumar & Akhai, 2022, Akhai & Khang, 2024, Tanwar & Akhai, 2017, Akhai, Thareja, & Singh, 2017, Akhai & Abbass, 2025, Akhai, 2024, Akhai & Wadhwa, 2024).

The quality of indoor air must also be considered. Ventilation and indoor air quality are achieved by carefully balancing the amount of outdoor air brought in, the use of filters, and the control of pollutants. Too little ventilation harms health, while too much may waste energy. Renewable energy targets are another important part of the design. Engineers estimate the expected annual output from solar panels or solar thermal systems, usually expressed either per square meter of building area or per unit of installed capacity (Akhai, Mala, & Jerin, 2021, Akhai, Mala, & Jerin, 2020; Tungnung, 2020).

Carbon emissions form another layer of measurement. The carbon intensity of the electricity or fuels used is calculated in kilograms of carbon dioxide equivalent per unit of energy Bukhary et al., 2020; Mannan & Al-Ghamdi, 2020; Santos et al., 2020). This shows how "clean" the energy supply is, and in some cases, marginal grid factors are considered to reflect the additional emissions caused by new loads on the grid. Water use is also linked to energy performance (Taneja, Bhatia, & Akhai, 2024, Taneja, Bhatia, & Akhai, 2025, Akhai & Taneja, 2025, Akhai & Taneja, 2024). This is known as the water–energy nexus, where hot water use, cooling tower operation, and opportunities for recovering waste heat all influence overall building efficiency.

In a net-zero energy building, the balance is simple in principle but complex in practice. Over the course of a year, the total renewable energy generated on-site, together with the useful heat that is recovered and reused, must be at least equal to the total energy demand of the building. Achieving this balance is not only about the total amount of energy but also about the timing of its use. Energy storage systems and smart controls help shift energy consumption to the right moments, reducing stress on the grid and allowing systems to be sized more appropriately (Verma, Kaur, Kaur, & Akhai, 2025, Kaur, Kaur, Bansal, Singh, Bhatia, & Akhai, 2024, Singh & Akhai, 2024). This dynamic balance is what makes net-zero performance possible in real-world conditions (Allen et al., 2020; Medved et al., 2021; Ren et al., 2021).

Table 1 presents performance metrics and targets for net-zero energy buildings, key energy balance indicators, literature review on metrics, comfort, and energy targets, and framework for sustainable building performance.

Table 1. Key indicators of energy balance in NZEB design

Metric / Target	Definition / Key Features	Importance in NZEB Design	Challenges / Considerations	Key References
Energy Use Intensity (EUI)	Annual energy use divided by total floor area (kWh/ m^2·year).	Enables fair comparison across different building sizes; key benchmark for efficiency.	Requires accurate energy modeling and monitoring; varies with occupancy and climate.	Apanavičienė et al., 2020; Crespi et al., 2020; Zeng et al., 2020
Peak Demand	Maximum energy demand at any point in time.	Determines size of transformers, chillers, HVAC, and other equipment; affects electricity tariffs (demand charges).	Highly variable; requires load forecasting and demand-response measures.	Singh & Maini, 2020; Zorn et al., 2020
Thermal Comfort	Indoor temperature and humidity maintained within acceptable ranges (e.g., 23–26 °C, 40–60% RH). Adaptive comfort allows wider ranges in naturally ventilated spaces.	Essential for occupant health, productivity, and satisfaction.	Varies by climate, culture, and occupant expectation; too strict standards increase energy use.	Tungnung, 2020
Indoor Air Quality (IAQ)	Balance between ventilation, filtration, and pollutant control.	Ensures healthy environment, supports productivity, prevents sick-building syndrome.	Too little ventilation harms health; too much wastes energy. Requires careful balancing.	Bukhary et al., 2020
Renewable Energy Yield	Expected annual output from PV, solar thermal, or other renewable systems.	Measures contribution to on-site generation; ensures supply-demand balance.	Dependent on location, orientation, shading, and system performance.	Medved et al., 2021
Carbon Intensity	Emissions factor of electricity/fuel used, in kg CO_2e per unit energy.	Tracks how "clean" the energy is; critical for net-zero carbon goals.	Grid emission factors vary; marginal vs. average factors can change results.	Mannan & Al-Ghamdi, 2020
Water–Energy Nexus	Link between energy use and water consumption (hot water, cooling towers, waste-heat recovery).	Helps optimize both energy and water performance simultaneously.	Complex interactions; requires system-wide modeling and monitoring.	Santos et al., 2020
Energy Balance (NZEB Target)	Annual renewable generation (plus recovered heat) ≥ annual building demand.	Core principle of net-zero energy design.	Achieving temporal balance is challenging; storage and smart controls are essential.	Allen et al., 2020; Ren et al., 2021

3. THERMO-MECHANICAL FUNDAMENTALS

Heat flows in buildings are mainly governed by three basic processes: conduction, convection, and radiation. These processes control how heat moves across the building envelope and within heating, ventilation, and air-conditioning (HVAC) systems. To manage these flows effectively, engineers use a set of key design levers that influence both comfort and energy performance(Kumar et al., 2020; Mishra et al., 2020).

One of the most important measures is the U-value, which describes how easily heat passes through a material or building element such as a wall, roof, or window. A lower U-value means better insulation and, as a result, smaller heating and cooling demands. Another crucial factor is the solar heat gain coefficient (SHGC), which tells us how much solar radiation passes through glass. The right value depends on climate: lower SHGC is desirable in hot regions to block excess heat, while higher SHGC is useful in cold climates to capture warmth. In mixed or composite climates, a balance is required (Bizoňová et al., 2020; Horiuchi & Yoshimura, 2020; Kalmár & Kalmár, 2020; Latifah & Rahadian, 2020).

The thermal mass of materials also plays a vital role. Heavy materials like concrete or layers containing phase change materials (PCMs) can absorb heat during the day and release it later. This dampens indoor temperature swings and shifts cooling needs away from the hottest hours of the day, helping reduce peak energy demand. Moisture control is another hidden but critical aspect. Water vapor can move through building assemblies by diffusion or air leakage, adding to latent cooling loads. Without proper air barriers and vapor retarders, this not only increases energy use but also risks long-term durability problems such as condensation and mold(Boardman et al., 2020; Kayll et al., 2020; Kumar et al., 2020).

Within HVAC systems, efficiency is often improved by using heat pumps, which move heat instead of generating it. The performance of a heat pump is measured by the coefficient of performance (COP) or by the seasonal COP, which averages efficiency over a year. Higher values mean the system uses less energy for the same amount of heating or cooling. Finally, the way energy is distributed within the building is equally important. Hydronic radiant systems, which use water pipes instead of air ducts, can reduce fan energy and provide good comfort even at slightly higher thermostat setpoints. Dedicated outdoor air systems (DOAS), when paired with energy recovery, supply fresh air efficiently while minimizing energy losses (Alessio et al., 2020; Bae et al., 2020; Hwang & Jeong, 2021).

These strategies collectively ensure that buildings not only reduce their heating and cooling loads but also distribute energy more effectively, which is essential for meeting net-zero goals.

4. RENEWABLE ENERGY OPTIONS FOR BUILDINGS

Renewable energy technologies form the backbone of net-zero energy buildings (NZEBs). They provide the clean energy needed to balance building demand while also reducing carbon emissions. The choice of technology depends on local climate, site conditions, and building type. Each option has unique strengths and limitations, but together they create a flexible toolkit for designers. Table 2 gives a concise yet comparative view of renewable energy technologies for NZEBs, highlighting technical features, strengths, limitations, and sources.

4.1 Photovoltaics (PV) and Building-Integrated PV (BIPV)

Rooftop solar photovoltaic panels are the most common renewable option because they are modular, reliable, and well understood. They can be added to most buildings with minimal disruption. However, when roof space is limited, additional strategies are needed. Building-integrated photovoltaics (BIPV) embed solar modules directly into façades, skylights, shading devices, or spandrel panels. This allows the building skin to serve a dual purpose: generating electricity while also acting as weather protection and an architectural feature. Although façades generally produce less electricity per square meter than rooftops, they can generate power during morning and afternoon hours. This helps flatten the building's demand curve and reduces reliance on the grid. System efficiency is influenced by tilt angle, shading, cell temperature, and inverter performance (Awan et al., 2020; Reffat & Ahmad, 2020; Rezk et al., 2020).

4.2 Photovoltaic-Thermal (PV-T) Systems

A further innovation is the photovoltaic-thermal system, which combines electricity generation with heat recovery. In these systems, PV cells are cooled by a circulating fluid. This improves the electrical efficiency of the PV module since cooler cells perform better, while the heat collected can be used for hot water or low-temperature radiant heating. In space-constrained sites, PV-T systems can be more effective than installing separate PV panels and solar thermal collectors because they deliver two outputs from the same area(Chaabane et al., 2020; Olczak et al., 2020; Singh, 2020).

4.3 Solar Thermal Systems

Traditional solar thermal technologies, such as flat-plate or evacuated-tube collectors, are efficient sources of hot water. They are especially suitable for applications with steady and significant hot water demand, such as hotels, hospitals, and university campuses. When properly designed, they can also support low-temperature space heating. Successful operation depends on stratified storage tanks, reliable controls, and hygienic measures to prevent risks such as Legionella bacteria(Guo et al., 2020; Smirnova et al., 2020; Yang & Svendsen, 2020).

4.4 Ground-Source Heat Pumps and Energy Piles

Ground-source heat pumps make use of the earth's stable underground temperatures to provide efficient heating and cooling. By drawing heat from the ground in winter and rejecting it in summer, these systems can achieve higher coefficients of performance than conventional systems. Civil engineering works offer additional opportunities through energy piles, where foundation piles are embedded with U-tubes to act as heat exchangers. This approach allows the structural elements of a building to serve a second function, providing energy exchange without requiring extra land. This is an excellent example of structural and mechanical systems working together for sustainability(Dabaieh & Serageldin, 2020; Jiang et al., 2020; Sridharan et al., 2020).

4.5 Ambient-Loop and District Energy Systems

Beyond individual buildings, shared energy systems can greatly improve efficiency. Ambient loops circulate water at moderate temperatures, typically between 15 and 30 °C, through a network of interconnected buildings. Each building then uses its own heat pump to draw heat from or reject heat into the loop. Waste heat from equipment such as data centers or supermarkets can be captured and reused by neighboring buildings. Because the loop operates at near-ambient temperatures, thermal losses are low, and the separation of generation from consumption allows greater flexibility and resilience(Kauko et al., 2020; Sommer et al., 2020; Wheatcroft et al., 2020).

4.6 Small Wind and Hybrid Systems

Wind energy can also play a role, though it is highly dependent on-site conditions. In dense urban areas, turbulence and shading often make small wind turbines ineffective. However, in open campuses, industrial estates, or coastal locations, wind can complement solar output. Hybrid systems that combine wind with PV, supported by shared inverters and battery storage, can smooth fluctuations in supply and provide a more stable renewable mix (Badge & Upadhyay, 2020; Madsen et al., 2020; Akhai et al., 2020).

4.7 Biomass and Biogas (Selective Use)

In some contexts, biomass or biogas can be a viable renewable option, especially for buildings with high, constant heating loads such as hospitals or large hostels. Modern biomass boilers or combined heat and power (CHP) units can provide reliable base-load heat while reducing reliance on fossil fuels (Yadav, Singh, Garg, & Akhai, 2024, Akhai, 2023, Akhai, 2024). However, their use requires strict safeguards. Air quality regulations must be met, robust emission control systems must be in place, and ash disposal must be properly managed. Additionally, the life-cycle carbon impacts must be carefully evaluated to ensure that the feedstock is sustainable and that the system does not create unintended environmental harm (Tiammee & Likasiri, 2020; Xie et al., 2020).

Table 2. Key renewable energy options, benefits, and challenges in building applications

Technology	Key Features	Benefits	Limitations / Challenges	Key References
Photovoltaics (PV) & Building-Integrated PV (BIPV)	Rooftop PV is modular, reliable, widely adopted; BIPV integrates solar modules into façades, skylights, spandrels, shading devices.	Generates electricity onsite; BIPV provides dual function (energy + building skin); reduces grid reliance; enables façade-based power generation during non-peak solar hours.	Limited roof area; lower efficiency for vertical façades; performance affected by tilt, shading, cell temp., and inverter quality.	Awan et al., 2020; Reffat & Ahmad, 2020; Rezk et al., 2020
Photovoltaic–Thermal (PV-T) Systems	Combines PV with thermal collectors by cooling PV cells with circulating fluid.	Improves PV electrical efficiency; provides hot water/low-temp heating; maximizes use of limited roof/facade area.	Higher system complexity; added cost; requires reliable thermal storage integration.	Chaabane et al., 2020; Olczak et al., 2020; Singh, 2020
Solar Thermal Systems	Flat-plate or evacuated-tube collectors for hot water & space heating.	High efficiency for steady hot-water demand (hospitals, hotels, campuses); supports low-temp heating.	Requires stratified storage & hygiene control (Legionella risk); best suited to buildings with continuous demand.	Guo et al., 2020; Smirnova et al., 2020; Yang & Svendsen, 2020

continued on following page

Table 2. Continued

Technology	Key Features	Benefits	Limitations / Challenges	Key References
Ground-Source Heat Pumps & Energy Piles	Uses stable subsurface temps; foundation piles with U-tubes act as exchangers.	Very high COP compared to conventional HVAC; integrates structural and mechanical functions; saves land space.	High upfront costs; site-specific soil/groundwater constraints; requires careful design to avoid interference.	Dabaieh & Serageldin, 2020; Jiang et al., 2020; Sridharan et al., 2020
Ambient-Loop & District Energy Systems	Shared low-temp water loops (15–30°C) across multiple buildings; integrates waste heat.	Reduces losses; flexible & resilient; recycles waste heat from data centers/ supermarkets; enables diversity in demand.	Requires multi-building coordination; infrastructure investment; complex ownership/management.	Kauko et al., 2020; Sommer et al., 2020; Wheatcroft et al., 2020
Small Wind & Hybrid Systems	Onsite wind turbines; often paired with PV & storage.	Complements solar (esp. in open/coastal areas); hybrid systems stabilize renewable mix; reduces intermittency.	Poor performance in dense urban areas due to turbulence; site-dependent wind availability.	Badge & Upadhyay, 2020; Madsen et al., 2020
Biomass & Biogas (Selective Use)	Biomass boilers or CHP; biogas from organic waste.	Provides reliable base-load heat; effective for large, continuous heating demand (hospitals, hostels).	Requires strict emissions control, sustainable feedstock sourcing, ash disposal; possible air-quality issues.	Tiammee & Likasiri, 2020; Xie et al., 2020

5. ENVELOPE INNOVATIONS WITH MECHANICAL IMPACT

The building envelope plays a central role in shaping energy performance. By carefully designing walls, roofs, windows, and shading elements, engineers can greatly reduce the energy required for heating and cooling. High-performance insulation and the use of thermal breaks help limit heat conduction through walls and structural elements, while also reducing the risk of condensation. New materials such as vacuum insulation panels and aerogels make it possible to achieve very low U-values even in thin layers, which is particularly useful in retrofit projects where space is limited (Chung et al., 2020; Johansson & Wahlgren, 2020).

Roofs and façades can also be adapted to minimize unwanted solar heat gains. Cool roofs and reflective surfaces reduce the amount of solar radiation absorbed, while green roofs add additional benefits. Through evapotranspiration and added thermal mass, green roofs help regulate temperatures, lower cooling demand, and provide stormwater management. Windows, often the weakest part of the envelope, are being transformed by dynamic glazing technologies. Electrochromic or thermochromic glass can change its properties depending on sunlight, adjusting the solar heat gain coefficient automatically throughout the day (Chen et al., 2020; Ciampi et al., 2020).

Thermal storage within the envelope is another area of innovation. Materials enhanced with phase change materials, such as specialized plasters or panels, can absorb heat during the day and release it at night, reducing peak loads and evening out temperature fluctuations. Shading devices, designed according to the path of the sun, allow natural daylight to enter without causing glare or overheating (Soudian et al., 2020; Verbruggen et al., 2020).

Each of these strategies reduces the load placed on the HVAC system. When the envelope performs well, the required size of heating and cooling equipment decreases. This creates a positive cycle: smaller systems mean lower capital costs, reduced noise, and lower ongoing energy demand. In this way, innovations in the envelope directly improve mechanical efficiency while supporting the larger goal of net-zero building design (Chen et al., 2020; Piselli et al., 2020).

6. HVAC STRATEGIES FOR NET-ZERO

Mechanical systems are at the heart of any net-zero building, since they determine how efficiently heating, cooling, and ventilation needs are met. Traditional HVAC systems often consume large amounts of energy, but new strategies and technologies now make it possible to provide comfort with far less energy use. Therefore, optimization of these systems is essential to maximize efficiency, minimize costs, and ensure long-term sustainability (Kumar, Wadhwa, Akhai, & Kaushik, 2024, Kanda, Akhai, & Bansal, 2013, Akhai, 2024, Kumar, Wadhwa, Akhai, & Kaushik, 2024). Table 3 discusses HVAC strategies for achieving net-zero building performance, energy-efficient approaches in NZEBs, innovative comfort and efficiency innovations, sustainable building design optimization strategies, and energy demand reduction pathways.

6.1 Heat Pumps as the Engine

Heat pumps form the central engine of many net-zero buildings. Air-source heat pumps are the simplest to install and have improved significantly in performance, while water-source and ground-source heat pumps can reach even higher efficiencies where site conditions allow. Modern designs use variable-speed compressors, which adjust their operation to match the actual load. Since buildings rarely operate at full load, this feature allows the system to remain highly efficient during part-load conditions, which dominate most of the year (Abdullah et al., 2020; Avan et al., 2020; Li et al., 2020).

6.2 Radiant Heating and Cooling

Instead of relying only on moving air, radiant heating and cooling systems use hydronic pipes embedded in floors, ceilings, or walls to transfer heat primarily by radiation. This approach allows chilled water to be supplied at higher temperatures and hot water at lower temperatures, which makes heat pumps more efficient. Radiant systems are usually paired with a dedicated outdoor air system (DOAS) that provides fresh air and manages humidity. Together, they deliver excellent comfort while significantly reducing fan power (Alessio et al., 2020; Bae et al., 2020; Singh, 2020).

6.3 Energy Recovery Ventilation and Dedicated Outdoor Air

Bringing in fresh air is essential for indoor quality but also adds to energy loads. Energy recovery systems solve this by capturing heat and sometimes moisture from exhaust air and transferring it to incoming fresh air. This reduces the additional heating or cooling needed. A dedicated outdoor air system further improves efficiency by handling ventilation separately from temperature control in the occupied zones, ensuring better indoor air quality and lower energy use (Chuah & Yang, 2020; Tran et al., 2020; Zhao et al., 2020).

6.4 Desiccants and Latent Control

In hot and humid climates, controlling moisture is as important as controlling temperature. Desiccant systems, using either solid desiccant wheels or liquid desiccant solutions, can efficiently remove moisture from incoming air. These systems can be regenerated using solar thermal energy or waste heat, making them especially attractive for facilities with high ventilation demands.

6.5 Thermal Storage

Thermal storage provides a way to align building loads with renewable energy production or utility pricing. Sensible storage, such as water tanks or concrete cores, stores energy in the form of temperature change, while latent storage, such as ice tanks or phase change materials, stores energy more densely. Storage allows buildings to shift cooling or heating loads to times when solar power is available or electricity tariffs are lower. Even modest storage capacity can reduce the required size of chillers and heat pumps while smoothing the building's interaction with the grid(Odufuwa et al., 2020; Wang et al., 2020; Yan et al., 2020).

6.6 Controls: From Schedules to Predictive Systems

Effective control strategies ensure that HVAC systems run only when needed and in the most efficient way. Simple approaches such as night temperature setbacks, adjusting supply-air temperature, and using economizers already save substantial energy. Advanced approaches go further by using predictive controls that rely on weather forecasts, occupancy patterns, and even expected solar generation. Known as model predictive control (MPC), these systems can pre-cool or pre-heat spaces, charge storage systems, and schedule equipment to reduce both energy costs and peak demand while maintaining comfort(Drgoňa et al., 2020; Fathi et al., 2020; Jain et al., 2020).

Table 3. HVAC strategies for achieving net-zero building performance

Strategy	Key Features / Mechanism	Benefits	Limitations / Considerations	Key References
Heat Pumps as the Engine	Air-source, water-source, and ground-source heat pumps; modern versions with variable-speed compressors adjust operation to match part-loads.	High efficiency compared to traditional systems; adaptable to site conditions; efficient under part-load (dominant operating mode).	Higher capital cost (esp. ground-source); performance depends on climate and ground conditions.	Abdullah et al., 2020; Avan et al., 2020; Li et al., 2020
Radiant Heating and Cooling	Hydronic pipes embedded in floors/ceilings/walls transfer heat mainly by radiation; paired with DOAS for ventilation and humidity control.	Higher efficiency due to elevated chilled-water and reduced hot-water temperatures; reduces fan energy; excellent comfort.	Higher installation complexity; careful design needed to avoid condensation issues.	Alessio et al., 2020; Bae et al., 2020; Singh, 2020
Energy Recovery Ventilation (ERV) & Dedicated Outdoor Air (DOAS)	ERV captures heat and moisture from exhaust air; DOAS handles ventilation separately from space conditioning.	Improves IAQ with less energy; reduces additional heating/ cooling loads; efficient ventilation.	Added system complexity; maintenance required for ERV devices.	Chuah & Yang, 2020; Tran et al., 2020; Zhao et al., 2020

continued on following page

Table 3. Continued

Strategy	Key Features / Mechanism	Benefits	Limitations / Considerations	Key References
Desiccants & Latent Control	Solid or liquid desiccants remove moisture; can be regenerated using solar thermal energy or waste heat.	Provides effective humidity control in hot-humid climates; reduces cooling loads; suitable for high ventilation demand facilities.	Additional system components; regeneration requires heat input; cost and space requirements may increase.	–
Thermal Storage	Sensible storage (water tanks, concrete cores) and latent storage (ice tanks, PCMs) shift heating/ cooling loads.	Aligns loads with renewable generation; reduces peak demand; decreases required chiller/heat pump size; improves grid interaction.	Storage adds upfront cost and space requirement; thermal losses possible over long durations.	Odufuwa et al., 2020; Wang et al., 2020; Yan et al., 2020
Controls (Schedules to Predictive Systems)	Range from simple setbacks and economizers to advanced Model Predictive Control (MPC) using forecasts and occupancy data.	Ensures systems run only when needed; MPC optimizes costs, demand peaks, and comfort; enables integration with renewables and storage.	Requires accurate forecasting and data; more complex design and commissioning; operator training essential.	Drgoňa et al., 2020; Fathi et al., 2020; Jain et al., 2020

7. ELECTRICAL INTEGRATION, MICROGRIDS, AND DEMAND RESPONSE

As buildings move toward net-zero operation, electrical systems must evolve to become smarter and more flexible. Traditional building electrical designs focused only on distributing power within the building, but net-zero buildings increasingly act as active participants in the wider grid. This requires advanced integration of renewable generation, storage, and demand-response strategies (Beucker & Hinterholzer, 2020; Mishra et al., 2020).

Smart inverters are a key technology in this transition. Unlike conventional inverters, which simply convert DC power from solar panels into usable AC power, smart inverters also provide grid-support functions. They can regulate voltage, help stabilize frequency, and ride through short disturbances without disconnecting. These features make renewable energy systems more reliable for both the building and the grid (Alhasnawi et al., 2020; Rangarajan et al., 2020; Tiwari, 2020).

Battery energy storage systems add another layer of capability. They help smooth the variability of solar generation, store excess electricity for use during cloudy periods or nighttime and provide backup during outages. Batteries also allow buildings to manage their demand more intelligently, reducing peak electricity use and saving costs on demand charges (Cook et al., 2020; Machlev et al., 2020; Samoita et al., 2020).

Beyond the building itself, new strategies such as building-to-grid and vehicle-to-grid integration are becoming important. In building-to-grid arrangements, buildings can export excess renewable energy to the grid when supply is high and draw from the grid when needed. Vehicle-to-grid integration goes further by treating fleets of electric vehicles as mobile storage units. During peak demand, vehicles can discharge power back to the grid or to the building and then recharge during periods of low demand or high renewable availability (Mohan et al., 2020; Sufyan et al., 2020).

Electrical integration also enhances resilience. By using critical load panels, buildings can prioritize essential services—such as life-safety systems, medical equipment, or communication infrastructure—during power outages. This ensures that even in emergencies, the most important functions of a building continue to operate. For modern civil infrastructure, this combination of resilience, flexibility, and smart control is no longer optional but increasingly a core requirement (Clark-Ginsberg et al., 2020; Mahabadi et al., 2021; Najafi et al., 2020).

8. RELIABILITY, DURABILITY, AND SAFETY OF THERMO-MECHANICAL SYSTEMS

For net-zero buildings to succeed in practice, their systems must not only be efficient but also reliable, durable, and safe throughout their lifespan. Thermo-mechanical systems operate under demanding conditions, and if they are not carefully designed, problems such as equipment failures, moisture damage, or indoor air quality issues can undermine performance and comfort (Novoseltsev et al., 2020; Riza et al., 2020; Tam & Agnew, 2020).

One of the major concerns is the effect of thermal cycling. As PV modules and other components expand and contract with temperature changes, solder joints may weaken and glazing seals can fail, leading to reduced performance and costly repairs. Moisture is another persistent threat, since water can enter through small gaps and cause corrosion in connectors, ducts, and other metallic parts. Poor management of condensate in HVAC systems may even result in biofilm growth, which compromises indoor air quality and poses health risks (Fořt et al., 2020; Kayll et al., 2020; Li et al., 2020).

To avoid these problems, engineers must pay special attention to moisture management in building envelopes. Continuous air barriers, correct placement of vapor retarders, and positive drainage on façades and roofs are essential strategies.

Material selection also plays a vital role, particularly in coastal or industrial regions where corrosion risks are high. In some cases, protective measures such as coatings or cathodic protection are necessary to extend system life (Arce-Recatalá et al., 2020; Ardi et al., 2020).

Safety is equally important. Fire safety measures include ensuring that PV and building-integrated PV modules meet appropriate fire classifications, maintaining roof setback zones for firefighting access, and designing shafts and risers with proper compartmentation to slow fire spread. Acoustics and vibration must also be considered, since poorly located heat pumps or pumps can create noise and vibrations that transfer through the building structure, reducing occupant comfort (Brink et al., 2020; Brzezińska & Bryant, 2020).

Finally, long-term maintainability is critical. Systems must be designed with easy access to filters, coils, valves, and isolation points so that maintenance tasks can be carried out efficiently. Clear operation and maintenance manuals, along with a plan for regular calibration of sensors, ensure that systems continue to operate as intended. Without these measures, even the most advanced building can fail to deliver on its net-zero promise (Guyot et al., 2020; Rodrigues et al., 2020).

9. TECHNO-ECONOMIC EVALUATION IN PLAIN TERMS

Designing a net-zero building is not only a technical challenge but also an economic one. For a project to be viable, its costs and benefits must be carefully analyzed over the entire life of the building. This involves looking beyond just the initial investment to also include operating expenses, maintenance needs, replacement cycles, and even the value of systems at the end of their service life (Hajare & Elwakil, 2020; Kiamili et al., 2020).

One widely used measure is the levelized cost of energy (LCOE) or levelized cost of heat (LCOH). This represents the average cost of each kilowatt-hour of electricity or heat delivered by a system over its lifetime. It takes into account the upfront capital cost, the ongoing costs of operation and maintenance, and applies a discount rate to reflect the time value of money. By reducing all costs to a single metric, LCOE or LCOH makes it possible to compare different technologies on a fair basis (Elkhayat et al., 2020; Loewen, 2020; Shen et al., 2020).

Another common approach is the payback period, which asks how many years it takes for the energy savings to equal the original investment. A simple payback ignores the time value of money, while a discounted payback includes it, giving a more realistic picture of financial performance. Although simple, these tools are often persuasive for decision-makers who want to see how quickly their investment will pay off(Lobo, 2023; Ветрова et al., 2020).

Because future conditions are uncertain, a sensitivity analysis is also important. This involves varying assumptions such as electricity tariffs, solar generation levels, equipment efficiency, or financing terms to see which factors have the biggest impact on results. Sensitivity analysis shows whether a project remains attractive under conservative or less favorable scenarios (Khadra et al., 2020; Sukor et al., 2020).

In practice, some measures prove beneficial in nearly every case. Steps such as tightening the building envelope, upgrading to heat pumps, and installing photovoltaic panels usually provide strong economic value regardless of future uncertainties. These "no-regrets" measures form a solid foundation for both financial and environmental performance, making them a reliable starting point for any net-zero project (Drzymała & Korzeniewska, 2020; Guno et al., 2020).

Designing a net-zero building requires balancing performance with cost. A project that saves energy but is too expensive will not be adopted widely. To evaluate projects, engineers and planners use simple but powerful financial tools such as levelized cost of energy, payback period, and sensitivity analysis (Hajare & Elwakil, 2020; Neupane et al., 2020; Ürge-Vorsatz et al., 2020).

Example: A Small Office NZEB Project

Imagine a small office building that plans to install a 100-kW rooftop photovoltaic (PV) system combined with a heat pump upgrade.

Step 1: Capital and Operating Costs

- PV system cost: 40,00,000 (about $50,000)
- Annual maintenance: 40,000 (about $500)
- Expected lifetime: 25 years
- Heat pump upgrade cost: 16,00,000 (about $20,000)
- Annual savings from heat pump efficiency: 1,60,000 (about $2,000)

Step 2: Energy Production and Savings

- PV output: 1,50,000 kWh per year (average 1,500 kWh/kW)
- Grid electricity price: 8 per kWh (about $0.10)
- Annual PV savings: 1,50,000 × 8 = 12,00,000 (about $15,000)
- Total annual savings (PV + heat pump): 13,60,000 (about $17,000)

Step 3: Simple Payback

- Total capital cost = 56,00,000
- Annual savings = 13,60,000
- Simple payback = 56,00,000 ÷ 13,60,000 ≈ 4.1 years

This means the system pays for itself in just over four years.

Step 4: Discounted Payback

If we apply a 6% discount rate (to reflect the time value of money), the discounted annual savings reduce slightly each year. In this case, the discounted payback extends to about 5 years, still quite attractive.

Step 5: Levelized Cost of Energy (LCOE)

To calculate LCOE for the PV system:

- Lifetime energy produced = 1,50,000 × 25 = 37,50,000 kWh
- Total lifetime cost = Capital cost + discounted O&M ≈ 40,00,000 + 5,00,000 = 45,00,000
- LCOE = 45,00,000 ÷ 37,50,000 ≈ 1.20 per kWh

This is far below the grid electricity price of 8 per kWh, showing the strong financial case.

Step 6: Sensitivity Analysis

- If electricity tariffs rise by 5% per year, the savings improve even further, shortening payback.
- If PV output is 10% lower due to weather or shading, the payback increases only slightly to about 4.5 years.
- Even with conservative assumptions, the project remains economically sound.

This simple example shows how techno-economic tools give a clear picture of project performance. The PV system and heat pump upgrade not only reduce emissions but also save money, with a short payback and very competitive LCOE. Such "no-regrets" measures are the cornerstone of net-zero building design, as they remain financially viable under a wide range of future scenarios.

10. LIFE-CYCLE ASSESSMENT (LCA) AND EMBODIED CARBON

Operational energy is only part of the picture. Embodied carbon from cement, steel, insulation, glass, and mechanical equipment can be significant. A balanced strategy:

1. Reduce operational energy via envelope and HVAC first (it also allows smaller equipment).
2. Choose lower-carbon materials (e.g., blended cements, recycled steel/aluminum, low-GWP insulation blowing agents).
3. Right-size equipment using realistic load models to avoid oversizing and excess embodied carbon.
4. Plan for circularity: modular plant rooms, standardized connections, and take-back schemes for PV modules and batteries.

System boundaries (A1–A3 products, A4 transport, A5 construction, B1–B7 use stage including energy and water, C1–C4 end-of-life) should be stated clearly (Alaux et al., 2021; Kiamili et al., 2020; Robati et al., 2021; Zeng et al., 2020).

When we think of energy in buildings, most attention goes to operational energy—the electricity and fuels used for heating, cooling, lighting, and equipment. However, a large share of a building's total climate impact comes from embodied carbon—the emissions associated with producing and transporting construction materials such as cement, steel, insulation, glass, and even the mechanical equipment itself. For a building to truly be sustainable, both operational and embodied carbon must be considered together(Arehart et al., 2020; Pan & Teng, 2021; Robati et al., 2021).

A balanced approach begins with reducing operational energy through efficient envelopes and HVAC systems. This not only lowers day-to-day emissions but also reduces the need for oversized mechanical equipment, which saves on embodied carbon as well. The choice of construction materials plays an important role. Blended cements, recycled steel and aluminum, and insulation materials with low global-warming-potential (GWP) blowing agents can significantly reduce embodied impacts (Arehart et al., 2020; Helal et al., 2020).

Right-sizing equipment is equally important. Oversized chillers, boilers, or pumps not only waste energy in operation but also add unnecessary embodied carbon in manufacturing and installation. Careful load modeling ensures that systems are large enough to meet demand but not larger than needed (Ha et al., 2020; Suamir et al., 2020; Timofeevsky et al., 2020).

Circularity is another growing priority. Modular plant rooms, standardized connections, and product take-back schemes for items such as photovoltaic modules and batteries extend the life cycle and reduce end-of-life waste. Life-cycle assessment methods provide a framework to evaluate these impacts, covering product stage (A1–A3), transport (A4), construction (A5), use stage including energy and water (B1–B7), and end-of-life processes (C1–C4). For clarity and transparency, it is essential to state which boundaries are being included in the analysis (Bjørn et al., 2020; Reinales et al., 2020; Rewlay-ngoen & Papong, 2020).

11. THREE CLIMATE ARCHETYPES: ILLUSTRATIVE PATHWAYS

To make the discussion more concrete, we can look at how the principles of net-zero design apply in three different climate contexts. These examples are simplified narratives, meant to illustrate design choices rather than provide universal prescriptions. The numbers are indicative and should always be recalibrated to local conditions.

11.1 Composite Climate Office (Example: North India, ~10,000 m^2 GFA)

In a composite climate with both hot summers and cold winters, the goal is to achieve an energy use intensity (EUI) below 45 kWh per square meter per year, while remaining net-zero on site. The envelope is designed with high levels of insulation: wall U-values of about 0.35 W/m^2·K, roof U-values around 0.25 W/m^2·K with cool reflective coatings and double-glazed low-E windows with U-values near 1.6 W/m^2·K and SHGC around 0.28. Airtightness is maintained at or below 3 m^3/h·m^2 at 50 Pa (Kosonen & Keskisaari, 2020; Kumar et al., 2020; Pont et al., 2020).

The HVAC system is built around water-to-water heat pumps connected to energy piles, with hydronic radiant slabs for space conditioning. A dedicated outdoor air system with total energy recovery ensures efficient ventilation, while supply air dew-point control handles latent loads. Phase change material–enhanced ceiling tiles help shave peak loads.

Renewable energy is provided by a dense 1.2 MWp rooftop PV system and 0.4 MWp of façade-integrated PV on the south and east sides, along with photovoltaic-thermal units for hot water in the cafeteria. Controls include model predictive control to anticipate afternoon peaks and monsoon humidity, demand-response systems to shift chiller operation into solar-rich hours, and chilled-water setpoint resets. With modeled annual PV generation of about 1.9 GWh against a building demand of 1.75 GWh, the building achieves net-positive performance with margin for system degradation and soiling.

11.2 Hot-Humid Hospital (24/7 Loads)

Hospitals in hot and humid climates face the challenge of continuous loads and strict reliability requirements. Latent load control is the top priority. A high-efficiency dedicated outdoor air system with desiccant-based dehumidification, regenerated by solar thermal energy, is used to manage humidity. Cooling is delivered by water-cooled heat pumps with heat rejection to a seawater loop where available, or by hybrid cooling towers. Infection control is enhanced by UV germicidal irradiation in the air-handling units. For resilience, battery storage is installed selectively to back up critical loads. A large PV system is deployed on the roof, sized to offset significant portions of the daytime cooling demand. Waste heat from medical equipment and laundry is also recovered to further improve efficiency(Final Project Report, A Novel, Renewable Energy Microgrid for a California Healthcare Facility, 2020; Lagrange et al., 2020; Sun et al., 2020).

11.3 Cold Climate School

In cold climates, heating and fresh air supply dominate energy use, but only during occupied hours. The building envelope is super-insulated with triple glazing to minimize heat loss. Ground-source heat pumps provide efficient heating, supported by high-effectiveness energy recovery systems in ventilation. Demand-controlled ventilation based on indoor CO_2 levels reduces unnecessary outdoor air intake. Additional recovery from spaces with high heat output, such as the gymnasium and cafeteria, further lowers heating needs. With careful design, the building can achieve an EUI below 30 kWh/m^2 per year. A combination of roof- and façade-mounted PV then brings the building to net-zero on an annual basis (Alajmi et al., 2020; Attia et al., 2020; Stamponi et al., 2021).

12. DESIGN WORKFLOW AND COORDINATION CHECKLIST

1. Define boundary and targets (NZEB type, carbon boundary, peak demand limits).
2. Climate and site analysis (sun paths, wind roses, soil temperatures).
3. Massing and orientation to balance daylight with solar gain control.
4. Envelope optimization (U, SHGC, airtightness) and early condensation analysis.
5. Preliminary load modeling (zonal peak and annual) including latent loads.
6. Renewable concept (roof/facade PV, PV-T, energy piles, solar thermal).
7. HVAC concept (heat pumps, radiant + DOAS, ERV, storage).
8. Electrical topology (smart inverters, BESS, critical loads, EV/V2G).

9. Controls strategy (MPC, demand response, fault detection & diagnostics).
10. Techno-economic and LCA (embodied vs operational trade-offs).
11. Detailed coordination (fire, acoustics, maintenance access, structural anchorage).
12. Commissioning and M&V (calibrate models, verify performance, train O&M).

13. COMMON BARRIERS AND PRACTICAL SOLUTIONS

The section discusses common barriers to solar energy installation, including limited roof area, high humidity, capital constraints, skills gap, and grid constraints. Solutions include extending roofs with BIPV, adding PV canopies, improving envelope, using performance-based contracts, addressing skills gaps, and addressing grid constraints like on-site storage, soft-start drives, load management, and coordination with utilities (Alboaouh & Mohagheghi, 2020; Aleem et al., 2020; Laaksonen et al., 2020).

1. Limited roof area: extend to façades with BIPV; add PV canopies over parking; improve envelope to lower EUI so less PV is needed.
2. High humidity: DOAS with energy recovery and desiccant assistance; airtightness upgrades.
3. Capital constraints: phase investments; start with envelope + high-COP heat pumps; add PV later; use performance-based contracts.
4. Skills gap: early contractor involvement; clear controls sequences; training for facilities staff; digital twin for operations.
5. Grid constraints: on-site storage, soft start drives, load management, and coordination with utilities for export limits.

14. WORKED-THROUGH, BACK-OF-THE-ENVELOPE CALCULATIONS

Although advanced software modeling is central to net-zero building design, simple calculations are still valuable. They allow engineers to sanity-check results and make sure that modeling outputs are realistic. These quick estimates also help decision-makers understand the scale of impacts without complex simulations. This section shows how small calculation exercises provide quick checks on glazing design, system efficiency, and renewable generation potential, reinforcing confidence in more detailed simulation outputs(Kamel et al., 2020; Manfren & Nastasi, 2020; O'Neill et al., 2020).

14.1 Cooling Load Impact of Window SHGC

Solar heat gain through glazing is a major contributor to cooling load. Consider a 10,000 m^2 office building with 40% window-to-wall ratio, giving about 4,000 m^2 of glazing. If the average solar radiation during peak hours is 500 W/m^2, the cooling impact of glazing depends strongly on the solar heat gain coefficient (SHGC). Reducing SHGC from 0.45 to 0.28 cuts transmitted solar energy by 0.17 per unit area. The resulting reduction in heat load is:

$$\Delta Q = 500 \times (0.45 - 0.28) \times 4{,}000$$

$$\Delta Q \approx 500 \times 0.17 \times 4{,}000$$

$$\Delta Q \approx 340 \text{ kW}$$

This means that by improving the glazing specification, the peak cooling requirement is reduced by 340 kW. This directly reduces the size of the chiller or heat pump needed, saving both capital cost and operating energy.

14.2 COP Improvement via Radiant Cooling

Radiant systems allow chilled-water supply temperatures to be higher than in conventional air systems. For instance, raising the chilled-water temperature from 7 °C to 12 °C reduces the lift that the heat pump must achieve. Since heat pumps operate more efficiently at smaller lifts, the coefficient of performance (COP) can improve by about 10–20%. In addition, radiant cooling reduces fan energy use, providing further savings. Together, these effects lower energy consumption while maintaining comfort (Brehm et al., 2021; Hwang & Jeong, 2021; Liu et al., 2020).

14.3 Annual PV Yield Estimate

Estimating photovoltaic generation can also be done with a simple approach. Suppose the roof and façade can accommodate 1.6 MWp of PV panels. If the local climate provides a typical yield of 1,200 kWh per kWp per year, the expected generation is:

$$1.6 \text{ MWp} \times 1{,}200 \text{ kWh/kWp·yr} = 1.92 \text{ GWh/year}$$

To make the estimate conservative, losses must be considered. With 10% system losses and 2% soiling due to dust or dirt, the actual annual yield would be about 1.67–1.73 GWh. This figure helps in sizing batteries and negotiating export agreements with the grid operator.

15. CONTROLS SEQUENCES THAT ACTUALLY WORK

Even the most efficient systems cannot deliver net-zero performance unless they are controlled properly. Control sequences translate design intent into daily operation, ensuring that equipment runs only when needed and at the right settings. A few proven strategies are widely applied in high-performance buildings (Kim & Braun, 2020; Sporr et al., 2020).

One is the use of economizer logic. When the outdoor air is cooler and less humid than the return air, the system can safely increase the proportion of outdoor air brought into the building. This reduces the need for mechanical cooling while still maintaining comfort and air quality. If outdoor conditions are not favorable, the system reverts to the minimum outdoor air requirement (Chuah & Yang, 2020; Ya-ming et al., 2020).

Another strategy is the chilled-water reset, which raises the chilled-water supply temperature whenever indoor zones are already meeting their setpoints with relatively low valve openings. For example, if zone valves are operating at 40% or less, this indicates that cooling demand is modest, and the chilled-water temperature can be increased. This improves heat-pump efficiency and reduces energy use (Khalilnejad et al., 2021; Park et al., 2020; Suamir et al., 2020).

Thermal storage dispatch adds further flexibility. When photovoltaic generation is higher than the building's immediate demand, and forecasts predict a high evening peak, the storage system can be charged during the day. Later, during peak tariff hours, the stored cooling or heating can be discharged, reducing costs and grid demand (Bandera et al., 2020; Wang et al., 2020).

Finally, fault detection ensures long-term reliability. If a heat pump's coefficient of performance drops below a threshold that has been adjusted for outdoor temperature and load conditions, the system raises an alarm. This allows issues to be investigated quickly before they waste energy or affect comfort (Boahen et al., 2020; Hu et al., 2020).

In brief:

- Economizer logic: If outdoor enthalpy $<$ return enthalpy and humidity within limits, increase outdoor air fraction; else revert to minimum.
- Chilled-water reset: Raise supply temperature when zones meet setpoints with valve positions $\leq$40% on average.
- Thermal storage dispatch: Charge when PV output $>$ building demand by a margin and weather forecast predicts evening peak; discharge during peak tariff windows.
- Fault detection: Alarm if heat-pump COP falls below dynamic threshold adjusted for outdoor temperature and load.

Together, these straightforward but powerful control strategies demonstrate that achieving net-zero is not only about advanced technology, but also about making sure systems are operated intelligently and consistently over time.

16. CASE NOTES ON CIVIL-SCALE INTEGRATION

Net-zero energy strategies often extend beyond the boundaries of a single building. Civil-scale integration allows infrastructure elements to provide additional energy functions, turning structural or urban systems into contributors to efficiency and resilience (Alajmi et al., 2020; Piselli et al., 2020).

One example is the use of energy piles. These are structural foundation piles embedded with thermal loops, allowing them to serve as heat exchangers. Their performance depends on factors such as pile spacing and grout conductivity, which affect thermal interference. While structural stability always remains the primary consideration, careful coordination ensures that thermal loops are integrated without conflicting with reinforcement bars or construction sequencing (Liu et al., 2020; Saaly et al., 2020).

Bridges and canopies can also support renewable energy generation through solar photovoltaic installations. To be effective and safe, these systems require lightweight, wind-rated mounting structures, along with careful attention to potential glare that might affect drivers or pedestrians. Electrical cabling must also account for structural expansion joints, ensuring long-term durability and safety (Abuzwidah et al., 2020; Dasaraju & Shahnasser, 2020; Prakash et al., 2020).

In tunnels and stations, waste heat from exhaust air can be captured using air-to-fluid heat exchangers. Instead of being lost, this recovered energy can be redirected into nearby ambient loops, supplying heating or cooling to adjacent buildings. This demonstrates how transport infrastructure can be part of a wider urban energy system (Dudkiewicz & Szałański, 2020; Ren et al., 2020; Tong et al., 2020).

Even water mains can serve as a valuable thermal resource. With cooperation from utilities, city water distribution systems may act as low-temperature heat reservoirs, enabling efficient operation of heat pumps. By using such networks, entire districts can benefit from shared renewable thermal energy without requiring additional land or large-scale new infrastructure (Arabkoohsar & Alsagri, 2020; Kauko et al., 2020).

Overall, these examples show how net-zero strategies can move beyond isolated buildings, creating synergies with civil engineering works to build more resilient and sustainable cities.

17. QUALITY ASSURANCE: COMMISSIONING AND MEASUREMENT & VERIFICATION (M&V)

High performance in net-zero buildings is not guaranteed by design alone; it must be verified and maintained through systematic quality assurance. This process begins with commissioning, which ensures that the systems installed in the building actually match the design intent. Commissioning involves checking that sensors are correctly located and calibrated, control sequences are implemented as intended, valves are installed in the right orientation, and all safety interlocks are tested. By carrying out these steps before handover, many problems can be prevented from appearing later during operation (Čech et al., 2020; Hu, 2020; O'Connor & Mock, 2020).

Once the building is occupied, measurement and verification (M&V) play a central role. Actual performance is compared with the design targets for energy use intensity (EUI) and indoor comfort. If the measured values begin to drift away from targets, the causes can be investigated, and corrective actions can be taken. M&V ensures that the building continues to perform as expected rather than gradually losing efficiency over time (Guyot et al., 2020; Manfren & Nastasi, 2020; Mavrigiannaki et al., 2020).

A new layer of capability comes from the use of digital twins. These are virtual models of the building that continuously update with live operational data. Digital twins allow engineers and operators to visualize performance in real time, identify issues more quickly, and test control strategies before applying them. They also support continuous optimization throughout the life of the building, helping to sustain net-zero performance year after year (Cui et al., 2020; Jafari et al., 2020).

18. CONCLUSIONS AND FUTURE PATHWAYS

Net-zero buildings are not a distant aspiration but an achievable outcome of rigorous engineering. They combine disciplined demand reduction, smart thermomechanical systems, and thoughtfully integrated renewable energy. High-performance envelopes, efficient heat pumps, radiant systems, energy recovery, thermal storage, and advanced controls together drive down energy use intensity (EUI) and reduce peak demand. The benefits extend beyond individual buildings. Civil-scale strategies such as energy piles, ambient thermal loops, and microgrids allow entire districts to share resources, recover waste heat, and build resilience. By pairing these design measures with life cycle thinking and strong commissioning practices, net-zero performance can be achieved and maintained. The path forward is clear: set explicit boundaries and performance targets, design the envelope and HVAC systems to minimize demand, integrate renewables intelligently, and verify results through commissioning and measurement. With these steps, net-zero buildings become not just slogans but real, functioning contributions to climate action and sustainable development.

The next generation of net-zero buildings will depend on emerging technologies that address both efficiency and sustainability. Heat pumps are evolving to use low global-warming-potential (GWP) refrigerants and to deliver higher temperature lifts, enabling them to provide domestic hot water without electric boosters. This improves both performance and climate compatibility.

Energy storage is also moving forward. Advanced phase change materials (PCMs) and thermochemical storage systems promise more compact and longer-duration storage. Unlike conventional systems that mainly balance daily loads, these innovations may align with weekly variations, providing deeper integration with renewable supply cycles.

In renewable integration, transparent or semi-transparent building-integrated photovoltaics (BIPV), as well as new perovskite-on-glass technologies, are expanding the possibilities of solar harvesting without sacrificing aesthetics or daylighting. Research is ongoing to improve their reliability and durability, which remain important hurdles before widespread adoption.

Controls will also become more intelligent. Artificial intelligence (AI) applied to model predictive control (MPC) can merge weather forecasts, occupancy patterns, tariff structures, and equipment health data. This allows simultaneous optimization of multiple assets—such as PV arrays, battery storage systems, heat pumps, and electric vehicles—in real time.

Finally, circularity in mechanical, electrical, and plumbing (MEP) systems will gain importance. Modular plant skids, remanufactured compressors, and standardized take-back programs for PV modules and batteries reduce embodied carbon and encourage closed-loop material cycles. These pathways show that net-zero design is not static but part of a continuous process of innovation.

REFERENCES

Abdullah, H., Ibrahim, O., Jaafar, M. N. M., Mohamad, M., Baharain, A., & Sulaiman, S. (2020). Energy Savings In a Multi-Circuit Water-Cooled Packaged Unit Air-Conditioning System. *Journal of Advanced Research in Fluid Mechanics and Thermal Sciences*, *76*(1), 39–53. DOI: 10.37934/arfmts.76.1.3953

Abuzwidah, M., Balwan, M., Abdel-Aty, M., & Al-Kateeb, G. (2020). Traffic Safety Benefits of Using Highway Lighting Technology: (Case Study: 4-Lanes Expressways in the Rural Areas). *IOP Conference Series. Materials Science and Engineering*, *811*(1), 12023. DOI: 10.1088/1757-899X/811/1/012023

Akhai, S. (2023). Biogas Plants for Sustainable Municipal Waste Management: A Brief Review. *Handbook of Research on Safe Disposal Methods of Municipal Solid Wastes for a Sustainable Environment*, 162-179. DOI: 10.4018/978-1-6684-8117-2.ch012

Akhai, S. (2024). Smart Home Environment for Society: An Overview of IoT-Enabled Air Conditioners. In *Modelling of Virtual Worlds Using the Internet of Things* (Vol. 1, pp. 74-87). CRC Press. DOI: 10.1201/9781003480181-3

Akhai, S. (2024). Trends and Environmental Impact of Paper Consumption: A Prognostic Scenario for the Indian Market by 2030-A Case Study. In *International Conference on Interdisciplinary Approaches in Civil Engineering for Sustainable Development* (Vol. 464, pp. 11-18). Springer Nature Singapore. DOI: 10.1007/978-981-97-0910-6_2

Akhai, S. (2024). Towards Trustworthy and Reliable AI: The Next Frontier. In *Explainable Artificial Intelligence (XAI) in Healthcare* (pp. 89-99). CRC Press. DOI: 10.1201/9781003426073-7

Akhai, S., & Abbass, M. (2025). Air Conditioning for Healthier Living. In *Managing the Health Risks of Climate Change* (pp. 135–154). IGI Global Scientific Publishing., DOI: 10.4018/979-8-3693-4183-4.ch006

Akhai, S., & Abbass, M. (2025). Toward Resilient Futures: The Role of AI-Driven Strategies in Climate Adaptation. In *Nexus of AI, Climatology, and Urbanism for Smart Cities* (pp. 325-340). IGI Global Scientific Publishing. DOI: 10.4018/979-8-3693-5918-1.ch012

Akhai, S., Bansal, S. A., & Singh, S. (2020). A critical review of thermal insulators from natural materials for energy saving in buildings. *Journal of Critical Reviews*, 7(19), 278–283.

Akhai, S., & Khang, A. (2024). Energy efficiency and human comfort: AI and IoT integration in hospital HVAC systems. *Medical Robotics and AI-Assisted Diagnostics for a High-Tech Healthcare Industry*, 93-108. DOI: 10.4018/979-8-3693-2105-8.ch007

Akhai, S., Mala, S., & Jerin, A. A. (2020). Apprehending Air Conditioning Systems in Context to COVID-19 and Human Health: A Brief Communication. *International Journal of Healthcare Education & Medical Informatics (ISSN: 2455-9199), 7*(1&2), 28-30. DOI: 10.24321/2455.9199.202004

Akhai, S., Mala, S., & Jerin, A. A. (2021). Understanding whether air filtration from air conditioners reduces the probability of virus transmission in the environment. *Journal of Advanced Research in Medical Science & Technology*, *8*(1), 36–41. DOI: 10.24321/2394.6539.202109

Akhai, S., Singh, V. P., & John, S. (2016). Human performance in industrial design centers with small unit air conditioning systems. *Journal of Advanced Research in Production Industrial Engineering*, *3*(2), 5–11.

Akhai, S., Srivastava, P., & Sharma, S. (2020). Developments in horizontal axis wind turbines-A brief review. *J. Crit. Rev*, *7*(19), 255–260.

Akhai, S., Srivastava, P., Sharma, V., & Bhatia, A. (2021). Investigating weld strength of AA8011-6062 alloys joined via friction-stir welding using the RSM approach. In *Journal of Physics: Conference Series* (Vol. 1950, No. 1, p. 012016). IOP Publishing. DOI: 10.1088/1742-6596/1950/1/012016

Akhai, S., & Taneja, T. (2024). Liquid Gold: Safeguarding Our Health Through Pure Drinking Water and Rigorous Water Testing. In *Natural Resource Management Issues in Human-Influenced Landscapes* (pp. 129-148). IGI Global. DOI: 10.4018/978-1-6684-7051-0.ch007

Akhai, S., & Taneja, T. (2025). The Critical Role of Water Quality: Health Impacts, Contaminants, and Sustainable Solutions for Environmental and Human Well-being. In *Smart Water Technology for Sustainable Management in Modern Cities* (pp. 101-116). IGI Global. DOI: 10.4018/979-8-3693-8074-1.ch005

Akhai, S., Thareja, P., & Singh, V. P. (2017). Assessment of Indoor Environment Health Sustenance in Air Conditioned Class Rooms. *Journal of Advanced Research in Civil and Environmental Engineering*, *4*(1&2), 1–9. DOI: 10.24321/2393.8307.201701

Akhai, S., & Wadhwa, A. S. (2024). Wireless networking solutions for climate control and comfort in modern air conditioning systems. In *Intelligent Networks* (pp. 165–181). CRC Press., DOI: 10.1201/9781003541363-11

Alajmi, A., Short, A., Ferguson, J., Poel, K. V., & Griffin, C. (2020). Detailed energy efficiency strategies for converting an existing office building to NZEB: A case study in the Pacific Northwest. *Energy Efficiency, 13*(6), 1089–1104. DOI: 10.1007/s12053-020-09861-9

Alaux, N., Hoxha, E., Saade, M. R. M., & Passer, A. (2021). Embodied greenhouse gas emissions reduction for structural elements in office buildings. *Journal of Physics: Conference Series, 2042*(1), 12165. DOI: 10.1088/1742-6596/2042/1/012165

Alboaouh, K., & Mohagheghi, S. (2020). Impact of Rooftop Photovoltaics on the Distribution System. *Journal of Renewable Energy, 1*, 1–23. Advance online publication. DOI: 10.1155/2020/4831434

Aleem, S. A., Hussain, S. M. S., & Ustun, T. S. (2020). A Review of Strategies to Increase PV Penetration Level in Smart Grids [Review of *A Review of Strategies to Increase PV Penetration Level in Smart Grids*]. *Energies, 13*(3), 636. Multidisciplinary Digital Publishing Institute. DOI: 10.3390/en13030636

Alessio, G., Emmi, G., Carli, M. D., & Zarrella, A. (2020). All-air system and radiant floor for heating and cooling in residential buildings: A simulation-based analysis. *Science and Technology for the Built Environment, 26*(10), 1397–1411. DOI: 10.1080/23744731.2020.1795515

Alhasnawi, B. N., Jasim, B. H., Issa, W., Anvari-Moghaddam, A., & Blaabjerg, F. (2020). A New Robust Control Strategy for Parallel Operated Inverters in Green Energy Applications. *Energies, 13*(13), 3480. DOI: 10.3390/en13133480

Allen, J., Halberstadt, A., Powers, J. T., & El-Farra, N. H. (2020). An Optimization-Based Supervisory Control and Coordination Approach for Solar-Load Balancing in Building Energy Management. *Mathematics, 8*(8), 1215. DOI: 10.3390/math8081215

Apanavičienė, R., Urbonas, R., & Fokaides, P. A. (2020). Smart Building Integration into a Smart City: Comparative Study of Real Estate Development. *Sustainability (Basel), 12*(22), 9376. DOI: 10.3390/su12229376

Arabkoohsar, A., & Alsagri, A. S. (2020). Thermodynamic analysis of ultralow-temperature district heating system with shared power heat pumps and triple-pipes. *Energy, 194*, 116918. DOI: 10.1016/j.energy.2020.116918

Arce-Recatalá, M., García-Morales, S., & Bossche, N. V. D. (2020). Quantifying Wind-driven Rain Intrusion – A Comparative Study on The Water Management Features of Different Types of Rear-Ventilated Facade Systems. *E3S Web of Conferences, 172*, 23007. DOI: 10.1051/e3sconf/202017223007

Ardi, S. Z., DİKER, B., & Bahri, P. A. (2020, January 1). Environmental Deterioration Factors in Metal Claddings and GFRC Panels Implemented on Facades: An Assessment through Two Cases in Istanbul. *XV International Conference on Durability of Building Materials and Components. eBook of Proceedings*. DOI: 10.23967/dbmc.2020.060

Arehart, J. H., Nelson, W. S., & Srubar, W. V.III. (2020). On the theoretical carbon storage and carbon sequestration potential of hempcrete. *Journal of Cleaner Production, 266*, 121846. DOI: 10.1016/j.jclepro.2020.121846

Attia, S., Shadmanfar, N., & Ricci, F. (2020). Developing two benchmark models for nearly zero energy schools. *Applied Energy, 263*, 114614. DOI: 10.1016/j.apenergy.2020.114614

Avan, A., Maleknejad, M., & Navi, K. (2020). High-speed energy efficient process, voltage and temperature tolerant hybrid multi-threshold 4:2 compressor design in CNFET technology. *IET Circuits, Devices & Systems, 14*(3), 357–368. DOI: 10.1049/iet-cds.2019.0105

Awan, A. B., Alghassab, M. A., Zubair, M., Bhatti, A. R., Uzair, M., & Abbas, G. (2020). Comparative Analysis of Ground-Mounted vs. Rooftop Photovoltaic Systems Optimized for Interrow Distance between Parallel Arrays. *Energies, 13*(14), 3639. DOI: 10.3390/en13143639

Badge, A. K., & Upadhyay, Mrs. M. (2020). Study on Hybrid Solar Wind Power System with PMSG for Power Enhancement Using AI Technique. *SMART MOVES JOURNAL IJOSCIENCE, 6*(1), 5. DOI: 10.24113/ijo-science.v6i1.252

Bae, S., Nam, Y., & Choi, J.-H. (2020). Comparative Analysis of System Performance and Thermal Comfort for an Integrated System with PVT and GSHP Considering Two Load Systems: Convective Heating and Radiant Floor Heating. *Energies, 13*(20), 5524. DOI: 10.3390/en13205524

Bandera, C. F., Pachano, J. E., Salom, J., Peppas, A., & Ruiz, G. R. (2020). Photovoltaic Plant Optimization to Leverage Electric Self Consumption by Harnessing Building Thermal Mass. *Sustainability (Basel), 12*(2), 553. DOI: 10.3390/su12020553

Beucker, S., & Hinterholzer, S. (2020). *Effects of ICT-Enabled Flexible Energy Consumption on the Reduction of CO2 Emissions in Buildings*. 1. DOI: 10.1145/3401335.3401336

Bizoňová, S., Katunský, D., & Bagoňa, M. (2020). Comparative analysis of selected glass systems by dynamic simulation using measured real environmental conditions. *E3S Web of Conferences, 172*, 19009. DOI: 10.1051/e3sconf/202017219009

Bjørn, A., Chandrakumar, C., Boulay, A., Doka, G., Fang, K., Gondran, N., Hauschild, M. Z., Kerkhof, A. C., King, H., Margni, M., McLaren, S. J., Mueller, C., Owsianiak, M., Peters, G., Roos, S., Sala, S., Sandin, G., Sim, S., Vargas-Gonzalez, M., & Ryberg, M. (2020). Review of life-cycle based methods for absolute environmental sustainability assessment and their applications. *Environmental Research Letters*, *15*(8), 83001. DOI: 10.1088/1748-9326/ab89d7

Boahen, S., Mensah, K., Nam, Y., & Choi, J. M. (2020). Fault Detection Methodology for Secondary Fluid Flow Rate in a Heat Pump Unit. *Energies*, *13*(11), 2974. DOI: 10.3390/en13112974

Boardman, C. R., Glass, S. V., & Zelinka, S. L. (2020). Moisture Redistribution in Full-Scale Wood-Frame Wall Assemblies: Measurements and Engineering Approximation. *Buildings (Basel, Switzerland)*, *10*(8), 141. DOI: 10.3390/buildings10080141

Brehm, J. K., Augustine, D. B., Ziviani, D., & Groll, E. A. (2021). Assessing the performance limits of a variable-speed residential heat pump system. *IOP Conference Series. Materials Science and Engineering*, *1180*(1), 12042. DOI: 10.1088/1757-899X/1180/1/012042

Brink, H. W., Loomans, M., Mobach, M. P., & Kort, H. S. M. (2020). Classrooms' indoor environmental conditions affecting the academic achievement of students and teachers in higher education: A systematic literature review [Review of *Classrooms' indoor environmental conditions affecting the academic achievement of students and teachers in higher education: A systematic literature review*]. *Indoor Air*, *31*(2), 405. Wiley. DOI: 10.1111/ina.12745

Brzezińska, D., & Bryant, P. (2020). Risk Index Method–A Tool for Sustainable, Holistic Building Fire Strategies. *Sustainability (Basel)*, *12*(11), 4469. DOI: 10.3390/su12114469

Bukhary, S., Batista, J. R., & Ahmad, S. (2020). Design Aspects, Energy Consumption Evaluation, and Offset for Drinking Water Treatment Operation. *Water (Basel)*, *12*(6), 1772. DOI: 10.3390/w12061772

Canale, L., Fazio, A. R. D., Russo, M., Frattolillo, A., & Dell'Isola, M. (2021). An Overview on Functional Integration of Hybrid Renewable Energy Systems in Multi-Energy Buildings. *Energies*, *14*(4), 1078. DOI: 10.3390/en14041078

Čech, M., Goubej, M., Sobota, J., & Visioli, A. (2020). Model-based system engineering in control education using HIL simulators. *IFAC-PapersOnLine*, *53*(2), 17302–17307. DOI: 10.1016/j.ifacol.2020.12.1812

Chaabane, M., Mhiri, H., & Bournot, P. (2020). Thermal performance evaluation and enhancement of a parabolic trough collector. *Journal of Renewable and Sustainable Energy*, *12*(4), 043702. Advance online publication. DOI: 10.1063/1.5145257

Chen, K. W., Teitelbaum, E., Meggers, F., Pantelic, J., & Rysanek, A. (2020). Exploring membrane-assisted radiant cooling for designing comfortable naturally ventilated spaces in the tropics. *Building Research and Information*, *49*(5), 483–495. DOI: 10.1080/09613218.2020.1847025

Chen, Z., Hammad, A. W. A., Kamardeen, I., & Haddad, A. (2020). *Optimizing Window Design on Residential Building Facades by Considering Heat Transfer and Natural Lighting in Non-tropical Regions of Australia*. DOI: 10.20944/preprints202010.0111.v1

Chuah, Y. K., & Yang, J. (2020). A Integrated Dedicated Outdoor Air System to Optimize Energy Saving. *Sustainability (Basel)*, *12*(3), 1051. DOI: 10.3390/su12031051

Chung, S.-Y., Sikora, P., Stephan, D., & Elrahman, M. A. (2020). The Effect of Lightweight Concrete Cores on the Thermal Performance of Vacuum Insulation Panels. *Materials (Basel)*, *13*(11), 2632. DOI: 10.3390/ma13112632 PMID: 32526963

Ciampi, G., Scorpio, M., Spanodimitriou, Y., Rosato, A., & Sibilio, S. (2020). Thermal model validation of an electric-driven smart window through experimental data and evaluation of the impact on a case study. *Building and Environment*, *181*, 107134. DOI: 10.1016/j.buildenv.2020.107134

Clark-Ginsberg, A., Rueda, I. A., Monken, J., Liu, J., & Chen, H. (2020). Maintaining critical infrastructure resilience to natural hazards during the COVID-19 pandemic: Hurricane preparations by US energy companies. *Journal of Infrastructure Preservation and Resilience*, *1*(1), 10. Advance online publication. DOI: 10.1186/s43065-020-00010-1 PMID: 34778806

Cook, J., Hotchkiss, E., Li, X., & Cruce, J. (2020). Planning for the storm: Considering renewable energy for critical infrastructure resilience. *Journal of Emergency Management (Weston, Mass.)*, *18*(4), 295. DOI: 10.5055/jem.2020.0475 PMID: 32804398

Crespi, G., Becchio, C., & Corgnati, S. P. (2020). Towards Post-Carbon Cities: Which retrofit scenarios for hotels in Italy? *Renewable Energy*, *163*, 950–963. DOI: 10.1016/j.renene.2020.08.023

Cui, L., Sun, H., Yang, Z., & Duan, H. (2020). The digital twin system of PERA company air-conditioning system. *IFAC-PapersOnLine*, *53*(5), 720–725. DOI: 10.1016/j.ifacol.2021.04.207

Dabaieh, M., & Serageldin, A. A. (2020). Earth air heat exchanger, Trombe wall and green wall for passive heating and cooling in premium passive refugee house in Sweden. *Energy Conversion and Management*, *209*, 112555. DOI: 10.1016/j.enconman.2020.112555

Dasaraju, S., & Shahnasser, H. (2020). Smart Solar-Powered Lighting System for Smart Cities. *International Journal of Computer and Electrical Engineering*, *12*(2), 83–92. DOI: 10.17706/IJCEE.2020.12.2.83-92

Delavar, H., & Sahebi, H. (2020). A sustainable mathematical model for design of net zero energy buildings. *Heliyon*, *6*(1), e03190. Advance online publication. DOI: 10.1016/j.heliyon.2020.e03190 PMID: 31956715

Drgoňa, J., Arroyo, J., Figueroa, I. C., Blum, D., Arendt, K., Kim, D., Ollé, E. P., Oravec, J., Wetter, M., Vrabie, D., & Helsen, L. (2020). All you need to know about model predictive control for buildings. *Annual Reviews in Control*, *50*, 190–232. DOI: 10.1016/j.arcontrol.2020.09.001

Drzymała, A., & Korzeniewska, E. (2020). Economic and Technical Aspects of a Hybrid Single-Family House Heating Based on Photovoltaic and Heat Pump Installation. *2020 IEEE Problems of Automated Electrodrive*. [PAEP]. *Theory into Practice*, *1*, 1–4. Advance online publication. DOI: 10.1109/PAEP49887.2020.9240906

Dudkiewicz, E., & Szałański, P. (2020). Overview of exhaust gas heat recovery technologies for radiant heating systems in large halls. *Thermal Science and Engineering Progress*, *18*, 100522. DOI: 10.1016/j.tsep.2020.100522

Elkhayat, Y. O., Ibrahim, M. G., Tokimatsu, K., & Ali, A. A. M. (2020). Multi-criteria selection of high-performance glazing systems: A case study of an office building in New Cairo, Egypt. *Journal of Building Engineering*, *32*, 101466. DOI: 10.1016/j.jobe.2020.101466

Fathi, S., Srinivasan, R., Kibert, C. J., Steiner, R., & Demirezen, E. M. (2020). AI-Based Campus Energy Use Prediction for Assessing the Effects of Climate Change. *Sustainability (Basel)*, *12*(8), 3223. DOI: 10.3390/su12083223

Final Project Report, A Novel, Renewable Energy Microgrid for a California Healthcare Facility. (2020). https://www.energy.ca.gov/sites/default/files/2021-06/CEC-500-2019-034.pdf

Fořt, J., Šál, J., Kočí, J., & Černý, R. (2020). Energy Efficiency of Novel Interior Surface Layer with Improved Thermal Characteristics and Its Effect on Hygrothermal Performance of Contemporary Building Envelopes. *Energies*, *13*(8), 2012. DOI: 10.3390/en13082012

Gangwar, G., Kaur, P., & Singh, I. P. (2020). A Study of Passive and Active Strategies through Case Studies for the Composite Climate Zone of India. *Civil Engineering and Architecture*, *8*(6), 1370–1389. DOI: 10.13189/cea.2020.080620

Guno, C. S., Agaton, C. B., Villanueva, R. O., & Villanueva, R. O. (2020). Optimal Investment Strategy for Solar PV Integration in Residential Buildings: A Case Study in The Philippines. *International Journal of Renewable Energy Development*, *10*(1), 79–89. DOI: 10.14710/ijred.2021.32657

Guo, L., Zhang, Y., Wang, Z., Zeng, J., Zhang, Y., & Zhang, Z. (2020). Parameter sensitivity analysis and optimization strategy research of enhanced geothermal system: A case study in Guide Basin, Northwestern China. *Renewable Energy*, *153*, 813–831. DOI: 10.1016/j.renene.2020.02.058

Guyot, D., Giraud, F., Simon, F., Corgier, D., Marvillet, C., & Tréméac, B. (2020). Detailed monitoring as an essential tool for achieving energy performance targets in operation conditions: the HIKARI case study. *E3S Web of Conferences, 172*, 22006. DOI: 10.1051/e3sconf/202017222006

Ha, J.-W., Cho, S., Kim, H., & Song, Y. (2020). Annual Energy Consumption Cut-Off with Cooling System Design Parameter Changes in Large Office Buildings. *Energies*, *13*(8), 2034. DOI: 10.3390/en13082034

Hajare, A., & Elwakil, E. (2020). Integration of life cycle cost analysis and energy simulation for building energy-efficient strategies assessment. *Sustainable Cities and Society*, *61*, 102293. DOI: 10.1016/j.scs.2020.102293

Helal, J., Stephan, A., & Crawford, R. H. (2020). The influence of structural design methods on the embodied greenhouse gas emissions of structural systems for tall buildings. *Structures (Oxford)*, *24*, 650–665. DOI: 10.1016/j.istruc.2020.01.026

Horiuchi, T., & Yoshimura, K. (2020). Solar heat gain coefficient and heat transmission coefficient of Al-doped ZnO thin-film coated low-emissivity glass. *Journal of the Ceramic Society of Japan*, *128*(4), 220–223. DOI: 10.2109/jcersj2.19217

Hu, M.-S. (2020). Design and development of a high-precision automatic safety valve testing system. *Advances in Mechanical Engineering*, *12*(4). Advance online publication. DOI: 10.1177/1687814020914733

Hu, Y., Yuill, D. P., Ebrahimifakhar, A., & Rooholghodos, A. (2020). An experimental study of the behavior of a high efficiency residential heat pump in cooling mode with common installation faults imposed. *Applied Thermal Engineering*, *184*, 116116. DOI: 10.1016/j.applthermaleng.2020.116116

Hurwitz, Z. L., Dubief, Y., & Almassalkhi, M. (2020). Economic efficiency and carbon emissions in multi-energy systems with flexible buildings. *International Journal of Electrical Power & Energy Systems, 123*, 106114. DOI: 10.1016/j.ijepes.2020.106114

Hussain, I., Khan, F., Ahmad, I., Khan, S., & Saeed, M. (2021). Power Loss Reduction via Distributed Generation System Injected in a Radial Feeder. *Mehran University Research Journal of Engineering and Technology, 40*(1), 160–168. DOI: 10.22581/muet1982.2101.15

Hwang, Y.-J., & Jeong, J. (2021). Energy Saving Potential of Radiant Floor Heating Assisted by an Air Source Heat Pump in Residential Buildings. *Energies, 14*(5), 1321. DOI: 10.3390/en14051321

Jafari, M. A., Zaidan, E., Ghofrani, A., Mahani, K., & Farzan, F. (2020). Improving Building Energy Footprint and Asset Performance Using Digital Twin Technology. *IFAC-PapersOnLine, 53*(3), 386–391. DOI: 10.1016/j.ifacol.2020.11.062

Jain, A., Smarra, F., Reticcioli, E., D'Innocenzo, A., & Morari, M. (2020). NeurOpt: Neural network based optimization for building energy management and climate control. *arXiv (Cornell University)*. /arXiv.2001.07831DOI: 10.48550

Jha, R., Lee, S., Iyengar, S., Hajiesmaili, M., Irwin, D., & Shenoy, P. (2020). *Emission-aware Energy Storage Scheduling for a Greener Grid*. DOI: 10.1145/3396851.3397755

Jiang, S., Li, X., Lyu, W., Wang, B., & Shi, W. (2020). Numerical investigation of the energy efficiency of a serial pipe-embedded external wall system considering water temperature changes in the pipeline. *Journal of Building Engineering, 31*, 101435. DOI: 10.1016/j.jobe.2020.101435

Johansson, P., & Wahlgren, P. (2020, January 1). Experiences from Interior Super Insulation of a Brick Wall from the 1800s. *XV International Conference on Durability of Building Materials and Components. eBook of Proceedings*. DOI: 10.23967/dbmc.2020.065

Kalmár, F., & Kalmár, T. (2020). Thermal Comfort Aspects of Solar Gains during the Heating Season. *Energies, 13*(7), 1702. DOI: 10.3390/en13071702

Kamel, E., Sheikh, S., & Huang, X. (2020). Data-driven predictive models for residential building energy use based on the segregation of heating and cooling days. *Energy, 206*, 118045. DOI: 10.1016/j.energy.2020.118045

Kanda, R., Akhai, S., & Bansal, R. (2013). Analysis of MOST technique for elimination of ideal time by synchronization of different lines. *International Journal of Research, 1*(4).

Kauko, H., Rohde, D., & Hafner, A. (2020). Local Heating Networks with Waste Heat Utilization: Low or Medium Temperature Supply? *Energies*, *13*(4), 954. DOI: 10.3390/en13040954

Kaur, P., Kaur, G., Bansal, J., Singh, A., Bhatia, B., & Akhai, S. (2024, September). A Review of Opportunities and Challenges of Incorporating Artificial Intelligence and Machine Learning to Improve the Reliability of Electrical Power Systems. In *2024 International Conference on Communication, Computing and Energy Efficient Technologies (I3CEET)* (pp. 493-498). IEEE. DOI: 10.1109/I3CEET61722.2024.10993874

Kayll, D., Defo, M., Moore, T. M., & Lacasse, M. (2020, January 1). A Methodology for Assessment of Building Assembly Air Leakage Moisture Response, Condensation Risk, and Expected Durability When Subjected to Projected Future Climate Loads. *XV International Conference on Durability of Building Materials and Components. eBook of Proceedings*. DOI: 10.23967/dbmc.2020.183

Khadra, A., Hugosson, M., Akander, J., & Myhren, J. A. (2020). Economic performance assessment of three renovated multi-family buildings with different HVAC systems. *Energy and Building*, *224*, 110275. DOI: 10.1016/j.enbuild.2020.110275

Khalilnejad, A., French, R. H., & Abramson, A. R. (2021). Evaluation of cooling setpoint setback savings in commercial buildings using electricity and exterior temperature time series data. *Energy*, *233*, 121117. DOI: 10.1016/j.energy.2021.121117

Kiamili, C., Hollberg, A., & Habert, G. (2020). Detailed Assessment of Embodied Carbon of HVAC Systems for a New Office Building Based on BIM. *Sustainability (Basel)*, *12*(8), 3372. DOI: 10.3390/su12083372

Kim, D., & Braun, J. E. (2020). Model predictive control for supervising multiple rooftop unit economizers to fully leverage free cooling energy resource. *Applied Energy*, *275*, 115324. DOI: 10.1016/j.apenergy.2020.115324

Kosonen, A., & Keskisaari, A. (2020). Zero-energy log house – Future concept for an energy efficient building in the Nordic conditions. *Energy and Building*, *228*, 110449. DOI: 10.1016/j.enbuild.2020.110449

Kumar, H., Wadhwa, A. S., Akhai, S., & Kaushik, A. (2024). Parametric analysis, modeling and optimization of the process parameters in electric discharge machining of aluminium metal matrix composite. *Engineering Research Express*, *6*(2), 025542. DOI: 10.1088/2631-8695/ad4ba9

Kumar, H., Wadhwa, A. S., Akhai, S., & Kaushik, A. (2024). Parametric optimization of the machining performance of Al-SiCp composite using combination of response surface methodology and desirability function. *Engineering Research Express*, *6*(2), 025505. DOI: 10.1088/2631-8695/ad38ff

Kumar, P., & Akhai, S. (2022). Effective energy management in smart buildings using VRV/VRF systems. In *Additive Manufacturing in Industry 4.0* (pp. 27–35). CRC Press., DOI: 10.1201/9781003360001-2

Kumar, R., Waghmare, A., Nayak, S., Paswan, M., & Achintya, . (2021). Design of self sustainable zero energy building. *Materials Today: Proceedings*, *46*, 6737–6742. DOI: 10.1016/j.matpr.2021.04.263

Kumar, R. V., N, S. K., Narsimha, K., K, S. R. R., & E, V. K. (2020). Spatial Day Light Autonomy and Energy Analysis of a Residential Building for Different Climatic Conditions and Window-to-Wall Ratios. *E3S Web of Conferences, 184*, 1117. DOI: 10.1051/e3sconf/202018401117

Laaksonen, H., Parthasarathy, C., Hafezi, H., Shafie-khah, M., Khajeh, H., & Hatziargyriou, N. (2020). Solutions to Increase PV Hosting Capacity and Provision of Services from Flexible Energy Resources. *Applied Sciences (Basel, Switzerland)*, *10*(15), 5146. DOI: 10.3390/app10155146

Lagrange, A., de Simón-Martín, M., González-Martínez, A., Bracco, S., & Rosales-Asensio, E. (2020). Sustainable microgrids with energy storage as a means to increase power resilience in critical facilities: An application to a hospital. *International Journal of Electrical Power & Energy Systems, 119*, 105865. DOI: 10.1016/j.ijepes.2020.105865

Latifah, N. L., & Rahadian, E. Y. (2020). Energy Saving Building Strategies through The Application of Solar Control Glass. *ELKOMIKA Jurnal Teknik Energi Elektrik Teknik Telekomunikasi & Teknik Elektronika*, *8*(2), 388. DOI: 10.26760/elkomika.v8i2.388

Li, X., Cui, T., Huang, K., & Ma, X. (2020). Optimization of load sharing for parallel compressors using a novel hybrid intelligent algorithm. *Energy Science & Engineering*, *9*(3), 330–342. DOI: 10.1002/ese3.821

Li, Y., Dang, X., Xia, C., Ma, Y., Ogura, D., & Hokoi, S. (2020). The Effect of Air leakage through the Air Cavities of Building Walls on Mold Growth Risks. *Energies*, *13*(5), 1177. DOI: 10.3390/en13051177

Liu, H., Zhao, B. Y., Zhang, Z., Li, H., Hu, B., & Wang, R. Z. (2020). Experimental validation of an advanced heat pump system with high-efficiency centrifugal compressor. *Energy, 213*, 118968. DOI: 10.1016/j.energy.2020.118968

Liu, R. Y. W., Sailer, E., Taborda, D. M. G., & Potts, D. M. (2020). Evaluating the impact of different pipe arrangements on the thermal performance of thermo-active piles. *E3S Web of Conferences, 205*, 5006. DOI: 10.1051/e3sconf/202020505006

Lobo, A. C. (2023). The Use of Time Value of Money in Making Investment Decision. SSRN *Electronic Journal*. DOI: 10.2139/ssrn.4338203

Loewen, J. (2020). LCOE is an undiscounted metric that inaccurately disfavors renewable energy resources. *The Electricity Journal, 33*(6), 106769. DOI: 10.1016/j.tej.2020.106769

Machlev, R., Zargari, N., Chowdhury, N. R., Belikov, J., & Levron, Y. (2020). A review of optimal control methods for energy storage systems - energy trading, energy balancing and electric vehicles [Review of *A review of optimal control methods for energy storage systems - energy trading, energy balancing and electric vehicles*]. *Journal of Energy Storage, 32*, 101787. Elsevier BV. DOI: 10.1016/j.est.2020.101787

Madsen, H. A., Zahle, F., Meng, F., Barlas, T., Rasmussen, F., & Rudolf, R. T. (2020). Initial performance and load analysis of the LowWind turbine in comparison with a conventional turbine. *Journal of Physics: Conference Series, 1618*(3), 32011. DOI: 10.1088/1742-6596/1618/3/032011

Mahabadi, Z., Varga, L., & Dolan, T. (2021). Network Properties for Robust Multilayer Infrastructure Systems: A Percolation Theory Review. *IEEE Access : Practical Innovations, Open Solutions, 9*, 135755–135773. DOI: 10.1109/ACCESS.2021.3116868

Manfren, M., & Nastasi, B. (2020). Parametric Performance Analysis and Energy Model Calibration Workflow Integration—A Scalable Approach for Buildings. *Energies, 13*(3), 621. DOI: 10.3390/en13030621

Mannan, M., & Al-Ghamdi, S. G. (2020). Environmental impact of water-use in buildings: Latest developments from a life-cycle assessment perspective [Review of *Environmental impact of water-use in buildings: Latest developments from a life-cycle assessment perspective*]. *Journal of Environmental Management, 261*, 110198. Elsevier BV. DOI: 10.1016/j.jenvman.2020.110198

Martinez-Bolaños, J. R., da Silva, V. O., Zucchi, M., Heideier, R. B., Relva, S. G., Saidel, M. A., & Fadigas, E. A. F. A. (2020). Performance Analysis of Topologies for Autonomous Hybrid Microgrids in Remote Non-Interconnected Communities in the Amazon Region. *Sustainability (Basel), 13*(1), 44. DOI: 10.3390/su13010044

Mavrigiannaki, A., Gobakis, K., Kolokotsa, D., Kalaitzakis, K., Pisello, A. L., Piselli, C., Gupta, R., Gregg, M., Laskari, M., Saliari, M., Assimakopoulos, M.-N., & Synnefa, A. (2020). Measurement and Verification of Zero Energy Settlements: Lessons Learned from Four Pilot Cases in Europe. *Sustainability (Basel)*, *12*(22), 9783. DOI: 10.3390/su12229783

Medved, S., Domjan, S., & Arkar, C. (2021). Contribution of energy storage to the transition from net zero to zero energy buildings. *Energy and Building*, *236*, 110751. DOI: 10.1016/j.enbuild.2021.110751

Mishra, M., Deshmukh, A., & sri, K. S. (2020). Influence Of Passive Techniques on Thermal Performance of Building in Moderate Zone. *International Journal of Innovative Technology and Exploring Engineering*, *9*(3), 360–365. DOI: 10.35940/ijitee.B7164.019320

Mishra, S., Frank, S., Petersen, A., Buechler, R., & Slovensky, M. (2020). Deep-Learning-Based, Multi-Timescale Load Forecasting in Buildings: Opportunities and Challenges from Research to Deployment. *arXiv (Cornell University)*. /arXiv.2008.05458DOI: 10.48550

Mohan, C. V. R., & Deepthi, R. (2020). SIMULATION MODEL OF ELECTRIC VEHICLE TO GRID TECHNOLOGY FOR BALANCING LOADS AND FREQUENCY REGULATIONS. *International Journal of Engineering Applied Sciences and Technology*, *5*(4), 203–208. DOI: 10.33564/IJEAST.2020.v05i04.028

Najafi, J., Peiravi, A., & Anvari-Moghaddam, A. (2020). Enhancing Integrated Power and Water Distribution Networks Seismic Resilience Leveraging Microgrids. *Sustainability (Basel)*, *12*(6), 2167. DOI: 10.3390/su12062167

Neupane, P., Afroze, D., & Phommasone, P. (2020). Cost Benefit Analysis of Green Building: A Case Study of Public Office Building in Nepal. *Saudi Journal of Engineering and Technology*, *5*(10), 382–391. DOI: 10.36348/sjet.2020.v05i10.005

Novoseltsev, B. P., Жерлыкина, M. H., & Garmonov, K. V. (2020). Maintenance of heat stability of the heating system work in a structure. *IOP Conference Series. Materials Science and Engineering*, *962*(3), 32003. DOI: 10.1088/1757-899X/962/3/032003

O'Connor, J. T., & Mock, B. (2020). Responsibilities and accountabilities for industrial facility commissioning and startup activities. *Construction Innovation*, *20*(4), 625–645. DOI: 10.1108/CI-09-2019-0094

O'Neill, R., Window, A., Kenway, S., & Dargusch, P. (2020). Integrated operational and life-cycle modelling of energy, carbon and cost for building façades. *Journal of Cleaner Production*, *286*, 125370. DOI: 10.1016/j.jclepro.2020.125370

Odufuwa, O. Y., Kusakana, K., & Numbi, B. P. (2020). Optimal energy management of Ice thermal energy storage-based air conditioning system for commercial buildings in real-time – A review based on POET framework [Review of *Optimal energy management of Ice thermal energy storage-based air conditioning system for commercial buildings in real-time – A review based on POET framework*]. *Journal of Physics Conference Series, 1577*(1), 12049. IOP Publishing. DOI: 10.1088/1742-6596/1577/1/012049

Olczak, P., Matuszewska, D., & Zabagło, J. (2020). The Comparison of Solar Energy Gaining Effectiveness between Flat Plate Collectors and Evacuated Tube Collectors with Heat Pipe: Case Study. *Energies, 13*(7), 1829. DOI: 10.3390/en13071829

Pan, W., & Teng, Y. (2021). A systematic investigation into the methodological variables of embodied carbon assessment of buildings. *Renewable & Sustainable Energy Reviews, 141*, 110840. DOI: 10.1016/j.rser.2021.110840

Park, M.-K., Lee, J.-M., Kang, W.-H., Kim, C., & Lee, K. H. (2020). Air-Handling-Unit Discharge Air Temperature Reset Based on Outdoor Air Temperature and Cooling Energy Performance in an Office Building. *Journal of Energy Engineering, 146*(3), 04020013. Advance online publication. DOI: 10.1061/(ASCE)EY.1943-7897.0000660

Piselli, C., Grazia, M. D., & Pisello, A. L. (2020). Combined Effect of Outdoor Microclimate Boundary Conditions on Air Conditioning System's Efficiency and Building Energy Demand in Net Zero Energy Settlements. *Sustainability (Basel), 12*(15), 6056. DOI: 10.3390/su12156056

Pont, U., Wöltzl, M., Schuß, M., Schober, P., & Mahdavi, A. (2020). Exploring novel solutions for incorporating vacuum glazing in new and existing window constructions. *E3S Web of Conferences, 172*, 24006. DOI: 10.1051/e3sconf/202017224006

Prakash, V. S. J., Pavithra, D., Nisha, R., & Nishanthini, P. (2020). Automatic street light intensity control. *International Journal of Advance Research, Ideas and Innovations in Technology, 6*(2), 258. https://www.ijariit.com/manuscripts/v6i2/V6I2-1169.pdf

Rangarajan, S. S., Sharma, J., & Sundarabalan, C. K. (2020). Novel Exertion of Intelligent Static Compensator Based Smart Inverters for Ancillary Services in a Distribution Utility Network-Review. *Electronics (Basel), 9*(4), 662. DOI: 10.3390/electronics9040662

Reffat, R. M., & Ahmad, R. M. (2020). Determination of optimal energy-efficient integrated daylighting systems into building windows. *Solar Energy, 209*, 258–277. DOI: 10.1016/j.solener.2020.08.086

Reinales, D., Zambrana-Vásquez, D., & de Guinoa, A. S. (2020). Social Life Cycle Assessment of Product Value Chains Under a Circular Economy Approach: A Case Study in the Plastic Packaging Sector. *Sustainability (Basel)*, *12*(16), 6671. DOI: 10.3390/su12166671

Ren, H., Sun, Y., Albdoor, A. K., Tyagi, V. V., Pandey, A. K., & Ma, Z. (2021). Improving energy flexibility of a net-zero energy house using a solar-assisted air conditioning system with thermal energy storage and demand-side management. *Applied Energy*, *285*, 116433. DOI: 10.1016/j.apenergy.2021.116433

Ren, M., Yuan, Y., Cao, X., Sun, L., & Jiang, F. (2020). Numercial analysis on the thermal performance of capillary heat exchange system in metro running tunnel. *Energy and Built Environment*, *1*(2), 207–214. DOI: 10.1016/j.enbenv.2020.01.003

Rewlay-ngoen, C., & Papong, S. (2020). Environmental impact assessment of a rotary compressor in Thailand based on life cycle assessment methodology. *SN Applied Sciences*, *2*(9), 1526. Advance online publication. DOI: 10.1007/s42452-020-03278-w

Rezk, H., Kanagaraj, N., & Al-Dhaifallah, M. (2020). Design and Sensitivity Analysis of Hybrid Photovoltaic-Fuel-Cell-Battery System to Supply a Small Community at Saudi NEOM City. *Sustainability (Basel)*, *12*(8), 3341. DOI: 10.3390/su12083341

Riza, I. W., Latief, Y., & Machfudiyanto, R. A. (2020). Enhancing safety and comfort through mechanical component maintenance performance on green building using BIM-based information system. *IOP Conference Series. Materials Science and Engineering*, *830*(2), 22041. DOI: 10.1088/1757-899X/830/2/022041

Robati, M., Oldfield, P., Nezhad, A. A., Carmichael, D. G., & Kuru, A. (2021). Carbon value engineering: A framework for integrating embodied carbon and cost reduction strategies in building design. *Building and Environment*, *192*, 107620. DOI: 10.1016/j.buildenv.2021.107620

Rodrigues, M. G., & Júnior, F. (2020). Execution Manual and Preventive Maintenance in Residential Works. *International Journal of Advanced Engineering Research and Science*, *7*(5), 304–313. DOI: 10.22161/ijaers.75.37

Saaly, M., Maghoul, P., & Holländer, H. (2020). Investigation of the effects of heat loss through below-grade envelope of buildings in urban areas on thermo-mechanical behaviour of geothermal piles. *E3S Web of Conferences*, *205*, 5010. DOI: 10.1051/e3sconf/202020505010

Samoita, D., Nzila, C., Østergaard, P. A., & Remmen, A. (2020). Barriers and Solutions for Increasing the Integration of Solar Photovoltaic in Kenya's Electricity Mix. *Energies*, *13*(20), 5502. DOI: 10.3390/en13205502

Santos, A. F., Gaspar, P. D., & de Souza, H. J. L. (2020). New HVAC Sustainability Index—TWI (Total Water Impact). *Energies*, *13*(7), 1590. DOI: 10.3390/en13071590

Satoła, D., Balouktsi, M., Lützkendorf, T., Wiberg, A. H., & Gustavsen, A. (2021). How to define (net) zero greenhouse gas emissions buildings: The results of an international survey as part of IEA EBC annex 72. *Building and Environment*, *192*, 107619. DOI: 10.1016/j.buildenv.2021.107619

Satoła, D., Wiberg, A. H., Singh, M., Babu, S., James, B., Dixit, M., Sharston, R., Grynberg, Y., & Gustavsen, A. (2022). Comparative review of international approaches to net-zero buildings: Knowledge-sharing initiative to develop design strategies for greenhouse gas emissions reduction. *Energy Sustainable Development. Energy for Sustainable Development*, *71*, 291–306. DOI: 10.1016/j.esd.2022.10.005

Sharma, V., & Akhai, S. (2019). Trends in Utilization of Coal Fly Ash in India: A Review. *Journal of Engineering Design and Analysis*, *2*(1), 12–16.

Shen, W., Chen, X., Qiu, J., Hayward, J. A., Sayeef, S., Osman, P., Meng, K., & Dong, Z. Y. (2020). A comprehensive review of variable renewable energy levelized cost of electricity [Review of *A comprehensive review of variable renewable energy levelized cost of electricity*]. *Renewable and Sustainable Energy Reviews*, *133*, 110301. Elsevier BV. DOI: 10.1016/j.rser.2020.110301

Singh, A. (2020). Evaluation of Performance of Photovoltaic Thermal Hybrid System. *International Journal of Engineering Research & Technology (Ahmedabad)*, *5*(5). Advance online publication. DOI: 10.17577/IJERTV9IS050559

Singh, M., & Maini, R. (2020). Various Electricity Load Forecasting Techniques with Pros and Cons. [IJRTE]. *International Journal of Recent Technology and Engineering*, *8*(6), 220–229. DOI: 10.35940/ijrte.F6997.038620

Singh, S., & Akhai, S. (2024, July). Electronics in precision control for mechanical engineering: A brief overview. In. AIP Conference Proceedings: Vol. 3121. *No. 1* (p. 020007). AIP Publishing LLC., DOI: 10.1063/5.0221464

Smirnova, E., Larionova, Y., & Larionov, A. (2020). Sewer system optimization in housing and communal services. *E3S Web of Conferences*, *157*, 2002. DOI: 10.1051/e3sconf/202015702002

Sommer, T., Sulzer, M., Wetter, M., Sotnikov, A., Mennel, S., & Stettler, C. (2020). The reservoir network: A new network topology for district heating and cooling. *Energy*, *199*, 117418. DOI: 10.1016/j.energy.2020.117418

Soudian, S., Berardi, U., & Laschuk, N. O. (2020, January 1). Effects of Long-Term UV Exposure on the Performance of Cement Plasters Integrated with Thermochromic Paint and PCMs for Building Facade Applications. *XV International Conference on Durability of Building Materials and Components. eBook of Proceedings*. DOI: 10.23967/dbmc.2020.163

Sporr, A., Zucker, G., & Hofmann, R. (2020). Automatically Creating HVAC Control Strategies Based on Building Information Modeling (BIM): Heat Provisioning and Distribution. *Energies*, *13*(17), 4403. DOI: 10.3390/en13174403

Sridharan, A., Mohan, P. K., Vellaiyan, P., & Ibrahim, M. T. (2020). An experimental investigation on the performance of novel hybrid heat exchanger over conventional shell and tube heat exchanger. *AIP Conference Proceedings*, *2240*, 40001. DOI: 10.1063/5.0011074

Stamponi, E., Lattanzi, N., Giorgini, F., Serpilli, F., Montelpare, S., & Moretti, E. (2021). Energy analysis, numerical simulations and intervention proposals for a NZEB industrial building: the "Loccioni Leaf Lab" case study. *E3S Web of Conferences*, *238*, 6004. DOI: 10.1051/e3sconf/202123806004

Suamir, I. N., Sudirman, S., Ardita, I. N., & Santanu, G. (2020). Experimental and numerical optimization on chilled water configuration for improving temperature performance and economic viability of a centralized chiller plant. *Journal of Physics: Conference Series*, *1450*(1), 12106. DOI: 10.1088/1742-6596/1450/1/012106

Sufyan, M., Rahim, N. A., Muhammad, M. A., Tan, C., Raihan, S. R. S., & Bakar, A. H. A. (2020). Charge coordination and battery lifecycle analysis of electric vehicles with V2G implementation. *Electric Power Systems Research*, *184*, 106307. DOI: 10.1016/j.epsr.2020.106307

Sukor, N. R., Shamsuddin, A. H., Mahlia, T. M. I., & Isa, M. F. M. (2020). Techno-Economic Analysis of CO2 Capture Technologies in Offshore Natural Gas Field: Implications to Carbon Capture and Storage in Malaysia. *Processes (Basel, Switzerland)*, *8*(3), 350. DOI: 10.3390/pr8030350

Sun, K., Specian, M., & Hong, T. (2020). Nexus of thermal resilience and energy efficiency in buildings: A case study of a nursing home. *Building and Environment*, *177*, 106842. DOI: 10.1016/j.buildenv.2020.106842

Tam, I. C. K., & Agnew, B. (2020). Thermal Systems—An Overview. *Energies, 14*(1), 175. DOI: 10.3390/en14010175

Taneja, T., Bhatia, B. S., & Akhai, S. (2024, December). Assessing Groundwater Quality in Rampura-Bathinda for Physicochemical Parameters & Heavy Metal Concentrations using Electronic Equipment. In *2024 1st International Conference on Sustainability and Technological Advancements in Engineering Domain (SUSTAINED)* (pp. 791-796). IEEE. DOI: 10.1109/SUSTAINED63638.2024.11073880

Taneja, T., Bhatia, B. S., & Akhai, S. (2025). Agricultural Resilience: Water Quality and Human Well-Being. *Emerging Smart Agricultural Practices Using Artificial Intelligence*, 1-23. DOI: 10.1002/9781394274277.ch1

Tanwar, N., & Akhai, S. (2017). Survey Analysis for Quality Control Comfort Management in Air Conditioned Classroom. *Journal of Advanced Research in Civil and Environmental Engineering*, *4*(1&2), 20–23. DOI: 10.24321/2393.8307.201702

Thareja, P., & Akhai, S. (2016). Processing Aluminum Fly Ash Composites via Parametric Analysis of Stir Casting. *Journal of Advanced Research in Manufacturing. Material Science & Metallurgical Engineering*, *3*(3&4), 21–28.

Thareja, P., & Akhai, S. (2017). Processing parameters of powder aluminium-fly ash P/M composites. *Journal of advanced research in manufacturing, material science & metallurgical engineering, 4*(3&4), 24-35.

Tiammee, S., & Likasiri, C. (2020). Sustainability in corn production management: A multi-objective approach. *Journal of Cleaner Production*, *257*, 120855. DOI: 10.1016/j.jclepro.2020.120855

Timofeevsky, A. L., Sulin, A. B., & Rubtsov, A. K. (2020). Calculation of the required coolant mass taking account the dynamics of the object heat load and technical characteristics of the chiller. *AIP Conference Proceedings*, *2296*, 30001. DOI: 10.1063/5.0027418

Tiwari, V. (2020). Inverter Control Performance Study on Solar PV Systems and Its Effect on Power Quality. *SMART MOVES JOURNAL IJOSCIENCE*, *6*(7). Advance online publication. DOI: 10.24113/ijoscience.v6i7.337

Tong, Z., Cao, T., Zang, G., Hu, S., Liu, G., & Wang, Y. (2020). Performance analysis of capillary front-end heat exchanger for subway tunnel. *Applied Thermal Engineering*, *174*, 115360. DOI: 10.1016/j.applthermaleng.2020.115360

Tran, V. V., Park, D., & Lee, Y. (2020). Indoor Air Pollution, Related Human Diseases, and Recent Trends in the Control and Improvement of Indoor Air Quality [Review of *Indoor Air Pollution, Related Human Diseases, and Recent Trends in the Control and Improvement of Indoor Air Quality*]. *International Journal of Environmental Research and Public Health, 17*(8), 2927. Multidisciplinary Digital Publishing Institute. DOI: 10.3390/ijerph17082927

Tungnung, K. (2020). Parametric strategy on passive design affordances in cool-humid climate zones of North-East India. *Solar Energy, 207*, 1–9. DOI: 10.1016/j.solener.2020.06.024

Ürge-Vorsatz, D., Khosla, R., Bernhardt, R., Chan, Y. C., Vérez, D., Hu, S., & Cabeza, L. F. (2020). Advances Toward a Net-Zero Global Building Sector. *Annual Review of Environment and Resources, 45*(1), 227–269. DOI: 10.1146/annurev-environ-012420-045843

Verbruggen, S., Hertoge, J., Delghust, M., Laverge, J., & Janssens, A. (2020). The Use of Solar Shading in a Nearly Zero-Energy Neighbourhood. *E3S Web of Conferences, 172*, 19003. DOI: 10.1051/e3sconf/202017219003

Verma, R., Kaur, K., Kaur, A., & Akhai, S. (2025, June). Demand response and smart grids in the renewable energy transition: A review. In. AIP Conference Proceedings: Vol. 3261. *No. 1* (p. 030001). AIP Publishing LLC., DOI: 10.1063/5.0258780

Wang, Y., Hu, C., Wu, B., Zhang, J., Zi, Z., & Kang, L. (2020). Matching Characteristic Research of Building Renewable Energy System Based on Virtual Energy Storage of Air Conditioning Load. *Energies, 13*(5), 1269. DOI: 10.3390/en13051269

Wehkamp, S., Schmeling, L., Vorspel, L., Roelcke, F., & Windmeier, K.-L. (2020). District Energy Systems: Challenges and New Tools for Planning and Evaluation. *Energies, 13*(11), 2967. DOI: 10.3390/en13112967

Wheatcroft, E., Wynn, H. P., Lygnerud, K., Bonvicini, G., & Leonte, D. (2020). The Role of Low Temperature Waste Heat Recovery in Achieving 2050 Goals: A Policy Positioning Paper. *Energies, 13*(8), 2107. DOI: 10.3390/en13082107

Wu, W., & Skye, H. M. (2021). Residential net-zero energy buildings: Review and perspective. *Renewable & Sustainable Energy Reviews, 142*, 110859. DOI: 10.1016/j.rser.2021.110859 PMID: 34413697

Xie, B., Qin, J., Wang, S., Li, X., Sun, H., & Chen, W. (2020). Adsorption of Phenol on Commercial Activated Carbons: Modelling and Interpretation. *International Journal of Environmental Research and Public Health, 17*(3), 789. DOI: 10.3390/ijerph17030789 PMID: 32012816

Ya-ming, F., Li, X., Zheng, M., Weng, R., & Tu, J. (2020). Numerical Study on Effects of Air Return Height on Performance of an Underfloor Air Distribution System for Heating and Cooling. *Energies, 13*(5), 1070. DOI: 10.3390/en13051070

Yadav, V., Singh, S., Garg, M. P., & Akhai, S. (2024). Effect of natural Sapindus mukorossi treatment process on bio-waste bagasse fibers for biocomposite fabrication and application purposes. *Industrial Crops and Products, 219*, 119066. DOI: 10.1016/j.indcrop.2024.119066

Yan, C., Wang, F., Yan, P., Shan, K., & Kosonen, R. (2020). A multi-timescale cold storage system within energy flexible buildings for power balance management of smart grids. *Renewable Energy, 161*, 626–634. DOI: 10.1016/j.renene.2020.07.079

Yang, X., & Svendsen, S. (2020). Improving the district heating operation by innovative layout and control strategy of the hot water storage tank. *Energy and Building, 224*, 110273. DOI: 10.1016/j.enbuild.2020.110273

Zeng, R., Chini, A. R., & Ries, R. (2020). Innovative design for sustainability. *Engineering, Construction, and Architectural Management, 28*(3), 747–764. DOI: 10.1108/ECAM-09-2019-0491

Zhao, T., Wang, J., Liu, C., Hua, P., Zhou, Y., & Zhao, Y. (2020). Comparative study of outdoor airflow requirement and distribution in multi-zone VAV system with different control strategies. *Science and Technology for the Built Environment, 27*(4), 489–508. DOI: 10.1080/23744731.2020.1819062

Zorn, C., Koks, E., & Eggimann, S. (2020). Demand side management impacts on electricity network vulnerability. *arXiv (Cornell University)*. /arXiv.2008.04954DOI: 10.48550

KEY TERMS AND DEFINITIONS

Battery Energy Storage System (BESS): Storage technology that stores electricity for later use, stabilizing energy supply.

Building-Integrated Photovoltaics (BIPV): Solar panels integrated into façades, roofs, or windows to serve both as building skin and power source.

Circularity: Approach where materials and systems are reused, recycled, or recovered to minimize waste.

Coefficient of Performance (COP): Efficiency measure of heat pumps, ratio of heating/cooling output to energy input.

Dedicated Outdoor Air System (DOAS): Ventilation system delivering conditioned fresh air independently of temperature control.

Energy Recovery Ventilation (ERV): System that recovers heat and/or moisture from exhaust air to pre-condition incoming air.

Energy Use Intensity (EUI): Annual energy use of a building per unit floor area (kWh/m^2·year).

Global Warming Potential (GWP): Measure of how much a gas contributes to global warming relative to CO_2.

Heating, Ventilation, and Air Conditioning (HVAC): Mechanical systems ensuring indoor thermal comfort and air quality.

Levelized Cost of Energy (LCOE): Average cost per unit of energy produced over a system's lifetime.

Life Cycle Assessment (LCA): Evaluation of environmental impacts across a product or system's entire life cycle.

Model Predictive Control (MPC): Advanced control strategy using forecasts to optimize building energy use.

Net-Zero Energy Building (NZEB): A building that generates as much renewable energy as it consumes annually.

Phase Change Material (PCM): Materials that absorb/release heat during phase transitions, used for thermal storage.

Photovoltaic-Thermal (PV-T): Hybrid systems that generate electricity and recover heat simultaneously.

Solar Heat Gain Coefficient (SHGC): Fraction of solar radiation admitted through windows or glazing.

Vehicle-to-Grid (V2G): Technology enabling electric vehicles to supply stored energy back to the grid.

Chapter 13
Smart Aesthetics of Sustainable Construction as a Synthesis of Biomimicry, Emotion, and Analytics

Oleksandra Humenna
http://orcid.org/0000-0002-3725-5272
Kyiv National Economic University Named After Vadym Hetman, Ukraine

Hanna Melehanych
Uzhhorod National University, Ukraine

Maryna Chaikovska
Odesa I.I. Mechnikov National University, Ukraine

ABSTRACT

This chapter introduces an integrated model of smart aesthetics as a foundation for sustainable transformation in the construction sector. Responding to ecological challenges, post-war recovery needs, and rising expectations for human-centered design, the model combines three complementary components: biomimicry, Kansei engineering, and the Analytic Hierarchy Process. Biomimicry provides naturally optimized solutions for energy efficiency and climate resilience, while Kansei engineering enhances psychological comfort, emotional safety, and user well-being. AHP ensures balanced decision-making by evaluating ecological, social, cultural, and economic criteria within reconstruction projects. The chapter emphasizes that sustainable development requires not only engineering efficiency but also emotional

DOI: 10.4018/979-8-3373-6167-3.ch013

and cultural intelligence embedded in the built environment. The proposed framework demonstrates practical value for schools, hospitals, residential areas, and public spaces, offering a pathway toward environmentally responsible and emotionally supportive urban environments.

INTRODUCTION: FOUNDATIONS OF SMART AESTHETICS IN CONSTRUCTION

In the 21st century, the construction industry is undergoing one of the most large-scale transformations in its entire history. A combination of global challenges – climate change, energy crises, increasing urbanization, digitalization, and rising societal expectations – has led to a situation in which traditional technocratic approaches are no longer capable of meeting the complex needs of society. According to the UNEP *Global Status Report for Buildings and Construction 2023* (United Nations Environment Programme, 2023), buildings and infrastructure account for nearly 37% of global CO_2 emissions. This figure clearly demonstrates that the evolution of construction is not merely an engineering issue but, above all, a civilizational one.

In accordance with the approved 2030 Agenda for Sustainable Development (United Nations, 2015), the world has 17 Sustainable Development Goals (SDGs), progress toward which is measured through 169 targets and indicators. Goals such as SDG 7 "Affordable and Clean Energy," SDG 8 "Decent Work and Economic Growth," SDG 9 "Industry, Innovation and Infrastructure," SDG 11 "Sustainable Cities and Communities," SDG 12 "Responsible Consumption and Production," and SDG 17 "Partnerships for the Goals" directly or indirectly include objectives related to the construction industry. In particular, progress under SDG 17 in Ukraine is measured, among other indicators, through the implementation of public–private partnership (PPP) projects, the vast majority of which are related to construction. Under SDG 11 "Sustainable Cities and Communities," one of the indicators is the coverage of territorial communities in Ukraine by affordable housing programs.

The author's interpretation of the concept of "sustainable development of the construction industry" is as follows: sustainable development in construction is an innovative and integrated approach to the design, implementation, and operation of construction projects, aimed at harmonizing environmental, economic, and social factors. It involves the use of renewable resources, minimization of negative environmental impact, the introduction of energy-efficient and circular technologies, and the creation of a comfortable, safe, and adaptive environment for people.

According to this definition, the implementation of sustainable development principles in the construction industry can be achieved through:

- the use of environmentally friendly building materials that reduce the carbon footprint;
- the development of energy-efficient technologies and the use of renewable energy sources;
- the introduction of a circular economy, in which materials are reused or recycled;
- optimization of design processes, enabling the reduction of resource consumption at all stages of the building life cycle;
- the carbon footprint assessment (using the Life Cycle Assessment methodology), economic comparison (Life Cycle Cost and total construction costs), and expert assessment (based on technical and ergonomic parameters).

The modern development of energy-efficient construction began after the energy crisis of thc 1970s, when a sharp increase in energy prices forced the United States and European countries to seek new solutions. The first energy-efficient buildings focused on thermal insulation, solar energy, and reducing heat losses (Melnyk, 2018):

- 1972, USA – construction of the first house with maximum thermal insulation and no windows on the northern facade;
- 1973–1979, Finland – development of the "ECONO-HOUSE" complex, combining thermal insulation with geothermal technologies;
- 1983, Sweden – introduction of standards regulating building heat losses (no more than 50–60 kWh/m^2 per year);
- 1980s, Germany – development of the "passive house" concept, in which heating energy consumption was reduced by 90% compared to standard buildings.

By the early 2010s, more than 15,000 passive houses had been built in Germany and Austria alone. Modern technologies already make it possible to create buildings that generate more energy than they consume.

According to international economic forecasts, the global construction market is expected to expand by USD 4.5 trillion over the next decade, reaching a total volume of over USD 15 trillion by 2030. This projected growth underscores the increasing relevance of integrated design models (such as smart aesthetics) that combine ecological optimisation, emotional ergonomics, and data-driven decision-making to shape sustainable and future-oriented development trajectories (PwC, 2025).

Overall, the evolution of the concept of sustainable development in construction has passed through several stages:

1. Awareness of the environmental impact of construction: over time, it became evident that traditional construction methods lead to environmental degradation, prompting the search for new approaches to minimize negative impacts on nature;
2. Development and implementation of "green" technologies: design and construction methods emerged that take environmental aspects into account, such as energy efficiency, the use of renewable energy sources, and environmentally friendly materials. This includes the implementation of green structures, such as green roofs and façades, which contribute to improved microclimates and building energy efficiency;
3. Integration of sustainable development principles into project management: modern construction involves a comprehensive approach in which sustainability principles are integrated at all stages of the building life cycle – from design to operation. This includes the use of environmentally friendly materials, energy-efficient technologies, and ensuring comfort for occupants (Pihul & Siomushkin, 2023). The implementation of LCA can identify and mitigate environmental impacts at the development stage, thereby contributing to the sustainability of the construction industry. The total environmental impact of a building is the result of the environmental loads that occur overthe life cycle of the building: the initial impact, the annualised recurrent impact and the deconstructedimpact. Initial impacts occur during the design and construction of a building, including projectmanagement activities, material use, the construction process and waste management. Annualised recurrent impacts are those that recur each year and result from the use of energy for heating, lighting, ventilation and cooling, as well as repairs and renovations over long periods of time. The third stage of deconstruction occurs during the demolition of the building and is the final stage (Delehan et al., 2024);
4. Development of a regulatory and standards framework: in many countries, including Ukraine, standards and regulations have been developed to govern requirements for energy efficiency and environmental performance of buildings. This encourages developers to incorporate sustainable development principles into their projects.

At the same time, the international analysis by UN-Habitat, *World Cities Report 2022* (UN-Habitat, 2022), emphasizes that cities of the future must not only function efficiently but also ensure psychological safety, comfort, a sense of belonging, and opportunities for interaction with nature. The global trend is moving toward a synthesis of the technological and humanitarian dimensions of architecture, creating a demand for a new approach – smart aesthetics of sustainable construction – where beauty, efficiency, resilience, and human emotions are combined into a single system.

This approach is based on the integration of three complementary methodological directions: biomimicry, Kansei engineering, and analytical decision-making methods (Saaty's AHP). Together, they form a synthesis of science, technology, and culture that makes it possible to create not merely buildings, but ecosystems capable of responding to environmental change, supporting human psychological comfort, reducing resource loads, and strengthening social cohesion.

The Ukrainian context gives this approach particular significance. According to UNDP (2023), in the process of post-war reconstruction the country has the opportunity to move away from outdated infrastructure models toward modern ones that place people, culture, and ecology at the center of every spatial decision. Therefore, the concept of smart aesthetics is not merely a theoretical idea, but a practically necessary strategic platform for future development.

INNOVATIVE APPROACHES AND SMART AESTHETICS

As early as 2018, the report of the Club of Rome highlighted the increasing vulnerability of cities, as major centres of human settlement, to natural disasters such as earthquakes, tsunamis, wildfires, and floods (Weizsäcker & Wijkman, 2018). This, accordingly, necessitates the development of new solutions in the planning of disaster-prevention infrastructure, for example, in the construction of dams and in maximising their resilience. In addition, the authors emphasize the need to design and implement projects for the urban regeneration of neglected cities in industrialised countries.

To address the global challenge of high carbon emissions in the construction sector, technologies aimed at reducing the negative environmental impact are being actively developed worldwide. For example, in the United States and Europe, hemp-based construction materials (hempcrete) are gaining popularity due to their capacity to absorb carbon dioxide, thereby ensuring a negative carbon footprint. Several countries, including the United States (specifically California, New York, and New Jersey) and France, have introduced strict regulations restricting the use of construction materials with high levels of carbon dioxide emissions, thereby stimulating the transition to environmentally sustainable alternatives.

Hemp-based construction materials, which are becoming increasingly widespread, are capable of absorbing tonnes of carbon dioxide even after being incorporated into completed buildings, thereby reducing greenhouse gas concentrations to low or even near-zero levels. Carbon credits and financial incentives may also contribute to making hemp-based buildings more attractive and economically feasible for both consumers and construction companies. Consequently, several municipal programmes in the United States promote the use of hemp raw materials for the production of construction materials with the purpose of reducing carbon dioxide emissions (BDO, 2024).

Green construction involves the use of environmentally friendly materials and technologies that contribute to lowering greenhouse gas emissions, ensuring the efficient use of energy and resources, and creating comfortable and safe residential and commercial buildings. The global trend towards environmental sustainability is compelling the construction industry to adapt to new standards, and Ukraine is also becoming part of this transformative process.

Innovations in the construction sector contribute to sustainable development through the use of energy-efficient materials (advanced thermal insulation solutions, environmentally friendly building materials), the implementation of digital technologies (Building Information Modelling, automated construction management systems), the development of modular construction, and the application of three-dimensional printing technologies, which reduce construction costs and decrease carbon dioxide emissions.

Table 1 presents the proposed innovative approaches and solutions aimed at promoting sustainable development in the construction sector.

Table 1. Innovative approaches for sustainable development of the construction

Nº	Elements of Innovation	Characteristics of the Proposed Innovative Approaches
1	Use of Advanced Materials	**Energy-efficient materials:** application of new-generation thermal insulation materials aimed at reducing energy consumption.
		Recycling: utilisation of secondary materials obtained from construction waste.
		Biomaterials: incorporation of natural-origin materials such as bamboo, straw, or fungal mycelium.
		Biomimicry in construction (architectural bionics): implementation of natural forms, processes, and systems in the development of materials, structural solutions, energy-efficient technologies, and resilient infrastructure.
2	Energy-efficient Solutions	Installation of solar panels and rainwater harvesting systems.
		Use of heat pumps and ventilation systems with heat recovery.
3	Smart Cities and Smart Buildings	Integration of "smart" technologies for the automation of building and urban infrastructure management
		Application of sensors and artificial intelligence for monitoring energy consumption and assessing the condition of buildings.
4	Modular Construction	Production of construction modules in factory conditions followed by on-site assembly.
		Reduction of construction timelines and minimisation of waste.
5	Digital Technologies	**Building Information Modelling:** creation of digital building models to optimise design, construction, and operational processes.
		Three-dimensional printing: use of additive technologies for the production of building components.
		Drones: monitoring of construction processes and assessment of building conditions through aerial imaging.

(Humenna, 2024)

Given the global climate challenges, Ukraine and other countries must integrate sustainable construction approaches to comply with international standards and ensure alignment with the United Nations Sustainable Development Goals. This also opens opportunities for attracting foreign investment and international assistance. Effective implementation of innovations in the construction sector will contribute to creating a comfortable and safe environment for people, improving the energy efficiency of buildings, and minimising environmental impact.

In this context, smart aesthetics, as understood by the authors, represents an integrated scientific and design concept for the formation of the architectural environment. It combines nature-oriented principles (biomimicry), psychological and emotional dimensions of human perception (neuroaesthetics and Kansei Engineering), and evidence-based methods for choosing technical solutions (multicriteria models, including the Analytic Hierarchy Process), with the aim of creating spaces that simultaneously ensure environmental efficiency, emotional comfort, functional adaptability, and social resilience.

This approach views aesthetics not merely as a category of "beauty" or style, but as an instrument of sustainable development, psychological well-being, and the safe organisation of human life processes. Within the framework of smart aesthetics, architectural form, materiality, lighting, tactile qualities, and spatial structures are designed based on measurable interactions between humans, nature, and technology. Aesthetic decisions are thus regarded as having a direct impact on health, behaviour, and community resilience.

Therefore, smart aesthetics constitutes a methodological framework in which scientific knowledge about cognitive reactions, the environment, resource efficiency, and social dynamics is integrated into design practices that create coherent, adaptive, and human-centred architectural spaces. This framework is particularly significant for shaping resilient infrastructure in the future.

BIOMIMICRY AS A BASIS FOR SUSTAINABLE DESIGN

One of the key components of smart aesthetics is architectural bionics (biomimicry), an approach based on the imitation of structures, behaviours, and processes found in living nature. In the construction sector, biomimicry involves the use of natural forms, processes, and systems for the development of materials, structural solutions, energy-efficient technologies, and resilient infrastructure. The core idea of biomimicry is to learn from nature rather than merely exploit its resources. Natural ecosystems possess billions of years of evolutionary experience in creating adaptive, resource-efficient, and durable structures.

This field has been actively advanced through the work of Benyus (2009), who emphasizes that nature has accumulated millions of years of "research and development"; therefore, its solutions are highly optimized in terms of resources, energy efficiency, and functionality.

The advantages of applying biomimicry in construction include the following (Humenna, 2024):

- energy efficiency – the use of natural methods of ventilation, lighting, and cooling;
- resilience – reduction of environmental impact through the use of environmentally friendly materials and technologies;
- innovation – the creation of unique and highly efficient design solutions;
- cost reduction – optimization of resources and reduction of operational expenses.

The OECD report *Nature-Based Solutions for Climate Resilience* (2023) (OECD, 2023) confirms that the application of nature-inspired structures enables an average reduction in building energy consumption by 18–32 percent, while operational costs can be reduced by 15–25 percent. This is explained by the fact that the forms and principles developed by nature are inherently adapted to changing environmental conditions.

Biomimicry is not a one-dimensional approach. It manifests itself at several interrelated levels:

- structural, including forms inspired by shells, carapaces, honeycombs, and skeletal frameworks;
- functional, encompassing natural systems for regulating temperature, humidity, and airflow;
- behavioural, involving adaptability, self-regulation, and responsiveness to external changes;
- ecosystem-based, reflecting the interconnection of elements embedded within broader ecological networks.

Global examples of successful biomimicry-based projects in construction include the following:

1. Buildings inspired by termite mounds – the Eastgate Centre business complex in Harare, Zimbabwe. This building employs a passive cooling system that

mimics the ventilation mechanisms of termite mounds. In nature, termites are able to maintain a stable internal temperature within their structures despite significant fluctuations in external temperatures. Similarly, the Eastgate Centre uses ventilation shafts to enable natural air circulation, thereby significantly reducing the need for mechanical air conditioning systems (CdT, n.d.).

2. Facades inspired by animal skin – the Institut du Monde Arabe in Paris. The building's facade imitates the functioning principle of the human eye pupil. It consists of metal panels equipped with diaphragms that automatically regulate the amount of daylight entering the interior spaces (French Moments, n.d.). This significantly reduces electricity consumption for both lighting and air conditioning.
3. Self-cleaning surfaces inspired by lotus leaves – the application of the "Lotus Effect" technology in paints and building coatings. This technology replicates the water- and dirt-repellent properties of lotus leaves, which result from their microscopic surface texture. For example, facade coatings such as StoLotusan reduce the need for frequent façade cleaning and maintenance (Schramm, n.d.).
4. Structures inspired by marine organism skeletons – the Gherkin Tower in London. The design of the tower is inspired by the strength and efficiency of marine sponge skeletons, particularly species such as *Euplectella.* This structural concept increases resistance to mechanical loads while reducing material use (Discover Walks, n.d.). Moreover, the aerodynamic form of the building decreases wind resistance and enhances overall energy efficiency.
5. Roof systems inspired by spider webs – the Allianz Arena stadium in Munich. Its façade and roof are constructed using ETFE polymer materials that mimic the lightness and strength of spider silk (Dreamstime, 2022). This solution reduces the overall weight of the structure while enabling high levels of natural daylight penetration.
6. Sustainable materials inspired by natural processes – bio-bricks developed by analogy with coral reefs. Bio-bricks are produced through bacterial processes that induce calcium carbonate formation, closely resembling coral growth mechanisms. This approach allows for the production of durable building materials without traditional kiln firing (The Raycat Solution, 2016). The technology offers an environmentally sustainable alternative to cement manufacturing by significantly reducing carbon dioxide emissions and enabling the reuse of waste materials as raw inputs.

Biomimicry is of particular importance for both contemporary and post-war Ukraine. At present, the application of biomimicry in Ukrainian construction remains in a developmental stage; however, several projects and initiatives have already begun to implement biomimicry principles or nature-inspired ecological design. Selected examples include the following (Humenna, 2024):

- Arena Lviv Stadium (Lviv) – the stadium roof is designed to optimise natural lighting and ventilation, thereby reducing energy consumption. Although this is not a direct example of biomimicry, the principle of using natural processes to reduce energy demand is conceptually aligned with biomimetic thinking.
- Podilski Tovtry National Nature Park – the Ecotourism Centre was developed with consideration of biomimicry principles, using local natural resources and emulating ecosystem functions to enhance energy efficiency. The roof structures are inspired by tree leaves, providing shading and natural cooling.
- Modular housing for internally displaced persons – several modular housing projects designed for internally displaced populations incorporate environmentally friendly materials resembling natural structures, such as timber and composite materials. These buildings apply principles of natural ventilation and thermal insulation observed in natural systems.
- Unit.City (Kyiv) – an innovation park integrating principles of sustainable and environmentally responsible construction. New residential and office buildings employ energy-efficient technologies such as natural ventilation systems and rainwater harvesting. Certain design elements are inspired by natural processes, including facades with glass systems that imitate thermal self-regulation mechanisms.

In conditions requiring autonomous, energy-efficient, and rapidly deployable buildings, natural models offer ready-made solutions. Examples include lotus-inspired schools, whose circular form enables continuous air circulation, natural daylighting, and a community-centred interior layout; coral-inspired medical hubs that allow for flexible modular expansion; and shell-inspired housing units that provide both physical and psychological protection. Such qualities are critically important for communities that have experienced the trauma of war, as confirmed by the *World Health Organization* (2023).

KANSEI ENGINEERING AND EMOTIONAL DESIGN

If biomimicry answers the question "how nature works", Kansei Engineering (KE) answers the question "how a person feels". It is a Japanese scientific approach initiated by Ishihara et al. (2019), which combines *kansei* (sensations, emotions) and *kougaku* (engineering) and translates users' emotional and psychological needs into specific engineering parameters.

Today, Kansei Engineering is widely applied across various industries, ranging from the automotive and furniture sectors to the design of gadgets, appliances, clothing, packaging, architecture, and interiors. Its influence is also expanding in the digital domain, particularly in the development of interfaces, websites, and mobile applications (Humenna, 2013). In construction, the primary objective of Kansei Engineering is the creation of spaces that evoke positive emotions, comfort, and satisfaction.

Several research (Weinberger, et al., 2021; Evans, 2003) demonstrates that emotionally comfortable spaces increase productivity by 12–18 percent, reduce anxiety levels by up to 30 percent, and enhance trust in reconstruction processes by a factor of 2.3.

Within contemporary sustainable construction, Kansei Engineering is evolving into an important social instrument, as it enables the creation of environments that not only comply with standards but also resonate with the inner emotional state of individuals.

Kansei Engineering in construction focuses on the creation of comfortable, aesthetically pleasing, and emotionally attractive spaces that influence people's mood, behavior, and overall well-being.

The main criteria considered in the development of buildings and architectural spaces include:

1. Emotional perception of space – a sense of comfort, coziness, and safety; harmony of form, color, and textures; balance between natural and artificial elements.
2. Sensory perception of the environment – lighting (natural and artificial, intensity, color temperature); acoustic comfort (noise reduction, pleasant sound effects); tactile sensations (material textures, surface warmth); aromas and air quality.
3. Functionality and ergonomics – logical spatial layout (intuitive navigation); ease of movement and accessibility for all users (inclusivity); compliance with the needs of residents or users.
4. Aesthetics and design – proportions of colors, forms, and textures; creation of an atmosphere appropriate to the function of the space (a working mindset for offices, relaxation for residential environments, and so forth); use of natural materials to achieve harmony.
5. Individual spatial customization – personalization of interiors according to user preferences; adaptive systems (smart technologies that adjust lighting, temperature, colors, and similar parameters).
6. Impact on the psychological state of individuals – anti-stress architectural solutions (biophilic design, natural colors, smooth forms); use of rhythmic and

harmonious elements for calm perception; consideration of social interaction (communication zones and spaces for solitude).

Equally important in Kansei Engineering is the use of cultural codes and the collective memory of communities. According to UNESCO's report (2025), spaces that reflect regional cultural codes can increase levels of social cohesion by 15–28 percent. This aspect is particularly critical for Ukraine, where post-war reconstruction architecture possesses significant therapeutic and healing potential (UNESCO, 2025).

Kansei Engineering in construction can be implemented across several key directions:

Residential construction:

1) Design of apartments and houses that takes into account the emotional perception of space.
2) Interiors that evoke a sense of coziness and calm through combinations of colors, textures, and lighting.
3) Development of comfortable communal areas in multi-apartment buildings (green zones, spaces for social interaction).

Office and commercial facilities:

1) Offices that enhance productivity through ergonomics, acoustic comfort, and appropriate lighting.
2) Retail spaces and restaurants that create positive emotional customer experiences through design, aromas, and music.

Public spaces and urban planning:

1) Development of urban environments that prioritize residents' psychological comfort.
2) Parks that promote relaxation and social interaction.
3) Transportation hubs (airports, railway stations) that minimize passenger stress through convenient navigation and thoughtful design.

Smart cities and Internet of Things technologies:

1) Application of artificial intelligence to create personalized and comfortable environments.

2) Interactive architectural elements that change their appearance or functionality depending on users' needs.

The Kansei Engineering method involves the use of so-called "kansei words" as a tool for formalizing and evaluating emotional responses and consumer preferences. These function as parameters that allow for more precise consideration of the emotional component of user perception. However, researchers note that developers of innovations often neglect the principle of user orientation. Their attention is mainly focused on solving technical problems rather than addressing the real needs of target audiences. In many cases, consumers are not regarded as key subjects in the product development process, which leads to reduced market demand. As a result, inventions remain unrealized despite their potential technological value. By contrast, a strategic orientation toward the consumer is a fundamental principle of innovative development that must be applied systematically. The formation of relevant consumer value, oriented not only toward functionality but also toward emotional perception, combined with attractive design and alignment with user expectations, creates the foundation for sustainable competitive advantage.

Kansei words (or "emotional triggers") are verbal markers that consumers associate with emotionally appealing characteristics of a product. These attributes contribute to a positive user experience and enhance the aesthetic and functional value of a product. They are terms that describe feelings, emotions, and associations arising during interaction with a space or architectural elements, for example:

- Comfortable and cozy: warm, calm, homelike, harmonious, enveloping, soft, natural, welcoming;
- Modern and stylish: minimalist, dynamic, technological, elegant, futuristic, innovative;
- Luxurious and aesthetic: exclusive, refined, expensive, sophisticated, elite;
- Functional and practical: logical, ergonomic, flexible, intuitive, smart;
- Natural and ecological: organic, natural, green, biophilic, environmentally friendly;
- Safe and reliable: solid, resilient, protected, durable, stable;
- Interactive and adaptive: smart, personalized, responsive, transformable.

The practical application of Kansei Engineering in construction depends on defined objectives. If the goal is to create an office space that motivates work, designers use kansei words such as energetic, stimulating, dynamic, technological, and organized. Bright color accents, appropriate lighting, ergonomic furniture, and open collaboration spaces are selected. To create a relaxing residential interior, kansei words such as warm, soft, natural, cozy, and calm are applied. Designers use pastel tones, textiles, wooden and stone surfaces, and warm lighting. Kansei Engineering enables construction projects to become not merely functional but also emotionally interactive with users.

Emotional engagement of stakeholders in design and planning processes is a key element of successful marketing strategies for construction companies, as it fosters deeper connections with consumers. This engagement goes beyond attention to include emotional reactions – preferably positive, though not necessarily so. This creates a dual effect: value creation for the client and profitability for the company. Kansei Engineering opens new possibilities in construction, enabling the creation of spaces that are not only functional but also emotionally attractive. Innovative materials, ergonomics, and smart technologies shape the future of construction as energy-efficient and oriented toward quality of life and urban comfort.

A distinct emerging direction is the integration of emotional feedback into digital design systems, referred to as "emotional digital twins". The *World Economic Forum report* (World Economic Forum, 2024) emphasizes that modeling users' behavioral responses enables optimization of spatial loads, shelter comfort, school lighting conditions, and internal circulation logic within buildings.

For Ukraine, this technology may become a key instrument in creating safe and supportive spaces for children and vulnerable population groups.

CULTURAL IDENTITY AS THE CORE OF SMART AESTHETICS

Cultural and social sustainability play a crucial role in ensuring the sustainable development of Ukraine's construction industry. They contribute to the creation of a harmonious environment that takes into account societal needs and preserves cultural heritage. In our view, cultural sustainability implies the integration of architectural solutions that reflect national traditions and specific features with modern technologies and materials. Social sustainability, in turn, involves the creation of infrastructure that meets the needs of various social groups and ensures accessibility, safety, and comfort.

The construction industry is currently developing at an extremely rapid pace due to the urgent need to restore destroyed infrastructure. However, alongside physical reconstruction, it is no less important to take into account the cultural dimension of the development of cities and communities. In this context, Corporate Cultural Responsibility (CCR) becomes an important component both in shaping the brands of construction companies and in their contribution to long-term sustainable development (Humenna & Bedii, 2022). As a concept, CCR emerges from the idea of Corporate Social Responsibility (CSR) of business and is only beginning to gain clarity in academic and professional discourse.

The concept of "Corporate Cultural Responsibility" should be understood as the collective responsibility of business or an organization for supporting, restoring, and developing key elements of the cultural identity of society – such as language, religion, art, creativity, media, and others. It is precisely these components that shape value orientations, which later transform into public expectations and demands addressed to companies and their brands (Humenna & Bedii, 2022).

Corporate cultural responsibility in the global construction industry represents both a strategic component of corporate branding and a substantive pillar of sustainable development. Developers who consciously integrate cultural values into construction projects not only strengthen their corporate reputation and brand differentiation but also contribute to the creation of inclusive, meaningful, and human-centered built environments. Through architecture and spatial design, construction companies can support cultural continuity, foster social cohesion, and enhance the overall quality of life in urban and regional contexts worldwide.

Every construction brand aspires to be perceived in line with its core values and strategic vision. Beyond communication strategies, branding in construction encompasses the development of semantic meaning and visual expression, including architectural language, identity systems, symbolic elements, and non-verbal cues embedded in the built environment. Within this framework, corporate cultural responsibility (CCR) plays a particularly significant role, as it connects economic objectives with social, cultural, and ethical dimensions of development.

The implementation of corporate cultural responsibility in the construction sector typically includes several key aspects:

- Architectural identity – the incorporation of culturally distinctive elements into architectural solutions, the preservation and adaptive reuse of historical heritage, and the integration of contemporary design with local traditions and context-sensitive aesthetics;
- Support for cultural and creative industries – collaboration with artists, designers, sculptors, and other creative professionals to shape public spaces, façades, murals, interior environments, and landscape architecture;
- Inclusive communication and information environment – the use of culturally respectful and locally appropriate language, symbols, and narratives in visual design, marketing, documentation, and corporate communication;
- Social and community integration – the design of residential and commercial facilities that include spaces for cultural activities, social interaction, and community development, fostering a sense of belonging and shared identity;
- Integration of material, environmental, and symbolic value – the creation of environments that are not only aesthetically refined and environmentally sustainable but also enriched with cultural and symbolic meaning. This includes

the use of natural and non-toxic materials, recycling and circular construction practices, ecological balance support, and the incorporation of elements reflecting local cultural, historical, or spiritual heritage.

In a global context, corporate cultural responsibility in construction contributes to more resilient, culturally sensitive, and sustainable cities, aligning economic performance with long-term social and cultural value creation.

The need to conceptualize the practice of CCR and integrate it into the professional environment is driven by the fact that, despite existing interaction between business and the cultural sphere, there is often a lack of genuine mutual understanding. The absence of a clearly articulated concept of CCR and adequate awareness of its significance results in the fact that potentially productive cooperation between these sectors does not yield the expected results (Czyzewska, 2018):

1. Business participates in cultural projects episodically and mostly formally – primarily in the form of sponsorship, which does not create a significant reputational or economic effect in the medium- and long-term perspective;
2. The cultural sector, receiving irregular sponsorship contributions, lacks resources for systematic development and does not fully utilize the potential of audience loyalty to a brand or company.

In fact, effective cooperation between business and culture is entirely possible and capable of generating mutual benefits (Table 2).

Table 2. Benefits of applying CCR principles for business and culture in the construction sector

Benefits	Business (construction sector)	Culture (societal and cultural sphere)	SDGs / ESG impact
Reputational	Construction companies that integrate cultural narratives into architectural design strengthen brand identity and differentiation. For example, *BIG* (Bjarke Ingels Group - https://big.dk/) systematically incorporates local cultural context into landmark projects, enhancing global brand recognition.	Cultural initiatives gain legitimacy and public trust through association with a reputable developer. For instance, the inclusion of public art or heritage preservation projects within residential developments elevates cultural initiatives from isolated acts to socially meaningful urban transformations.	*SDG 11 – Sustainable Cities and Communities* (cultural heritage protection; inclusion of culture in urban spaces) *ESG – Social / Governance* **(social responsibility, community engagement, cultural awareness).**
Informational	Developers engage creative communities (architects, artists, urbanists) who generate user-driven narratives and visual content. Projects such as *Hudson Yards* (USA) or *UNIT. City* (Kyiv) stimulate extensive media coverage and UGC, amplifying brand visibility through cultural discourse rather than traditional advertising (Gorjian, 2025; Chen et al., 2022).	Cultural actors gain access to broader, non-traditional audiences through high-traffic construction projects. Murals, sculptures, and cultural events embedded in housing or mixed-use developments attract residents, visitors, and online audiences well beyond conventional cultural venues.	*SDG 11 – Sustainable Cities and Communities* (inclusive cultural-urban environment) *ESG – Social* **(community engagement, cultural value creation, increased transparency and visibility).**
Economic	Cultural integration increases emotional attachment and loyalty, positively affecting sales velocity and long-term asset value. Projects like *King's Cross redevelopment* (UK) demonstrate that culturally enriched environments command higher real estate value and sustained commercial interest (Centre for Cities, 2022). In Ukraine, culturally branded residential complexes show stronger presales and community retention.	Cultural projects benefit economically through increased visibility, attendance, and funding opportunities. Collaboration with developers provides artists and cultural institutions with stable financing, expanded reach, and long-term presence in urban space rather than one-time events or exhibitions.	*SDG 8 – Decent Work and Economic Growth* (job creation, mixed-use development, economic revitalization) *SDG 11 – Sustainable Cities and Communities* *ESG – Economic / Social* **(investment in cultural infrastructure, long-term cultural-economic synergy).**

The introduction of CCR in construction contributes to:

- the formation of a cultural urban environment: unique architecture oriented toward cultural identity strengthens local culture and attracts tourists;
- the enhancement of corporate reputation: developers who integrate cultural responsibility gain public trust, which increases the loyalty of customers and partners;

- economic benefits: the use of cultural features in construction creates added value for projects, increases their profitability, and contributes to the expansion of the client base.

The integration of cultural elements into construction projects occurs primarily through heritage preservation initiatives. In particular, some construction companies adapt old buildings to modern needs while preserving their architectural value – for example, the restoration of historic façades combined with modern residential technologies. In practice, however, there are also opposite (negative) cases of deliberate or accidental destruction of historical and architectural heritage sites by developers.

The incorporation of cultural elements into residential complexes takes place through the use of Ukrainian ethnomotifs in the design of courtyards, entrances, and parks (for example, decorating facades with traditional ornaments or using street names honoring Ukrainian artists). Some construction companies cooperate with local artists and designers to create unique architectural solutions that reflect national identity and cultural heritage. This may include the use of traditional ornaments in facade design or the creation of public spaces featuring elements of Ukrainian art. At times, construction companies also participate in financing or organizing cultural events such as festivals, exhibitions, or educational programs aimed at promoting Ukrainian culture and art. The integration of spaces for exhibitions, performances, and workshops within residential areas fosters community development and local culture, strengthening ties with the community and increasing cultural awareness.

Overall, cultural sustainability contributes to the preservation of cultural identity and enhances the attractiveness of buildings for local residents and tourists, while social sustainability improves quality of life and social cohesion, as it encompasses the construction of housing, public spaces, educational institutions, and healthcare facilities.

In different regions of Ukraine, this may take the form of:

- Carpathian wooden structures in schools and cultural centers,
- Polissian water-oriented façades and natural materials,
- Podillian geometric motifs in the interiors of public buildings.

Such integration transforms architecture into a carrier of memory and an instrument of social therapy.

According to the *UNESCO World Heritage: Reconstruction and Recovery Report* (UNESCO, 2023), infrastructure reconstruction becomes more effective and resilient when it reproduces local symbols, ornaments, and traditional materials. Cultural codes do not merely embellish the environment – they restore a sense of home and historical continuity, thereby strengthening psychological resilience.

AHP AS A TOOL FOR SUSTAINABLE DECISION-MAKING

Targeted programmes for sustainable development in construction emerged as a response to global challenges, particularly environmental, economic, and social problems, and to the necessity of transforming the construction industry. These programmes make it possible to optimise energy consumption, reduce carbon dioxide emissions, improve quality of life, and foster the development of innovative technologies.

Targeted sustainable development programmes in the construction sector represent comprehensive governmental, international, or corporate initiatives aimed at implementing environmentally friendly, energy-efficient, and socially responsible approaches in construction. Such programmes are usually developed based on the principles of sustainable development and align with the United Nations Sustainable Development Goals in the areas of climate protection, energy efficiency, and urban development.

It is expedient to present an interpretation of programme implementation levels through the prism of high value added (HVA) in production and social spheres (Table 3).

Table 3. Targeted programmes in construction with high value added

Level	Sphere	High Value-Added Programmes
Global	Production	Participation in global sustainable manufacturing initiatives (ISO, green manufacturing)
	Social	Orientation towards the UN Sustainable Development Goals: decent work, inclusiveness, education
National	Production	State programmes supporting manufacturers of energy-efficient materials, incentives for Building Information Modelling
	Social	"Energy Home" programme, investments in social housing, schools, hospitals
Regional	Production	Regional industrial parks, support for small and medium-sized enterprises in construction
	Social	Regional insulation programmes, inclusive social housing
Local	Production	Modular and energy-independent buildings, reconstruction using Building Information Modelling
	Social	Implementation of energy-efficient solutions at the community level, participation in participatory budgeting

The main objectives of the modified construction project management model within targeted sustainable development programmes include:

- ensuring timely and cost-effective implementation of construction projects;
- minimisation of infrastructural and financial risks through the introduction of digital technologies and analytical forecasting methods;

- integration of resource-efficient construction practices in order to reduce negative environmental impacts;
- optimisation of multifactor resource management with regard to socio-economic priorities of sustainable development;
- enhancement of control at all stages of programme implementation, ensuring transparency and compliance with international standards.

Modern construction requires precise and well-substantiated decision-making, particularly when balancing environmental, economic, technological, and social criteria. For this reason, the Analytic Hierarchy Process developed by Saaty (1980) constitutes an important tool in shaping the concept of smart aesthetics.

According to the McKinsey Global Institute report *Infrastructure Productivity Report* (2023), the application of the AHP method reduces planning error risks by 27 percent, decreases budget overruns by 19–22 percent, and accelerates decision-making processes by up to 30 percent (McKinsey Global Institute, 2023).

This method enables the construction of a hierarchy of criteria, which in the context of smart aesthetics primarily include:

- environmental performance (emissions, materials, energy efficiency);
- emotional sensitivity and comfort (Kansei Engineering metrics);
- functional resilience (adaptability, autonomy);
- socio-cultural relevance;
- life-cycle economic costs.

Such a multicriteria model forms the basis for the creation of an electronic decision-support dashboard for construction companies, local authorities, and international donors.

Stages of Constructing the Saaty Hierarchy for a Creative and Innovative Model of Sustainable Development in Construction (using materials **(Humenna, 2025)).**

1. Construction of the hierarchical structure, which includes three levels:
   Level 1: Identification of the main objective;
   Level 2: Selection of key evaluation criteria;
   Level 3: Determination of alternatives or strategies.
2. Formation of pairwise comparison matrices, where experts assess the relative importance of each criterion using the Saaty scale (from 1 to 9) and construct pairwise comparison matrices.

3. Calculation of weighting coefficients, involving matrix normalisation, computation of the priority eigenvector, and consistency assessment.
4. Selection of the optimal strategy, based on the calculated weighting coefficients.

The proposed Analytic Hierarchy Process method makes it possible to calculate and determine the optimal strategy for sustainable development in the construction sector.

1. Defining the Objective and Constructing the Three-Level Hierarchical Structure

Level 1: Main objective – selection of the optimal sustainable development strategy for the construction sector.

Level 2: Key evaluation criteria for innovative and creative approaches:

- ✓ economic sustainability – cost reduction, increased profitability, new business models (E);
- ✓ environmental efficiency – use of environmentally friendly materials, energy efficiency, waste minimisation (Ec);
- ✓ social responsibility – occupational safety, comfort for residents, community development (S);
- ✓ innovativeness – level of technological change, implementation of digital solutions (Building Information Modelling, Internet of Things) (In);
- ✓ creativity – unconventional architectural solutions, flexibility and adaptability of buildings (Cr).

Level 3: Alternatives or innovation implementation strategies:

- ✓ Green construction (GC) – introduction of environmentally friendly materials and energy-saving technologies;
- ✓ Smart city (SC) – digitalisation of construction and implementation of Internet of Things technologies;
- ✓ Modular construction (MC) – use of prefabricated modular structures to reduce costs;
- ✓ Circular economy (CE) – reuse of materials and minimisation of construction waste;
- ✓ Building Information Modelling (BIM) – digital transformation of the construction industry.

2. Construction of the Pairwise Comparison Matrix for Criteria

Using the Saaty scale from 1 to 9, experts evaluate the importance of each criterion relative to the others: the higher the value, the stronger the preference of one criterion over another (Table 4).

Table 4. Pairwise comparison matrix of criteria according to the saaty scale

Criteria	E	Ec	S	In	Cr
E	1	1/3	2	1/2	3
Ec	3	1	4	2	5
S	1/2	¼	1	1/3	2
In	2	½	3	1	4
Cr	1/3	1/5	½	1/4	1

3. Matrix Normalisation

To determine the weights of the criteria, each matrix element is divided by the sum of values in its column (Table 5).

Table 5. Normalised matrix of criteria values

Criteria	E	Ec	S	In	Cr	Row Sum	Average Value (Criterion Weight)
E	0,14	0,15	0,17	0,15	0,20	0,81	0,16
Ec	0,43	0,45	0,34	0,31	0,33	1,86	0,37
S	0,07	0,11	0,08	0,05	0,13	0,44	0,09
In	0,29	0,22	0,25	0,31	0,27	1,34	0,27
Cr	0,07	0,07	0,17	0,08	0,07	0,46	0,09

The resulting criterion weights are:

- environmental efficiency (Ec) = 0,37;
- innovativeness (In) = 0,27;
- economic sustainability (E) = 0,16;
- social responsibility (S) = 0,09;
- creativity (Cr) = 0,09.

4. Construction of Alternative Comparison Matrices (for Each Criterion)

For each criterion, alternatives are evaluated; for example, ecological efficiency is assessed below. The calculated alternative weights are based on logical reasoning, prevailing trends, and expert assessments (Table 6).

Table 6. Evaluation of alternatives for Environmental Efficiency (Ec)

Alternative	GC	SC	MC	CE	BIM
GC	1	3	4	2	2
SC	1/3	1	2	½	1
MC	1/4	1/2	1	1/3	1/2
CE	1/2	2	3	1	1
BIM	1/2	1	2	1	1

After normalisation, the resulting weights are:
Green construction = 0,33;
Smart city = 0,15;
Modular construction = 0,09;
Circular economy = 0,21;
Building Information Modelling = 0,22.
Analogous calculations are conducted for all other criteria.

5. Calculation of the Overall Priority of Alternatives

The weights of the strategies are determined by multiplying them by the weights of the criteria and summing the results (Table 7).

Green Construction (GC):

$$0{,}20 \times 0{,}16 + 0{,}33 \times 0{,}37 + 0{,}25 \times 0{,}09 + 0{,}30 \times 0{,}27 + 0{,}25 \times 0{,}09 =$$

$$= 0{,}032 + 0{,}122 + 0{,}0225 + 0{,}081 + 0{,}0225 = 0{,}28$$

Smart City (SC):

$$0{,}30 \times 0{,}16 + 0{,}15 \times 0{,}37 + 0{,}30 \times 0{,}09 + 0{,}40 \times 0{,}27 + 0{,}35 \times 0{,}09 =$$

$$= 0{,}048 + 0{,}0555 + 0{,}027 + 0{,}108 + 0{,}0315 = 0{,}27$$

Modular Construction (MC):

$$0{,}25 \times 0{,}16 + 0{,}09 \times 0{,}37 + 0{,}20 \times 0{,}09 + 0{,}20 \times 0{,}27 + 0{,}20 \times 0{,}09 =$$

$$= 0{,}040 + 0{,}0333 + 0{,}018 + 0{,}054 + 0{,}018 = 0{,}16$$

Circular Economy (CE):

$$0{,}25 \times 0{,}16 + 0{,}21 \times 0{,}37 + 0{,}25 \times 0{,}09 + 0{,}10 \times 0{,}27 + 0{,}10 \times 0{,}09 =$$

$$= 0{,}040 + 0{,}0777 + 0{,}0225 + 0{,}027 + 0{,}009 = 0{,}18$$

BIM:

$$0{,}40 \times 0{,}16 + 0{,}22 \times 0{,}37 + 0{,}30 \times 0{,}09 + 0{,}45 \times 0{,}27 + 0{,}35 \times 0{,}09 =$$

$$= 0{,}064 + 0{,}0814 + 0{,}027 + 0{,}1215 + 0{,}0315 = 0{,}325$$

Table 7. Final calculation of the priority of each alternative using the Saaty method

Alternative	E (0.16)	Ec (0.37)	S (0.09)	In (0.27)	Cr (0.09)	Total Weight	Rank
BIM	0,40	0,22	0,30	0,45	0,35	0,325	1
GC	0,20	0,33	0,25	0,30	0,25	0,28	2
SC	0,30	0,15	0,30	0,40	0,35	0,27	3
CE	0,25	0,21	0,25	0,10	0,10	0,18	4
MC	0,25	0,09	0,20	0,20	0,20	0,16	5

Thus, the implementation of BIM technologies ranks first as the most effective strategy for sustainable development, primarily due to high scores on the criteria of innovation, economic resilience, and social responsibility. Green construction and the Smart City concept also demonstrate high priority, while modular construction appears to be the least effective alternative in this assessment.

Accordingly, for the sustainable development of the construction sector, it is recommended to prioritize targeted programs focused on BIM, green construction, and smart city initiatives.

The application of the Saaty method makes it possible to substantiate the selection of a sustainable development strategy for the construction industry by accounting for economic, environmental, social, and technological aspects. The optimal model may combine several approaches (particularly Smart City and Green Construction) which receive the greatest weights under the criteria of environmental performance and innovation.

PRACTICAL VALUE OF THE SMART AESTHETICS MODEL

The comprehensive model of smart aesthetics operates according to a cyclical–hierarchical logic of interaction in which each component influences the others.

Stage 1 – Generation of Natural Concepts (Biomimicry)

Forms, functions, and behavioural patterns derived from nature constitute the conceptual basis of a project.

Stage 2 – Emotional and Sensory Adaptation (Kansei Engineering)

Natural forms are evaluated for their emotional alignment with user needs; textures, colours, and the atmospheric characteristics of the space are refined accordingly.

Stage 3 – Analytical Optimization (AHP)

At this stage, all previous groups of design decisions are "weighted" through a system of ecological, psychological, technical, and economic criteria.

Stage 4 – Adjustment of Decisions

AHP may reveal that certain natural forms are economically unfeasible or that emotionally comfortable materials involve excessively high CO_2 emissions.

This initiates a feedback loop that affects previous stages: the biomimetic prototype may be reconsidered; emotional parameters may be revised; the overall concept is adjusted until an optimal balance is achieved.

The following table presents potential criteria and their estimated weights in sustainable construction projects (Table 8).

Table 8. Multicriteria AHP model for sustainable construction projects

Group of Criteria	Sub-criteria	Weight (assumption)	Relevance for Smart Aesthetics
Environmental factors	Emissions, energy efficiency, water conservation	0.35	Ensures nature-oriented design
Emotional factors	Colour, acoustics, comfort, cultural codes	0.25	Ensures psychological resilience
Functional factors	Flexibility, modularity, risk adaptation	0.20	Enables responsiveness to environmental changes
Economic factors	Life cycle, materials, maintenance	0.15	Enhances affordability and feasibility
Social impact	Community cohesion, accessibility, universality	0.05	Strengthens community-oriented design

The key strength of the smart aesthetics model lies precisely in its integrative nature:

- without biomimicry, a project loses its natural optimality;
- without Kansei engineering, the space cannot provide psychological or cultural comfort;
- without AHP, decisions lack sufficient rational justification.

The interaction of these elements allows for the creation of future-oriented environments that:

- conserve resources (through natural models);
- support mental health (through sensory ergonomics);
- acquire long-term value (through optimization mechanisms).

The concept of smart aesthetics – combining biomimicry, Kansei engineering, and the Analytic Hierarchy Process (AHP) – possesses profound practical significance in the context of contemporary global climate challenges, infrastructure modernization, and the need to create psychologically safe environments. Internationally, this model evolves as a response to planetary ecological threats and urban pressures, whereas in Ukraine it emerges as a strategic foundation for post-war reconstruction. Its multidimensionality lies in the simultaneous integration of natural laws (biomimicry), users' emotional needs (Kansei engineering), and rational justification of decisions (AHP), thus forming a unique platform for sustainable, human-centred, and economically efficient design.

Given the extensive environmental damage caused by the construction industry (particularly its high carbon emissions) the integration of natural models of thermal and airflow regulation, as proposed by biomimicry, becomes a key factor in achieving climate goals. A striking example is the Eastgate Centre in Harare, whose ventilation system was modelled on termite mounds. This approach enabled a reduction of energy consumption by over 90% compared to conventional buildings (Benyus, 2009). At the same time, the application of AHP would enable a mathematical verification of the optimality of such solutions by assessing the balance among costs, efficiency, and technical risks – an especially important consideration during the planning of large-scale infrastructure projects.

An equally important component of global practical value is psychological comfort, which is often underestimated at the level of engineering standards. Kansei engineering demonstrates that emotional compatibility of the spatial environment directly affects users' health, behaviour, and productivity.

According to World Health Organization (2023), well-designed environments can reduce stress levels by 30%, accelerate cognitive recovery, and foster a sense of calm and security.

Key neuroaesthetic parameters include smooth forms, biophilic elements, rhythmic compositions, and natural textures – all of which are critically important in territories undergoing reconstruction where people have experienced loss, fear, and destruction (Table 9).

Table 9. Neuroaesthetic indicators and their impact on psychophysiological state

Neuroaesthetic Factor	Description	Impact on the Individual	Source
Smoothness of forms	Soft, organic lines, absence of sharp transitions	Reduction of anxiety, enhanced sense of safety	(World Health Organization, 2023)
Biophilic elements	Plants, water motifs, natural materials	Increased productivity, emotional stabilisation	(UN-Habitat, 2022)
Rhythmicity of composition	Repetitive elements, symmetry, harmonic structures	Improved cognitive focus	(Weinberger, et al., 2021)
Natural lighting	Maximisation of daylight, soft-spectrum illumination	Regulation of circadian rhythms	(World Economic Forum, 2024)
Wood and clay textures	Warm, natural tactile materials	Reduction of sensory overload	(UNESCO, 2025)

In particular, within the series of Maggie's Centres in the United Kingdom – designed with the use of warm natural materials, soft lighting, and spatial openness – research conducted by UCL (2020) recorded a 48% reduction in patient stress levels (Ishihara et al., 2019; Weinberger et al., 2021). These findings confirm that architecture has significant therapeutic potential, which fully aligns with the principles of smart aesthetics.

Such synergy between natural and emotional design approaches proves especially effective in contemporary urbanism. For example, the Masdar City project in the

UAE applies biomimetic models of air movement and building orientation based on natural logic, reducing urban cooling energy demand by more than 50% (UN-Habitat, 2022). The integration of this approach with AHP tools enables optimal selection of materials, modelling of transportation flows, and balancing of shaded and sunlit zones according to climatic conditions.

Furthermore, the integrated system of smart aesthetics demonstrates its applicability in large-scale infrastructure megaprojects, such as the reconstruction of the Port of Rotterdam in the Netherlands. Here, soft nature-derived forms of breakwaters (coral-inspired biomimicry) were used alongside consideration of the emotional perception of public spaces (Kansei component), and the application of AHP enabled the optimal selection of materials and structural solutions (OECD, 2023). The outcome is a system that simultaneously:

- protects the coastline from storms,
- stimulates the restoration of marine ecosystems,
- enhances the region's tourist attractiveness,
- optimises budgetary expenditures.

These examples demonstrate that the smart aesthetics model has a universal character, because it not only combines various methods but creates a synergy between natural, psychological, and technical parameters.

In the national (Ukrainian) context, the practical value of this model becomes particularly significant due to the large-scale post-war reconstruction of the country. According to the World Bank's *Ukraine Rapid Damage and Needs Assessment 2023* (European Union, United Nations, Government of Ukraine, & World Bank, 2023), the total damage to infrastructure and housing stock has exceeded USD 411 billion, creating a unique opportunity for the implementation of innovative approaches. The smart aesthetics model enables the construction of new environments that not only compensate for physical losses but also support the emotional and social recovery of the population.

The most illustrative area of application is the creation of new schools. Children who have experienced war require safe, predictable, and psychologically supportive environments (World Health Organization, 2023). Biomimetic forms, such as the "lotus" structural model, can ensure natural lighting, geometric smoothness, and energy efficiency. Kansei engineering allows the colour palette, textures, and acoustic properties to be adapted to children's needs, reducing anxiety levels. AHP, in this context, is indispensable for selecting economically and technically optimal materials, considering safety requirements, installation speed, and community budget constraints.

Equally important is the application of the smart aesthetics model to the creation of new medical facilities. Hospitals designed according to biomimetic principles – for example, the "root-system hospital" with a branched structural layout – naturally reduce sensory noise, simplify navigation, and create private zones essential for individuals experiencing post-traumatic stress (UNESCO, 2025). The Kansei component contributes to an atmosphere of trust and calmness, while AHP optimises the interaction among engineering, economic, and social requirements.

Another promising application involves residential districts based on coral reef structures. These configurations ensure natural microclimates, social cohesion, and spatial compactness – especially important in communities formed after mass population displacement (World Bank, 2024). Biomimetic forms deliver efficiency, Kansei engineering ensures psychological comfort, and AHP provides rational decision-making.

An equally important direction for Ukraine is the creation of next-generation shelters. Modern shelters must not only provide physical protection but also reduce anxiety. The biomimetic form of the turtle shell creates a dome-like protective space, and the Kansei approach enables the use of warm colours, noise reduction, and clear navigation. AHP ensures optimal selection of materials, air filtration systems, and structural schemes.

The integration of smart aesthetics principles into the construction sector generates substantial benefits at both global and national scales. In the global context, the model contributes to enhanced environmental performance, improved user well-being, and greater long-term resilience of built environments. At the national level – particularly in countries undergoing rapid reconstruction – smart aesthetics supports psychological recovery, promotes community cohesion, and ensures cost-efficient, human-centred development (Table 10).

Table 10. Benefits of the smart aesthetics model in construction at the global and national levels

Component	Global Benefit	National Benefit (Ukraine)
Biomimicry	Energy efficiency, adaptability	Autonomous buildings in wartime risk conditions
Kansei engineering	Psychological comfort	Healing war trauma, supporting children's well-being
AHP	Rational decision-making	Optimisation of reconstruction budgets

Finally, the smart aesthetics model has strong potential for the development of tourism infrastructure as a tool of cultural diplomacy. For instance, projects in the Carpathian region can combine biomimicry of mountain landscapes, Hutsul cultural codes, and AHP for the analysis of tourist flows and transport logistics. This creates a balance between nature conservation, regional development, and cultural identity.

All these examples demonstrate that the smart aesthetics model can transform not only individual buildings but entire cities and regions. Its strength lies in balancing nature, human experience, and technology. Globally, it promotes climate neutrality and innovation. Nationally, it becomes a platform for rebuilding Ukraine in a way that ensures infrastructure not only fulfils functional needs but also restores psychological resilience, cultural identity, and confidence in the future.

CONCLUSIONS AND RECOMMENDATIONS

The concept of *smart aesthetics* emerges as an integrated methodological framework that unites three complementary approaches – biomimicry, Kansei engineering, and the Analytic Hierarchy Process (AHP). Biomimetic principles provide naturally optimized architectural solutions, as demonstrated by such innovations as the Eastgate Centre or the Lotus Effect, and align with global trends confirming the capacity of biomimicry to reduce energy consumption and improve microclimatic performance. Kansei engineering enriches this framework by shaping environments that respond to psychological needs, lowering anxiety levels, enhancing emotional comfort, and influencing productivity and social resilience. AHP, in turn, ensures rational decision-making by balancing ecological, economic, social, and cultural criteria; empirical results confirm its ability to reduce project risks by 19–27% and validate the prioritization of BIM, green construction, and smart-city models.

Taken together, these insights demonstrate that sustainable construction requires not only engineering efficiency but also a nuanced understanding of human experience. Psychological safety, cultural identity, and emotional stability have become as crucial as energy efficiency and material optimisation. This is particularly significant for Ukraine, where reconstruction must simultaneously restore infrastructure, strengthen collective memory, and support community resilience. Within this context, Corporate Cultural Responsibility (CCR) functions not merely as a reputational instrument but as a mechanism of economic and cultural cooperation that generates shared value for developers and cultural institutions.

Thus, *smart aesthetics* provides an integrative pathway for the development of architectural environments in which natural optimisation, emotional sensitivity, and strategically justified resource management form the foundation of sustainable, human-centred spaces capable of supporting society through periods of transformation and recovery.

REFERENCES

BDO. (2024). *Green technologies – new opportunities in construction.* https://www.bdo.ua/uk-ua/insights-2/information-materials/2024/green-technologies-new-opportunities-in-construction

Benyus, J. M. (2009). *Biomimicry: Innovation inspired by nature.* HarperCollins. https://elmoukrie.com/wp-content/uploads/2021/11/janine-m-benyus-biomimicry_-innovation-inspired-by-nature-harper-perennial-2002-2.pdf

CdT. (n.d.). *The Eastgate Centre in Harare, Zimbabwe.* https://www.cdt.ch/prodotti/hub/the-eastgate-centre-in-harare-zimbabwe-286733

Centre for Cities. (2022). *Making places: The role of regeneration in levelling up (Case study: King's Cross).* https://www.centreforcities.org/reader/making-places/learning-from-kings-cross-regeneration/

Chen, K., Lin, H., You, S., & Han, Y. (2022). Review of the impact of urban parks and green spaces on residence prices in the environmental health context. *Frontiers in Public Health, 10*, 993801. DOI: 10.3389/fpubh.2022.993801 PMID: 36159304

Czyzewska, C. (2018). Should corporate cultural responsibility replace cultural sponsorship? https://www.linkedin.com/pulse/should-corporate-cultural-responsibility-replace-cyrielle-czy%C5%BCewska/

Delehan, S., Vilčeková, S., Melehanych, H., Krídlová Burdová, E., & Khorolskyi, A. (2024). A comparative assessment of the capabilities and success of the wood construction industry in Slovakia and Ukraine based on life cycle assessment certification standards. *Frontiers in Environmental Science, 12*, 1319823. Advance online publication. DOI: 10.3389/fenvs.2024.1319823

Discover Walks. (n.d.). *Top facts about the Gherkin in London.* https://www.discoverwalks.com/blog/london/top-facts-about-the-gherkin-in-london/

Dreamstime. (2022, August). *The Allianz Arena football stadium exterior aerial view.* https://www.dreamstime.com/aerial-view-allianz-arena-munich-germany-august-football-stadium-image124543606

European Union, United Nations, Government of Ukraine, & World Bank. (2023). *Ukraine – Third rapid damage and needs assessment: February 2022. December 2023.* https://documents.worldbank.org/en/publication/documents-reports/documentdetail/099021324115085807

Evans, G. W. (2003). The built environment and mental health. *Journal of Urban Health : Bulletin of the New York Academy of Medicine, 80*(4), 536–555. DOI: 10.1093/jurban/jtg063 PMID: 14709704

French Moments. (n.d.). *Arab World Institute of Paris*. https://frenchmoments.eu/arab-world-institute-of-paris/

Gorjian, M. (2025). *Greening schoolyards and urban property values: A systematic review of geospatial and statistical evidence*. arXiv. /arXiv.2507.19934DOI: 10.48550

Humenna, O. (2024). Innovative approaches in construction in the context of Ukraine's recovery based on the principles of sustainable development. *Ways to Improve Construction Efficiency, 2*(53), 289–301. DOI: 10.32347/2707-501x.2024.53(2).289-301

Humenna, O., & Bedii, N. (2022). Corporate cultural responsibility as a component of company branding in Ukraine. *NaUKMA Research Papers. Economics, 7*(1), 41–47. DOI: 10.18523/2519-4739.2022.7.1.41-47

Humenna, O. V. (2013). Kansei engineering in the marketing system of innovative products. *Economic Bulletin of NTUU KPI, 10*, 311–315. https://ela.kpi.ua/server/api/core/bitstreams/52c90934-7a51-4e7d-9772-0c04773fd8ac/content

Humenna, O. V. (2025). *Formation of a creative-innovative ecosystem for sustainable development of the construction industry: Theory, methodology, practice* (Doctoral dissertation, specialty 08.00.03 – Economics and Management of the National Economy). Kyiv.

Ishihara, S., Nagamachi, M., & Tsuchiya, T. (2019). Development of a Kansei engineering artificial intelligence sightseeing application. In *Advances in Affective and Pleasurable Design* (pp. 312–322). DOI: 10.1007/978-3-319-94944-4_34

McKinsey Global Institute. (2023). *Infrastructure productivity report*. https://www.mckinsey.com/capabilities/operations/our-insights/infrastructure-productivity

Melnyk, L. H. (2018). *Narodzhennia sostenovoi ekonomiky: dosvid YeS i praktyka Ukrainy v svitli III i IV promyslovykh revoliutsii* [The birth of the sustain economy: EU experience and Ukraine's practice]. Universytetska Knyha.

OECD. (2023). *Nature-based solutions for climate resilience*. https://www.oecd.org/en/topics/nature-based-solutions-and-climate-resilient-infrastructure.html

Pihul, O., & Siomushkin, V. (2023). Integration of sustainable development concepts and project management in modern housing construction models. *Economics and Society, 53*. Advance online publication. DOI: 10.32782/2524-0072/2023-53-64

PwC. (2025, November). *What's next in engineering and construction*. https://www.pwc.com/us/en/industries/industrial-products/library/engineering-and-construction-trends.html

Saaty, T. L. (1980). *The analytic hierarchy process*. McGraw-Hill.

Schramm. (n.d.). *StoLotusan – The facade paint with the Lotus Effect*. https://www.schramm.de/en/stolotusan-the-facade-paint-with-the-lotus-effect/

The Raycat Solution. (2016, November 27). *Biobricks and how to start hacking DNA*. http://www.theraycatsolution.com/news/2016/11/27/biobricks-and-how-to-start-hacking-dna

UN-Habitat. (2022). *World cities report 2022: Envisioning the future of cities*. https://unhabitat.org/wcr/

UNDP. (2023). From ruins to resilience on the road to recovery in Ukraine. https://www.undp.org/eurasia/blog/ruins-resilience-road-recovery-ukraine

UNESCO. (2023). World Heritage: Reconstruction and Recovery. https://whc.unesco.org/en/review/86

UNESCO. (2025). *The UNESCO Global Report on Cultural Policies*. https://www.unesco.org/en/culture/global-report/global-report-cultural-policies?hub=150011

United Nations. (2015). *Transforming our world: The 2030 Agenda for Sustainable Development*. https://www.un.org/ga/search/view_doc.asp?symbol=A/RES/70/1

United Nations Environment Programme. (2023). *Global status report for buildings and construction 2023*. https://www.unep.org/resources/report/global-status-report-buildings-and-construction

Weinberger, A. B., Christensen, A. P., Coburn, A., & Chatterjee, A. (2021). Psychological responses to buildings and natural landscapes. *Journal of Environmental Psychology*, *77*, 101676. Advance online publication. DOI: 10.1016/j.jenvp.2021.101676

Weizsäcker, E. U., & Wijkman, A. (2018). *Come on! Capitalism, short-termism, population and the destruction of the planet*. Springer. (Ukrainian edition: Sammit-Knyha, 2019) DOI: 10.1007/978-1-4939-7419-1

World Bank. (2024). Resilient Housing Guidance for the Maldives. https://documents.worldbank.org/en/publication/documents-reports/documentdetail/099554312202436138

World Economic Forum. (2024). *Reimagining Real Estate: A Framework for the Future*. https://reports.weforum.org/docs/WEF_Reimagining_Real_Estate_2024.pdf

World Health Organization. (2023). Supporting Mental Health and Community Well-Being. https://ukraine.iom.int/sites/g/files/tmzbdl1861/files/documents/2024-03/iom-ukraine-mhpss-activities-overview-july-december-2023_1.pdf

Compilation of References

Abbas, A., Adeel, M., Akbar, G., Hayyat, S., & Din, G. M. (2025). Integrating Digital Twin and Nature Based Solutions for Climate Resilient Urban Design: A Predictive Framework for Future Ready Residential Neighborhoods in the Anthropocene. *EJSMT*, *1*(5), 1–14. DOI: 10.59324/ejsmt.2025.1(5).01

Abbas, M., & Shafiee, M. (2018). Structural health monitoring and determination of surface defects in large metallic structures using ultrasonic guided waves. *Sensors (Basel)*, *18*(11), 3958. https://www.mdpi.com/1424-8220/18/11/3958

Abbas, N., Yousaf, M., Akbar, M., Saeed, M. A., Huali, P., & Hussain, Z. (2022). An Experimental Investigation and Computer Modeling of Direct Tension Tensile Test of Reinforced Concrete Cylinder. *Inventions (Basel, Switzerland)*, *7*(3), 77. DOI: 10.3390/inventions7030077

Abbass, M., Abbas, U., Jafri, R., Arif, S. M., & Akhai, S. (2024). *AI and Machine Learning Applications in Sustainable Smart Cities.*, DOI: 10.4018/979-8-3693-6740-7.ch001

Abbass, M., & Akhai, S. (2025a). Eco-friendly geopolymer bricks synthesized from alkali-activated sugarcane bagasse ash and rice husk ash. *Industrial Crops and Products*, *233*, 121410. DOI: 10.1016/j.indcrop.2025.121410

Abbass, M., & Akhai, S. (2025b). Exploring the sustainability potential of geopolymer concrete with coal bottom ash and basalt fibres. *Multiscale and Multidisciplinary Modeling. Experiments and Design*, *8*(3), 199. DOI: 10.1007/s41939-025-00794-3

Abdullah, H., Ibrahim, O., Jaafar, M. N. M., Mohamad, M., Baharain, A., & Sulaiman, S. (2020). Energy Savings In a Multi-Circuit Water-Cooled Packaged Unit Air-Conditioning System. *Journal of Advanced Research in Fluid Mechanics and Thermal Sciences*, *76*(1), 39–53. DOI: 10.37934/arfmts.76.1.3953

Abedi, M., Rawai, N. M., Fathi, M. S., & Mirasa, A. K. (2014). Cloud computing as a construction collaboration tool for precast supply chain management. *Jurnal Teknologi (Sciences & Engineering)*, *70*(7).

Aboutaha, R. S., Engelhardt, M. D., Jirsa, J. O., & Kreger, M. E. (2010). Numerical study on the seismic retrofitting of reinforced concrete columns using rectified steel jackets. *ACI Special Publications, SP-160*(2), 59–72. DOI: 10.14359/1546

Abuzwidah, M., Balwan, M., Abdel-Aty, M., & Al-Kateeb, G. (2020). Traffic Safety Benefits of Using Highway Lighting Technology: (Case Study: 4-Lanes Expressways in the Rural Areas). *IOP Conference Series. Materials Science and Engineering, 811*(1), 12023. DOI: 10.1088/1757-899X/811/1/012023

Adithya, K., Girimurugan, R., Vairavel, M., Jawahar, M., Surya, K., Tamilselvan, M., & Thiyagarajan, P. (2020). Structural and thermal research of steam turbine blades by finite element method. *International Journal of Innovative Technology and Exploring Engineering*, *9*(5), 2275–2278. Advance online publication. DOI: 10.35940/ijitee.E2495.039520

Adnan, & Anas, M. (2025). Geopolymer concrete as a sustainable alternative to OPC. *Journal of Umm Al-Qura University for Engineering and Architecture*. DOI: 10.1007/s43995-025-00155-8

Aftimi, S., Kerroum, Y., Guenbour, A., Bellaouchou, A., Idrissi, H., Boulif, R., Semlal, N., Warad, I., & Zarrouk, A. (2024). New insights on the corrosion–erosion behavior of 904L stainless steel in phosphoric acid containing impurities. *Journal of Bio- and Tribo-Corrosion, 10*(1), 13. Advance online publication. DOI: 10.1007/s40735-024-00818-4

Agrawal, A., Sanghai, S. S., & Dabhekar, K. (2021). A review on analysis and design of precast structures. *International Journal of Scientific Research in Science, Engineering and Technology*, *8*(2), 345–350. DOI: 10.32628/IJSRSET218267

Ahmad, A. (2025). *Opposition-based NSGA-III framework for multi-objective optimization of retrofitting projects: Balancing time, cost, quality, energy, safety, and environmental impact*. Asian Journal of Civil Engineering. DOI: 10.1007/S42107-025-01568-5

Ahmad, S., Soetanto, R., & Goodier, C. (2019). Lean approach in precast concrete component production. *Built Environment Project and Asset Management, 9*(3), 457–470. DOI: 10.1108/BEPAM-02-2018-0051

Ahmed, H., Shahzada, K., & Farooq, M. (2021). Performance-based seismic assessment of capacity enhancement of building infrastructure and its cost-benefit evaluation. International Journal of Disaster Risk Reduction. https://www.sciencedirect.com/science/article/pii/S2212420921003071

Ahmed, H., Zahedi, A., & Sanchez, L. (2022). Condition assessment of ASR-affected reinforced concrete columns after nearly 20 years in service. Construction and Building Materials. https://www.sciencedirect.com/science/article/pii/S0950061822022292

Ahmed, H. (2025). Seismic retrofitting of RC buildings using a performance-based approach for risk resilience and vulnerability assessment. *Buildings (Basel, Switzerland), 15*(8), 1333. https://www.mdpi.com/2075-5309/15/8/1333

Ahmed, H., & Lee, Y. (2025). Performance-based seismic retrofitting and resilience assessment for RC buildings considering direct and indirect losses. *Advances in Structural Engineering, 28*(12), 2237–2259. DOI: 10.1177/13694332251325919

Ajayi, S. O., Oyedele, L. O., Bilal, M., Akinade, O. O., Alaka, H. A., & Owolabi, H. A. (2017). Critical management practices influencing on-site waste minimization in construction projects. *Waste Management, 59*, 330–339. DOI: 10.1016/j.wasman.2016.10.040 PMID: 27839690

Akbar, A., Njoroge, J. M., Lee, S., Chang, Y., & Kwon, S. (2023). CNN-Based Automatic Mobile Reporting System and Quantification for the Concrete Crack Size of the Precast Members of OSC Construction. *Engineering Proceedings, 36*(1), 15.

Akbar, G., Hayyat, S., Bilal, M., & Hameed, A. (2025). Sustainable innovations in ultra-high-performance concrete for smart and resilient structures: A survey. European Journal of Applied Science. *Engineering and Technology, 3*(5), 182–196. DOI: 10.59324/EJASET.2025.3(5).15

Akhai, S. (2023). Biogas Plants for Sustainable Municipal Waste Management: A Brief Review. *Handbook of Research on Safe Disposal Methods of Municipal Solid Wastes for a Sustainable Environment*, 162-179. DOI: 10.4018/978-1-6684-8117-2.ch012

Akhai, S. (2024). Smart Home Environment for Society: An Overview of IoT-Enabled Air Conditioners. In *Modelling of Virtual Worlds Using the Internet of Things* (Vol. 1, pp. 74-87). CRC Press. DOI: 10.1201/9781003480181-3

Akhai, S. (2024). Towards Trustworthy and Reliable AI. In *Explainable Artificial Intelligence (XAI) in Healthcare* (pp. 89–99). CRC Press. DOI: 10.1201/9781003426073-7

Akhai, S. (2024). Trends and Environmental Impact of Paper Consumption: A Prognostic Scenario for the Indian Market by 2030-A Case Study. In *International Conference on Interdisciplinary Approaches in Civil Engineering for Sustainable Development* (Vol. 464, pp. 11-18). Springer Nature Singapore. DOI: 10.1007/978-981-97-0910-6_2

Akhai, S., & Abbass, M. (2025). Toward Resilient Futures: The Role of AI-Driven Strategies in Climate Adaptation. In *Nexus of AI, Climatology, and Urbanism for Smart Cities* (pp. 325-340). IGI Global Scientific Publishing. DOI: 10.4018/979-8-3693-5918-1.ch012

Akhai, S., & Khang, A. (2024). Energy efficiency and human comfort: AI and IoT integration in hospital HVAC systems. *Medical Robotics and AI-Assisted Diagnostics for a High-Tech Healthcare Industry*, 93-108. DOI: 10.4018/979-8-3693-2105-8.ch007

Akhai, S., & Taneja, T. (2024). Liquid Gold: Safeguarding Our Health Through Pure Drinking Water and Rigorous Water Testing. In *Natural Resource Management Issues in Human-Influenced Landscapes* (pp. 129-148). IGI Global. DOI: 10.4018/978-1-6684-7051-0.ch007

Akhai, S., & Taneja, T. (2025). The Critical Role of Water Quality: Health Impacts, Contaminants, and Sustainable Solutions for Environmental and Human Well-being. In *Smart Water Technology for Sustainable Management in Modern Cities* (pp. 101-116). IGI Global. DOI: 10.4018/979-8-3693-8074-1.ch005

Akhai, S., Mala, S., & Jerin, A. A. (2020). Apprehending Air Conditioning Systems in Context to COVID-19 and Human Health: A Brief Communication. *International Journal of Healthcare Education & Medical Informatics (ISSN: 2455-9199)*, *7*(1&2), 28-30. DOI: 10.24321/2455.9199.202004

Akhai, S., Srivastava, P., Sharma, V., & Bhatia, A. (2021). Investigating weld strength of AA8011-6062 alloys joined via friction-stir welding using the RSM approach. In *Journal of Physics: Conference Series* (Vol. 1950, No. 1, p. 012016). IOP Publishing. DOI: 10.1088/1742-6596/1950/1/012016

Akhai, S., & Abbass, M. (2025). Air Conditioning for Healthier Living. In *Managing the Health Risks of Climate Change* (pp. 135–154). IGI Global Scientific Publishing., DOI: 10.4018/979-8-3693-4183-4.ch006

Akhai, S., Bansal, S. A., & Singh, S. (2020). A critical review of thermal insulators from natural materials for energy saving in buildings. *Journal of Critical Reviews*, *7*(19), 278–283.

Akhai, S., Mala, S., & Jerin, A. A. (2021). Understanding whether air filtration from air conditioners reduces the probability of virus transmission in the environment. *Journal of Advanced Research in Medical Science & Technology*, *8*(1), 36–41. DOI: 10.24321/2394.6539.202109

Akhai, S., Singh, V. P., & John, S. (2016). Human performance in industrial design centers with small unit air conditioning systems. *Journal of Advanced Research in Production Industrial Engineering*, *3*(2), 5–11.

Akhai, S., Srivastava, P., & Sharma, S. (2020). Developments in horizontal axis wind turbines-A brief review. *J. Crit. Rev*, *7*(19), 255–260.

Akhai, S., Thareja, P., & Singh, V. P. (2017). Assessment of Indoor Environment Health Sustenance in Air Conditioned Class Rooms. *Journal of Advanced Research in Civil and Environmental Engineering*, *4*(1&2), 1–9. DOI: 10.24321/2393.8307.201701

Akhai, S., & Wadhwa, A. S. (2024). Wireless networking solutions for climate control and comfort in modern air conditioning systems. In *Intelligent Networks* (pp. 165–181). CRC Press., DOI: 10.1201/9781003541363-11

Aktan, A. E. (2009). Structural health monitoring: What is the prescription? Mechanics Research Communications. https://www.sciencedirect.com/science/article/pii/S0093641308001122

Alajmi, A., Short, A., Ferguson, J., Poel, K. V., & Griffin, C. (2020). Detailed energy efficiency strategies for converting an existing office building to NZEB: A case study in the Pacific Northwest. *Energy Efficiency*, *13*(6), 1089–1104. DOI: 10.1007/s12053-020-09861-9

Alaux, N., Hoxha, E., Saade, M. R. M., & Passer, A. (2021). Embodied greenhouse gas emissions reduction for structural elements in office buildings. *Journal of Physics: Conference Series*, *2042*(1), 12165. DOI: 10.1088/1742-6596/2042/1/012165

Al-Bazi, A., & Dawood, N. (2010). Developing crew allocation system for the precast industry using genetic algorithms. *Computer-Aided Civil and Infrastructure Engineering*, *25*(8), 581–595. DOI: 10.1111/j.1467-8667.2010.00666.x

Alboaouh, K., & Mohagheghi, S. (2020). Impact of Rooftop Photovoltaics on the Distribution System. *Journal of Renewable Energy*, *1*, 1–23. Advance online publication. DOI: 10.1155/2020/4831434

Aldebei, F., & Dombi, M. (2021). Mining the built environment: Telling the story of urban mining. *Buildings (Basel, Switzerland)*, *11*(9), 388. DOI: 10.3390/buildings11090388

Aleem, S. A., Hussain, S. M. S., & Ustun, T. S. (2020). A Review of Strategies to Increase PV Penetration Level in Smart Grids [Review of *A Review of Strategies to Increase PV Penetration Level in Smart Grids*]. *Energies*, *13*(3), 636. Multidisciplinary Digital Publishing Institute. DOI: 10.3390/en13030636

Alessio, G., Emmi, G., Carli, M. D., & Zarrella, A. (2020). All-air system and radiant floor for heating and cooling in residential buildings: A simulation-based analysis. *Science and Technology for the Built Environment*, *26*(10), 1397–1411. DOI: 10.1080/23744731.2020.1795515

Alhasnawi, B. N., Jasim, B. H., Issa, W., Anvari-Moghaddam, A., & Blaabjerg, F. (2020). A New Robust Control Strategy for Parallel Operated Inverters in Green Energy Applications. *Energies*, *13*(13), 3480. DOI: 10.3390/en13133480

Ali-Benyahia, K., Sbartaï, Z. M., Breysse, D., Ghrici, M., & Kenai, S. (2019). Improving the Non-Destructive Assessment of Concrete Strength in the Field: Influence of the Selection Process of Core Placement on the Assessment Quality for Single and Combined NDT Techniques. *Construction & Building Materials*, *195*, 613–622. DOI: 10.1016/j.conbuildmat.2018.10.032

Ali, E., Mansour, A., Mohammed Abdelkader, E., Elshaboury, N., & Zayed, T. (2025). Digital twin for climate resilience: Transforming smart cities for a sustainable future. *The International Archives of the Photogrammetry, Remote Sensing and Spatial Information Sciences*, *48*, 139–145. DOI: 10.5194/isprs-archives-XLVIII-G-2025-139-2025

Ali, G. E., & Magalhaes, R. (2008). Barriers to implementing e-learning: A Kuwaiti case study. *International Journal of Training and Development*, *12*(1), 36–53. DOI: 10.1111/j.1468-2419.2007.00294.x

Ali, M., Issayev, G., Shehab, E., & Sarfraz, S. (2022). A critical review of 3D printing and digital manufacturing in construction engineering. *Rapid Prototyping Journal*, *28*(7), 1312–1324. DOI: 10.1108/RPJ-07-2021-0160

Alkhamis, M., Ghasemi, M., & Ghasemi, A. (2018). Performance-based optimum retrofitting design of concrete bridge piers. Jordan Journal of Civil Engineering. https://www.academia.edu/download/85695491/paper.pdf

Allan, M., Rajabifard, A., & Foliente, G. (2024). Climate resilient urban regeneration and SDG 11–stakeholders' view on pathways and digital infrastructures. *International Journal of Digital Earth*, *17*(1), 2385076. DOI: 10.1080/17538947.2024.2385076

Allen, J., Halberstadt, A., Powers, J. T., & El-Farra, N. H. (2020). An Optimization-Based Supervisory Control and Coordination Approach for Solar-Load Balancing in Building Energy Management. *Mathematics*, *8*(8), 1215. DOI: 10.3390/math8081215

Al-Sabah, S., Sourav, S. N. A., & McNally, C. (2021). The post-installed screw pull-out test: Development of a method for assessing in-situ concrete compressive strength. *Journal of Building Engineering*, *33*, 101658. DOI: 10.1016/j.jobe.2020.101658

Alves, R., Lousada, S., Cabezas, J., & Gómez, J. M. N. (2022). The importance of the strategic urban rehabilitation plan in the sustainable development of the municipality of Machico. *Sustainability*, *14*(24), 16816. DOI: 10.3390/SU142416816

Alves, R., Lousada, S., Cabezas, J., & Gómez, J. M. N. (2023). Local housing strategy: Analysis of importance and implementation in Machico municipality, Madeira. *Land (Basel)*, *12*(9), 1778. DOI: 10.3390/LAND12091778

Ameri, F., & Dutta, D. (2005). Product Lifecycle Management: Closing the Knowledge Loops. *Computer-Aided Design and Applications*, *2*(5), 577–590. DOI: 10.1080/16864360.2005.10738322

Anbuchezian, A., & Vairavel, M. (2022). Steel fibre impacts on mechanical performance and toughness of steel fibre reinforced high strength concrete after normal and hygrothermal curing (SFRHSC). *NeuroQuantology : An Interdisciplinary Journal of Neuroscience and Quantum Physics*, *20*(5), 1135–1143.

Andriulaitytė, I., & Valentukevičienė, M. (2020). Circular economy in buildings. *Construction Optimisation and Energy Potential*, *9*(2), 23–29.

Antunes, M., Santos, R. L., Pereira, J., Rocha, P., Horta, R. B., & Colaço, R. (2021). Alternative Clinker Technologies for Reducing Carbon Emissions in Cement Industry: A Critical Review. *Materials (Basel)*, *15*(1), 209. DOI: 10.3390/ma15010209 PMID: 35009355

Anwar, G. A., Dong, Y., & Zhai, C. (2020). Performance-based probabilistic framework for seismic risk, resilience, and sustainability assessment of reinforced concrete structures. *Advances in Structural Engineering*, *23*(7), 1454–1472. DOI: 10.1177/1369433219895363

Apanavičienė, R., Urbonas, R., & Fokaides, P. A. (2020). Smart Building Integration into a Smart City: Comparative Study of Real Estate Development. *Sustainability (Basel)*, *12*(22), 9376. DOI: 10.3390/su12229376

Aprile, A., Spacone, E., & Limkatanyu, S. (2001). Role of Bond in RC Beams Strengthened with Steel and FRP Plates. *Journal of Structural Engineering*, *127*(12), 1445–1452. DOI: 10.1061/(ASCE)0733-9445(2001)127:12(1445)

Arabkoohsar, A., & Alsagri, A. S. (2020). Thermodynamic analysis of ultralow-temperature district heating system with shared power heat pumps and triple-pipes. *Energy*, *194*, 116918. DOI: 10.1016/j.energy.2020.116918

Aragón, A., & Alberti, M. G. (2024). Limitations of machine-interpretability of digital EPDs used for a BIM-based sustainability assessment of construction assets. *Journal of Building Engineering*, *96*, 110418. DOI: 10.1016/j.jobe.2024.110418

Araújo, C. S., Ferreira, E. D. A. M., & Costa, D. B. (2025). System for tracking and managing metallic formworks adopted in cast-in-place concrete wall systems by integrating IoT and BIM. *Construction Innovation*, *25*(4), 1315–1341. DOI: 10.1108/CI-06-2023-0127

Arce-Recatalá, M., García-Morales, S., & Bossche, N. V. D. (2020). Quantifying Wind-driven Rain Intrusion – A Comparative Study on The Water Management Features of Different Types of Rear-Ventilated Facade Systems. *E3S Web of Conferences, 172*, 23007. DOI: 10.1051/e3sconf/202017223007

Ardi, S. Z., DİKER, B., & Bahri, P. A. (2020, January 1). Environmental Deterioration Factors in Metal Claddings and GFRC Panels Implemented on Facades: An Assessment through Two Cases in Istanbul. *XV International Conference on Durability of Building Materials and Components. eBook of Proceedings.* DOI: 10.23967/dbmc.2020.060

Arditi, D., Ergin, U., & Günhan, S. (2000). Factors affecting the use of precast concrete systems. *Journal of Architectural Engineering*, *6*(3), 79–86. DOI: 10.1061/(ASCE)1076-0431(2000)6:3(79)

Arehart, J. H., Nelson, W. S., & Srubar, W. V.III. (2020). On the theoretical carbon storage and carbon sequestration potential of hempcrete. *Journal of Cleaner Production, 266*, 121846. DOI: 10.1016/j.jclepro.2020.121846

Arfa, F. H., Zijlstra, H., Lubelli, B., & Quist, W. (2022). Adaptive Reuse of Heritage Buildings: From a Literature Review to a Model of Practice. *The Historic Environment, 13*(2), 148–170. DOI: 10.1080/17567505.2022.2058551

Arganiz, P., & Muslim, F. (2024). Development of a prototype system integration model for RFID technology with the internet of things and its implementation to improve precast concrete material management in Indonesia. *Journal of Project Management (Canada)*, *9*(3), 213–226. DOI: 10.5267/j.jpm.2024.5.001

Arı, A. C. (2023). Kompozit Malzemelerin Tahribatsız Muayene Yöntemlerinden Olan Ultrasonik Test ile Ölçülmesi Üzerine Yapılan Çalışmaların İncelenmesi. *Uluslararası Mühendislik Tasarım Ve Teknoloji Dergisi*, *4*(1-2), 1–10.

Arif, F., Ayub, L., Khan, F. A., & Khan, A. U. R. (2025). Conceptual Framework for IoT-Based Structural Health Monitoring of RC Bridges for Maintenance Decision-Making. *Construction Technologies and Architecture, 15*, 19–26. DOI: 10.4028/p-q8nkQW

Arooj, M., & Yadav, G. (2017). Comparison of Destructive and Non-Destructive Testing of Concrete - A Review. [IJERT]. *International Journal of Engineering Research & Technology (Ahmedabad)*, *6*(11), 62–67.

Arthur, C., Onuche, P. U. O., Yusuf, M., Animashaun, T., Boansi, S. O., Adewale, K., & Bart-Addison, C. E. (2025). Decarbonizing Concrete: Ultra-Low-Carbon Pathways and Policy Integration. *Scientific. Journal of Engineering Technology*, *2*(1), 134–142. DOI: 10.69739/sjet.v2i1.669

Arya, A., Gunarani, G. I., Rathinakumar, V., Sharma, A., Pati, A. K., & Sethi, K. C. (2024). NSGA-III based optimization model for balancing time, cost, and quality in resource-constrained retrofitting projects. *Asian Journal of Civil Engineering*, *25*(7), 5613–5625. DOI: 10.1007/S42107-024-01133-6

Arya, A., & Sharma, R. L. (2022). Strategies for green building rating in India: A comparison of LEED and GRIHA criteria. *Materials Today: Proceedings, 57*, 2311–2316. DOI: 10.1016/j.matpr.2022.01.100

Asadi, E., da Silva, M., Antunes, C., & Dias, L. (2012). A multi-objective optimization model for building retrofit strategies using TRNSYS simulations, GenOpt and MATLAB. Building and Environment. https://www.sciencedirect.com/science/article/pii/S0360132312001217

Asamoah, R. O., Ankrah, J. S., Offei-Nyako, K., & Tutu, E. O. (2016). Cost Analysis of Precast and Cast-in-Place Concrete Construction for Selected Public Buildings in Ghana. *Journal of Construction Engineering, 2016*(1), 8785129. DOI: 10.1155/2016/8785129

Asdrubali, F., Grazieschi, G., & Gandola, D. M. (2025). The Role of Environmental Product Declarations in the Decarbonization of Building Materials and Components. *Energies, 18*(5), 1308. DOI: 10.3390/en18051308

Askar, R., Karaca, F., Bragança, L., & Gervásio, H. (2024). The role of BIM in supporting circularity: A conceptual framework for developing BIM-based circularity assessment models in buildings. In International Conference "Coordinating Engineering for Sustainability and Resilience" (pp. 649–658). Springer Nature.

Asmara, Y. P. (2023). Concrete reinforcement degradation and rehabilitation: Damages, corrosion and prevention. In *Engineering materials*. Springer. DOI: 10.1007/978-981-99-5933-4_8

ASTM. (2002). *Standard Practice for Predicting Concrete Strength by the Maturity Method. C 1074-98*. American Society for Materials Testing.

ASTM. (2007). *Standard Test Method for Tensile Strength of Hardened Concrete, C 900 – 06*. American Society for Materials Testing.

Ateş, E. (2020). Deprem ve Mikrotremor Kayıtları Kullanılarak Elde Edilen Yatay/Düşey Spektral Oranların Karşılaştırılması: Amasya Örneği. *Türk Deprem Araştırma Dergisi*, 2(2), 160–175. DOI: 10.46464/tdad.746926

Ateş, E., & Uyanık, O. (2019). Jeofizik yöntemler ile yer ve yapı etkileşimi. *Süleyman Demirel Üniversitesi Fen Bilimleri Enstitüsü Dergisi, 23*, 46–60. DOI: 10.19113/sdufenbed.430906

Ateş, E., & Uyanık, O. (2023). Farklı derinliklerde ölçülen ivme kayıtlarının karşılaştırılması. *Türk Deprem Arastirma Dergisi, 5*(2), 215–232.

Ateş, E., & Uyanık, O. (2025). Comparison of soil-structure interaction obtained from spectral ratio methods applied to earthquake and microtremor records. *Journal of Applied Geophysics, 234*, 105645. DOI: 10.1016/j.jappgeo.2025.105645

Atiş, C., Abdılahı, A., İlkentapar, S., Durak, U., & Karahan, O. (2022). Flexural Tensile Strength Measurement and Determination for Kerbstones with Unsymmetrical Section. *Süleyman Demirel Üniversitesi Fen Bilimleri Enstitüsü Dergisi, 26*(1), 63–68. DOI: 10.19113/sdufenbed.943602

Attia, S., Santos, M. C., Al-Obaidy, M., & Baskar, M. (2021). Leadership of EU member States in building carbon footprint regulations and their role in promoting circular building design. *IOP Conference Series. Earth and Environmental Science, 855*(1), 012023. DOI: 10.1088/1755-1315/855/1/012023

Attia, S., Shadmanfar, N., & Ricci, F. (2020). Developing two benchmark models for nearly zero energy schools. *Applied Energy, 263*, 114614. DOI: 10.1016/j.apenergy.2020.114614

Avan, A., Maleknejad, M., & Navi, K. (2020). High-speed energy efficient process, voltage and temperature tolerant hybrid multi-threshold 4:2 compressor design in CNFET technology. *IET Circuits, Devices & Systems, 14*(3), 357–368. DOI: 10.1049/iet-cds.2019.0105

Awan, A. B., Alghassab, M. A., Zubair, M., Bhatti, A. R., Uzair, M., & Abbas, G. (2020). Comparative Analysis of Ground-Mounted vs. Rooftop Photovoltaic Systems Optimized for Interrow Distance between Parallel Arrays. *Energies, 13*(14), 3639. DOI: 10.3390/en13143639

Azevedo, A. C., Lemos, S., Delgado, J. M. P. Q., Silva, F. A. N., & Sousa, C. A. P. (2023). Non-destructive and Destructive Tests to Drive Corrective Intervention Procedure of Concrete Elements. In Delgado, J. M. P. Q. (Ed.), *Building Pathologies: Experimental Campaigns and Numerical Procedures. Building Pathology and Rehabilitation, 25*. Springer. DOI: 10.1007/978-3-031-17061-4_5

Azman, M. N. A., Ahamad, M. S. S., Majid, T. A., & Shah, M. N. S. A. (2012). A study of precast concrete in Malaysia. *Concrete (London), 46*(10), 50–52.

Babiak, K., & Thibault, L. (2009). Challenges in Multiple Cross-Sector Partnerships. *Nonprofit and Voluntary Sector Quarterly, 38*(1), 117–143. DOI: 10.1177/0899764008316054

Bachmann, H., & Steinle, A. (2011). *Precast concrete structures*. Ernst & Sohn.

Badge, A. K., & Upadhyay, Mrs. M. (2020). Study on Hybrid Solar Wind Power System with PMSG for Power Enhancement Using AI Technique. *SMART MOVES JOURNAL IJOSCIENCE, 6*(1), 5. DOI: 10.24113/ijo-science.v6i1.252

Bae, S., Nam, Y., & Choi, J.-H. (2020). Comparative Analysis of System Performance and Thermal Comfort for an Integrated System with PVT and GSHP Considering Two Load Systems: Convective Heating and Radiant Floor Heating. *Energies, 13*(20), 5524. DOI: 10.3390/en13205524

Baghalzadeh Shishehgarkhaneh, M., Keivani, A., Moehler, R. C., Jelodari, N., & Roshdi Laleh, S. (2022). Internet of Things (IoT), Building Information Modeling (BIM), and Digital Twin (DT) in Construction Industry: A Review, Bibliometric, and Network Analysis. *Buildings (Basel, Switzerland)*, *12*(10), 1503. DOI: 10.3390/buildings12101503

Bahadur, P. S., Srivastava, R., & Srivastava, V. (2023). Review on the Development of Smart Materials, Including Shape Memory Alloys, Characterization, and Recent Applications. *Modeling, Characterization, and Processing of Smart Materials*, 107-139.

Bain, A., Shenoy, M., Ashton, W., & Chertow, M. (2010). Industrial symbiosis and waste recovery in an Indian industrial area. *Resources, Conservation and Recycling*, *54*(12), 1278–1287. DOI: 10.1016/j.resconrec.2010.04.007

Bajare, D., Zsembinszki, G., Rizaov, D., Tambovceva, T., Cudečka-Purina, N., Atstāja, D., & Cavdar, A. D. (2024). Circular Value Chain Management—Barriers and Opportunities. In *Circular Economy Design and Management in the Built Environment: A Critical Review of the State of the Art* (pp. 667–735). Springer Nature Switzerland.

Bakker, C. A., Mugge, R., Boks, C., & Oguchi, M. (2021). Understanding and managing product lifetimes in support of a circular economy. *Journal of Cleaner Production*, *279*, 123764. DOI: 10.1016/j.jclepro.2020.123764

Bakker, C., Wang, F., Huisman, J., & Den Hollander, M. (2014). Products that go round: Exploring product life extension through design. *Journal of Cleaner Production*, *69*, 10–16. DOI: 10.1016/j.jclepro.2014.01.028

Balamuralikrishnan, R., AlHarthy, M. N., & Krishnamurthy, B. S. (2017). Destructive and Non-Destructive Testing Methods for Condition Monitoring of Concrete Members. [JMEST]. *Journal of Multidisciplinary Engineering Science and Technology*, *4*(6), 7405–7409.

Bandera, C. F., Pachano, J. E., Salom, J., Peppas, A., & Ruiz, G. R. (2020). Photovoltaic Plant Optimization to Leverage Electric Self Consumption by Harnessing Building Thermal Mass. *Sustainability (Basel)*, *12*(2), 553. DOI: 10.3390/su12020553

Banerjee, A., & Nayaka, R. R. (2022). A comprehensive overview on BIM-integrated cyber physical system architectures and practices in the architecture, engineering and construction industry. *Construction Innovation*, *22*(4), 727–748. DOI: 10.1108/CI-02-2021-0029

Banihashemi, S., Meskin, S., Sheikhkhoshkar, M., Mohandes, S. R., Hajirasouli, A., & LeNguyen, K. (2024). Circular economy in construction: The digital transformation perspective. *Cleaner Engineering and Technology*, *18*, 100715. DOI: 10.1016/j.clet.2023.100715

Bao, Y., Chen, Z., Wei, S., Xu, Y., Tang, Z., & Li, H. (2019). The state of the art of data science and engineering in structural health monitoring. Engineering. https://www.sciencedirect.com/science/article/pii/S2095809918308026

Bao, Z., Lu, W., Chi, B., Yuan, H., & Hao, J. (2019). Procurement innovation for a circular economy of construction and demolition waste: Lessons learnt from Suzhou, China. *Waste Management*, *99*, 12–21. DOI: 10.1016/j.wasman.2019.08.031 PMID: 31454595

Bapat, H., Sarkar, D., & Gujar, R. (2021). Application of integrated fuzzy FCM-BIM-IoT for sustainable material selection and energy management of metro rail station box project in western India. *Innovative Infrastructure Solutions*, *6*(2), 73. DOI: 10.1007/s41062-020-00431-7

Baran, E., Mahamid, M., Baran, M., Kurtoglu, M., & Torra-Bilal, I. (2021). Performance of a moment resisting beam-column connection for precast concrete construction. *Engineering Structures*, *246*, 113005. DOI: 10.1016/j.engstruct.2021.113005

Barski, M., Kędziora, P., & Muc, A. (2014). Structural health monitoring methods in machine design and operation. Archive of Mechanical Engineering. DOI: 10.2478/meceng-2014-0037

Bašková, R., Struková, Z., & Kozlovská, M. (2019, August). Construction cost saving through adoption of IoT applications in concrete works. In *International Conference Current Issues of Civil and Environmental Engineering Lviv-Košice–Rzeszów* (pp. 452-459). Cham: Springer International Publishing.

Bastidas-Arteaga, E., Sánchez-Silva, M., Chateauneuf, A., Ribas Silva, M., & Sánchez-Silva, M. (2008). Coupled reliability model of biodeterioration, chloride ingress and cracking for reinforced concrete structures. *Structural Safety*, *30*(2), 110–129. DOI: 10.1016/j.strusafe.2006.09.001

BDO. (2024). *Green technologies – new opportunities in construction*. https://www.bdo.ua/uk-ua/insights-2/information-materials/2024/green-technologies-new-opportunities-in-construction

Belsky, M., Eastman, C., Sacks, R., Venugopal, M., Aram, S., & Yang, D. (2014). Interoperability for precast concrete building models. *PCI Journal*, *59*(2), 144–155. DOI: 10.15554/pcij.03012014.144.155

Benachio, G. L. F., Freitas, M. D. C. D., & Tavares, S. F. (2020). Circular economy in the construction industry: A systematic literature review. *Journal of Cleaner Production*, *260*, 121046. DOI: 10.1016/j.jclepro.2020.121046

Benachio, G. L. F., Freitas, M. D. C. D., & Tavares, S. F. (2021). Interactions between Lean Construction Principles and Circular Economy Practices for the Construction Industry. *Journal of Construction Engineering and Management*, *147*(7), 04021068. DOI: 10.1061/(ASCE)CO.1943-7862.0002082

Ben-Daya, M., Hassini, E., & Bahroun, Z. (2019). Internet of things and supply chain management: A literature review. *International Journal of Production Research*, *57*(15-16), 4719–4742. DOI: 10.1080/00207543.2017.1402140

Benyus, J. M. (2009). *Biomimicry: Innovation inspired by nature*. HarperCollins. https://elmoukrie.com/wp-content/uploads/2021/11/janine-m-benyus-biomimicry_-innovation-inspired-by-nature-harper-perennial-2002-2.pdf

Bertino, G., Kisser, J., Zeilinger, J., Langergraber, G., Fischer, T., & Österreicher, D. (2021). Fundamentals of Building Deconstruction as a Circular Economy Strategy for the Reuse of Construction Materials. *Applied Sciences (Basel, Switzerland)*, *11*(3), 939. DOI: 10.3390/app11030939

Beucker, S., & Hinterholzer, S. (2020). *Effects of ICT-Enabled Flexible Energy Consumption on the Reduction of CO2 Emissions in Buildings*. 1. DOI: 10.1145/3401335.3401336

Bharath, S. R., Lavanya, N., Bharath Kumar, H. B., Chaitra, R. K., & Dandautiya, R. (2021). Utilization Potential of Iron Ore Tailing Waste in Various Applications. In *Advances in Energy and Environment: Select Proceedings of TRACE 2020* (pp. 125-134). Singapore: Springer Singapore. DOI: 10.1007/978-981-33-6695-4_12

Bhargava, K., Ghosh, A., Mori, Y., & Ramanujam, S. (2007). Corrosion-induced bond strength degradation in reinforced concrete: Analytical and empirical models. Nuclear Engineering and Design. https://www.sciencedirect.com/science/article/pii/S0029549307001021

Bhogayata, A., & Sata, A. (2025). *Introduction to Internet of Things in Civil Engineering: Theory and Practice*. Springer. DOI: 10.1007/978-981-97-9450-8

Bijivemula, S., & Wanjari, S. (2018). Comparative Study on in Situ and Laboratory Testing of Concrete Cores with Different Height to Diameter Ratios. *Journal of Testing and Evaluation*. Advance online publication. DOI: 10.1520/JTE20170373

Birgani, Y. T., & Yazdandoost, F. (2016). Resilience in urban drainage risk management systems. *Proceedings of the Institution of Civil Engineers -. Water Management*, *169*(1), 3–16. DOI: 10.1680/wama.14.00043

Bishko, A. V., Samokrutov, A. A., & Shevaldykin, V. G. (2008). Ultrasonic Echo-Pulse Tomography of Concrete Using Shear Wave Low-Frequency Phased Antenna Arrays. *17th World Conference on Non-Destructive Testing*, 25–28 October 2008, Shanghai, China.

Bishr, H. A., Al-Amoudi, O. S. B., Basunbul, I. A., & Al-Sulaimani, G. J. (1995). Assessment of concrete compressive strength using the Lok test. *Construction & Building Materials*, *9*(4), 227–237. DOI: 10.1016/0950-0618(95)00025-B

Bizoňová, S., Katunský, D., & Bagoňa, M. (2020). Comparative analysis of selected glass systems by dynamic simulation using measured real environmental conditions. *E3S Web of Conferences, 172*, 19009. DOI: 10.1051/e3sconf/202017219009

Bjørn, A., Chandrakumar, C., Boulay, A., Doka, G., Fang, K., Gondran, N., Hauschild, M. Z., Kerkhof, A. C., King, H., Margni, M., McLaren, S. J., Mueller, C., Owsianiak, M., Peters, G., Roos, S., Sala, S., Sandin, G., Sim, S., Vargas-Gonzalez, M., & Ryberg, M. (2020). Review of life-cycle based methods for absolute environmental sustainability assessment and their applications. *Environmental Research Letters*, *15*(8), 83001. DOI: 10.1088/1748-9326/ab89d7

Blagojević, B., Vasilevska, L., Anđelković, D., Bogojević, A., & Lousada, S. (2023). A framework for assessing nature-based urban stormwater management solutions: A preliminary spatial analysis approach applied to Southeast Serbia. *Water (Basel)*, *15*(20), 3604. DOI: 10.3390/W15203604

Blumensaat, F., Sivertsen, L., & Schaar, E. (2025). The value of urban drainage systems data: Facts, discussions and future perspectives. *Journal of Hydroinformatics*.

Boahen, S., Mensah, K., Nam, Y., & Choi, J. M. (2020). Fault Detection Methodology for Secondary Fluid Flow Rate in a Heat Pump Unit. *Energies*, *13*(11), 2974. DOI: 10.3390/en13112974

Boardman, C. R., Glass, S. V., & Zelinka, S. L. (2020). Moisture Redistribution in Full-Scale Wood-Frame Wall Assemblies: Measurements and Engineering Approximation. *Buildings (Basel, Switzerland)*, *10*(8), 141. DOI: 10.3390/buildings10080141

Boateng, E. A., Asibey, M. O., Cobbinah, P. B., Adutwum, I. O., & Blija, D. K. (2023). Enabling nature-based solutions: Innovating urban climate resilience. *Journal of Environmental Management*, *332*, 117433. DOI: 10.1016/j.jenvman.2023.117433 PMID: 36738724

Bonaldo, E., Barros, J. A. O., & Lourenco, P. B. (2005). Bond Characterization Between Concrete Substrate and Repair SFRC Using Tensile Testing. *International Journal of Adhesion and Adhesives*, *25*(6), 463–474. DOI: 10.1016/j.ijadhadh.2005.01.002

Bondarenko, S., Lebedev, M., Strokova, V., Markova, I., Kozhukhova, N., Lukyanenko, N., & Potapov, D. (2025). Interaction Mechanisms in «Portland Cement—Functional Polymer Mineral Additives» Binder Produced by Different Methods. *Materials (Basel)*, *18*(13), 3178. DOI: 10.3390/ma18133178 PMID: 40649666

Bonev, M., Wörösch, M., & Hvam, L. (2015). Utilizing platforms in industrialized construction: A case study of a precast manufacturer. *Construction Innovation, 15*(1), 84–106. DOI: 10.1108/CI-04-2014-0023

Bora, D. S. (2016). Tests and Acceptance Criteria for Concrete Based on Compressive Strength. *National Technical Seminar on Innovation, Project Delivery and Management in Infrastructure Technology for Sustainable Development*, 133–141.

Bouchard, K. M. (2007). A performance-based approach to retrofitting unreinforced masonry structures for seismic loads [Master's thesis, Massachusetts Institute of Technology]. MIT DSpace. https://dspace.mit.edu/handle/1721.1/38944

Bovio, F. L., Bovio, G. L., Brencich, A., Cassini, G., Pera, D., & Riotto, G. (2014). A new pull-out technique for in-place estimation of concrete compressive strength. *Advances in Materials Science and Engineering, 6*(1), 1–8. DOI: 10.1155/2014/984341

Bramston, D., & Maycroft, N. (2014). Designing with Waste. In *Materials Experience* (pp. 123–133). Elsevier., DOI: 10.1016/B978-0-08-099359-1.00009-6

Brehm, J. K., Augustine, D. B., Ziviani, D., & Groll, E. A. (2021). Assessing the performance limits of a variable-speed residential heat pump system. *IOP Conference Series. Materials Science and Engineering, 1180*(1), 12042. DOI: 10.1088/1757-899X/1180/1/012042

Brencich, A., Bovolenta, R., Ghiggi, V., Pera, D., & Redaelli, P. (2020). Rebound hammer test: An investigation into its reliability in applications on concrete structures. *Advances in Materials Science and Engineering, 2020*(1), 6450183. DOI: 10.1155/2020/6450183

Breysse, D., Larget, M., Sbartai, Z. M., Lataste, J. F., & Balayssac, J. P. (2011). Quality of NDT measurements and accuracy of concrete physical properties quantitative assessment. *European Journal of Environmental and Civil Engineering, 15*(4), 619–632. DOI: 10.1080/19648189.2011.9693351

Brink, H. W., Loomans, M., Mobach, M. P., & Kort, H. S. M. (2020). Classrooms' indoor environmental conditions affecting the academic achievement of students and teachers in higher education: A systematic literature review [Review of *Classrooms' indoor environmental conditions affecting the academic achievement of students and teachers in higher education: A systematic literature review*]. *Indoor Air, 31*(2), 405. Wiley. DOI: 10.1111/ina.12745

Brzev, S., & Guevara-Perez, T. (2010). *Precast concrete construction.* British Columbia Institute of Technology.

Brzezińska, D., & Bryant, P. (2020). Risk Index Method–A Tool for Sustainable, Holistic Building Fire Strategies. *Sustainability (Basel), 12*(11), 4469. DOI: 10.3390/su12114469

Bukhari, S. A. S., Shafi, I., Ahmad, J., Villar, S. G., Villena, E. G., Khurshaid, T., & Ashraf, I. (2025). Review of flood monitoring and prevention approaches: A data analytic perspective. *Natural Hazards*, *121*(5), 5103–5128. DOI: 10.1007/s11069-024-07050-w

Bukhary, S., Batista, J. R., & Ahmad, S. (2020). Design Aspects, Energy Consumption Evaluation, and Offset for Drinking Water Treatment Operation. *Water (Basel)*, *12*(6), 1772. DOI: 10.3390/w12061772

Bullen, P. A. (2007). Adaptive reuse and sustainability of commercial buildings. *Facilities*, *25*(1/2), 20–31. DOI: 10.1108/02632770710716911

Bungey, J. H., & Madandoust, R. (1992). Factors influencing pull-off tests on concrete. *Magazine of Concrete Research*, *44*(158), 21–30. DOI: 10.1680/macr.1992.44.158.21

Busch, J., Dawson, D., & Roelich, K. (2017). Closing the low-carbon material loop using a dynamic whole system approach. *Journal of Cleaner Production*, *149*, 751–761. DOI: 10.1016/j.jclepro.2017.02.166

Caldas, L. R., Silva, M. V., Silva, V. P., Carvalho, M. T. M., & Toledo Filho, R. D. (2022). How different tools contribute to climate change mitigation in a circular building environment? A systematic literature review. *Sustainability (Basel)*, *14*(7), 3759. DOI: 10.3390/su14073759

Canale, L., Fazio, A. R. D., Russo, M., Frattolillo, A., & Dell'Isola, M. (2021). An Overview on Functional Integration of Hybrid Renewable Energy Systems in Multi-Energy Buildings. *Energies*, *14*(4), 1078. DOI: 10.3390/en14041078

Carbonara, N., & Pellegrino, R. (2018). Fostering innovation in public procurement through public private partnerships. *Journal of Public Procurement*, *18*(3), 257–280. DOI: 10.1108/JOPP-09-2018-016

Carbonara, N., & Pellegrino, R. (2020). The role of public private partnerships in fostering innovation. *Construction Management and Economics*, *38*(2), 140–156. DOI: 10.1080/01446193.2019.1610184

Carcassi, O. B., Minotti, P., Habert, G., Paoletti, I., Claude, S., & Pittau, F. (2022). Carbon Footprint Assessment of a Novel Bio-Based Composite for Building Insulation. *Sustainability (Basel)*, *14*(3), 1384. DOI: 10.3390/su14031384

Carino, N. J., & Lew, H. S. (2001). The Maturity Method: From Theory to Practice, Construction Congress and Exposition, May 21-23, 2001, Washington, DC, American Society of Civil Engineers, Reston, Virginia, Peter C. Chang, Editor, 2001, 19 p.

Cassagnabère, F., Mouret, M., Escadeillas, G., Broilliard, P., & Bertrand, A. (2010). Metakaolin, a solution for the precast industry to limit the clinker content in concrete: Mechanical aspects. *Construction & Building Materials*, *24*(7), 1109–1118. DOI: 10.1016/j.conbuildmat.2009.12.032

Caterino, N., Iervolino, I., Manfredi, G., & Cosenza, E. (2008). Multi-criteria decision making for seismic retrofitting of RC structures. *Journal of Earthquake Engineering*, *12*(4), 555–583. DOI: 10.1080/13632460701572872

Caterino, N., Iervolino, I., Manfredi, G., & Cosenza, E. (2009). Comparative analysis of multi-criteria decision-making methods for seismic structural retrofitting. *Computer-Aided Civil and Infrastructure Engineering*, *24*(6), 432–445. DOI: 10.1111/J.1467-8667.2009.00599.X

CdT. (n.d.). *The Eastgate Centre in Harare, Zimbabwe*. https://www.cdt.ch/prodotti/hub/the-eastgate-centre-in-harare-zimbabwe-286733

Čech, M., Goubej, M., Sobota, J., & Visioli, A. (2020). Model-based system engineering in control education using HIL simulators. *IFAC-PapersOnLine*, *53*(2), 17302–17307. DOI: 10.1016/j.ifacol.2020.12.1812

Çelik, S. B., Çobanoğlu, İ., & Gireson, K. (2025). Prediction and classification of flexural strength of rock materials used as dimension stones. *Bulletin of Engineering Geology and the Environment*, *84*(1), 44. DOI: 10.1007/s10064-024-04070-5

Centre for Cities. (2022). *Making places: The role of regeneration in levelling up (Case study: King's Cross)*. https://www.centreforcities.org/reader/making-places/learning-from-kings-cross-regeneration/

Çetin, M., Günen, A., Kalkandelen, M., & Karakaş, M. S. (2021). Microstructural, wear and corrosion characteristics of boronized AISI 904L superaustenitic stainless steel. *Vacuum*, *187*, 110145. DOI: 10.1016/j.vacuum.2021.110145

Chaabane, M., Mhiri, H., & Bournot, P. (2020). Thermal performance evaluation and enhancement of a parabolic trough collector. *Journal of Renewable and Sustainable Energy*, *12*(4), 043702. Advance online publication. DOI: 10.1063/1.5145257

Chang, S., Castro-Lacouture, D., & Yamagata, Y. (2020). Decision support for retrofitting building envelopes using multi-objective optimization under uncertainties. Journal of Building Engineering. https://www.sciencedirect.com/science/article/pii/S2352710219319825

Chang, C., & Habib, A. (2023). Life-cycle cost analysis of ultra-high-performance concrete in retrofitting techniques for ABC projects. *Third International Interactive Symposium on Ultra-High Performance Concrete*. DOI: 10.21838/UHPC.16694

Chang, M., Maguire, M., & Sun, Y. (2019). Stochastic modeling of bridge deterioration using classification tree and logistic regression. *Journal of Infrastructure Systems, 25*(1). Advance online publication. DOI: 10.1061/(ASCE)IS.1943-555X.0000466

Chang, R., Zhang, N., & Gu, Q. (2023). A review on mechanical and structural performances of precast concrete buildings. *Buildings (Basel, Switzerland), 13*(7), 1575. DOI: 10.3390/buildings13071575

Chazdon, R. L., & Guariguata, M. R. (2016). Natural regeneration as a tool for large-scale forest restoration in the tropics: Prospects and challenges. *Biotropica, 48*(6), 716–730. DOI: 10.1111/btp.12381

Chen, G., Xu, P., Mi, G., & Zhu, J. (2019). Compressive strength and cracking of composite concrete in hot-humid environments based on microscopic quantitative analysis. Construction and Building Materials. https://www.sciencedirect.com/science/article/pii/S0950061819318392

Chen, H., Jiang, Y., & Miao, G. (2022). Structural performance deterioration of corroding reinforced concrete columns in marine environments. Ocean Engineering. https://www.sciencedirect.com/science/article/pii/S0029801822014718

Chen, A. S., Leandro, J., & Djordjević, S. (2016). Modelling sewer discharge via displacement of manhole covers during flood events using 1D/2D SIPSON/P-DWave dual drainage simulations. *Urban Water Journal, 13*(8), 830–840. DOI: 10.1080/1573062X.2015.1041991

Chen, C. (2023). Advantages and barriers of modular construction method in constructing buildings. *Proceedings of the Institution of Civil Engineers: Infrastructure and Construction, 176*(2), 75–84.

Chen, C., & Tiong, R. (2024). Solutions to Address the Low-Capacity Utilization Issue in Singapore's Precast Industry. *Information (Basel), 15*(8), 458. DOI: 10.3390/info15080458

Cheng, Y., Lin, F., Wang, W., & Zhang, J. (2022). Vision-based trajectory monitoring for assembly alignment of precast concrete bridge components. *Automation in Construction, 140*, 104350. DOI: 10.1016/j.autcon.2022.104350

Chen, H., & Nepal, J. (2015). Stochastic modelling and lifecycle performance assessment of bond strength of corroded reinforcement in concrete. *Structural Engineering and Mechanics, 54*(2). Advance online publication. DOI: 10.12989/sem.2015.54.2.000

Chen, J. H., Ha, N. T. T., Tai, H. W., & Chang, C. A. (2020). The willingness to adopt the Internet of Things (IoT) conception in Taiwan's construction industry. *Journal of Civil Engineering and Management, 26*(6), 534–550. DOI: 10.3846/jcem.2020.12639

Chen, K. W., Teitelbaum, E., Meggers, F., Pantelic, J., & Rysanek, A. (2020). Exploring membrane-assisted radiant cooling for designing comfortable naturally ventilated spaces in the tropics. *Building Research and Information*, *49*(5), 483–495. DOI: 10.1080/09613218.2020.1847025

Chen, K., Lin, H., You, S., & Han, Y. (2022). Review of the impact of urban parks and green spaces on residence prices in the environmental health context. *Frontiers in Public Health*, *10*, 993801. DOI: 10.3389/fpubh.2022.993801 PMID: 36159304

Chen, W. F., & Drucker, D. C. (1969). Constitutive Relationships for Concrete. *Journal of the Engineering Mechanics Division.*

Chen, Y., Ning, Y., Chen, X., Xuan, W., & Zhu, X. (2023). Experimental and Numerical Investigation on Triaxial Compressive Behavior of Steam-Cooked Recycled Aggregate Concrete. *Buildings (Basel, Switzerland)*, *13*(2), 334. DOI: 10.3390/buildings13020334

Chen, Z., Hammad, A. W. A., Kamardeen, I., & Haddad, A. (2020). *Optimizing Window Design on Residential Building Facades by Considering Heat Transfer and Natural Lighting in Non-tropical Regions of Australia*. DOI: 10.20944/preprints202010.0111.v1

Chien, H., Hori, K., & Saito, O. (2022). Urban commons in the techno-economic paradigm shift: An information and communication technology-enabled climate-resilient solutions review. *Environment and Planning. B, Urban Analytics and City Science*, *49*(5), 1389–1405. DOI: 10.1177/23998083211066324

Chisari, C., & Bedon, C. (2017). Performance-based design of FRP retrofitting of existing RC frames by means of multi-objective optimisation. *Bollettino di Geofisica Teorica ed Applicata*, *58*(4), 377–394. DOI: 10.4430/bgta0202

Choi, H. K., Choi, Y. C., & Choi, C. S. (2013). Development and testing of precast concrete beam-to-column connections. *Engineering Structures*, *56*, 1820–1835. DOI: 10.1016/j.engstruct.2013.07.021

Chojnacka, K., & Widera, B. (2025). Nature-Based Solutions and Other Critical Aspects of Urban Innovations for Green Transition. *European Review (Chichester, England)*, *33*(4), 1–10. DOI: 10.1017/S1062798725100343

Chopade, R., Pathak, P., Chougule, R., Bothra, N., Mane, N., & Minde, P. (2021). Analysis of Safety Management in Precast Construction Projects. *Turkish Journal of Computer and Mathematics Education*, *12*(6), 5225–5236.

Chuah, Y. K., & Yang, J. (2020). A Integrated Dedicated Outdoor Air System to Optimize Energy Saving. *Sustainability (Basel)*, *12*(3), 1051. DOI: 10.3390/su12031051

Chung, S.-Y., Sikora, P., Stephan, D., & Elrahman, M. A. (2020). The Effect of Lightweight Concrete Cores on the Thermal Performance of Vacuum Insulation Panels. *Materials (Basel), 13*(11), 2632. DOI: 10.3390/ma13112632 PMID: 32526963

Ciampi, G., Scorpio, M., Spanodimitriou, Y., Rosato, A., & Sibilio, S. (2020). Thermal model validation of an electric-driven smart window through experimental data and evaluation of the impact on a case study. *Building and Environment, 181*, 107134. DOI: 10.1016/j.buildenv.2020.107134

Clark-Ginsberg, A., Rueda, I. A., Monken, J., Liu, J., & Chen, H. (2020). Maintaining critical infrastructure resilience to natural hazards during the COVID-19 pandemic: Hurricane preparations by US energy companies. *Journal of Infrastructure Preservation and Resilience, 1*(1), 10. Advance online publication. DOI: 10.1186/s43065-020-00010-1 PMID: 34778806

Cole, A., Ameri Sianaki, O., & Shee, H. (2024, July). Conceptualizing a Digital Twin Architecture for Enhanced Control in Precast Concrete Production. In *International Conference on Complex, Intelligent, and Software Intensive Systems* (pp. 261-272). Cham: Springer Nature Switzerland. DOI: 10.1007/978-3-031-70011-8_24

Cook, J., Hotchkiss, E., Li, X., & Cruce, J. (2020). Planning for the storm: Considering renewable energy for critical infrastructure resilience. *Journal of Emergency Management (Weston, Mass.), 18*(4), 295. DOI: 10.5055/jem.2020.0475 PMID: 32804398

Cornet, M. (2019). The circular cornerstone: Applying a circular economy through design for disassembly to the development of Amstel III.

Costa, A., Proverbio, M., & Smith, I. F. (2017, November). Cyber civil infrastructure and IoT for cities. In *2017 IEEE Symposium Series on Computational Intelligence (SSCI)* (pp. 1-8). IEEE. DOI: 10.1109/SSCI.2017.8280815

Crespi, G., Becchio, C., & Corgnati, S. P. (2020). Towards Post-Carbon Cities: Which retrofit scenarios for hotels in Italy? *Renewable Energy, 163*, 950–963. DOI: 10.1016/j.renene.2020.08.023

Cruz, V. F., & Góes, T. M. (2021). Contribuição do BIM para o desenvolvimento da economia circular no ambiente construído: Uma revisão sistemática da literatura.

Cui, L., Sun, H., Yang, Z., & Duan, H. (2020). The digital twin system of PERA company air-conditioning system. *IFAC-PapersOnLine, 53*(5), 720–725. DOI: 10.1016/j.ifacol.2021.04.207

Czyzewska, C. (2018). Should corporate cultural responsibility replace cultural sponsorship? https://www.linkedin.com/pulse/should-corporate-cultural-responsibility-replace-cyrielle-czy%C5%BCewska/

D'Oca, S., Ferrante, A., Ferrer, C., Pernetti, R., Gralka, A., Sebastian, R., & Op 't Veld, P. (2018). Technical, Financial, and Social Barriers and Challenges in Deep Building Renovation: Integration of Lessons Learned from the H2020 Cluster Projects. *Buildings (Basel, Switzerland)*, *8*(12), 174. DOI: 10.3390/buildings8120174

Dabaieh, M., & Serageldin, A. A. (2020). Earth air heat exchanger, Trombe wall and green wall for passive heating and cooling in premium passive refugee house in Sweden. *Energy Conversion and Management*, *209*, 112555. DOI: 10.1016/j.enconman.2020.112555

Daehn, K., Basuhi, R., Gregory, J., Berlinger, M., Somjit, V., & Olivetti, E. A. (2021). Innovations to decarbonize materials industries. *Nature Reviews. Materials*, *7*(4), 275–294. DOI: 10.1038/s41578-021-00376-y

Dagur, A. H., Kartha, A. A., Subodh, M. A., Vishnu, C., Arun, D., Vijay Kumar, M. G., Abraham, W. S., Chatterjee, A., & Abraham, J. (2017). Microstructure, mechanical properties and biocorrosion behavior of dissimilar welds of AISI 904L and UNS S32750. *Journal of Manufacturing Processes*, *30*, 27–40. DOI: 10.1016/j.jmapro.2017.09.001

Dal Lago, B., Del Galdo, M., & Bisi, D. (2022). Tests and design of welded-bar angle connections of precast floor elements. *Journal of Advanced Concrete Technology*, *20*(2), 43–56. DOI: 10.3151/jact.20.43

Dalal, S., & Dalal, P. (2021). Strength, deformation and fragility assessment of reinforced concrete moment resisting frame designed by force-based design and performance-based plastic design. Structures. https://www.sciencedirect.com/science/article/pii/S2352012420306706

Dara, H. M., Raut, A., Adamu, M., Ibrahim, Y. E., & Ingle, P. V. (2024). Reducing non-value added (NVA) activities through lean tools for the precast industry. *Heliyon*, *10*(7), e29148. DOI: 10.1016/j.heliyon.2024.e29148 PMID: 38633654

Dasaraju, S., & Shahnasser, H. (2020). Smart Solar-Powered Lighting System for Smart Cities. *International Journal of Computer and Electrical Engineering*, *12*(2), 83–92. DOI: 10.17706/IJCEE.2020.12.2.83-92

Dave, B., Kubler, S., Främling, K., & Koskela, L. (2016). Opportunities for enhanced lean construction management using Internet of Things standards. *Automation in Construction*, *61*, 86–97. DOI: 10.1016/j.autcon.2015.10.009

David, C. G., Kremer, J., Ashwini, M., Kloft, H., & Goseberg, N. (2025). Digital fabrication of Hybrid Nature-based Solutions as new opportunity for coastal climate change adaptation. *Cambridge Prisms: Coastal Futures*, *3*, e24. DOI: 10.1017/cft.2025.10014 PMID: 41307040

David, M., Derigent, W., Loubet, G., Takacs, A., & Dragomirescu, D. (2024). Communicating materials: Communicating concrete development for construction industry. *IEEE Transactions on Industrial Informatics, 20*(4), 6983–6994. DOI: 10.1109/TII.2024.3352261

Davoudi, R., Miller, G., & Kutz, J. (2018). Data-driven vision-based inspection for reinforced concrete beams and slabs: Quantitative damage and load estimation. Automation in Construction. https://www.sciencedirect.com/science/article/pii/S0926580518303686

de Almeida Filho, R. R., & da Cruz, F. M. (2022). Organizational factors: A case study of the cost-benefit of implementing control measures in precast structure constructions. *Research. Social Development, 11*(1), e33711124981–e33711124981.

De Anício, S. O. (2022). Connections between circular economy and ecological economics. *Sustainability and Climate Change, 15*(3), 162–165. DOI: 10.1089/scc.2022.0013

De Belie, N., Gruyaert, E., Al-Tabbaa, A., Antonaci, P., Baera, C., Bajare, D., Darquennes, A., Davies, R., Ferrara, L., Jefferson, T., Litina, C., Miljevic, B., Otlewska, A., Ranogajec, J., Roig-Flores, M., Paine, K., Lukowski, P., Serna, P., Tulliani, J., & Jonkers, H. M. (2018). A Review of Self-Healing Concrete for Damage Management of Structures. *Advanced Materials Interfaces, 5*(17), 1800074. Advance online publication. DOI: 10.1002/admi.201800074

de Carvalho, A. R., da Silva Calderón-Morales, B. R., Júnior, J. C. B., de Oliveira, T. M., & Silva, G. J. B. (2023). Proposition of geopolymers obtained through the acid activation of iron ore tailings with phosphoric acid. *Construction & Building Materials, 403*, 133078. DOI: 10.1016/j.conbuildmat.2023.133078

de Oliveira, J., Schreiber, D., & Jahno, V. D. (2024). Circular economy and buildings as material banks in mitigation of environmental impacts from construction and demolition waste. *Sustainability (Basel), 16*(12), 5022. DOI: 10.3390/su16125022

Delavar, H., & Sahebi, H. (2020). A sustainable mathematical model for design of net zero energy buildings. *Heliyon, 6*(1), e03190. Advance online publication. DOI: 10.1016/j.heliyon.2020.e03190 PMID: 31956715

Delehan, S., Vilčeková, S., Melehanych, H., Krídlová Burdová, E., & Khorolskyi, A. (2024). A comparative assessment of the capabilities and success of the wood construction industry in Slovakia and Ukraine based on life cycle assessment certification standards. *Frontiers in Environmental Science, 12*, 1319823. Advance online publication. DOI: 10.3389/fenvs.2024.1319823

Demir, T., Ulucan, M., & Alyamaç, K. E. (2023). Development of Combined Methods Using Non-Destructive Test Methods to Determine the In-Place Strength of High-Strength Concretes. *Processes (Basel, Switzerland), 11*(3), 673. DOI: 10.3390/pr11030673

Deng, K., Cui, Y., Deng, Q., Liu, R., Chen, Z., & Wang, S. (2025). Multi-objective optimization of urban residential envelope structures in cold regions of China based on performance and economic efficiency. *Buildings (Basel, Switzerland), 15*(13), 2365. https://www.mdpi.com/2075-5309/15/13/2365

Deng, P., Sun, Y., Liu, Y., & Song, X. (2019). Revised Rebound Hammer and Pull-Out Test Strength Curves for Fiber-Reinforced Concrete. *Advances in Civil Engineering*, (2), 1–12.

Dhiman, S., Seema, S., & Akhai, S. (2025). Performance evaluation of OPC 43 concrete incorporating micro silica and recycled ceramic tile waste as partial cement and coarse aggregate replacements. *Asian Journal of Civil Engineering*. DOI: 10.1007/s42107-025-01504-7

Didenko, N. I., Klochkov, Y. S., & Skripnuk, D. F. (2018). Ecological Criteria for Comparing Linear and Circular Economies. *Resources, 7*(3), 48. DOI: 10.3390/resources7030048

Dilakshan, S., Rathnasinghe, A. P., & Seneviratne, L. D. I. P. (2021, July). Potential of internet of things (Iot) in the construction industry. In *Proceedings The 9th World Construction Symposium| July* (p. 446).

Dilakshan, S., Rathnasinghe, A. P., & Seneviratne, L. D. I. P. (2021). An Investigation on the Readiness of the Sri Lankan Construction Industry towards Internet of Things (IoT): Conceptual Guideline. In *International Conference on Business Management*-2020. DOI: 10.31357/icbm.v17.5194

Ding, H., Huang, H., Liu, Z., & Wang, F. (2025). Digital aggregate for concrete smart management and lifecycle traceability from gamma-dicalcium silicate carbonation. *Construction & Building Materials, 472*, 140869. DOI: 10.1016/j.conbuildmat.2025.140869

Ding, H., Li, M., Zhong, R. Y., & Huang, G. Q. (2023). Multistage self-adaptive decision-making mechanism for prefabricated building modules with IoT-enabled graduation manufacturing system. *Automation in Construction, 148*, 104755. DOI: 10.1016/j.autcon.2023.104755

Ding, K. W., & Zhang, X. (2012). Literature review of precast concrete frame structures in civil engineering. *Advanced Materials Research, 568*, 3–6. DOI: 10.4028/www.scientific.net/AMR.568.3

Ding, S., Hu, H., Xu, F., Chai, Z., & Wang, W. (2024). Blockchain-based security-minded information-sharing in precast construction supply chain management with scalability, efficiency and privacy improvements. *Automation in Construction, 168*, 105698. DOI: 10.1016/j.autcon.2024.105698

Discover Walks. (n.d.). *Top facts about the Gherkin in London.* https://www.discoverwalks.com/blog/london/top-facts-about-the-gherkin-in-london/

Disfani, M. M., Arulrajah, A., Bo, M. W., & Sivakugan, N. (2012). Environmental risks of using recycled crushed glass in road applications. *Journal of Cleaner Production, 20*(1), 170–179. DOI: 10.1016/j.jclepro.2011.07.020

do Amaral Cassilha, S., & Tavares, S. F. (2024). Circularidade na governança do ambiente construído: Instrumentos políticos e atores. *Encontro Nacional de Tecnologia do Ambiente Construído, 20*, 1–13. DOI: 10.46421/entac.v20i1.6095

Dolati, S. S. K., & Mehrabi, A. (2021, April). Review of available systems and materials for splicing prestressed-precast concrete piles. [). Elsevier.]. *Structures (Oxford), 30*, 850–865. DOI: 10.1016/j.istruc.2021.01.029

Drage, J., & Kennedy, G. (2020). Building a Low-Cost, Internet-of-Things, Real-Time Groundwater Level Monitoring Network. *Ground Water Monitoring and Remediation, 40*(4), 67–73. DOI: 10.1111/gwmr.12408

Dreamstime. (2022, August). *The Allianz Arena football stadium exterior aerial view.* https://www.dreamstime.com/aerial-view-allianz-arena-munich-germany-august-football-stadium-image124543606

Drgoňa, J., Arroyo, J., Figueroa, I. C., Blum, D., Arendt, K., Kim, D., Ollé, E. P., Oravec, J., Wetter, M., Vrabie, D., & Helsen, L. (2020). All you need to know about model predictive control for buildings. *Annual Reviews in Control, 50*, 190–232. DOI: 10.1016/j.arcontrol.2020.09.001

Drzymała, A., & Korzeniewska, E. (2020). Economic and Technical Aspects of a Hybrid Single-Family House Heating Based on Photovoltaic and Heat Pump Installation. *2020 IEEE Problems of Automated Electrodrive.* [PAEP]. *Theory into Practice, 1*, 1–4. Advance online publication. DOI: 10.1109/PAEP49887.2020.9240906

Dudkiewicz, E., & Szałański, P. (2020). Overview of exhaust gas heat recovery technologies for radiant heating systems in large halls. *Thermal Science and Engineering Progress, 18*, 100522. DOI: 10.1016/j.tsep.2020.100522

Dutta, C., Kumar, J., Das, T., & Saha, S. (2021). Recent advancements in the development of sensors for structural health monitoring at high-temperature environment: A review. IEEE Sensors Journal. https://ieeexplore.ieee.org/abstract/document/9415648/

Eberhardt, L. C. M., Birgisdottir, H., & Birkved, M. (2019). Potential of circular economy in sustainable buildings. In IOP Conference Series: Materials Science and Engineering (Vol. 471, Article 092051). IOP Publishing. DOI: 10.1088/1757-899X/471/9/092051

Echeverria-Valiente, E., Garcia-Alvarado, R., Celis-D'Amico, F., & Saelzer-Fuica, G. (2019). Integrated design experiences for energy-efficient housing in Chile. *Construction Innovation, 19*(2), 236–255. DOI: 10.1108/CI-05-2017-0042

Ekin, N. (2020). Betonların elastik modül hesabında Poisson oranının önemi. *Çanakkale Onsekiz Mart Üniversitesi Fen Bilimleri Enstitüsü Dergisi, 6*, 239–252.

Ekin, N. (2021a). Relationship between amplitude anisotropy and compressive strength of reinforced concrete depending on curing conditions. *Indian Journal of Engineering and Materials Sciences, 28*(3), 286–299.

Ekin, N. (2021b). Concrete anisotropy estimated from ultrasonic signal amplitudes. *Materials Testing, 63*(12), 1164–1173. DOI: 10.1515/mt-2021-0053

Ekin, N. (2025a). The relationships between ultrasonic P and S wave velocities and resistivity in reinforced concrete. *Construction & Building Materials, 479*, 141475. DOI: 10.1016/j.conbuildmat.2025.141475

Ekin, N. (2025b). Beton Dayanimi Tahmininde İkili Ve Çoklu Doğrusal Regresyon Analizlerinin Karşilaştirilmasi. *SDU Mühendislik Bilimleri Ve Tasarım Dergisi, 13*(1), 64–77. DOI: 10.21923/jesd.1572342

Ekin, N. (2025c). Yüksek dayanımlı betonlarda uluslararası standartlardaki elastik modüllerin karşılaştırılması. *Süleyman Demirel Üniversitesi Fen Bilimleri Enstitüsü Dergisi, 29*(2), 312–326. DOI: 10.19113/sdufenbed.1651363

Ekin, N., & Uyanık, O. (2021a). *Comparison of static and dynamic elastic moduli in concrete: Effects of compressive strength, curing conditions and reinforcement.* Iranian Journal of Science and Technology, Transactions of Civil Engineering., DOI: 10.1007/s40996-020-00513-7

Ekin, N., & Uyanık, O. (2021b). Anisotropy of reinforced concrete from geophysical methods. *Science of Sintering, 53*(3), 323–345. DOI: 10.2298/SOS2103323E

Elkhayat, Y. O., Ibrahim, M. G., Tokimatsu, K., & Ali, A. A. M. (2020). Multi-criteria selection of high-performance glazing systems: A case study of an office building in New Cairo, Egypt. *Journal of Building Engineering, 32*, 101466. DOI: 10.1016/j.jobe.2020.101466

Ellingwood, B. R. (2005). Risk-informed condition assessment of civil infrastructure: State of practice and research issues. *Structure and Infrastructure Engineering, 1*(1), 7–18. DOI: 10.1080/15732470412331289341

Elliott, K. S. (2000). Research and development in precast concrete framed structures. *Progress in Structural Engineering and Materials*, *2*(4), 405–428. DOI: 10.1002/pse.65

Elze, M., Müller, S., & Günther, V. (2022). Industrielle modulare Konstruktion mit niedrigem CO_2-Fußabdruck–EDGE Suedkreuz Berlin. *Bautechnik*, *99*(S2), 100–108. DOI: 10.1002/bate.202200077

Ercan A. (2001). Yapı Jeofiziği IV-Beton Esnekliği ve Bozukluklar. *Dünya İnşaat Dergisi*. 18, 2001(12), 70-73.

Ercan A. (2002). Beton İçi Bozukluklar, Çatlak ve Kırıklar. *Dünya İnşaat Dergisi*, 18(2002-01), 66-71

European Union, United Nations, Government of Ukraine, & World Bank. (2023). *Ukraine – Third rapid damage and needs assessment: February 2022. December 2023*. https://documents.worldbank.org/en/publication/documents-reports/documentdetail/099021324115085807

Evans, G. W. (2003). The built environment and mental health. *Journal of Urban Health : Bulletin of the New York Academy of Medicine*, *80*(4), 536–555. DOI: 10.1093/jurban/jtg063 PMID: 14709704

Fallah, A., Vafaei, M., & Alih, S. C. (2020). A review of seismic response of precast structures. *International Journal of Engineering Research & Technology (Ahmedabad)*, *13*(10), 2694–2705. DOI: 10.37624/IJERT/13.10.2020.2694-2705

Fang, M., Zhou, R., Ke, W., Tian, B., Zhang, Y., & Liu, J. (2022). Precast system and assembly connection of cement concrete slabs for road pavement: A review. [english edition]. *Journal of Traffic and Transportation Engineering*, *9*(2), 208–222. DOI: 10.1016/j.jtte.2021.10.003

Fang, Y., & Ng, S. T. (2019). Genetic algorithm for determining the construction logistics of precast components. *Engineering, Construction, and Architectural Management*, *26*(10), 2289–2306. DOI: 10.1108/ECAM-09-2018-0386

Farahani, A., Naderpour, H., & Kheyroddin, G. (2023). Developing a fuzzy expert system for diagnosing chemical deterioration in reinforced concrete structures. *Applied Sciences (Basel, Switzerland)*, *13*(18), 10372. https://www.mdpi.com/2076-3417/13/18/10372

Faris, N., Zayed, T., & Faris, A. (2026). Review of condition rating and deterioration modeling approaches for concrete bridges. *Buildings (Basel, Switzerland)*, *15*(2), 219. https://www.mdpi.com/2075-5309/15/2/219

Farouk, H., & Abdelsabour, I. (2019). DFD (design for disassembly) framework: An approach to enhance the design construction process. *Journal of Engineering Sciences*, *47*(6), 883–898.

Farrar, C., & Lieven, N. (2007). Damage prognosis: The future of structural health monitoring. Philosophical Transactions of the Royal Society A: Mathematical, Physical and Engineering Sciences, 365(1851), 623–632. DOI: 10.1098/RSTA.2006.1927

Farrar, C., & Worden, K. (2012). Structural health monitoring: A machine learning perspective. Wiley. https://books.google.com/books?hl=en&lr=&id=2w_sp6lersUC

Fathi, S., Srinivasan, R., Kibert, C. J., Steiner, R., & Demirezen, E. M. (2020). AI-Based Campus Energy Use Prediction for Assessing the Effects of Climate Change. *Sustainability (Basel)*, *12*(8), 3223. DOI: 10.3390/su12083223

Fawad, M., Salamak, M., Poprawa, G., Koris, K., Jasinski, M., Lazinski, P., Piotrowski, D., Hasnain, M., & Gerges, M. (2023). Automation of structural health monitoring system of a bridge using BIMification approach and BIM-based finite element model development. *Scientific Reports*, *13*, 13215. DOI: 10.1038/s41598-023-40355-7

Fernandes, I., Broekmans, M. A. T. M., & Noronha, F. (2009). Petrography and geochemical analysis for the forensic assessment of concrete damage. In *Criminal and environmental soil forensics* (pp. 163–180). Springer. DOI: 10.1007/978-1-4020-9204-6_11

Ferriz-Papi, J. A., Lee, A., & Alhawamdeh, M. (2024). Examining the challenges for circular economy implementation in construction and demolition waste management: A comprehensive review using systematic methods. *Buildings (Basel, Switzerland)*, *14*(5), 1237. DOI: 10.3390/buildings14051237

Final Project Report, A Novel, Renewable Energy Microgrid for a California Healthcare Facility. (2020). https://www.energy.ca.gov/sites/default/files/2021-06/CEC-500-2019-034.pdf

Fořt, J., Šál, J., Kočí, J., & Černý, R. (2020). Energy Efficiency of Novel Interior Surface Layer with Improved Thermal Characteristics and Its Effect on Hygrothermal Performance of Contemporary Building Envelopes. *Energies*, *13*(8), 2012. DOI: 10.3390/en13082012

Fragiadakis, M., & Papadrakakis, M. (2008). Performance-based optimum seismic design of reinforced concrete structures. *Earthquake Engineering & Structural Dynamics*, *37*(6), 825–844. DOI: 10.1002/EQE.786

French Moments. (n.d.). *Arab World Institute of Paris*. https://frenchmoments.eu/arab-world-institute-of-paris/

Fu, C., & Li, Y. (2023). Steel–concrete bond deterioration in reinforced concrete tension members due to primary cracks. Structures. https://www.sciencedirect.com/science/article/pii/S2352012423009852

Fuller, D. A., & Ottman, J. A. (2004). Moderating unintended pollution: The role of sustainable product design. *Journal of Business Research*, *57*(11), 1231–1238. DOI: 10.1016/S0148-2963(02)00446-0

Gallardo, C. A., Granja, A. D., Picchi, F. A., & Folch, A. T. (2006). Stabilization and standardization of a precast production process. In *Proceedings of 14th Annual Conference of the International Group for Lean Construction.* Santiago, Chile (Vol. 205).

Gangwar, G., Kaur, P., & Singh, I. P. (2020). A Study of Passive and Active Strategies through Case Studies for the Composite Climate Zone of India. *Civil Engineering and Architecture*, *8*(6), 1370–1389. DOI: 10.13189/cea.2020.080620

Gao, M. Y., Han, J., Yang, Y., Tiong, R. L., Zhao, C., & Han, C. (2024). BIM-based and IoT-driven smart tracking for precast construction dynamic scheduling. *Journal of Construction Engineering and Management*, *150*(9), 04024117. DOI: 10.1061/JCEMD4.COENG-14498

Garcia-Monzon, H. (2003). Maturity of Concrete Strength Development, Texas Tech University Graduate School of Civil Engineering, M.S. Thesis.

Gardner, P., Liu, X., & Worden, K. (2020). On the application of domain adaptation in structural health monitoring. Mechanical Systems and Signal Processing. https://www.sciencedirect.com/science/article/pii/S088832701930771X

Gelowitz, M. D. C., & McArthur, J. J. (2017). Comparison of type III environmental product declarations for construction products: Material sourcing and harmonization evaluation. *Journal of Cleaner Production*, *157*, 125–133. DOI: 10.1016/j.jclepro.2017.04.133

Ghayeb, H. H., Sulong, N. R., Razak, H. A., Mo, K. H., Ismail, Z., Hashim, H., & Gordan, M. (2022). A review of the seismic performance behaviour of hybrid precast beam-to-column connections. *Archives of Civil and Mechanical Engineering*, *23*(1), 35. DOI: 10.1007/s43452-022-00558-7

Gibbons, L. V. (2020). Regenerative—The New Sustainable? *Sustainability (Basel)*, *12*(13), 5483. DOI: 10.3390/su12135483

Gomez-Cardenas, C. (2015). *Development of New Tools to Optimize Non-Destructive Testing Campaigns in Reinforced Concrete Structures, Durable Construction.* Université Paul Sabatier – Toulouse III.

González-Mejía, R., Berrone, L., & Gualtieri, S. (2024). Decentralized vs. Centralized Urban Drainage Systems: A Comparative Analysis of Resilience and Flow Management. *Water (Basel)*, *17*(10), 1536.

Grattan-Bellew, P. E. (2001). Petrographic and technological methods for evaluation of concrete aggregates. In *Handbook of analytical techniques in concrete science and technology* (pp. 63–104). Elsevier. DOI: 10.1016/B978-081551437-4.50005-9

Grussing, M. N., Liu, L. Y., Uzarski, D. R., El-Rayes, K., & El-Gohary, N. (2016). Discrete Markov approach for building component condition, reliability, and service-life prediction modeling. *Journal of Performance of Constructed Facilities*, *30*(5). Advance online publication. DOI: 10.1061/(ASCE)CF.1943-5509.0000865

Gucunski, N., La, H., Dinh, K., & Kee, M. K. (2023). Advancing condition assessment of reinforced concrete bridge elements through automation, visualization, and improved interpretation of multi-NDE technology. Materials Evaluation. https://www.researchgate.net/profile/Mahsa-Sanei/publication/368352480_Automated_Geometric_Quality_Inspection_of_Rebar_Layout_using_RGBD_Data/links/64d996af66f0e0067d9316af/Automated-Geometric-Quality-Inspection-of-Rebar-Layout-using-RGBD-Data.pdf#page=58

Güemes, A., Fernandez-Lopez, A., & Pozo, A. R. (2020). Structural health monitoring for advanced composite structures: A review. *Journal of Composites Science*, *4*(1), 13. https://www.mdpi.com/2504-477X/4/1/13

Guerra, B. C., Shahi, S., Mollaei, A., Skaf, N., Weber, O., Leite, F., & Haas, C. (2021). Circular economy applications in the construction industry: A global scan of trends and opportunities. *Journal of Cleaner Production*, *324*, 129125. DOI: 10.1016/j.jclepro.2021.129125

Guerrini, G., Restrepo, J. I., Vervelidis, A., & Massari, M. (2015). Self-centering precast concrete dual-steel-shell columns for accelerated bridge construction: Seismic performance, analysis, and design. *Report No. PEER*, *2015*, 13.

Gu, H., & Liu, Q. (2022). Updating deterioration models of reinforced concrete structures in carbonation environment using in-situ inspection data. *Structure and Infrastructure Engineering*, *18*(2), 266–277. DOI: 10.1080/15732479.2020.1841246

Güngör, B., Agibayeva, A., Karaca, F., Askar, R., Giarma, C., Rosado, L., ... & Bragança, L. (2024). Circularity Tools and Frameworks for New Buildings. *Circular Economy Design and Management in the Built Environment: A Critical Review of the State of the Art*, 431-458.

Guno, C. S., Agaton, C. B., Villanueva, R. O., & Villanueva, R. O. (2020). Optimal Investment Strategy for Solar PV Integration in Residential Buildings: A Case Study in The Philippines. *International Journal of Renewable Energy Development*, *10*(1), 79–89. DOI: 10.14710/ijred.2021.32657

Guo, H., Dong, Y., & Gu, X. (2019). Durability assessment of reinforced concrete structures considering global warming: A performance-based engineering and experimental approach. *Construction & Building Materials*. Advance online publication. DOI: 10.1016/j.conbuildmat.2019.117251

Guo, L., Zhang, Y., Wang, Z., Zeng, J., Zhang, Y., & Zhang, Z. (2020). Parameter sensitivity analysis and optimization strategy research of enhanced geothermal system: A case study in Guide Basin, Northwestern China. *Renewable Energy, 153*, 813–831. DOI: 10.1016/j.renene.2020.02.058

Gupta, A. (2019). Accelerating circularity in built-environment through active procurement: An aggregated assessment framework to make sustainable choices while using secondary materials in the early design phase.

Gupta, A. K. (2014). *Non-Destructive Testing of Bridges*. Indian Institute of Civil Railway Engineering.

Guyot, D., Giraud, F., Simon, F., Corgier, D., Marvillet, C., & Tréméac, B. (2020). Detailed monitoring as an essential tool for achieving energy performance targets in operation conditions: the HIKARI case study. *E3S Web of Conferences, 172*, 22006. DOI: 10.1051/e3sconf/202017222006

Hagnell, M. K., & Åkermo, M. (2019). The economic and mechanical potential of closed loop material usage and recycling of fibre-reinforced composite materials. *Journal of Cleaner Production, 223*, 957–968. DOI: 10.1016/j.jclepro.2019.03.156

Ha, J.-W., Cho, S., Kim, H., & Song, Y. (2020). Annual Energy Consumption Cut-Off with Cooling System Design Parameter Changes in Large Office Buildings. *Energies, 13*(8), 2034. DOI: 10.3390/en13082034

Hajare, A., & Elwakil, E. (2020). Integration of life cycle cost analysis and energy simulation for building energy-efficient strategies assessment. *Sustainable Cities and Society, 61*, 102293. DOI: 10.1016/j.scs.2020.102293

Hampel, T., & Curbach, M. (2000). Behavior of high performance concrete under multiaxial loading. In *Proc., PCI/FHWA/FIB Int. Symp. on High Performance Concrete* (pp. 185–196). Precast/Prestressed Concrete Institute.

Han, C., & Ye, H. (2018, August). A novel IoT-Cloud-BIM based intelligent information management system in building industrialization. In *International Conference on Construction and Real Estate Management 2018* (pp. 72-77). Reston, VA: American Society of Civil Engineers. DOI: 10.1061/9780784481721.008

Hanif, A., Kim, Y., Parthasarathy, P., Usman, M., & Li, Z. (2019). Flexural fatigue behavior of lightweight ferrocement: Experimental investigation & numerical modeling. 3291–3302.

Hanifehzadeh, M., Aryan, H., Gencturk, B., & Alipour, A. (2021). Structural response of steel jacket-UHPC retrofitted reinforced concrete columns under blast loading. *Materials (Basel), 14*(6), 1521. https://www.mdpi.com/1996-1944/14/6/1521

Han, P., Qiao, G., Guo, B., Li, D., & Ou, J. (2022). Investigation of low-frequency stray current induced corrosion in reinforced concrete infrastructure in high-speed railway power supply system. *International Journal of Electrical Power & Energy Systems, 134*, 107436. DOI: 10.1016/j.ijepes.2021.107436

Han, Y., Yang, Z., Ding, T., & Xiao, J. (2021). Environmental and economic assessment on 3D printed buildings with recycled concrete. *Journal of Cleaner Production, 278*, 123884. DOI: 10.1016/j.jclepro.2020.123884

Hasan, M. (2024). Deterioration prediction of concrete bridge components using artificial intelligence and stochastic methods [Doctoral dissertation, RMIT University]. RMIT Research Repository. https://research-repository.rmit.edu.au/articles/thesis/Deterioration_prediction_of_concrete_bridge_components_using_artificial_intelligence_and_stochastic_methods/27591186

Heisel, F., Schlesier, K., & Hebel, D. E. (2019). Prototypology for a circular building industry: The potential of re-used and recycled building materials. In IOP Conference Series: Earth and Environmental Science (Vol. 225, Article 012023). IOP Publishing.

Helal, J., Sofi, M., & Mendis, P. (2015). Non-Destructive Testing of Concrete: Review of Methods. *Electronic Journal of Structural Engineering, 14*(1), 97–105. DOI: 10.56748/ejse.141931

Helal, J., Stephan, A., & Crawford, R. H. (2020). The influence of structural design methods on the embodied greenhouse gas emissions of structural systems for tall buildings. *Structures (Oxford), 24*, 650–665. DOI: 10.1016/j.istruc.2020.01.026

Hendrickson, C., & Horvath, A. (2000). Resource Use and Environmental Emissions of U.S. Construction Sectors. *Journal of Construction Engineering and Management, 126*(1), 38–44. DOI: 10.1061/(ASCE)0733-9364(2000)126:1(38)

He, R., & Lu, N. (2023). Unveiling the dielectric property change of concrete during hardening process by ground penetrating radar with the antenna frequency of 1.6 GHz and 2.6 GHz. *Cement and Concrete Composites, 144*, 105279. DOI: 10.1016/j.cemconcomp.2023.105279

Hewes, J. T. (2013). Analysis of the state of the art of precast concrete bridge substructure systems.

Hieber, D. G., Wacker, J. M., Eberhard, M. O., & Stanton, J. F. (2005). State-of-the-art report on precast concrete systems for rapid construction of bridges.

Hola, J., & Runkiewicz, L. (2018). Methods and Diagnostic Techniques Used to Analyze the Technical Condition of Reinforced Concrete Structures. *Struct. Environ.*, *10*(4), 309–337. DOI: 10.30540/sae-2018-030

Holton, I. R. (2009). *Developing a sector sustainability strategy for the UK precast concrete industry*. Loughborough University.

Horiuchi, T., & Yoshimura, K. (2020). Solar heat gain coefficient and heat transmission coefficient of Al-doped ZnO thin-film coated low-emissivity glass. *Journal of the Ceramic Society of Japan*, *128*(4), 220–223. DOI: 10.2109/jcersj2.19217

Hossain, M. I., & Islam, Md. S., & Islam, Md. M. (2024). Evaluating concrete compressive strength using non-destructive test techniques. In *Proceedings of the 7th International Conference on Civil Engineering for Sustainable Development 2024* (pp. 1–10). Khulna University of Engineering & Technology.

Hridoy, S. M., Injamamul Haque Protyai, M., Rashid, A., & Abu Sayeed, M. (2025). Marshall property analysis of an improved bituminous mix obtained by mixing plastic waste with bitumen. *South African Journal of Chemical Engineering*, *53*, 40–48. DOI: 10.1016/j.sajce.2025.04.003

Huang, W., Quan, W., Miao, X., An, Y., & Sun, W. (2023). Study on the compressive mechanical properties of mixed recycled aggregate concrete with silane modified. *Journal of Building Engineering*, *75*, 106950. DOI: 10.1016/j.jobe.2023.106950

Hu, H. T., Fu, C. H., Zhan, Q. W., Zhang, X., Ge, Y. M., & Zhou, J. L. (2024). Study on improving the crack self-healing width and depth based on microbial mineralization modified by Ca–Al bimetallic hydroxides. *Journal of Building Engineering*, *91*, 109598. DOI: 10.1016/j.jobe.2024.109598

Hu, J., Zhang, S., Lin, J., Yu, Y., Yu, W., & Huang, Y. (2025). Towards Sustainable High-Performance Materials: Tailoring the Mechanical Properties of Bamboo-Based Fiber Composites through Controlled Processing. *Chemical Engineering Journal*, *512*, 162331. DOI: 10.1016/j.cej.2025.162331

Hu, M.-S. (2020). Design and development of a high-precision automatic safety valve testing system. *Advances in Mechanical Engineering*, *12*(4). Advance online publication. DOI: 10.1177/1687814020914733

Humenna, O. V. (2013). Kansei engineering in the marketing system of innovative products. *Economic Bulletin of NTUU KPI, 10*, 311–315. https://ela.kpi.ua/server/api/core/bitstreams/52c90934-7a51-4e7d-9772-0c04773fd8ac/content

Humenna, O. V. (2025). *Formation of a creative-innovative ecosystem for sustainable development of the construction industry: Theory, methodology, practice* (Doctoral dissertation, specialty 08.00.03 – Economics and Management of the National Economy). Kyiv.

Humenna, O. (2024). Innovative approaches in construction in the context of Ukraine's recovery based on the principles of sustainable development. *Ways to Improve Construction Efficiency*, *2*(53), 289–301. DOI: 10.32347/2707-501x.2024.53(2).289-301

Humenna, O., & Bedii, N. (2022). Corporate cultural responsibility as a component of company branding in Ukraine. *NaUKMA Research Papers. Economics*, *7*(1), 41–47. DOI: 10.18523/2519-4739.2022.7.1.41-47

Hurwitz, Z. L., Dubief, Y., & Almassalkhi, M. (2020). Economic efficiency and carbon emissions in multi-energy systems with flexible buildings. *International Journal of Electrical Power & Energy Systems*, *123*, 106114. DOI: 10.1016/j.ijepes.2020.106114

Hussain, I., Khan, F., Ahmad, I., Khan, S., & Saeed, M. (2021). Power Loss Reduction via Distributed Generation System Injected in a Radial Feeder. *Mehran University Research Journal of Engineering and Technology*, *40*(1), 160–168. DOI: 10.22581/muet1982.2101.15

Hu, X., Shen, J., & Ma, Z. (2024, November). Prefabricated Buildings for Smart Cities: A Literature Survey. In *2024 Twelfth International Conference on Advanced Cloud and Big Data (CBD)* (pp. 302-307). IEEE. DOI: 10.1109/CBD65573.2024.00061

Hu, X., Yang, Z., Wu, M., Wang, F., Yuan, H., & Jiang, Y. (2024). Evaluation of performance and healing effect of two-component biological self-healing mortar. *Journal of Sustainable Cement-Based Materials*, *13*(3), 462–475. DOI: 10.1080/21650373.2023.2282678

Hu, Y., Yuill, D. P., Ebrahimifakhar, A., & Rooholghodos, A. (2020). An experimental study of the behavior of a high efficiency residential heat pump in cooling mode with common installation faults imposed. *Applied Thermal Engineering*, *184*, 116116. DOI: 10.1016/j.applthermaleng.2020.116116

Hwang, Y.-J., & Jeong, J. (2021). Energy Saving Potential of Radiant Floor Heating Assisted by an Air Source Heat Pump in Residential Buildings. *Energies*, *14*(5), 1321. DOI: 10.3390/en14051321

Ilugbusi, B. S., Adisa, O., Obi, O., Awonuga, K. F., Adelekan, O. A., Asuzu, O. F., & Ndubuisi, N. L. (2024). Urban resilience to climate change: A review of adaptation strategies and infrastructure innovations. *Ecofeminism and Climate Change*, *5*(1), 18–23. DOI: 10.26480/efcc.01.2024.18.23

Incorvaja, D., Celik, Y., Petri, I., & Rana, O. (2022). Circular economy and construction supply chains. In *2022 IEEE/ACM International Conference on Big Data Computing, Applications and Technologies (BDCAT)* (pp. 92–99). IEEE. DOI: 10.1109/BDCAT56447.2022.00019

Ingham, J. P. (2009). Application of petrographic examination techniques to the assessment of fire-damaged concrete and masonry structures. Materials Characterization. https://www.sciencedirect.com/science/article/pii/S1044580308003173

Ishihara, S., Nagamachi, M., & Tsuchiya, T. (2019). Development of a Kansei engineering artificial intelligence sightseeing application. In *Advances in Affective and Pleasurable Design* (pp. 312–322). DOI: 10.1007/978-3-319-94944-4_34

Islam, F. S. (2025). Artificial intelligence-driven optimization of nature-based carbon sequestration: A scalable architecture for urban climate resilience. *International Journal of Environment and Climate Change*, *15*(7), 252–277. DOI: 10.9734/ijecc/2025/v15i74928

Ismail, Z. A. B. (2021). Hybrid intelligent vehicle system for managing construction supply chain in precast concrete building construction projects. *World Journal of Engineering*, *18*(4), 538–546. DOI: 10.1108/WJE-08-2020-0361

Ismail, Z. A. B. (2022). Thermal comfort practices for precast concrete building construction projects: Towards BIM and IOT integration. *Engineering, Construction, and Architectural Management*, *29*(3), 1504–1521.

Iyer-Raniga, U., Erasmus, P., Huovila, P., & Maity, S. (2019). Circularity in the built environment: A focus on India. In *International business, trade and institutional sustainability* (pp. 739–755). Springer.

Jadhav, O., Minde, P., Yadhav, A., & Gaidhankar, D. (2023). A review of emerging trends & advances in construction technology in the Indian scenario. *Materials Today: Proceedings*, *77*, 897–904. DOI: 10.1016/j.matpr.2022.12.041

Jafari, A., & Valentin, V. (2018). Selection of optimization objectives for decision-making in building energy retrofits. Building and Environment. https://www.sciencedirect.com/science/article/pii/S0360132317305954

Jafari, M. A., Zaidan, E., Ghofrani, A., Mahani, K., & Farzan, F. (2020). Improving Building Energy Footprint and Asset Performance Using Digital Twin Technology. *IFAC-PapersOnLine*, *53*(3), 386–391. DOI: 10.1016/j.ifacol.2020.11.062

Jaillon, L., & Poon, C. S. (2007, May). Advantages and limitations of precast concrete construction in high-rise buildings: Hong Kong case studies. In *Proceedings of the CIB World Building Congress* (pp. 14-17).

Jaillon, L., & Poon, C. S. (2008). Sustainable construction aspects of using prefabrication in dense urban environment: A Hong Kong case study. *Construction Management and Economics*, *26*(9), 953–966. DOI: 10.1080/01446190802259043

Jaillon, L., & Poon, C. S. (2009). The evolution of prefabricated residential building systems in Hong Kong: A review of the public and the private sector. *Automation in Construction, 18*(3), 239–248. DOI: 10.1016/j.autcon.2008.09.002

Jain, A., Smarra, F., Reticcioli, E., D'Innocenzo, A., & Morari, M. (2020). NeurOpt: Neural network based optimization for building energy management and climate control. *arXiv (Cornell University).* /arXiv.2001.07831DOI: 10.48550

Jang, Y., Son, J., & Hwang, S. (2022). Requirements analysis for development of off-Site construction project management system: Focusing on precast concrete construction. *Buildings (Basel, Switzerland), 12*(10), 1499. DOI: 10.3390/buildings12101499

Jayanthi, N., Raja, K. T., Wadhwa, G., Shneka, K., & Swathi, R. (2021). IoT based-civil labour safety monitoring system in construction site. *Turkish Journal of Computer and Mathematics Education, 12*(9), 1723–1728.

Jedidi, M. (2018). Evaluation of Concrete Quality Using Destructive and Non-Destructive Testing. *MOJ Civil Engineering, 4*(4), 219–223. DOI: 10.15406/mojce.2018.04.00122

Jha, A. K., Bloch, R., & Lamond, J. (2011). *Cities and Flooding: A Guide to Integrated Urban Flood Risk Management for the 21st Century.*

Jha, R., Lee, S., Iyengar, S., Hajiesmaili, M., Irwin, D., & Shenoy, P. (2020). *Emission-aware Energy Storage Scheduling for a Greener Grid.* DOI: 10.1145/3396851.3397755

Jia, G., & Gardoni, P. (2019). Stochastic life-cycle analysis: Renewal-theory life-cycle analysis with state-dependent deterioration stochastic models. *Structure and Infrastructure Engineering, 15*(8), 1001–1014. DOI: 10.1080/15732479.2019.1590424

Jia, G., Gardoni, P., Trejo, D., & Masad, V. (2021). Stochastic modeling of deterioration and time-variant performance of reinforced concrete structures under joint effects of earthquakes, corrosion, and ASR. *Journal of Structural Engineering, 147*(2). Advance online publication. DOI: 10.1061/(ASCE)ST.1943-541X.0002884

Jia, G., Tabandeh, A., & Gardoni, P. (2017). Life-cycle analysis of engineering systems: Modeling deterioration, instantaneous reliability, and resilience. In *Life-cycle of engineering systems* (pp. 465–494). Springer. DOI: 10.1007/978-3-319-52425-2_20

Jiang, L., Wu, M., Du, F., Chen, D., Xiao, L., Chen, W., Du, W., & Ding, Q. (2024). State-of-the-Art Review of Microcapsule Self-Repairing Concrete: Principles, Applications, Test Methods, Prospects. *Polymers, 16*(22), 3165. DOI: 10.3390/polym16223165 PMID: 39599256

Jiang, S., Li, X., Lyu, W., Wang, B., & Shi, W. (2020). Numerical investigation of the energy efficiency of a serial pipe-embedded external wall system considering water temperature changes in the pipeline. *Journal of Building Engineering*, *31*, 101435. DOI: 10.1016/j.jobe.2020.101435

Jiang, Y., Li, M., Li, M., Liu, X., Zhong, R. Y., Pan, W., & Huang, G. Q. (2022). Digital twin-enabled real-time synchronization for planning, scheduling, and execution in precast on-site assembly. *Automation in Construction*, *141*, 104397. DOI: 10.1016/j.autcon.2022.104397

Jiang, Y., Zhao, D., Wang, D., & Xing, Y. (2019). Sustainable Performance of Buildings through Modular Prefabrication in the Construction Phase: A Comparative Study. *Sustainability (Basel)*, *11*(20), 5658. DOI: 10.3390/su11205658

Jiménez-Rivero, A., De Guzmán-Báez, A., & García-Navarro, J. (2017). Enhanced On-Site Waste Management of Plasterboard in Construction Works: A Case Study in Spain. *Sustainability (Basel)*, *9*(3), 450. DOI: 10.3390/su9030450

Johansson, P., & Wahlgren, P. (2020, January 1). Experiences from Interior Super Insulation of a Brick Wall from the 1800s. *XV International Conference on Durability of Building Materials and Components. eBook of Proceedings*. DOI: 10.23967/dbmc.2020.065

Joseph, H. S., Devaraj, A. R., Priscila, A., Selvalakshmi, V., & Rameshkumaar, V. P. (2026). Advancing Sustainable Civil Engineering Through Intelligent Motion Control: Construction Equipment and Infrastructure Development. In *Methodologies and Applications of Intelligent Motion Control Systems* (pp. 105-132). IGI Global Scientific Publishing.

Joyce, J., Chang, N. B., Harji, R., & Ruppert, T. (2018). Coupling infrastructure resilience and flood risk assessment via copulas analyses for a coastal green-grey-blue drainage system under extreme weather events. *Environmental Modelling & Software*, *100*, 82–103. DOI: 10.1016/j.envsoft.2017.11.008

Kadioğlu, S. (2010). Description of buried archaeological remains from the Temple of Augustus in Ankara, Türkiye, using a new 3D visualization technique of the underground radar dataset. *Near Surface Geophysics*, *8*(5), 397–406. DOI: 10.3997/1873-0604.2010028

Kadioğlu, S., Kadioğlu, M., & Kadioğlu, Y. K. (2013b). Identification of buried archaeological remains in the ancient city of Nysa in Aydın- Türkiye, using ground penetrating radar, polarized microscope, and confocal Raman spectroscopy methods. *Journal of Archaeological Science*, *40*(10), 3569–3583. DOI: 10.1016/j.jas.2013.04.001

Kadioğlu, S., Kadioğlu, Y. K., & Akyol, A. A. (2011). Monitoring buried remains in the Zeynel Bey tomb in Hasankeyf ancient city, Türkiye, with transparent 3D semi-bird's-eye view of underground radar data. *Journal of Geophysics and Engineering*, *8*(3). Advance online publication. DOI: 10.1088/1742-2132/8/3/S07

Kadioğlu, S., Kadioğlu, Y. K., Catapano, I., & Soldovieri, F. (2013a). Ground penetrating radar and microwave tomography for security management of cultural heritage site: Miletus Ilyas Bey Mosque (Türkiye). *Journal of Geophysics and Engineering*, *10*(6), 064007. DOI: 10.1088/1742-2132/10/6/064007

Kairu, W. M., Otieno, M., Olanrewaju, O. A., & Mwero, J. M. (2025). Application of Schmidt hammer and ultrasonic pulse velocity for structural integrity assessment in water dams. *Journal of Infrastructure Preservation and Resilience*, *5*(1), 11. Advance online publication. DOI: 10.1186/s43065-025-00123-5

Kaja, N., & Jauswal, A. (2021). Review of precast concrete technology in India. [IJERT]. *International Journal of Engineering Research & Technology (Ahmedabad)*, *10*(06), 867–872.

Kale, S., & Arditi, D. (2006). Diffusion of ISO 9000 certification in the precast concrete industry. *Construction Management and Economics*, *24*(5), 485–495. DOI: 10.1080/01446190600601594

Kalmár, F., & Kalmár, T. (2020). Thermal Comfort Aspects of Solar Gains during the Heating Season. *Energies*, *13*(7), 1702. DOI: 10.3390/en13071702

Kamel, E., Sheikh, S., & Huang, X. (2020). Data-driven predictive models for residential building energy use based on the segregation of heating and cooling days. *Energy*, *206*, 118045. DOI: 10.1016/j.energy.2020.118045

Kampli, G., Chickerur, S., & Chitawadagi, M. (2024). IoT system implementation for real-time concrete strength prediction: Experimental design, variance evaluation, cost analysis, and implementation ease. *Innovative Infrastructure Solutions*, *9*(7), 260. DOI: 10.1007/s41062-024-01586-3

Kanda, R., Akhai, S., & Bansal, R. (2013). Analysis of MOST technique for elimination of ideal time by synchronization of different lines. *International Journal of Research*, *1*(4).

Karmakar, S., Dauji, S., Kshirsagar, S. S., Saini, S. K., Bhargava, K., & Mahapatra, K. (2021). Estimation of present-day strength of concrete for a 40-year-old building from non-destructive tests: A case study. *Journal of Asian Concrete Federation*, *7*(2), 17–32. DOI: 10.18702/acf.2021.12.7.2.17

Karthik, M., Mander, J., & Hurlebaus, S. (2020). Simulating behaviour of large reinforced concrete beam-column joints subject to ASR/DEF deterioration and influence of corrosion. Engineering Structures. https://www.sciencedirect.com/science/article/pii/S0141029620303606

Karunarathna, S., Linforth, S., Kashani, A., Liu, X., & Ngo, T. (2021). Effect of recycled rubber aggregate size on fracture and other mechanical properties of structural concrete. *Journal of Cleaner Production*, *314*, 128230. DOI: 10.1016/j.jclepro.2021.128230

Katiyar, A., & Kumar, P. (2022). Automated construction progress monitoring of prefabricated concrete structures using building information modeling and internet of things. *International Journal on Recent and Innovation Trends in Computing and Communication*, *10*(7), 14–18. DOI: 10.17762/ijritcc.v10i7.5560

Kauko, H., Rohde, D., & Hafner, A. (2020). Local Heating Networks with Waste Heat Utilization: Low or Medium Temperature Supply? *Energies*, *13*(4), 954. DOI: 10.3390/en13040954

Kaur, P., Kaur, G., Bansal, J., Singh, A., Bhatia, B., & Akhai, S. (2024, September). A Review of Opportunities and Challenges of Incorporating Artificial Intelligence and Machine Learning to Improve the Reliability of Electrical Power Systems. In *2024 International Conference on Communication, Computing and Energy Efficient Technologies (I3CEET)* (pp. 493-498). IEEE. DOI: 10.1109/I3CEET61722.2024.10993874

Kaya, Y., & Safak, E. (2015). Real-time analysis and interpretation of continuous data from structural health monitoring systems. *Bulletin of Earthquake Engineering*, *13*(3), 917–934. DOI: 10.1007/S10518-014-9642-9

Kayll, D., Defo, M., Moore, T. M., & Lacasse, M. (2020, January 1). A Methodology for Assessment of Building Assembly Air Leakage Moisture Response, Condensation Risk, and Expected Durability When Subjected to Projected Future Climate Loads. *XV International Conference on Durability of Building Materials and Components. eBook of Proceedings*. DOI: 10.23967/dbmc.2020.183

Kehl, R. J., & Carrasquillo, R. L. (1998). *Investigation of the use of match cure technology in the precast concrete industry* (No. FHWA/TX-01/1714-2). University of Texas at Austin. Center for Transportation Research.

Kennedy, D., Habel, K., & FitzGerald, G. (2023). Ultra-high-performance concrete column jacket retrofit for the Mission Bridge. *Third International Interactive Symposium on Ultra-High Performance Concrete*. DOI: 10.21838/UHPC.16730

Khadra, A., Hugosson, M., Akander, J., & Myhren, J. A. (2020). Economic performance assessment of three renovated multi-family buildings with different HVAC systems. *Energy and Building*, *224*, 110275. DOI: 10.1016/j.enbuild.2020.110275

Khairunnisa, M., Latief, Y., Hidayawanti, R., Machfudiyanto, R. A., & Riantini, L. S. (2025, February). Innovation of technology application in the manufacturing of spun pile products at precast construction company with a risk approach and ISO 56002 to increase competitiveness. In. AIP Conference Proceedings: Vol. 3169. *No. 1* (p. 040047). AIP Publishing LLC. DOI: 10.1063/5.0255371

Khalilnejad, A., French, R. H., & Abramson, A. R. (2021). Evaluation of cooling setpoint setback savings in commercial buildings using electricity and exterior temperature time series data. *Energy*, *233*, 121117. DOI: 10.1016/j.energy.2021.121117

Khare, R. K., Maniyar, S. R., Uma, S. R., & Bidwai, V. B. (2011). Seismic performance and design of precast concrete building structures: An overview. *Journal of Structural Engineering*, *38*(3), 272–284.

Khudair, A. W. A., Naser, A. F., Aldhalemi, A. A., & Mahmood, H. F. (2025). International Journal of Transport Development and Integration. *Integration (Amsterdam)*, *9*(1), 151–162.

Khurshid, K., Danish, A., Salim, M. U., Bayram, M., Ozbakkaloglu, T., & Mosaberpanah, M. A. (2023). An in-depth survey demystifying the Internet of Things (IoT) in the construction industry: Unfolding new dimensions. *Sustainability (Basel)*, *15*(2), 1275. DOI: 10.3390/su15021275

Kiamili, C., Hollberg, A., & Habert, G. (2020). Detailed Assessment of Embodied Carbon of HVAC Systems for a New Office Building Based on BIM. *Sustainability (Basel)*, *12*(8), 3372. DOI: 10.3390/su12083372

Kibar, H., & Öztürk, T. (2015). Determination of Concrete Quality by Destructive and Non-Destructive Methods. *Computers and Concrete*, *15*(3), 473–484. DOI: 10.12989/cac.2015.15.3.473

Kim, S., Choi, I., Lee, K., & Song, J. (2025). A rapid decision-making process of retrofit scheme for seismically vulnerable reinforced concrete school buildings using simple information. Structures. https://www.sciencedirect.com/science/article/pii/S2352012425009889

Kim, D., & Braun, J. E. (2020). Model predictive control for supervising multiple rooftop unit economizers to fully leverage free cooling energy resource. *Applied Energy*, *275*, 115324. DOI: 10.1016/j.apenergy.2020.115324

Kim, K. G., & Choi, H. S. (2021). Planning instruments for climate smart and wise cities: a spatial, green and digital deal approach. In *Planning Climate Smart and Wise Cities: A Multidisciplinary Approach* (pp. 3–97). Springer International Publishing.

Kim, S., & Kim, S.-A. (2020). Framework for Designing Sustainable Structures through Steel Beam Reuse. *Sustainability (Basel)*, *12*(22), 9494. DOI: 10.3390/su12229494

Kim, T., Kim, Y. W., & Cho, H. (2020). Dynamic production scheduling model under due date uncertainty in precast concrete construction. *Journal of Cleaner Production*, *257*, 120527. DOI: 10.1016/j.jclepro.2020.120527

Kiong, N. B., & Akasah, Z. (2012). An overview of precast concrete system for building maintenance: Malaysian perspective. *International Journal of Engineering Science and Advanced Technology*, *2*(6), 1684–1689.

Kokorin, A. S., Koval, S. V., & Emelyanovich, A. A. (n.d.). Quality Management In Precast Concrete Industry: Problems And Perspectives. *European Proceedings of Social and Behavioural Sciences*.

Kolbeck, L., Kovaleva, D., Manny, A., Stieler, D., Rettinger, M., Renz, R., Tošić, Z., Teschemacher, T., Stindt, J., Forman, P., Borrmann, A., Blandini, L., Stempniewski, L., Stark, A., Menges, A., Schlaich, M., Albers, A., Lordick, D., Bletzinger, K.-U., & Mark, P. (2023). Modularisation Strategies for Individualised Precast Construction—Conceptual Fundamentals and Research Directions. *Designs*, *7*(6), 143. DOI: 10.3390/designs7060143

Köliö, A., Pakkala, T., Annila, P., & Lahdensivu, J. (2014). Possibilities to validate design models for corrosion in carbonated concrete using condition assessment data. Engineering Structures. https://www.sciencedirect.com/science/article/pii/S0141029614003940

Konradsen, F., Hansen, K. S. H., Ghose, A., & Pizzol, M. (2024). Same product, different score: How methodological differences affect EPD results. *The International Journal of Life Cycle Assessment*, *29*(2), 291–307. DOI: 10.1007/s11367-023-02246-x

Korper, A. K., Patrício, L., Holmlid, S., & Witell, L. (2020). Service design as an innovation approach in technology startups: A longitudinal multiple case study. *Creativity and Innovation Management*, *29*(2), 303–323. DOI: 10.1111/caim.12383

Kosonen, A., & Keskisaari, A. (2020). Zero-energy log house – Future concept for an energy efficient building in the Nordic conditions. *Energy and Building*, *228*, 110449. DOI: 10.1016/j.enbuild.2020.110449

Kosse, S., Forman, P., Stindt, J., Hoppe, J., König, M., & Mark, P. (2023, June). Industry 4.0 enabled modular precast concrete components: a case study. In *International RILEM Conference on Synergising expertise towards sustainability and robustness of CBMs and concrete structures* (pp. 229-240). Cham: Springer Nature Switzerland. DOI: 10.1007/978-3-031-33211-1_21

Kosse, S., Hagedorn, P., Maibaum, J., & König, M. (2024, August). Digital twins for precast concrete: advancing environmental analysis through integrated life cycle assessment. In *International Conference on Computing in Civil and Building Engineering* (pp. 613-627). Cham: Springer Nature Switzerland.

Kostadino, E. (2023). *Prediction Of U-Ditch Raw Material Inventories In.* Multiguna Precast Mandiri Using A Web-Based Single Exponential Smoothing Method.

Krause, K., & Hafner, A. (2022). Resource Efficiency in the Construction Sector: Material Intensities of Residential Buildings—A German Case Study. *Energies, 15*(16), 5825. DOI: 10.3390/en15165825

Kumar, R. V., N, S. K., Narsimha, K., K, S. R. R., & E, V. K. (2020). Spatial Day Light Autonomy and Energy Analysis of a Residential Building for Different Climatic Conditions and Window-to-Wall Ratios. *E3S Web of Conferences, 184*, 1117. DOI: 10.1051/e3sconf/202018401117

Kumar, H., Wadhwa, A. S., Akhai, S., & Kaushik, A. (2024). Parametric analysis, modeling and optimization of the process parameters in electric discharge machining of aluminium metal matrix composite. *Engineering Research Express, 6*(2), 025542. DOI: 10.1088/2631-8695/ad4ba9

Kumar, H., Wadhwa, A. S., Akhai, S., & Kaushik, A. (2024). Parametric optimization of the machining performance of Al-SiCp composite using combination of response surface methodology and desirability function. *Engineering Research Express, 6*(2), 025505. DOI: 10.1088/2631-8695/ad38ff

Kumar, P., & Akhai, S. (2022). Effective energy management in smart buildings using VRV/VRF systems. In *Additive Manufacturing in Industry 4.0* (pp. 27–35). CRC Press., DOI: 10.1201/9781003360001-2

Kumar, P., Srijan, S., Aggarwal, V., & Walia, B. S. (2025). Influence of Mineral Composition and Microstructural Features on the Tensile Failure Mechanisms of Granite and Marble. *Journal of Mines. Metals and Fuels, 73*(8), 2515–2537. DOI: 10.18311/jmmf/2025/48211

Kumar, P., Srijan, S., Aggarwal, V., & Walia, B. S. (2025). The unique modified seesaw test (MST) instrument for the direct assessment of rock's tensile strength utilizing ring-shaped specimens. *Sadhana, 50*(3), 123. DOI: 10.1007/s12046-025-02803-0

Kumar, P., & Walia, B. S. (2025). An analytical–numerical framework for progressive contact analysis in radially compressed direct disc tests. *Journal of Earth System Science, 134*(4), 202. DOI: 10.1007/s12040-025-02650-x

Kumar, R., Waghmare, A., Nayak, S., Paswan, M., & Achintya, . (2021). Design of self sustainable zero energy building. *Materials Today: Proceedings, 46*, 6737–6742. DOI: 10.1016/j.matpr.2021.04.263

Kumar, S. (2022). *Evolving Role of Sustainable Procurement in Coming Decades.*, DOI: 10.1007/978-3-031-08258-0_7

Kupfer, H. B. (1973). *Das Verhalten des Betons unter mehrachsiger Kurzzeitbelastung unter besonderer Berücksichtigung der zweiachsigen Beanspruchung* (Schriftenreihe des DAfStb, Heft 229). Berlin: Ernst & Sohn.

Kupfer, H. B., Hilsdorf, H. K., & Rusch, H. (1969). Behavior of Concrete under Biaxial Stresses. *ACI Journal.*

Kurama, Y. C., Sritharan, S., Fleischman, R. B., Restrepo, J. I., Henry, R. S., Cleland, N. M., Ghosh, S. K., & Bonelli, P. (2018). Seismic-resistant precast concrete structures: State of the art. *Journal of Structural Engineering, 144*(4), 03118001. DOI: 10.1061/(ASCE)ST.1943-541X.0001972

Laaksonen, H., Parthasarathy, C., Hafezi, H., Shafie-khah, M., Khajeh, H., & Hatziargyriou, N. (2020). Solutions to Increase PV Hosting Capacity and Provision of Services from Flexible Energy Resources. *Applied Sciences (Basel, Switzerland), 10*(15), 5146. DOI: 10.3390/app10155146

Lagrange, A., de Simón-Martín, M., González-Martínez, A., Bracco, S., & Rosales-Asensio, E. (2020). Sustainable microgrids with energy storage as a means to increase power resilience in critical facilities: An application to a hospital. *International Journal of Electrical Power & Energy Systems, 119*, 105865. DOI: 10.1016/j.ijepes.2020.105865

Langston, C., Wong, F. K. W., Hui, E. C. M., & Shen, L.-Y. (2008). Strategic assessment of building adaptive reuse opportunities in Hong Kong. *Building and Environment, 43*(10), 1709–1718. DOI: 10.1016/j.buildenv.2007.10.017

Lassandro, P., Capotorto, S., & Mammone, V. (2025). Nature-Based Solutions: Green and Smart Façade with an Innovative Cultivation System for Sustainable Buildings and More Climate-Resilient Cities. *Sustainability (Basel), 17*(10), 4580. DOI: 10.3390/su17104580

Latifah, N. L., & Rahadian, E. Y. (2020). Energy Saving Building Strategies through The Application of Solar Control Glass. *ELKOMIKA Jurnal Teknik Energi Elektrik Teknik Telekomunikasi & Teknik Elektronika, 8*(2), 388. DOI: 10.26760/elkomika.v8i2.388

Lee, E. H., Lee, Y. S., Joo, J. G., Jung, D., & Kim, J. H. (2016). Flood reduction in urban drainage systems: Cooperative operation of centralized and decentralized reservoirs. *Water (Switzerland), 8*(10).

Lehmann, S. (2011). Optimizing Urban Material Flows and Waste Streams in Urban Development through Principles of Zero Waste and Sustainable Consumption. *Sustainability (Basel), 3*(1), 155–183. DOI: 10.3390/su3010155

Leising, E., Quist, J., & Bocken, N. (2018). Circular Economy in the building sector: Three cases and a collaboration tool. *Journal of Cleaner Production, 176*, 976–989. DOI: 10.1016/j.jclepro.2017.12.010

Lei, Y., & Zhang, Z. (2013). Review of physical based monitoring techniques for condition assessment of corrosion in reinforced concrete. *Mathematical Problems in Engineering, 2013*. Advance online publication. DOI: 10.1155/2013/953930

Lemian, D., & Bode, F. (2025). Digital twins in the building sector: Implementation and key features. In E3S Web of Conferences (Vol. 550, Article 05004). EDP Sciences.

Lencis, U., Udris, A., Kara De Maeijer, P., Korjakins, A., & Zvejnieks, E. (2025). A Comparison of Different Types of Pull-Off Testing and Splitting Methods for Determining the Tensile Strength of Concrete. *Buildings (Basel, Switzerland), 15*(7), 1068. DOI: 10.3390/buildings15071068

Leuzzo, A. (2022, May). A Methodology Toward a Just, Digital and Ecological Transition for Resilient and Sustainable Scenarios: Innovations in Urban Water Management. In *International Symposium: New Metropolitan Perspectives* (pp. 332-356). Cham: Springer International Publishing.

Lewrick, M., & Raeside, R. (2010). Transformation and change process in innovation models: Start-up and mature companies. *International Journal of Business Innovation and Research, 4*(6), 515. DOI: 10.1504/IJBIR.2010.035711

Liang, J., Cao, Z., Zu, Q., Huang, H., & Guo, S. (2024). Production sequencing and layout optimization of precast concrete components under mold resource constraints. *Buildings (Basel, Switzerland), 14*(10), 3173. DOI: 10.3390/buildings14103173

Li, C. Z., Hong, J., Xue, F., Shen, G. Q., Xu, X., & Luo, L. (2016). SWOT analysis and Internet of Things-enabled platform for prefabrication housing production in Hong Kong. *Habitat International, 57*, 74–87. DOI: 10.1016/j.habitatint.2016.07.002

Li, C. Z., Hong, J., Xue, F., Shen, G. Q., Xu, X., & Mok, M. K. (2016). Schedule risks in prefabrication housing production in Hong Kong: A social network analysis. *Journal of Cleaner Production, 134*, 482–494. DOI: 10.1016/j.jclepro.2016.02.123

Li, C. Z., Xue, F., Li, X., Hong, J., & Shen, G. Q. (2018). An Internet of Things-enabled BIM platform for on-site assembly services in prefabricated construction. *Automation in Construction, 89*, 146–161. DOI: 10.1016/j.autcon.2018.01.001

Ligabue, V., Pampanin, S., & Savoia, M. (2018). Seismic performance of alternative risk-reduction retrofit strategies to support decision making. *Bulletin of Earthquake Engineering, 16*(7), 3001–3030. DOI: 10.1007/S10518-017-0291-7

Li, J., & Zhao, K. (2025). Sustainable Development Advantages of Cross-Laminated Timber (CLT) and Cross-Laminated Bamboo and Timber (CLBT). *Materials (Basel), 18*(21), 4913. DOI: 10.3390/ma18214913 PMID: 41227871

Li, K., Gan, V. J., Li, M., Gao, M. Y., Tiong, R. L., & Yang, Y. (2024). Automated generative design and prefabrication of precast buildings using integrated BIM and graph convolutional neural network. *Developments in the Built Environment, 18*, 100418. DOI: 10.1016/j.dibe.2024.100418

Li, M., Zou, D., Li, Y., & Tong, L. (2023). Effect of cooling rate on pitting corrosion behavior of 904L austenitic stainless steel in a simulated flue gas desulfurization solution. *Metals and Materials International, 29*(3), 730–747. DOI: 10.1007/s12540-022-01255-z

Lima, T. A. R. M., da Mota, I. T. S., & Junior, C. M. M. (2023). Blockchain e contratos inteligentes na construção civil 4.0. Revista Multidisciplinar do Nordeste Mineiro, 11(1).

Lima, C. M., Ferreira, E. D. A. M., & Calmon, J. L. (2021). BIM and IoT integration to support management processes: An application for precast concrete systems. *The Journal of Modern Project Management, 9*(2).

Li, R. W., Cao, D. S., Wu, H., & Wang, D. F. (2021). Collapse analysis and damage assessment of a typical simply supported double-pier RC bridge under truck collision. *Structures (Oxford), 33*, 3222–3238. DOI: 10.1016/j.istruc.2021.06.041

Liu, J., Hang, M., Jiang, M., Song, H., & Zhang, X. (2024). Performance and mechanism of triethanolamine-triisopropanolamine corrosion inhibitor to deterioration of reinforced concrete under the coupling effect of freeze-thaw. Journal of Building Engineering. https://www.sciencedirect.com/science/article/pii/S2352710224026342

Liu, R. Y. W., Sailer, E., Taborda, D. M. G., & Potts, D. M. (2020). Evaluating the impact of different pipe arrangements on the thermal performance of thermo-active piles. *E3S Web of Conferences, 205*, 5006. DOI: 10.1051/e3sconf/202020505006

Liu, C., Zhang, F., & Zhang, H. (2020). Comparative analysis of off-site precast concrete and cast-in-place concrete in low-carbon built environment. *Fresenius Environmental Bulletin, 29*(3), 1804–1812.

Liu, H., Zhao, B. Y., Zhang, Z., Li, H., Hu, B., & Wang, R. Z. (2020). Experimental validation of an advanced heat pump system with high-efficiency centrifugal compressor. *Energy, 213*, 118968. DOI: 10.1016/j.energy.2020.118968

Liu, L., Li, X., Xia, G., Jin, J., & Chen, G. (2016). Spatial fuzzy clustering approach to characterize flood risk in urban storm water drainage systems. *Natural Hazards, 83*(3), 1469–1483. DOI: 10.1007/s11069-016-2371-4

Liu, T., Zhu, C., Tong, Z., & Luo, D. (2024, June). Research on smart manufacturing technology of reinforcing bars for bridge precast concrete structure. *Journal of Physics: Conference Series*, *2775*(1), 012028. DOI: 10.1088/1742-6596/2775/1/012028

Liu, X., Li, Y., & Zhao, X. (2022). Impacts of urbanization on flood risk and stormwater management: A review of trends and solutions. *Environmental Earth Sciences*, *81*(7), 310–326.

Liu, Y., Tong, W., Li, Q., Yao, F., Li, Y., Li, H. X., & Huang, J. (2022). Study on complexity of precast concrete components and its influence on production efficiency. *Advances in Civil Engineering*, *2022*(1), 9926547. DOI: 10.1155/2022/9926547

Liu, Z. S., Liu, S. N., Wang, S. W., Zhao, Y. H., & Yu, J. Q. (2018). Information solution of precast building process based on low power wide area Internet of Things. *Construction Technology*, *16*(47), 117–122.

Liu, Z., Takasu, K., Koyamada, H., & Suyama, H. (2022). A study on engineering properties and environmental impact of sustainable concrete with fly ash or GGBS. *Construction & Building Materials*, *316*, 125776. DOI: 10.1016/j.conbuildmat.2021.125776

Li, X., Cui, T., Huang, K., & Ma, X. (2020). Optimization of load sharing for parallel compressors using a novel hybrid intelligent algorithm. *Energy Science & Engineering*, *9*(3), 330–342. DOI: 10.1002/ese3.821

Li, Y., Dang, X., Xia, C., Ma, Y., Ogura, D., & Hokoi, S. (2020). The Effect of Air leakage through the Air Cavities of Building Walls on Mold Growth Risks. *Energies*, *13*(5), 1177. DOI: 10.3390/en13051177

Li, Z., Delsaute, B., Lu, T., Kostiuchenko, A., Staquet, S., & Ye, G. (2021). A comparative study on the mechanical properties, autogenous shrinkage and cracking proneness of alkali-activated concrete and ordinary Portland cement concrete. *Construction & Building Materials*, *292*, 123418. DOI: 10.1016/j.conbuildmat.2021.123418

Li, Z., Shen, G. Q., & Xue, X. (2014). Critical review of the research on the management of prefabricated construction. *Habitat International*, *43*, 240–249. DOI: 10.1016/j.habitatint.2014.04.001

Lobo, A. C. (2023). The Use of Time Value of Money in Making Investment Decision. SSRN *Electronic Journal*. DOI: 10.2139/ssrn.4338203

Lockrey, S., Nguyen, H., Crossin, E., & Verghese, K. (2016). Recycling the construction and demolition waste in Vietnam: Opportunities and challenges in practice. *Journal of Cleaner Production*, *133*, 757–766. DOI: 10.1016/j.jclepro.2016.05.175

Loewen, J. (2020). LCOE is an undiscounted metric that inaccurately disfavors renewable energy resources. *The Electricity Journal, 33*(6), 106769. DOI: 10.1016/j.tej.2020.106769

Lopes, Y., Vannali, L., & Jose, V. (2016). Prediction of Concrete Compressive Strength by Non-Destructive Testing Method: A Case Study. *Open Journal of Civil Engineering, 06*(04), 503–515. DOI: 10.4236/ojce.2016.64043

Lorusso, A., & Celenta, G. (2023, May). Internet of Things in the construction industry: A general overview. In *International Conference "New Technologies, Development and Applications"* (pp. 577-584). Cham: Springer Nature Switzerland. DOI: 10.1007/978-3-031-31066-9_65

Loto, R. T., Loto, C. A., & Fajobi, M. (2020). Corrosion resistance of GX4CrNiMo16-5-1 martensitic, 316L austenitic, and 904L austenitic stainless steels subjected to high temperature variation. In Ikhmayies, S., Li, B., Carpenter, J. S., de Oliveira, D. L. C., Monteiro, F., Anyalebechi, P. N., Sagüés, A. A., & Solanki, K. N. (Eds.), *Characterization of minerals, metals, and materials 2020* (pp. 261–267). Springer., DOI: 10.1007/978-3-030-36628-5_24

Loures, L., Castanho, R. A., Gómez, J. M. N., Lousada, S. A., Fernández-Pozo, L., Cabezas, J., & Loures, A. L. C. V. (2019). Impactos socioculturais da cooperação transfronteiriça no espaço europeu. Fronteira: Journal of Social. *Technological and Environmental Science, 8*(3), 292–312. DOI: 10.21664/2238-8869.2019V8I3.P292-312

Lousada, S., Cabezas, J., Castanho, R. A., & Gómez, J. M. N. (2022). Land-use changes in insular urban territories: A retrospective analysis from 1990 to 2018. The case of Madeira Island—Ribeira Brava. *Sustainability, 14*(24), 16839. DOI: 10.3390/SU142416839

Lousada, S., Gonçalves, L., & Atmaca, A. (2022). Hydraulic planning in insular urban territories: The case of Madeira Island—São Vicente. *Water (Basel), 14*(1), 112. DOI: 10.3390/W14010112

Lucena, J. R. P. G., Carabbacan, A. P. I., Manggapis, F. F., Kumar, S. D., & Dela Cruz, O. G. (2024). Strength, Serviceability, and Anchorage of Fiber-Reinforced Polymer and Carbon Fiber Composite Prestressing Tendons. *IOP Conference Series. Earth and Environmental Science, 1326*(1), 012047. DOI: 10.1088/1755-1315/1326/1/012047

Lu, L. (2024). Optimal Replacement Ratio of Recycled Concrete Aggregate Balancing Mechanical Performance with Sustainability: A Review. *Buildings (Basel, Switzerland), 14*(7), 2204. DOI: 10.3390/buildings14072204

Luo, L., Wei, H., Lin, Z., Wu, J., Wang, W., & Sun, Y. (2025). Multi-objective optimal energy-efficient retrofit determination using hybrid urban building energy model: Considering uncertainties between models. *Building Simulation, 18*(1), 183–206. DOI: 10.1007/S12273-024-1206-6

Lu, Y., Liu, Z., Li, S., & Hu, J. (2018). Axial compression behavior of hybrid fiber reinforcedconcrete filled steel tube stub column. *Construction & Building Materials, 174*, 96–107. DOI: 10.1016/j.conbuildmat.2018.04.089

Mabrouk, M., Mahran, M. G. N., Saleh, D., & AbouKorin, S. A. A. (2025). Digital Twin Cities for Climate-Resilient Urban Planning: A Systematic Review of Frameworks, Applications, and Future Perspectives. *Advance in Sustainability, 5*(2).

Machlev, R., Zargari, N., Chowdhury, N. R., Belikov, J., & Levron, Y. (2020). A review of optimal control methods for energy storage systems - energy trading, energy balancing and electric vehicles [Review of *A review of optimal control methods for energy storage systems - energy trading, energy balancing and electric vehicles*]. *Journal of Energy Storage, 32*, 101787. Elsevier BV. DOI: 10.1016/j.est.2020.101787

Madandoust, R., & Kazemi, M. (2017). Numerical Analysis of Rupture Test Method in Concrete. *Construction & Building Materials, 151*, 487–493. DOI: 10.1016/j.conbuildmat.2017.05.126

Madireddy, H., Naganathan, S., & Mahalingam, B. (2021). A review on precast concrete construction. *Sustainable Practices and Innovations in Civil Engineering: Select Proceedings of SPICE 2021*, 77-113.

Madsen, H. A., Zahle, F., Meng, F., Barlas, T., Rasmussen, F., & Rudolf, R. T. (2020). Initial performance and load analysis of the LowWind turbine in comparison with a conventional turbine. *Journal of Physics: Conference Series, 1618*(3), 32011. DOI: 10.1088/1742-6596/1618/3/032011

Mahabadi, Z., Varga, L., & Dolan, T. (2021). Network Properties for Robust Multilayer Infrastructure Systems: A Percolation Theory Review. *IEEE Access : Practical Innovations, Open Solutions, 9*, 135755–135773. DOI: 10.1109/ACCESS.2021.3116868

Maheskumar, P., Girimurugan, R., Sivaraman, G., Purushothaman, S., & Vairavel, M. (2021). Thermal analysis on regenerative cooled liquid propellant rocket thrust chamber using ANSYS. *IOP Conference Series. Materials Science and Engineering, 1059*(1), 012068. Advance online publication. DOI: 10.1088/1757-899X/1059/1/012068

Mainini, A. G., Poli, T., Speroni, A., Cavaglià, M., & Blanco Cadena, J. D. (2024). *The Relevance of Performance-Based Design Within Early-Stage Design.*, DOI: 10.1007/978-3-031-75298-8_2

Malek, J., & Kaouther, M. (2014). Destructive and Non-Destructive Testing of Concrete Structures. *Jordan Journal of Civil Engineering*, *8*, 432–441.

Malhotra, V. M. (1975). Evaluation of Tensile Test for Determining the Strength of In-Situ Concrete. *Matériaux Et Constructions*, *8*(43), 19–31. DOI: 10.1007/BF02538954

Mallick, R. (2025). Smart Cities and Climate Resilience. In *Community Resilience and Climate Change Challenges: Pursuit of Sustainable Development Goals (SDGs)* (pp. 401–412). IGI Global Scientific Publishing.

Ma, M., Tam, V. W. Y., Le, K. N., & Li, W. (2020). Challenges in current construction and demolition waste recycling: A China study. *Waste Management*, *118*, 610–625. DOI: 10.1016/j.wasman.2020.09.030 PMID: 33010692

Mamin, A. N., Avdeev, K. V., Zenin, S. A., & Bammatov, A. A. (2025). Failure mechanisms of reinforced concrete bridges. *Concrete and Reinforced Concrete*, *627*(2), 5–15. DOI: 10.37538/0005-9889-2025-2(627)-5-15

Mancini, T., & Bolzoni, F. (2003). Resistivity measurements as a tool for assessment of durability of concrete structures. *Construction & Building Materials*, *17*(3), 193–199. DOI: 10.1016/S0950-0618(02)00056-0

Manfren, M., & Nastasi, B. (2020). Parametric Performance Analysis and Energy Model Calibration Workflow Integration—A Scalable Approach for Buildings. *Energies*, *13*(3), 621. DOI: 10.3390/en13030621

Mangialardo, A., & Micelli, E. (2018). Rethinking the Construction Industry Under the Circular Economy: Principles and Case Studies. In Bisello, A., Vettorato, D., Laconte, P., & Costa, S. (Eds.), *Smart and Sustainable Planning for Cities and Regions* (pp. 333–344). Springer International Publishing., DOI: 10.1007/978-3-319-75774-2_23

Mannan, M., & Al-Ghamdi, S. G. (2020). Environmental impact of water-use in buildings: Latest developments from a life-cycle assessment perspective [Review of *Environmental impact of water-use in buildings: Latest developments from a life-cycle assessment perspective*]. *Journal of Environmental Management, 261*, 110198. Elsevier BV. DOI: 10.1016/j.jenvman.2020.110198

Mariak, A., Kurpinska, M., & Wilde, K. (2019). Maturity Curve for In-Situ Strength Prediction of High Performance Concrete. *MATEC Web of Conferences, 262*, 06007. DOI: 10.1051/matecconf/201926206007

Martinez-Bolaños, J. R., da Silva, V. O., Zucchi, M., Heideier, R. B., Relva, S. G., Saidel, M. A., & Fadigas, E. A. F. A. (2020). Performance Analysis of Topologies for Autonomous Hybrid Microgrids in Remote Non-Interconnected Communities in the Amazon Region. *Sustainability (Basel)*, *13*(1), 44. DOI: 10.3390/su13010044

Martins, R., Carmo, R. D., Costa, H., & Júlio, E. (2023). A review on precast structural concrete walls and connections. *Advances in Structural Engineering*, *26*(14), 2600–2620. DOI: 10.1177/13694332231191073

Martiradonna, S., Lombillo, I., & Fatiguso, F. (2019, October). Performance Monitoring of Refurbished Buildings through Innovative Precast Concrete Modules. In *2019 IEEE International Conference on Systems, Man and Cybernetics (SMC)* (pp. 953-957). IEEE. DOI: 10.1109/SMC.2019.8914472

Masoumi, F., Akgül, F., & Mehrabzadeh, A. (2013). Development of an optimised condition assessment plan for common reinforced concrete bridges in Estonia. *IACSIT International Journal of Engineering and Technology*, *5*, 708–711. DOI: 10.7763/IJET.2013.V5.647

Maurya, S., & Madan, S. K. (2023, December). Precast Concrete Technology: A Catalyst for Sustainable Urbanization. In *International Conference on Green Energy and Sustainable Technology* (pp. 229-245). Singapore: Springer Nature Singapore.

Maurya, S., & Madan, S. K. (2025). Precast Concrete Technology: A Catalyst. *Advances in Green Energy Technologies: Proceedings of ICGEST 2023,* Volume 2, *1314*, 229.

Mavrigiannaki, A., Gobakis, K., Kolokotsa, D., Kalaitzakis, K., Pisello, A. L., Piselli, C., Gupta, R., Gregg, M., Laskari, M., Saliari, M., Assimakopoulos, M.-N., & Synnefa, A. (2020). Measurement and Verification of Zero Energy Settlements: Lessons Learned from Four Pilot Cases in Europe. *Sustainability (Basel)*, *12*(22), 9783. DOI: 10.3390/su12229783

Ma, Z., Liu, Y., & Li, J. (2023). Review on automated quality inspection of precast concrete components. *Automation in Construction*, *150*, 104828. DOI: 10.1016/j.autcon.2023.104828

Mazurkiewicz, A., & Poteralska, B. (2017). Technology Transfer Barriers and Challenges Faced by R&D Organisations. *Procedia Engineering*, *182*, 457–465. DOI: 10.1016/j.proeng.2017.03.134

McCann, D. M., & Forde, M. C. (2001). Review of NDT Methods in the Evaluation of Concrete and Masonry Structures. *NDT & E International*, *34*(2), 71–84. DOI: 10.1016/S0963-8695(00)00032-3

McKinsey Global Institute. (2023). *Infrastructure productivity report*. https://www.mckinsey.com/capabilities/operations/our-insights/infrastructure-productivity

Medina, P., & Gomez, J. (2022). Reinforced concrete long-term deterioration prediction for the implementation of a bridge management system. Materials Today: Proceedings. https://www.sciencedirect.com/science/article/pii/S2214785322005995

Medved, S., Domjan, S., & Arkar, C. (2021). Contribution of energy storage to the transition from net zero to zero energy buildings. *Energy and Building*, *236*, 110751. DOI: 10.1016/j.enbuild.2021.110751

Melnyk, L. H. (2018). *Narodzhennia sostenovoi ekonomiky: dosvid YeS i praktyka Ukrainy v svitli III i IV promyslovykh revoliutsii* [The birth of the sustain economy: EU experience and Ukraine's practice]. Universytetska Knyha.

Miano, A. (2017). Performance based assessment and retrofit for existing RC structures [Doctoral dissertation]. https://tesidottorato.depositolegale.it/handle/20.500.14242/142579

Miano, A., Sezen, H., Jalayer, F., & Prota, A. (2017). Performance based comparison of different retrofit methods for reinforced concrete structures. In *Proceedings of the 6th International Conference on Computational Methods in Structural Dynamics and Earthquake Engineering* (Vol. 1, pp. 1515–1535). DOI: 10.7712/120117.5510.17150

Micic, T., Stojic, D., & Velimirovic, N. (2013). Stochastic process model for timber-concrete composite beam deterioration. In Safety, reliability, risk and life-cycle performance of structures and infrastructures: Proceedings of the 11th International Conference on Structural Safety and Reliability (pp. 5127–5133). CRC Press. DOI: 10.1201/B16387-745

Miller, D., Ho, N. M., Talebian, N., & Javanbakht, Z. (2023). Real-time monitoring of early-age compressive strength of concrete using an IoT-enabled monitoring system: An investigative study. *Innovative Infrastructure Solutions*, *8*(2), 75. DOI: 10.1007/s41062-023-01043-7

Miller, J. D., & Hutchins, M. (2017). The impacts of urbanisation and climate change on urban flooding and urban water quality: A review of the evidence concerning the United Kingdom. *Journal of Hydrology. Regional Studies*, *12*(January), 345–362. DOI: 10.1016/j.ejrh.2017.06.006

Miller, S., Davis, A., & Brown, T. (2021). Sustainable urban drainage systems for flood risk management and water quality improvement. *Urban Water Journal*, *18*(5), 450–463.

Min, H., & Junyu, L. (2013, August). An intelligent system for precast concrete element manufacturing management based on RFID technology. In *2013 IEEE International Conference on Green Computing and Communications and IEEE Internet of Things and IEEE Cyber, Physical and Social Computing* (pp. 1138-1143). IEEE. DOI: 10.1109/GreenCom-iThings-CPSCom.2013.198

Mirza, S., & Amleh, L. (2002). *Sustainable infrastructure and civil engineering education*. National Guidelines for Environmental Sustainability.

Mishra, A., Bajpai, S., & Bhadwal, S. (2025). Nature-based solutions for climate-resilient development: a technical, policy, and governance perspective. In *Handbook of nature-based solutions to mitigation and adaptation to climate change* (pp. 1831–1854). Springer Nature Switzerland. DOI: 10.1007/978-3-031-34967-6_159

Mishra, M., Deshmukh, A., & sri, K. S. (2020). Influence Of Passive Techniques on Thermal Performance of Building in Moderate Zone. *International Journal of Innovative Technology and Exploring Engineering*, *9*(3), 360–365. DOI: 10.35940/ijitee.B7164.019320

Mishra, S., Singh, S., & Srivastava, R. (2017). Green Building a Sustainable Approach to Save Energy. *International Journal of Engineering and Management Research*, *7*(3), 14–20.

Mitra, G., Jain, K., & Bhattacharjee, B. (2010). Condition assessment of corrosion-distressed reinforced concrete buildings using fuzzy logic. *Journal of Performance of Constructed Facilities*, *24*(6), 562–570. DOI: 10.1061/(ASCE)CF.1943-5509.0000137

Mohammadinia, A., Wong, Y. C., Arulrajah, A., & Horpibulsuk, S. (2019). Strength evaluation of utilizing recycled plastic waste and recycled crushed glass in concrete footpaths. *Construction & Building Materials*, *197*, 489–496. DOI: 10.1016/j.conbuildmat.2018.11.192

Mohammed, H. A. (2018). Variety of Precast Concrete. *Presentation presented to Al-Furat Al-Awsat Technical University-Technical Collage of Al-Mussaib-Session Presentation, Iraq.*

Mohan, C. V. R., & Deepthi, R. (2020). SIMULATION MODEL OF ELECTRIC VEHICLE TO GRID TECHNOLOGY FOR BALANCING LOADS AND FREQUENCY REGULATIONS. *International Journal of Engineering Applied Sciences and Technology*, *5*(4), 203–208. DOI: 10.33564/IJEAST.2020.v05i04.028

Molavi, J., & Barral, D. L. (2016). A Construction Procurement Method to Achieve Sustainability in Modular Construction. *Procedia Engineering*, *145*, 1362–1369. DOI: 10.1016/j.proeng.2016.04.201

Moré, F. B., Galindro, B. M., & Soares, S. R. (2022). Assessing the completeness and comparability of environmental product declarations. *Journal of Cleaner Production*, *375*, 133999. DOI: 10.1016/j.jclepro.2022.133999

Mori, Y. (1993a). Methodology for reliability based condition assessment: Application to concrete structures in nuclear plants. https://books.google.com/books?hl=en&lr=&id=t9dIT8kTGZkC

Mori, Y. (1993b). Reliability-based condition assessment and life prediction of concrete structures [Doctoral dissertation]. ProQuest Dissertations & Theses. https://search.proquest.com/openview/d9f3f6e7e18337612fd70b5089cfce17/1

Moroni, I., Arruda, A., & Araujo, K. (2015). The Design and Technological Innovation: How to Understand the Growth of Startups Companies in Competitive Business Environment. *Procedia Manufacturing*, *3*, 2199–2204. DOI: 10.1016/j.promfg.2015.07.361

Morris, I., Kumar, V., & Glisic, B. (2020). Predicting material properties of concrete from ground-penetrating radar attributes. *Structural Health Monitoring*, *20*(6), 2791–2812.

Motia, K., Kumar, R., & Akhai, S. (2024). *AI and Smart Manufacturing*., DOI: 10.4018/979-8-3693-6720-9.ch001

Moufti, S., Zayed, T., & Dabous, S. (2014). Defect-based condition assessment of concrete bridges: Fuzzy hierarchical evidential reasoning approach. *Transportation Research Record: Journal of the Transportation Research Board*, *2431*(1), 88–96. DOI: 10.3141/2431-12

Mugume, S. N., & Butler, D. (2017). Evaluation of functional resilience in urban drainage and flood management systems using a global analysis approach. *Urban Water Journal*, *14*(7), 727–736. DOI: 10.1080/1573062X.2016.1253754

Müller, E. (2020). Regenerative development as natural solution for sustainability. In Sarmiento, F. O., & Frolich, L. M. (Eds.), *The Elgar Companion to Geography, Transdisciplinarity and Sustainability*. Edward Elgar Publishing., DOI: 10.4337/9781786430106.00021

Mulrow, J. S., Derrible, S., Ashton, W. S., & Chopra, S. S. (2017). Industrial Symbiosis at the Facility Scale. *Journal of Industrial Ecology*, *21*(3), 559–571. DOI: 10.1111/jiec.12592

Muralidhar, P., & JaiSai, T. (2024, August). Kavach: A Technological and Strategic Framework for Railway Safety in India on Automatic Train Protection (ATP) Practices. In *International Conference on Construction, Real Estate, Infrastructure & Project Management* (pp. 276-284). Singapore: Springer Nature Singapore.

Muralidhar, P., & JaiSai, T. (2025). Framework for Railway Safety in India on Automatic Train Protection (ATP) Practices. *Recent Trends and Advances in Construction, Real Estate, Infrastructure and Project Management (CRIP) Sector: Global Perspective*, 276.

Murari, S. S., & Joshi, A. M. (2017). Precast construction methodology in construction industry. *Available at SSRN* 3496019.

Mustapha, F., Aris, K., & Wardi, N. (2012). Structural health monitoring for composite structure undergoing tensile and thermal testing. Journal of Vibroengineering. https://search.ebscohost.com/login.aspx?direct=true&profile=ehost&scope=site&authtype=crawler&jrnl=13928716&AN=79998236

Nadeem, A., Memon, S., & Lo, T. Y. (2013). Qualitative and quantitative analysis and identification of flaws in the microstructure of fly ash and metakaolin blended high performance concrete after exposure to elevated temperatures. Construction and Building Materials. https://www.sciencedirect.com/science/article/pii/S0950061812007246

Nair, P. S., Gupta, R., Agrawal, V., & Thomas, B. S. (2023). A critical review on the effectiveness of microbial concentrations for enhancing self-healing in cement concrete and mortar. *Materials Today: Proceedings*. Advance online publication. DOI: 10.1016/j.matpr.2023.03.546

Najafi, J., Peiravi, A., & Anvari-Moghaddam, A. (2020). Enhancing Integrated Power and Water Distribution Networks Seismic Resilience Leveraging Microgrids. *Sustainability (Basel)*, *12*(6), 2167. DOI: 10.3390/su12062167

Najjar, M. K., Figueiredo, K., Evangelista, A. C. J., Hammad, A. W. A., Tam, V. W. Y., & Haddad, A. (2022). Life cycle assessment methodology integrated with BIM as a decision-making tool at early-stages of building design. *International Journal of Construction Management*, *22*(4), 541–555. DOI: 10.1080/15623599.2019.1637098

Nanyam, V. N., Basu, R., Sawhney, A., Vikram, H., & Lodha, G. (2017). Implementation of precast technology in India–opportunities and challenges. *Procedia Engineering*, *196*, 144–151. DOI: 10.1016/j.proeng.2017.07.184

Naraharisetty, V., Talari, V. S., Neridu, S., Kalapatapu, P., & Pasupuleti, V. D. K. (2021, August). Cloud architecture for IoT based bridge monitoring applications. In *2021 international conference on emerging techniques in computational intelligence (ICETCI)* (pp. 39-42). IEEE. DOI: 10.1109/ICETCI51973.2021.9574044

Nasir, M. H. A., Genovese, A., Acquaye, A. A., Koh, S. C. L., & Yamoah, F. (2017). Comparing linear and circular supply chains: A case study from the construction industry. *International Journal of Production Economics*, *183*, 443–457. DOI: 10.1016/j.ijpe.2016.06.008

Nasr, A., Honfi, D., & Larsson Ivanov, O. (2022). Probabilistic analysis of climate change impact on chloride-induced deterioration of reinforced concrete considering Nordic climate. *Journal of Infrastructure Preservation and Resilience*, *3*(1). Advance online publication. DOI: 10.1186/S43065-022-00053-6

Nassar, R.-U.-D., Soroushian, P., & Sufyan-Ud-Din, M. (2021). Long-term field performance of concrete produced with powder waste glass as partial replacement of cement. *Case Studies in Construction Materials*, *15*, e00745. DOI: 10.1016/j.cscm.2021.e00745

Nepomuceno, M. C., & Bernardo, L. F. (2019). Evaluation of self-compacting concrete strength with non-destructive tests for concrete structures. *Applied Sciences (Basel, Switzerland)*, *9*(23), 5109. DOI: 10.3390/app9235109

Neupane, P., Afroze, D., & Phommasone, P. (2020). Cost Benefit Analysis of Green Building: A Case Study of Public Office Building in Nepal. *Saudi Journal of Engineering and Technology*, *5*(10), 382–391. DOI: 10.36348/sjet.2020.v05i10.005

Neville, A. M. (2012). *Properties of Concrete*. Prentice Hall.

Nguanthaisong, M., Terakulsatit, B., Setwong, W., & Phoovasawat, C. (2025). Deterioration of reinforced concrete structures due to chloride exposure in water: A case study, Thailand. *International Journal of GEOMATE, 28*(126), 1–7. DOI: 10.21660/2025.126.4713

Nguyen, H. Q., Radhakrishnan, M., Huynh, T. T. N., Baino-Salingay, M. L., Ho, L. P., Van der Steen, P., & Pathirana, A. (2017). Water Quality Dynamics of Urban Water Bodies during Flooding in Can Tho City, Vietnam. *Water (Basel)*, *9*(4), 260. DOI: 10.3390/w9040260

Nicał, A., & Anysz, H. (2020). The quality management in precast concrete production and delivery processes supported by association analysis. *International Journal of Environmental Science and Technology*, *17*(1), 577–590. DOI: 10.1007/s13762-019-02597-9

Nilimaa, J., & Nilforoush, R. (2023). A Direct Tensile Strength Testing Method for Concrete from Existing Structures. *CivilEng*, *4*(1), 333–344. DOI: 10.3390/civileng4010019

Novak, J., Kohoutková, A., Křístek, V., & Vodička, J. (2017, September). Precast concrete pavement–systems and performance review. *IOP Conference Series. Materials Science and Engineering*, *236*(1), 012030. DOI: 10.1088/1757-899X/236/1/012030

Novoseltsev, B. P., Жерлыкина, M. H., & Garmonov, K. V. (2020). Maintenance of heat stability of the heating system work in a structure. *IOP Conference Series. Materials Science and Engineering*, *962*(3), 32003. DOI: 10.1088/1757-899X/962/3/032003

O'Connor, J. T., & Mock, B. (2020). Responsibilities and accountabilities for industrial facility commissioning and startup activities. *Construction Innovation*, *20*(4), 625–645. DOI: 10.1108/CI-09-2019-0094

O'Hegarty, R., & Kinnane, O. (2020). Review of precast concrete sandwich panels and their innovations. *Construction & Building Materials, 233*, 117145. DOI: 10.1016/j.conbuildmat.2019.117145

O'Neill, R., Window, A., Kenway, S., & Dargusch, P. (2020). Integrated operational and life-cycle modelling of energy, carbon and cost for building façades. *Journal of Cleaner Production, 286*, 125370. DOI: 10.1016/j.jclepro.2020.125370

Obunguta, F., & Matsushima, K. (2022). Optimal pavement management strategy development with a stochastic model and its practical application to Ugandan national roads. *The International Journal of Pavement Engineering, 23*(7), 2405–2419. DOI: 10.1080/10298436.2020.1857759

Odufuwa, O. Y., Kusakana, K., & Numbi, B. P. (2020). Optimal energy management of Ice thermal energy storage-based air conditioning system for commercial buildings in real-time – A review based on POET framework [Review of *Optimal energy management of Ice thermal energy storage-based air conditioning system for commercial buildings in real-time – A review based on POET framework*]. *Journal of Physics Conference Series, 1577*(1), 12049. IOP Publishing. DOI: 10.1088/1742-6596/1577/1/012049

OECD. (2023). *Nature-based solutions for climate resilience*. https://www.oecd.org/en/topics/nature-based-solutions-and-climate-resilient-infrastructure.html

Ogunmakinde, O. E., Egbelakin, T., & Sher, W. (2022). Contributions of the circular economy to the United Nations sustainable development goals through sustainable construction. *Resources, Conservation and Recycling, 178*, 106023. DOI: 10.1016/j.resconrec.2021.106023

Oh, J.-W., Park, K.-S., Kim, H. S., Kim, I., Pang, S.-J., Ahn, K.-S., & Oh, J.-K. (2023). Comparative CO2 emissions of concrete and timber slabs with equivalent structural performance. *Energy and Building, 281*, 112768. DOI: 10.1016/j.enbuild.2022.112768

Olczak, P., Matuszewska, D., & Zabagło, J. (2020). The Comparison of Solar Energy Gaining Effectiveness between Flat Plate Collectors and Evacuated Tube Collectors with Heat Pipe: Case Study. *Energies, 13*(7), 1829. DOI: 10.3390/en13071829

Olidis, C., Swan, D. J., Saeed, A., Mellerski, R. C., & Hammons, M. I. (2010). Precast slab literature review report: Repair of rigid airfield pavements using precast concrete panels-a state-of-the-art review.

Oliveira, M. D. P. S. L., De Oliveira, E. A., & Fonseca, A. M. (2021). Strategies to promote circular economy in the management of construction and demolition waste at the regional level: A case study in Manaus, Brazil. *Clean Technologies and Environmental Policy, 23*(9), 2713–2725. DOI: 10.1007/s10098-021-02197-7

Omar, O. (2025). *Navigating Nearly Zero-Energy Strategies for Urban Climate Change Adaptation and Mitigation.*, DOI: 10.1007/978-981-96-1521-6_6

Omar, T., & Nehdi, M. (2018). Condition assessment of reinforced concrete bridges: Current practice and research challenges. *Infrastructures, 3*(3), 36. https://www.mdpi.com/2412-3811/3/3/36

Öncü, Z. (2025). Estimation of concrete strength by using artificial intelligence (AI) methods. *Nondestructive Testing and Evaluation,* •••, 1–18. DOI: 10.1080/10589759.2025.2536660

Ormondroyd, G. A., Spear, M. J., & Skinner, C. (2016). The Opportunities and Challenges for Re-use and Recycling of Timber and Wood Products Within the Construction Sector. In Kutnar, A., & Muthu, S. S. (Eds.), *Environmental Impacts of Traditional and Innovative Forest-based Bioproducts* (pp. 45–103). Springer Singapore., DOI: 10.1007/978-981-10-0655-5_3

Osei-Tutu, S., Ayarkwa, J., Osei-Asibey, D., Nani, G., & Afful, A. E. (2023). Barriers impeding circular economy (CE) uptake in the construction industry. *Smart and Sustainable Built Environment, 12*(4), 892–918. DOI: 10.1108/SASBE-03-2022-0049

Ostachowicz, W., & Güemes, A. (Eds.). (2013). *New trends in structural health monitoring* (Vol. 542). Springer Vienna. DOI: 10.1007/978-3-7091-1390-5

Ou, J., & Li, H. (2010). Structural health monitoring in mainland China: Review and future trends. *Structural Health Monitoring, 9*(3), 219–231. DOI: 10.1177/1475921710365269

Oyebamiji, H. O., Adewuyi, A. Y., Halidu, A., Togo, B., Alebiosu, Y. T., Atilola, M. O., Eromosele, E. I., & Atanda, A. A. (2025). Integrating circular economy in retrofit: Global practices in sustainable construction. *Journal of Management, and Development Research, 2*(2), 10–18. DOI: 10.69739/jmdr.v2i2.811

Oyeniran, K. A., Oyeniran, O. H., & Oladunmoye, K. M. (2025). Artificial Intelligence, Innovation, and Safety in Biotechnology Research. In *Cases on AI Innovations in Occupational Health and Safety* (pp. 73–98). IGI Global Scientific Publishing. DOI: 10.4018/979-8-3693-9301-7.ch004

Özçelik, E., & Gelişli, K. (2018). BETONARME YAPILARDA JEOFİZİK YÖNTEMLERLE BETON KALİTESİ VE YAPI DONATI DURUMUNUN ARAŞTIRILMASI. *Uygulamalı Yerbilimleri Dergisi, 17*(1), 1–11. DOI: 10.30706/uybd.433735

Öziçer, S. (2016). Jeofizik Yöntemler ile Riskli Yapıların Belirlenmesi ve İzmir Örneği. *SDÜ Fen Bilimleri Enstitüsü Jeofizik Müh. Anabilim dalı.* s:93

Öziçer, S., & Uyanık, O. (2017). Beton Dayanımının Yerinde P Dalga Hızından Belirlenmesi ve İzmir Örneği. *SDU International Journal of Technological Sciences*, *9*(1), 1–16.

Pacheco, R., Ordóñez, J., & Martínez, G. (2012). Energy efficient design of building: A review. *Renewable & Sustainable Energy Reviews*, *16*(6), 3559–3573. DOI: 10.1016/j.rser.2012.03.045

Padalu, P., & Singh, M. (2024). An overview of performance-based seismic design framework for reinforced concrete frame buildings. *Iranian Journal of Science and Technology. Transaction of Civil Engineering*, *48*(2), 635–667. DOI: 10.1007/S40996-023-01217-4

Padhiary, M. (2025). The convergence of deep learning, IoT, sensors, and farm machinery in agriculture. In *Designing sustainable internet of things solutions for smart industries* (pp. 109–142). IGI Global.

Palacio, J. P., Pablo, A. A., Tingatinga, E. J., & Mata, W. L. (2013). Structural performance assessment and retrofit of reinforced concrete buildings under seismic loads. In *Proceedings of the Structures Congress 2013* (pp. 2044–2055). ASCE. DOI: 10.1061/9780784412848.179

Pampanin, S. (2005). Emerging solutions for high seismic performance of precast/prestressed concrete buildings. *Journal of Advanced Concrete Technology*, *3*(2), 207–223. DOI: 10.3151/jact.3.207

Pampanin, S., & Archer, U. (2011). Performance-based seismic retrofit of existing reinforced concrete frame buildings using fibre-reinforced polymers: Challenges and solutions. *Structural Engineering International*, *21*(3), 260–270. DOI: 10.2749/101686611X13049248220041

Panagiotidou, M., Aye, L., & Rismanchi, B. (2021). Optimisation of multi-residential building retrofit, cost-optimal and net-zero emission targets. Energy and Buildings. https://www.sciencedirect.com/science/article/pii/S0378778821006691

Pan, W., & Teng, Y. (2021). A systematic investigation into the methodological variables of embodied carbon assessment of buildings. *Renewable & Sustainable Energy Reviews*, *141*, 110840. DOI: 10.1016/j.rser.2021.110840

Parida, S. (2009). Vaccination against foot-and-mouth disease virus: Strategies and effectiveness. *Expert Review of Vaccines*, *8*(3), 347–365. DOI: 10.1586/14760584.8.3.347 PMID: 19249976

Park, J. S. (2005). Opportunity recognition and product innovation in entrepreneurial hi-tech start-ups: A new perspective and supporting case study. *Technovation*, *25*(7), 739–752. DOI: 10.1016/j.technovation.2004.01.006

Park, M.-K., Lee, J.-M., Kang, W.-H., Kim, C., & Lee, K. H. (2020). Air-Handling-Unit Discharge Air Temperature Reset Based on Outdoor Air Temperature and Cooling Energy Performance in an Office Building. *Journal of Energy Engineering*, *146*(3), 04020013. Advance online publication. DOI: 10.1061/(ASCE)EY.1943-7897.0000660

Pedro, D., De Brito, J., & Evangelista, L. (2017). Mechanical characterization of high performance concrete prepared with recycled aggregates and silica fume from precast industry. *Journal of Cleaner Production*, *164*, 939–949. DOI: 10.1016/j.jclepro.2017.06.249

Pérez-Sánchez, M., Ramos, H. M., Franco, A. B., & López-Jiménez, P. A. (2017). Urban Floods Adaptation and Sustainable Drainage Measures. *Fluids (Basel, Switzerland)*, *2*(4), 61. DOI: 10.3390/fluids2040061

Piccardo, C., & Hughes, M. (2022). Design strategies to increase the reuse of wood materials in buildings: Lessons from architectural practice. *Journal of Cleaner Production*, *368*, 133083. DOI: 10.1016/j.jclepro.2022.133083

Pierzchlewicz, D., Woźniak, A., & Widera, B. (2025). Recent research on circular architecture: A literature review of 2021–2024 on circular strategies in the built environment. *Sustainability (Basel)*, *17*(17), 7580. DOI: 10.3390/su17177580

Pihul, O., & Siomushkin, V. (2023). Integration of sustainable development concepts and project management in modern housing construction models. *Economics and Society*, *53*. Advance online publication. DOI: 10.32782/2524-0072/2023-53-64

Piselli, C., Grazia, M. D., & Pisello, A. L. (2020). Combined Effect of Outdoor Microclimate Boundary Conditions on Air Conditioning System's Efficiency and Building Energy Demand in Net Zero Energy Settlements. *Sustainability (Basel)*, *12*(15), 6056. DOI: 10.3390/su12156056

Polat, G. (2010). Precast concrete systems in developing vs. industrialized countries. *Journal of Civil Engineering and Management*, *16*(1), 85–94. DOI: 10.3846/jcem.2010.08

Pont, U., Wöltzl, M., Schuß, M., Schober, P., & Mahdavi, A. (2020). Exploring novel solutions for incorporating vacuum glazing in new and existing window constructions. *E3S Web of Conferences*, *172*, 24006. DOI: 10.1051/e3sconf/202017224006

Poon, C. S., Yu, A. T. W., & Ng, L. H. (2001). On-site sorting of construction and demolition waste in Hong Kong. *Resources, Conservation and Recycling*, *32*(2), 157–172. DOI: 10.1016/S0921-3449(01)00052-0

Potemans, A. (2017). Modular building in a circular economy [Master's thesis, Delft University of Technology].

Pour-Ali, S., Tavangar, R., & Hejazi, S. (2023). Effect of micro-shot peening on microstructure and fluoride-induced corrosion performance of AISI 904L super-austenitic stainless steel. *Materials Today. Communications*, *36*, 106682. DOI: 10.1016/j.mtcomm.2023.106682

Pradhan, P., Verma, M. S., & Sahu, M. S. (2025). Optimization of Cast-In-Situ and Precast Concrete Methods Through BIM and Structural Simulations. *Optimization*.

Prakash, V. S. J., Pavithra, D., Nisha, R., & Nishanthini, P. (2020). Automatic street light intensity control. *International Journal of Advance Research, Ideas and Innovations in Technology, 6*(2), 258. https://www.ijariit.com/manuscripts/v6i2/V6I2-1169.pdf

Prakashvel, J., Umarani, C., & Sathiskumar, K. (2023). State-of-the-art review on performance-based retrofitting of structures. In *Lecture Notes in Civil Engineering* (Vol. 330, pp. 289–302). Springer. DOI: 10.1007/978-981-99-1604-7_22

Praveen, G., kashi Vishwanath, N., Sree, C. C., & Rohith, D. (2025). Climate-Resilient Urban Planning. *International Journal of Advances in Agricultural Science and Technology*, *12*(3), 226–234.

Priddy, L. P., Bly, P. G., & Flintsch, G. W. (2013). *Review of precast portland cement concrete panel technologies for use in expedient portland cement concrete airfield pavement repairs* (No. 13-2956).

Priya, P. K., & Neamitha, M. (2018). A review on precast concrete. *IACSIT International Journal of Engineering and Technology*, *5*(1).

Pucinotti, R., & De Lorenzo, R. A. (2005). In situ non-destructive testing: The steel and concrete resistance assessment of "ancient" r/c structures. *In Structural Studies, Repairs and Maintenance of Heritage Architecture* IX (p. 355).

PwC. (2025, November). *What's next in engineering and construction*. https://www.pwc.com/us/en/industries/industrial-products/library/engineering-and-construction-trends.html

Qiao, M., Ma, Y. X., Fung, T. C., & Tan, K. H. (2024). Experimental study on exterior precast concrete beam-to-CECFST column joints with wet connections under cyclic loading. *Journal of Building Engineering*, *88*, 109109. DOI: 10.1016/j.jobe.2024.109109

Qi, B., Razkenari, M., Costin, A., Kibert, C., & Fu, M. (2021). A systematic review of emerging technologies in industrialized construction. *Journal of Building Engineering*, *39*, 102265. DOI: 10.1016/j.jobe.2021.102265

Qiu, T., Chen, X., Su, D., Chen, K., He, Q., Zheng, Z., Zhou, W., & Wang, Y. (2024). LEGO®-inspired decision-making model for automatic construction of precast diaphragm walls. *Automation in Construction, 166*, 105667. DOI: 10.1016/j.autcon.2024.105667

Rahim, A. A., Hamid, Z. A., Zen, I. H., Ismail, Z., & Kamar, K. A. M. (2012). Adaptable housing of precast panel system in Malaysia. *Procedia: Social and Behavioral Sciences, 50*, 369–382. DOI: 10.1016/j.sbspro.2012.08.042

Rahman, M., & Sobuz, H. R. (2018, February). Comparative study of IPS & PPVC precast system—A case study of public housing buildings project in Singapore. In *Proceedings of the 4th International Conference on Civil Engineering for Sustainable Development (ICCESD 2018), KUET, Khulna, Bangladesh* (pp. 9-11).

Rajala, R., Hakanen, E., Mattila, J., Seppälä, T., & Westerlund, M. (2018). How Do Intelligent Goods Shape Closed-Loop Systems? *California Management Review, 60*(3), 20–44. DOI: 10.1177/0008125618759685

Raj, B., Jayakumar, T., & Rao, B. P. C. (1995). Non-Destructive Testing and Evaluation for Structural Integrity. *Sadhana: Academy Proceedings in Eng. The Sciences, 20*, 5–38.

Raj, B., & Venkatraman, B. (2013). Overview of Diagnostic and Monitoring Methods and Techniques. In Czichos, H. (Ed.), *Handbook of Technical Diagnostics* (pp. 43–68). Springer Verlag. DOI: 10.1007/978-3-642-25850-3_4

Ramos, N. M. M., Simoes, M. L., Delgado, J. M. P. Q., & Freitas, V. P. (2011). Applicability of Tensile Testing: Lessons from a Large Sample of In Situ Tests. *International Conference on Durability of Building Materials and Components, 12–15 April 2011, XII DBMC, Porto, Portugal.*

Rangarajan, S. S., Sharma, J., & Sundarabalan, C. K. (2020). Novel Exertion of Intelligent Static Compensator Based Smart Inverters for Ancillary Services in a Distribution Utility Network-Review. *Electronics (Basel), 9*(4), 662. DOI: 10.3390/electronics9040662

Rani, J., Guru, R., Singh, J., & Fangueiro, R. School of Engineering, University of Minho, Guimaraes, Portugal, Santhanam, S., & Department of Textile Technology, Park College of Engineering and Technology, Coimbatore, Tamilnadu, India. (2024). Applying Giloy Stems Dye to Silk and Lyocell Fabrics to Improve Their Antimicrobial and Fastness Properties Using Bio-Mordant. *Textile & Leather Review, 7*, 1197–1215. DOI: 10.31881/TLR.2024.117

Rani, L. Hurmat, Kaur, B., & Arora, R. (2025). Role of AI in Developing Smart Urban Cities and Resilience to Climate Change. In *Climate Resilience: Impact of Quantum Computing and Artificial Intelligence on Urban Planning* (pp. 187-208). Cham: Springer Nature Switzerland.

Rani, J. (2025). Cultural and Social Norms Supporting Sustainability. In Mokdad, M. (Ed.), *Social System Reforms to Achieve Global Sustainability* (pp. 43–78). IGI Global., DOI: 10.4018/979-8-3373-1280-4.ch003

Rani, J., Guru, R., & Santhanam, S. (2025). Case Studies on the Use of AI in Literary Research and Academic Studies. In Asrifan, A. (Ed.), (pp. 1–36). Advances in Computational Intelligence and Robotics. IGI Global Scientific Publishing., DOI: 10.4018/979-8-3373-6330-1.ch001

Rani, J., Guru, R., Singh, J., & Santhanam, S. (2024). Eco-Dyeing and Functional Finishing of Cotton Fabric Using a Natural Colour Derived From Lotus Seed: Enhanced Fastness Properties with Chitosan. *Textile & Leather Review*, *7*, 1039–1060. DOI: 10.31881/TLR.2024.099

Rani, J., Santhanam, S., Guru, R., & Seidu, R. K. (2025). Sustainable Finishing Processes: Enhancing Performance with Minimal Impact. In Jiang, S.-X., Seidu, R. K., & Tawiah, B. (Eds.), *Advances in Textile Materials and Processing Techniques for Sustainability* (pp. 161–199). Springer Nature Singapore., DOI: 10.1007/978-981-95-0469-5_6

Rashed, G. M., Mohrez, W. A., Abd Allah, M., & Ghayad, I. M. (2020). Corrosion behavior of super austenitic stainless steel 904L weldment in sulfuric acid media used for leaching application. *Key Engineering Materials*, *835*, 384–391. . DOI: 10.4028/www.scientific.net/KEM.835.384

Rasmussen, F., Andersen, C., Wittchen, A., Hansen, R., & Birgisdóttir, H. (2021). Environmental Product Declarations of Structural Wood: A Review of Impacts and Potential Pitfalls for Practice. *Buildings (Basel, Switzerland)*, *11*(8), 362. DOI: 10.3390/buildings11080362

Reed, B. (2007). Shifting from 'sustainability' to regeneration. *Building Research and Information*, *35*(6), 674–680. DOI: 10.1080/09613210701475753

Reffat, R. M., & Ahmad, R. M. (2020). Determination of optimal energy-efficient integrated daylighting systems into building windows. *Solar Energy*, *209*, 258–277. DOI: 10.1016/j.solener.2020.08.086

Reichenbach, S., & Kromoser, B. (2021). State of practice of automation in precast concrete production. *Journal of Building Engineering*, *43*, 102527. DOI: 10.1016/j.jobe.2021.102527

Reinales, D., Zambrana-Vásquez, D., & de Guinoa, A. S. (2020). Social Life Cycle Assessment of Product Value Chains Under a Circular Economy Approach: A Case Study in the Plastic Packaging Sector. *Sustainability (Basel)*, *12*(16), 6671. DOI: 10.3390/su12166671

Ren, H., Sun, Y., Albdoor, A. K., Tyagi, V. V., Pandey, A. K., & Ma, Z. (2021). Improving energy flexibility of a net-zero energy house using a solar-assisted air conditioning system with thermal energy storage and demand-side management. *Applied Energy*, *285*, 116433. DOI: 10.1016/j.apenergy.2021.116433

Ren, M., Yuan, Y., Cao, X., Sun, L., & Jiang, F. (2020). Numercial analysis on the thermal performance of capillary heat exchange system in metro running tunnel. *Energy and Built Environment*, *1*(2), 207–214. DOI: 10.1016/j.enbenv.2020.01.003

Resan, S. F., Chassib, S. M., Zemam, S. K., & Madhi, M. J. (2020). New approach of concrete tensile strength test. *Case Studies in Construction Materials*, *12*, e00347. DOI: 10.1016/j.cscm.2020.e00347

Rewlay-ngoen, C., & Papong, S. (2020). Environmental impact assessment of a rotary compressor in Thailand based on life cycle assessment methodology. *SN Applied Sciences*, *2*(9), 1526. Advance online publication. DOI: 10.1007/s42452-020-03278-w

Rezk, H., Kanagaraj, N., & Al-Dhaifallah, M. (2020). Design and Sensitivity Analysis of Hybrid Photovoltaic-Fuel-Cell-Battery System to Supply a Small Community at Saudi NEOM City. *Sustainability (Basel)*, *12*(8), 3341. DOI: 10.3390/su12083341

RILEM bulletin. (1987). Materials and Structures, 20, 79–80.

Riza, I. W., Latief, Y., & Machfudiyanto, R. A. (2020). Enhancing safety and comfort through mechanical component maintenance performance on green building using BIM-based information system. *IOP Conference Series. Materials Science and Engineering*, *830*(2), 22041. DOI: 10.1088/1757-899X/830/2/022041

Robati, M., Oldfield, P., Nezhad, A. A., Carmichael, D. G., & Kuru, A. (2021). Carbon value engineering: A framework for integrating embodied carbon and cost reduction strategies in building design. *Building and Environment*, *192*, 107620. DOI: 10.1016/j.buildenv.2021.107620

Robles, K., Yee, J., & Kee, S. (2022). Electrical resistivity measurements for non-destructive evaluation of chloride-induced deterioration of reinforced concrete: A review. *Materials (Basel)*, *15*(8), 2725. https://www.mdpi.com/1996-1944/15/8/2725

Rodrigues, M. G., & Júnior, F. (2020). Execution Manual and Preventive Maintenance in Residential Works. *International Journal of Advanced Engineering Research and Science*, *7*(5), 304–313. DOI: 10.22161/ijaers.75.37

Rodrigues, R., Gaboreau, S., Gance, J., Ignatiadis, I., & Betelu, S. (2021). Reinforced concrete structures: A review of corrosion mechanisms and advances in electrical methods for corrosion monitoring. *Construction & Building Materials*. Advance online publication. DOI: 10.1016/j.conbuildmat.2020.121240

Rothschild, K. A. (2025). An Analysis of Engineering Roles in Achieving System Integrity.

Rucka, M. (2020). Special Issue: Non-Destructive Testing of Structures. *Materials (Basel), 13*(21), 4996. DOI: 10.3390/ma13214996 PMID: 33171907

Rui, Y., & Sen, Z. (2018). The review of problems in precast construction activities. *The Journal of Social Sciences Research*, 173-176.

Ruiter, H., De Feijter, F., & Wagensveld, K. (2021). Management Control and Business Model Innovation in the Context of a Circular Economy in the Dutch Construction Industry. *Sustainability (Basel), 14*(1), 366. DOI: 10.3390/su14010366

Ryang, C. K., Hyuk, K. K., & Sup, W. Y. (2004). New Precast Technologies for Building in Japan. *Magazine of the Korea Concrete Institute, 16*(5), 20–25.

Saaly, M., Maghoul, P., & Holländer, H. (2020). Investigation of the effects of heat loss through below-grade envelope of buildings in urban areas on thermo-mechanical behaviour of geothermal piles. *E3S Web of Conferences, 205*, 5010. DOI: 10.1051/e3sconf/202020505010

Saaty, T. L. (1980). *The analytic hierarchy process*. McGraw-Hill.

Sabbağ, N. (2016). Determination of Concrete Quality by Geophysical Methods. PhD Thesis, *Süleyman Demirel University Geophysical Engineering*, Isparta, Türkiye, 172 p.

Sabbağ, N., & Uyanık, O. (2016). Time-Dependent Variation of Seismic Velocities in Low-Strength Concrete. [TOJSAT]. *The Online Journal of Science and Technology, 6*(4), 49–57.

Sabbağ, N., & Uyanık, O. (2017). Prediction of Reinforced Concrete Strength by Ultrasonic Velocities. *Journal of Applied Geophysics, 141*, 13–23. DOI: 10.1016/j.jappgeo.2017.04.005

Sabbağ, N., & Uyanık, O. (2018a). Determination of Reinforced Concrete Strength by Electrical Resistivity Depending on Curing Conditions. *Journal of Applied Geophysics, 155*, 13–25. DOI: 10.1016/j.jappgeo.2018.03.007

Sabbağ, N., & Uyanık, O. (2018b). Comparison of Static and Dynamic Elastic Parameters of Saturated Concretes. *Afyon Kocatepe University Journal of Science and Engineering Sciences, 18*(3), 1181–1189. DOI: 10.5578/fmbd.67588

Sabbağ, N., & Uyanık, O. (2018c). Determination of Reinforced Dry Concrete Strength by Electrical Resistivity Method. *Pamukkale University Journal of Engineering Sciences, 24*(6), 1230–1236. DOI: 10.5505/pajes.2017.76702

Sabbağ, N., & Uyanık, O. (2020). Determination of the anisotropy of reinforced concretes by the resistivity method. *Pamukkale University Journal of Engineering Sciences, 26*(3), 572–580. DOI: 10.5505/pajes.2019.00008

Sacks, R., Eastman, C. E., & Lee, G. (2003). Process improvements in precast concrete construction using top-down parametric 3-D computer modeling. *PCI Journal, 48*(3), 46–55. DOI: 10.15554/pcij.05012003.46.55

Sacks, R., Eastman, C. M., & Lee, G. (2004). Process model perspectives on management and engineering procedures in the precast/prestressed concrete industry. *Journal of Construction Engineering and Management, 130*(2), 206–215. DOI: 10.1061/(ASCE)0733-9364(2004)130:2(206)

Sádaba, J., Luzarraga, A., & Lenzi, S. (2025). Designing for climate adaptation: A case study integrating nature-based solutions with urban infrastructure. *Urban Science (Basel, Switzerland), 9*(3), 74. DOI: 10.3390/urbansci9030074

Sadeghi, F., Zhu, X., Li, J., & Rashidi, M. (2021). Integrated remote sensing technologies for condition assessment of concrete bridges. *Remote Sensing, 13*(17), 3377. DOI: 10.3390/RS13173377

Sadeghi, H., & Singh, R. M. (2023). Driven precast concrete geothermal energy piles: Current state of knowledge. *Building and Environment, 228*, 109790. DOI: 10.1016/j.buildenv.2022.109790

Sadrolodabaee, P., Claramunt, J., Monserrat-López, A., Aguado, A., & de la Fuente, A. (2025). Low-carbon cementitious paving systems with ladle furnace slag and fine recycled aggregate: A multi-scale experimental study. *Construction & Building Materials, 490*, 142547. DOI: 10.1016/j.conbuildmat.2025.142547

Sæthre, S. A., Torgersen, G. S., & Gjevestad, J. G. O. (2021). Assessing the Dilemma of Traditional Flood Protection Measures versus Maintaining River Ecosystems: A Case Study of the River Lærdal in Norway. *Water (Basel), 13*(14), 1884. DOI: 10.3390/w13141884

Sajid, S., & Çakmak, L. (2019). Impulse response test for condition assessment of concrete: A review. Construction and Building Materials. https://www.sciencedirect.com/science/article/pii/S0950061819306658

Salman, M., Najid, S. V. P., Ali, U., Khairoowala, Z., & Siddiquee, T. A. R. (2025). Building for Tomorrow: Climate-Resilient Infrastructure Strategies for Urban Sustainability. In *Urbanization, Climate Change, and Health: Integrating Strategies for Sustainable and Resilient Cities* (pp. 291–311). Springer Nature Switzerland. DOI: 10.1007/978-3-031-99132-5_14

Salmeron, M., Criner, N., Zhang, X., Dyke, S. J., Ramirez, J. A., Wogen, B. E., & Rearick, A. (2022). Predictive analytics for holistic lifecycle modeling of concrete bridge decks with construction defects. *Journal of Bridge Engineering*, *29*(9). Advance online publication. DOI: 10.1061/JBENF2.BEENG-6636

Samii, R., Van Wassenhove, L. N., & Bhattacharya, S. (2002). An Innovative Public–Private Partnership: New Approach to Development. *World Development*, *30*(6), 991–1008. DOI: 10.1016/S0305-750X(02)00015-3

Samoita, D., Nzila, C., Østergaard, P. A., & Remmen, A. (2020). Barriers and Solutions for Increasing the Integration of Solar Photovoltaic in Kenya's Electricity Mix. *Energies*, *13*(20), 5502. DOI: 10.3390/en13205502

Samson, D., & Omoniyi, T. M. (2014). Correlation between Non-Destructive and Destructive Tests of Concrete. *International Journal of Engineering and Science Invention*, *3*(9), 12–17.

Sanchez, L., Fournier, B., Drimalas, T., Bastien, J., Mitchell, D., & Noel, M. (2020). Semi-quantitative condition assessment of concrete distress through the damage rating index. *Cement*, *1–2*, 100001. DOI: 10.1016/J.CEMENT.2020.100001

Sangoremi, A. A., Abosede, O. O., & Adeleke, A. E. (2025). A review on circular economy: Precursor to achieving global sustainability. *Asian Research Journal of Current Science*, *10*(3), 309–324.

Santos, A. F., Gaspar, P. D., & de Souza, H. J. L. (2020). New HVAC Sustainability Index—TWI (Total Water Impact). *Energies*, *13*(7), 1590. DOI: 10.3390/en13071590

Santos, P., Cervantes, G. C., Zaragoza-Benzal, A., Byrne, A., Karaca, F., Ferrández, D., Salles, A., & Bragança, L. (2024). Circular material usage strategies and principles in buildings: A review. *Buildings (Basel, Switzerland)*, *14*(1), 281. DOI: 10.3390/buildings14010281

Sarkar, D., Dhaneshwar, D., & Raval, P. (2023). Automation in monitoring of construction projects through BIM-IoT-blockchain model. *Journal of The Institution of Engineers (India): Series A*, *104*(2), 317-333.

Sarkar, D., Patel, H., & Dave, B. (2022). Development of integrated cloud-based Internet of Things (IoT) platform for asset management of elevated metro rail projects. *International Journal of Construction Management*, *22*(10), 1993–2002. DOI: 10.1080/15623599.2020.1762035

Satoła, D., Balouktsi, M., Lützkendorf, T., Wiberg, A. H., & Gustavsen, A. (2021). How to define (net) zero greenhouse gas emissions buildings: The results of an international survey as part of IEA EBC annex 72. *Building and Environment*, *192*, 107619. DOI: 10.1016/j.buildenv.2021.107619

Satoła, D., Wiberg, A. H., Singh, M., Babu, S., James, B., Dixit, M., Sharston, R., Grynberg, Y., & Gustavsen, A. (2022). Comparative review of international approaches to net-zero buildings: Knowledge-sharing initiative to develop design strategies for greenhouse gas emissions reduction. *Energy Sustainable Development. Energy for Sustainable Development, 71*, 291–306. DOI: 10.1016/j.esd.2022.10.005

Sberna, A., Deb, A., Di Trapani, F., & Calvi, J. (2025). Reliability-based seismic retrofitting design methodology for non-ductile reinforced concrete frame structures. Probabilistic Engineering Mechanics. https://www.sciencedirect.com/science/article/pii/S0266892025000906

Schaarschmidt, M., Westerkamp, C., Hennewig, A., Pieper, D., Speckmann, H., & Bisle, W. (2015). Structural health monitoring: Overview on technologies under development. In Structural Health Monitoring 2015 (Vol. 2, pp. 1181–1188). DOI: 10.12783/SHM2015/148

Scheer, S., Santos, A., Quevedo, J. R. S., Filho, J. M., & Boiça, S. M. R. (2005). IT as a Success Factor for a Lean Construction System in a Precast Concrete Construction Industry. In *Computing in Civil Engineering (2005)* (pp. 1-12). DOI: 10.1061/40794(179)160

Schipper, H. R. (2015). Double-curved precast concrete elements.

Schramm. (n.d.). *StoLotusan – The facade paint with the Lotus Effect*. https://www.schramm.de/en/stolotusan-the-facade-paint-with-the-lotus-effect/

Seidu, S., Chan, D. W., Edwards, D. J., & Owusu-Manu, D. G. (2025). A systematic review of green infrastructure and ecosystem-based technologies for climate resilience: Development trends, research outlook and future projections. *Environmental Research Letters*, *20*(8), 083001. DOI: 10.1088/1748-9326/ade730

Selcuk, M. (2015). Investigation of Mechanical Properties of Concrete by Triaxial Compression Test, Master Thesis, ITU.

Senthilvel, M., Varghese, K., & Ramesh Babu, N. (2016). Building information modeling for precast construction: a review of research and practice. In *Construction Research Congress 2016* (pp. 2250-2259). DOI: 10.1061/9780784479827.224

Seo, J., Hu, J., & Lee, J. (2016). Summary review of structural health monitoring applications for highway bridges. *Journal of Performance of Constructed Facilities*, *30*(4). Advance online publication. DOI: 10.1061/(ASCE)CF.1943-5509.0000824

Sepasgozar, S., Khan, A., Smith, K., Romero, J., Shen, X., Shirowzhan, S., Li, H., & Tahmasebinia, F. (2023). BIM and Digital Twin for Developing Convergence Technologies as Future of Digital Construction. *Buildings (Basel, Switzerland)*, *13*(2), 441. DOI: 10.3390/buildings13020441

Sethi, K. C., Prajapati, U., Parihar, A., Gupta, C., Shrivastava, G., & Sharma, K. (2024). Development of optimization model for balancing time, cost, and environmental impact in retrofitting projects with NSGA-III. *Asian Journal of Civil Engineering, 25*(7), 5123–5136. DOI: 10.1007/S42107-024-01102-Z

Shahbazi, S., Wiktorsson, M., Kurdve, M., Jönsson, C., & Bjelkemyr, M. (2016). Material efficiency in manufacturing: Swedish evidence on potential, barriers and strategies. *Journal of Cleaner Production, 127*, 438–450. DOI: 10.1016/j.jclepro.2016.03.143

Shao, C., Gu, C., Meng, Z., & Han, Y. (2019). A data-driven approach based on multivariate copulas for quantitative risk assessment of concrete dam. *Journal of Marine Science and Engineering*, *7*(10), 353. https://www.mdpi.com/2077-1312/7/10/353

Sharma, K. M., & Srivastava, R. (2025). Risk mitigation and safety assurance in mechanical and civil engineering projects. *International Journal for Research in Applied Science and Engineering Technology, 13*(9), 714–719. DOI: 10.22214/ijraset.2025.74100

Sharma, V., & Akhai, S. (2019). Trends in Utilization of Coal Fly Ash in India: A Review. *Journal of Engineering Design and Analysis*, *2*(1), 12–16.

Shehu, M. O., Almeida, R., & Mahapatra, K. (2025). Opportunities for digital tracking technologies in the precast concrete sector in Sweden. *Frontiers in Built Environment, 11*, 1566784. DOI: 10.3389/fbuil.2025.1566784

Sheik, S., Tirumalla, A., Gurrala, A. K., & Mohammed, R. (2022). Effect of microstructural morphology on corrosion susceptibility of austenitic and super austenitic stainless steels. *Materials Today: Proceedings, 66*, 514–518. DOI: 10.1016/j.matpr.2022.06.113

Shen, W., Chen, X., Qiu, J., Hayward, J. A., Sayeef, S., Osman, P., Meng, K., & Dong, Z. Y. (2020). A comprehensive review of variable renewable energy levelized cost of electricity [Review of *A comprehensive review of variable renewable energy levelized cost of electricity*]. *Renewable and Sustainable Energy Reviews, 133*, 110301. Elsevier BV. DOI: 10.1016/j.rser.2020.110301

Sherfudeen, A. P., Kumar, N., Raghavan, N., Pillai, R. G., & Kalidindi, S. N. (2016). Promoting precast concrete for affordable housing–An overview on promotional policies worldwide and challenges and possibilities in India. *Indian Concrete Journal, 90*(5), 13–25.

Shrestha, B., & Giri, O. P. (2023). Study on concrete compressive strength through destructive and non-destructive testing. *Int J Sci Math Technol Learn, 31*(2), 336–356.

Shukla, B. K., Yadav, H., Singh, S., Verma, S., Shukla, A. K., & Sharma, C. (2025). Sustainable engineering of recycled aggregate concrete: Structural performance and environmental benefits under circular economy frameworks. Construction Materials, 5(3), Article 67.

Sieffert, Y., Huygen, J. M., & Daudon, D. (2014). Sustainable construction with repurposed materials in the context of a civil engineering–architecture collaboration. *Journal of Cleaner Production*, *67*, 125–138. DOI: 10.1016/j.jclepro.2013.12.018

Singh, A. (2020). Evaluation of Performance of Photovoltaic Thermal Hybrid System. *International Journal of Engineering Research & Technology (Ahmedabad)*, *5*(5). Advance online publication. DOI: 10.17577/IJERTV9IS050559

Singhal, S., Chourasia, A., Chellappa, S., & Parashar, J. (2019). Precast reinforced concrete shear walls: State of the art review. *Structural Concrete*, *20*(3), 886–898. DOI: 10.1002/suco.201800129

Singh, M., & Maini, R. (2020). Various Electricity Load Forecasting Techniques with Pros and Cons. [IJRTE]. *International Journal of Recent Technology and Engineering*, *8*(6), 220–229. DOI: 10.35940/ijrte.F6997.038620

Singh, N., & Rani, J. (2024). Experimental Retail: Conceptualizing Immersive and Memorable Shopping Journeys. In Rajput, S., Malik, F. A., Mahajan, S., & Javed, A. (Eds.), (pp. 31–50). Advances in Logistics, Operations, and Management Science. IGI Global., DOI: 10.4018/979-8-3693-7620-1.ch003

Singh, R. P., & Raghubanshi, A. S. (2020). 'Green building' movement in India: Study on institutional support and regulatory support. In *Urban Ecology* (pp. 435–455). Elsevier., DOI: 10.1016/B978-0-12-820730-7.00022-7

Singh, R., Gehlot, A., Akram, S. V., Gupta, L. R., Jena, M. K., Prakash, C., Singh, S., & Kumar, R. (2021). Cloud manufacturing, internet of things-assisted manufacturing and 3D printing technology: Reliable tools for sustainable construction. *Sustainability (Basel)*, *13*(13), 7327. DOI: 10.3390/su13137327

Sirivivatnanon, V., Xue, C., & Khatri, R. (2023). Long-term reinforcement corrosion in low carbon concrete with a high volume of SCMs exposed to NaCl solutions and field marine environment. *Construction & Building Materials*, *393*, 132071. DOI: 10.1016/j.conbuildmat.2023.132071

Skrzypczak, I. (2023). Statistical quality inspection methodology in production of precast concrete elements. *Materials (Basel)*, *16*(1), 431. DOI: 10.3390/ma16010431 PMID: 36614770

Smirnova, E., Larionova, Y., & Larionov, A. (2020). Sewer system optimization in housing and communal services. *E3S Web of Conferences*, *157*, 2002. DOI: 10.1051/e3sconf/202015702002

Smith, R. M. (2015). "Green" building in India: A comparative and spatial analysis of the LEED-India and GRIHA rating systems. *Asian Geographer, 32*(2), 73–84. DOI: 10.1080/10225706.2015.1020065

Soda, P. R. K., Chakravarthi, E. K., Mogal, A., & Mini, K. M. (2022). Statistical and experimental investigation on self-healing of microcracks in cement mortar by encapsulation of calcite precipitating bacteria into expanded perlite. *Construction & Building Materials, 342*, 127985. DOI: 10.1016/j.conbuildmat.2022.127985

Sohn, H., Farrar, C., Hemez, F., Shunk, D., Stinemates, D., & Nadler, B. (1999). A review of structural health monitoring literature: 1996–2001. IEE Colloquium on Condition Monitoring Machinery, External Structures and Health, 34, 3–6. DOI: 10.1049/IC:19990185

Solanki, A., & Sarkar, D. (2022, August). A Framework of Internet of Things–Cloud–Building Information Modeling Based Life Cycle Management of Precast Components in Prefabricated Buildings. In *International Conference Innovation in Smart and Sustainable Infrastructure* (pp. 671-681). Singapore: Springer Nature Singapore.

Sommer, T., Sulzer, M., Wetter, M., Sotnikov, A., Mennel, S., & Stettler, C. (2020). The reservoir network: A new network topology for district heating and cooling. *Energy, 199*, 117418. DOI: 10.1016/j.energy.2020.117418

Sommese, F. (2024). Nature-based solutions to enhance urban resilience in the climate change and post-pandemic era: A taxonomy for the built environment. *Buildings (Basel, Switzerland), 14*(7), 2190. DOI: 10.3390/buildings14072190

Song, Y.-K., Jo, Y.-H., Lim, Y.-J., Cho, S.-Y., Yu, H.-C., Ryu, B.-C., Lee, S.-I., & Chung, C.-M. (2013). Sunlight-Induced Self-Healing of a Microcapsule-Type Protective Coating. *ACS Applied Materials & Interfaces, 5*(4), 1378–1384. DOI: 10.1021/am302728m PMID: 23373694

Soudian, S., Berardi, U., & Laschuk, N. O. (2020, January 1). Effects of Long-Term UV Exposure on the Performance of Cement Plasters Integrated with Thermochromic Paint and PCMs for Building Facade Applications. *XV International Conference on Durability of Building Materials and Components. eBook of Proceedings*. DOI: 10.23967/dbmc.2020.163

Sporr, A., Zucker, G., & Hofmann, R. (2020). Automatically Creating HVAC Control Strategies Based on Building Information Modeling (BIM): Heat Provisioning and Distribution. *Energies, 13*(17), 4403. DOI: 10.3390/en13174403

Sree, P. U., Kumar, K. N., Kumar, G. R., Manohar, K., Sampath, G., & Sumfoniya, J. (2025). AI-Artificial Intelligence Powered Autonomous Vehicle Control System. *International Journal of Computer Technology and Electronics Communication, 8*(1), 10055–10061.

Sridhar, K. V., Yamini, & Cheran. (2023, June). State-of-the-art review on precast composite sandwich slab panels. In *5TH International Conference On Innovative Design, Analysis & Development Practices In Aerospace & Automotive Engineering: I-DAD'22* (Vol. 2766, No. 1, p. 020079). AIP Publishing LLC.

Sridharan, A., Mohan, P. K., Vellaiyan, P., & Ibrahim, M. T. (2020). An experimental investigation on the performance of novel hybrid heat exchanger over conventional shell and tube heat exchanger. *AIP Conference Proceedings, 2240*, 40001. DOI: 10.1063/5.0011074

Srivastava, R., & Thakur, A. (2025). The Foundational Principles of FDM: An Acronym Pivotal t Its Operation and Its Wide Range of Applications. In *Modeling, Analysis, and Control of 3D Printing Processes* (pp. 355-380). IGI Global Scientific Publishing.

Srivastava, R., Thakur, A., & Rana, A. (2024). The Basic Understanding of Nano Materials: A Review. *Innovations in Nanomaterials-Based Corrosion Inhibitors*, 1-20.

Srivastava, R., Arun, J., & Patel, D. K. (2019, April). Amalgamating the Service Quality Aspect in Supply Chain Management. In *2019 International Conference on Automation, Computational and Technology Management (ICACTM)* (pp. 63-67). IEEE. DOI: 10.1109/ICACTM.2019.8776839

Srivastava, V. K., Maurya, L., & Mehra, E. R. (2014). Detection of noise in high pass Butterworth IIR Filter using MATLAB. *IJERT, 3*(2), 1232–1235.

Stamponi, E., Lattanzi, N., Giorgini, F., Serpilli, F., Montelpare, S., & Moretti, E. (2021). Energy analysis, numerical simulations and intervention proposals for a NZEB industrial building: the "Loccioni Leaf Lab" case study. *E3S Web of Conferences, 238*, 6004. DOI: 10.1051/e3sconf/202123806004

Stanley, J. (1995). *Nondestructive Testing Handbook 9: Special Nondestructive Testing Methods*. American Society for Nondestructive Testing.

Stefanidou, S., & Kappos, A. (2021). Fragility-informed selection of bridge retrofit scheme based on performance criteria. Engineering Structures. https://www.sciencedirect.com/science/article/pii/S0141029621001267

Straatman, R., & Vambersky, J. N. J. A. (2001). Precast construction and environment. *Structural Concrete, 2*(2), 93–98. DOI: 10.1680/stco.2001.2.2.93

Sua-iam, G., & Makul, N. (2024). Potential future direction of the sustainable production of Precast concrete with recycled concrete aggregate: A critical review. *Engineering and Science, 28*, 1075.

Suamir, I. N., Sudirman, S., Ardita, I. N., & Santanu, G. (2020). Experimental and numerical optimization on chilled water configuration for improving temperature performance and economic viability of a centralized chiller plant. *Journal of Physics: Conference Series, 1450*(1), 12106. DOI: 10.1088/1742-6596/1450/1/012106

Sudarsan, R., Fenves, S. J., Sriram, R. D., & Wang, F. (2005). A product information modeling framework for product lifecycle management. *Computer Aided Design, 37*(13), 1399–1411. DOI: 10.1016/j.cad.2005.02.010

Sufyan, M., Rahim, N. A., Muhammad, M. A., Tan, C., Raihan, S. R. S., & Bakar, A. H. A. (2020). Charge coordination and battery lifecycle analysis of electric vehicles with V2G implementation. *Electric Power Systems Research, 184*, 106307. DOI: 10.1016/j.epsr.2020.106307

Sukor, N. R., Shamsuddin, A. H., Mahlia, T. M. I., & Isa, M. F. M. (2020). Techno-Economic Analysis of CO2 Capture Technologies in Offshore Natural Gas Field: Implications to Carbon Capture and Storage in Malaysia. *Processes (Basel, Switzerland), 8*(3), 350. DOI: 10.3390/pr8030350

Sun, K., Specian, M., & Hong, T. (2020). Nexus of thermal resilience and energy efficiency in buildings: A case study of a nursing home. *Building and Environment, 177*, 106842. DOI: 10.1016/j.buildenv.2020.106842

Sunny, K. T., Korra, N. N., Muthukumaran, V., & Arivazhagan, B. (2022). Corrosion studies and thermogravimetric analysis performed on activated tungsten inert gas and hot wire tungsten inert gas welded super austenitic stainless steel AISI 904L. *Proceedings of the Institution of Mechanical Engineers. Part C, Journal of Mechanical Engineering Science, 236*(20), 10528–10540. DOI: 10.1177/09544062221106120

Swadhi, R. (2025). Innovative strategies for widespread adoption in a climate-smart future: Scaling up agroforestry. In *Agroforestry for a Climate-Smart Future* (pp. 473–496). IGI Global Scientific Publishing. DOI: 10.4018/979-8-3693-8282-0.ch016

Syed, A., & Sonparote, R. (2020). A review of precast concrete pavement technology. *The Baltic Journal of Road and Bridge Engineering, 15*(4), 22–53. DOI: 10.7250/bjrbe.2020-15.493

Takada, T. (2020). Study of CIM and IoT-Simulation for Cost Performance Analysis-(Cost Management). , (2), 4-25.

Talaat, A. A., Emad, A., Tarek, A., Masbouba, M., Essam, A., & Kohail, M. (2021). Factors affecting the results of concrete compression testing: A review. *Ain Shams Engineering Journal, 12*(1), 205–221. DOI: 10.1016/j.asej.2020.07.015

Tam, I. C. K., & Agnew, B. (2020). Thermal Systems—An Overview. *Energies, 14*(1), 175. DOI: 10.3390/en14010175

Tam, V. W. Y., & Hao, J. J. L. (2019). Adaptive reuse in sustainable development. *International Journal of Construction Management, 19*(6), 509–521. DOI: 10.1080/15623599.2018.1459154

Taneja, T., Bhatia, B. S., & Akhai, S. (2024, December). Assessing Groundwater Quality in Rampura-Bathinda for Physicochemical Parameters & Heavy Metal Concentrations using Electronic Equipment. In *2024 1st International Conference on Sustainability and Technological Advancements in Engineering Domain (SUSTAINED)* (pp. 791-796). IEEE. DOI: 10.1109/SUSTAINED63638.2024.11073880

Taneja, T., Bhatia, B. S., & Akhai, S. (2025). Agricultural Resilience: Water Quality and Human Well-Being. *Emerging Smart Agricultural Practices Using Artificial Intelligence*, 1-23. DOI: 10.1002/9781394274277.ch1

Tan, Q., Zhu, H., Yang, S., Yang, X., & Ou, L. (2024). Precast assembled road paving technology: Progress and prospects. *Materials (Basel), 17*(10), 2245. DOI: 10.3390/ma17102245 PMID: 38793312

Tan, S., Poon, J., Chan, R., & Chia, D. (2020). Retrofitting of reinforced concrete beam-column via steel jackets against close-in detonation. *International Journal of Engineering, 33*(5), 770–783. DOI: 10.5829/ije.2020.33.05b.08

Tanwar, N., & Akhai, S. (2017). Survey Analysis for Quality Control Comfort Management in Air Conditioned Classroom. *Journal of Advanced Research in Civil and Environmental Engineering, 4*(1&2), 20–23. DOI: 10.24321/2393.8307.201702

Tapia, C., & Padgett, J. (2016). Multi-objective optimisation of bridge retrofit and post-event repair selection to enhance sustainability. *Structure and Infrastructure Engineering, 12*(1), 93–107. DOI: 10.1080/15732479.2014.995676

Tarussov, A., Vandry, M., & De La Haza, A. (2013). Condition assessment of concrete structures using a new analysis method: Ground-penetrating radar computer-assisted visual interpretation. Construction and Building Materials. https://www.sciencedirect.com/science/article/pii/S0950061812003819

Tayabji, S., Ye, D., & Buch, N. (2013). Precast concrete pavements: Technology overview and technical considerations. *PCI Journal, 58*(1), 112–128. DOI: 10.15554/pcij.01012013.112.128

Teisserenc, B., & Sepasgozar, S. (2021). Adoption of Blockchain Technology through Digital Twins in the Construction Industry 4.0: A PESTELS Approach. *Buildings (Basel, Switzerland), 11*(12), 670. DOI: 10.3390/buildings11120670

Thakuri, P. K., Alkki, L., & Aarikka-Stenroos, L. (2024). Digital technologies enabling component reuse in circular value chains: Using digital twin, internet of things and robots in construction and manufacturing sectors. *R & D Management.*

Thareja, P., & Akhai, S. (2017). Processing parameters of powder aluminium-fly ash P/M composites. *Journal of advanced research in manufacturing, material science & metallurgical engineering, 4*(3&4), 24-35.

Thareja, P., & Akhai, S. (2016). Processing Aluminum Fly Ash Composites via Parametric Analysis of Stir Casting. *Journal of Advanced Research in Manufacturing. Material Science & Metallurgical Engineering, 3*(3&4), 21–28.

The Raycat Solution. (2016, November 27). *Biobricks and how to start hacking DNA*. http://www.theraycatsolution.com/news/2016/11/27/biobricks-and-how-to-start-hacking-dna

Thériault, F., Noël, M., & Sanchez, L. (2022). Simplified approach for quantitative inspections of concrete structures using digital image correlation. Engineering Structures. https://www.sciencedirect.com/science/article/pii/S0141029621018113

Tiammee, S., & Likasiri, C. (2020). Sustainability in corn production management: A multi-objective approach. *Journal of Cleaner Production, 257*, 120855. DOI: 10.1016/j.jclepro.2020.120855

Tian, M. (2021, November). Monitoring Method of Vertical Stress on Precast Concrete of Prefabricated Building. In *International Conference on Forthcoming Networks and Sustainability in the IoT Era* (pp. 220-227). Cham: Springer International Publishing.

Tibbetts, C. M., Riding, K. A., & Ferraro, C. C. (2021). A critical review of the testing and benefits of permeability-reducing admixtures for use in concrete. *Cement, 6*, 100016. DOI: 10.1016/j.cement.2021.100016

Tikalsky, P. J. (2006). Overview of Non-Destructive Testing (NDT) Material Evaluation Methods, *NSF/RILEM Workshop*, July 10-14, 2006, Prague, Czech Republic.

Timofeevsky, A. L., Sulin, A. B., & Rubtsov, A. K. (2020). Calculation of the required coolant mass taking account the dynamics of the object heat load and technical characteristics of the chiller. *AIP Conference Proceedings, 2296*, 30001. DOI: 10.1063/5.0027418

Tingsanchali, T. (2012). Urban flood disaster management. *Procedia Engineering, 32*, 25–37. DOI: 10.1016/j.proeng.2012.01.1233

Tiwari, V. (2020). Inverter Control Performance Study on Solar PV Systems and Its Effect on Power Quality. *SMART MOVES JOURNAL IJOSCIENCE, 6*(7). Advance online publication. DOI: 10.24113/ijoscience.v6i7.337

Tomek, R. (2017). Advantages of precast concrete in highway infrastructure construction. *Procedia Engineering, 196*, 176–180. DOI: 10.1016/j.proeng.2017.07.188

Tong, Z., Cao, T., Zang, G., Hu, S., Liu, G., & Wang, Y. (2020). Performance analysis of capillary front-end heat exchanger for subway tunnel. *Applied Thermal Engineering*, *174*, 115360. DOI: 10.1016/j.applthermaleng.2020.115360

Torgersen, G., Bjerkholt, J. T., & Lindholm, O. G. (2014). Addressing flooding and SUDS when improving drainage and sewerage systems-A comparative study of selected Scandinavian cities. *Water (Basel)*, *6*(4), 839–857. DOI: 10.3390/w6040839

Tozan, B., Hoxha, E., Stapel, E. B., & Birgisdóttir, H. (2025). Prioritizing low-carbon materials and reducing material quantities to mitigate embodied GHG emissions in buildings. *Sustainable Cities and Society*, *131*, 106770. DOI: 10.1016/j.scs.2025.106770

Trach, R., Moshynskyi, V., Chernyshev, D., & Bushuyev, O. (2022). Modeling the quantitative assessment of the condition of bridge components made of reinforced concrete using ANN. *Sustainability*, *14*(23), 15779. https://www.mdpi.com/2071-1050/14/23/15779

Tran, V. V., Park, D., & Lee, Y. (2020). Indoor Air Pollution, Related Human Diseases, and Recent Trends in the Control and Improvement of Indoor Air Quality [Review of *Indoor Air Pollution, Related Human Diseases, and Recent Trends in the Control and Improvement of Indoor Air Quality*]. *International Journal of Environmental Research and Public Health, 17*(8), 2927. Multidisciplinary Digital Publishing Institute. DOI: 10.3390/ijerph17082927

Tripathi, A., Kaushal, I., Revolti, A., Benedetti, L., & Dallasega, P. (2025). Digital Twin Driven Quality Control in Smart Mobile Factories: A Framework in Precast Manufacturing. *Procedia CIRP*, *134*, 801–806. DOI: 10.1016/j.procir.2025.02.194

Trubina, N., Leindecker, G., Askar, R., Karanafti, A., Gómez-Gil, M., Blázquez, T., . . . Bragança, L. (2024, May). Digital technologies and material passports for circularity in buildings: an in-depth analysis of current practices and emerging trends. In *International Conference" Coordinating Engineering for Sustainability and Resilience"* (pp. 690-699). Cham: Springer Nature Switzerland. DOI: 10.1007/978-3-031-57800-7_64

Tsoka, S., & Tsikaloudaki, K. (2024). Design for circularity, design for adaptability, design for disassembly. In *Circular economy design and management in the built environment: A critical review of the state of the art* (pp. 257–272). Springer Nature.

Tungnung, K. (2020). Parametric strategy on passive design affordances in cool-humid climate zones of North-East India. *Solar Energy*, *207*, 1–9. DOI: 10.1016/j.solener.2020.06.024

Turai, V. R., & Patil, A. R. (2022). Review on precast concrete technology vs. cast-in-place concrete. *International Journal of Civil Engineering*, *9*(3), 27–32.

Turkish Ready Mixed Concrete Association (THBB). (2020). *Izmir Earthquake Investigation Report*. Earthquake Investigation Reports.

U.S. Environmental Protection Agency (EPA). (2021). *Combined Sewer Overflow (CSO) Control Policy Implementation Guidance*. Office of Water.

U.S. Environmental Protection Agency (EPA). (2025). *Low Impact Development Technical Guidance and Regulatory Compliance*. Office of Water.

UNDP. (2023). From ruins to resilience on the road to recovery in Ukraine. https://www.undp.org/eurasia/blog/ruins-resilience-road-recovery-ukraine

UNESCO. (2023). World Heritage: Reconstruction and Recovery. https://whc.unesco.org/en/review/86

UNESCO. (2025). *The UNESCO Global Report on Cultural Policies*. https://www.unesco.org/en/culture/global-report/global-report-cultural-policies?hub=150011

UN-Habitat. (2022). *World cities report 2022: Envisioning the future of cities*. https://unhabitat.org/wcr/

United Nations Environment Programme. (2023). *Global status report for buildings and construction 2023*. https://www.unep.org/resources/report/global-status-report-buildings-and-construction

United Nations. (2015). *Transforming our world: The 2030 Agenda for Sustainable Development*. https://www.un.org/ga/search/view_doc.asp?symbol=A/RES/70/1

Unlu, D. (2010). Rapid bridge construction technology: precast elements for substructures.

Ürge-Vorsatz, D., Khosla, R., Bernhardt, R., Chan, Y. C., Vérez, D., Hu, S., & Cabeza, L. F. (2020). Advances Toward a Net-Zero Global Building Sector. *Annual Review of Environment and Resources*, *45*(1), 227–269. DOI: 10.1146/annurev-environ-012420-045843

Uyanık, O. (1999). Kayaçlarda Sismik Hızlar ve Kayma Direncinin İncelenmesi. *52. Türkiye Jeoloji Kurultayı*, 63-70, Ankara.

Uyanık, O. (2014). Jeofizik ve Klasik Yöntemlerle Yapı İncelemeleri (Yapı Jeofiziği). JFMO Eğitim Yayınları No:19 ISBN:978-605-01-0643-5.

Uyanık, O. (1991). *Kaya Mekaniği ve Jeofizik Laboratuar Parametrelerinin İlişkilendirme Açısından Önemi*. Dokuz Eylül Üniv. Müh.-Mim. Fak. Jeofizik Müh. Böl.

Uyanık, O. (2012). Determination of Concrete Strength from Seismic Velocities. *Geophysical Bulletin*, *23*(70), 25–30.

Uyanık, O. (2015). Deprem ağır hasar alanlarının önceden belirlenmesi ve şehir planlaması için makro ve mikro bölgelendirmelerin önemi. *Süleyman Demirel Üniversitesi Fen Bilimleri Enstitüsü Dergisi, 19*(2), 24–38.

Uyanık, O., & Çatlıoğlu, B. (2015). Determination of Density from Seismic Velocities. *Jeofizik, 17*(1-2), 3–15.

Uyanık, O., Çatlıoğlu, B., Sabbağ, N., Öncü, Z., & Uyanık, N. A. (2012a). Determination of Concrete Quality of Reinforced Concrete Structures in Urban Transformation Projects from Seismic Ultrasonic Velocities. *1st Earth Sciences Symposium, SDU*, Türkiye.

Uyanık, O., Çatlıoğlu, B., Sabbağ, N., Öncü, Z., & Uyanık, N. A. (2012b). Relationships Between Rock Properties and Seismic Ultrasonic Velocities. *1st Earth Sciences Symposium, SDU*, Türkiye.

Uyanık, O., Kaptan, K., Gülay, F. G., & Tezcan, S. (2011). Determination of Concrete Strength by Non-Destructive Ultrasonic Method. *Construction World, 184*, 55–58.

Uyanık, O., Sabbağ, N., Çatlıoğlu, B., Uyanık, N. A., & Öncü, Z. (2012c). Determination of Fracture and Fissure Indexes of Rocks from Seismic Ultrasonic Velocities. *1st Earth Sciences Symposium, SDU*, Türkiye.

Uyanık, O., Sabbağ, N., Uyanık, N. A., & Öncü, Z. (2019). Estimation of Mechanical and Physical Properties of Some Sedimentary Rocks from Ultrasonic Velocities. *Bulletin of Engineering Geology and the Environment, 78*(8), 6003–6016. DOI: 10.1007/s10064-019-01501-6

Uyanık, O., Şenli, G., & Çatlıoğlu, B. (2012d). Determination of Concrete Quality of Buildings by Non-Destructive Geophysical Methods. *International Construction Congress 2012, ICONC2012-SDÜ.*

Uyanık, O., Şenli, G., & Çatlıoğlu, B. (2013). Determination of Concrete Quality of Buildings by Non-Destructive Geophysical Methods. *SDÜ International Journal of Technology Sciences, 5*(2), 156–165.

Uyanık, O., & Tezcan, S. (2012). Determination of Concrete Strength by Ultrasonic Method. *Geophysical Bulletin, 23*(70), 41–45.

Uygunoğlu, T., & Topçu, İ. B. (2020). Nesnelerin internetinin (iot) inşaat mühendisliğindeki rolü: rfid uygulamalari. *International Journal of 3D Printing Technologies and Digital Industry, 4*(3), 270-277.

Vagdatli, T., Petroutsatou, K., Panetsos, P., Barmpa, Z., & Fragkakis, N. (2024). Probabilistic pre-estimation of life-cycle costs of road bridges using dynamic Bayesian networks. *Structure and Infrastructure Engineering*. Advance online publication. DOI: 10.1080/15732479.2024.2396603

Vaghefi, K., Ahlborn, T. M., Harris, D. K., & Brooks, C. N. (2015). Combined imaging technologies for concrete bridge deck condition assessment. *Journal of Performance of Constructed Facilities*, *29*(4). Advance online publication. DOI: 10.1061/(ASCE)CF.1943-5509.0000465

Vairavel, M., Vinod Kumar, T., Chandrasekaran, M., & Vivek, P. (2018). Evaluation of tensile and thermal behavior analysis of nano carbon fiber reinforced polymer composite. *International Journal of Engineering & Technology, 7*(3.27), 177–180. DOI: 10.14419/ijet.v7i3.27.17750

Valinejadshoubi, M., Moselhi, O., Bagchi, A., & Salem, A. (2021). Development of an IoT and BIM-based automated alert system for thermal comfort monitoring in buildings. *Sustainable Cities and Society*, *66*, 102602. DOI: 10.1016/j.scs.2020.102602

Van Dinter, R., Tekinerdogan, B., & Catal, C. (2022). Predictive maintenance using digital twins: A systematic literature review. *Information and Software Technology*, *151*, 107008. DOI: 10.1016/j.infsof.2022.107008

Varnavina, A., Sneed, L., Khamzin, A., & Torgashov, E. (2017). An attempt to describe a relationship between concrete deterioration quantities and bridge deck condition assessment techniques. Journal of Applied Geophysics. https://www.sciencedirect.com/science/article/pii/S0926985117304883

Vasilca, I.-S., Nen, M., Chivu, O., Radu, V., Simion, C.-P., & Marinescu, N. (2021). The Management of Environmental Resources in the Construction Sector: An Empirical Model. *Energies*, *14*(9), 2489. DOI: 10.3390/en14092489

Vasiliev, D., Stevens, R., & Bornmalm, L. (2025, February). Nature-Based Solutions and Smart Cities. In *International Congress on Information and Communication Technology* (pp. 183-192). Singapore: Springer Nature Singapore.

Vasudevan, G., & Tait, D. (2023). Bayesian model for sustainable, resilient, and cost-effective concrete mixtures. *Sustainable and Resilient Infrastructure*, *8*(3), 325–339. DOI: 10.1080/23789689.2023.2188348

Vatin, N. I., Korsun, V. I., Velichkin, V. Z., Kozinets, G., Zhuvak, O., & Rybakov, V. A. (2018). Precast-monolithic reinforced concrete beam-slabs technology with claydit blocks. *Construction of Unique Buildings and Structures*, *70*(7), 43–59.

Verbruggen, S., Hertoge, J., Delghust, M., Laverge, J., & Janssens, A. (2020). The Use of Solar Shading in a Nearly Zero-Energy Neighbourhood. *E3S Web of Conferences, 172*, 19003. DOI: 10.1051/e3sconf/202017219003

Vettriselvan, R., Sathya, M., & Velmurugan, T. (2018). Productivity and profitability mechanical engineering entrepreneurs: Business perspective. *International Journal of Mechanical Engineering and Technology*, *9*(8), 758–765.

Vishnukumar, M., Muthupandi, V., & Jerome, S. (2023). Microstructural characteristics, mechanical properties and corrosion performance of super austenitic stainless steel 904L produced by wire arc additive manufacturing. *Materials Today. Communications*, *35*, 105801. DOI: 10.1016/j.mtcomm.2023.105801

Vogel, R., Nix, S., & Thompson, A. (2023). Urban flood management in the context of climate change: The role of sustainable drainage systems. *Climate Risk Management*, *36*(4), 62–75.

Vulevic, A., Castanho, R. A., Gómez, J. M. N., Lousada, S., Loures, L., Cabezas, J., & Fernández-Pozo, L. (2021). Cross-border cooperation and adaptation to climate change in Western Balkans Danube Area. In *Advances in spatial science* (pp. 289–308). Springer. DOI: 10.1007/978-3-030-72124-4_14

Walker, H., Lane, D., & Stutzman, P. (2006). Petrographic methods of examining hardened concrete: A petrographic manual. https://rosap.ntl.bts.gov/view/dot/38050

Wang, B., Zhong, S., Lee, T. L., Fancey, K. S., & Mi, J. (2020). *Nondestructive Testing and Evaluation of Composite Materials/Structures: A State-of-the-Art Review*. Advances in Mechanical Engineering., DOI: 10.1177/1687814020913761

Wang, J., Shi, W., Xiang, S., & Ballinger, R. G. (2021). Study of the corrosion behaviour of sensitized 904L austenitic stainless steel in Cl^- solution. *Corrosion Science*, *181*, 109234. DOI: 10.1016/j.corsci.2020.109234

Wang, L., Zhao, Y., & Yin, X. (2023). Precast production scheduling in off-site construction: Mainstream contents and optimization perspective. *Journal of Cleaner Production*, *405*, 137054. DOI: 10.1016/j.jclepro.2023.137054

Wang, P., Lai, H., Li, W., Ke, L., Wu, H., Li, W., & Leung, C. K. Y. (2025). Mechanical properties and environmental impact assessments of GFRP rebar reinforced limestone calcinated clay cement (LC3) concrete. *Construction & Building Materials*, *472*, 140788. DOI: 10.1016/j.conbuildmat.2025.140788

Wang, S., Mei, P., Chang, L., Wu, C., Chen, S., Chen, Q., & Li, G. (2024). Evaluation of the corrosion resistance of 904L composite plate in a high-temperature and high-pressure gas field environment. *Processes (Basel, Switzerland)*, *12*(11), 2372. Advance online publication. DOI: 10.3390/pr12112372

Wang, X., Cui, F., Cui, L., & Jiang, D. (2023). Research on a multi-objective optimization design for the durability of high-performance fiber-reinforced concrete based on a hybrid algorithm. *Coatings*, *13*(12), 2054. https://www.mdpi.com/2079-6412/13/12/2054

Wang, X., Nguyen, M., Foliente, G., & Ye, L. (2007). An approach to modelling concrete bridge condition deterioration using a statistical causal relationship based on inspection data. *Structure and Infrastructure Engineering*, *3*(1), 3–15. DOI: 10.1080/15732470500103682

Wang, X., Wang, S., Song, X., & Han, Y. (2020). IoT-based intelligent construction system for prefabricated buildings: Study of operating mechanism and implementation in China. *Applied Sciences (Basel, Switzerland)*, *10*(18), 6311. DOI: 10.3390/app10186311

Wang, Y., Hu, C., Wu, B., Zhang, J., Zi, Z., & Kang, L. (2020). Matching Characteristic Research of Building Renewable Energy System Based on Virtual Energy Storage of Air Conditioning Load. *Energies*, *13*(5), 1269. DOI: 10.3390/en13051269

Wang, Y., Thangasamy, V. K., Tiong, R. L., & Zhang, L. (2022). Improved workflow for precast element design based on BIM and lean construction. *Journal of Construction Engineering and Management*, *148*(8), 04022065. DOI: 10.1061/(ASCE)CO.1943-7862.0002316

Wang, Y., Yuan, Z., & Sun, C. (2018). Research on assembly sequence planning and optimization of precast concrete buildings. *Journal of Civil Engineering and Management*, *24*(2), 106–115. DOI: 10.3846/jcem.2018.458

Wang, Z., & Hu, H. (2017). Improved precast production–scheduling model considering the whole supply chain. *Journal of Computing in Civil Engineering*, *31*(4), 04017013. DOI: 10.1061/(ASCE)CP.1943-5487.0000667

Wang, Z., Hu, H., Gong, J., Ma, X., & Xiong, W. (2019). Precast supply chain management in off-site construction: A critical literature review. *Journal of Cleaner Production*, *232*, 1204–1217. DOI: 10.1016/j.jclepro.2019.05.229

Wang, Z., Wang, T., Hu, H., Gong, J., Ren, X., & Xiao, Q. (2020). Blockchain-based framework for improving supply chain traceability and information sharing in precast construction. *Automation in Construction*, *111*, 103063. DOI: 10.1016/j.autcon.2019.103063

Wang, Z., & Xu, H. (2024). Scheduling redundancy optimization for precast production to enhance disruption management in prefabricated construction. *Computers & Industrial Engineering*, *192*, 110251. DOI: 10.1016/j.cie.2024.110251

Wang, Z., Zhang, Q., Yang, B., Wu, T., Lei, K., Zhang, B., & Fang, T. (2021). Vision-based framework for automatic progress monitoring of precast walls by using surveillance videos during the construction phase. *Journal of Computing in Civil Engineering*, *35*(1), 04020056. DOI: 10.1061/(ASCE)CP.1943-5487.0000933

Warszawski, A., Avraham, M., & Carmel, D. (1984). Utilization of precast concrete elements in building. *Journal of Construction Engineering and Management, 110*(4), 476–485. DOI: 10.1061/(ASCE)0733-9364(1984)110:4(476)

Webster, M. D. (2007). Structural Design for Adaptability and Deconstruction: A Strategy for Closing the Materials Loop and Increasing Building Value. *New Horizons and Better Practices*, 1–6. DOI: 10.1061/40946(248)27

Wehkamp, S., Schmeling, L., Vorspel, L., Roelcke, F., & Windmeier, K.-L. (2020). District Energy Systems: Challenges and New Tools for Planning and Evaluation. *Energies, 13*(11), 2967. DOI: 10.3390/en13112967

Weinberger, A. B., Christensen, A. P., Coburn, A., & Chatterjee, A. (2021). Psychological responses to buildings and natural landscapes. *Journal of Environmental Psychology, 77*, 101676. Advance online publication. DOI: 10.1016/j.jenvp.2021.101676

Weizsäcker, E. U., & Wijkman, A. (2018). *Come on! Capitalism, short-termism, population and the destruction of the planet*. Springer. (Ukrainian edition: Sammit-Knyha, 2019) DOI: 10.1007/978-1-4939-7419-1

Wheatcroft, E., Wynn, H. P., Lygnerud, K., Bonvicini, G., & Leonte, D. (2020). The Role of Low Temperature Waste Heat Recovery in Achieving 2050 Goals: A Policy Positioning Paper. *Energies, 13*(8), 2107. DOI: 10.3390/en13082107

Whitehurst, E. A. (1951). Soniscope Testing of Concrete Structures. *Journal of the American Concrete Institute. Proceedings, 47*, 443–444.

Wijenayake, A. (2025). Climate Change and Urban Resilience: Strategies for Sustainable Cities. *International Journal of Academic Research in Business & Social Sciences, 15*(4), 247–261. DOI: 10.6007/IJARBSS/v15-i4/25093

Wilkinson, M. J., & Dixon, S. (2016). A cost-benefit analysis model for the retrofit of sustainable urban drainage systems towards improved flood risk mitigation. *International Journal of Building Pathology and Adaptation, 31*(5), 654–668. DOI: 10.1108/IJBPA-05-2015-0017

Wilson, S. (2025). *Strategies to Mitigate Contractor Safety Noncompliance in the Construction Industry* (Doctoral dissertation, Walden University).

Wilson, S., Adu-Duodu, K., Li, Y., Sham, R., Wang, Y., Solaiman, E., . . . Rana, O. (2023, October). Tracking material reuse across construction supply chains. In *2023 IEEE 19th International Conference on e-Science (e-Science)* (pp. 1-4). IEEE. DOI: 10.1109/e-Science58273.2023.10254935

Wilson, M., Paschen, J., & Pitt, L. (2022). The circular economy meets artificial intelligence (AI): Understanding the opportunities of AI for reverse logistics. *Management of Environmental Quality, 33*(1), 9–25. DOI: 10.1108/MEQ-10-2020-0222

Winkler, H. (2011). Closed-loop production systems—A sustainable supply chain approach. *CIRP Journal of Manufacturing Science and Technology*, *4*(3), 243–246. DOI: 10.1016/j.cirpj.2011.05.001

Wong, L. S. (2022). Durability Performance of Geopolymer Concrete: A Review. *Polymers*, *14*(5), 868. DOI: 10.3390/polym14050868 PMID: 35267691

Wong, R. W., & Loo, B. P. (2022). Sustainability implications of using precast concrete in construction: An in-depth project-level analysis spanning two decades. *Journal of Cleaner Production*, *378*, 134486. DOI: 10.1016/j.jclepro.2022.134486

Won, J. J., Hak, L. K., & Mun, A. J. (2004). Recent domestic precast concrete slab technologies. *Magazine of the Korea Concrete Institute*, *16*(5), 16–19.

Wood, L. C., Duong, L. N. K., & Wang, J. X. (2022). Business Process Improvement for Sustainable Technologies Investments in Construction: A Configurational Approach. *Sustainability (Basel)*, *14*(9), 5697. DOI: 10.3390/su14095697

World Bank. (2024). Resilient Housing Guidance for the Maldives. https://documents.worldbank.org/en/publication/documents-reports/documentdetail/099554312202436138

World Economic Forum. (2024). *Reimagining Real Estate: A Framework for the Future.* https://reports.weforum.org/docs/WEF_Reimagining_Real_Estate_2024.pdf

World Health Organization. (2023). Supporting Mental Health and Community Well-Being. https://ukraine.iom.int/sites/g/files/tmzbdl1861/files/documents/2024-03/iom-ukraine-mhpss-activities-overview-july-december-2023_1.pdf

Wu, D., Liang, S., Shen, M., Guo, Z., Zhu, X., & Sun, C. (2019). Experimental estimation of seismic properties of new precast shear wall spatial structure model. *Engineering Structures*, *183*, 319–339. DOI: 10.1016/j.engstruct.2018.12.096

Wu, W., & Skye, H. M. (2021). Residential net-zero energy buildings: Review and perspective. *Renewable & Sustainable Energy Reviews*, *142*, 110859. DOI: 10.1016/j.rser.2021.110859 PMID: 34413697

Xia, W., Bai, E., Xu, J., & Liu, G. (2021). Experimental Study on the Strength and Deformation Characteristics of Concrete under True Triaxial Compression after Sulfate Attack. *Advances in Civil Engineering, 5548313*, 15 pages.

Xiao, J., Jiang, S., & Li, Y. (2024). Seismic Performance of Prefabricated Reinforced Concrete Structure Beam Joints Based on the Internet of Things. *Journal of Testing and Evaluation*, *52*(3), 1789–1803. DOI: 10.1520/JTE20230027

Xiao, S., & Fomin, N. I. (2024). Development and systematic review of connection techniques for RC Precast structural elements: Beam and Column Connection. *Heliyon*, *10*(16), e35886. DOI: 10.1016/j.heliyon.2024.e35886 PMID: 39224382

Xiaosheng, T., & Hamzeh, F. (2020, July). Precast concrete building construction process comparison. In *Proceedings of the Annual Conference of the International Group for Lean Construction,* Berkeley, CA, USA (pp. 6-12).

Xie, B., Qin, J., Wang, S., Li, X., Sun, H., & Chen, W. (2020). Adsorption of Phenol on Commercial Activated Carbons: Modelling and Interpretation. *International Journal of Environmental Research and Public Health, 17*(3), 789. DOI: 10.3390/ijerph17030789 PMID: 32012816

Xu, D., Zhang, X., He, X., Liu, C., Liu, Z., Zhang, X., Zhao, M., & Li, X. (2024). Mechanism and evaluation method of stress corrosion susceptibility of 904L stainless steel with optimized structure in seawater. *Corrosion Science, 229*, 111865. DOI: 10.1016/j.corsci.2024.111865

Xu, J., Zhang, Q., Teng, Y., & Pan, W. (2023). Integrating IoT and BIM for tracking and visualising embodied carbon of prefabricated buildings. *Building and Environment, 242*, 110492. DOI: 10.1016/j.buildenv.2023.110492

Yadav, D. P., Patel, D. N., & Morkos, B. W. (2018). Development of Product Recyclability Index Utilizing Design for Assembly and Disassembly Principles. *Journal of Manufacturing Science and Engineering, 140*(3), 031015. DOI: 10.1115/1.4038515

Yadav, V., Singh, S., Garg, M. P., & Akhai, S. (2024). Effect of natural Sapindus mukorossi treatment process on bio-waste bagasse fibers for biocomposite fabrication and application purposes. *Industrial Crops and Products, 219*, 119066. DOI: 10.1016/j.indcrop.2024.119066

Yamamoto, N. (2009). Condition indicators for the assessment of local and spatial deterioration of concrete structures. In *Condition assessment of aged structures*. CRC Press. DOI: 10.1201/9781439832554.CH6

Ya-ming, F., Li, X., Zheng, M., Weng, R., & Tu, J. (2020). Numerical Study on Effects of Air Return Height on Performance of an Underfloor Air Distribution System for Heating and Cooling. *Energies, 13*(5), 1070. DOI: 10.3390/en13051070

Yan, C., Wang, F., Yan, P., Shan, K., & Kosonen, R. (2020). A multi-timescale cold storage system within energy flexible buildings for power balance management of smart grids. *Renewable Energy, 161*, 626–634. DOI: 10.1016/j.renene.2020.07.079

Yang, S., Xu, Z., & Bian, Z. (2020). Prediction on compressive strength concrete using modified pull-off testing method (MPTM). *Construction and Building Materials, 250*, 118834, ISSN 0950-0618.

Yang, B., Liu, B., Xiao, J., Zhang, B., Wang, Z., & Dong, M. (2021). A novel construction scheduling framework for a mixed construction process of precast components and cast-in-place parts in prefabricated buildings. *Journal of Building Engineering, 43*, 103181. DOI: 10.1016/j.jobe.2021.103181

Yang, X., & Svendsen, S. (2020). Improving the district heating operation by innovative layout and control strategy of the hot water storage tank. *Energy and Building*, *224*, 110273. DOI: 10.1016/j.enbuild.2020.110273

Yang, Z., Tang, D., Li, D., Chen, T., Liu, J., Qi, H., & Zhou, J. (2025). Automatic dimensional quality inspection system for regular precast concrete elements based on 3D structure lighting scan technology. *Expert Systems with Applications*, 129149.

Yao, Y., Tam, V. W., Wang, J., Le, K. N., & Butera, A. (2025). Deep reinforcement learning-based schedule optimization for parallel precast production. *Engineering, Construction, and Architectural Management*, *32*(13), 1–34. DOI: 10.1108/ECAM-03-2025-0429

Yaqoob, S., & Silfwerbrand, J. (2023). Rapid repair of concrete pavements using Precast technology. In *14th International Symposium on Concrete Roads,* Krakow, Poland, June 25-28, 2023.

Yardim, Y. (2018). Review of research on the application of ferrocement in composite precast slabs. *Periodica Polytechnica. Civil Engineering*, *62*(4), 1030–1038. DOI: 10.3311/PPci.11737

Ye, X., Kosse, S., Hagedorn, P., & König, M. (2025). Workflow automation using digital twins and smart contracts: a decentralised framework for precast concrete production.

Yee, P. T. L., Adnan, A. B., Mirasa, A. K., & Rahman, A. B. A. (2011). Performance of IBS precast concrete beam-column connections under earthquake effects: A literature review. *American Journal of Engineering and Applied Sciences*, *4*(1), 93–101. DOI: 10.3844/ajeassp.2011.93.101

Ye, M., Li, L., Pei, B., Yoo, D. Y., Li, H., & Zhou, C. (2024). A critical review on shear performance of joints in precast Ultra-High-Performance Concrete (UHPC) segmental bridges. *Engineering Structures*, *301*, 117224. DOI: 10.1016/j.engstruct.2023.117224

Ye, X., Zhou, Y., Guo, H., & Luo, Z. (2024). A computer vision-based approach to automatically extracting the aligning information of precast structural components. *Automation in Construction*, *164*, 105478. DOI: 10.1016/j.autcon.2024.105478

Yıldız, A., & Daloğlu, A. T. (2021). Determination of Concrete Strength in Reinforced Concrete Buildings by Non-Destructive Methods. *Gazi University Journal of Science Part C*, *9*(3), 783–794.

Yılmaz, D. G. (2023). Nature-based Solutions for climate-resilient cities: A proposal of a model for successful implementation. *Journal of Design for Resilience in Architecture and Planning*, *4*(2), 189–203. DOI: 10.47818/DRArch.2023.v4i2092

Yohanna, A., Latief, Y., Hidayawanti, R., Armyn, R., & Machfudiyanto, L. S. R. (2023). Innovation manufacturing method of precast concrete for spun pile products at precast construction company in Indonesia with a risk approach and ISO 56002 innovation process to increase competitiveness.

Yong, Y. P., Lee, S. J., Chang, Y. H., Lee, K. H., Kwon, S. W., Cho, C. S., & Chung, S. W. (2023). Object detection and distance measurement algorithm for collision avoidance of precast concrete installation during crane lifting process. *Buildings (Basel, Switzerland), 13*(10), 2551. DOI: 10.3390/buildings13102551

Yousif Q.A. (2018). Beton Yaşının Basınç Dayanımı Üzerine Olan Etkisinin Ultrases Yöntemi ile Belirlenmesi. *Dicle Üniversitesi Fen Bilimler Enstitüsü YL Tezi, Fizik Anabilim Dalı*, Diyarbakır, 50s.

Yuan, H., Lu, W., & Jianli Hao, J. (2013). The evolution of construction waste sorting on-site. *Renewable & Sustainable Energy Reviews, 20*, 483–490. DOI: 10.1016/j.rser.2012.12.012

Yu, H., Ma, Z., & Chen, G. (2024). *SUDS as a Paradigm Shift: Global Trends and Policy Drivers for Sustainable Urban Drainage*. Challenges and Innovations in Urban Drainage Systems.

Yu, J., Liu, D., & Zhang, Z. (2022). Prediction analysis of recycled aggregate concrete with ceramic waste powder under freeze-thaw conditions based on impact-echo method and Grey-Markov model. *Frontiers in Materials, 9*, 1060294. DOI: 10.3389/FMATS.2022.1060294/FULL

Yu, Z., Wei, Y., & Cao, Z. (2019, December). The Design and Application of Information Management System Based on IOT for Precast Concrete Segments. *IOP Conference Series. Earth and Environmental Science, 371*(3), 032056. DOI: 10.1088/1755-1315/371/3/032056

Zameeruddin, M., & Sangle, K. K. (2021). Performance-based seismic assessment of reinforced concrete moment resisting frame. Journal of King Saud University—Engineering Sciences. https://www.sciencedirect.com/science/article/pii/S1018363920302245

Zamen, S., & Dackermann, U. (2020). Fractal analysis of nonlinear ultrasonic waves in phase-space domain as a quantitative method for damage assessment of concrete structures. NDT & E International. https://www.sciencedirect.com/science/article/pii/S0963869519306334

Zeng, R., Chini, A. R., & Ries, R. (2020). Innovative design for sustainability. *Engineering, Construction, and Architectural Management, 28*(3), 747–764. DOI: 10.1108/ECAM-09-2019-0491

Zeng, X., Ogunseitan, O. A., Nakamura, S., Suh, S., Kral, U., Li, J., & Geng, Y. (2022). Reshaping global policies for circular economy. *Circular Economy*, *1*(1), 100003. DOI: 10.1016/j.cec.2022.100003

Zhang, Y., Gao, L., & Bian, W. (2020). Mechanical Performance of Concrete Made with Recycled Aggregates from Concrete Pavements, *Advances in Materials Science and Engineering*, 2020, 5035763, 8 pages. DOI: 10.1155/2020/5035763

Zhang, C., Liu, S., & Yang, H. (2021). Urban flooding and water quality management: A case study on the limitations of traditional drainage systems. *Journal of Water and Climate Change*, *12*(3), 411–425.

Zhang, R., Feng, X., Mou, Z., & Zhang, Y. (2023). Green optimization for precast production rescheduling based on disruption management. *Journal of Cleaner Production*, *420*, 138406. DOI: 10.1016/j.jclepro.2023.138406

Zhang, R., Zhang, Y., Li, A., & Yang, T. Y. (2022). Experimental study on a new type of precast beam-column joint. *Journal of Building Engineering*, *51*, 104252. DOI: 10.1016/j.jobe.2022.104252

Zhang, W., Kang, K., & Zhong, R. Y. (2021). A cost evaluation model for IoT-enabled prefabricated construction supply chain management. *Industrial Management & Data Systems*, *121*(12), 2738–2759. DOI: 10.1108/IMDS-12-2020-0735

Zhang, W., & Nicholson, C. (2016). A multi-objective optimization model for retrofit strategies to mitigate direct economic loss and population dislocation. *Sustainable and Resilient Infrastructure*, *1*(3–4), 123–136. DOI: 10.1080/23789689.2016.1254995

Zhang, X., Zhang, J., Feng, Q., & Luo, Q. (2025). Are all non-kaolinitic phases detrimental to LC3 performance? Effects of mineral type, particle size, and mechanochemical activation. *Construction & Building Materials*, *500*, 144182. DOI: 10.1016/j.conbuildmat.2025.144182

Zhang, Y., & Pan, W. (2021). Reducing construction waste through modular construction. In *International symposium on advancement of construction management and real estate* (pp. 339–347). Springer.

Zhao, J., Hou, S., Shi, D., Fan, J., Wu, T., Sun, W., & Wu, G. (2024). Damage assessment of precast reinforced concrete beam-column connections using AE technology. *Case Studies in Construction Materials*, *20*, e03069. DOI: 10.1016/j.cscm.2024.e03069

Zhao, L., Chen, X., & Lee, H. (2024). Stakeholder perceptions and barriers to SUDS adoption in Asian cities: A comparative study. *Urban Climate*, *40*, 123–135.

Zhao, L., Liu, Z., & Mbachu, J. (2019). Development of intelligent prefabs using IoT technology to improve the performance of prefabricated construction projects. *Sensors (Basel), 19*(19), 4131. DOI: 10.3390/s19194131 PMID: 31554246

Zhao, T., Wang, J., Liu, C., Hua, P., Zhou, Y., & Zhao, Y. (2020). Comparative study of outdoor airflow requirement and distribution in multi-zone VAV system with different control strategies. *Science and Technology for the Built Environment, 27*(4), 489–508. DOI: 10.1080/23744731.2020.1819062

Zhao, Y., Cao, C., & Liu, Z. (2022). A framework for prefabricated component hoisting management systems based on digital twin technology. *Buildings (Basel, Switzerland), 12*(3), 276. DOI: 10.3390/buildings12030276

Zhao, Y., Cao, C., Liu, Z., & Mu, E. (2021). Intelligent control method of hoisting prefabricated components based on internet-of-things. *Sensors (Basel), 21*(3), 980. DOI: 10.3390/s21030980 PMID: 33540530

Zhao, Y., & Zhang, Y. (2023). A Review on Hydration Process and Setting Time of Limestone Calcined Clay Cement (LC3). *Solids, 4*(1), 24–38. DOI: 10.3390/solids4010003

Zheng, Y., Guo, Z., Guan, D., & Zhang, X. (2018). Parametric study on a novel grouted rolling pipe splice for precast concrete construction. *Construction & Building Materials, 166*, 452–463. DOI: 10.1016/j.conbuildmat.2018.01.182

Zhong, R. Y., Peng, Y., Xue, F., Fang, J., Zou, W., Luo, H., Thomas Ng, S., Lu, W., Shen, G. Q. P., & Huang, G. Q. (2017). Prefabricated construction enabled by the Internet-of-Things. *Automation in Construction, 76*, 59–70. DOI: 10.1016/j.autcon.2017.01.006

Zhou, P., Li, X., & Zhang, Y. (2024). Stormwater runoff characterization and adaptation of best management practices under urbanization and climate change scenarios. *Journal of Hydrology (Amsterdam), 634*, 130985.

Zhu, Z. Q., Yong, Y. P., Lee, S. J., Chang, Y. H., & Kwon, S. W. (2021). Vision-based precast concrete management plan in off-site construction site: Using PC member quality grades. In *ISARC. Proceedings of the International Symposium on Automation and Robotics in Construction* (Vol. 38, pp. 197-203). IAARC Publications.

Zirps, M., Lepech, M., Savage, S., Michel, A., Stang, H., & Geiker, M. (2020). Probabilistic design and management of sustainable concrete infrastructure using multi-physics service life models. *Frontiers in Built Environment, 6*. Advance online publication. DOI: 10.3389/FBUIL.2020.00072

Zou, X., Teng, J., De Lorenzis, L., & Xia, S. (2007). Optimal performance-based design of FRP jackets for seismic retrofit of reinforced concrete frames. Composites Part B: Engineering. https://www.sciencedirect.com/science/article/pii/S1359836806001673

zu Castell-Rüdenhausen, M., Wahlström, M., Astrup, T. F., Jensen, C., Oberender, A., Johansson, P., & Waerner, E. R. (2021). Policies as drivers for circular economy in the construction sector in the Nordics. *Sustainability (Basel)*, *13*(16), 9350. DOI: 10.3390/su13169350

Zub, C.-I., Stratan, A., Dogariu, A., Vulcu, C., & Dubina, D. (2018). Performance based seismic assessment of buildings for evaluation of retrofitting systems efficiency. In *Springer natural hazards* (pp. 373–387). Springer. DOI: 10.1007/978-3-319-74724-8_25

About the Contributors

Sérgio António Neves Lousada holds an International PhD. in Civil Engineering – Hydraulics, at the University of Minho (UMinho), Portugal and an International PhD. in Sustainable Territorial Development at the University of Extremadura (UEx), Spain. He teaches at the University of Madeira (UMa), field of Hydraulics, Environment and Water Resources and Construction. Furthermore, he collaborates with the Research Group on Environment and Spatial Planning (MAOT) - University of Extremadura (UEx), Badajoz, Spain; VALORIZA - Research Center for the Enhancement of Endogenous Resources, Polytechnic Institute of Portalegre (IPP), Portugal; CITUR-Madeira - Centre for Tourism Research, Development and Innovation, Madeira, Portugal; CERIS - Civil Engineering Department of University of Aveiro, Aveiro, Portugal; OSEAN - Outermost Regions Sustainable Ecosystem for Entrepreneurship and Innovation, Madeira, Portugal. Moreover, he holds and International Interuniversity Master's Degree in Ports and Coasts Engineering at the University of Las Palmas de Gran Canaria (ULPGC) and at the University of La Laguna (ULL).

***

Mahapara Abbass is a distinguished academician and researcher in the Department of Civil Engineering at Maharishi Markandeshwar (Deemed to be University), renowned for her groundbreaking contributions to structural engineering and sustainable construction materials. With an impressive portfolio of publications in prestigious SCI and Scopus-indexed journals, including Multiscale Multidisciplinary Modeling, Experimentation and Design, and Lecture Notes in Civil Engineering, Dr. Abbass has also presented her work at esteemed conferences and authored/co-authored book chapters in critically acclaimed volumes, such as Advances in Geotechnics and Structural Engineering. Her expertise in geopolymer concrete, sustainable construction practices, and innovative materials has been recognized through publications in esteemed platforms like Elsevier, cementing her prominent role in advancing the frontiers of civil engineering.

Shalom Akhai has established himself as a prominent figure in the field of engineering, leveraging over 17 years of academic, research, and technical experience to make significant contributions. With a robust portfolio of over 100+ publications in prestigious journals, conferences and book chapters, Dr. Akhai demonstrates a commitment to advancing knowledge through scholarly work. Additionally, he has authored/co-authored several technical textbooks, further solidifying his expertise and dedication to education in the field. Beyond academia, Dr. Akhai's impact extends to real-world applications through the establishment of IGNIS Technical Solutions, a consultancy firm to translate his knowledge into real-world applications. His skills have resulted in the successful development of over 50 patents+, underscoring his innovative approach and influence in shaping the engineering landscape.

Ali Biglarifadafan is a Senior Lecturer in Civil Engineering within the Faculty of Engineering at the University of Golestan. He/She earned a Ph.D. in Civil Engineering (Structural Engineering) from the University of Glasgow in 2014, supported by an elite scholarship from the MSRT. His/Her doctoral research was conducted under the supervision of Professors P. Harrison, N. Bicanic, C. Pearce, and Z. Guo, contributing to the field of computational mechanics. This was followed by a postdoctoral research fellowship in Geotechnical Engineering at the University of Warwick. Appointed as an Assistant Professor at Golestan University in 2015 and subsequently promoted to Senior Lecturer, his/her primary research and teaching agenda focuses on sustainable and resilient design methodologies. Dr. Biglari currently leads a research team of doctoral and postgraduate students investigating a Resilient Seismic Design framework for structures and geo-structures, employing integrated traditional and artificial intelligence (AI) techniques. In addition to research and teaching, Dr. Biglari has held significant administrative and leadership roles, including Course Coordinator, Head of Subject, Head of the In

Maryna Chaikovska, D.Sc. (Economics), Professor, is a Ukrainian scholar specializing in digital transformation, responsible innovation, marketing management, IT project management, and sustainable development. She holds a Doctor of Economic Sciences degree with a focus on marketing IT project management and has over thirty-five years of scientific, academic, and professional experience. Prof. Chaikovska serves as Professor at the Department of Marketing and Business Administration, Odesa I.I. Mechnikov National University, where she teaches courses in IT project management, IT entrepreneurship, marketing of the

IT industry, business information systems, automation of business processes. Prof. Chaikovska has participated in numerous international academic programs supported by USAID, SEUME, EERS, DAAD, AccEnt, SEA-EU, and Erasmus+, including research training at Kiel University (Germany) and Visiting Professor positions at Alma Mater Studiorum – University of Bologna (Italy), University of Cádiz (Spain), International Hellenic University (Greece), and Wroclaw University of Science and Technology (Poland). She is the author of more than 250 scholarly works, textbooks, and original methodologies in IT project management, innovation, and business modeling, with indexed profiles in Scopus, Google Scholar, and ORCID. Her expertise bridges research, education, and practical application in responsible marketing, Industry 5.0, and the sustainable development of societal systems.

Nevbahar Ekin currently works at the Department of Geophysical Engineering, Süleyman Demirel University in Isparta-Turkey. She completed her PhD in "Süleyman Demirel Univesity (Turkey)" under supervision of Assoc. Prof. Dr. Osman Uyanık. Her researches are in Counstruction Geophysics, Electrical Prospecting, Soil dynamic, soil mechanic, Geotechnics. She is involved to a number of different national project.

Oleksandra Humenna, D.Sc. (Economics), Associate Professor, is a Ukrainian scholar specializing in sustainable development, creative economy, project management, and cultural and creative industries. She holds a Doctor of Economic Sciences degree in the field of economics and management of the national economy and has over twenty years of academic and administrative experience. Dr. Humenna currently serves as Professor at the Department of Innovative Business Technologies of the Vadym Hetman Kyiv National Economic University, teaching courses in project management and sustainable business development. Her professional background includes leadership positions such as Acting President and Dean of the Faculty of Economics at the National University of Kyiv-Mohyla Academy, Rector of the Salvador Dalí Academy of Contemporary Art, and Honorary Professor at ISMA University (Latvia). She has participated in numerous international academic programs, including research training at Indiana University (USA), Cornell University, and Berkeley Executive Leadership Academy. Dr. Humenna has been involved in multiple international and national projects supported by USAID, UNDP, the U.S. Embassy in Ukraine, NAWA, Erasmus+, and the British Council. She is the author of more than 130 scholarly and educational publications, with indexed profiles in Scopus, Web of Science, Google Scholar.

Ashifa Kariveliparambil is a faculty member in the Faculty of Health Sciences at Istanbul Gelisim University, Türkiye. She is an interdisciplinary scholar working across social work, public health, community development, and gender studies, with growing expertise in AI and data analytics for social research. Her work focuses on women's health, disaster resilience, social capital, and rights-based practice. She has published widely in Scopus and WoS indexed journals, led national and EU-funded projects, and holds a patented intervention model integrating social, health, and data-driven approaches.

Parveen Kumar, an Assistant Professor in the Department of Civil Engineering at Maharishi Markandeshwar (Deemed to be University), Mullana–Ambala, is an emerging academician and researcher with a distinctive portfolio spanning rock mechanics, geotechnical engineering, sustainable materials, and environmental education. His research contributions include advanced analytical–numerical and experimental studies on tensile behaviour of rock materials, notably the development and calibration of the Modified Seesaw Test (MST) for direct tensile assessment of ring-shaped rock specimens, published in reputed Springer journals such as Sādhanā (Academy Proceedings in Engineering Sciences). In parallel, he has co-developed innovative testing and engineering solutions, including patented designs for a direct tensile test machine for rock and a fingerprint- and camera-based smart door handle, reflecting his strong inclination toward practical, industry-oriented innovation. Beyond rock and soil mechanics, Parveen Kumar has co-authored several books and articles on environmental education, sustainable development, rural technology, and composite materials for sustainability, including A Textbook of Environmental Education (Exploring Environmental Issues from Local to Global Perspectives) and volumes on planning for sustainable development and rural/community technologies. His publication record, comprising more than a dozen research papers, books, and patents, is complemented by recognition as an IGBC Accredited Faculty, underscoring his engagement with green building and sustainability agendas. Through this diverse body of work, Parveen Kumar continues to contribute to the advancement of civil engineering, with a particular emphasis on innovative testing methods, sustainable infrastructure, and environmentally responsible development.

I**M. Vairavel,** M.E., Ph.D., PDF, is an Associate Professor in the Department of Mechanical Engineering at Dhanalakshmi Srinivasan College of Engineering, Coimbatore, Tamil Nadu, India. He serves as the Chief Editor of Kalaivani Publications. He holds two Ph.Ds.—the first (2017) from the Indian School of Science, Gorgand, and the second (2021) in welding from Vels Institute of Technology—and is currently pursuing postdoctoral research. With 123 journal articles and 19 honors, his expertise spans robotics, AI, neural networks, RSM, and solar desalination. From 2010 to 2014, he worked as a steel metal designer at CRAFTSMAN AUTOMATION PVT LTD. He later served as Assistant Professor at various institutions, including Annapoorna Engineering College. He is also active in publishing, having edited and published over 52 books. His contributions include 9 Indian and 3 international patents in solar energy, Rs 1.4 crore in research funding, and extensive CFD-based design projects. He continues to mentor students and professionals in academic and industrial sectors.

Hanna Melehanych, PhD (Political Science), Associate Professor, is a Ukrainian scholar specializing in political institutions, democratic governance, sustainable development, and cross-border cooperation. She is an Associate Professor at the Department of International Studies and Public Communications of Uzhhorod National University and serves as Director of the Sustainable Development Centre of the university. As.prof. Melehanych has extensive experience in academic, project, and expert work, participating in numerous international initiatives supported by ERASMUS+, the Visegrad Fund, IFES, the International Organization for Migration, and UNDP. Her professional portfolio includes roles as project manager, communication coordinator, strategic planning expert, and contributor to regional development strategies. She has been actively engaged in educational mobility programmes, expert accreditation activities, and training programs related to democracy, climate resilience, and public governance. As.prof. Melehanych is a member of the editorial boards of several scientific journals and participates in the Transcarpathian coalition "Women, Peace, Security." Her research and teaching interests focus on sustainable development policy, democratic participation, public communications, and institutional capacity building.

Ziya Öncü is a researcher in the field of geophysical engineering with a particular focus on artificial intelligence applications in construction materials and structural analysis. His academic work centers on the integration of data-driven models into traditional engineering practices, aiming to improve the accuracy of concrete strength estimation and enhance the sustainability of construction processes. He completed his undergraduate studies in Geophysical Engineering in Turkey, where he cultivated a strong foundation in structural design and material behavior. Building on this background, he has participated in several research projects exploring machine learning, nondestructive testing, and performance optimization in engineering applications. He has authored and co-authored academic papers presented at national and international conferences and has contributed to peer-reviewed journals in the fields of construction materials, artificial intelligence, and nondestructive testing. His research interests include sustainable infrastructure development, predictive modeling, and the digital transformation of engineering practices. In addition to his research activities, he is actively engaged in academic collaborations and professional development initiatives. He has taken part in workshops and seminars focused on advanced computational methods, sustainability in construction, and the future role of AI in engineering.

R. Vettriselvan is an Associate Professor at AMET University, specializes in HRM and Marketing. He is acting as a mentor at Saraswathi Institute of Medical Sciences, Hapur. He served as a Review Board member of the National Council for Higher Education, Malawi; Head of the Department of Commerce and Management Studies at St. Eugene University, Zambia; and Director of Research and Publication at St. John the Baptist University, Malawi. He has published 31 books and 100+ articles in peer-reviewed journals and edited book chapters. Under his guidance, two have completed and seven are pursuing their PhD.

Sripriya Ranganathan working as Associate Professor Department of Electrical Engineering, Saveetha Engineering College, Chennai, India

Jyoti Rani is an accomplished Assistant Professor at the School of Fashion Design and Technology, with over two years of experience in academia and research. Her work focuses on innovative and sustainable practices in fashion and textiles, making her a prominent figure in the field. Her research areas include vegan leather, natural dyeing, consumer behavior, artificial intelligence (AI) in fashion, and merchandising, reflecting her commitment to addressing contemporary challenges in the industry. Dr. Rani has published 28 research papers in reputed, peer-reviewed journals, all indexed in Scopus, demonstrating the high quality and impact of her work. Additionally, she has contributed 10 book chapters, showcasing her interdisciplinary expertise and ability to connect traditional practices with modern advancements. Her research often bridges the gap between sustainability and technology, particularly in areas like vegan leather and natural dyeing, which are critical to reducing the environmental footprint of the fashion industry.One of her standout achievements is securing 1 patent in fiber technology, highlighting her innovative approach to material science and its applications

Chadetrik Rout, a distinguished academician and researcher in the Department of Civil Engineering at Maharishi Markandeshwar (Deemed to be University), Mullana, Ambala, has established a strong reputation in the domain of environmental engineering and sustainable infrastructure. Currently serving as an Associate Professor, he brings more than eighteen years of teaching and research experience, with core expertise spanning groundwater quality assessment, water resources management, environmental impact analysis, and the mitigation of pollution from heavy metals and industrial effluents. Dr. Rout has authored and co-authored over sixty-five research and review papers in reputed national and international journals and conference proceedings, complemented by contributions to scholarly books and textbooks on environmental education and rural and sustainable technologies. His recent works include advanced reviews on geopolymers, composite and bio-based materials, and environmentally sustainable construction practices, published with outlets such as STM Journals and other peer-reviewed platforms. Collectively, his research portfolio reflects a sustained commitment to developing environmentally responsible solutions for civil engineering applications and to promoting sustainability-focused education for the next generation of engineers.

Nurten Ayten Uyanık currently works at the Vocational School of Tech. Sciences, Isparta University of Applied Sciences in Isparta-Turkey. She completed her PhD in "Süleyman Demirel Univesity (Turkey)" under supervision of Prof. Dr. İskender Akkurt. Her researches are in Natural Radioactivity, Magnetic Prospecting, Soil dynamic, Geotechnics. She is involved to a number of different national project.

Osman Uyanık currently works at the Department of Geophysical Engineering, Süleyman Demirel University in Isparta-Turkey. He completed his PhD in "Dokuz Eylül Univesity (Turkey)" under supervision of Prof. Dr. A. Güngör Taktak. His researches are in Seismic Refraction, Soil dynamic, Soil mechanic, Geotechnics, Counstruction Geophysics, Natural Radioactivity. He is involved to a number of different national project.

Vijay Singh Khunga working as Associate Professor, Department of Community Medicine, Saraswathi Institute of Medical Sciences, Hapur, India

V. Savitha Assistant Professor, Department of Electrical Engineering, Saveetha Engineering College, Chennai

Index

A

Adaptive Infrastructure Systems 373
adaptive reuse 40, 55, 83, 97, 98, 100, 104, 447
AISI 904L 149, 150, 151, 152, 153, 154, 155, 158, 162, 164, 165, 166, 167
Analytic Hierarchy Process 433, 439, 452, 453, 458, 462, 465

B

BIM 33, 36, 39, 40, 48, 49, 50, 52, 54, 55, 56, 57, 58, 62, 64, 71, 84, 91, 96, 97, 103, 175, 177, 182, 186, 210, 213, 215, 216, 217, 218, 219, 220, 221, 226, 227, 237, 239, 240, 241, 242, 243, 244, 245, 287, 288, 323, 421, 426, 428, 453, 455, 456, 462
Biomimicry 433, 437, 438, 439, 440, 441, 442, 457, 458, 460, 461, 462, 463
building-integrated photovoltaics 316, 391, 411, 431

C

Cast in Situ 173
Circular Economy 12, 15, 39, 40, 41, 42, 43, 44, 45, 47, 49, 50, 51, 53, 54, 56, 58, 59, 61, 62, 64, 65, 66, 67, 69, 71, 73, 75, 78, 87, 88, 89, 92, 95, 96, 98, 99, 100, 101, 102, 103, 104, 105, 255, 302, 314, 426, 435, 453, 455, 456
Civil Engineering Practice 364, 365, 369, 375, 376, 377, 379
Climate-Resilient Infrastructure 1, 3, 362, 365, 367, 373, 377, 378, 381, 384
Concrete 2, 4, 5, 6, 7, 8, 9, 10, 11, 12, 13, 14, 15, 16, 17, 18, 19, 20, 21, 22, 23, 24, 25, 26, 27, 28, 29, 30, 31, 33, 34, 35, 36, 37, 38, 47, 55, 67, 74, 80, 81, 82, 84, 85, 87, 100, 101, 107, 108, 109, 110, 111, 112, 113, 115, 116, 117, 118, 119, 120, 121, 122, 124, 125, 126, 127, 128, 129, 130, 131, 132, 133, 134, 136, 137, 138, 139, 140, 141, 142, 143, 144, 145, 146, 148, 166, 169, 170, 171, 172, 173, 174, 175, 184, 185, 186, 187, 188, 189, 190, 191, 192, 193, 194, 195, 196, 197, 198, 199, 200, 201, 202, 203, 205, 206, 208, 209, 210, 211, 212, 213, 214, 215, 216, 217, 218, 219, 220, 221, 222, 223, 224, 225, 226, 227, 228, 229, 230, 231, 232, 233, 234, 235, 236, 237, 238, 239, 240, 241, 242, 243, 244, 245, 246, 247, 248, 249, 251, 254, 257, 258, 260, 262, 263, 264, 265, 266, 267, 269, 271, 272, 274, 275, 276, 278, 279, 280, 282, 283, 284, 285, 289, 295, 297, 298, 299, 303, 304, 305, 314, 315, 317, 319, 320, 321, 322, 323, 324, 325, 326, 327, 328, 329, 330, 331, 332, 374, 390, 396, 398, 404, 417
Construction 2, 4, 8, 12, 13, 25, 28, 29, 33, 34, 36, 37, 38, 39, 40, 41, 42, 43, 44, 45, 46, 47, 48, 49, 50, 51, 52, 53, 54, 55, 56, 57, 58, 59, 60, 61, 62, 63, 64, 65, 66, 67, 69, 70, 71, 72, 73, 74, 78, 80, 81, 82, 83, 84, 85, 86, 87, 88, 89, 91, 92, 95, 96, 97, 98, 99, 100, 101, 102, 103, 104, 105, 108, 112, 136, 137, 138, 139, 142, 143, 145, 146, 148, 169, 170, 171, 172, 173, 174, 175, 176, 177, 178, 179, 180, 181, 182, 183, 184, 185, 186, 187, 188, 189, 190, 191, 192, 193, 194, 196, 197, 198, 210, 211, 212, 213, 214, 215, 216, 217, 218, 219, 220, 221, 222, 223, 224, 225, 226, 227, 228, 229, 230, 231, 232, 233, 234, 235, 237, 238, 239, 240, 241, 242, 243, 244, 245, 246, 250, 271, 285, 291, 308, 310, 319, 321, 322, 324, 327, 329, 330, 334, 342, 343, 344, 374, 375, 379, 383, 386, 403, 404, 409, 424, 431, 433, 434, 435, 436, 437, 438, 439, 440, 441, 442, 443, 444, 445, 446, 447, 448, 449, 450, 451,

452, 453, 454, 455, 456, 457, 458, 460, 461, 462, 463, 464, 465
Corrosion Resistance 28, 149, 150, 151, 153, 162, 163, 165, 166, 168, 303
Cultural Identity 446, 447, 449, 450, 462

D

Degradation Modeling 253, 294
demand response 398, 406, 430
destructive testing 108, 109, 110, 111, 112, 116, 125, 133, 134, 135, 136, 137, 139, 140, 141, 142, 143, 144, 148, 176, 210, 253, 271
Digital Twins 39, 40, 49, 51, 66, 67, 96, 104, 177, 182, 216, 221, 229, 241, 245, 257, 315, 362, 363, 366, 367, 368, 372, 373, 374, 376, 379, 380, 386, 410, 446
DRX 150, 152, 155, 165

E

embodied carbon 2, 4, 15, 29, 44, 47, 50, 58, 77, 210, 221, 245, 251, 255, 305, 306, 316, 317, 403, 411, 421, 425, 426
energy piles 190, 385, 392, 394, 404, 405, 409, 410

F

Flood Risk Reduction 346, 347, 348, 352, 353

G

Geopolymers 7, 8, 10, 11, 214, 238
Green Stormwater Infrastructure 335, 342, 345, 353

H

High-Temperature Performance 19, 149
HVAC 385, 387, 389, 390, 394, 395, 397, 399, 403, 404, 405, 411, 413, 421, 427, 428, 432

I

Industry 4.0 71, 92, 104, 179, 182, 183, 216, 241, 422
IoT 51, 71, 92, 97, 177, 185, 186, 213, 214, 215, 217, 218, 219, 220, 221, 222, 223, 224, 225, 228, 229, 230, 231, 232, 233, 234, 235, 236, 237, 238, 239, 240, 241, 242, 243, 244, 245, 246, 285, 286, 367, 368, 373, 412, 413
ISO 9001 195, 196, 198, 199, 200, 203, 204, 205, 206, 207, 208, 209, 210, 211, 212

K

Kansei Engineering 433, 437, 439, 442, 443, 444, 445, 446, 452, 457, 458, 459, 460, 461, 462, 464

L

Life-Cycle Assessment 2, 48, 313, 385, 403, 404, 423
Low-Carbon Concrete 5

M

Manufacturing 2, 36, 75, 97, 101, 103, 104, 105, 150, 166, 167, 169, 170, 174, 177, 179, 182, 196, 198, 200, 202, 203, 214, 216, 217, 220, 222, 225, 226, 227, 228, 230, 231, 239, 241, 242, 243, 244, 250, 403, 422, 429, 441, 451
material efficiency 69, 76, 78, 92, 103, 306
Material Reuse 44, 45, 48, 54, 62, 67, 89
Microstructure 130, 149, 151, 152, 166, 167, 327
modular construction 39, 40, 45, 46, 55, 56, 57, 64, 67, 70, 101, 235, 438, 453, 455, 456
Monitoring Techniques 279, 325

N

Nature-Based Solutions 316, 362, 364, 366, 369, 371, 373, 376, 379, 381, 382, 383, 384, 440, 464
Net-Zero Energy Building 92, 388, 432
Neuroaesthetics 439
non-destructive testing 108, 109, 110, 111, 112, 125, 133, 134, 135, 136, 137, 139, 140, 141, 142, 143, 144, 148, 176, 210

P

Performance-Based Retrofitting 328
phase change materials 390, 395, 396, 411
Precast 7, 9, 11, 16, 139, 169, 170, 171, 172, 173, 174, 175, 176, 177, 178, 179, 180, 181, 182, 183, 184, 185, 186, 187, 188, 189, 190, 191, 192, 193, 194, 195, 196, 197, 198, 199, 200, 201, 202, 203, 204, 205, 206, 207, 208, 209, 210, 211, 212, 213, 214, 215, 216, 217, 218, 219, 220, 221, 222, 223, 224, 225, 226, 227, 228, 229, 230, 231, 232, 233, 234, 235, 236, 237, 238, 239, 240, 241, 242, 243, 244, 245, 246, 304
Precast Concrete 169, 170, 171, 172, 173, 184, 185, 186, 187, 188, 189, 190, 191, 192, 193, 194, 195, 196, 197, 198, 199, 200, 201, 203, 205, 206, 208, 209, 210, 211, 212, 213, 214, 215, 216, 217, 218, 219, 220, 221, 222, 223, 224, 225, 226, 227, 228, 229, 230, 231, 232, 233, 234, 235, 236, 237, 238, 240, 241, 242, 243, 244, 245, 246

Q

Quality 3, 4, 11, 12, 13, 44, 45, 46, 48, 62, 75, 80, 85, 90, 92, 104, 108, 112, 115, 118, 119, 125, 126, 127, 136, 140, 141, 143, 145, 146, 170, 171, 172, 173, 174, 175, 176, 177, 179, 181, 182, 188, 191, 193, 195, 196, 197, 198, 199, 200, 201, 202, 203, 204, 205, 206, 207, 208, 209, 210, 211, 212, 216, 218, 219, 220, 222, 223, 224, 227, 230, 233, 234, 235, 243, 244, 246, 263, 271, 272, 281, 284, 319, 320, 321, 324, 333, 337, 338, 340, 341, 342, 344, 346, 347, 349, 352, 353, 355, 356, 357, 358, 359, 368, 369, 370, 371, 373, 374, 375, 376, 386, 388, 389, 393, 394, 396, 399, 408, 410, 413, 429, 430, 432, 443, 446, 447, 450, 451

R

Recycled Aggregate 12, 13, 14, 15, 35, 37, 67, 115, 138, 331
recycling 19, 41, 42, 43, 50, 52, 59, 61, 62, 66, 69, 70, 72, 74, 76, 78, 80, 81, 82, 84, 87, 89, 92, 96, 98, 99, 101, 102, 182, 311, 438, 448
renewable energy 48, 74, 75, 80, 88, 385, 386, 387, 388, 389, 391, 393, 396, 398, 399, 404, 405, 409, 410, 414, 416, 417, 418, 419, 423, 427, 430, 431, 432, 435, 436
renewable materials 45, 69, 86, 95
resistivity 7, 8, 10, 11, 109, 110, 111, 125, 128, 129, 130, 132, 134, 139, 142, 144, 253, 261, 276, 283, 284, 285, 328
reuse 39, 40, 41, 43, 44, 45, 46, 47, 48, 50, 52, 53, 54, 55, 58, 59, 62, 67, 69, 71, 72, 73, 74, 75, 77, 78, 80, 81, 83, 84, 87, 89, 95, 97, 98, 100, 102, 104, 171, 225, 244, 317, 441, 447, 453

S

schmidt 110, 111, 116, 117, 120, 121, 124, 141, 275
SDGs 40, 41, 54, 56, 383, 434, 449
Self-Healing Concrete 2, 20, 24, 315
SEM 149, 150, 153, 163, 164, 322
semi-destructive testing 110, 116
Stainless Steel 149, 150, 151, 152, 153, 154, 155, 158, 159, 162, 164, 165, 166, 167, 168
Structural Health Monitoring 142, 213,

225, 227, 228, 230, 237, 247, 248, 252, 253, 266, 270, 273, 274, 277, 279, 280, 281, 285, 286, 313, 314, 319, 320, 322, 323, 324, 325, 327, 328, 329, 330
Supplementary Cementitious Materials 1, 5
Sustainability 2, 15, 18, 19, 20, 29, 33, 36, 40, 42, 44, 47, 48, 51, 53, 54, 55, 56, 60, 64, 65, 66, 67, 70, 73, 74, 75, 76, 80, 81, 83, 89, 90, 91, 92, 96, 98, 99, 100, 101, 102, 103, 105, 172, 174, 177, 178, 179, 181, 182, 183, 186, 192, 205, 209, 210, 211, 212, 216, 220, 224, 230, 231, 235, 236, 240, 241, 243, 244, 248, 250, 255, 257, 304, 306, 308, 314, 315, 316, 318, 319, 320, 326, 330, 362, 364, 370, 372, 374, 376, 380, 383, 384, 385, 386, 392, 395, 411, 414, 415, 416, 417, 418, 421, 423, 424, 425, 426, 427, 429, 431, 436, 438, 446, 450, 453, 454
sustainable construction 42, 66, 69, 82, 84, 96, 100, 104, 181, 243, 433, 436, 439, 443, 457, 458, 462
Sustainable Development 35, 40, 48, 54, 56, 62, 66, 70, 78, 88, 89, 90, 96, 104, 137, 140, 190, 319, 383, 411, 412, 427, 433, 434, 435, 436, 438, 439, 446, 447, 451, 452, 453, 456, 457, 464, 465
Sustainable Urban Design 362, 367, 369, 371
Sustainable Urban Drainage Systems 335, 338, 339, 340, 341, 342, 345, 347, 353, 355, 357, 358, 359

T

TGA 149, 150, 152, 153, 154, 165
Thermal Stability 149, 150, 153, 165

U

ultrasonic 13, 108, 109, 110, 111, 112, 113, 116, 121, 124, 125, 126, 127, 130, 133, 134, 137, 139, 141, 144, 145, 146, 148, 176, 210, 252, 268, 269, 270, 271, 274, 319, 332
uniaxial compressive strength 110, 112, 113, 134, 147, 148
Urban Resilience 345, 367, 369, 370, 382, 384

W

Waste Reduction 15, 57, 58, 78, 87, 92
Water Quality Improvement 346, 347, 349, 352, 353, 357

X

XRD 150, 153, 159

Implementation and Evaluation of Green Materials in Technology Development:

Emerging Research and Opportunities

M.T. Mastura
Universiti Teknikal Malaysia Melaka, Malaysia

S. M. Sapuan
Universiti Putra Malaysia, Malaysia

A volume in the Advances in Environmental Engineering and Green Technologies (AEEGT) Book Series

IGI Global
DISSEMINATOR OF KNOWLEDGE

Published in the United States of America by
IGI Global
Engineering Science Reference (an imprint of IGI Global)
701 E. Chocolate Avenue
Hershey PA, USA 17033
Tel: 717-533-8845
Fax: 717-533-8661
E-mail: cust@igi-global.com
Web site: http://www.igi-global.com

Library of Congress Cataloging-in-Publication Data

Names: Mastura, M. T., 1985- editor. | Sapuan, S. M., editor.
Title: Implementation and evaluation of green materials in technology development: emerging research and opportunities / M.T. Mastura and S.M. Sapuan, editors.
Description: Hershey, PA : Engineering Science Reference, 2020. | Includes bibliographical references and index. | Summary: “”This book explores the theoretical, conceptual, and practical uses of environmentally friendly materials in technology development”--Provided by publisher”--Provided by publisher.
Identifiers: LCCN 2019027710 (print) | LCCN 2019027711 (ebook) | ISBN 9781799813743 (hardcover) | ISBN 9781799813750 (paperback) | ISBN 9781799813767 (ebook)
Subjects: LCSH: Technology--Environmental aspects. | Materials--Environmental aspects. | Sustainable engineering.
Classification: LCC T173.8 .E93 2020 (print) | LCC T173.8 (ebook) | DDC 600--dc23
LC record available at https://lccn.loc.gov/2019027710
LC ebook record available at https://lccn.loc.gov/2019027711

This book is published in the IGI Global book series Advances in Environmental Engineering and Green Technologies (AEEGT) (ISSN: 2326-9162; eISSN: 2326-9170)

British Cataloguing in Publication Data
A Cataloguing in Publication record for this book is available from the British Library.

Advances in Environmental Engineering and Green Technologies (AEEGT) Book Series

ISSN:2326-9162
EISSN:2326-9170

Editor-in-Chief: Sang-Bing Tsai, University of Electronic Science and Technology of China Zhongshan Institute, China; Ming-Lang Tseng, Lunghwa University of Science and Technology, Taiwan; and Yuchi Wang, University of Electronic Science and Technology of China Zhongshan Institute, China

MISSION

Growing awareness and an increased focus on environmental issues such as climate change, energy use, and loss of non-renewable resources have brought about a greater need for research that provides potential solutions to these problems. Research in environmental science and engineering continues to play a vital role in uncovering new opportunities for a "green" future.

The **Advances in Environmental Engineering and Green Technologies (AEEGT)** book series is a mouthpiece for research in all aspects of environmental science, earth science, and green initiatives. This series supports the ongoing research in this field through publishing books that discuss topics within environmental engineering or that deal with the interdisciplinary field of green technologies.

COVERAGE

- Air Quality
- Waste Management
- Sustainable Communities
- Electric Vehicles
- Green Transportation
- Industrial Waste Management and Minimization
- Green Technology
- Contaminated Site Remediation
- Radioactive Waste Treatment
- Water Supply and Treatment

The Advances in Environmental Engineering and Green Technologies (AEEGT) Book Series (ISSN 2326-9162) is published by IGI Global, 701 E. Chocolate Avenue, Hershey, PA 17033-1240, USA, www.igi-global.com. This series is composed of titles available for purchase individually; each title is edited to be contextually exclusive from any other title within the series. For pricing and ordering information please visit http://www.igi-global.com/book-series/advances-environmental-engineering-green-technologies/73679. Postmaster: Send all address changes to above address.

Titles in this Series

For a list of additional titles in this series, please visit:
https://www.igi-global.com/book-series/advances-environmental-engineering-green-technologies/73679

Modern Techniques for Agricultural Disease Management and Crop Yield Prediction
N. Pradeep (Bapuji Institute of Engineering and Technology, India) Sandeep Kautish (Lord Buddha Education Foundation, Nepal & Asia Pacific University, Malaysia) C.R. Nirmala (Bapuji Institute of Engineering and Technology, India) Vishal Goyal (Punjabi University, Patiala, India) and Sonia Abdellatif (Seoul National University, South Korea)
Engineering Science Reference • © 2020 • 291pp • H/C (ISBN: 9781522596325) • US $195.00 (our price)

Novel Advancements in Electrical Power Planning and Performance
Smita Shandilya (Sagar Institute of Research, Technology and Science, India) Shishir Kumar Shandilya (Vellore Institute of Technology, India) Tripta Thakur (Maulana Azad National Institute of Technology, India) and Atulya K. Nagar (Liverpool Hope University, UK)
Engineering Science Reference • © 2020 • 388pp • H/C (ISBN: 9781522585510) • US $225.00 (our price)

Global Food Politics and Approaches to Sustainable Consumption Emerging Research and Opportunities
Luke Amadi (University of Port Harcourt, Nigeria) and Fidelis Allen (University of Port Harcourt, Nigeria)
Engineering Science Reference • © 2020 • 242pp • H/C (ISBN: 9781799801252) • US $185.00 (our price)

Fuzzy Expert Systems and Applications in Agricultural Diagnosis
A.V. Senthil Kumar (Hindusthan College of Arts and Science, India) and M. Kalpana (Tamil Nadu Agricultural University, India)
Engineering Science Reference • © 2020 • 335pp • H/C (ISBN: 9781522591757) • US $245.00 (our price)

Handbook of Research on Environmental and Human Health Impacts of Plastic Pollution
Khursheed Ahmad Wani (Government Degree College Bijbehara, India) Lutfah Ariana (Indonesian Institute of Sciences, Indonesia) and S.M. Zuber (Government Degree College Bijbehara, India)
Engineering Science Reference • © 2020 • 600pp • H/C (ISBN: 9781522594529) • US $285.00 (our price)

IGI Global
DISSEMINATOR OF KNOWLEDGE

701 East Chocolate Avenue, Hershey, PA 17033, USA
Tel: 717-533-8845 x100 • Fax: 717-533-8661
E-Mail: cust@igi-global.com • www.igi-global.com

Table of Contents

Preface viii

Chapter 1
Green Material for Fused Filament Fabrication: A Review 1
Mastura Mohammad Taha, Universiti Teknikal Malaysia Melaka, Malaysia
Ridhwan Jumaidin, Universiti Teknikal Malaysia Melaka, Malaysia
Nadlene M. Razali, Universiti Teknikal Malaysia Melaka, Malaysia
Syahibudil Ikhwan Abdul Kudus, Universiti Teknikal Malaysia Melaka, Malaysia

Chapter 2
Starch Cellulosic Bio-Composites: A Sustainable and Multifunctional Material for Green Technology 28
M. J. Halimatul, Universiti Putra Malaysia, Malaysia
S. M. Sapuan, Universiti Putra Malaysia, Malaysia
N. Julkapli, Universiti Malaya
M. Jawaid, Universiti Putra Malaysia, Malaysia
M. R. Ishak, Universiti Putra Malaysia, Malaysia
Mohammad Taha Mastura, Universiti Teknikal Malaysia Melaka, Malaysia

Chapter 3
Green Materials in Hybrid Composites for Automotive Applications: Green Materials 56
N. M. Norizan, Universiti Putra Malaysia, Malaysia
A. Atiqah, Universiti Tenaga Nasional, Malaysia
M. N. M. Ansari, Universiti Tenaga Nasional, Malaysia
M. Rahmah, Universiti Teknologi MARA, Malaysia

Chapter 4
Modifications of Biodegradable Thermoplastic Starch (TPS) From Sago Starch via Cross-Linking Method 77
Yusriah Lazim, Malaysian Institute of Chemical Bio-Engineering Technology, Universiti Kuala Lumpur, Malaysia
Abdul Baith Abu Hanafi, Malaysian Institute of Chemical Bio-Engineering Technology, Universiti Kuala Lumpur, Malaysia
Mohd Syazwan Adura, Malaysian Institute of Chemical Bio-Engineering Technology, Universiti Kuala Lumpur, Malaysia
Siti Afifah Muda, Malaysian Institute of Chemical Bio-Engineering Technology, Universiti Kuala Lumpur, Malaysia
Lily Suhaila Yacob, Malaysian Institute of Chemical Bio-Engineering Technology, Universiti Kuala Lumpur, Malaysia
Ahmad Marzio Mohd Yusof, Malaysian Institute of Chemical Bio-Engineering Technology, Universiti Kuala Lumpur, Malaysia

Chapter 5
Non-Edible Oil Biodiesel Production via Microwave Irradiation Technologies Using Waste-Heterogeneous Catalyst Derived From Natural Calcium Oxide 92
Mahanum Mohd Zamberi, Universiti Teknikal Malaysia Melaka, Malaysia
Farid Nasir Ani, Universiti Teknologi Malaysia, Malaysia

Chapter 6
Optimizing the Mercerisation Effect on the Mode I Fracture Toughness of Bambusa Vulgaris Bamboo Using Surface Response Method 112
Mohamad Zaki Hassan, Universiti Teknologi Malaysia, Malaysia
Siti Amni Roslan, Universiti Teknologi Malaysia, Malaysia
Zainudin A. Rasid, Universiti Teknologi Malaysia, Malaysia
Mohd Zuhri Mohamed Yusoff, Universiti Putra Malaysia, Malaysia
S. M. Sapuan, Universiti Putra Malaysia, Malaysia
Firdaus Muhammad-Sukki, Robert Gordon University, UK

Chapter 7
Investigation of Recycled Acrylonitrile Butadiene Styrene for Additive Manufacturing 130
Shajahan Bin Maidin, Universiti Teknikal Malaysia Melaka, Malaysia
Zulkeflee Abdullah, Universiti Teknikal Malaysia Melaka, Malaysia
Ting Kung Hieng, Universiti Teknikal Malaysia Melaka, Malaysia

Related Readings 155

About the Contributors 168

Index 173

Preface

Regarding on the examination of scientist from the University of Exeter and Society for the Protection of Turtles (Cyprus) on green turtles, they found that green turtles are more likely to swallow a bunch of plastics that resembles their natural diet of sea grass (University of Exeter, 2019). In another case, scientist from University of Tasmania found a serious problem that related to plastic debris that found in the seabirds (University of Tasmania, 2019). These two cases have shown serious issues on plastic pollution that really happen around the world. The impact is not only on the health and physiology on the affected animals but also on the whole eco-system. The findings on those two animals also are a sign of the disruption eco-system that caused by the serious plastic pollution. Therefore, it is our responsibility as human beings that able to think and create the solutions that could save the whole creatures on Earth. One of the ways is implementation of green materials or biodegradable materials could at least reduce the percentage of possible plastic pollution. With the new researches and development of new technologies, the chance to save the environment and the whole eco-system could be achieved. However, there are always challenges in every solution. There are a few questions that need to be answered in order to implement the green materials as solution of environmental issues effectively. Therefore, it is very important to evaluate how good is green materials as substitute materials in terms of their performance to be comparable with the conventional materials. This would require investigation on the raw materials properties and enhancement techniques in order to meet up the standard that already set by the conventional materials. Some of the end users that have little knowledge in technical information of green materials would doubt about the substitute materials and will question their technical ability. Consequently, efforts by the government to ask community to change their interest would be useless and unsuccessful. Through this book, hidden benefits of green materials are demonstrated through the researches that performed successfully by the authors. Moreover, to convince manufacturer to change their direction to green materials would require serious actions in all perspective. People would not easily accept the changes without knowing or obtaining the benefits of the changes. In particular, changes in industry would require commitments in all perspective and this would be the challenges for the industry players to implement the green materials in their scopes. The main concern is how this implementation would beneficial to them financially as they are already in stable position in the market. Generally, raw green materials are considered cheaper as they are easily obtained and readily available with lower cost of maintenance compared with the metal and plastic types of materials. However, cost of processing would raise the total cost of the end products that would put a risk for the industry player in the market. There would be a question on asking for the reason of changes by putting a risk on their business. Therefore, adequate techniques and technology should be in parallel with the demands from all stakeholders of green materials to ensure the changes of direction from conventional materials to green materials will not damage the industry financially. This book presented the studies from different researchers that have one common goal which is to promote implementation of green materials in manufacturing, product and innovation, automotive industry and others by explaining the advantages and challenges of the implementation. Opportunities of the green materials to be applied are demonstrated together with the recommended techniques and technologies that explained in every chapter.

Furthermore, environmentally friendly products have brought an attention to all stages of group of people in worldwide. The consciousness has brought to many efforts and policies that require commitment and cooperation from all. Currently, green materials could be found in most manufacturing industry as one of the efforts to support eco-friendly policies and campaigns. Moreover, interest on the research and development based on the green materials has increased as this type of material has a lot of advantages in terms of the economy, technical and environment. Green materials as defined by some researchers are materials that environmentally friendly which mean have lowest negative impact towards environment and would give stability towards eco-system during the pre and post processing of the materials. Economically, this type of material is considered cheaper as the raw materials could be found either as a waste or naturally found either it is from plant or animals (Monteiro, Lopes, Ailton Silva Ferreira, & Nascimento, 2009). For bio-materials, the process of extraction is simple and requires lowest energy consumption as they are naturally ready to be used with some simple modifications or treatment. For example, natural fibres are extracted either from the bast, leaf, seeds, fruits, or stalk. In order to improve the properties of the raw fibres, fibre treatments are required. Researchers suggested the natural fibres are treated by alkaline solution, silane or seawater to remove the impurities and enhance the interfacial bonding of the fibres with the matrix materials as a result for good composite material (Pickering, Efendy, & Le, 2016). In comparison with the steel production, production of natural fibres are almost zero emission and carbon footprint ((Mastura, Sapuan, Mansor, & Nuraini,

2016). This is what we are looking for in the new era of any industry where less negative impact on the environment is required. Therefore, implementation of this type of green materials has become an interest of researchers to widen the potential of green materials and convince the industry players about the advantages of the green materials in product and innovation. Moreover, green materials like natural fibres are proven as substitute materials for synthetic materials such as carbon fibres and glass fibres. The material properties of the natural fibres are comparable with those synthetic fibres with a few modifications and combination with other type of materials (Joshi, Drzal, Mohanty, & Arora, 2004). Natural fibre could be the reinforced materials for the hybrid composites to reduce the percentage of the polymer and make the resultant composite material more reliable. There are a lot of studies have been performed by the researchers on the material properties of the natural fibre based composite materials that exhibit excellent performance (Fazita et al., 2017). By all the advantages of green materials, opportunities to be implemented in various industries should be exposed widely.

In addition, in the context of green materials, the focus is also not on the natural fibres only. As mentioned previously, green materials also include materials that come from the waste. Recycled materials could consider as green materials as they support sustainability and give less negative impact on the environment generally. Recycled materials such as thermoplastic polymer also have reached attention from several researchers for their potential to be implemented in many product innovation and technology (Mathijsen, 2016; Nayak, Dixit, & K K, 2012; Kesarwani, Jahan, & Kesarwani, 2015). Implementation of recycled materials could greatly reduce environmental impact on pollution. However, there are challenges in implementation of recycled materials especially for polymer materials. This type of materials already being processed primarily and requires secondary process in order to improve their properties that might be degraded during the primary process. In Stoof and Pickering (2017) study also, they had mentioned about the challenges in processing recycled materials and suggested a new technique that could greatly impact the opportunities on this type of materials to be implemented in many application area. Despite the challenges of processing recycled materials, the positive impact of this type of material towards environment should not be forgotten. New technology development would lead to various research opportunities and product innovation. Moreover, implementation of waste materials or recycled materials such as plastics could reduce the issues on the waste management that concern every human being in this planet. The issues affect the whole eco-system either on or under the ground and sea. If there are more technologies that developed and exposed the recyclability of materials, the waste problem could be solved significantly and people will start realize the way to turn the waste into something that valuable.

Moreover, implementation of green, waste and recycle materials on the production of biofuels also has been increased the opportunities in research and technology development particularly in automotive industry. Generally, potential feedstocks for the production of biofuels could be obtained from non-edible oil, edible oil, animal fat, and plant oil as these feedstock are considered as cheap and abundant materials (Manaf et al., 2019). This type of material has great potential to be implemented as it significantly contribute to less pollution and harmful gaseous towards eco-system. However, even though non-edible oil has high free fatty acids which may leads to some serious drawbacks; it is still often selected by researchers due to the issue of food versus fuel. When Indonesia as one of the largest producer of palm oil in the world has making a step ahead implemented the B20 (20% palm methyl ester and 80% petroleum diesel) it shows that the future of biofuels is promising. This will not only strengthened the palm oil industry but also boost up the country's economy. Unlike natural fibres, production of bio-diesel requires more stages in order to obtain the desirable properties that could meet up with the application requirements. Therefore, this challenge has been part of motivation of researchers in this field to discover more techniques and technologies. Properties of the bio-diesel are most likely depending on its feedstock. Therefore, selection of the feedstocks and the processing techniques play an important role in implying potential of bio-diesel in various applications.

In order to fulfil the requirements based on green materials characteristic, several technologies are developed according to the researchers' finding. This would widen the opportunities of green materials to be applied in more area. Application of green materials are not only focus on the end product but it also focus on the process of the green materials that requires development of new technique and this would indirectly improve some material processing and characterization techniques. Prior to implementation of green materials, industry players have to identify the possible techniques that would not degrade the good material properties and find a way to improve them. In additive manufacturing technology, employment of green materials leads to many new techniques and knowledge that widen the opportunities of this type of material. In comparison with the traditional manufacturing process, additive manufacturing also known as environmentally friendly process where the consumption rate of the materials are higher and less waste materials are produced (Mellor, Hao, & Zhang, 2014). Moreover, parts development time also has been reduced that makes this type of manufacturing process more preferable (Thompson et al., 2016). This technology has attracted many major industries particularly automotive industry. Therefore, more technologies will be developed to implement natural fibre based materials from additive manufacturing process for automotive components.

In this book, readers would discover the potential of green materials that could be implemented in various applications. It is expected to widen the perspective of the readers about the benefits and challenges of green materials. In this new era of technologies, topic on the sustainability has brought too many attentions from all industrial players and researchers where the elements of the sustainability could give significant positive impact on all perspective. Based on the three important pillars of sustainability which are people, economy and environment, application of green materials seems compliance with the sustainable goal in product innovation and technology. Moreover, successful of implementation of green materials that demonstrated in this book would lead to more efforts and further actions from government level by being more supportive and creating more policies that could solve many environmental issues. This would motivate and attract researchers to discover more knowledge and develop more technology hands-in hands with industry.

In a summary, implementation of green materials has been proven as one of the method towards sustainability goals. Challenges and issues that occur during the whole lifecycle of the materials should be discussed and solved in order to enhance the potential and opportunities of the green materials to be applied in various areas. As demonstrated in all the chapters in this book, the importance of discovering a new techniques and technologies would open much wider range of doors for green materials. This continuous effort should be supported by exposing the challenges and solutions that could convince all stages of community about the impact of implementation of green materials.

ORGANIZATION OF THE BOOK

The book is organized into seven chapters. A brief description of each of the chapters follows:

Chapter 1 reviews the potential of green materials in additive manufacturing process particularly in fused filament fabrication. The authors discussed the characteristics of the natural fibre composites that compatible with the process parameters that required in fused filament fabrication. In overall, this chapter shows the compatibility of natural fibres to be implemented as materials for fused filament fabrication technology and how the technology could contribute to positive environmental impact.

Chapter 2 specifically presents material properties of starch cellulose bio-composites. Nanocellulose or lignocellulose fibre is used as reinforced materials for starch matrix and its mechanical properties and water uptake is reviewed based on past researchers. The potential of this bio-composites is discussed as these type of materials could be apply in food packaging industry as they are environmental friendly and non-toxic materials.

Chapter 3 reviews the employment of green materials in hybrid composites in automotive manufacturing. The authors explain about the classification of green materials and their characteristics that make them suitable for the automotive application. This chapter also discusses the challenges of this type of materials in automotive manufacturing and suggests a few approaches to overcome them.

Chapter 4 particularly discusses the biodegradable thermoplastic starch (TPS) from sago starch that requires some modifications to enhance the properties of the material. The authors suggested cross-linking method to be applied by using cross-linking agents such as organic acids. This chapter also shows techniques in cross-linking that should be included in the process for the effective results.

Chapter 5 addresses the techniques in producing non-edible oil bio-diesel using microwave irradiation technologies. This technique is explained in this chapter by emphasizing the employment of waste heterogeneous catalyst. This chapter also shows the great potential of the microwave irradiation technologies to produce the eco-friendly oil diesel economically, sustainable and productive.

Chapter 6 presents an analysis of Mercerisation effect on the mode I fracture toughness of bambusa vulgaris bamboo using Surface Response Method. This chapter also addresses the potential of natural plant fibres as alternative fibres to substitute synthetic fibres in polymer composites. However, the authors suggested some fibre modification based on their analysis on the mode I fracture toughness of the bamboo fibres.

Chapter 7 investigates the mechanical properties of 100% recycled waste Acrylonitrile Butadiene Styrene (ABS) that commonly applied as feedstock materials in additive manufacturing. Based on the investigation, the authors suggest that there is a potential for the recycled waste ABS in additive manufacturing that could contribute to positive environmental impact and sustainability. However, some materials modification or reinforcement is needed to enhance the performance of recycled waste ABS.

M. T. Mastura
Universiti Teknikal Malaysia Melaka, Malaysia

S. M. Sapuan
Universiti Putra Malaysia, Malaysia

REFERENCES

Fazita, M. R. N., Nurnadia, M. J., Khalil, H. P. S. A., Haa, M. K. M., Fizree, H. M., & Suraya, N. L. M. (2017). Green Biocomposites, 199–224. doi:10.1007/978-3-319-49382-4

Joshi, S. V., Drzal, L. T., Mohanty, A. K., & Arora, S. (2004). Are natural fiber composites environmentally superior to glass fiber reinforced composites? *Composites. Part A, Applied Science and Manufacturing*, *35*(3), 371–376. doi:10.1016/j.compositesa.2003.09.016

Kesarwani, P., Jahan, S., & Kesarwani, K. (2015). *Composites: Classification and its manufacturing process*. Academic Press.

Manaf, I. S. A., Embong, N. H., Khazaai, S. N. M., Rahim, M. H. A., Yusoff, M. M., Lee, K. T., & Maniam, G. P. (2019). A review for key challenges of the development of biodiesel industry. *Energy Conversion and Management*, *185*, 508–517. doi:10.1016/j.enconman.2019.02.019

Mastura, M. T., Sapuan, S. M., Mansor, M. R., & Nuraini, A. A. (2016). Environmentally conscious hybrid bio-composite material selection for automotive anti-roll bar. *International Journal of Advanced Manufacturing Technology*, 1–17. doi:10.100700170-016-9217-9

Mathijsen, D. (2016). Composites made from recycled thermoplastics and natural fibers: A new philosophy for the future. Reinforced Plastics, 60(3), 142–145. doi:10.1016/j.repl.2016.04.068

Mellor, S., Hao, L., & Zhang, D. (2014). Additive manufacturing: A framework for implementation. *International Journal of Production Economics*, *149*, 194–201. doi:10.1016/j.ijpe.2013.07.008

Monteiro, S. N., Lopes, F. P. D., Ailton Silva Ferreira, A., & Nascimento, D. C. O. (2009). Natural-Fiber Polymer-Matrix Composites: Cheaper, Tougher, and Environmentally Friendly. Characterization and Analysis, 61(1).

Nayak, S. K., Dixit, G., & K K, A. (2012). Sisal fiber (SF) reinforced recycled polypropylene (RPP) composites. International Journal of Plastics Technology, 16(2), 150–165. doi:10.1007/s12588-012-9042-4

Pickering, K. L., Efendy, M. G. A., & Le, T. M. (2016). A review of recent developments in natural fibre composites and their mechanical performance. *Composites. Part A, Applied Science and Manufacturing*, *83*, 98–112. doi:10.1016/j.compositesa.2015.08.038

Stoof, D., & Pickering, K. (2017). 3D Printing of Natural Fibre Reinforced Recycled Polypropylene. Processing and Fabrication of Advanced Materials, 668–691.

Thompson, M. K., Moroni, G., Vaneker, T., Fadel, G., Campbell, R. I., Gibson, I., … Martina, F. (2016). Design for Additive Manufacturing: Trends, opportunities, considerations, and constraints. CIRP Annals - Manufacturing Technology, 65(2), 737–760. doi:10.1016/j.cirp.2016.05.004

University of Exeter. (2019). Green turtles eat plastic that looks like their food. ScienceDaily. Retrieved from www.sciencedaily.com/releases/2019/08/190809085731.htm

University of Tasmania. (2019, July 30). Non-lethal impacts of seabirds' plastic ingestion. ScienceDaily. Retrieved August 22, 2019 from www.sciencedaily.com/releases/2019/07/190730092638.htm

Chapter 1
Green Material for Fused Filament Fabrication:
A Review

Mastura Mohammad Taha
Universiti Teknikal Malaysia Melaka, Malaysia

Ridhwan Jumaidin
Universiti Teknikal Malaysia Melaka, Malaysia

Nadlene M. Razali
Universiti Teknikal Malaysia Melaka, Malaysia

Syahibudil Ikhwan Abdul Kudus
Universiti Teknikal Malaysia Melaka, Malaysia

ABSTRACT

Fused filament fabrication (FFF) has been developed in additive manufacturing technology as a fast and simple manufacturing process in product design. Advantage of the process such as flexibility in terms of the materials employment has attracted many researchers to develop new materials for the feed stock filament in the heat extrusion process of FFF. Green materials or bio-composites materials have been found in FFF and successfully commercialized in the market. However, a deep research should have been performed prior the application because of the unique characteristics of the material itself. The challenge for the researchers to develop bio-composite materials as the filament in FFF technology is to determine the right composition of the composites with the right thermal, mechanical, and rheological properties. Therefore, in this study, a review has been conducted to highlight the important requirements of the process and materials. Green materials such as bio-composites have a great potential in the FFF technology and could improve the sustainability impact.

DOI: 10.4018/978-1-7998-1374-3.ch001

INTRODUCTION

Dematerialization is defined as reduction in quantity of materials used and/or the quantity of waste generated (Cleveland & Ruth, 1998). This principle included in sustainable development where it could reduce waste materials and environmental effects. Technology in manufacturing process has been developed into several techniques that support sustainability with dematerialization principle. One of the current manufacturing process technologies that optimize the usage of materials is additive manufacturing. Additive manufacturing technologies process the 3D model data and joining the materials layer upon layer according to the desired shape and geometry. This technique is opposed the principle in traditional machining where the materials are removed part upon part for desired shape and generates material waste after the process. Consequently, waste management technologies are required and consume more energy and cost. Hence, additive manufacturing is considered as one of the initiative for supporting sustainable product development.

Additive manufacturing technologies include powder bed fusion (polymers and metals, e.g.: Laser Sintering, Laser Melting), binder jetting, vat polymerization (e.g.: Stereolitography), material jetting and material extrusion (e.g.: Fused Filament Fabrication). Among these technologies, Fused Filament Fabrication or FFF (commonly referred to as Fused Filament Fabrication or FFF by Stratasys) is found to be compatible with composite materials that filled with natural s (Bikas, Stavropoulos, & Chryssolouris, 2016). The process utilizes materials with lower melting point by extruding the molten filament through movable circular nozzle. Generally, the filament is heated up to 1°C above its melting point and solidifies right after extrusion. In comparison with the other types of additive manufacturing technologies, FFF requires less cost with high speed of production due to its simplicity (Frăţilă & Rotaru, 2017). However, poor surface finish is produced and secondary manufacturing process is needed for the good quality of product finishing.

What Is Fused Filament Fabrication (FFF)?

Fused Filament Fabrication (FFF) is a type of material extrusion technology, in which, a process where a string of thermoplastic filament is dispensed through a heated nozzle or orifice and melting it in the process, in *X, Y* and *Z* (ISO/ASTM International, 2016). The principle of FFF is based on chemistry, thermal energy and layer manufacturing technology (Chua & Leong, 2015). The FFF printer deposits the filament material on a horizontal build platform, where the material cools down, and the material laying process repeats layer-by-layer until it forms a solid, three-dimensional part. The first commercialised FFF system that used material extrusion process was developed and trademarked by Stratasys Ltd. (Caffrey et al., 2016). Figure 1 shows a diagrammatic illustration of FFF process.

FFF machines are operated automatically by the printer's hardware and firmware. There are four primary stages of how FFF works, they are (a) the generation of digital 3D model from computer-aided design (CAD) software, (b) converting the 3D model to a tetrahedral mesh file, generally Standard Triangle Language (STL) format were used, (c) splitting the tetrahedral mesh into layers and producing the tool-path using a program known as a 'slicer', and (d) using computer numerical control (CNC) to move the extrusion head to produce parts (Steuben et al., 2015). Figure 2 illustrates the steps of FFF printing process.

To get an accurate printed part, FFF system, where they are also known as desktop 3D printers, allow users to adjust several process parameters such as layer height, shell thickness, part infills, print speed, temperature of the nozzle and the build platform, and the control of the cooling fan. The most important printer characters that one need to know is the size of build platform and layer height. FFF system offers variety of build platform size, ranging from 100x100x100 to 1000x1000x1000 mm. However, on average, desktop 3D printers usually offer around 200x200x200 mm build size. Layer height has significant effects on the printed part where it will determine the smoothness of part surface. A typical layer height is generally between 50 to 400 microns (0.05 to 0.4 mm). Different printers offer different maximum resolutions of layer height. Smaller layer height and nozzle diameters produces smoother surfaces and make the part produced more precise.

Figure 1. Principle of FFF (Additively, 2017)

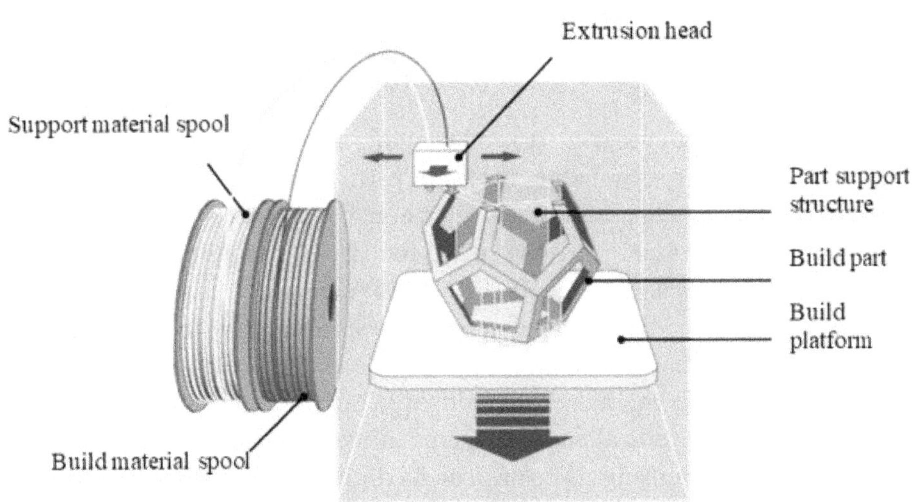

Figure 2. The FFF printing process (Steuben et al., 2015)

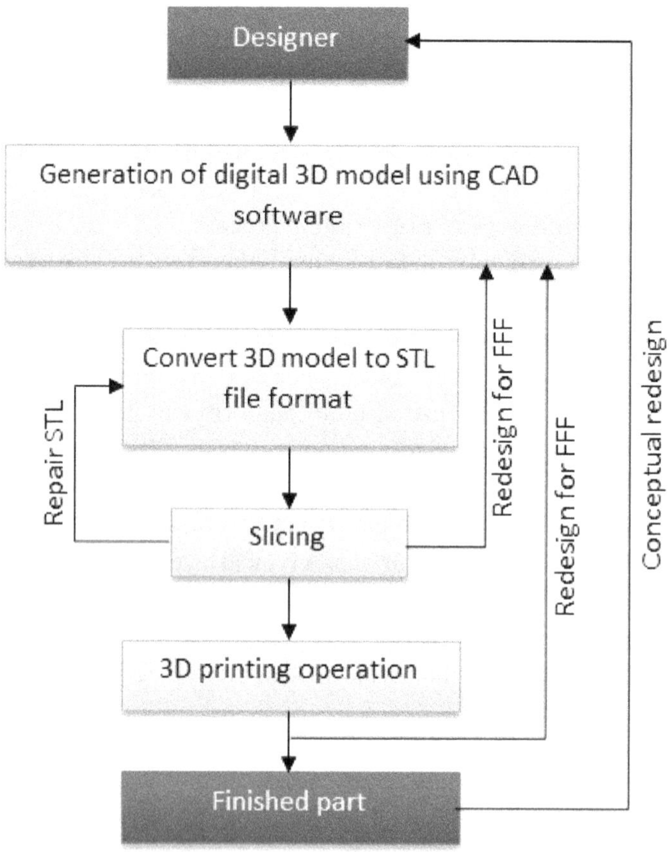

Printed part sometimes is at risk of warping and shrinkage. This is due to temperature variations surrounding the build part. To prevent this from happen, some of the solution is to print raft and brim at the base layer of the part to make the printed part more stable and has strong foundation during build up process. Thermoplastic materials such as ABS, PLA and PA are widely used because it has good mechanical properties, good temperature resistance, easy to print with, and suitable for end-use prints (Redwood et al., 2018). Users also need to consider support structure when there is overhangs and bridges in the part design. Design features like angled walls and vertical edges are some of the examples of overhangs. Any design features that overhangs is below than 45 degrees require support structures.

Application of FFF system for industrial and public use has been raised dramatically in past years as the rapid production of physical models is the key driven for the FFF market. Study shows that majority of 3D printing users, including business and individuals, owned FFF 3D printers, which contributes 75% of total market purchases and ownership (see Figure 3). The reason behind is mainly due to the price of FFF system because it is cheaper as well as easier to handle to compare with other type of 3D printing machines. On the other hand, with the usage trends of metal 3D printing increasing, the use of plastic materials remained in the lead as the most material used by the users (see Figure 4). Plastic still have a lot of advantages as it is the most convenient material for making prototypes and part production. Based on this, it can be seen that FFF system is still the first 3D printing technology that users mostly used.

Figure 3. Ownership of 3D printers type in 2018 (Sculpteo, 2018)

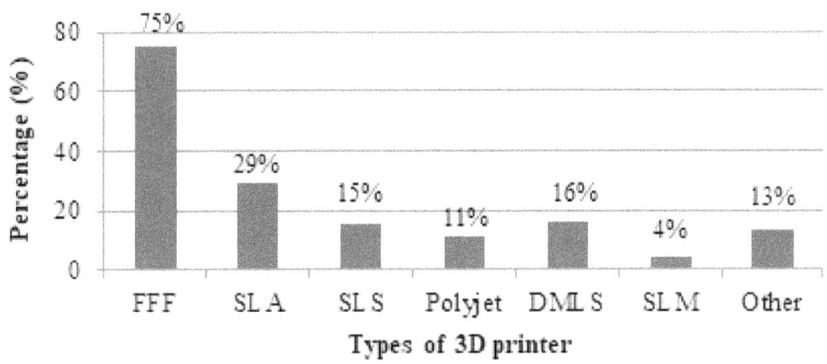

Figure 4. The use of 3D printing materials in 2018 (Sculpteo, 2018)

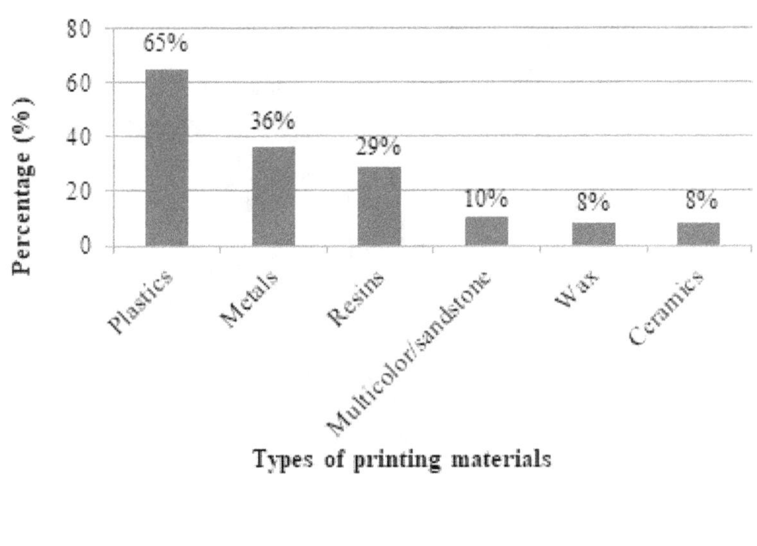

Application of Fused Filament Fabrication (FFF)

FFF could create accurate and working models for individual requirements with lower cost within shorter turnaround times. Products with low mechanical strength and have unique geometries that is difficult to be made by traditional manufacturing processes such as subtractive methods is suitable to be made through FFF system (Melenka et al., 2015). Models produced by FFF can be applied in a variety of uses. The most common applications of FFF is creating functional parts for low-volume production parts. Range of general applications of FFF includes prototypes for presentation model, research and education use, direct parts production, and artistic items. Many users were motivated by the 'do-it-yourself' (DIY) movement and used FFF system to replicate existing items, create customise and personalise objects, repair broken parts, and make household items (Abdul Kudus, 2017). Figure 5 shows the example of 3D printed customise camera mounting fabricated using FFF.

In engineering application, FFF commonly used to make patterns for investment casting, electronic enclosures and housings, form and fit testing, and customisation of jig and fixtures (Redwood et al., 2018). For example, Volkswagen Autoeurope embraced these advantages by creating highly complex designs of jigs, fixtures and other manufacturing aids such as wheel protection jig (see Figure 6). Through FFF, the geometric design of the jig can be tailored to match specific requirements at minimal extra cost. The development time also reduced from 56 to 10 days. Figure 5 and 6 shows the example of 3D printed parts fabricated using FFF.

Figure 5. Example of 3D printed part fabricated using FFF (Chris, 2018)

Figure 6. 3D printed wheel protection jig fabricated using FFF (de Vries, 2017).

Advantages and Disadvantages of FFF

Application of FFF for industrial and public use has been raised dramatically in past years as the rapid production of physical models from FFF is the key driven for the FFF market. FFF is quite popular among hobbyist where they can successfully extrude the molten filament through simple heated nozzle like pen and "print" the 2D layers one on top of another until the part is complete. Moreover, the simple printing kit can be found in cheaper price and can be handled by young user. On the other hand, various types of products have been made through FFF techniques. These include products of biomedical, electronics and toys and rapid prototyping. This technique could create accurate and working models for individual requirements with lower cost within shorter turnaround times. Therefore, product such as in biomedical disciplines is really suitable to be made through FFF where mass-customisation is possible (Stansbury & Idacavage, 2016). Moreover, the aforementioned advantages of FFF show the capability of this technique to produce simple conductive thermoplastic composite for electronic device (Leigh, Bradley, Purssell, Billson, & Hutchins, 2012).

The other advantage of FFF is its ability to produce complex geometry or complex structure based on the generated STL drawing file. Any shape and geometry of the model could be produced from the thermal extrusion of FFF process. The melting filament will be extruded through the nozzle and start to be distributed layer by layer according to the image data from STL file. This would be the ideal environmentally friendly manufacturing process since minimal waste is generated and almost all the feedstock materials are utilized during the process. It is found that the manufacturing technology could reduce the amount of waste by 10% to 20% in general (Ngo et.al, 2018). However, some of the geometry could not be produced accurately and properly due to some limitation of the product structure. Therefore, it would require supporting materials as an external support that can be removed after printing. Mohammed (2017) suggested recycling the waste of usable thermoplastic filament in order to reduce the burden of increased landfill would be viable. This strategy would open more opportunities for FFF to be applied in various product backgrounds and would lower environment resources. In addition, no tooling is required for FFF that leads to zero tooling waste, less production ramp-up time and expense (Holmström, Partanen, Tuomi, & Walter, 2010). Table 1 shows the general advantages of FFF application.

Although FFF provides several great advantage to design and product development, it also has several restrictions that limit its application. Table 2 below shows the general limitations of FFF application.

Table 1. Advantages of FFF system, adapted and adjusted from Chua and Leong (2015) and Redwood et al. (2018)

Advantages	Descriptions
Fabrication of functional parts	FFF able to fabricate parts with material that are similar to actual moulded product.
Minimal wastage	The FFF process builds part directly by extruding molten materials that needed to build the part and its support.
Ease of support removal	Support structures generated can be easily snapped off or washed-away in water-based solution.
Ease of material change	Build materials can be changed instantly when they are running low.
Ease of operation	FFF machines are generally can be handled by lay users. It requires less manufacturing engineering skills.
Large range of materials available	The most popular system for prototyping parts. Thus, wide range of thermoplastic materials with different colors is available in the market.

Table 2. Limitations of FFF system, adapted and adjusted from Abdul Kudus (2017) and Chua and Leong (2015)

Limitations	Descriptions
Restricted accuracy	Part built from FFF process have restricted accuracy due to the shape of material used, i.e. the filament form.
Slow process	Building speed is restricted by the extrusion rate of the build material from the nozzle. As the plastics viscosity is high, the build process cannot be sped up.
Unpredictable shrinkage	Shrinkages and warping are difficult to predict. Users need to experimentally adjust the process parameters of the machine.
Build-up space	The build space of FFF machines sets a physical limit to product dimensions.
Build quality	The quality of the produced parts that related to accuracy, surface finish, strength in the Z direction, and material density still needs improvement.

Impact of FFF Towards Environment in Terms of Material

In contrary with subtraction principle in traditional manufacturing process, additive manufacturing is found to be more sustainable with regards to material and energy consumption, waste management and environmental impact. Generally, materials that are used in FFF are developed in commercially available wide range of non-toxic thermoplastic such as acrylonitrile butadiene styrene (ABS) and polylactic acid (PLA). Both of these materials are safe to be used as long as the operation temperature is not exceeded their melting temperature to avoid the fumes that would cause some health problems (Frăţilă & Rotaru, 2017). However, ABS that contains styrene is likely to emit high volatile organic compounds (VOC) (Azimi et. al, 2016). Volatile organic compounds include emission of styrene, caprolactam, and lactide from thermal decomposition of thermoplastic filaments and inhaled by individuals within the surrounding area of the printing apparatus. Ultrafine particles also emitted in a form of VOC from printing process of ABS and PLA in 3D printing environment. Both of the emissions would cause health problem to FFF users such as asthma, allergies and could be transmitted through respiratory tract. Appropriate action should be taken to reduce the emission during the operation of FFF. It is suggested to install high-flow spot ventilation system and the 3D printer is enclosed with high efficiency particulate air (HEPA) filter (Azimi, Fazli, & Stephens, 2017; Kwon et al., 2017).

Another goal for the sustainable manufacturing is to reduce flux consumption that includes electrical, materials and fluids during the whole process. Particularly, FFF only requires quantity of materials that would be enough for the production and would stop operation during the completion of the fabrication. In contrary, most of traditional manufacturing or subtraction manufacturing process are prepared with excessive amount of materials and requires pre and post processing such as pre-heating and curing. Consequently, less energy efficiency is found and more electricity is consumed during the manufacturing process. Thus, layer-by-layer principle of additive manufacturing process consumes less energy and more beneficial in terms of environmental perspective (Kreiger & Pearce, 2013). Moreover, in comparison with other additive manufacturing such as SLA and Polyjet printing, filament of FFF has low energy consumption during the build process (Li et al., 2017).

FFF is based on the thermal extrusion process where it is highly depending on thermal properties of the filament materials to assure the compatibility of the filament materials. Both ABS and PLA exhibit an excellence melt characteristics and high viscosity materials that ideally for thermal extrusion process in FFF (Bourell et al., 2017). The printed products from FFF have an issue with regards to its strength and other material properties (Singh, Ramakrishna, & Singh, 2017). There are many extended studies have been conducted by researchers and they found the potential of FFF produce market-ready products. In a result, pure thermoplastic filament has to overcome this limitation by adding reinforced materials such as metal and s (Ning et. al, 2015). However there are a lot of efforts and time are needed to successfully commercialize the reinforced filament for FFF that compatible with FFF process parameters and constraints. In this study, a comprehensive review is conducted to obtain the potential and challenges of reinforced filaments of FFF from past researchers.

EMPLOYMENT OF GREEN MATERIALS FOR FUSED FILAMENT FABRICATION

Fused filament fabrication is one of the most interesting techniques in additive manufacturing due to the relatively simple process to produce a three-dimensional product from digital drawing. However, there are limitations on the range of materials available for the FFF. Therefore, researchers have increased the effort to develop alternative materials available for the FFF process. Among these materials, green materials could be one of the most promising due to the low cost, easily available, and renewable for future use.

The green materials could be referred to materials derived from natural resources such as biopolymer which could be derived from starch, protein, carrageenan, agar, alginate, etc. (Jumaidin, Sapuan, Jawaid, & Ishak, 2017; Jumaidin et.al, 2016a, 2016b; Rhim, 2012). One of the most important advantages of this material is the biodegradable characteristics which is the main drawback for the synthetic materials. The biodegradable characteristics are highly important for the sustainable development since it will produce less impact to the environment upon disposal. It is known that the petroleum based polymer has caused severe problems to the environment due to the non-biodegradable characteristics, which led to the accumulation of plastic waste in the landfill, river, and sea. This phenomenon has affected the ecosystem especially for the wildlife which often suffers after consuming the plastic waste. Hence, the applications of green materials are highly essential especially in the new technology such as FFF in order to create an established platform for the application of these materials in this technology.

Biopolymer

Earlier, Lam et al., (2002) has developed a tissue scaffold using 3D printing method. In their study, the scaffold was produced by using the mixture of corn starch, dextran, and gelatine (Lam, Mo, Teoh, & Hutmacher, 2002). These materials were mixed according to 50:30:20 ratio of corn starch: dextran: gelatine. The printed materials were undergone post-processing in order to enhance the strength of the material and to increase the resistance of the product against water uptake. The post-processing method includes drying the product at 100°C for 1 hour after printing. According to Lam et al (2002), the printed product has become strong enough to be manually handled after drying. Another treatment was conducted on the product via infiltration with different amount of copolymer solutions derived from 75% poly (L-lactide) acid and 25% poly-caprolactone. Another batch of samples was infiltrated by the same method, but after drying, the scaffolds were soaked in deionised water for 10 minute and dried at 100 °C. The treated product were undergone several testing such as water uptake, differential scanning calorimetry, mechanical test, and porosity analysis. The results show that the products were subjected to shrinkage in the range of 2 to 12% after the processing. The treatment of the composites using copolymer has successfully increase the resistance of the material to water absorption. The samples that subjected to 3-days water-soaking were lasted without much visible surface degradation and the Scanning Electron Microscope (SEM) image shows "undegraded" and uniform interior. Lam et al (2002) attributed the phenomena to the uniform infiltration of copolymer into the specimen. The copolymer coats the microstructure of the specimens both externally and internally. In terms of mechanical properties, the post-processing treatment was reported to improve the yield strength and stiffness of the product. The microporosity also decreased after infiltration since the micropores and void has been infiltrated by the copolymer. Figure 7 shows the 3D printed scaffolds from green material.

Another type of green material that was used for 3D printing is poly (lactic acid) (PLA). PLA is an interesting materials which is extensively utilized for biodegradable and renewable thermoplastic application, and have great potential as alternative for conventional petrochemical-based polymers (Rasal, Janorkar, & Hirt, 2010). In

Figure 7. 3D printed scaffolds from corn starch, dextran, and gelatine (Lam et al., 2002)

general, PLA is produced by converting sugar or starch using bacterial fermentation or petrochemical route. This biopolymer have been used for resorbable structures and prosthetic devices due its bio-resorbability and biocompatibility (Rasal et al., 2010). Apart from that, this biopolymer is categorized under thermoplastic material which made this material suitable as a 3D printing material. PLA is also known for its renewability, biodegradability, transparency, processability, and mechanical properties.

In a previous study, a group of researchers investigates the fatigue behaviour of PLA parts processed by FFF (Afrose et.al, 2016). Afrose et al. (2016) investigates the effect of part build orientation on the tensile fatigue properties of PLA material. Three different build orientations (X, Y and 45°) were used for this study. In static condition, the results shows that the parts processed in Z build orientation exhibit the higher tensile stress, which is in the range of 60% until 64% of the injection moulded PLA samples. Meanwhile, in cyclic condition, the part in 45° shows higher fatigue life than the other orientations. The results reported in these authors have provided a significant finding on the performance of 3D printed materials under cyclic loading. Figure 8 and 9 shows the build orientation and deposited toolpath pattern on the tensile samples.

On the other hand, high cost is a major constraint for PLA as a green material that has confined the wide application of these materials in the market. Therefore, researchers have introduced alternative approach to reduce the cost of this green

Figure 8. Build orientation of tensile samples (Afrose et al., 2016)

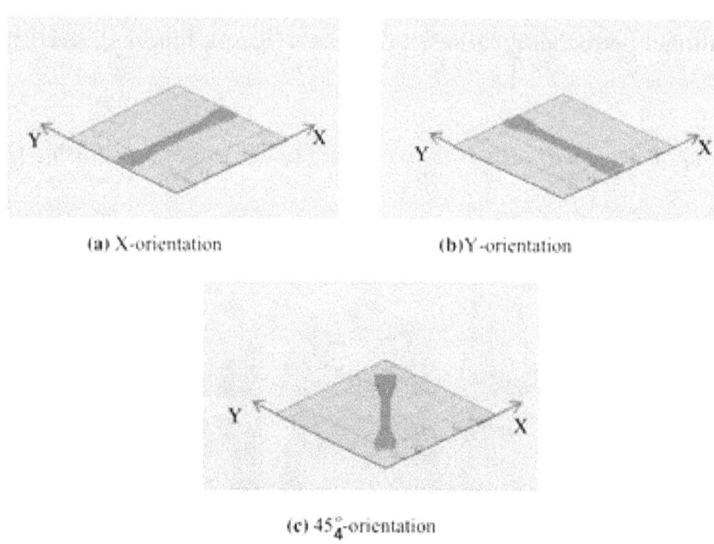

Figure 9. Deposited toolpath pattern on the specimen in three build orientations (Afrose et al., 2016)

(a) X-orientation (b) Y-orientation (c) 45°-orientation

material by mixing the PLA with natural fibre. This approach will transform the biopolymer into bio-composites which has potential to reduce the costs of the material while improving the mechanical properties as well.

Another study conducted by (Kuo et al., 2016) has reported the blending of starch with acrylonitrile-butadiene-styrene copolymers (ABS). In general, the polymer blend consists of starch, plasticizers (i.e. glycerol and water), enzyme (i.e. a-iso-amylase), ABS (U200B), compatibilizer (i.e. styrene-maleic anhy-dride copolymer; SMA), impact modifier (i.e. methyl- methacrylate butadiene styrene; MBS), and pigments (i.e. titanium white (TiO^2) and carbon black (CB)). The polymer blend was compounded using twin-screw extruder to produce filament with 1.75 mm of diameter. The experimental results show that the polymer blend shows superior physical properties than the commercial ABS filament in terms of mechanical properties, thermal resistance, flowability, and emissions of volatile organic compounds (VOCs). The authors attributed this finding to the introduction of SMA as the compatibilizer. Regardless of the weak nature of starch, the findings reported by this study has proven that modification of conventional polymer (i.e. ABS) with green material (i.e starch) has successfully improved the properties of the materials while increasing the environmental friendly aspect of the material at the same time. Hence, this might trigger more study on the utilization of green materials in 3D printing technology.

Natural Fibre Composite

The incorporation of natural fibre into biopolymer might help to reduce the cost of expensive biopolymer such as PLA by replacing the portion with natural fibre which commonly exists as agriculture processing waste. A study conducted by Matsuzaki et al., (2016) developed a method for 3D printing of PLA reinforced with continuous carbon fibre or twisted yarn of natural fibre (Matsuzaki et al., 2016). The technique used for this study is quite unique where the PLA filament was separately supplied to the 3D printer and the fibres were impregnated with the filament within the heated

Figure 10. Schematic of the 3D printer head used to produce continuous fibre reinforced polymer composites using in-nozzle impregnation based on FFF (Matsuzaki et al., 2016)

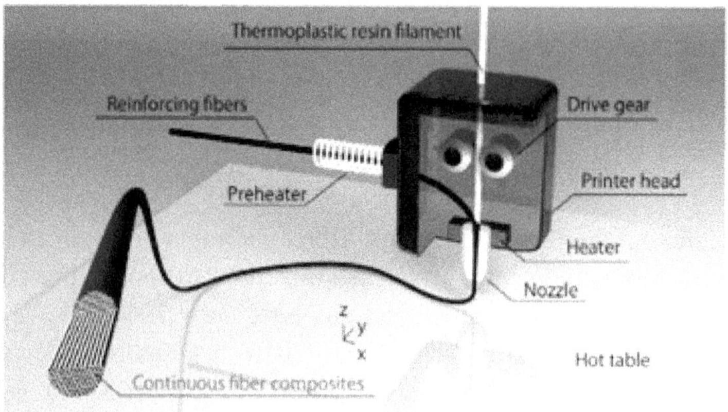

nozzle of the printer immediately before printing. Figure 10 shows the schematic of the 3D printer head used to produce continuous fibre reinforced polymer composites using in-nozzle impregnation based on FFF.

The results show that the incorporation of jute fibre has successfully improved the mechanical properties of the PLA matrix (i. e tensile strength, tensile modulus, and tensile strain). The authors attributed the properties improvement to the orientation of the fibre. The fibres were aligned in longitudinal direction which is according to the loading direction during tensile test. Hence, jute fibres were aligned along the longitudinal direction, which was chosen as the loading direction, and no fibres were aligned in the transverse (y) or thickness (z) directions in the gage region of the tensile test specimens. Therefore, the mechanical properties of the tensile specimen are orthotropic and maximized along the loading direction. Figure 11 shows the 3D printed jute fibre reinforced PLA composites. In addition, the authors claimed that the 3D printed product produced using this method is superior to the conventional 3D printing method.

In general, the findings reported from the previous study shows the feasibility of employing natural resources i. e biopolymer and natural fibre in 3D printing technology. The potential of these materials might be significant for cost reduction while adding environmental friendly aspect on the final product as well. Hence, more studies should be carried out on the modification of conventional 3D printing materials using bio-based resources in order to increase the potential of other bio-based materials in this technology.

Figure 11. 3D printed jute fibre reinforced PLA composites (Matsuzaki et al., 2016)

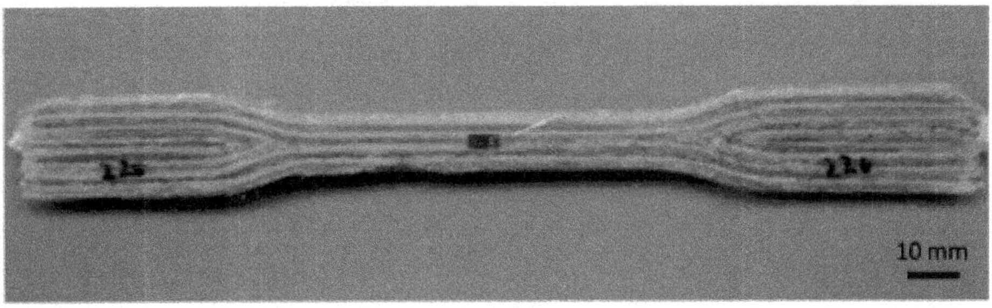

MATERIAL REQUIREMENT FOR FFF

Selection of suitable material for 3D printing is no longer a straight forward process. Radically new materials have also been considered and regularly used (Castles et al. 2016). Previously, PLA and ABS have been used widely as the two main polymers for 3D printing. However, it was found out that their selection reason was not mostly due to feasibility reasons. Therefore, other polymers should be considered to replace their dominance in the industry.

Nowadays, most of popular material for FFF is polymers and composites such as; PLA, ABS, Polyethylene terephthalate (PET), Nylon, and Thermoplastic Polyurethane (TPU). Each material has different specific properties. The usage of the material is depending on the applications. The materials are generally classified into three categories; mechanical performance, process / fabrication requirement and visual quality. The choice of material really depends on product types, cost and applications. Other important criteria to make decision of material usage is ease of printing, visual quality, maximum stress, elongation at breaks, impact resistance, impact resistance, layer adhesion, and heat resistance.

Mechanical Properties

Composites (metal and polymer with filler) material is widely used as the based material for the 3D printing product. Recently, researchers were using polymer based material due to the good characteristic in term of cost reduction. Polymers have low melting point, T_m and for liquid types are widely used in 3D printing industry because of their low weight, low cost and processing flexibility (Raut et. al., 2014). The advantages of using 3D printer in fabricating samples is it can be able to produce the complex shape of design with good quality of surface finish (Wang et. al., 2017). However, polymer product by 3D printer has a lot of challenges especially on the

Table 3. Criteria to be consider for material used in FFF

Criteria	Specific characteristic
Ease of printing	Bed adhesion, maximum printing speed, flow accuracy, ease to feed into the printer
Visual Quality	Visual appearance after printing; in of the surface finish of the material
Maximum Stress	Maximum stress the object can withstand under a certain load
Elongation at break	Maximum length the object has been stretched before breaking
Impact resistance	Energy needed to break an object with a sudden impact.
Heat resistance	Maximum temperature the object can sustain before softening and deforming
Layer adhesion	The adhesion between layers of material is. It is linked to "isotropy" (uniformity in all directions): the better the layer adhesion, the more isotropic the object will be.

stability; functionality and lack mechanical strength make the applications limited. Solutions to this problem, researchers were actively studies on combining various materials for achieving desired mechanical and functional properties. Therefore, in recent years, development of composite materials that are compatible with the available printers has attracted great attentions.

In polymer composites engineering, particle fillers are widely used to improve the mechanical properties of printed product. Particles are easy to be mixed with polymers, especially in powder form and further to be extruded into printable filaments for FFF process. Table 4 summarizes various particle reinforced polymer materials used for FFF and the properties improvements of resulting composites. The crucial part in the 3D printing of particle reinforced composites is to improve tensile strength and storage modulus. For examples, by adding glass beads (Athreya et. al., 2006), iron or copper particles (Nikzad et. al., 2011) can improved that tensile properties. Besides that, wear resistance can be improve by adding aluminium and aluminium oxide (Al_2O_3) (Boparai et. al., 2015). From the findings, the powder particle is focused on the metal powder and ceramics. Less number of literatures published on the natural s. This might be due to the instability of the .

Duigou et. al. (2016) carried out an experiment in order to evaluate the mechanical properties of wood reinforced bio-composites prepared by FFF. From their findings, mechanical properties of the composites were dependent on the printing orientation (0 or 90^0) due to anisotropy and the printing width (overlapping of filaments). The microstructure analysis is relatively high porosity as shown in Figure 12. This phenomenon contributed to the damage mechanisms; increase the water absorption and swelling of the samples. However, this type of microstructures achieved by printing could be used to produce bio-composites with a faster moisture-induced bending response compared to compressed samples.

Table 4. The effect of filler to based material for FFF

Types of filler	Improvement - effect of fillers	Reference
Iron/ABS, and Copper/ABS	Improved storage modulus and thermal conductivity, reduced coefficient of thermal expansion	(Nikzad et. al., 2011b)
BaTiO3/ABS Thermoplastic	Improved and tuneable dielectric permittivity, periodic and graded structures were printed	(Castles et al., 2016)
Thermoplastic elastomer/ABS	Reduced anisotropy of printed components	(Torrado et. al., 2014)
Short glass fibre/ABS	improve the strength of an ABS filament at the expense of reduced flexibility and handleability	(Zhong et. al., 2001)

Stoof & Pickering (2017) conducted a research on the using of waste thermoplastic materials blended with natural s. They used hemp and harakeke s for the fillers and all the samples went through tensile test in order to evaluate the mechanical properties of the samples fabricated by using 3D printer. The s were controlled in the form of short with length of 8mm. Comparing these results to those of plain polypropylene (PP) filament with the natural composites, showed improvements in tensile strength and Young's modulus of 77% and 275% respectively. Figure 13 shows one the findings results. Furthermore, the shrinkage issue showed improvement for composites sample compare to the neat PP samples as shown in Figure 14.

Thermal Properties

Thermal effect one of the important criteria needs to be considering in choosing the best material which can suit the FFF application. Thermogravimetric Analysis (TGA) and Differential Scanning Calorimetry (DSC) is the example method to identify the glass transition temperature and the exact temperature for FFF process. The DSC results can give the range of temperature in FFF setup which is allow for the correct decision to be made when selecting deposition temperatures by setting the upper and lower limits on suitable fabrication temperatures.

A lot of studies have been done on the thermal characterization of the polymer material that used in FFF in order the to obtain the glass transition temperature and melting temperature for the process set-up (Ajinjeru et al., 2016; Elbadawi, 2018; Jagenteufel et al., 2017; Valášek et. al., 2012). For thermoplastics material, fabrication temperature is a crucial indicator where it need to obtain the appropriate value to setup the processing parameter to produce the target strength and modulus of the product.

Figure 12. 3D printing of wood bio-composites and the cross section of the samples which high porosity found in the samples (Le Duigou et al., 2016)

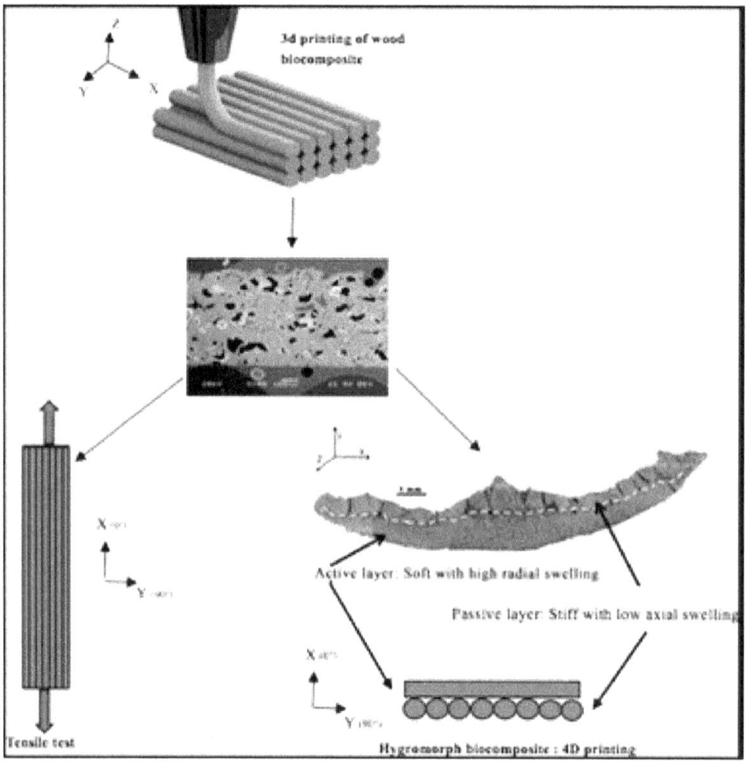

Figure 13. Tensile strength of post-consumer PP/Harakeke and PP/hemp composites (Stoof & Pickering, 2017)

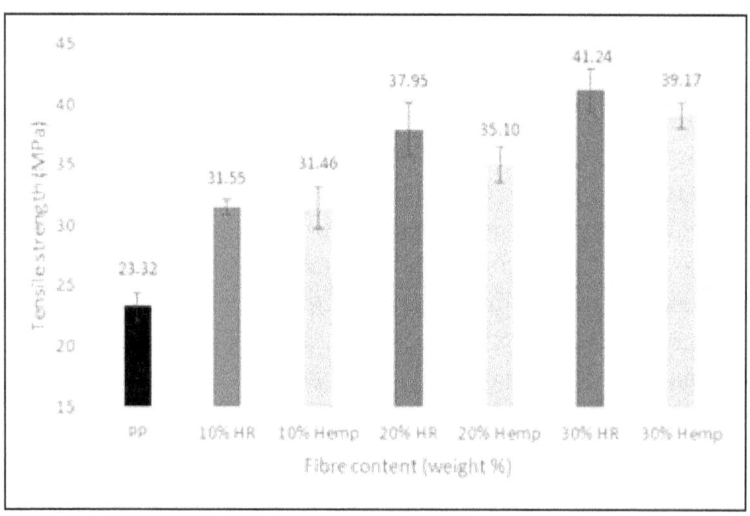

Figure 14. a) Plain polypropylene shrinkage sample b) 30 wt% hemp shrinkage sample (Stoof & Pickering, 2017)

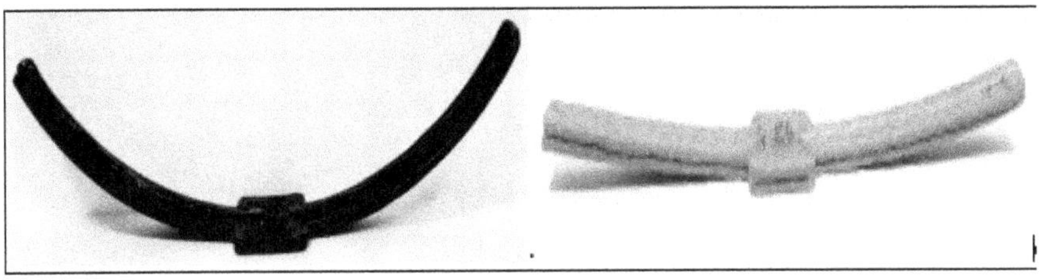

Rheological Properties

Rheological properties affect high product quality to the polymer fabrication. The rheology properties need to understand well before starts the fabrication especially in addictive manufacturing. The characterization data is reliable and provide extensive physical information. Based on the literature review, a few studies have highlighted the rheological effect on the mechanical and physical properties of the product on the FFF printing process such as those from Caroline A. Murphy & Collins (2016) and McIlroy & Olmsted (2017). Thus, an area which is vital to be explored is research on rheology of polymer during FFF application in order to understand how its melting characteristics influence the final printing result.

Rheological measurements of the material can be done by using rheometer Physica MCR 301(Jagenteufel et al., 2017). For a small-amplitude oscillatory shear were performed on a Rheostress RC501 (Arsad et. al., 2011) and other equipment can be used is Discovery Hybrid Rheometer-2 (DHR-2) (Ajinjeru et al., 2016).

The viscosity of the polymer material is influenced by the introducing filler material. The viscosity of neat polymer is consistent compared to the polymer blended with reinforced material. Arsad et.al studied the effect of mechanical and rheological properties of polyamides and ABS blended with short glass. The viscosity of the material increase with incorporation and uniformity of short glass blended with the polymer. However, introducing filler slightly decrease the viscosity of the material (Arsad et al., 2011).

The optimum viscosity set up of the material needs to identify earlier before the fabrication process. The consequent of high viscosity is the filament buckling (Elbadawi, 2018) as illustrated in Figure 15. The filament will acts as the piston that drives the extrusion/ FFF process. If the filament did not flow at the optimum rate it can create a backpressure to the filament, and will cause buckling to the filament. Pressure and stress applied in the extrusion process must be controlled not exceed the critical stress limit, σ_c of the filament to avoid buckling. It is important since the product quality is concerned, this is not desirable.

Figure 15. Illustration depicting buckling and no buckling conditions(Elbadawi, 2018)

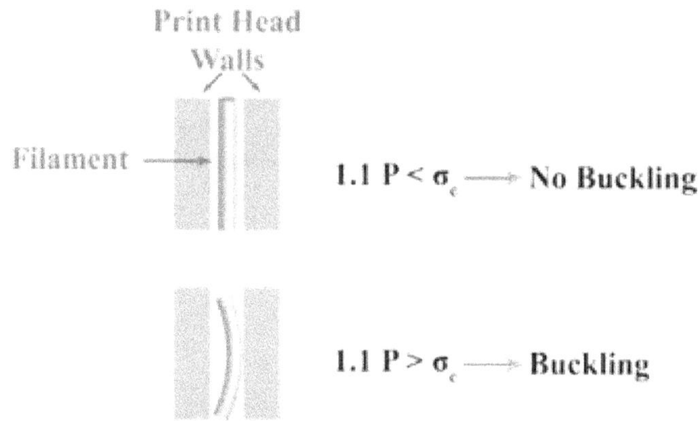

PROCESS REQUIREMENT FOR FFF

Pre-Processing

Readily available filaments could be purchased with regards to the machine of FFF and its producer. Some of the commercial filaments compatible for any type of 3D printers but some of the producers recommend using only their original "machine based custom-made" filament which is very expensive. Consequently, the end users try to develop their own filament which is low cost and able to achieve to the desired properties. Moreover, reinforced materials are added in preparation of filament FFF to enhance the properties of thermoplastic filament. Preparation of this type of filament FFF consists of compounding and palletizing the materials before extrusion process for filament. Stoof and Pickering (2017) had prepared 10, 20, 30, 40 and 50% hemp and harakeke reinforced polypropylene filament using twin screw extruder and spooling machine to keep the dimensional consistency of the filament. Kariz et al. (2018) prepared 1.75 mm wood-PLA filament from pallet form and were extruded using a single-screw filament. Šafka et al. (2016) incorporated coir s into the ABS by compounding these two materials by granulation process and two-screw extruder which is advisable for natural s reinforced thermoplastic composites. It will increase the effectivity and homogeneity and quality of the final material.

Post-Processing

Post-processing usually requires secondary manufacturing process for fabrication part to archive the design requirement and meet some particular standards and specifications. Fused Filament Fabrication as one type of the additive manufacturing is known as near-zero material waste that requires minimum post-processing procedure. However, extrusion of the filament through FFF technology requires secondary manufacturing process due to its limitation in terms of geometrical and surface quality (Boschetto, Bottini, & Veniali, 2016). Surface quality of fabrication parts implies dimension accuracy that becomes an issue for some manufacturing process. This would cause damages in terms of the cost and time for in design process. Therefore, manufacturers will find the optimum process parameter to improve the surface quality and dimension accuracy. In another way, Boschetto and Bottini (2015) proposed an integration between FFF with barrel finishing where a geometrical model of the profile under the action of machining is proposed. This model will predict the surface morphology by reducing the expensive trial and error phase in the post-processing operation. In another study, Boschetto and Bottini (2016) found that layer thickness and deposition angle influence surface quality of parts. Manufacturers are require to find the optimum value of these two parameter to ensure the parts that will eventually functioned and assembled as per planned.

Furthermore, post-processing procedure is not only involved machining but it is also involving some chemical process where the unwanted materials in the part are removing using chemical procedure. Gibson, Rosen, and Stucker (2015) mentioned that post-processing for additive manufacturing can be performed by solvent bathing, sand blasting, sealing parts, electrochemical polishing, bonding and gluing. Type of the material of the filament also determines type of post-processing that required for the fabricated parts. Ceramic filaments are not easy to be extruded in the 3D printer especially in terms of accuracy of the printing and layer-by-layer appearance. Therefore, fabrication of complex shapes is followed by sintering to produce attractive ceramics parts (Ngo et al., 2018).

Although FFF produces almost zero material waste, the appearance of the fabricated parts requires post-processing operation to improve its surface quality and dimension accuracy. There are two types of post-processing operation which is chemical and physical post-processing. These two types of process are mainly determined by their types of materials. Metal or polymer based materials are mostly undergo some machining to remove excessive materials or supporting materials. This includes cutting and surface finishing. For metal based materials, heat treatment also involves in the post-processing procedure to improve the morphology of the fabricated parts. Ceramic based parts require sintering for better appearance of the fabricated parts. On the other hand, there are several methods to reduce or predict the surface quality issues on the fabricated parts and avoid the post-processing operations. Optimizing the process parameters, for example deposition angle and layer thickness would improve the surface quality of the parts. It is important for the manufacturer to carefully determine the process phase to avoid cost and time damage.

CONCLUSION

In a conclusion, FFF allows flexibility in materials for the feed filament to increase the level of the technology of additive manufacturing. In consideration of sustainability and environmentally friendly, green materials are employed in the heat extrusion technique to fabricate parts by using 3D printer. This type of material is mostly plant based fibres that blended in the polymer as composites with unique material characteristics. It is a bit challenging for heat extrusion to process the composite filaments as the materials are limited in terms of the thermal and rheological processing. It is important to the manufacturer to determine the optimum process parameter to meet the fibre based material characteristics to produce high quality parts. Therefore, several criteria are requires to consider the green materials in FFF. However, the potential of green materials in FFF will greatly give positive impact towards all perspective including manufacturing industry, environment and social.

REFERENCES

Abdul Kudus, S. I. (2017). *The Value Of Personalised Consumer Product Design Facilitated Through Additive Manufacturing Technology*. Loughborough University.

Additively. (2017). Fused deposition modeling. Author.

Afrose, M. F., Masood, S. H., Iovenitti, P., Nikzad, M., & Sbarski, I. (2016). Effects of part build orientations on fatigue behaviour of FDM-processed PLA material. Progress in Additive Manufacturing, 1(1–2), 21–28. doi:10.1007/s40964-015-0002-3

Ajinjeru, C., Kishore, V., Chen, X., Lindahl, J., Sudbury, Z., Hassen, A. A., ... Duty, C. (2016). The Influence of Rheology on Melt Processing Conditions of Amorphous Thermoplastics for Big Area Additive Manufacturing (BAAM). *Solid Freeform Fabrication, 2016*, 754–761.

Arsad, A., Rahmat, A. R., Hassan, A., & Iskandar, S. N. (2011). Mechanical and rheological characterization of PA6 and ABS blends-with and without short glass fibre. *Journal of Applied Sciences (Faisalabad), 11*(13), 2313–2319. doi:10.3923/jas.2011.2313.2319

Athreya, S. R., & Suman Das, K. K. (2006). Processing and properties of glass bead particulate-filled functionally graded Nylon-11 composites produced by selective laser sintering. *Materials Science and Engineering A*, *437*(2), 226–234. doi:10.1016/j.msea.2006.07.112

Azimi, P., Fazli, T., & Stephens, B. (2017). Predicting Concentrations of Ultrafine Particles and Volatile Organic Compounds Resulting from Desktop 3D Printer Operation and the Impact of Potential Control Strategies. *Journal of Industrial Ecology*, *21*(S1), S107–S119. doi:10.1111/jiec.12578

Azimi, P., Zhao, D., Pouzet, C., Crain, N. E., & Stephens, B. (2016). Emissions of ultrafine particles and volatile organic. *Environmental Science & Technology*, *50*(3), 1260–1268. doi:10.1021/acs.est.5b04983

Bikas, H., Stavropoulos, P., & Chryssolouris, G. (2016). Additive manufacturing methods and modeling approaches: A critical review. *International Journal of Advanced Manufacturing Technology*, *83*(1–4), 389–405. doi:10.1007/s00170-015-7576-2

Boparai, K., Singh, R., & Singh, H. (2015). Comparison of tribological behaviour for Nylon6-Al-Al 2 O 3 and ABS parts fabricated by fused deposition modelling. *Virtual and Physical Prototyping*, *10*(2), 59–66. doi:10.1080/17452759.2015.103 7402

Boschetto, A., & Bottini, L. (2015). Surface improvement of fused deposition modeling parts by barrel finishing. *Rapid Prototyping Journal*, *21*(6), 686–696. doi:10.1108/RPJ-10-2013-0105

Boschetto, A., & Bottini, L. (2016). Design for manufacturing of surfaces to improve accuracy in Fused Deposition Modeling. *Robotics and Computer-integrated Manufacturing*, *37*, 103–114. doi:10.1016/j.rcim.2015.07.005

Boschetto, A., Bottini, L., & Veniali, F. (2016). Finishing of Fused Deposition Modeling parts by CNC machining. *Robotics and Computer-integrated Manufacturing*, *41*, 92–101. doi:10.1016/j.rcim.2016.03.004

Bourell, D., Kruth, J. P., Leu, M., Levy, G., Rosen, D., Beese, A. M., & Clare, A. (2017). Materials for additive manufacturing. CIRP Annals - Manufacturing Technology, 66(2), 659–681. doi:10.1016/j.cirp.2017.05.009

Caffrey, T., Wohlers, T., & Campbell, R. I. (2016). *Wohlers Report 2016*. Fort Collins, CO: Wohlers Associates.

Castles, F., Isakov, D., Lui, A., Lei, Q., Dancer, C. E. J., Wang, Y., & Janurudin, J. M. (2016). *Microwave dielectric characterisation of 3D-printed BaTiO 3 / ABS polymer composites*. Nature Publishing Group; doi:10.1038rep22714

Chris, L. (2018). *BetaFPV 75x Customizable Cam Mount*. Academic Press.

Chua, C. K., & Leong, K. F. (2015). *3D Printing and Additive Manufacturing: Principles and Applications* (4th ed.). Singapore: World Scientific Publishing.

Cleveland, C. J., & Ruth, M. (1998). Indicators of Dematerialization and the Materials Intensity of Use. *Journal of Industrial Ecology*, *2*(3), 15–50. doi:10.1162/jiec.1998.2.3.15

de Vries, C. (2017). *Volkswagen Autoeuropa: Maximizing Production Efficiency with 3D Printed Tools and Fixtures*. Academic Press.

Elbadawi, M. (2018). *Polymeric Additive Manufacturing: The Necessity and Utility of Rheology*. Polymer Rheology; doi:10.5772/intechopen.77074

Frăţilă, D., & Rotaru, H. (2017). Additive manufacturing – a sustainable manufacturing route. MATEC Web of Conferences, 94, 03004. 10.1051/matecconf/20179403004

Gibson, I., Rosen, D., & Stucker, B. (2015). Additive Manufacturing Technologies: 3D Printing and Direct Digital Manufacturing. *Johnson Matthey Technology Review*, *59*(3), 498.

Holmström, J., Partanen, J., Tuomi, J., & Walter, M. (2010). Rapid manufacturing in the spare parts supply chain: Alternative approaches to capacity deployment. *Journal of Manufacturing Technology Management*, *21*(6), 687–697. doi:10.1108/17410381011063996

ISO/ASTM International. (2016). *ISO/ASTM 52900: Standard Terminology for Additive Manufacturing Technologies*. ISO/ASTM International; doi:10.1520/F2792-12A.2

Jagenteufel, R., Hofstaetter, T., Kamleitner, F., Pedersen, D., Tosello, G., & Hansen, H. (2017). Rheology of high melt strength polypropylene for additive manufacturing. Advanced Materials Letters, 8(6), 712–716. doi:10.5185/amlett.2017.1450

Jumaidin, R., Sapuan, S. M., Jawaid, M., Ishak, M. R., & Sahari, J. (2016a). Characteristics of Thermoplastic Sugar Palm Starch/Agar Blend: Thermal, Tensile, and Physical Properties. *International Journal of Biological Macromolecules*, *89*, 575–581. doi:10.1016/j.ijbiomac.2016.05.028

Jumaidin, R., Sapuan, S. M., Jawaid, M., Ishak, M. R., & Sahari, J. (2016b). Effect of seaweed on physical properties of thermoplastic sugar palm starch/agar composites. *Journal of Mechanical Engineering Science, 10*(3), 2214–2225. doi:10.15282/jmes.10.3.2016.1.0207

Jumaidin, R., Sapuan, S. M., Jawaid, M., Ishak, M. R., & Sahari, J. (2017). Effect of Agar on Flexural, Impact, and Thermogravimetric Properties of Thermoplastic Sugar Palm Starch. *Current Organic Synthesis, 14*(2), 200–205. doi:10.2174/1570179413666160921110732

Kariz, M., Sernek, M., Obućina, M., & Kuzman, M. K. (2018). Effect of wood content in FDM filament on properties of 3D printed parts. *Materials Today Communications, 1*, 135–140. doi:10.1016/j.mtcomm.2017.12.016

Kreiger, M., & Pearce, J. M. (2013). Environmental impacts of distributed manufacturing from 3-D printing of polymer components and products. *Materials Research Society Symposium Proceedings*, 1492, 85–90. 10.1557/opl.2013.319

Kuo, C. C., Liu, L. C., Teng, W. F., Chang, H. Y., Chien, F. M., Liao, S. J., ... Chen, C. M. (2016). Preparation of starch/acrylonitrile-butadiene-styrene copolymers (ABS) biomass alloys and their feasible evaluation for 3D printing applications. *Composites. Part B, Engineering, 86*, 36–39. doi:10.1016/j.compositesb.2015.10.005

Kwon, O., Yoon, C., Ham, S., Park, J., Lee, J., Yoo, D., & Kim, Y. (2017). Characterization and Control of Nanoparticle Emission during 3D Printing. *Environmental Science & Technology, 51*(18), 10357–10368. doi:10.1021/acs.est.7b01454

Lam, C. X. F., Mo, X. M., Teoh, S. H., & Hutmacher, D. W. (2002). Scaffold development using 3D printing with a starch-based polymer. *Materials Science and Engineering C, 20*(1-2), 49–56. doi:10.1016/S0928-4931(02)00012-7

Le Duigou, A., Castro, M., Bevan, R., & Martin, N. (2016). 3D printing of wood biocomposites: From mechanical to actuation functionality. *Materials & Design, 96*, 106–114. doi:10.1016/j.matdes.2016.02.018

Leigh, S. J., Bradley, R. J., Purssell, C. P., Billson, D. R., & Hutchins, D. A. (2012). A Simple, Low-Cost Conductive Composite Material for 3D Printing of Electronic Sensors. *PLoS One, 7*(11), e49365. doi:10.1371/journal.pone.0049365

Li, Y., Linke, B. S., Voet, H., Falk, B., Schmitt, R., & Lam, M. (2017). Cost, sustainability and surface roughness quality – A comprehensive analysis of products made with personal 3D printers. *CIRP Journal of Manufacturing Science and Technology, 16*, 1–11. doi:10.1016/j.cirpj.2016.10.001

Matsuzaki, R., Ueda, M., Namiki, M., Jeong, T. K., Asahara, H., & Horiguchi, K. (2015, December). Three-dimensional printing of continuous-fibre composites by in-nozzle impregnation. *Scientific Reports*, *6*, 1–7. doi:10.1038rep23058

McIlroy, C., & Olmsted, P. D. (2017). Deformation of an Amorphous Polymer during the Fused-Filament-Fabrication Method for Additive Manufacturing. *Journal of Rheology (New York, N.Y.)*, *61*(2), 379–397. doi:10.1122/1.4976839

Melenka, G. W., Schofield, J. S., Dawson, M. R., & Carey, J. P. (2015). Evaluation of Dimensional Accuracy and Material Properties of the MakerBot 3D Desktop Printer. *Rapid Prototyping Journal*, *21*(5), 618–627. doi:10.1108/RPJ-09-2013-0093

Murphy & Collins. (2016). Microcrystalline Cellulose Reinforced Polylactic Acid Biocomposite Filaments for 3D Printing Caroline. *Polymer Composites*, *116*(1), 37–43. doi:10.1002/pc

Mohammed, M. I., Mohan, M., Das, A., Johnson, D., Singh, M., Badwal, P., . . . Gibson, I. (2017). A low carbon footprint approach to the reconstitution of plastics into 3D-printer filament for enhanced waste reduction. KnE Engineering, 2(2), 234. doi:10.18502/keg.v2i2.621

Ngo, T. D., Kashani, A., Imbalzano, G., Nguyen, K. T. Q., & Hui, D. (2018). Additive manufacturing (3D printing): A review of materials, methods, applications and challenges. *Composites. Part B, Engineering*, *143*, 172–196. doi:10.1016/j.compositesb.2018.02.012

Nikzad, M., Masood, S. H., & Sbarski, I. (2011). Thermo-mechanical properties of a highly filled polymeric composites for Fused Deposition Modeling. *Materials & Design*, *32*(6), 3448–3456. doi:10.1016/j.matdes.2011.01.056

Ning, F., Cong, W., Qiu, J., Wei, J., & Wang, S. (2015). Additive manufacturing of carbon fi ber reinforced thermoplastic composites using fused deposition modeling. *Composites. Part B, Engineering*, *80*, 369–378. doi:10.1016/j.compositesb.2015.06.013

Rasal, R. M., Janorkar, A. V., & Hirt, D. E. (2010). Poly(lactic acid) modifications. Progress in Polymer Science (Oxford), 35(3), 338–356. doi:10.1016/j.progpolymsci.2009.12.003

Raut, S., Jatti, V. S., Khedkar, N. K., & Singh, T. P. (2014). Investigation of the effect of built orientation on mechanical properties and total cost of FDM parts. *Procedia Materials Science*, *6*, 1625–1630. doi:10.1016/j.mspro.2014.07.146

Redwood, B., Schöffer, F., & Garret, B. (2018). The 3D Printing Handbook: Technologies, Design and Applications. Amsterdam: 3D Hubs B.V.

Rhim, J.-W. (2012). Physical-mechanical properties of agar/κ-carrageenan blend film and derived clay nanocomposite film. *Journal of Food Science*, *77*(12), N66–N73. doi:10.1111/j.1750-3841.2012.02988.x

Šafka, J., Ackermann, M., Bobek, J., Seidl, M., Habr, J., & Běhálek, L. (2016). Use of Composite Materials for FDM 3D Print Technology. *Materials Science Forum*, *862*, 174–181. www.scientific.net/MSF.862.174 doi:10.4028/www.scientific.net/MSF.862.174

Sculpteo. (2018). The State of 3D Printing. Villejuif, France: Author.

Singh, S., Ramakrishna, S., & Singh, R. (2017). Material issues in additive manufacturing: A review. *Journal of Manufacturing Processes*, *25*, 185–200. doi:10.1016/j.jmapro.2016.11.006

Stansbury, J. W., & Idacavage, M. J. (2016). 3D printing with polymers: Challenges among expanding options and opportunities. *Dental Materials*, *32*(1), 54–64. doi:10.1016/j.dental.2015.09.018

Steuben, J., Van Bossuyt, D. L., & Turner, C. J. (2015). Design for Fused Filament Fabrication Additive Manufacturing. International Design Engineering Technical Conferences & Computers and Information in Engineering Conference. doi:10.1115/DETC2015-46355

Stoof, D., & Pickering, K. (2017). 3D Printing of Natural Reinforced Recycled Polypropylene. *Proceedings of the Processing and Fabrication of Advanced Materials*, *25*, 668–691.

Torrado, A. R., David, P., & Wicker, R. B. (2014). Fracture Surface Analysis of 3D-Printed Tensile Specimens of Novel ABS-Based Materials. *Journal of Failure Analysis and Prevention*, *14*(3), 343–353. doi:10.1007/s11668-014-9803-9

Valášek, P., Müller, M., & Proshlyakov, A. (2012). Effect of sedimentation on the final hardness of polymeric particle composites. *Research in Agricultural Engineering*, *58*(3), 92–98. doi:10.17221/5/2011-RAE

Wang, X., Jiang, M., Zhou, Z., Gou, J., & Hui, D. (2017). 3D printing of polymer matrix composites : A review and prospective. *Composites. Part B, Engineering*, *110*, 442–458. doi:10.1016/j.compositesb.2016.11.034

Zhong, W., Li, F., Zhang, Z., Song, L., & Li, Z. (2001). *Short fibre reinforced composites for fused deposition modeling*. Academic Press.

Chapter 2
Starch Cellulosic Bio-Composites:
A Sustainable and Multifunctional Material for Green Technology

M. J. Halimatul
Universiti Putra Malaysia, Malaysia

M. Jawaid
Universiti Putra Malaysia, Malaysia

S. M. Sapuan
Universiti Putra Malaysia, Malaysia

M. R. Ishak
Universiti Putra Malaysia, Malaysia

N. Julkapli
Universiti Malaya

Mohammad Taha Mastura
Universiti Teknikal Malaysia Melaka, Malaysia

ABSTRACT

This review discusses the effect of nanocellulose or lignocellulosic fibers introduced in starch films. A concise comparison in the mechanical and water absorption properties of the nanocellulose-reinforced starch originated from different plant species was made. It was found that most of these studies prefer solution casting as popular fabrication technique. Studies found nanocellulose generates positive results on mechanical and water uptake properties. The increment in tensile strength was reported between 1.08 to slightly higher than 2-fold while water uptake was decreased between 1.14 to 1.19-fold. In addition, the fibers also serves well as a reinforcement material for starch matric although not as competent as nanocellulose. Discussion on improvement in mechanical, water uptake, thermal, and biodegradation of lignocellulosic fiber-reinforced starch was presented in this chapter. This review also emphasizes potential uses of nanocellulose reinforced starch composite as a smart food packaging and bio-carrier in bio-delivery system where it contributes considerably to a better life.

DOI: 10.4018/978-1-7998-1374-3.ch002

INTRODUCTION

Bio-composite complete the return of material used and energetic value back to the nature cycles when its production appropriately taking into account sustainable energy and chemical consumption starts from selection of raw materials, processing, service life of the product, up to waste management procedure (Vilaplana, Strömberg, & Karlsson, 2010). These efforts are important due to the fact that uncontrolled industrial development had imposed negative impact on the planet we live in. Some experts argue earth geology has been switched to a new geological era known as Anthropocene, where environmental equilibrium enjoyed by the earth and many lives inside it for as long as 10000 years were disturbed by - mostly human activity (Crutzen, 2002). Although the industrial revolution has led to economic prosperity and rise in living standard, but it also at the same time has imposed industrial pollution cost whenever earth's fragile nature is compromised. Therefore, in dealing with this issue, a concept inspired by the nature has been introduced to be infused into business management which is known as cradle-to-cradle concept (McDonough, Braungart, Anastas, & Zimmerman, 2003). This concept will give unimaginable benefit if material and energetic products and processes successfully optimized.

One of the many commercialized bio-composites is bio-composite made from cellulosic plant fiber. Driven by the increase in the price of timber resources, concern for the environment, competition and the availability of new sources of fiber neither from agricultural biomass nor wood residue, all these has led to increased demand for cellulosic bio-composite. Cellulosic bio-composite can be divided into two categories, namely fabricated from wood or non-wood resources. Both of them are made up by cellulose chains which are linear polysaccharides consist of 10,000 glucopyranose units (in woody plant) linked together by β-1-4-linkages, depicted by **Figure 1.**

Figure 1. Chemical structure of cellulose; the arrows point to the cellobiose molecule which is the basic unit of cellulose (Habibi, Lucia, & Rojas, 2010)

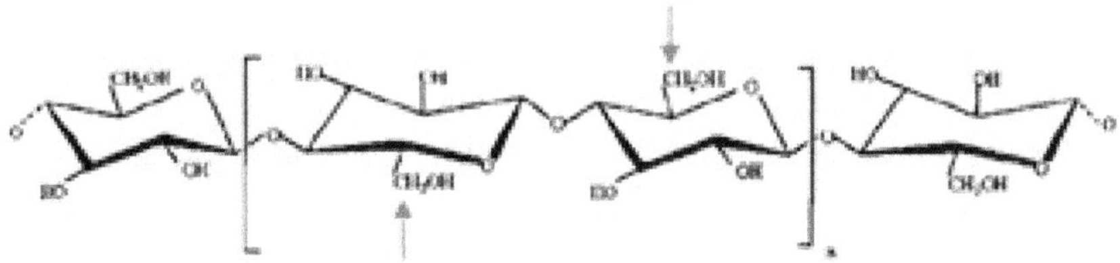

These linear polymers align together to produce elementary fibril that further pack into microfibrils which are in turn assembled into fibers. The microfibrils range between 2 to 20 nm in diameter and length up to micrometer in size; they consist of alternating crystalline and amorphous domain. Crystalline domain is where cellulose chains are arranged in highly ordered while amorphous domain consist disordered cellulose chains (Moon, Martini, Nairn, Simonsen, & Youngblood, 2011). **Table 1** demonstrated some example of wood and non-wood based cellulosic bio-composites for different properties and applications where the improvement on certain properties of polymeric materials has been recorded.

STARCH BIOPOLYMER: PROPERTIES AND LIMITATION

Starch is one of the most studied and promising agro-resources for the production of biodegradable polymers as matrices for bio-composite applications. Starch, derivate from variety of botanical sources (cereals, legumes, and tubers), is widely available raw material for use in biodegradable materials. Indeed, starch is a material with attractive combination of attributes such as low cost, abundant availability and thermoplastic behavior, besides the fact they are biodegradable, recyclable and renewable (Wu & Hakkarainen, 2015). Moreover, due to isotropic property, scentless, tasteless, colorless and besides it is non-toxic to human and ecology starch is potentially used for raw material in the manufacture of bioplastic (Flores, Famá, & Rojas, 2007). However, starches have some drawbacks, such as naturally very brittle because of the existence of a strong cohesive bonds between the polymer molecules (Maran & Sivakumar, 2013), poor melting process ability, high water solubility, difficulty of processing and brittleness, causing the need for a plasticizer to make them suitable for engineering applications. Plasticizer such as water, glycerol, sorbitol and acetic acid is added to the starch to increase the flexibility and extensibility of starch by reducing intermolecular forces between the adjacent polymer chains (Romero-Bastida & Bello-Pérez, 2005). The transformation of native starch to this thermoplastic starch occurs due to the disintegration of molecular order within the starch granules when high temperatures, mechanical stirring and plasticizers are added which subsequently caused the starch granules to swell and burst resulting the releasing of starch polymers contained therein. These polymers form hydrogen bonds with plasticizers and thus prevent them from realigning into crystallite formation again.

Natural starch is normally made into thermoplastic by processing through the use of various industrial techniques, such as injection, extrusion or use of a batch mixer connected to a torque rheometer starch absorbs moisture when exposed to humidity and has inferior but useful mechanical properties compared to synthetic

Table 1. List of lignocellulosic fiber reinforced composites and its properties

Lignocellulosic fiber reinforced composites	Properties	Reference
Wood reinforced composites: Maple wood flour (30 wt% fiber) reinforced polylactic acid composite	Mechanical properties Flexural modulus = 8.9 GPa Flexural strength = 116.6 ± 2.4 MPa Tensile strength = 63.3 ± 5.8 MPa Tensile modulus = 5.3 ± 0.8 GPa Impact strength = 23.2 ± 2.9 J/m Thermomechanical properties T_g = 77°C Storage modulus at 40°C = 10 GPa	(Huda, Drzal, Misra, & Mohanty, 2006)
Wood fiber reinforced maleated polypropylene composites	Mechanical properties Tensile strength = 50% increment with 50 wt% fiber loading Impact strength = 20% increment with 30 wt% fiber loading Thermomechanical properties Damping index = 60% decrement with 50 wt% fiber loading	(Bledzki, Faruk, & Huque, 2002)
Saw dust reinforced high density polypropylene	Elastic modulus 2.3 GPa for jack pine fiber composite 2.2 GPa for black spruce composite	(Bouafif, Koubaa, Perré, & Cloutier, 2009)
Wood flour reinforced maleated polyprolylene	Mechanical properties Stiffness = 4GPa	(Dányádi et al., 2007)
Non-wood reinforced composites: Henequen fiber- HDPE (treated with silane)	Mechanical properties Tensile strength = 30% increment Flexural strength = ~13% increment Shear strength = 25% increment	(Ashori, 2008)
Sugar palm fiber (alkali treated) reinforced epoxy composites	Flexural modulus = 6948 MPa	(Bachtiar, Sapuan, & Hamdan, 2010)
Pineapple fiber reinforced polycarbonate composites	Mechanical properties Young's modulus = 1960 MPa Thermomechanical properties Onset temperature for thermal degradation = 270°C	(Threepopnatkul, Kaerkitcha, & Athipongarporn, 2009)
Hybrid banana/kenaf reinforced	Tensile strength = 139 MPa Flexural strength = 172.2 MPa	(Alavudeen, Rajini, Karthikeyan, Thiruchitrambalam, & Venkateshwaren, 2015)
Coconut shell powder reinforced ultra-high molecular weight polyethylene	Compressive strength = 50% decrease with 40 wt% fiber loading	(Pradhan, Dwarakadasa, & Reucroft, 2004)
Banana fiber reinforced epoxy composites	Ultimate tensile strength = 45.57 MPa Flexural strength = 73.58 MPa Impact strength = 6.95 kJ/m^2	(Maleque, Belal, & Sapuan, 2007)

polymeric materials. Furthermore, the nature of starch makes them as an excessive hydrophilicity material. One of the methods to overcome these negative attributes, as well as to improve the strength properties in general, is to incorporate starch by abundant and relatively inexpensive lignocellulosic materials, such as hard and soft wood fibers or other plant fibers including jute, coconut, banana, cane bagasse and others The resulting materials are termed "bio-composites" or "green" composites, which are considered to be totally biodegradable.

STARCH BIO-COMPOSITES: PROPERTIES

Starch/Cellulose Bio-Composites

To improve the properties of starch while maintaining its bio properties, research laboratories have developed a strategy to incorporate starch matrix with the cellulosic filler. In this case, two different types of cellulosic materials have be taken into consideration which is microcellulose and nanocellulose. Several studies and applications have demonstrated the interest of using cellulosic fillers as reinforcement in starch matrices. This incorporation show an improvement of water resistance by reinforcing plasticized starch films with cellulose microfibrils. This behavior is related to the hydrophobic character of the cellulose fillers in comparison to starch hydrophilic property.

Starch/Microcellulose Bio-Composites

Incorporation of microcellulose into stract matrix is one of the promising attempt was made to improve the thermomechanical properties and to decrease the water sensitivity of starch-based systems, preserving the biodegradability of the material. An alternative way to palliate these restrictions is the use of microcellulose as natural and biodegradable filler. Microcellulose are gaining attention as a reinforcing phase due to its low density, a highly reduced wear of the processing machinery, and a relatively reactive surface may be mentioned as attractive properties, together with their abundance and low price (Siró & Plackett, 2010). In addition, the recycling by combustion of cellulose composites is easier in comparison with inorganic filler systems.

At the low contents of microcellulose derivatives (<9 wt%), the greater values tensile strength of the starch bio-composite obtained. The values of water vapor decreased with the increasing of microcellulose derivatives (Avérous, Fringant, & Moro, 2001b; Dufresne & Vignon, 1998; Ma, Chang, & Yu, 2008). This behavior is related to the highly crystalline 'hydrophobic' character of the micro cellulose fibers in comparison to starch hydrophilic property. In addition, these authors demonstrated an improved thermal stability due to a higher and longer thermal resistance of micro cellulose fibers. Indeed, the starch/microcellulose also demonstrated a homogenous distribution with strong interfacial bonding between the composites components (Ma et al., 2008). In other study demonstrate that the dynamic mechanical analysis shows for starch composites a strong evolution of the main relaxation temperature, which can be linked to the existence of cellulose–starch interactions resulting in a decrease of starch chain mobility (Figure 2).

Starch/Nanocellulose Bio-Composites

The addition of plasticizer into the starch produce a stretchable and flexible film layer, but mechanically low, has barrier properties with high water uptake. To overcome the problems, nanocellulose is added to the mixture of starch and plasticizer, Nanocellulose can be obtained by two methods, namely through biological synthesis by using bacteria (bottom-up method) or extraction from plants (top-down method). It has high strength and stiffness besides low weight with valuable and sustainable properties as it is derived from renewable sources. In addition, it decomposes naturally and bio-compatible, thus making it the best candidate for reinforcing the thermoplastic starch.

In literature, it is difficult to determine the true strength and modulus of cellulose crystals. Hence, theoretical calculations and numerical simulations have been used to estimate the axial modulus of crystalline cellulose. Based on theoretical calculations, axial modulus of cellulose crystal is of 58 GPa where the value is close to the value owned by Kevlar (60 GPa). In addition, the numerical simulation shows cellulose crystals have a higher axial modulus of 180 GPa compared to Kevlar which shows 125 GPa. Tensile strength of nanocellulose shows the theoretical predictions of 0.3 to 22 GPa. This high value is due to extended chair conformation shown by cellulose crystal, as well as the high density of covalent bonds per cross-sectional area in addition to the high inter and intra molecular hydrogen bonding sites (Hamad, 2008) (Xiong, Zhang, Tian, Zhou, & Lu, 2012).

Mechanical Properties

Though identical in chemical composition, the mechanical properties and structure of nanocellulose differ from those of plant cellulose. The comparison between nanocellulose with cellulose fiber was acquired that nanocellulose provides a better properties, such as mechanical, barrier and morphology at the same percentage of filler (Y Chen, Liu, Chang, Cao, & Anderson, 2009). Due to at a low loading level, nanocellulose was dispersed evenly in the starch matrix, the tensile strength increased from 3.15 to 10.98 MPa when nanocellulose content went from 0 to 5 wt.% (Peter

Figure 2. Storage modulus and tan δ curves (DMTA) for starch, starch–10%microcellulose and starch–18%microcellulose (Avérous, Fringant, & Moro, 2001a)

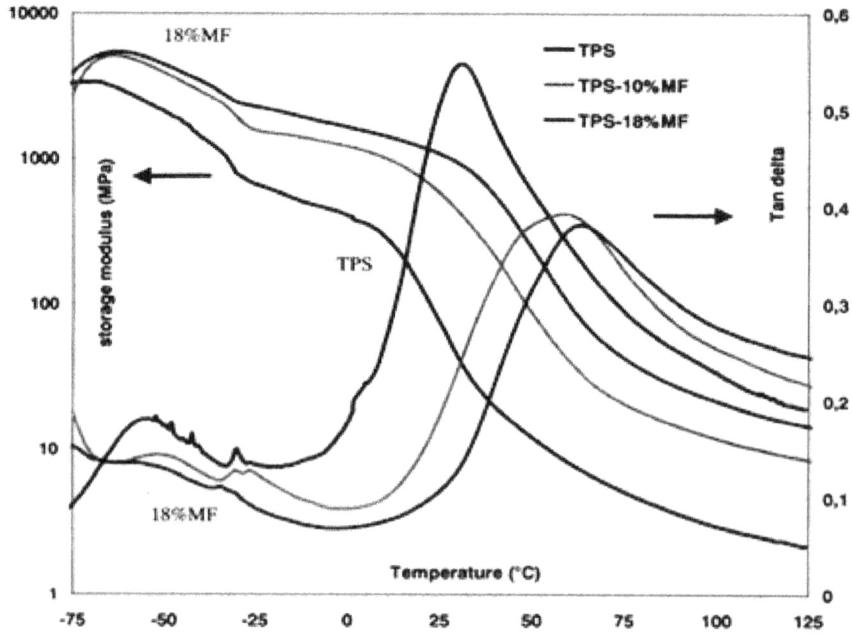

et.al,2010). The increase in tensile strength also proved by many researches such as an increment of 1.28-fold for thermoplastic wheat starch film reinforced 2 wt% of oil palm empty fruit bunches nanocellulose (M. Salehudin, Salleh, Mamat, & Muhamad, 2014), 1.9-fold for pea starch film reinforced 10 wt% hull pea nanocellulose (Y Chen et al., 2009), 1.08-fold for potato starch film reinforced 0.5 wt% nanocellulose (Arun, Kumar, & Sreekala, 2012b) and more than doubling of the strength of the modified potato film reinforced 0.5 wt% sisal nanocellulose (J. Morán, Vázquez, & Cyras, 2013) when compared to their respective neat starch films. Beside tensile strength, Young modulus also increased by 30 and 17 fold (with 5% nanocellulose), while the elongation at break was reduced from 144% to 24% and 48% with increasing nanocellulose content, respectively for the starch bio-composites (Martins et al., 2009; Wan et al., 2009). The increase in mechanical properties are driven by good compatibility between these two types of polysaccharides in addition to high number of active sites on nanocellulose surfaces which then allow more hydrogen bonds to be formed with the starch matrix. Hydrogen bonds formed between nanocellulose leads to a dense percolating cellulosic network that is fundamental for the reinforcing effects. Table 2 below shows some of the mechanical properties of nanocellulose reinforced thermoplastic starch film derived from different mixing formula.

Water Absorption Properties

Film resistance against water invasion is one of the most important properties for food packaging applications, particularly those involving high water activity food. According to Tongdeesoontorn (2011) water insolubility is a prerequisite to improve product intensity and improve moisture barrier property as well as overall shelf-life. Thermoplastic starch film is highly hydrophilic in nature indicated by high water diffusion rate (Karimi, Dufresne, & Tahir, 2014) but nanocellulose addition to thermoplastic starch film has been successful in lowering these properties. The films water uptake is decreased by 1.19-fold when added with 10 wt% nanocellulose (Y Chen et al., 2009), similarly a 1.92-fold decrease when added with 70 wt% nanocellulose (Svagan, Hedenqvist, & Berglund, 2009), by a 1.5-fold decrease when added with 10wt% nanocellulose (Karimi et al., 2014) and by 1.14-fold decrease when added with 2 wt% nanocellulose (M. Salehudin & Salleh, 2013) compared to their respective neat films. Moreover, water vapor permeability decreased from 5.75 × 10−10 to 3.43×10^{-10} g m^{-1} s^{-1} Pa^{-1} (Chang, Jian, Zheng, Yu, & Ma, 2010).

Starch/Cellulosic Fibers Bio-Composites

Different kinds of fibers have been investigated in blends for formulation of bio-composites. The incorporation of these fibers as fillers or for reinforcement into blends has received increased attention, particularly for price-driven and high volume applications.

Examples of Starch/Cellulosic Fibers Bio-Composites

Starch/Kapok Fibers Bio-Composites

Kapok (Ceiba pentandra) is obtained from the fruit of the kapok tree and is yellowish-brown and slightly lustrous in appearance. The fiber is light, very buoyant, resilient and resistant to water. Kapok fibers are abundantly natural microtubes with round tube structure and have been widely utilized as stuffing, buoyancy material and oil-absorbing material. Indeed, Kapok fiber has high oil absorbency characteristics but exhibits good water repellency because of its waxy surface and consequently make it as the potential natural non wood fiber to starch bio-composites (Mathew & Dufresne, 2002). This phenomenon was explained by the more hydrophobic properties of Kapok fibers due to its high crystallinity and by the strong interfacial adhesion.

Where denoted as below,

A= neat starch film
B = thermoplastic starch compressed with ramie mat

Table 2. Mechanical properties of nanocellulose reinforced thermoplastic starch film derived from different mixing formula

Starch type	Nanocellulose origin	Nanocellulose extraction process	Nanocellulose properties	Fabrication technique	Mechanical properties			Testing method	Reference
Pea	Pea hull	Acid hydrolysis	7 to 12 nm in diameter, 240 to 400 nm long and 10 wt% content inside the film	Solution casting	**Properties**	**Neat starch film**	**Nanocellulose added film**	Universal testing machine	(Yun Chen, Liu, Chang, Cao, & Anderson, 2009)
					Tensile strength (MPa)	4.1	7.8		
					Elongation at break (%)	30.1	62.2		
Wheat	Oil palm empty fruit bunches fiber	Acid hydrolysis	1 to 100 nm in size, 2 wt% casting content inside the film	Solution casting	**Properties**	**Neat starch film**	**Nanocellulose added film**	Universal testing machine according to ASTM D638-03	(Harfiz et al., 2013)
					Tensile strength (MPa)	3.66	4.68		
					Elongation at break (%)	1.79	1.19		
Potato	Commercial microcrystalline cellulose	High pressure homogenization	392 ± 36 nm in size, 1 wt% content inside the film	Solution casting	**Properties**	**Nanocellulose added film**	**Nanocellulose gum arabic added film**	Texture analyser according to ASTM D882-91	(Vigneshwaran, Ammayappan, & Huang, 2011)
					Tensile strength (MPa)	3.27 ± 1.1	4.79 ± 1.3		
					Elongation at break (%)	22.8 ± 2.9	36.6 ± 3.8		
Wheat	Oil palm empty fruit bunches fiber	Acid hydrolysis	1 to 100 nm in size, 2 wt% casting content inside the film	Solution casting	**Properties**	**Chitosan added film**	**Nanocellulose and chitosan added film**	Universal testing machine according to ASTM D638-03	(M. H. Salehudin, Salleh, Mamat, & Muhamad, 2014)
					Tensile strength (MPa)	3.96	5.25		
					Elongation at break (%)	30.27	25.71		

continues on following page

Table 2. Continued

Starch type	Nanocellulose origin	Nanocellulose extraction process	Nanocellulose properties	Fabrication technique	Mechanical properties							Testing method	Reference
Potato	Commercial nanofiber		0.5 wt% content inside the film	Hand lay-up and compression molding	Properties	A	B	C	D	E	F	Universal testing machine according to ASTM 882	(Arun, Kumar, & Sreekala, 2012a)
					Tensile strength (MPa)	6.5	8.5	7.0	104	3.5	108		
					Elongation at break (%)	2.4	19.8	3.2	21.7	6.5	21.7		
					Young's modulus (Mpa)	20	215	60	250	30	456		
Potato (acetylated potato and maleated potato)	Sisal fiber	Acid hydrolysis	30.9 ± 12.5 nm in diameter, 5 wt% content inside the film	Solution casting	Elastic modulus starch film	1-fold increase compared to neat potato						Universal testing machine Instron according to ASTM D638	(J. I. Morán, Vázquez, & Cyras, 2013)

C= nanocellulose added to thermoplastic starch
D = nanocellulose added to thermoplastic starch compressed with ramie mat
E = nanocellulose and polycaprolactone added to thermoplastic starch
F = nanocellulose and polycaprolactone added to thermoplastic starch compressed with ramie

Compared to the starch thermal resistance of the bio-composite improved as a consequence of the kapok fiber reinforcement (Jutarat Prachayawarakorn, Chaiwatyothin, Mueangta, & Hanchana, 2013). The latter hindered the post-processing crystallization of starch significantly; hence influenced the changes in the properties and the embrittlement of the material.

Starch/Jute Fibers Bio-Composites

Jute (Corchorus olitorius) is a bast fiber, obtained from the stem of the jute plant. Jute has a reasonable strength. Jute is mainly used for the production of low-value products, such as ropes and sackings. It is also used for carpet backing materials and home interiors such as furnishing fabrics and wall hangings. Incorporation of jute fiber demonstrates perform a new hydrogen bond as detected by the infrared peak shift. These hydrogen bonds are responsible for mechanical properties of the starch matrix reinforced by jute fibers were significantly improved. In addition, water absorption of the starch matrix was clearly reduced by the incorporation of the jute fibers up to 30 wt % (Jutarat Prachayawarakorn et al., 2013).

Starch/Luffa Fibers Bio-Composites

Luffa fiber provides good qualities as it is cheap, abundant and non-toxic. Improvement of the tensile and water sorption properties of starch film by using luffa fiber has been reported (Shen, Min Xie, Huang, Zhou, & Ruan, 2012). In general, the tensile strength of the starch/luffa fiber composite with 10 wt.% of luffa fiber had a twofold increase compared to un-filled film (Kaewtatip & Thongmee, 2012). However, in most of studies, the water absorption of the starch/luffa fiber composites decreased significantly when the luffa fiber contents increased.

Starch/Coir Fibers Bio-Composites

Coir is one of abundant, versatile, renewable, cheap, and biodegradable non wood fiber used for making a wide variety of products. Coir has also been tested as filler or reinforcement in different composite materials. Furthermore, it represents an additional agro industrial nonfood feedstock (agro industrial and food industry waste) that should be considered as feedstock for the formulation of Eco compatible composite materials (Harish, Michael, Bensely, Lal, & Rajadurai, 2008).

Due to this promising stability, coir has been selected as one of the reinforcement agent used for the starch bio-composites. It is anticipated that the Young's modulus of starch films were increased by about 75% with addition of 15wt% of coir (Rosa et al., 2009).

Similarly, observed values were ~373.5 MPa and ~176 MPa for Young modulus and 10–11 MPa and 3 MPa for tensile strength for starch/coir bio-composites containing 0–30% coir fibers respectively (Lomelí Ramírez et al., 2011).

The surface treatment of coir such as acid/base, hydrophilic/hydrophobic treatment contributed to a better interface of starch/coir bio-composites **(Figure 3).** Those treatments removed fatty-deposits and tyloses leading to the formation of a rough surface with pits, resulting in mechanical anchorage and reduced pullout and gaps between the fiber and matrix in composite materials. For example, the alkali treatment resulted in a higher amount of starch exposed on the coir fibers surface, thereby increasing the number of possible reaction sites. In advanced, addition of co-matrix such as polyethylene glycol (Iovino, Zullo, Rao, Cassar, & Gianfreda, 2008), gluten (Muensri, Kunanopparat, Menut, & Siriwattanayotin, 2011), ethylene vinyl alcohol and acrylamide are recorded to improve the biodegradability, antimicrobial activity, light irradiation resistance and flow ability. This material can be used for different fields from pharmaceutical and medical device to industrial packaging, but its utilization is rather limited due to high production cost.

Starch/Ramie Bio-Composites

Ramie or China grass (Boehmeria nivea (L.) Gaud.) is a perennial herbaceous plant of the Urticaceous family. This crop is mainly planted in China and other Asian countries including Philippines and India. Ramie fibers have been used as a textile fiber for centuries due to their excellent fiber characteristics. Ramie fibers are very fine and silk-like, naturally white in color, and have a high luster (Arun et al., 2012b). Other advantages of it are good resistance to bacteria, mildew, and insect attack. The fibers are stable in alkaline media and not harmed by mild acids (Ochi, 2006).

It recorded that, the starch/ramie bio-composites, conditioned at 50% relative humidity, increases, respectively, in both tensile strength and Young's modulus from 2.8 MPa for starch film to 6.9 MPa and from 56 MPa for starch film to 480 MPa with increasing ramie fibers content from 0 to 40 wt% (**Figure 4**)

Figure 3. Scanning electron micrographs of (a–d) coir fiber subjected to different treatments (a: untreated; b: washed; c: mercerized; d: bleached) and their composites (e–f: starch/EVOH and untreated coir; g–h: starch/EVOH and washed coir; i–j: starch/EVOH and mercerized coir; k–l: starch/EVOH and blenched coir). Scale bars: a–d (in d), 50 μm; e, g, i and k (in k), 1 mm; f, h, j and l (in l), 100 μm (Rosa et al., 2009)

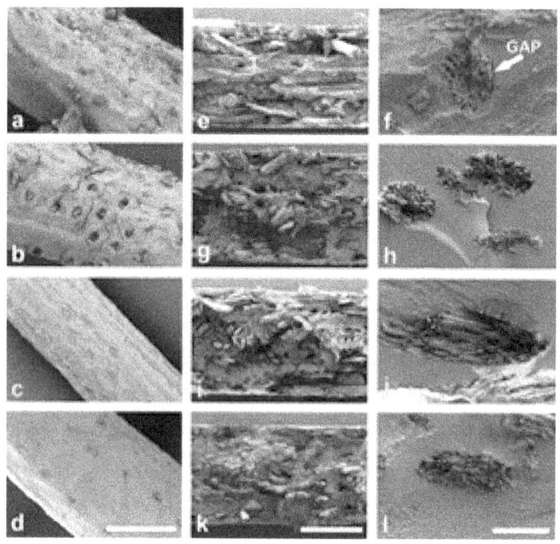

Properties of Starch/Cellulosic Fibers Bio-Composites

Mechanical Properties

Fibers increase the mechanical properties of thermoplastic starch. Morsyleide et.al (2008) reported an increase of nearly 50% tensile modulus of coconut fiber reinforced starch composites compared to plain film. However, as the addition of fiber exceeds 15 wt%, probability of fiber-fiber bonded to each other are higher then, agglomerates and ultimately cause decrease in stress transfer efficiency from matrix to fiber (Rosa et al., 2009). In addition, decrement of mechanical properties as fiber loadings exceeding 15 wt% were also recorded on cotton fiber reinforced composite starch (Jutarat Prachayawarakorn, Ruttanabus, & Boonsom, 2011). At lower fiber loadings of 10 wt%, tensile strength was doubled but as fiber loading was increase to 15 wt% no improvement was observed. Further increments in fiber loading caused mechanical properties begin to decline. However, Belhassen et.al (2008) found that the mechanical properties continue to increase linearly until the fiber loadings achieved 30 wt%. The addition of 15 wt% alfa fiber into thermoplastic starch matric increases Young modulus 6.5 times, while the addition of 30 wt% alfa fiber increases 11.8 times compared to neat starch composite (Belhassen et al., 2009).

Figure 4. The dynamic mechanical behaviors of both storage modulus (E') and loss factor (tan δ) as a function of temperature for starch film and starch/ramie bio-composites (Lu, Weng, & Cao, 2006).

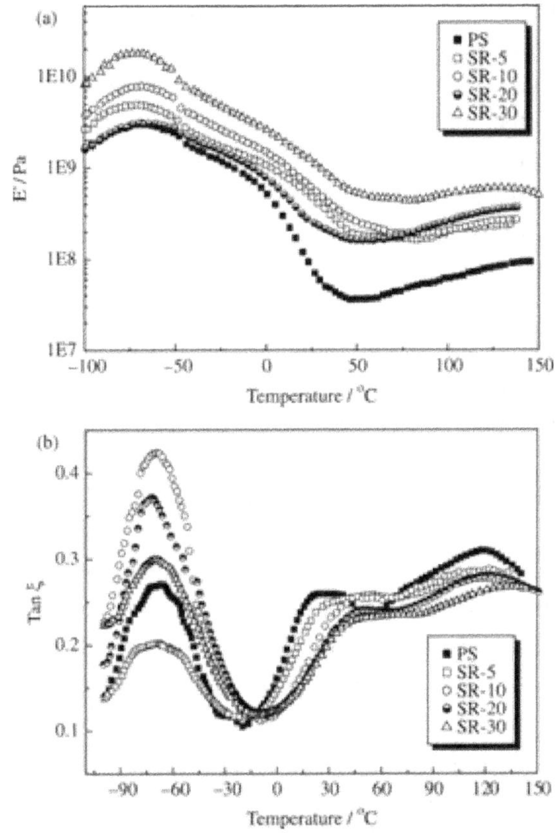

In addition, other study found that fiber loadings of 30 wt% is the optimum value for the lignocellulosic fiber reinforced starch composites where coconut fiber increase tensile strength to approximately 170% compared to the neat starch composite (Corradini, Imam, Agnelli, & Mattoso, 2009). Sisal and jute reinforced starch composites were also showed a similar trend, where they increase tensile strength 154% and 100% respectively with 30 wt% fiber loading. On the other hand, it was found that at lower fiber loading of 10 wt% sisal and jute fiber, increment of 100% and 54% was observed in tensile strength compared to neat starch composite (Torres, Arroyo, & Gomez, 2007). The difference between the degree of reinforcement of sisal and jute fiber in starch matrices is due to differences in chemical composition and morphology of the fibers (Corradini et al., 2009). These differences were also reported as sisal reinforced starch composite exhibit better mechanical properties compared to hemp reinforced starch composite. Besides, such differences were also observed in impact properties as sisal, jute and Cabuya fiber reinforced starch was compared (Torres et al., 2007).

Furthermore, flexural modulus also report increase of 5-fold when 15 wt% alfa fiber was introduced to the composite and 9.8-fold as 30 wt% alfa fiber was used (Belhassen et al., 2009). However, the opposite trend was reported on coconut fiber reinforced starch composite as fiber loading increased from 5 wt% to 15 wt% (Rosa et al., 2009). It is also reported that increasing fiber loading of sisal and hemp lowered flexibility properties. This occur due to fiber nature that is more rigid and stiff compared to starch therefore, addition of fiber degrades matrix ductile behavior and restrain matrix from deform (Gironès et al., 2012a). Moreover, this rigidity also lowers elongation at break of alfa fiber reinforced starch composite as much as 2.5 times compared to neat starch (Belhassen et al., 2009). The same observation was reported on cotton fiber reinforced starch composite. Good compatibility between matrix and reinforcement phase was characterized by FTIR with many hydrogen bonds formed between both phases (Jutarat Prachayawarakorn et al., 2011) and thus contribute to some degree of reinforcement in thermoplastic starch matrix (Teixeira et al., 2012).

Properties of the lignocellulosic reinforced starch composite is influenced by many factors such as the type of starch (Jutarat Prachayawarakorn et al., 2011), fiber and processing conditions such as temperature and duration (Torres et al., 2007). Difference in starch origins significantly affect the composite because of difference in types of starch composition, ratio of amylose to amylopectin as well as the size and shape of starch granules. High amylose content in the starch produces a stiffer composite. Therefore, rice starch that has higher amylose content compared to waxy starch produce a higher maximum load and Young modulus in cotton fiber reinforced starch composites (Jutarat Prachayawarakorn et al., 2011). Besides starch, processing condition also plays crucial role. Impact properties of 3 types of composite namely cabuya, jute and sisal reinforced starch composites show the highest result for composite processed in 130°C followed by 150 ° C and 170°C. On the other hand, as temperature was constant, processing duration of 45 minutes produce the highest impact properties followed by 35 and 25 minute Furthermore, the type and amount of plasticizer also reported affect the mechanical properties of lignocellulosic reinforced starch composite where mixture of ethylene glycol and glycerol provide the highest ultimate tensile strength followed glycerine and water while propylene glycol and chitosan produce composites of similar ultimate tensile strength. Furthermore, it was also reported that decrease in tensile strength of the composites when glycerol content increases. This is because increase in glycerol content increases amount of bubbles formed, caused by water evaporation during the gelation process. These bubbles eventually formed voids that become a weak point in the composites (Torres et al., 2007).

For starch foam, increment in plasticizer content enhances radial expansion ratio of the foam yet an opposite effect produce as lignocellulosic fiber amount is increase. Guan and Hanna (2004b) reported plasticizer content (ethanol) caused a lot of vapor generates during foaming process thus decrease bond formation between the polymer chains. This consequently reduces elastic properties of starch foam. The reduction cause cell walls thinner and even some were broken during foaming process due to low expansion of starch foam. Therefore, starch foam mixed lignocellulosic fiber is denser compared to neat starch foam and reported to have high mechanical properties as well. In contrast, insufficient plasticizer content cause starch to be expelled out from fiber-fiber matrix due to high bonding within the fibers (Guan & Hanna, 2004). The addition of fiber reduces starch foam strength by 3.75 fold when 54.5% corn fiber was added to the formulation. This is because, addition of fibers in the system causes the cell walls to collapse during foaming process because of abrasion effect (Cinelli, Chiellini, Lawton, & Imam, 2006a). Moreover, unmelted fiber prevent cell expansion thus forms clusters of shrunken cell (Guan & Hanna, 2004). Not only the strength of the foam decreased, flexural modulus also shows similar trend because fibers that interferes in interior foam foaming process cause the foam to be easily bent during the 3-point bending tests (Guan & Hanna, 2004).

Water Absorption

It was reported that increase in fiber loading reduce equilibrium water content of the composite. Equilibrium water content that achieved after 3-days of condition records 380% water content for neat starch composite and reduce to 305% as 10 wt% cotton fiber was introduced (J. Prachayawarakorn, Sangnitidej, & Boonpasith, 2010). Similar results were also reported where addition of corn cob fiber lowers water uptake of starch foam. However, the composite requires increase in both fiber and polyvinyl alcohol simultaneously in order to obtain such effects. Despite water uptake, fiber properties also influence diffusion coefficient. According to Vallejos et al. (2011a), decrease in diffusion coefficient was observed in lignocellulosic fiber reinforced starch composites compared to neat starch as fiber absorbs much less water compared to starch (Vallejos et al., 2011b). However, a contrary result reported on flax fiber reinforced starch where increase in diffusion coefficient from 2.7 x 10^{-8} m^2/h for 10 wt% flax fiber loading to 5.0 x $10^{-8} m^2/h$ for 30 wt% (Cañigueral et al., 2009). This is due to flax fiber addition increase surface roughness that increases surface specific thus enhance interaction with the external medium. The research shows increase in fiber content affect water sorption kinetics but not amount of water adsorbed at equilibrium. This is because fibers were incorporated in the system in dry condition therefore it reaches equilibrium with 12% water content. While this value falls within the range of samples humidity condition, it does not provide any changes to the composite equilibrium water content as latex polymer that was also introduced to the system, led to faster retrogradation of amylose chains thus increase crystallinity index of the composites (Gironès et al., 2012b).

The type of starch also plays important role in influencing water absorption. It was reported that cassava starch composite absorbs more water compared to corn starch composite when cotton fiber used as reinforcement material (Lomelí Ramírez et al., 2011). The difference in water absorption also reported in rice starch compare to waxy starch composites. Waxy starch composite that contains high amylopectin content reaches equilibrium water content within 4 days compared to 7 days for rice starch composite (Jutarat Prachayawarakorn et al., 2011). Beside starch, type of fiber also affects water uptake properties of lignocellulosic reinforced starch composites. Cellulose fiber reinforced starch composite has lower equilibrium moisture content compared to corn cob fiber reinforced starch composite. This is because starch prefer cellulose over corn cob fiber for bonding (Guan & Hanna, 2004).

Moreover, plasticizer that is also an important substituent in lignocellulosic reinforced starch composites however, does not affect the water absorption properties of composites (De Carvalho, Curvelo, & Agnelli, 2002a). Usually, when the plasticizer content increases and excessive, it will be trapped in the system without chemical bonding and thus easily perform bonding with surrounding water molecules due to hydroscopic nature of the plasticizer. However, when the fiber is introduced into the mixture of starch and plasticizer, excessive plasticizer content is no longer affecting water uptake because fibers bind with these excess plasticizer cause plasticizer content almost the same in all samples. In addition, fiber also obstructing the absorption of water by forming a network that prevents matrix from expands.

Crystallinity

Addition of fiber to the system shows new signal at $2\Theta = 16.6\,°$ and $22.5\,°$, which correspond to the crystalline portion of the fiber (Gironès et al., 2012b). Other study shows crystalline peak of cotton fiber reinforced starch at $2\Theta = 13.1\,°$, $18.0\,°$, $20.0\,°$, $22.5\,°$ and $25.7\,°$ in which these peaks are between neat crystalline starch film peak, $2\Theta = 13.1$ and 20.0 with cotton fiber peak, $2\Theta = 18.0\,°$, $22.5\,°$ and $25.7\,°$. This shows the crystallinity of the composites were successfully enhanced by addition of fiber (Gironès et al., 2012b). According Teixera et al, (2012), extruded fiber reinforced thermoplastic starch produce 3 types of crystalline structures which are V_H, V_A and the residue of crystalline type of A and B. Crystalline pattern V_H produce when amylose and plasticizer forms complex which, for cassava bagasse reinforced starch composite reported at $2\Theta = 20.4$. On the other hand, V_A pattern is formed during extrusion and the crystalline lattice is contracted and contains less water compared to V_H. The third pattern is corresponding to residue of type A and B crystalline obtained due to imperfect melting during the extrusion process (Teixeira et al., 2012).

Morphology

Micrograph of tensile fracture surface of alfa fiber reinforced starch composite shows fibers dispersed well in starch matrix while most of the fiber surface covered by the matrix. However, ductile tearing in matrix showed some fiber pull-out (Belhassen et al., 2009). Phase compatibility between starch matrix with fiber is also seen in rice and waxy starch composites where 2mm fiber wetted perfectly and well embedded in starch matrix. In contrast, the use of longer fiber (5mm) shows more cracks and voids image (Jutarat Prachayawarakorn et al., 2013). Other than the length of fiber, the impurities content also affect adhesion between starch and fiber. For instance, coconut fiber reinforced starch composites show poor adhesion with matrix even good wetting was observed. This is because, coconut fiber contains tyloses containing impurities, wax, fatty substances and globular protrusion (Rosa et al., 2009). In addition, coconut fibrils covered with a layer of wax aliphatic called cuticle which prevents the formation of hydrogen bonds between starch and fiber (Satyanarayana, Sukumaran, Mukherjee, Pavithran, & Pillai, 1990). Therefore, tensile fractured surface shows many voids as fibers pulled –out from matrix due to weak interface (Rosa et al., 2009). Beside impurities, lignin also prevents the formation of hydrogen bonds between starch and fiber. Research shows that the use of bleached kraft fiber that has undergone delignification process in starch composite shows image of fiber breakage results in excellent adhesion between starch and fiber. In addition to the delignification process, alkali treatment also removes some lignin together with impurities, parenchyma cells and hemicellulose, thus change the smooth wax coated fiber surface into rough surface. This makes the fiber surface area increases besides defibrillation that occur cause fiber to be more visible (De Carvalho, Curvelo, & Agnelli, 2002b). On the other hand, SEM image of the starch foam shows a more regular cell structure when 1 wt% fiber is introduced while the foam with higher fiber loadings, 2 wt% and 3 wt% fiber, cause accumulation of fibers near the surface of the foam (Carr, Parra, Ponce, Lugão, & Buchler, 2006). There is research shows that adding fiber to starch foam cause interference in the formation of foam cells and as corn fiber loading exceeds 50 wt%, the foaming process stop (Cinelli, Chiellini, Lawton, & Imam, 2006b).

Biodegradation

The study of biodegradation properties of lignocellulosic fiber reinforced starch composite found fiber successfully slow down decomposition process. Initially, in the early days of soil burial test, an increase in weight of the composites was observed. The increase is due to absorption of water from the soil. After then, the weight of the samples decrease until sample decomposed after 18 days (Jutarat Prachayawarakorn et al., 2013). The slowing down of decomposition rate because fiber helps composites to be more resistant against micro-organisms mitigating hydrolytic enzyme from matrix (De Carvalho et al., 2002b).

STARCH BIO-COMPOSITES: POTENTIAL APPLICATION

Smart Food Packaging

Globalization in food supply increased demand for food that is not only convenient but, more fresh and safe. Therefore, food packaging technology plays significant role in meeting customer requirements. Research on active packaging, specifically made from nanocellulose reinforced starch composites produces bioactive packaging that has a great potential to improve food quality and safety during food storage.

The addition of anti-microbial substances such as chitosan (M. Salehudin et al., 2014), cinnamaldehyde (A. C. De Souza, Dias, Sousa, & Tadini, 2014), natamycin and nicin (Ollé Resa, Jagus, & Gerschenson, 2014) into the thermoplastic starch reinforced nanocellulose showed effective inhibition against micro-organism replication. Chitosan had mitigated *Bacillus subtilis* bacteria from replicate productively as positively charged amino groups on chitosan polymer reacted with negatively charged bacterial membrane, causing the leakage of proteinaceous and other intracellular constituents. At the same time, the positively charged chitosan formed a good interphase with sulphate group found on the nanocellulose surface (M. Salehudin et al., 2014). The sulphate group exists after nanocellulose extraction process through acid hydrolysis by using sulphuric acid. Moreover, negative charged sulphate groups also provide negative electric double layer repulsion between the nanocellulose to ensure homogenous dispersion in the starch matrix (Bondeson, Mathew, & Oksman, 2006).

Besides chitosan, cinnamaldehyde also increase anti-microbe properties of the thermoplastic starch where it was impregnated on the composite by supercritical solvent impregnated technique. The film composites show significant decrease in equilibrium water vapor sorption and water vapor permeability (A. C de Souza, Dias, Sousa, & Tadini, 2014). Although anti-microbes provide anti-microbial property for the film, sometimes it may interfere with other film properties shown in natamycin and nicin film. The additions of these anti-microbes have degraded the mechanical strength of the thermoplastic starch reinforced nanocellulose film observed through reduction in young modulus.

Beside natamycin and nicin, yerba mate was also introduced in nanocellulose reinforced starch as valuable additive (Jaramillo, González Seligra, Goyanes, Bernal, & Famá, 2015). It showed a similar effect as natamycin and nicin where it exhibits plasticizing effect on nanocellulose reinforced starch film thus reduced the mechanical strength of the film. However, result in antioxidant activity shows that, yerba mate had largely inhibited the growth of hydroxyl radicals where addition of 5% of this material yields the best effect. Furthermore, study on benefit of red propolis as antioxidant in nanocellulose reinforced starch film also exhibit a good result (Costa, Druzian, Machado, De Souza, & Guimaraes, 2014). This active film was excellent in providing protection for butter along 60 days of storage where efficiency of antioxidant properties was evaluated through a low value of peroxide in food (butter). Interestingly, red propolis was not only serves as an excellent antioxidant additive but also acts as anti-microbes as well. Test on anti-microbe properties found, thermoplastic starch containing 0.5% and 0.7% of nanocellulose and red propolis respectively showed 1 log reduction in quantity of microbes on 4,12,20 and 28 days compared to control film. The decline was associated to phenolic compounds inside red propolis that tend to inhibit bacterial RNA polymerase, destroys cell membrane thus stop the growth of bacteria.

Bio-Delivery System

Great synergy between nanocellulose and starch provides outstanding protection for good bacteria, namely *Bacillus coagulans* along its way to the intestines (Khorasani & Shojaosadati, 2017). The good bacteria were used to fight against bad bacteria in human body. Nanocellulose in this research was not only acts as a barrier and matrix entrapment for the bacteria, but also as a source of dietary fiber for them. Starch used in this study was resistant starch which had been modified to prevent it hydrolyzed by α-amylase enzyme in the mouth. Result shown, incorporating starch to nanocellulose-pectin co-polymer had successfully increases the survivability of bacteria from 75% to 97.37% in simulated gastrointestinal condition. Apart from good bacteria, nanocellulose-starch cooperation also provides great protection to colon specific drug namely methotrexate (Meneguin et al., 2017). This drug is important in treating colon diseases such as colorectal cancer, Crohn's disease and ulcerative colitis. By manipulating the content of nanocellulose, drug release rate can be controlled. Nanocellulose with high hydrophilicity and swell ability, had facilitated drug to diffuse out through the starch. The film shows the drug dissolution was 80% within 150 minutes. Such increment in drug release rate could overcome problem of oral bioavailability faced by poorly water soluble drug such as methotraxate.

CONCLUSION

Nowadays, research and development on bioplastics is a very attractive proposition because of the increasing costs of alternatives made from fossil sources and the environmental pollution by synthetic plastics that are now causing worldwide environmental problems. A renewable bioplastic material such as starch has a huge advantage because it is an abundant low cost resource, has good biodegradability and has suitable properties for various usages. Starch is one of the attractive biopolymer for wide range of thin film applications. The properties of starch can be improved by incorporation of various types of natural filler including wood, non-wood and ceramic. Different natural filler such as micro winceyette fiber, sugarcane fiber, cotton fiber, eucalyptus fiber, cassava bagasse cellulose, wheat straw fiber and sisal fiber have been used for preparing starch composites. These natural fillers can have improved tensile strength, thermal stability and water resistance compared to TPS. Moreover, incorporated natural filler can hinder the re-crystallization of starch because of the strong interaction between the natural filler and starch.

REFERENCES

Alavudeen, A., Rajini, N., Karthikeyan, S., Thiruchitrambalam, M., & Venkateshwaren, N. (2015). Mechanical properties of banana/kenaf fiber-reinforced hybrid polyester composites: Effect of woven fabric and random orientation. *Materials & Design*, *66*, 246–257. doi:10.1016/J.MATDES.2014.10.067

Arun, S., Kumar, K. A. A., & Sreekala, M. S. (2012). Fully biodegradable potato starch composites: Effect of macro and nano fiber reinforcement on mechanical, thermal and water-sorption characteristics. International Journal of Plastics Technology. doi:10.100712588-012-9026-4

Ashori, A. (2008). Wood-plastic composites as promising green-composites for automotive industries! *Bioresource Technology*, *99*(11), 4661–4667. doi:10.1016/j.biortech.2007.09.043 PubMed

Avérous, L., Fringant, C., & Moro, L. (2001). Plasticized starch–cellulose interactions in polysaccharide composites. *Polymer*, *42*(15), 6565–6572. doi:10.1016/S0032-3861(01)00125-2

Bachtiar, D., Sapuan, S. M., & Hamdan, M. M. (2010). Flexural properties of alkaline treated sugar palm fiber reinforced epoxy composites. *International Journal of Automotive and Mechanical Engineering*, *1*(June), 79–90. doi:10.15282/ijame.1.2010.7.0007

Belhassen, R., Boufi, S., Vilaseca, F., López, J. P., Méndez, J., Franco, E., ... Mutjé, P. (2008, November). … Mutjé, P. (2009). Biocomposites based on Alfa fibers and starch-based biopolymer. *Polymers for Advanced Technologies*, *20*(12), 1068–1075. doi:10.1002/pat.1364

Bledzki, A. K., Faruk, O., & Huque, M. (2002). Physico-mechanical studies of wood fiber reinforced composites. *Polymer-Plastics Technology and Engineering*, *41*(3), 435–451. doi:10.1081/PPT-120004361

Bondeson, D., Mathew, A., & Oksman, K. (2006). Optimization of the isolation of nanocrystals from microcrystalline cellulose by acid hydrolysis. *Cellulose (London, England)*, *13*(2), 171–180. doi:10.1007/s10570-006-9061-4

Bouafif, H., Koubaa, A., Perré, P., & Cloutier, A. (2009). Effects of fiber characteristics on the physical and mechanical properties of wood plastic composites. *Composites. Part A, Applied Science and Manufacturing*, *40*(12), 1975–1981. doi:10.1016/j.compositesa.2009.06.003

Cañigueral, N., Vilaseca, F., Méndez, J. A., López, J. P., Barberà, L., & Puig, J. … Mutjé, P. (2009). Behavior of biocomposite materials from flax strands and starch-based biopolymer. *Chemical Engineering Science*. doi:10.1016/j.ces.2009.02.006

Carr, L. G., Parra, D. F., Ponce, P., Lugão, A. B., & Buchler, P. M. (2006). Influence of Fibers on the Mechanical Properties of Cassava Starch Foams. *Journal of Polymers and the Environment*, *14*(2), 179–183. doi:10.1007/s10924-006-0008-5

Chang, P. R., Jian, R., Zheng, P., Yu, J., & Ma, X. (2010). Preparation and properties of glycerol plasticized-starch (GPS)/cellulose nanoparticle (CN) composites. *Carbohydrate Polymers*, *79*(2), 301–305. doi:10.1016/j.carbpol.2009.08.007

Chen, Y., Liu, C., Chang, P. R., Cao, X., & Anderson, D. P. (2009). Bionanocomposites based on pea starch and cellulose nanowhiskers hydrolyzed from pea hull fibre: Effect of hydrolysis time. *Carbohydrate Polymers*. doi:10.1016/j.carbpol.2008.11.030

Cinelli, P., Chiellini, E., Lawton, J. W., & Imam, S. H. (2006). Foamed articles based on potato starch, corn fibers and poly(vinyl alcohol). *Polymer Degradation & Stability*, *91*(5), 1147–1155. doi:10.1016/j.polymdegradstab.2005.07.001

Corradini, E., Imam, S. H., Agnelli, J., & Mattoso, L. H. C. (2009). Effect of coconut, sisal and jute fibers on the properties of starch/ gluten/glycerol Matrix. *Journal of Polymers and the Environment*, *17*(1), 1–9. doi:10.1007/s10924-009-0115-1

Costa, S. S., Druzian, J. I., Machado, B. A. S., De Souza, C. O., & Guimaraes, A. G. (2014). Bi-functional biobased packing of the cassava starch, glycerol, licuri nanocellulose and red propolis. *PLoS One*, *9*(11), e112554. doi:10.1371/journal.pone.0112554 PubMed

Crutzen, P. J. (2002). Geology of mankind. *Nature*, *415*(6867), 23. doi:10.1038/415023a PubMed

Dányádi, L., Janecska, T., Szabó, Z., Nagy, G., Móczó, J., & Pukánszky, B. (2007). Wood flour filled PP composites: Compatibilization and adhesion. *Composites Science and Technology*, *67*(13), 2838–2846. doi:10.1016/j.compscitech.2007.01.024

De Carvalho, A. J. F., Curvelo, A. A. S., & Agnelli, J. A. M. (2002). Wood pulp reinforced thermoplastic starch composites. *International Journal of Polymeric Materials*, *51*(7), 647–660. doi:10.1080/714975803

de Souza, A., Dias, A., Sousa, H., & Tadini, C. (2014). Impregnation of cinnamaldehyde into cassava starch biocomposite films using supercritical fluid technology for the development of food active packaging. *Carbohydrate Polymers*, *102*, 830–837. doi:10.1016/j.carbpol.2013.10.082 PubMed

De Souza, A. C., Dias, A. M. A., Sousa, H. C., & Tadini, C. C. (2014). Impregnation of cinnamaldehyde into cassava starch biocomposite films using supercritical fluid technology for the development of food active packaging. *Carbohydrate Polymers*, *102*, 830–837. doi:10.1016/j.carbpol.2013.10.082 PubMed

Dufresne, A., & Vignon, M. R. (1998). Improvement of Starch Film Performances Using Cellulose Microfibrils. *Macromolecules*, *31*(8), 2693–2696. doi:10.1021/ma971532b

Flores, S., Famá, L., Rojas, A., Goyanes, S., & Gerschenson, L. (2007). Physical properties of tapioca-starch edible films: Influence of filmmaking and potassium sorbate. *Food Research*, *40*(2), 257–265. doi:10.1016/j.foodres.2006.02.004

Gironès, J., López, J. P., Mutjé, P., Carvalho, J. F., Curvelo, A. A. S., & Vilaseca, F. (2012). Natural fiber-reinforced thermoplastic starch composites obtained by melt processing. *Composites Science and Technology*, *72*(7), 858–863. doi:10.1016/j.compscitech.2012.02.019

Guan, J., & Hanna, M. A. (2004). Functional properties of extruded foam composites of starch acetate and corn cob fiber. *Industrial Crops and Products*, *19*(3), 255–269. doi:10.1016/j.indcrop.2003.10.007

Habibi, Y., Lucia, L. A., & Rojas, O. J. (2010). Cellulose Nanocrystals: Chemistry, Self-Assembly, and Applications. d, 3479–3500. doi:10.1021/cr900339w

Hamad, W. (2008). On the Development and Applications of Cellulosic Nanofibrillar and Nanocrystalline Materials. Canadian Journal of Chemical Engineering. doi:10.1002/cjce.5450840501

Harfiz, M., Eraricar, S., Nur, S., Mamat, H., Ida, P., & Muhamad, I. (2013). Empty Fruit Bunch (EFB) Cellulose Nanofiber as Filler f or Starch Based Biopolymer Reinforcement. Presentation at the "Advanced food processing and quality management," 1–6.

Harish, S., Michael, D. P., Bensely, A., Lal, D. M., & Rajadurai, A. (2008). Mechanical property evaluation of natural fiber coir composite. *Materials Characterization*, *60*(1), 44–49. doi:10.1016/j.matchar.2008.07.001

Huda, M. S., Drzal, L. T., Misra, M., & Mohanty, A. K. (2006). Wood-fiber-reinforced poly(lactic acid) composites: Evaluation of the physicomechanical and morphological properties. *Journal of Applied Polymer Science*, *102*(5), 4856–4869. doi:10.1002/app.24829

Iovino, R., Zullo, R., Rao, M. A., Cassar, L., & Gianfreda, L. (2008). Biodegradation of poly(lactic acid)/starch/coir biocomposites under controlled composting conditions. *Polymer Degradation & Stability*, *93*(1), 147–157. doi:10.1016/j.polymdegradstab.2007.10.011

Jaramillo, C. M., González Seligra, P., Goyanes, S., Bernal, C., & Famá, L. (2015). Biofilms based on cassava starch containing extract of yerba mate as antioxidant and plasticizer. Starch/Staerke. doi:10.1002/star.201500033

Kaewtatip, K., & Thongmee, J. (2012). Studies on the structure and properties of thermoplastic starch/luffa fiber composites. *Materials & Design*, *40*, 314–318. doi:10.1016/j.matdes.2012.03.053

Karimi, S., Dufresne, A., & Tahir, P. (2014). Biodegradable starch-based composites: Effect of micro and nanoreinforcements on composite properties. *Journal of Materials*.

Khorasani, A. C., & Shojaosadati, S. A. (2017). Starch- and carboxymethylcellulose-coated bacterial nanocellulose-pectin bionanocomposite as novel protective prebiotic matrices. *Food Hydrocolloids*, *63*, 273–285. doi:10.1016/j.foodhyd.2016.09.002

Lomelí Ramírez, M. G., Satyanarayana, K. G., Iwakiri, S., De Muniz, G. B., Tanobe, V., & Flores-Sahagun, T. S. (2011). Study of the properties of biocomposites. Part I. Cassava starch-green coir fibers from Brazil. *Carbohydrate Polymers*, *86*(4), 1712–1722. doi:10.1016/j.carbpol.2011.07.002

Lu, Y., Weng, L., & Cao, X. (2006). Morphological, thermal and mechanical properties of ramie crystallites - Reinforced plasticized starch biocomposites. *Carbohydrate Polymers*, *63*(2), 198–204. doi:10.1016/j.carbpol.2005.08.027

Ma, X., Chang, P. R., & Yu, J. (2008). Properties of biodegradable thermoplastic pea starch/carboxymethyl cellulose and pea starch/microcrystalline cellulose composites. *Carbohydrate Polymers*, *72*(3), 369–375. doi:10.1016/j.carbpol.2007.09.002

Maleque, M. A., Belal, F. Y., & Sapuan, S. M. (2007). Mechanical properties study of pseudo-stem banana fiber reinforced epoxy composite. *Arabian Journal for Science and Engineering*.

Maran, J., & Sivakumar, V. (2013). Development of model for barrier and optical properties of tapioca starch based edible films. Carbohydrate. doi:10.1016/j.carbpol.2012.09.069

Martins, I. M. G., Magina, S. P., Oliveira, L., Freire, C. S. R., Silvestre, A. J. D., Neto, C. P., & Gandini, A. (2009). New biocomposites based on thermoplastic starch and bacterial cellulose. *Composites Science and Technology*, *69*(13), 2163–2168. doi:10.1016/j.compscitech.2009.05.012

Mathew, A. P., & Dufresne, A. (2002). Morphological Investigation of Nanocomposites from Sorbitol Plasticized Starch and Tunicin Whiskers. *Biomacromolecules*, *3*(3), 609–617. doi:10.1021/bm0101769 PubMed

McDonough, W., Braungart, M., Anastas, P. T., & Zimmerman, J. B. (2003). Peer Reviewed: Applying the Principles of Green Engineering to Cradle-to-Cradle Design. *Environmental Science & Technology*, *37*(23), 434A–441A. doi:10.1021/es0326322 PubMed

Meneguin, A. B., Ferreira Cury, B. S., dos Santos, A. M., Franco, D. F., Barud, H. S., & da Silva Filho, E. C. (2017). Resistant starch/pectin free-standing films reinforced with nanocellulose intended for colonic methotrexate release. *Carbohydrate Polymers*, *157*, 1013–1023. doi:10.1016/j.carbpol.2016.10.062 PubMed

Moon, R. J., Martini, A., Nairn, J., Simonsen, J., & Youngblood, J. (2011). Cellulose nanomaterials review: Structure, properties and nanocomposites. *Chemical Society Reviews*, *40*. doi:10.1039/c0cs00108b

Morán, J. I., Vázquez, A., & Cyras, V. P. (2013). Bio-nanocomposites based on derivatized potato starch and cellulose, preparation and characterization. *Journal of Materials Science*, *48*(20), 7196–7203. doi:10.1007/s10853-013-7536-x

Muensri, P., Kunanopparat, T., Menut, P., & Siriwattanayotin, S. (2011). Effect of lignin removal on the properties of coconut coir fiber/wheat gluten biocomposite. *Composites. Part A, Applied Science and Manufacturing*, *42*(2), 173–179. doi:10.1016/j.compositesa.2010.11.002

Ochi, S. (2006). Development of high strength biodegradable composites using Manila hemp fiber and starch-based biodegradable resin. *Composites. Part A, Applied Science and Manufacturing*, *37*(11), 1879–1883. doi:10.1016/j.compositesa.2005.12.019

Ollé Resa, C. P., Jagus, R. J., & Gerschenson, L. N. (2014). Effect of natamycin, nisin and glycerol on the physicochemical properties, roughness and hydrophobicity of tapioca starch edible films. *Materials Science and Engineering C*, *40*, 281–287. doi:10.1016/j.msec.2014.04.005 PubMed

Prachayawarakorn, J., Chaiwatyothin, S., Mueangta, S., & Hanchana, A. (2013). Effect of jute and kapok fibers on properties of thermoplastic cassava starch composites. *Materials & Design*, *47*, 309–315. doi:10.1016/j.matdes.2012.12.012

Prachayawarakorn, J., Ruttanabus, P., & Boonsom, P. (2011). Effect of Cotton Fiber Contents and Lengths on Properties of Thermoplastic Starch Composites Prepared from Rice and Waxy Rice Starches. *Journal of Polymers and the Environment*, *19*(1), 274–282. doi:10.1007/s10924-010-0273-1

Prachayawarakorn, J., Sangnitidej, P., & Boonpasith, P. (2010). Properties of thermoplastic rice starch composites reinforced by cotton fiber or low-density polyethylene. *Carbohydrate Polymers*, *81*(2), 425–433. doi:10.1016/j.carbpol.2010.02.041

Pradhan, S. K., Dwarakadasa, E. S., & Reucroft, P. J. (2004). Processing and characterization of coconut shell powder filled UHMWPE. *Materials Science and Engineering A*, *367*(1-2), 57–62. doi:10.1016/j.msea.2003.09.065

Romero-Bastida, C., & Bello-Pérez, L. (2005). Physicochemical and microstructural characterization of films prepared by thermal and cold gelatinization from non-conventional sources of starches. Carbohydrate. doi:10.1016/j.carbpol.2005.01.004

Rosa, M. F., Sen Chiou, B., Medeiros, E. S., Wood, D. F., Mattoso, L. H. C., Orts, W. J., & Imam, S. H. (2009). Biodegradable composites based on starch/EVOH/glycerol blends and coconut fibers. *Journal of Applied Polymer Science*. doi:10.1002/app.29062

Salehudin, M., & Salleh, E. (2013). Empty Fruit Bunch (EFB) Cellulose Nanofiber as Filler for Starch Based Biopolymer Reinforcement. 2013 CIGR Section VII.

Salehudin, M. H., Salleh, E., Mamat, S. N. H., & Muhamad, I. I. (2014). Starch based Active Packaging Film Reinforced with Empty Fruit Bunch (EFB) Cellulose Nanofiber. *Procedia Chemistry*, *9*, 23–33. doi:10.1016/j.proche.2014.05.004

Satyanarayana, K. G., Sukumaran, K., Mukherjee, P. S., Pavithran, C., & Pillai, S. G. K. (1990). Natural fibre-polymer composites. *Cement and Concrete Composites*, *12*(2), 117–136. doi:10.1016/0958-9465(90)90049-4

Shen, J., & Xie, M. Y., Huang, X., Zhou, S., & Ruan, D. (2012). Mechanical properties of luffa sponge. Journal of the Mechanical Behavior of Biomedical Materials, 15, 141–152. doi:10.1016/j.jmbbm.2012.07.004 PubMed

Siró, I., & Plackett, D. (2010). Microfibrillated cellulose and new nanocomposite materials: A review. *Cellulose (London, England)*, *17*(3), 459–494. doi:10.1007/s10570-010-9405-y

Svagan, A., Hedenqvist, M., & Berglund, L. (2009). Reduced water vapour sorption in cellulose nanocomposites with starch matrix. Composites Science. doi:10.1016/j.compscitech.2008.11.016

Teixeira, E. de M., Curvelo, A. A. S., Corrêa, A. C., Marconcini, J. M., Glenn, G. M., & Mattoso, L. H. C. (2012). Properties of thermoplastic starch from cassava bagasse and cassava starch and their blends with poly (lactic acid). *Industrial Crops and Products*, *37*(1), 61–68. doi:10.1016/j.indcrop.2011.11.036

Threepopnatkul, P., Kaerkitcha, N., & Athipongarporn, N. (2009). Effect of surface treatment on performance of pineapple leaf fiber-polycarbonate composites. *Composites. Part B, Engineering*, *40*(7), 628–632. doi:10.1016/j.compositesb.2009.04.008

Tongdeesoontorn, W. (2011). Effect of carboxymethyl cellulose concentration on physical properties of biodegradable cassava starch-based films. Chemistry Central. doi:10.1186/1752-153X-5-6

Torres, F. G., Arroyo, O. H., & Gomez, C. (2007, March). Processing and Mechanical Properties of Natural Fiber Reinforced Thermoplastic Starch Biocomposites. *Journal of Thermoplastic Composite Materials*, *20*(2), 207–223. doi:10.1177/0892705707073945

Vallejos, M. E., Curvelo, A. A. S., Teixeira, E. M., Mendes, F. M., Carvalho, A. J. F., Felissia, F. E., & Area, M. C. (2011b). Composite materials of thermoplastic starch and fibers from the ethanol–water fractionation of bagasse. *Industrial Crops and Products*, *33*(3), 739–746. doi:10.1016/j.indcrop.2011.01.014

Vigneshwaran, N., Ammayappan, L., & Huang, Q. (2011). Effect of Gum arabic on distribution behavior of nanocellulose fillers in starch film. Applied Nanoscience. doi:10.1007/s13204-011-0020-5

Vilaplana, F., Strömberg, E., & Karlsson, S. (2010). Environmental and resource aspects of sustainable biocomposites. Polymer Degradation and Stability. doi:10.1016/j.polymdegradstab.2010.07.016

Wan, Y. Z., Luo, H., He, F., Liang, H., Huang, Y., & Li, X. L. (2009). Mechanical, moisture absorption, and biodegradation behaviours of bacterial cellulose fibre-reinforced starch biocomposites. *Composites Science and Technology*, *69*(7–8), 1212–1217. doi:10.1016/j.compscitech.2009.02.024

Wu, D., & Hakkarainen, M. (2015). Recycling PLA to multifunctional oligomeric compatibilizers for PLA/starch composites. *European Polymer Journal*, *64*, 126–137. doi:10.1016/j.eurpolymj.2015.01.004

Xiong, R., Zhang, X., Tian, D., Zhou, Z., & Lu, C. (2012). Comparing microcrystalline with spherical nanocrystalline cellulose from waste cotton fabrics. *Cellulose (London, England)*, *19*(4), 1189–1198. doi:10.1007/s10570-012-9730-4

Chapter 3
Green Materials in Hybrid Composites for Automotive Applications:
Green Materials

N. M. Norizan
Universiti Putra Malaysia, Malaysia

A. Atiqah
https://orcid.org/0000-0003-0583-7486
Universiti Tenaga Nasional, Malaysia

M. N. M. Ansari
Universiti Tenaga Nasional, Malaysia

M. Rahmah
Universiti Teknologi MARA, Malaysia

ABSTRACT

The increasing trend of using bio-based fibre, also known as natural fibre, provides many benefits for long-term environmental preservation. In the biocomposites group, green composites are a specific class whereby the bio-based polymer is reinforced with natural fibre. The current review deals with the advance of green materials in hybrid composites for automotive applications. The variation of green materials such as natural fibres is developed to be used as hybrid green materials as reinforcing materials in composites. There are many works done by another researcher that showed the improvement of utilizing the green materials of composites. The application and challenges of having green materials in composites for automotive applications are also presented.

DOI: 10.4018/978-1-7998-1374-3.ch003

INTRODUCTION

One of the main reasons to be called green composite is the composites that consisted of bio fibre or natural fibre as fibre reinforcing in the material composite. There is a different type of fibres used in composite materials such as particle, continuous, and discontinuous fibre reinforcement. The role of reinforcing fibre in the composite is to carry the applied load and stress and conveys enhanced the mechanical performance, as well as the stiffness and strength to the composites. Moreover, the essential of green materials in the composites is the biodegradable polymer that acts as a continuous phase or well knows as a matrix in distinguishing the demanding properties of the green composites. The benefits of having biodegradable polymers are can reduce the emissions of synthetic polymer in landfill that leads to preserving the composite against environmental and chemical attack. Besides, this biodegradable polymer can hold and binds the biofibre together and transfer the load. (Karim & Afrin, 2017; Thakur, Thakur, & Gupta, 2014). In another definition by Koronis et al. (2013), composite materials are classified as green composites is mainly due to their renewable and natural-based resources, degradable, and sustainable. These green material composites can be easily disposed of without harming the environment. While, the definition of natural-based material is a product made from renewable agriculture and forest products, including crops and crop by products and its residues. The general scheme of green composites comprising natural fibre reinforcement with bio-derived resin will produce green composites or biocomposites, as shown in Figure 1.

The application of green material composites is the replacement of synthetic fibre-glass and steel in automotive parts. The primary market identified that those green materials installed in automotive are the trim parts in-cabin linings, parcel

Figure 1. The general rule of green composites; adapted from (Saba, Jawaid, Sultan, & Alothman, 2017)

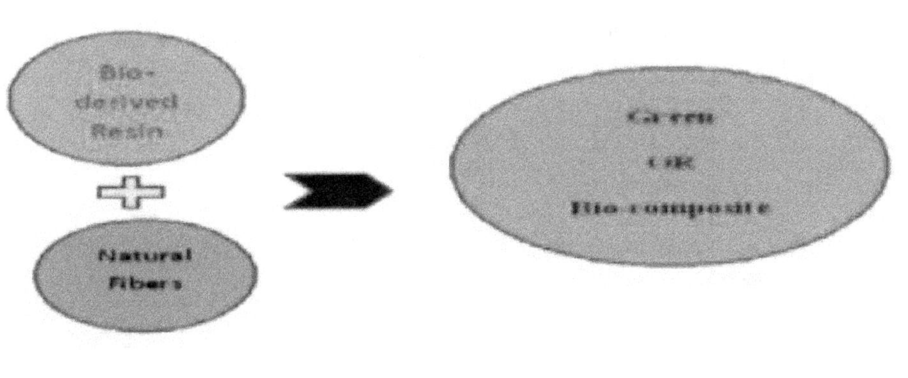

shelves, seat cushions, door panels, and dashboards. For instance, the conventional of asbestos fibres are replaced with using flax fibre in car disk brakes, and the use of hybrid kenaf, i.e., glass-reinforced epoxy composites, as car bumper beams (Bismarck, Baltazar-Y-Jimenez, & Sarikakis, 2006; Davoodi et al., 2010). Further market penetration of green composites will only occur only when their products can be rendered cost-effectiveness and competitive to the present molded parts used on many vehicles. According to Ahmad et al. (2015), the variation of natural fibre causes the difference in the mechanical properties of the fibre itself. This could affect the design, and quality aspect of the bio fibre reinforced composites. Thus, the lacking of the properties limits the potential application of flax fibre is often employed for poor composite implementation.

GREEN MATERIALS COMPOSITES FOR AUTOMOTIVE APPLICATIONS

It is essential to select green material to develop environment-friendly from renewable resources and reduce the dependence of petroleum-based chemicals. Currently, lignocellulose fibres are another option for the most widely utilized in composites materials. The potential of using natural fibres such as kenaf, sisal, sugar palm, flax, banana, etc. in the industry mainly in automotive application is due to the lower cost and high stiffness per weight than glass fibre. Moreover, the long term effect of using this natural fibre preserve the environment since it gives less thermally cycled rather than use synthetic fibre materials. Table 1 and Table 2 shows the green hybrid material for automotive applications and the automotive part that using natural fibre.

Table 1. Green hybrid materials for automotive applications

Type of matrix	Type of natural fibres	Applications	Refs
PP	non-woven/woven kenaf	Car door pocket	(Yuhazri et al., 2018)
Wheat straw	Coconut fibre/wheat husk	Automotive	(Sharma, Rao, & Krupashankara, 2018)
PLA	kenaf/coir	Indoor structural applications	(Yusoff, Takagi, & Nakagaito, 2016)
PLA	bamboo/coir	Indoor structural applications	(Yusoff et al., 2016)
PLA	Kenaf/bamboo/coir	Indoor structural applications	(Yusoff et al., 2016)
Epoxy	Kenaf fibre/ wooden stick	Elliptical tubes	(Alkateb, Sapuan, Leman, Jawaid, & Ishak, 2018)

Table 2. The automotive part that was using natural fibres

Type of natural fibres	Automotive makers	Automotive part	Refs
Soy foams	Ford	Car seat	(John & Thomas, 2008)
Wheat form	Ford	flex crossover injection moulded storage bins	(Andresen, Demuth, Lange, Stoick, & Pruszko, 2012)
100% castor bean oil derived nylon-11	Ford	Fuel tank tubes	(Holbery & Houston, 2006)
Jute/plastics, flax/sisal/epoxy	Mercedez Benz and Daimler Chrysler	Interior door panels	(La Mantia & Morreale, 2011; Marsh, 2003)
Flax/polyester	Mercedez Benz and Daimler Chrysler	Setra series engine	(Alkbir, Sapuan, Nuraini, & Ishak, 2016)
Sisal	Mercedez Benz and Daimler Chrysler	rear panel shelves	(Mohanty, Misra, & Drzal, 2002)

MECHANICAL PROPERTIES OF GREEN COMPOSITES

The mechanical properties of recent developments of green composites can be examined by several factors. Among of them are properties of the matrix and fibre reinforcement, aspect ratio of fibre reinforcement, orientation and geometry of the fibres, amount of fibre loading in the composites (fibre volume and fibre weight fraction), the interphase of matrix-fibre (Mallick, 2007; Mohd Nurazzi et al., 2017; Singh, Afrin, & Karim, 2017).

The mechanical properties of composite material are essential is due to these properties can determine the performance and application of the composite material. The most common mechanical properties examined are tensile, flexural, and impact properties. Table 3 and Table 4 shows the tensile properties of the thermosetting polymer matrix and selected natural fibres and synthetic fibres.

Table 3. Mechanical properties of thermosetting resins (Holbery & Houston, 2006; Iijima, Tochimoto, & Tomoi, 1991)

Type of Resin	Tensile strength (MPa)	Tensile Modulus (MPa)
Phenolic	10	375
Unsaturated polyester	40 to 90	2000 to 4500
Vinyl ester	69 to 83	3100 to 3800
Epoxy	35 to 100	3000 to 6000

Table 4. Tensile properties of selected natural fibres and synthetic fibres (Koronis et al., 2013; Ku, Wang, Pattarachaiyakoop, & Trada, 2011; Norizan, Abdan, Salit, & Mohamed, 2017; Satyanarayana, Arizaga, & Wypych, 2009)

Fibres	Tensile strength (MPa)	Elongation at break (%)	Young Modulus (GPa)
Sugar Palm	15.5 to 290	5.7 to 28.0	0.5 to 3.37
Bagasse	290	-	17
Bamboo	140 to 230	-	11 to 17
Flax	345 to 1035	2.7 to 3.2	27.6
Hemp	690	1.6 to 4	70
Jute	393 to 773	1.5 to 1.8	26.5
Kenaf	930	1.6	53
Sisal	511 to 535	2.0 to 2.5	9.4 to 22
Ramie	560	2.5 to 3.8	24.5
Pineapple	400 to 627	14.5	1.44
Coir	175	30	4 to 6
E-Glass	2000 to 3500	0.5	70
S-Glass	4570	2.8	86
Aramid	3000 to 3150	3.3 to 3.7	63.0 to 67.0

One of the most crucial of mechanical properties is tensile properties that involved of tensile strength and modulus. This tensile test is to evaluate the capability of a material to resist the load or force that tends to pull it apart and the expanse of its deformation before the material is breaking. Moreover, the tensile modulus displays the stiffness of a material and can be obtained from the stress-strain curve. In the formulation of tensile modulus, the force applied to produce deformation in a unit area of a test specimen and is the ratio of applied load to the original cross-sectional area is called Stress. While the strain is the change in the length per unit of the original length (Dl/l). Tensile strength is the maximum stress that a specimen can withstand during a tension test. Tensile Modulus also knew as Young's modulus is a measure of the stiffness of the material and is the ratio of stress to corresponding strain below the proportional limit of a material. Elongation at break of a material is the percentage increases in its length due to the applied tensile load up to its breaking point.

Both bio-fibre reinforcement and polymer matrix are essential in enhancing the mechanical properties of the composites. General principle is ruled in the composites concept are the matrix are not just protecting the fibre reinforcement from damage due to abrasion, but it also acts as load supporting constituents to cohere the fibre reinforcement together and to preserve the composites (Callister Jr & Rethwisch, 2012). Thus, by substituting the fibres in impregnated of polymer matrix, this could be lead the good resultant mechanical properties of composites.

Fibre Reinforcements

The principal function of fibre reinforcements is to deliver as the reinforcing materials and to withstand the load, limit deformation and improve the mechanical properties of the composites (Le Duigou, Pillin, Bourmaud, Davies, & Baley, 2008). This fibre reinforcement can provide the strength and stiffness of the composites. The critical parameter in a composite material is the fibre aspect ratio (length/diameter ratio). The increasing of fibre aspect ratio in fibre reinforcements leads to enhance mechanical properties of composites.

The fibre aspect ratio is crucial to study the fracture properties of the composites. In the discontinuous fibre reinforced composites or known as short fibre composites, the critical fibre length is needed to fulfil the full stressed condition in composites. When comparing the fibre lengths, the shorter length of fibre size lead to debonding of the fibre and matrix interphase. Conversely, the longer of the fibre than exceptional critical length, the fibres are better to withstand under applied load and will improve the properties of the composites.

When the increase of fibre length than the diameter itself, the most exceptional strength and stiffness will happen, even though the short fibre is limited by the length of fibre, but they promote excellent properties for composites. This could be agreed by another researcher that studied the content of fibre volume fraction in composites (Kumar, Yakabu, & Anandjiwala, 2010; Petinakis et al., 2009). When the increase of fibre contents, the strength and stiffness increases until beyond a particular volume fraction of fibre. Above beyond the optimum of fibre content, no proportional increase of the strength in composite materials.

Failure occurs when sufficient energy is built up within the material, and that energy is then dissipated in the form of cracks. As more and more energy is added to the material, the crack propagates through the material until failure occurs. As a crack begins to propagate through a matrix, it hits a reinforcement which forces the crack to change direction to go around the is. The crack must traverse a larger distance, which means more energy is being absorbed by the composite. The energy dissipated is proportional to the distance travelled by the crack (Figure 2).

According to Thomas and King (1993), the damage can spread differently during the fracture occurs in the composites. The fracture mechanism is occurring when the combination between durable fibre or matrix with brittle matrix, this will cause of stress concentration at crack trip that will lead to crack propagate due to fibre debonding between fibre and matrix. As shown in Figure 3. Since the main

Figure 2. Comparison of crack propagation between matrix only and composite (Thomas & King, 1993)

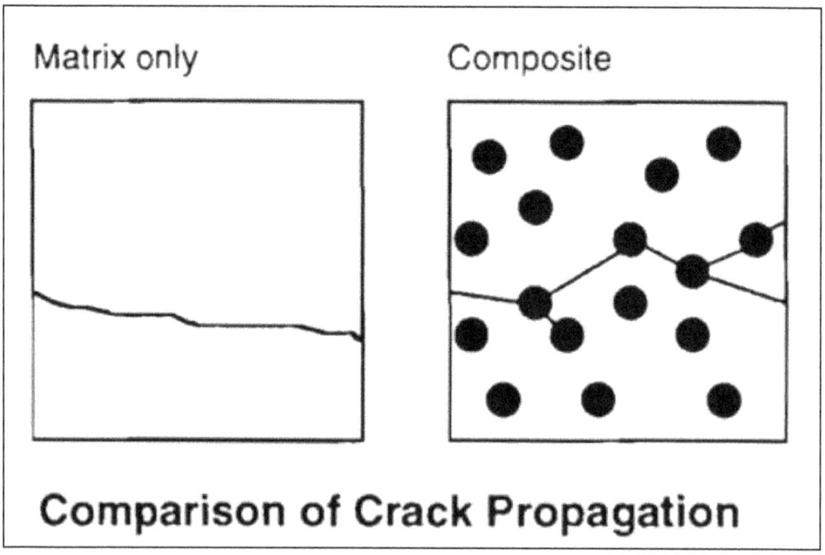

role of fibre reinforcement is to promote toughening properties, in another mode, the crack can yield around the matrix and shear failure can occur in the interphase region between fibre and matrix. This will be allowing the unloaded fibre to shrink back into the matrix.

Effect of Fibre Loading

Generally, increased the fibre loading will result in a significant increase of composites stiffness, as tensile and impact strengths are enhanced through the addition of natural fibres. Until now, a lot of studies have been proved to investigate the natural fibre content of optimum value within the composites to enhance the mechanical performance of the composites. According to Shalwan and Yousif (2013), there is no universal value for natural fibre content (weight) and fibre volume fraction where the higher tensile properties could be achieved. This is because, there is possible type-specific optimum loading for demonstrating good tensile strength for each fibre type. This can be correlated with chemical composition of natural fibre properties and their inherent strength, interfacial adhesion, and physical properties characteristics. Furthermore, several researchers have commonly agreed at higher 50% by volume fraction, the fibres tend to form aggregations which result in the weakening of the interfacial area and debonding between the fibres and matrix (Brahim & Cheikh,

Figure 3. Failure propagation (a) brittle cracking of matrix (high interfacial strength), (b) shear yielding of matrix, (c) interfacial failure and (d) propagation of fracture involving propagation of (a) and (c) (Hancox, 1986; Prakash & Viswanthan, 2019; Thomas & King, 1993)

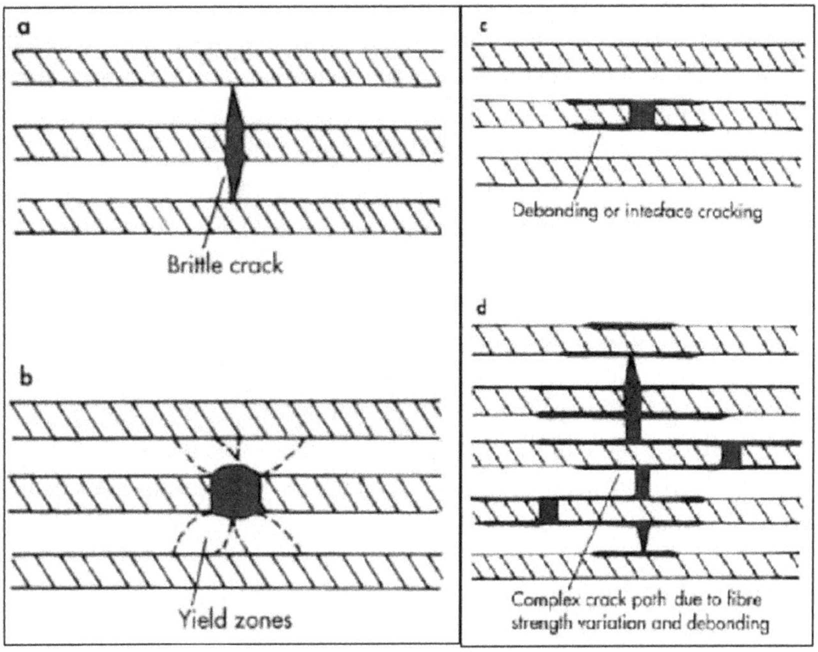

2007; George, Joseph, Boudenne, & Thomas, 2010; Venkateshwaran, Elayaperumal, & Sathiya, 2012; Zhai et al., 2006).

According to Hanafee et al. (2019), the increasing volume fraction of fibre means greater shear produced between the fibre and matrix. A possible resulting greater force that can break the specimens. Further flexural test carried out with higher fibre loading result the higher flexural strength. Conversely, the greater of fibre volume fraction at above 50%, the lower mechanical properties revealing in the composites. The reason is the fibres are not insufficient been absorbing by matrix loading that leads fibre-matrix debonding, increased sample brittleness and thus makes the specimen composite are failed at the beginning of flexural test. This will affect the overall mechanical performance of the composites. In another study, the researchers have found that the maximum mechanical properties (tensile, impact, flexural and compression properties) of variation sugar palm yarn reinforced unsaturated polyester resin was achieved at 30 wt.% and 40 wt. % (Nurazzi, Khalina, Sapuan, & Rahmah, 2018).

This could be stated that the tensile modulus is prone to dependent on the fibre properties while the tensile strength was more sensitive to the matrix. Taken together, to improve the mechanical properties, the composites should have a stable interface fibre-matrix bonding, fibre wetting in the matrix region, low stress concentration and high aspect ratio that can evaluate the tensile modulus.

Additionally, an optimum bonding level between fibre and matrix is essential to achieve the tensile and impact properties. The factors such as fibre pull-out, degree of adhesion and mechanism to absorb energy can influence the impact strength of the composites. The attributions mostly differ with the formulation according to the rules of mixtures and increase linearly with composition. Nevertheless, the linear dependence of the fibre content does not hold more than 80% of the fibre, probably due to poor adhesion between fibre reinforcement and matrix.

Specifically, materials with high specific strength and high specific stiffness are likely to have individual merit in applications in which weight will be a critical factor. As can be seen in Figure 4, E-glass is clearly shown the highest specific strength but is outperformed by other natural fibre such as hemp, kenaf and ramie. When comparing both charts, no optimum fibre that outperforms. A reasonable choice to select among the fibres, hemp is stronger than ramie fibre performance that similar and still stiffer than E-glass fibre.

Hybridization of Fibres

The development of hybrid composite materials based on the reinforcement of two or more fibres in a single matrix remains in its infancy. Research has revealed that the behavior of hybrid composites appears to be simply a weighted sum of

Figure 4. Average specific strength and stiffness on selected fibres and synthetic fibres (Koronis et al., 2013)

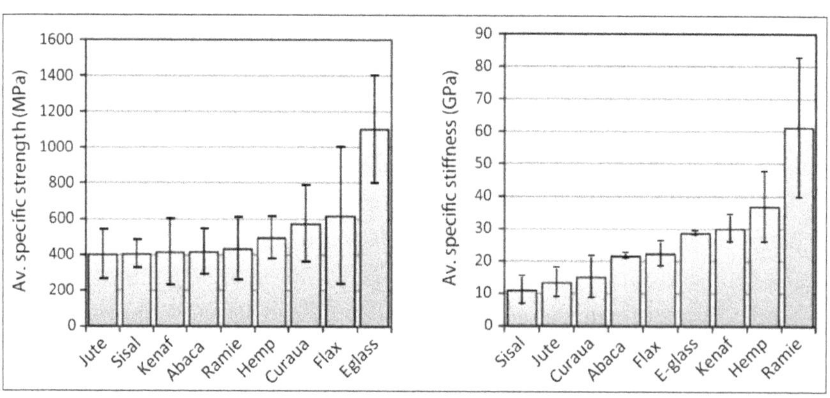

the individual components in which there is a more favorable balance between the advantages and disadvantages inherent in any composite material (Mishra et al., 2003)(Mishra et al., 2003)(Mishra et al., 2003). In other words, the beneficial aspect of hybridization of different fibres is that a particular fibre could complement the limitations of another fibre. As a result, a balance of mechanical performance and cost reduction for engineering applications could be achieved (Thwe & Liao, 2002; Venkata Reddy, Venkata Naidu, & Shobha Rani, 2008).

The effect of hybridization between natural fibres and synthetic fibres in a single matrix has been studied by previous researchers (AlMaadeed, Kahraman, Noorunnisa Khanam, & Al-Maadeed, 2013; Atiqah, Maleque, Jawaid, & Iqbal, 2014; Goud & Rao, 2011; Nunna, Chandra, Shrivastava, & Jalan, 2012; Saba, Tahir, & Jawaid, 2014; Shalwan & Yousif, 2013; Venkata Reddy et al., 2008). It has been found that the effect of this hybridization with optimized loadings has shown a promising impact on the improvement of the physical and mechanical properties. According to Atiqah et al. (2014). the limitation of fibre loading in hybrid composites is generally shown at a maximum of 50%. Figure 5 shows the schematic diagram of the hybridisation of natural fibres (sugar palm yarn fibre) with synthetic fibres (glass fibre mat).

Natural fibres to matrix adhesion play an important role in the reinforcement of the composites. When the stress is applied, stress is not applied directly to the fibres but the matrix. In order to develop composite with excellent mechanical properties (ultimate strength but toughness), the load must be transferred effectively from the

Figure 5. Schematic diagram of hybridization and sequence of fibre between the natural fibre (sugar palm yarn fibre) and synthetic fibre (glass fibre mat) (Nurazzi et al., 2018)

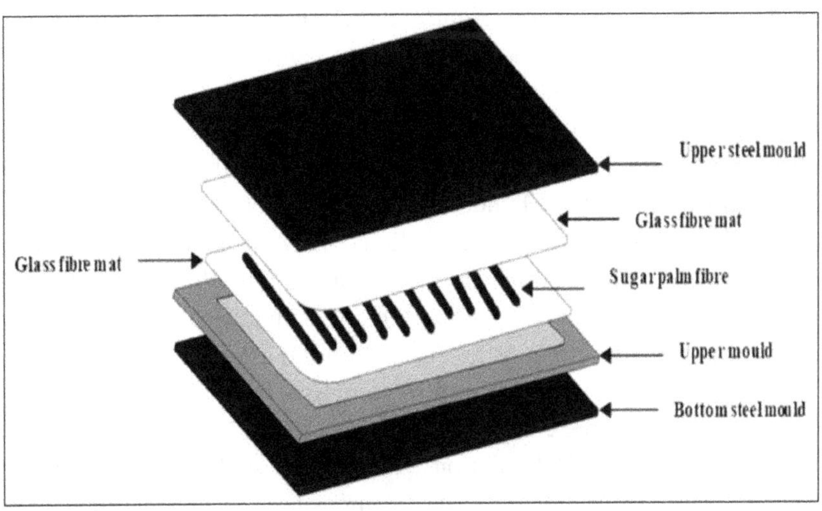

matrix to the fibres. This recommends good adhesion as well interaction between the fibres and the matrix such as efficient and robust fibre matrix bonding. This can be controlled by either surface treatment applied to the filler or by the use of additives such as coupling agents.

Other than the risks as mentioned above, an additional risk of using natural fibres is the poor compatibility between the fibres and the polymer matrix, which results in a non-uniform dispersion formation of fibres within the matrix, and thus generates poor interfacial bonding properties (Zin et al., 2019). Most polymers, especially thermosetting, are non-polar (hydrophobic, water-repelling) substances, which is not compatible with polar ('hydrophilic, water-attracting) natural fibres. Therefore, this causes poor adhesion between the fibre and the matrix interaction surfaces. Hence leads to the deterioration of mechanical properties of the composites.

Fibre Improvement

Many studies can be found on various composite systems in achieving desired mechanical properties by improving the stress transfer mechanism through the improvement of the adhesion between the natural fibre and the matrix. Chemical or pretreatment methods of the fibre will chemically modify the surface, clean the fibre surface as in Figure 5, reduce the moisture absorption process and increase the surface unevenness (Edeerozey, Akil, Azhar, & Ariffin, 2007). Some of the important industrial methods that are currently employed includes the mercerisation, acetylation, etherification, peroxide treatment, benzoylation, graft copolymerisation, acrylation, maleic anhydride, titanate treatment, permanganate treatment, sodium chlorite treatment, plasma treatment, isocyanate treatment and use of a coupling agent such as silane treatment used for natural fibres to improve the fibre-matrix interfacial bonding in composites (Kalia, Kaith, & Kaur, 2009). Figure 6 shows the schematic mechanism of the effect of alkaline treatment on the untreated natural fibre.

Bachtiar et al. (2010) studied the mechanical properties arising from the effects of different concentrations of sodium hydroxide (NaOH) on sugar palm reinforced epoxy composites. These fibres were treated with the alkaline solution by soaking with 0.25 M and 0.5 M of NaOH under 1 hour, 4 hours, and 8 hours of soaking time, respectively. The ultimate tensile strength (49.87 MPa) was achieved at 0.25 M NaOH solution with 1 hour soaking time, an improvement of 16.4% from untreated fibre composite. The tensile modulus (3.85 GPa) at this condition gave the increase of 13.6% from untreated fibre composite. The maximum flexural strength was achieved at 0.25 M NaOH solution with 1 hour of soaking time by 96.71 MPa, improving by 24.41% from untreated fibre composite. However, the maximum flexural modulus was achieved at 0.5 M NaOH solution with 4 hours soaking time by 6.90 GPa, improving by 148% from untreated fibre composite. The higher impact strength

Figure 6. (a) Surface morphology of untreated and (b) treated single sugar palm fibre (Norizan et al., 2017)

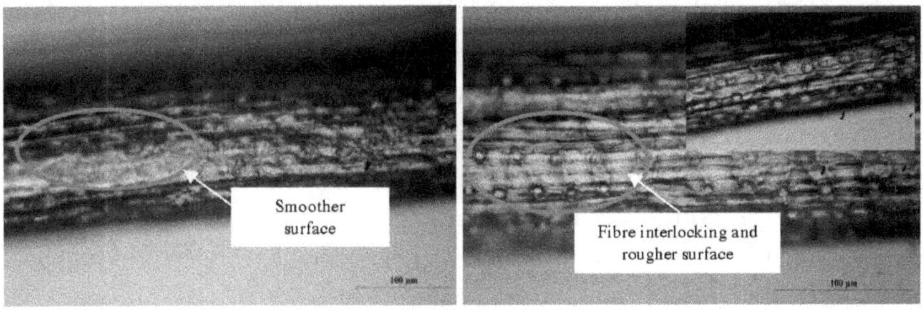

(a) (b)

of treated sugar palm fibre was achieved at 0.5 M NaOH solution with 8 hours soaking time by 60 J/m with improving by 12.85% from untreated fibre composite. The improvement in these properties might be due to the reduction of hydrophilic properties within the sugar palm fibres after being treated with NaOH. The treatment was also thought to have facilitated better interfacial bonding between the sugar palm fibre and the epoxy matrix.

Figure 7. Schematic mechanism of the untreated and treated natural fibre using alkaline treatment (Mohanty, Misra, & Drzal, 2001)

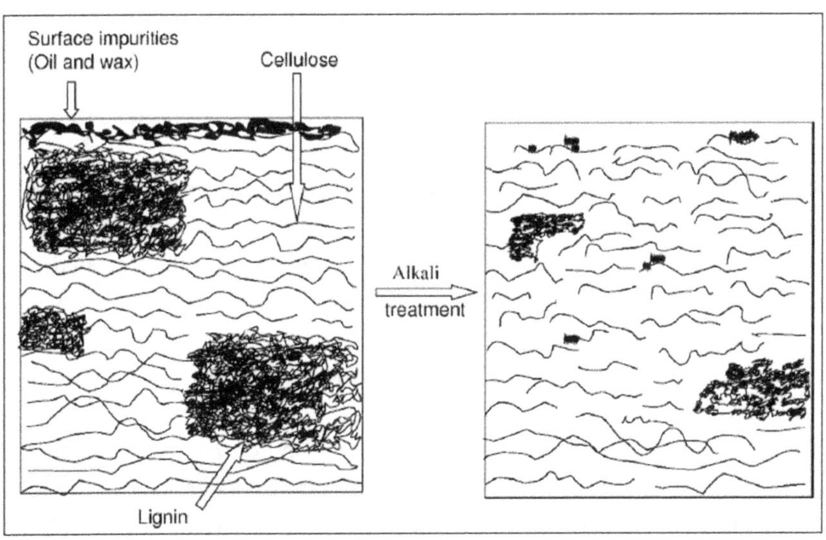

Good dispersion implies that fibres particles are separated from each other (there are no clumps and agglomerates) and each of the fillers is surrounded by the matrix. Insufficient fillers dispersion usually due to the inhomogeneous mixture of resin-rich areas and fibre rich domain. This repulsive is due to matrix is not well attached to fibre-rich areas (clumps) are susceptible to microcracking. Microcracks contribute to inferior mechanical properties of composites. It is therefore important to ensure homogeneous fibre dispersion in order to achieve maximum strength and performance of the composite materials. Two major factors affect the extent of fibre dispersion, fibre-particle interaction such as strong hydrogen bonding between the fibre and particle and fibre length (Thakur & Singha, 2010). The same factors also account for the tendency for some fibres, such as natural fibres to agglomerate during mixing. The fibre length used in the preparation of composites is critical. It should not be too long; otherwise, the fibres may get entangled with each other. This will cause problems with fibres to offer effective reinforcement in the mechanical performance of the composites.

Fibre Geometry

The geometry of the fibres in a composite is also important, since fibres have their highest mechanical properties along their lengths, rather than across their widths (Thakur & Singha, 2010). The orientation of the fibres in the composite plays an important role in imparting isotropic or anisotropic properties to a composite. It shows that the 0° (longitudinal) fibre orientation indicated tensile and flexural properties compared to the composite with 45° and 90° fibre orientation. This was because composites with 0° orientation possessed long and continuous sugar palm yarn fibres that resisted and transferred the tensile forces that were aligned perpendicular to the direction of force, whereas the other two orientations (45° and 90°) contained short and discontinuous fibres which were related to the fibre end effects (Tungjitpornkull & Sombatsompop, 2009).

Since the load was subjected at the middle of the composites (Figure 7(a)) until failure occurred, the incorporation of long and continuous fibres in the composites acted as carriers of the load, and the stress was transferred from the matrix along the fibres. This led to a productive and uniform stress distribution. However, for the 45° and 90° orientations, the short and discontinuous (fibre end effects) fibres when the force was applied at the middle detracted from the effectiveness of the uniform stress distribution.

This can be observed in the 90° fibre orientation composites. For 90° of fibre orientation or the transverse direction, cracks could occur at any place in the composite that was not supported by the fibres and the areas without fibres. The propagation of cracks near the fibre ends, and the constraining effect of the fibres in crack propagation perpendicular or transverse to the loading direction also occurred. Figure 8 (a), (b), (c), and (d) shows the schematic diagram of composites with different fibre orientations.

According to Mortazavian and Fatemi (2015), the significantly lower mechanical properties of composites in the 45° and 90° (transverse) orientation compared to 0° (longitudinal) fibre orientation would be due to the strain limit of the fibres. In the longitudinal direction, fibres are highly aligned with loading direction. Therefore, the fibres would have a low capacity to deal with strain, resulting in a low ductility composite. As the average number of fibres with angles more massive than 0° concerning the loading direction increases, the failure is more dominated by the matrix and interfacial properties. Therefore, matrix straining is more dominant in the 45° and 90° directions. A lower ductility was observed for the 90° angle or transverse direction.

Figure 8 (a) Schematic diagram of the flexural test and (b) the mechanism of 0°, (c) 45° and (d) 90° orientation during force applied (Mortazavian & Fatemi, 2015)

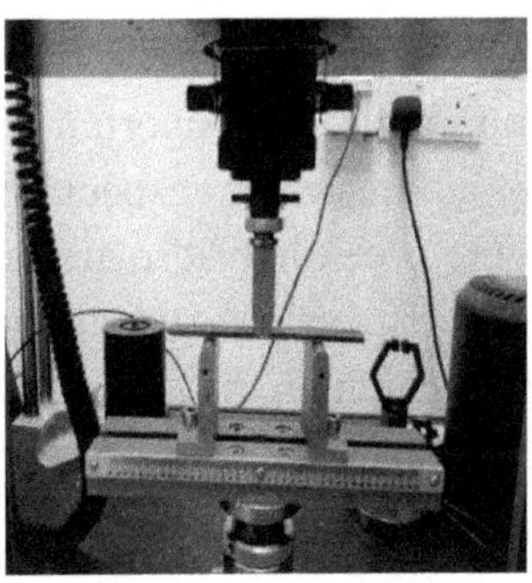

(a)

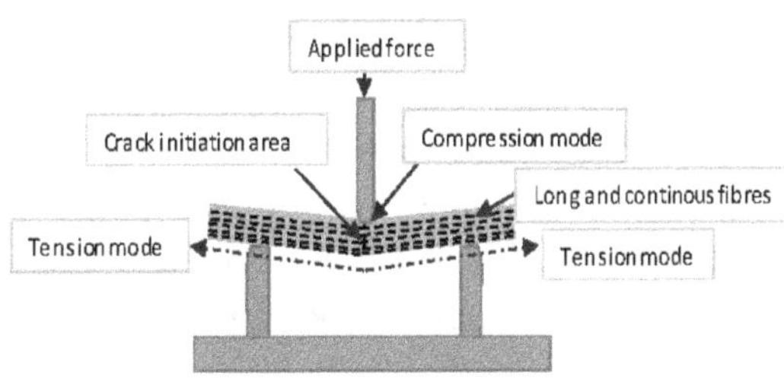

(b)

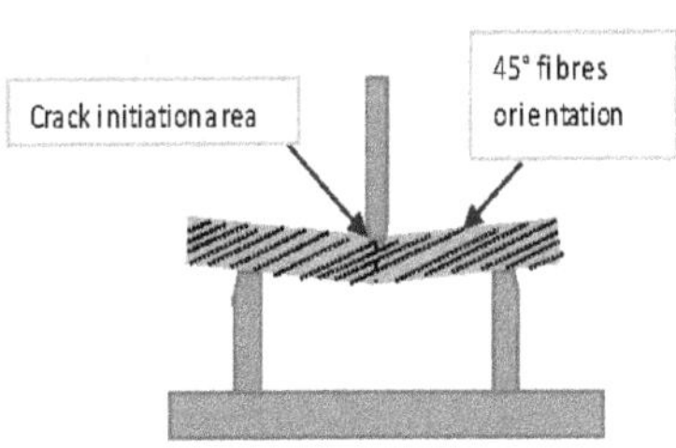

(c)

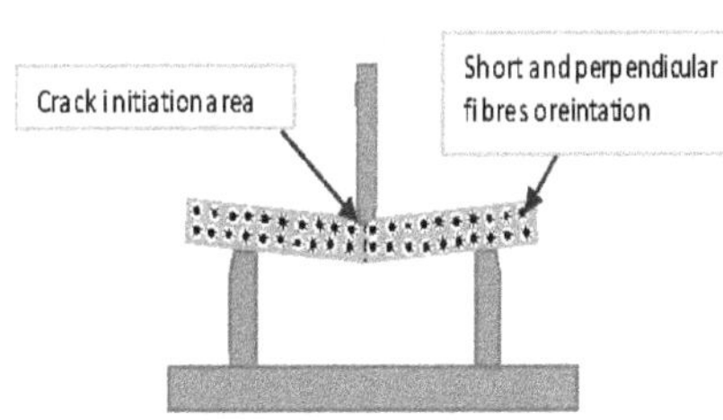

(d)

CHALLENGES

Even though many researchers prove that using green hybrid material for automotive applications. There is also a drawback of using those materials, the properties and their structure is the challenge to comply with the automotive requirement. For instance, high thermomechanical performance for short and long term period through testing impact, creep, and fatigue testing exposure to corrosive chemical and fluids. It is precaution steps that the industry should cover for establishing an uncertain venture than a well-established structure. Besides, the automotive manufacturer has to be patient and consider new capital for developing a whole new infrastructure that affects the manufacturing cost until the finished product. There are also need to investigate on the properties of agriculture resources during processing such as lower moisture resistance, lower thermal stability and other variances make it difficult to have new composite whereby the steps of designing the uniform structure, prone to fast microbial and environmental degradation after disposal.

CONCLUSION

From the review done, the properties of the hybrid green material are important to improve the performance the composites. The proper study need to be done especially for the automotive applications. The benefits of having the green hybrid material promote lightweight automotive part and better life cycle of disposal. This could be reducing the greenhouse gas emission and the increasing research on the feasibility of green material in automotive applications

ACKNOWLEDGMENT

Authors would like to gratitude to Tenaga Nasional Berhad Malaysia, through a seed fund research grant (Project No: U-TS-RD-18-10), Institute of Power Engineering, Department of Mechanical Engineering, and Uniten R&D Sdn. Bhd., Universiti Tenaga Nasional for the support to publish these project findings.

REFERENCES

Ahmad, F., Choi, H. S., & Park, M. K. (2015). A review: Natural fiber composites selection in view of mechanical, light weight, and economic properties. *Macromolecular Materials and Engineering, 300*(1), 10–24. doi:10.1002/mame.201400089

Alkateb, M., Sapuan, S. M., Leman, Z., Jawaid, M., & Ishak, M. R. (2018). Crushing behavior of kenaf fiber/wooden stick reinforced epoxy hybrid" green" composite elliptical tubes. *Polimery*, 63.

Alkbir, M. F. M., Sapuan, S. M., Nuraini, A. A., & Ishak, M. R. (2016). Fibre properties and crashworthiness parameters of natural fibre-reinforced composite structure: A literature review. *Composite Structures, 148*, 59–73. doi:10.1016/j.compstruct.2016.01.098

AlMaadeed, M., Kahraman, R., Noorunnisa Khanam, P., & Al-Maadeed, S. (2013). Characterization of untreated and treated male and female date palm leaves. *Materials & Design, 43*, 526–531. doi:10.1016/j.matdes.2012.07.028

Andresen, C., Demuth, C., Lange, A., Stoick, P., & Pruszko, R. (2012). Biobased automobile parts investigation. A Report Developed for the USDA Office of Energy Policy and New Uses.

Atiqah, A., Maleque, M. A., Jawaid, M., & Iqbal, M. (2014). Development of kenaf-glass reinforced unsaturated polyester hybrid composite for structural applications. *Composites. Part B, Engineering, 56*, 68–73. doi:10.1016/j.compositesb.2013.08.019

Bachtiar, D., Sapuan, S. M., & Hamdan, M. M. (2008). The effect of alkaline treatment on tensile properties of sugar palm fibre reinforced epoxy composites. *Materials & Design, 29*(7), 1285–1290. doi:10.1016/j.matdes.2007.09.006

Bachtiar, D., Sapuan, S. M., & Hamdan, M. M. (2009). The influence of alkaline surface fibre treatment on the impact properties of sugar palm fibre-reinforced epoxy composites. *Polymer-Plastics Technology and Engineering, 48*(4), 379–383. doi:10.1080/03602550902725373

Bachtiar, D., Sapuan, S. M., & Hamdan, M. M. (2010). Flexural properties of alkaline treated sugar palm fibre reinforced epoxy composites. *International Journal of Automotive and Mechanical Engineering, 1*(1), 79–90. doi:10.15282/ijame.1.2010.7.0007

Ben Brahim, S., & Ben Cheikh, R. (2007). Influence of fibre orientation and volume fraction on the tensile properties of unidirectional Alfa-polyester composite. *Composites Science and Technology*, *67*(1), 140–147. doi:10.1016/j.compscitech.2005.10.006

Bismarck, A., Baltazar-Y-Jimenez, A., & Sarikakis, K. (2006). Green composites as panacea? Socio-economic aspects of green materials. *Environment, Development and Sustainability*, *8*(3), 445–463. doi:10.1007/s10668-005-8506-5

Callister, W. D. Jr, & Rethwisch, D. G. (2012). *Fundamentals of materials science and engineering: an integrated approach*. John Wiley & Sons.

Davoodi, M. M., Sapuan, S. M., Ahmad, D., Ali, A., Khalina, A., & Jonoobi, M. (2010). Mechanical properties of hybrid kenaf/glass reinforced epoxy composite for passenger car bumper beam. *Materials & Design*, *31*(10), 4927–4932. doi:10.1016/j.matdes.2010.05.021

Edeerozey, A. M. M., Akil, H. M., Azhar, A. B., & Ariffin, M. I. Z. (2007). Chemical modification of kenaf fibers. *Materials Letters*, *61*. doi:10.1016/j.matlet.2006.08.006

George, G., Joseph, K., Boudenne, A., & Thomas, S. (2010). Recent advances in green composites. Key Engineering Materials. doi:10.4028/www.scientific.net/KEM.425.107

Goud, G., & Rao, R. N. (2011). The effect of alkali treatment on dielectric properties of Roystonea regia/epoxy composites. *International Journal of Polymer Analysis and Characterization*, *16*(4), 239–250. doi:10.1080/1023666X.2011.570039

Hancox, N. L. (1986). Composite materials. Pt. 3. Matrices for composite materials. *Metals and Materials*, *2*(7), 435.

Holbery, J., & Houston, D. (2006). *Natural-Fiber-Reinforced Polymer Composites in Automotive Applications*. Academic Press.

Iijima, T., Tochimoto, T., & Tomoi, M. (1991). Modification of epoxy resins with poly (aryl ether ketone) s. *Journal of Applied Polymer Science*, *43*(9), 1685–1692. doi:10.1002/app.1991.070430911

John, M. J., & Thomas, S. (2008). Biofibres and biocomposites. *Carbohydrate Polymers*, *71*(3), 343–364. doi:10.1016/j.carbpol.2007.05.040

Kalia, S., Kaith, B. S., & Kaur, I. (2009). Pretreatments of natural fibers and their application as reinforcing material in polymer composites—A review. *Polymer Engineering and Science*, *49*(7), 1253–1272. doi:10.1002/pen.21328

Karim, Z., & Afrin, S. (2017). Bacterial cellulose: Preparation and characterization. In *Cellulose-Reinforced Nanofibre Composites* (pp. 327–340). Elsevier; doi:10.1016/B978-0-08-100957-4.00014-0.

Koronis, G., Silva, A., & Fontul, M. (2013). Green composites: A review of adequate materials for automotive applications. *Composites. Part B, Engineering*, *44*(1), 120–127. doi:10.1016/j.compositesb.2012.07.004

Ku, H., Wang, H., Pattarachaiyakoop, N., & Trada, M. (2011). Composites : Part B A review on the tensile properties of natural fiber reinforced polymer composites. doi:10.1016/j.compositesb.2011.01.010

Kumar, R., Yakabu, M. K., & Anandjiwala, R. D. (2010). Effect of montmorillonite clay on flax fabric reinforced poly lactic acid composites with amphiphilic additives. *Composites. Part A, Applied Science and Manufacturing*, *41*(11), 1620–1627. doi:10.1016/j.compositesa.2010.07.012

La Mantia, F. P., & Morreale, M. (2011). Green composites: A brief review. *Composites. Part A, Applied Science and Manufacturing*, *42*(6), 579–588. doi:10.1016/j.compositesa.2011.01.017

Le Duigou, A., Pillin, I., Bourmaud, A., Davies, P., & Baley, C. (2008). Effect of recycling on mechanical behaviour of biocompostable flax/poly (l-lactide) composites. *Composites. Part A, Applied Science and Manufacturing*, *39*(9), 1471–1478. doi:10.1016/j.compositesa.2008.05.008

Mallick, P. K. (2007). Fiber-reinforced composites: materials, manufacturing, and design. CRC press. doi:10.1201/9781420005981

Marsh, G. (2003). Next step for automotive materials. *Materials Today*, *4*(6), 36–43. doi:10.1016/S1369-7021(03)00429-2

Mishra, S., Mohanty, A. K., Drzal, L. T., Misra, M., Parija, S., Nayak, S. K., & Tripathy, S. S. (2003). Studies on mechanical performance of biofibre/glass reinforced polyester hybrid composites. *Composites Science and Technology*, *63*(10), 1377–1385. doi:10.1016/S0266-3538(03)00084-8

Mohanty, A. K., Misra, M., & Drzal, L. T. (2001). Surface modifications of natural fibers and performance of the resulting biocomposites: An overview. *Composite Interfaces*, *8*(5), 313–343. doi:10.1163/156855401753255422

Mohanty, A. K., Misra, M., & Drzal, L. T. (2002). Sustainable Bio-Composites from renewable resources: Opportunities and challenges in the green materials world. *Journal of Polymers and the Environment*, *10*(1–2), 19–26. doi:10.1023/A:1021013921916

Mohd Nurazzi, N., Khalina, A., Sapuan, S. M., Dayang Laila, A., Rahmah, M., & Hanafee, Z. (2017). A Review: Fibres, Polymer Matrices and Composites. *Pertanika Journal of Science & Technology*, *25*(4).

Mortazavian, S., & Fatemi, A. (2015). Effects of fiber orientation and anisotropy on tensile strength and elastic modulus of short fiber reinforced polymer composites. *Composites. Part B, Engineering*, *72*, 116–129. doi:10.1016/j.compositesb.2014.11.041

Norizan, M. N., Abdan, K., Salit, M. S., & Mohamed, R. (2017). Physical, mechanical and thermal properties of sugar palm yarn fibre loading on reinforced unsaturated polyester composites. *The Journal of Physiological Sciences; JPS*, *28*(3), 115–136.

Nunna, S., Chandra, P. R., Shrivastava, S., & Jalan, A. K. (2012). A review on mechanical behavior of natural fiber based hybrid composites. *Journal of Reinforced Plastics and Composites*, *31*(11), 759–769. doi:10.1177/0731684412444325

Nurazzi, N. M., Khalina, A., Sapuan, S. M., & Rahmah, M. (2018). Development of sugar palm yarn/glass fibre reinforced unsaturated polyester hybrid composites. Materials Research Express, 5(4), 45308. doi:10.1088/2053-1591/aabc27

Oliver-Ortega, H., Julian, F., Espinach, F. X., Tarrés, Q., Ardanuy, M., & Mutjé, P. (2019). Research on the use of lignocellulosic fibers reinforced bio-polyamide 11 with composites for automotive parts: Car door handle case study. *Journal of Cleaner Production*, *226*, 64–73. doi:10.1016/j.jclepro.2019.04.047

Petinakis, E., Yu, L., Edward, G., Dean, K., Liu, H., & Scully, A. D. (2009). Effect of matrix–particle interfacial adhesion on the mechanical properties of poly (lactic acid)/wood-flour micro-composites. *Journal of Polymers and the Environment*, *17*(2), 83–94. doi:10.1007/s10924-009-0124-0

Prakash, V. R. A., & Viswanthan, R. (2019). Fabrication and Characterization of Echinoidea spike Particles and Kenaf Natural Fibre-Reinforced Azadirachta-Indica Blended Epoxy Multi-hybrid Bio Composite. *Composites. Part A, Applied Science and Manufacturing*.

Saba, N., Jawaid, M., Sultan, M. T. H., & Alothman, O. Y. (2017). Green biocomposites for structural applications. Green Energy and Technology. doi:10.1007/978-3-319-49382-4_1

Saba, N., Tahir, P. M., & Jawaid, M. (2014). A review on potentiality of nano filler/natural fiber filled polymer hybrid composites. *Polymers*, *6*(8), 2247–2273. doi:10.3390/polym6082247

Satyanarayana, K. G., Arizaga, G. G. C., & Wypych, F. (2009). Biodegradable composites based on lignocellulosic fibers—An overview. *Progress in Polymer Science*, *34*(9), 982–1021. doi:10.1016/j.progpolymsci.2008.12.002

Shalwan, A., & Yousif, B. F. (2013). In state of art: Mechanical and tribological behaviour of polymeric composites based on natural fibres. *Materials & Design*, *48*, 14–24. doi:10.1016/j.matdes.2012.07.014

Sharma, A., Rao, N. N., & Krupashankara, M. S. (2018). Development of eco-friendly and biodegradable Bio composites. [). Proceedings. doi:10.1016/j.matpr.2018.06.490]. *Materials Today*, *5*, 20987–20995.

Singh, A. A., Afrin, S., & Karim, Z. (2017). Green composites: Versatile material for future. In *Green Biocomposites* (pp. 29–44). Springer; doi:10.1007/978-3-319-49382-4_2.

Thakur, V. K., & Singha, A. S. (2010). Mechanical and water absorption properties of natural fibers/polymer biocomposites. *Polymer-Plastics Technology and Engineering*, *49*(7), 694–700. doi:10.1080/03602551003682067

Thakur, V. K., Thakur, M. K., & Gupta, R. K. (2014). Review: Raw natural fiber–based polymer composites. *International Journal of Polymer Analysis and Characterization*, *19*(3), 256–271. doi:10.1080/1023666X.2014.880016

Thomas, M. P., & King, J. E. (1993). Effect of thermal and mechanical processing on tensile properties of powder formed 2124 aluminium and 2124 Al-SiCp metal matrix composite. *Materials Science and Technology*, *9*(9), 742–753.

Thwe, M. M., & Liao, K. (2002). Effects of environmental aging on the mechanical properties of bamboo–glass fiber reinforced polymer matrix hybrid composites. *Composites. Part A, Applied Science and Manufacturing*, *33*(1), 43–52. doi:10.1016/S1359-835X(01)00071-9

Tungjitpornkull, S., & Sombatsompop, N. (2009). Processing technique and fiber orientation angle affecting the mechanical properties of E-glass fiber reinforced wood/PVC composites. *Journal of Materials Processing Technology*, *209*(6), 3079–3088. doi:10.1016/j.jmatprotec.2008.07.021

Venkata Reddy, G., Venkata Naidu, S., & Shobha Rani, T. (2008). Impact properties of kapok based unsaturated polyester hybrid composites. *Journal of Reinforced Plastics and Composites*, *27*(16–17), 1789–1804. doi:10.1177/0731684407087380

Venkateshwaran, N., Elayaperumal, A., & Sathiya, G. K. (2012). Prediction of tensile properties of hybrid-natural fiber composites. *Composites. Part B, Engineering*, *43*(2), 793–796. doi:10.1016/j.compositesb.2011.08.023

Yuhazri, M. Y., Amirhafizan, M. H., Abdullah, A., Yahaya, S. H., Lau, S. T. W., & Kamarul, A. M. (2018). Kenaf fibre composites as promising green-composites for automotive car door map pocket application. *International Journal of Mechanical and Mechatronics Engineering*, *18*(2), 15–21. Retrieved from https://www.scopus.com/inward/record.uri?eid=2-s2.0-85046781426&partnerID=40&md5=ffa672a37730c237d8135a84f490c390

Yusoff, R. B., Takagi, H., & Nakagaito, A. N. (2016). Tensile and flexural properties of polylactic acid-based hybrid green composites reinforced by kenaf, bamboo and coir fibers. *Industrial Crops and Products*, *94*, 562–573. doi:10.1016/j.indcrop.2016.09.017

Zhai, L., Berg, M. C., Cebeci, F. C., Kim, Y., Milwid, J. M., Rubner, M. F., & Cohen, R. E. (2006). Patterned superhydrophobic surfaces: Toward a synthetic mimic of the Namib Desert beetle. *Nano Letters*, *6*(6), 1213–1217. doi:10.1021/nl060644q PubMed

Zin, M. H., Abdan, K., & Norizan, M. N. (2019). The effect of different fiber loading on flexural and thermal properties of banana/pineapple leaf (PALF)/glass hybrid composite. In Structural Health Monitoring of Biocomposites, Fibre-Reinforced Composites and Hybrid Composites (pp. 1–17). Elsevier.

Chapter 4
Modifications of Biodegradable Thermoplastic Starch (TPS) From Sago Starch via Cross-Linking Method

Yusriah Lazim
Malaysian Institute of Chemical Bio-Engineering Technology, Universiti Kuala Lumpur, Malaysia

Abdul Baith Abu Hanafi
Malaysian Institute of Chemical Bio-Engineering Technology, Universiti Kuala Lumpur, Malaysia

Mohd Syazwan Adura
Malaysian Institute of Chemical Bio-Engineering Technology, Universiti Kuala Lumpur, Malaysia

Siti Afifah Muda
Malaysian Institute of Chemical Bio-Engineering Technology, Universiti Kuala Lumpur, Malaysia

Lily Suhaila Yacob
Malaysian Institute of Chemical Bio-Engineering Technology, Universiti Kuala Lumpur, Malaysia

Ahmad Marzio Mohd Yusof
Malaysian Institute of Chemical Bio-Engineering Technology, Universiti Kuala Lumpur, Malaysia

ABSTRACT

The increasing amount of synthetic plastic waste has contributed to environmental problems worldwide. As an alternative to synthetic plastics, thermoplastic starch (TPS) has been used for many applications, especially packaging application, since TPS provides good biodegradation and ease of disposal and reducing the consumption of petroleum. However, TPS tends to experience loss in mechanical properties and stability due to retrogradation of the starch. Modification of starch is often carried out to overcome this shortcoming in TPS. TPS can be made of various type of starch from various plant sources such as corn, cassava, rice, and potato.

DOI: 10.4018/978-1-7998-1374-3.ch004

Sago starch is a plant-based starch that originates from the sago palm tree. The chapter aims to give a short literature overview on the sago starch, issues related to starch and thermoplastic starch, modification of sago starch via crosslinking method, properties of crosslinked sago starch, and limitation and opportunities of modification of sago starch via crosslinking method.

INTRODUCTION

Thermoplastic starch (TPS) is one of promising material to replace common synthetic polymer in packaging such as low density polyethylene (LDPE) and polypropylene (PP) sheets and film. Demand for starch is high in packaging applications due to the good tensile and gas barrier properties, and can be obtained from various renewable plant-based sources such as rice, potato, corn and sago. Starch in its native form is brittle and the hydrophilic nature of starch will results in poor moisture barrier properties and high water sensitivity of the starch-based film, as compared to synthetic polymer. Plasticizer such as glycerol and sorbitol are added into native starch to form TPS with improved flexibility and good barrier properties. However, TPS film will experience gradual loss in flexibility and its strength due to retrogradation which will affect quality of TPS, especially on its mechanical strength after storage period. Various chemical modification techniques have been developed to overcome these disadvantages. Chemical modification such as acid modification, cross-linking, esterification and etherification are among the options. This chapter will report on the current knowledge of TPS made from sago starch, pertaining to its properties with chemical modification via crosslinking.

BACKGROUND

Starch

Starch can be obtained from different parts of plants such as in seeds, roots, stem, tubers, leaves and fruits (Alcazar & Meireles, 2015). Commonly in granular forms, the granules can be in spherical, oval, polygonal and elongated shapes, of varying size. Starch is a homo-polysaccharide produced by a combination of units of glucose, stored in the carbohydrates in plants, which amylose and amylopectin. According to the (Osorio-Dıaz et al., 2002), amylose is formed by 5-600 glucose units that linked by α-(1,4) glycoside bonds and it is referred to a linear polysaccharide. The linear structure of amylose contributes to the strength and flexible structure of starch granules. The structure of helix with six glucose units make it place in the core of granules and soluble in water with chemical formula of ($C_6H_{10}O_5$). Meanwhile, amylopectin is a branched molecule α-(1,6) produced by thousands of glucose units that make it prevent from the formation of helix structure. Therefore, a semi-crystalline structure is the arrangement between amylose and amylopectin forming a matrix of starch granules with alternating crystalline and amorphous that known as the growth ring in superior plant starch (Jenkins, Cameron, & Donald, 1993).

Starch can be obtained from various sources such as wheat, barley, sorghum, potato, cassava and rice (Alcazr- & Meireles, 2015). Starches from different resources have different in structure through chain length distribution in crystalline structure and amylopectin, the size distribution of granules, shape, amylose and lipid content (Ahmad et al., 1999).

Sago Starch

Sago starch is one of starch type extracted from pith inside the stem of sago palm (Metroxylon sagu) tree (Figure 1). Sago palm is classified under tropical crops in Southeast Asia, which prefer wet growing conditions such as in peat swamps (Jong, 1995, Karim et al., 2008).

Sago is an important food resource especially to the people in Sarawak, Malaysia; a rural areas where sago is one of the main food staple for the locals, and also been use in the production sago flour, sago pearl, noodles, jellies and soups (Mohamed et al., 2008). Despite Papua New Guinea has the biggest sago palm area with high genetic variation of sago palm tree, Malaysia is known as one of the top leading sago starch-producing countries in the world as sago palm plantations are being established as cash crop and has become one of the important contribution to export

Figure 1. Sago palm trees (Adapted from: Karim et al., 2008)

revenue of Malaysia (Karim et al., 2008). Sago palm plantations are planted mainly in Sarawak (90% cultivation area), and some minor plantations can be found along the coastal belts of Johor i.e. in Batu Pahat and Muar (Widodo & Hassan, 2014; Ahmad et al., 2011).

Sago starch granules is reported to exhibit a temple bell-like oval shape (Figure 2) (Okazaki, 2018). According to report by Karim et al. (2008), sago starch granules have a broad size range, between 10 and 50 μm in diameter with an average granule diameter of 32 μm. Sago starch granules are generally bigger than those of rice (3 to 10 μm), corn (5 to 20 μm), wheat (22 to 36 μm), or cassava (5 to 25 μm), but smaller than those of potato (15 to 85 μm). The starch granules are partially crystalline and insoluble in water at room temperature. Sago starch has been used as main ingredient in various traditional food products, and also has been utilized as thickening agent in food industries. Sago starch also has a great potential to be used in non-food applications such as in production of biodegradable packaging, green composites, ethanol, and fuel alcohol (Abd- Aziz, 2002, Karim et al., 2008). Ahmad and co-workers (1999) have reported that the physicochemical properties of sago starch are approximately similar to other commercial starches, and the amylose content as well as gelatinisation temperatures are close to that of corn starch, suggesting that sago starch could be commercially exploited.

Thermoplastic Starch (TPS)

TPS is obtained by mixing native starch in the presence of plasticizer (water, glycerol, etc.) under elevating heat and shear using common plastic film and sheet processing such as extrusion and injection moulding (Guilbert et al., 1997). When starch granules undergoes a spontaneous de-structuring process under heat and presence of water, the starch will turns into a homogeneous melt known as thermoplastics starch (TPS), which possesses the thermoplastic attributes (Zdrahala, 1997). This process requires gelatinization stage, where gelatinization is referring to the term "de-structured starch" i.e. disruption of the molecular structure of its granular state which effect in the loss order and crystallinity.

Figure 2. Sago starch granules (Adapted from: Okazaki, 2018)

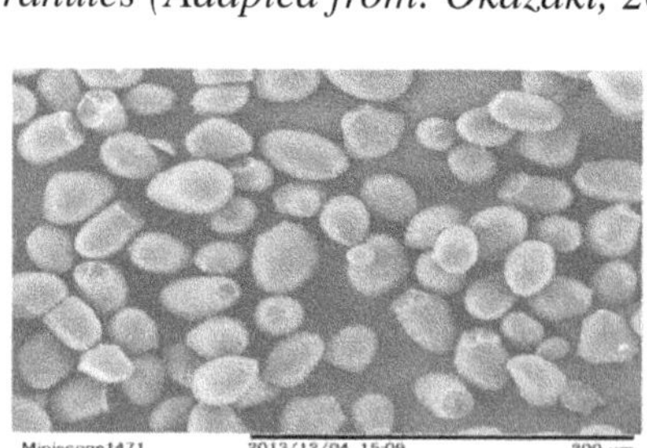

According to Shanks & Kong (2012), stability of TPS can be enhanced with the addition of hydrophilic polymers as the polymer will form hydrogen bonding with starch molecules to help stabilize the amorphous structure in starch after gelatinization stage. This will reduce the potential of recrystallization of starch molecules or known as retrogradation. Addition of additives such as plasticizer, crosslinking agents, texturizing agents and antimicrobial agents can alter the properties of TPS, which can widen the potential of TPS to be used in various applications (Zhang et al., 2014).

Starch is poised to establish an even stronger role in the manufacture of sustainable plastics and other bio-products largely because it is abundant, renewable, and inexpensive. Strategies for improving the properties of starch-based plastics such as blending starch with other polymers, using starch in composite materials, and using starch as a fermentation feedstock to make other biopolymers have been successful in developing viable replacements for petroleum based plastics. The prospects for starch in the packaging sector continue to become brighter as the market for sustainable plastics drives further innovation and development (Sulaiman, Manut, & Nur Firdaus, 2009).

Commercialisation of thermoplastic starch and starch-based polymers has been advancing. Because of the inalienable biodegradability of starch materials they are united generally to bundling as sheet or flims that can be subsequentially formed by thermoforming into custom bundling necessities. Starch are most suited for bundling of dry items, generally there will be a dispersion exchange and harmony set up between the TPS bundle and its substance. TPS foams are suited for damping effects to ensure delicate items. TPS solve the issue of transfer of bundling materials since they are biodegradable into ecologically cordial pieces, contingent upon the idea of non-starch segments of TPS structures (Zhang, Rempel, & McLaren, 2014). However, the crosslink starch also have found in many useful application such as in paper, textile and adhesive industry.

Retrogradation in TPS

Retrogradation is one of the major issue for TPS. During plasticization of starch into TPS, heating of starch with the aid of plasticizer causing the semi-crystalline starch to become more amorphous, forming viscous starch melt. With time, starch molecules tend to return to its original semi-crystalline structures, due to the re-crystallization of amylopectin or crystallization of amylose to form new V-type crystal. This will results in TPS being less flexible and gradually turns brittle, and losses its strength, hence, not suitable for packaging applications (Ashori et al., 2018).

Retrogradation is referred to as the long-term recrystallization of amylopectin because the reversible recrystallization of amylose is slower. Retrogradation can affect the performance of the thermoplastic such as its mechanical properties, texture, and colour due to the changes of the crystallinity. Retrogradation will affect the quality of TPS film in term of its mechanical properties, and its shelf-life (Zhang & Rempel, 2012).

Retrogradation is often occur in starch plasticized with glycerol compared to starch plasticized with other type of plasticizer (Shi et al., 2007). Attempts are made to substitute the use of glycerol with amide group based plasticizers to reduce retrogradation, but this type of plasticizers are prohibited in applications related to food packaging, biomedical and pharmaceutical due to the amide group-based plasticizer is known toxic (Shi et al., 2007). Therefore, glycerol is still favourable to be used as plasticizer in starch plastication, and modification of starch has been introduced an alternative option to reduce the tendency of retrogradation to occur in starch plasticized with glycerol.

Modification of Sago Starch TPS Via Cross-Linking

Modification of starch has been subject of interest to improve melt-processablilty of starch, improving the mechanical properties and thermal resistance of starch-based product. Modified starch is also one of the ways to overcome some of starch shortcomings and also introduce other functional properties to make starch more compatible for particular applications. Among the properties that can be altered via modification are hydrophobicity, pasting properties, thermal stability and resistance to retrogradation (Ojogbo, 2019). Starch modification could be via physical or chemical approaches. Chemical modifications methods in starch include esterification, etherification, oxidation, acetylation, hydrolysis, silyation, grafting and crosslinking, where reactions are targeted at the available hydroxyl groups on the B-D glucopyranosyl units of starch (Mohamed et al., 2017, Alcazar-Alay & Meireles, 2015).

Crosslinking is an important type of chemical modification of starch. It is a substitution reaction where the reagent is a chemical that contains two or more functional groups that able to react with the starch hydroxyl groups. Cross-linking in a starch happened when either linear or branched molecular chains are interconnected covalently (Alcazar-Alay & Meireles, 2015). Intermolecular cross-linking reactions increase the average molecular weight of the starch and provide chemical bridges between different molecules, which reduces the swelling in liquid water or at high relative humidity (Olsson, Hedenqvist, Johansson, & Järnström, 2013). All weak hydrogen bonds between starch molecular chains will be replaced by stronger covalent and more permanent.

Cross-linking is important to improve film formation and mechanical properties such as tensile strength and tensile modulus of the TPS film. Study by Seligra *et al* in 2016 stated that the crosslinked starch will increase the integrity, reduced the crystallinity and retrogradation process. The cross-linkages will form the covalent bond and act as "spot weld" to reinforce the molecular structure. The advantages of this type of crosslinking starch is it will improve the intermolecular interaction of the starch and provides stable granules.

Crosslinking for food grade-starch can be achieved by introduction of reagents, such as epicholorhydrin, sodium trimetaphospate, phosporyl chloride and sodium tripolyphospate, in the starch to form ether or ester bonds at hydroxyl group sites in starch molecules (Singh & Nath, 2012; Woo & Seib, 2002; Singh et al., 2007). Other chemicals such as malonic acid, poly(ethylene)glycol, dicyandiamide, borax, glyoxal, boric acid, adipic acid and citric acid have also been introduced as crosslinking agents in thermoplastic starch (Seligra et al., 2016; Mohamed et al., 2017; Yin et al., 2005). Use of polycarboxylic acid such as adipic acid, malonic acid and citric acid have been used in starch crosslinking as these this type of acid consists of two carboxylic groups which allows crosslinking at hydroxyl sites in starch to occur (Ackar et al., 2015, Reddy & Yang, 2010).

Citric acid is the common poly-carboxylic acids used for crosslinking agent. Citric acid is often used in starch modification because it is less harmful compared to other crosslinking agent for starch. Citric acid is a naturally occurring organic acid with three carboxylic groups and is generally classified as a safe food additive (Menzel et al., 2013). Hydroxyl group from the starch molecules have the potential to be cross-linked with the polyfunctional carboxylic acids.

The interaction could take place between the carboxylic group citric acids and the hydroxyl group of starch due to the multi-carboxylic structure. This interaction can increase the water resistibility due to the reducing available hydroxyl group of starch and it also can form stronger hydrogen bonds with the hydroxyl groups of starch molecule. Hence, the recrystallization and retrogradation can be prevented (Ghanbarzadeh, Almasi, & Entezami, 2011). Crosslinked starch has been observed to exhibit reduction in retrogradation rate and increase in gelatinization temperature owing to decrease in chains mobility of the starch due to formation of intermolecular bridges among the molecular chains (Liu et al., 1999).

Properties of Cross-Linked TPS From Sago Starch

Mohamed and co-workers in their studies on the effect of different crosslinking agents on the properties of glycerol-sago starch system, have concluded that introduction of borax and malonic acid as crosslinking agents in sago starch had improved swelling percentage, tensile strength and Young's modulus of TPS made from sago starch as compared to glyoxal, poly(ethylene)glycol and dicyandiamide. The findings indicate that effect of crosslinking in sago starch is more prominent when malonic acid and borax are used as crosslinking agents.

Singh and Nath (2012) in their investigation on the potential of using phosphoryl chloride as crosslinking reagent in sago starch have reported that an increase in crosslink reagent concentration resulted in higher degree of substitution due to more reacting sites are available for crosslinking in sago starch. Crosslinks induced by phosphoryl chloride also improved thermal stability of sago starch where the onset degradation temperature of crosslinked sago starch was higher than the native sago starch in the thermogravimetric analysis (TGA) curves.

Studies on crosslinked sago starch/polyvinyl alcohol (PVA) blend TPS film have been carried out by Ismail et al., (2014). Sago starch was initially crosslinked using sodium trimetaphosphate (STMP) and blended with PVA to form TPS film. Based on their observations, crosslinked sago starch have better interaction and good compatibility with PVA than unmodified sago starch, which contributed to improved tensile strength, reduced water absorption as well as decrease in water vapour transmission of the TPS film. However, the crosslinked sago starch/polyvinyl alcohol (PVA) blend TPS film experienced reduction in elongation at break. The lowering in water absorption and water vapour transmission of crosslinked sago starch/PVA film could be attributed to the phosphate groups replacing hydroxyl sites in sago starch, causing an increase in hydrophobicity of the film. Addition of 1 wt.% silica as filler in the crosslinked sago starch/PVA film also contributed to the increment in the tensile properties of the film.

Baith (2018) has conducted studies on the dynamic mechanical properties of sago starch crosslinked using citric acid (CA) (Figure 3). The author has reported that crosslinked TPS sago film with 5 wt.% citric acid exhibited higher storage modulus compare to the TPS sago film without citric acid. The storage modulus was observed to decrease for TPS crosslinked with 10 wt.% and 15 wt.% citric acid. The increasing of storage modulus of TPS/CA 5 wt.% explains that the TPS and citric acid could have produce crosslink at 5% CA content. In other word, it has a higher crosslinking density at 5 wt.% citric acids compare to the TPS/CA 10 wt. % and TPS/CA 15 wt.%. Higher degree of crosslinking will also contribute to the increase the storage modulus. However, for 10 wt.% and 15 wt.% citric acid, the result showed some reduction which could be due to the plasticizing effect imparted by CA in TPS. Furthermore, the possibility for formation of citrate esters and esterification of glycerol with citric acid as accordance to Shi et al (2007) to form citrate ester will reduce the storage modulus.

Similar trend was observed for loss modulus of the TPS sago film where TPS/CA 5wt.% shows higher loss modulus compare to the TPS/CA 0 wt.%. Loss modulus was found to decrease for TPS/CA 10 wt.% and TPS/CA 15 wt.%. This could be attributed to higher energy is needed to break the polymer chain due to the high degree of crosslink occurred in TPS/CA 5 wt.%. TPS without citric acid exhibited higher tan delta. Increasing tan delta means that the material has more energy dissipation

Figure 3. DMA results for crosslinked TPS sago film showing (a) storage modulus, (b) loss modulus, (c) tan delta of TPS with different citric acid content (Adapted from: Baith, 2018).

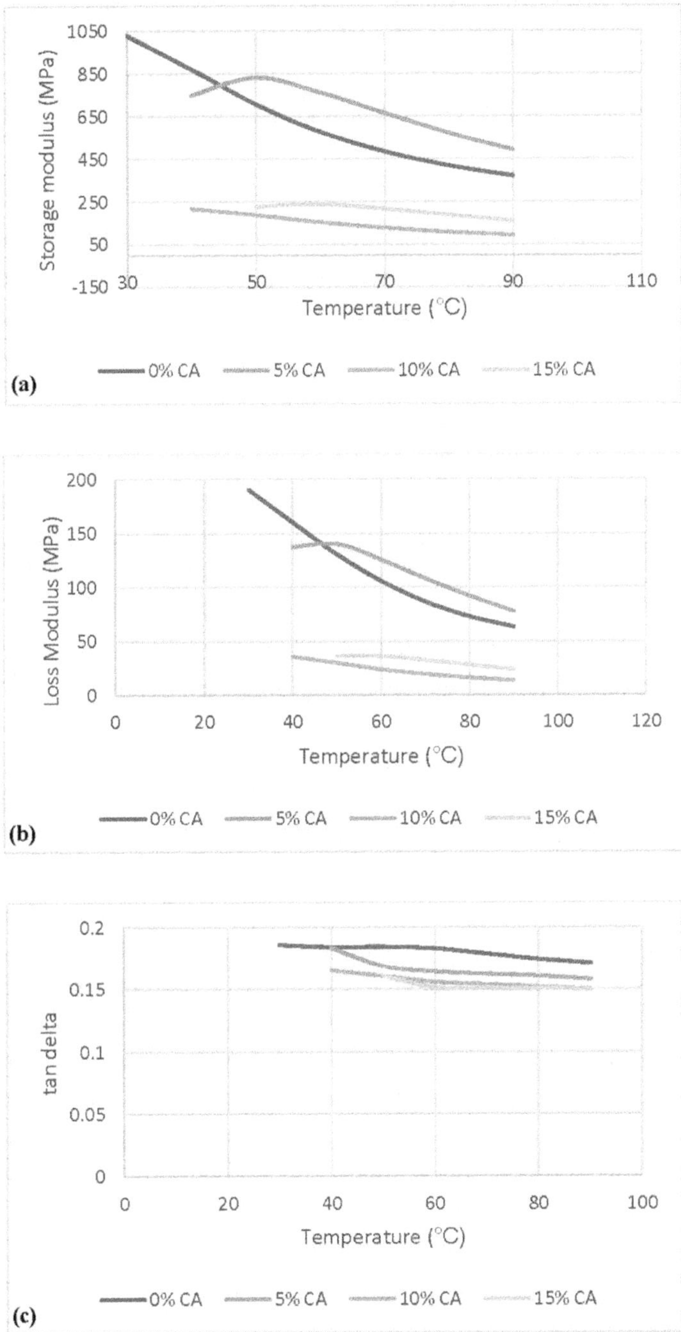

potential. So, the greater the tan delta, the more dissipative the material is. The increment of citric acid will diminished the value of tan delta as the crosslinking happen impose restriction against molecular motion of polymer chain (Heitor Luiz Ornaghi Junior et al, 2015). This statement can be related to the Tg values of TPS in the tan delta graph where Tg for TPS without citric acid was observed to be lower than the Tg of that of TPS/CA 5 wt.% of citric acids, suggesting that the crosslinking effect is at the highest when 5 wt.% citric acid is used to crosslink TPS sago starch.

Limitation and Opportunity of Cross-Linked Sago Starch

Modification of sago starch via crosslinking has been carried out to improve the mechanical properties (strength and stiffness), increase moisture resistance of TPS by increasing the hydrophobicity of sago starch, low swelling percentage which allows TPS sago to be exposed in aqueous environment. However, most of current reagents and method used in crosslinking process are costly, toxic, or insufficient to provide desired improvements in the TPS (Reddy & Yang, 2010).

Various works have been carried out to investigate the efficiency of low-cost crosslinking agent such as organic acids to be used in starch crosslinking process (Reddy & yang, 2010; Menzel et al., 2013; Dastidar & Netravali, 2012). They have concluded that organic acids such as malonic acid and citric acid, provide high degree of substitution in starch, and also reduce the swelling power of starch with increase of organic acid content in starch due to higher degree of crosslinking. These findings suggest that organic acids are good crosslinking agent at low cost. Furthermore, both sago starch and organic acid are of low-cost materials and easily available, which will attract further exploration works using these materials for starch modification.

Crosslinking of sago starch is indeed a versatile process. According to Ackar et al. (2015), crosslinking of starch is not restricted to single route or specific method. Crosslinking of starch can be modified by varying the selection of crosslinking agents, reaction temperature and time, catalyst, and concentration of crosslinking agents. They also reported that the complexity level of the modification stage can be altered by changing any reaction parameters in order to obtain desired properties in the crosslinked starch, which widen opportunity for further research on modification of starch, especially via crosslinking method. Since some of crosslinking agents are toxic, careful selection of selected material and reagents used during crosslinking stage is crucial when dealing with TPS for food and medical related applications such as in TPS based food packaging, food additives/ingredients and pharmaceutical products. This is to ensure the crosslinked TPS products are safe to be used and will not posed any health threats to the consumer.

Sago starch is reported to have the potential to be used as a base for dual-modification such as for hydroxypropylation and crosslinking process (Wattanachant et al., 2002). The authors compared the suitability of sago starch with various type of starch such as maize and tapioca starch in term of swelling power. It was observed from the report that sago starch exhibit higher swelling power compared to maize and tapioca starch owing to its bigger granule size, indicating that sago starch suits to be further use in dual-modification. Dual-modification allows combination of different modifications such as crosslinking-annealing or crosslinking-acetylation to produce modified starch with improved properties. Therefore, properties of crosslinked sago starch will be further altered to yield desired properties.

CONCLUSION

Sago starch and TPS film made from sago starch can be crosslinked using various crosslinking agents such as organic acids. Crosslinked sago starch and its TPS exhibit improved mechanical properties, increase to water resistance, lower the rate of retrogradation in TPS, and reduced swelling percentage of sago starch. The crosslinking process can be modified by varying the reaction parameters such as reagent concentration, reaction time and temperature, and type of crosslinking agent and catalyst, to obtain improvements in various properties. Selection of crosslinking agents and catalyst should be carefully done especially for TPS products used in food-related applications to ensure safety of the consumers.

REFERENCES

Ačkar, D., Babić, J., Jozinović, A., Miličević, B., Jokić, S., Miličević, R., ... Šubarić, D. (2005). Starch modification by organic acids and their derivatives: A review. *Molecules (Basel, Switzerland)*, *20*(10), 19554–19570. doi:10.3390/molecules201019554 PMID:26516831

Ahma, F. B., Williams, P. A., Doublier, J.-L., Durand, S., & Buleon, A. (1999). Physico-chemical characterization of sago starch. *Carbohydrate Polymers*, *38*(4), 361–370. doi:10.1016/S0144-8617(98)00123-4

Ahmad, F. B., Williams, P. A., Doublier, J. L., Durand, S., & Buleon, A. (1999). Physico-chemical characterisation of sago starch. *Carbohydrate Polymers*, *38*(4), 361–370. doi:10.1016/S0144-8617(98)00123-4

Ahmad, Z., Anuar, H., & Yusof, Y. (2011). The study of biodegradable thermoplastics sago starch. *Key Engineering Materials, 471-472*, 397–402. doi:10.4028/www.scientific.net/KEM.471-472.397

Alcazar-Alay, S. C., & Meireles, M. A. A. (2015). Physicochemical properties, modifications and applications of starches from different botanical sources. *Food Science and Technology (Campinas), 35*(2), 215–236. doi:10.1590/1678-457X.6749

Ashori, A., Shakeri, A., Tabarsa, T., Mashkour, M., & Ghanbari, A. (2018). Preparation and characterization of thermoplastic starch and cellulose nanofibers as green nanocomposites: Extrusion processing. *International Journal of Biological Macromolecules, 112*, 442–447. doi:10.1016/j.ijbiomac.2018.02.007 PMID:29410268

Baith, A. H. (2018). *Modification of thermoplastic starch via crosslinking* (Unpublished final year project dissertation). Universiti Kuala Lumpur, Malaysia.

Dastidar, G., & Netravali, A. N. (2012). Green crosslinking of native starches with malonic acid and their properties. *Carbohydrate Polymers, 90*(4), 1620–1628. doi:10.1016/j.carbpol.2012.07.041 PMID:22944425

Ghanbarzadeh, B., Almasi, H., & Entezami, A. A. (2011). Improving the barrier and mechanical properties of corn starch-based edible films: Effect of citric acid and carboxymethyl cellulose. *Industrial Crops and Products, 33*(1), 229–235. doi:10.1016/j.indcrop.2010.10.016

Guilbert, S., Cuq, B., & Gontard, N. (1997). Recent innovations in edible and/or biodegradable packaging materials. *Food Additives and Contaminants, 14*(6), 741–751. doi:10.1080/02652039709374585 PMID:9373537

Ismail, H., & Zaaba, N. F. (2014). Effect of unmodified and modified sago starch on properties of (sago starch)/silica/PVA plastic films. *Journal of Vinyl and Additive Technology, 20*(3), 185–192. doi:10.1002/vnl.21344

Jenkins, P. J., Cameron, R. E., & Donald, A. M. (1993). A Universal Feature in the Structure of Starch Granules from Different Botanical Sources. *Stärke, 45*(12), 417–420. doi:10.1002tar.19930451202

Jong, F. S. (1995). *Research for the development of sago palm (Metroxylon sagu Rottb.) cultivation in Sarawak, Malaysia* (Unpublished doctoral dissertation). Wageningen Agriculture University.

Karim, A. A., Tie, A. P.-L., Manan, D. M. A., & Zaidul, I. S. M. (2008). Starch from the Sago (Metroxylon sagu) Palm Tree: Properties, prospects and challenges as a new industrial source of food and other uses. *Comprehensive Reviews in Food Science and Food Safety*, *7*(3), 215–228. doi:10.1111/j.1541-4337.2008.00042.x

Liu, H., Ramsden, L., & Corke, H. (1999). Physical properties of cross-linked and acetylated normal and waxy rice starch. *Starch*, *51*(7), 249–252. doi:10.1002/(SICI)1521-379X(199907)51:7<249::AID-STAR249>3.0.CO;2-O

Menzel, C., Olsson, E., Plivelic, T. S., Andersson, R., Johansson, C., Kuktaite, R., ... Koch, K. (2013). Molecular structure of citric acid cross-linked starch films. *Carbohydrate Polymers*, *96*(1), 270–276. doi:10.1016/j.carbpol.2013.03.044 PMID:23688480

Mohamed, R., Nurazzi, N. M. N., Siti Aisyah, M. I., & Mohd Fauzi, F. (2017). Swelling and tensile properties of starch glycerol system with various crosslinking agents. *IOP Conference Series: Material Science and Engineering*, 223. 10.1088/1757-899X/223/1/012059

Ojogbo, E. C. (2019). *Starch modification for sustainable and functional material applications* (Unpublished master dissertation). University of Waterloo, Ontario, Canada.

Olsson, E., Hedenqvist, M. S., Johansson, C., & Järnström, L. (2013). Influence of citric acid and curing on moisture sorption, diffusion and permeability of starch films. *Carbohydrate Polymers*, *94*(2), 765–772. doi:10.1016/j.carbpol.2013.02.006 PMID:23544631

Ornaghi, H. L., Jr., Zattera, A. J., & Amico, S. C. (2015). Dynamic Mechanical Properties and Correlation With Dynamic Fragility of Sisal Reinforced Composites. *Polymer Composites*, 161–166. doi:10.1002/pc

Osorio-Díaz, P., Bello-Pérez, L. A., Agama-Acevedo, E., Vargas-Torres, A., Tovar, J., & Paredes-López, O. (2002). In vitro digestibility and resistant starch content of some industrialized commercial beans (Phaseolus vulgaris L.). *Food Chemistry*, *78*(3), 333–337. doi:10.1016/S0308-8146(02)00117-6

Reddy, N., & Yang, Y. (2010). Citric acid cross-linking of starch films. *Food Chemistry*, *118*(3), 702–711. doi:10.1016/j.foodchem.2009.05.050

Seligra, P. G., Medina Jaramillo, C., Famá, L., & Goyanes, S. (2016). Biodegradable and non-retrogradable eco-films based on starch-glycerol with citric acid as crosslinking agent. *Carbohydrate Polymers*, *138*, 66–74. doi:10.1016/j.carbpol.2015.11.041 PMID:26794739

Shanks, R., & Kong, I. (2012). *Thermoplastic starch, thermoplastic elastomers*. InTech. Available from: http://www.intechopen.com/bookd/thermoplastic-elastomers/thermoplastic-starch

Shi, R., Zhang, Z., Liu, Q., Han, Y., Zhang, L., Chen, D., & Tian, W. (2007). Characterization of citric acid/glycerol co-plasticized thermoplastic starch prepared by melt blending. *Carbohydrate Polymers*, *69*(4), 748–755. doi:10.1016/j.carbpol.2007.02.010

Singh, A. V., & Nath, L. K. (2012). Synthesis and evaluation of physicochemical properties of cross-linked sago starch. *International Journal of Biological Macromolecules*, *50*(1), 14–18. doi:10.1016/j.ijbiomac.2011.09.003 PMID:21971581

Singh, J., Kaur, L., & McCarthy, O. J. (2007). Factors influencing the pysico-chemical, morphological, thermal and rheological properties of some chemically modified starches for food applications-a review. *Food Hydrocolloids*, *21*(1), 1–22. doi:10.1016/j.foodhyd.2006.02.006

Sulaiman, S., Manut, A., & Nur Firdaus, A. R. (2009). Agriculture Applications. *ICIMT* '09. *International Conference, IEEE*, 513–516.

Widodo, R. T., & Hassan, A. (2014). Compression and mechanical properties of directly compressible pregelatinized sago starches. *Powder Technology*, *269*, 15–21. doi:10.1016/j.powtec.2014.08.039

Woo, K. S., & Seib, P. A. (2002). Cross-linked resistant starch: Preparation and properties. *Cereal Chemistry*, *79*(6), 819–825. doi:10.1094/CCHEM.2002.79.6.819

Yin, Y., Li, J., Liu, Y., & Li, Z. (2005). Starch crosslinked with poly(vinyl alcohol) by boric acid. *Journal of Applied Polymer Science*, *96*(4), 1394–1397. doi:10.1002/app.21569

Zdrahala, R. J. (1997). R & 1 Consulting International. Thermoplastic Starch Revisited. Structure/Property Relationship For "Dialed-In" Biodegradability, *Macromol*, 113–121.

Zhang, Y., & Rempel, C. (2012). *Retrogradation and antiplastication of thermoplastic starch, Thermoplastic Elastomers*. In Tech. Available from: http://www.intechopen.com/books/thermoplastic-elastomers/retrogradation-and-antiplastication-of-thermoplastic-starch

Zhang, Y., Rempel, C., & Liu, Q. (2014). Thermoplastic starch processing and characteristics-a review. *Critical Reviews in Food Science and Nutrition*, *54*(10), 1353–1370. doi:10.1080/10408398.2011.636156 PMID:24564592

Zhang, Y., Rempel, C., & McLaren, D. (2014). Thermoplastic Starch. In J. H. Han (Ed.), *Innovations in Food Packaging* (pp. 391–412). Academic Press. doi:10.1016/B978-0-12-394601-0.00016-3

Chapter 5
Non-Edible Oil Biodiesel Production via Microwave Irradiation Technologies Using Waste-Heterogeneous Catalyst Derived From Natural Calcium Oxide

Mahanum Mohd Zamberi
Universiti Teknikal Malaysia Melaka, Malaysia

Farid Nasir Ani
Universiti Teknologi Malaysia, Malaysia

ABSTRACT

Biodiesel production or synthesis by non-edible oils has been introduced recently due to its potential to overcome the problems associated with first generation of feedstock, especially the competition between food and transport biofuel. The production of this non-edible resource could be done without major investment and benefit many parties involved: agricultural sector, job creator, biodiversity, and many other benefits. Microwave-assisted technique has been found as one of the methods that has the potential to be a highlight due to its ability to reduce the cost production and produce higher quality biodiesel as well. This method has been identified to enhanced and accelerate the transesterification process in order to obtain higher yields in the biodiesel synthesis. Continuous process in the other hand seems to be more significant when compared to batch solution. This review is an attempt to gather and summarize the existing literature and study the natural heterogeneous catalyst as one of the green and economic methods using microwave irradiation method.

DOI: 10.4018/978-1-7998-1374-3.ch005

INTRODUCTION

Almost in every country worldwide, there are biodiesel activities, either in small or massive scales; in research stage or in industrial stage. Biodiesel production or synthesis by non-edible oils have been introduced recently due to its potential to overcome the problems associated with 1st generation of feedstock especially the competition between food and transport biofuel. The production of this non-edible resources could be done without major investments and benefited to many parties involved; agricultural sector, job creator, biodiversity and many other benefits.

Microwave assisted technique has been found as one of the method that have the potential to be highlighted due to its ability to reduce the cost production and produce higher quality biodiesel as well. This method has been identified to enhance and accelerate the transesterification process in order to obtain the higher yields in the biodiesel synthesis. Other than that, this method is also being used widely in oil extraction of bioactive compounds. Continuous process on the other hand seems to be more significant when compared to batch solution. This review is an attempt to gather and summarize the existing literature and study on the natural heterogeneous catalyst as one of the green and economical method using microwave irradiation method.

BIODIESEL

Biodiesel is an alternative fuel that can be used for conventional diesel engines made from a very wide range of feedstock that have been chemically transformed into alkyl esters with the aid of alcohol and with or without catalyst. Methanol is the most preferred alcohol compared to ethanol due to its chemical and physical advantages such as high volatility, better conversion and simpler to find (Idris, 2016).

Most of the biodiesel fuel manufactured from renewable, non-petroleum based sources range from animal fats to vegetable oils, algae and waste vegetable oils (Demirbas, 2017). In addition, the properties of these feedstocks vary widely from those of petroleum diesel fuel especially in the areas of viscosity, atomization and the coking of engine components. It is reported that most of the plant oils have viscosities that can be as much as 20 times higher than that of fossil fuel. This problem can be overcome by adapting the engine to the fuel or vice versa.

The most critical challenge encountered by biodiesel producer is to minimizing the cost of biodiesel production which is depending on the raw materials such as alcohol, oil feedstocks and catalyst as a booster (Zamberi et al., 2011). Utilizing edible vegetable oil as potential biodiesel feedstocks for example soybean oil in the United States and palm oil in the Southeast Asia has significantly increase because of its abundant agricultural production (Abbaszaadeh et al., 2012). However, these type of feedstocks has experienced several disadvantages over the past few years such as increases in feedstock price, low heating value and higher NOx emission compare to conventional diesel fuel (Ashraful et al., 2014). In order to overcome all these associated problems with edible oil, non-edible oil was introduced as an effective way to boost up the biodiesel production industries.

Non Edible Oil Feedstock

The demand of non-edible oil as feedstock in biodiesel production leads to explore alternative sources from non-edible oil seeds especially in developing countries such as Indonesia, Vietnam, Malaysia and South Africa due to the great concern of food demand when opting edible plants as raw feedstock (Zamberi et al., 2016). Non-edible oil plants can easily be grown in waste lands that are not suitable for food crops and the cost of cultivation is much lower because these plants can still sustain reasonably high yield without intensive care (Fatah et al., 2012). Most of these plants such as jatropha and rubber seeds and are also self-sowing crops and require almost no fertilizer input. The switch to these non-edible feedstock poses a positive impact on employment generation in rural and undeveloped areas.

The source for biodiesel production normally is influenced by the availability in each region or country. Factors such as agricultural policies, subsidies or government support, tax incentives, research and development fundings are the main instruments that can ensure the sustainable feedstock. The origin of the feedstock also determines the properties of the produced biodiesel and the type of the process to be used. Normally the low cost feedstock comprises substantial quantities of free fatty acids (FFA), that may contribute to a complex and high cost of production process.

This type of feedstock is well recognized as edible oil and being widely used in their origin due to the higher production and lower in cost. However, some Asian countries such as Malaysia, Vietnam, Thailand and Indonesia starts to explore for non-edible seed oil. *Atadashi et al., (2012) reported that the most appropriate non-edible feedstock due to fatty acids profiles were karanja, polanga (calophyllum inophyllum. L) and Jatropha Curcas. Utilizing non-edible vegetable oils as a diesel fuel offers many advantages such as higher combustion efficiency, renewability, lower sulfur and having aromatic content but the downside to this is that their high content of free fatty acids (FFA) which leads to higher production cost (Zahan et al., 2018).*

Jatropha Curcas

Jatropha curcas has been actively promoted as major biodiesel feedstock in tropical and subtropical regions throughout the Africa, India and South East Asia (Gmunder et al., 2012). Jatropha is emerging as the most preferred as it can produce high oil yield per hectare, easy propagation, hardiness, drought endurance, fast growth and can adapt to various climate conditions making it one of the most preferred choice among the biodiesel producer (Divakara et al., 2010). Even though the seed is highly poisonous, the oil extracted contain high viscous oil (more than 30% oil) that can be used for making candles and soap, as a diesel- paraffin substitute or extender (Mustafa, 2012). This plant can grow well in all types of soils and water and at the same time require very little irrigation which make it as a potential sustainable biofuel (Santos & Silva, 2016).

Hevea Brasiliensis (Rubber)

Hevea brasiliensis, well known as rubber tree is belonging to the family of Euphorbiaceae. It initially grew up in the Amazon rain forest mainly in Brazil. The tree is the primary source of rubber (known as latex) and produces 99% of world's natural rubber (Atabani et al., 2013). Latex from rubber tree can be utilized to produce various useful products such as tyre, gloves and hairband. Malaysia, Thailand, Indonesia, India and China are the main producer of natural rubber in the world. Malaysia Rubber Board has reported that Malaysia has an estimated acreage of 1,083, 480 hectares of rubber plantation as of 2018 (Malaysia Rubber Statistic, 2018). Hence, based on an estimated average production of 800 kg to 1200 kilograms of rubbers seeds per ha/year (Siriwardene and Nugara, 1972), the projected annual production of rubber seeds in Malaysia would be around 1 million metric tons where most of these rubber seeds are normally regarded as waste because they are not currently uses in daily life. Rubber seeds mostly weighing from 2 to 4 grams contain 50 to 60% of oil and kernel contains 40 to 50% of brown colour oil (Atabani et al., 2013). Ramadhas et al., (2005) reported that rubber seed oil consists of 18.9% saturated fatty acid (palmitic and stearic acids) and 80.5% unsaturated fatty acid (mainly of oleic, linoleic and linoleic acids). Saturated fatty acids will increase the cloud point, cetane number and improve oxidation stability. Moreover, Ikwuagwu et al., (2000) reported that rubber biodiesel has closer fuel properties with diesel fuel except slightly higher in viscosity.

Calophyllum Inophyllum L. (Polanga)

Alexanderian laurel or sometimes acknowledge as oil nut tree (*Calophyllum inophyllum*) commonly known as Polanga, belongs to the Clusiaceae family. The physical size of the tree is large and can grow up to 35 metre tall. It is best grows in deep soil or on exposed sea sands. It can easily be found in multiple origins including the East Africa, India, South East Asia and Australia. The trees often bear fruits throughout the year. Oleic, linoleic, stearic and palmitic acid are the main compounds that can be found in the seed oil (Orwa et al., 2009). The seed oil has identified very oily seeds, with up to 75% yield. (Venkanna & Venkataramana, 2009). Its biodiesel is compatible with diesel and possesses better lubrication capability (Fadhlullah et al., 2015).

Waste Cooking Oil (WCO)

A huge amount of waste vegetable oil (WVO) generated by the restaurants, fast food outlets, and food processing industries every day and everywhere around the world. Instead of let the WVO becomes a disposal burden to government, it is actually can be recycled to be an alternative raw material in biodiesel production (Zamberi, 2011). However, large amount of free fatty acids (FFA), water content, high viscosity and saponification number in WVO lead to high production cost. Transesterification process cannot be carried out directly. A two-step transesterification was utilized in order to reduce the FFA level of the WVO.

FATTY ACID COMPOSITION

Free fatty acids (FFAs) content in vegetable oils depends on its types. FFA content is the most important property to determine the efficiency of the process to produce biodiesel. The fatty acid composition and distribution of some non-edible oils are generally aliphatic compounds with a carboxyl group at the end of a straight-chain. The largest fatty acids methyl ester compound found in chromatography analysis are normally C16 and C18 acids. However some feedstocks contain significant amounts of fatty acids other than the typical C16 and C18 acids (Cherng & Yi, 2012). The oil having high content of oleic acid (C18) tend to withstand the oxidation level and therefore more suitable for process requiring high oxidative stability at high temperatures as in biodiesel and bio-lubricants applications (Bassie, 2008). The fatty acids composition of common non-edible oil feedstock is presented in Table 1.

CHARACTERIZATION OF THE BIODIESEL PROPERTIES

The properties of the non-edible biodiesel is vary depending on their chemical and fatty acids compositions. In order to full filled all the criteria that has been list out by the two main biodiesel standards testing, European Union (EN 14214) and American standards (ASTM D6751), all biodiesel producers need to ensure that their production compliance with all the standard requirement. Standards were developed to regulate and limit the amount of contaminants in the fuel/biodiesel which could

Table 1. Typical fatty acid composition of various non-edible oil feedstock (Atabani et al., 2013 and Roschat et al., 2017)

Fatty Acid	Typical Non-Edible Fatty Acid Composition (wt%)				
	Jatropha curcas L.	**Rubber seed oil [a*]**	**Calophyllum inophyllum. L**	**Palm oil**	**Waste vegetable oil**
C12:0 Lauric	-	-	-	-	-
C14:0 Myristic	1.4	-	-	-	-
C16:0 Palmitic	15.6	9.1	14.6	73.0	12.01
C16:1 Palmitoleic	-	-	2.5	-	-
C18:0 Stearic	9.7	5.6	19.96	5.0	3.1
C18:1 Oleic	40.8	24.0	35.57	18.0	21.2
C18:2 Linoleic	32.1	46.2	26.33	4.0	55.2
C18:3 Linolenic	-	14.2	0.27	-	5.9
C20.0 Arachidic	0.4	0.9	0.94	-	-
[a*]Roschat et al., 2017					

affect the quality of the biodiesel and cause problems towards the engine used. The properties of biodiesel include flash point, density, cetane number, viscosity, cloud point, pour point, acid value, calorific value, ash content and cold flow properties as tabulated in Table 2.

TRANSESTERIFICATION PROCESS FOR BIODIESEL PRODUCTION

There are various techniques to produce biodiesel production that have been reported such as transesterification, pyrolysis, micro-emulsification, and direct use and blending method. The most common method that been used is transesterification. The heart of biodiesel chemistry is the process of transforming the vegetable oil esters into the biodiesel esters by adding alcohol (methanol or ethanol), to form fatty acid alkyl esters and glycerol. At this stage, transesterification process is used in order to reduce the high viscosity and at the same time increase the volatility of the vegetable oils as well as animal fats into the desired level (Demirbas, 2005). Transesterification can be performed via catalytic or non-catalytic reaction using different heating systems. Base, acid and enzyme are the common catalyst used in the transesterification process. Transesterification is a reversible process which transformed the triglycerides into lower molecules of fatty acid esters. It is a chemical

Table 2. ASTM D6751 and EN 14214 properties

Properties	Unit	Test Method		Test limit	
		ASTM	EN	ASTM D6751	EN 14214
Kinematic viscosity at 40°C	mm^2/s	D 445	EN ISO 3104	1.9-6.0	3.5-5.0
Density at 15°C	kg/m^3	D1298	EN ISO 3675/12185	880	860-900
Calorific value	MJ//kg	-	EN 14214	-	35
Flash point	°C	D 93	ISO DIS 3679	Min. 100-170	>120
Pour point	°C	D 97	-	-15 to 16	-
Cloud point	°C	D 2500	-	-3 to 12	-
Oxidation stability hours, 110°C	-	D 675	EN 14112	Min. 3	Min. 6
Cetane number	-	D 613	EN ISO 5165	Min. 47	Min. 51
Acid value	mg KOH/g	D 664	EN 14104	Max.05	Max.0.5
Water content	%v	D 2709	EN ISO 12937	Max. 0.005	-
Carbon residue (100%)	m/m	D 4530	EN ISO 10370	Max. 0.05	Max. 0.03
Sulphated ash	% m/m	D 874	EN ISO 3987	0.02	0.02

reaction of vegetable oil or fats with the aid of alcohol in the presence of catalyst (strong base or strong acid) to produce new chemical compounds known as methyl esters. The transesterification reaction is presented in Figure 1.

MICROWAVE ROLE IN BIODIESEL PRODUCTION

Microwave technique has been utilized extensively to enhance the transesterification process of biodiesel production. A separation and purification process is carried out to produce a biodiesel which gives a good yield and at the same time meet up

Figure 1. Reaction for oil transesterification (Marchetti et al., 2007)

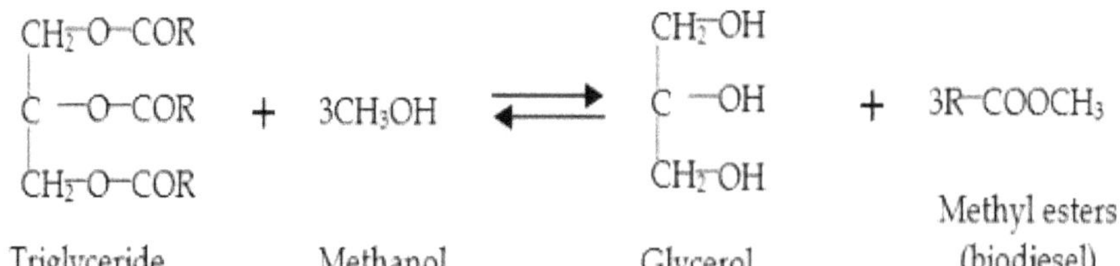

all the standard specifications. In the conventional heating method that is being used in the transesterification process, the three modes of heat transfer; conduction, convection and radiation is utilized to transfer the heat energy from the surface of the reactor to the reaction mixture (Motasemi & Ani, 2012). As a result, a large amount of energy is needed to heat the mixture which makes this method inefficient. This heating technique has reported the increased in reaction time and developed higher thermal gradients and at the same time produced non-uniform samples temperatures. In order to control such problems, other method must be incorporated to minimize these critical parameters and finally reduce the production cost.

As an alternative techniques for biodiesel production, microwave technologies offers several other advantages over the conventional heating method such as higher heating rates, non-contact heating between the heating source and heated material, energy transfer instead of heat transfer (penetrative radiation), reduced thermal gradients, fast start up and stopping and reverse thermal effect, reduced equipment size and waste, more effective heating, increased production (higher quality and yield product) and elimination of process step (A. Tangy *et al., 2017).* The comparison between heating methods for biodiesel production was summarized and tabulated in Table 3 (Gude et al., 2013). The use of microwave has been introduced in the past 20 years especially in the food processing industry. However, the application of microwave heating technology to the potential areas such as waste processing, mineral processing and other potential commercial developments is a relatively new development. Another major difficulty is scaling up the size of the microwave from laboratory scale to the industrial level.

Batch vs Continuous Microwave Transesterification Process

Batch operations is well recognized for small scale production because it is more energy efficient whereas the continuous flow process performs efficiently for a large volumes production and at the same time reducing the cost of production. In continuous flow process the mixture is continuously pumped and heated in microwave cavity which makes the process much faster and energy efficient than the conventional and batch process. However there can be also few limitations of this continuous process such as safety issues especially when dealing with the domestic microwave where malfunction or rupture could be disastrous, high microwave power will lead to damage the organic molecules (triglycerides) and the most important drawback is the issue on low penetration depth of microwave radiation into the absorbing material when scaling up the capacity and size from laboratory scale process to industrial production plant (Vyas AP et al., 2010).

For these reasons research groups have focused towards the continuous flow microwave reactors where the reacting mixture flows through a microwave reactor with a small cross-sectional area, solving the problems of penetration depth and

Table 3. Comparison between heating methods for biodiesel production

Characteristic/ parameter	Conventional heating	Supercritical heating	Microwave heating
Reaction time	Long (1-2 hr)	Short (<1 hr)	Very short (0.05 -0.1 hr)
Reaction temperature (°C)	40-100	250-400	40-100
Reaction pressure	Atmospheric	High pressure 35-60 MPa	Atmospheric
Catalyst required	Yes	No	Yes/No
Heat Lost	High	Moderate	Low
Form of energy	Electrical energy converted to thermal energy	Electrical energy converted to thermal energy	Electrical energy applied to microwaves
Process efficiency	Low	Moderate	High
Catalyst removal	Yes	No	Yes
Soap removal	Yes	No	Yes
Advantages	Simple operation, use of low energy source	Short reaction time, easy product separation	Short reaction time, cleaner products, and energy efficient
Limitations	High energy requirements	High capital cost, pressure vessel safety	May not be efficient with feedstock containing solids

high pressure reaction vessels. There have been several studies in the literature reporting on the biodiesel production of non-edible oils via continuous microwave transesterification process however the laboratory investigations on this method was very limited. Millano et al., (2018) investigated the production of biodiesel from waste palm oil using 800 Watt domestic microwave in a continuous transesterification reaction using sodium hydroxide (NaOH) as catalyst. A 97% FAME was obtained with optimum condition of 12:1 molar ratio of ethanol to oil, 30% catalyst and 30 seconds reaction time. Effects of other variables such as molar ratio of ethanol to oil, catalyst concentration and reaction time was also being observed. Most of the results were amazing because they achieved almost 100% conversion from different feed stock (coconut, rice bran and used cooking oil).

Barnard et al., (2007) conducted both batch and continuous microwave irradiation for transesterification process using a multimode microwave apparatus (CEM MARS) and the energy consumption by both techniques were observed. The continuous system was identified to be more energy efficient as compared to batch system and the conventional method. When a high power microwave device is used over a very small sample it leads to inefficiency. The study also showed that with a flow rate of 2 L/min the system possesses 97.9% overall conversion and 98.9% overall conversion with a flow rate of 7.2 L/min. With the same objective, Hernando et al., (2007) performed the experiment of transesterification of rapeseed or soybean oil under a small scale batch microwave reactor and microwave flow system. A very high yield, 97% was obtained in 1 minute in both techniques. It was shown that using modern commercial devices allows a straight forward transfer of the process from batch to stop flow.

In a different study performed by Encinar et al., (2011) the effect of temperature, molar ratio of methanol to oil and catalyst concentration of soybeans under the microwave irradiation continuous transesterification was examined. The effect of temperature ranging from 50 °C to 110 °C, methanol to oil molar ratio from 3:1 up to 12:1 and catalyst concentration varying from 0.5%, 1.0% and 1.5% in weight of KOH was investigated. A good conversion which is almost 99% was obtained with 12:1 ratio of methanol to oil, 1.0% catalyst concentration at reaction temperature of 70°C. The performance of FAME production was also made between the microwave irradiation technique and conventional heating and it was observed that the microwave irradiation was more efficient due to shorter reaction time (Encinar et al., 2011).

Tippayawong & Sittisun (2012) selected microwave technique via flow process to investigate the transesterification performance of Jatropha oil with the aid of alkaline catalyst, sodium methoxide ($NaOCH_3$). The optimum performance was observed at an excess molar ratio of oil to methanol 1:6 and 1 wt% catalyst loading with a very short reaction time less than 1 minutes. Yield of 96% was obtained and it can be concluded that the continuous transesterification process is simple, practical, reduced energy cost and time saving .

Existing continuous microwave assisted production of biodiesel technology design is still under extensive research around the world and has not been developed to date (Gude et al., 2013). Most of the researchers particularly from industries running the experiments using microwave heating on a large scale, often use different frequencies that can increase penetration depths through solvent and solids. The option to change the frequency may provide alternative solutions to the problem of microwave scale-up.

The concern for scale up development and performance studies have been worked out even though the process is quite impossible due to many limitations. Issues on suitable and special material such as Teflon which required a special design makes it impossible for the establishment on larger-scale reactors. The development of multimode reactors, where 6 to 12 samples can be heated paralleled is also unable to overcome the existing weakness (Bálint et al., 2019).

FROM HOMOGENEOUS TO HETEROGENEOUS CATALYTIC PROCESS

One of the major concerned among biodiesel producer in commercializing biodiesel is the cost of the feedstock and the production process. Due to this problem, many researchers are force to come out with other alternatives to reduce the cost of the biodiesel production. One of the possible alternatives is using heterogeneous catalysts transesterification method from waste sources which have simple product separation and purification process (Janaun, 2010). Even though homogeneous catalyst is said to be dominate the biodiesel industries due to their simple usage, very high yield production under mild conditions and shorter reaction time however this method still have a few drawbacks especially in large scale production . Among them are higher in costs production as the process involve washing and purification steps and problem with catalyst separation due to high solubility in the solution. More over high amounts of water are required in washing and resulting waste water treatment process which contributed to overall process cost.

Heterogeneous catalyst act in a different phase from the reaction mixture, usually as a solid. Heterogeneous catalytic reaction converts triglycerides into methyl ester and glycerol as a by-product slowly but produced biodiesel in a very feasible economic way due to the reusability of catalyst for both batch and continuous process and cost effective due to easy separation (Sakai et al., 2009). Base and solid acid catalyst are two types of heterogeneous catalyst that being used in biodiesel production. The solid base is found to be more active compared to solid acid catalyst which required relatively shorter reaction time and lower reaction temperature (Hara, 2009).

Heterogeneous Catalyst From Waste Calcium Oxide for Biodiesel Production

Review on the production of biodiesel from various types of feedstock especially from natural sources coming from unwanted or waste sources has been reported intensively in recent years. Most of these catalysts reported have a high potential to overcome the limitations of using homogeneous catalyst especially on the cost of production. The reusability factor is one of the major contributions of utilizing heterogeneous catalysts which is proven to reach up to 10 times, at the same time can reduce the purification process and the amount of water used. Calcium oxide (CaO) catalyst is among the most promising heterogeneous catalyst that can be classified having an eco-friendly characteristic and value added to the green biodiesel process (Zamberi et al., 2016; Yusuff et al., 2017). Apart from that, CaO heterogeneous catalyst is reported can withstand high temperatures, ease of recycling and its performance for biodiesel production is comparable to several homogeneous catalyst (Kouzu & Hidaka, 2012).

Preparation of CaO catalyst derived from waste shells is known as calcination process. Most of these waste shells consists of calcium carbonate ($CaCO_3$). The shells need to be calcined at very high temperature (700 °C to 1000 °C) to transform the calcium carbonate ($CaCO_3$) of the waste shells to calcium oxide (CaO). Usually, calcination temperature above 900 °C was selected because at and above these temperatures mostly an optimized CaO will be produced. The heterogeneous catalysts were characterized by various type of method such as X-ray diffraction (XRD), X-ray fluorescence (XRF), scanning electron microscopy (SEM), and the Brunauer-Emmett-Teller (BET) method.

Heterogeneous CaO From Waste Shells

Waste shells of mollusc which includes of snails, clams and tusk shells are often easily found in the sea, on the land and in freshwater. Boey et al., (2009) investigated on the use of waste mud crab shells (*Scylla serata*) contains mainly CaO as a catalyst to transesterified the palm olein with 0.5:1 methanol to oil ratio, catalyst amount of 5% at 65°C reaction temperature. The calcined catalyst that was reported can be reused more than three times several times without much reduction in its effectiveness. This finding is consistent with the findings by the same author, which used waste cockle shells (*Anadora granosa*) as a catalyst to in the transesterification process of chicken fat (Boey et al., 2011). Both of the finding showed that the conversion of methyl ester was 98.8% and 99.4% at about 3 hours' reaction time.

Xie et al., (2009) performed experimental investigations on the transesterification of non-edible rapeseed oil with biont (turtle) shell as catalysts. The catalyst was prepared by a tri-step procedure: incomplete carbonization at 500 °C, impregnation in potassium fluoride (KF), for 6 hours at 25 wt% and thermal activation at 300 °C. The yield product reached 97.5% with 3 hour reaction and 3 wt% catalyst. The heterogeneous biont catalyst poses a higher catalytic activity compared with conventional solid base catalyst. The above finding is consistent with the study by Yang et al. (2009). They examined the performance of shrimp shells that was being prepared in the similar process as Xie et al., (2009) except for the carbonation temperature of 450 °C and activation temperature of 250 °C. The production yield was recorded of 89.1% in 3 hours with 2.5 wt% catalyst.

Another studies by A. Buasri et al., (2013), derived CaO from waste mussel, cockles and scallops which were prepared by calcination at 700 °C to 1000 °C for 4 hours to transforms the materials to calcium oxide. The catalyst characterization was determined by XRD, XRF, SEM and BET. The calcined catalysts were further utilized as heterogeneous catalyst to transesterify palm oil with methanol. A high yield conversion of 95% was achieved for all shell derived catalysts with 3 hours' reaction time, 65 °C reaction temperature and catalyst loading of 10 wt%.

The same researchers also performed an experiments via microwave irradiation method to produced quality biodiesel from high free fatty acids (FFA) of Jatropha curcas oil (Buasri et al., 2015). This time they have used waste oyster and Pyramidella shells in their investigation. Same procedure was run as in previous studies. The optimum yield of nearly 93% was achieved for both shells catalysts with shorter reaction time, 5 minutes, microwave power 800 Watt, catalyst concentration of 4wt% and 15:1 methanol to oil molar ratio. The catalysts was used up to 3 times with low activity loss.

Heterogeneous CaO From Egg Shells

Calcium oxide derived from waste egg shells contained about 85 to 95% of calcium carbonate and 5 to 15% other components such as magnesium carbonate ($MgCO_3$), phosphate, organic matter and a small amount of metal in dry egg shells. Recently, Kara et al., (2019) analysed the activity of CaO derived from egg shells for the transesterification of waste vegetable oil. The yield reached maximum when the higher rate of methanol molar ratio was above 9 with 3 wt% catalyst loading. The interesting part is, it can be recycled up to 8 times. Study done by Ngadi et al., (2016) proved that the CaO derived from egg shells waste was comparable with commercial CaO. The optimum condition was achieved when the reaction temperature was 65°C for 3 hours reaction time with 6:1 methanol to oil molar ratio. Buasri et al., (2013) performed transesterification of palm oil using CaO derived from waste duck egg shells and the highest yields were obtained at 92.92 and 94.49% respectively.

Khemthong et al., (2012) performed transesterification process of palm oil using CaO from industrial eggshell wastes via microwave irradiation method. The maximum yield of FAME was 96.7% under 4 minutes of reaction time with 900 Watt microwave power, methanol to oil molar ratio of 18:1 and catalyst loading of 15%. The results also shows the ability of the microwave to produce biodiesel for only 4 minutes compared to conventional method which will take long reaction time. The research study by Bharadwaj et al., (2019) also found similar good performance of egg shells with yield more than 99% in transesterification of very high viscous rubber seeds oils.

Heterogeneous CaO From Natural Rocks

One of the natural source of calcium are coming from dolomite rocks which is of low cost and possess high basicity. Calcium carbonate and magnesium carbonate are the major elements that can be found in dolomite rocks (Marwaha et al., (2018). Ngamcharussrivichai et al., (2010) in their investigation on the performance on the use of modified dolomites for palm kernel oil transesterification in the conditions of

15:1 methanol to oil molar ratio for 3 hours at 333 K. The FAME yield almost reached 100% using 10 wt% catalyst amount. They also prepared 6 wt% of dolomite catalyst for transesterification of palm kernel oil with methanol at 333 K for 3 hours with 30:1 methanol to oil molar ratio. A total of 98% yield of methyl ester was obtained. The used of cement clinker (limestone) loaded with 21 wt% potassium hydroxide (KOH) as a eco-friendly catalyst have been examined by Soni & Agarwal (2013) for the soybean oil transesterification reaction with methanol.. It was reported the catalyst shows high catalytic performance and the yield can reached almost 85% after 1 hour of reaction at 65 °C with a 6:1 molar ratio of methanol to oil.

Heterogeneous CaO From Bones

Rohu fish (*Labeo rohita*) bone was being developed to be a low cost yet potential heterogeneous catalyst for the production of biodiesel from soybean oil. With the aid of response surface methodology (RSM) the maximum yield of fatty acid methyl esters reached 97.73% under the optimal parameter condition of methanol to oil molar ratio of 6.27:1 with calcination temperature at 997.42 °C and catalyst amount 1.01 wt%. Reusability of the catalyst could be reemployed up to 6 times indicating a promising low-cost catalyst for biodiesel production (Chakraborty et al., 2011). A study by Nisar et al., (2017) identified the performance of waste animal bones (cow and sheep) as heterogeneous catalyst that were collected from butcher shop supported with potassium hydroxide catalyst to convert the Jatropha oil into methyl ester. Catalyst loading and reaction temperature plays an important role to the conversion process. The maximum biodiesel yield of 96.01% was obtained under 6wt% of catalyst, 9:1 oil to methanol molar ratio with 70 ± 3 °C reaction temperature.

CONCLUSION

From manual to advance, different biodiesel production processes are providing distinct specifications in capacity, cost, feedstock type, process flow and production scale. Each of these shows particular preferences and limitations in biodiesel production. Microwave assisted technique has been found as one of the method that have the potential to be highlighted due to its ability to reduce the production cost and produce the higher quality biodiesel as well. This method has been identified to enhance and accelerate the transesterification process in order to obtain the higher yields in the biodiesel synthesis. The utilization of this method with other advanced techniques in the transesterification reaction have been proven to enhance the reaction rates and has been implemented to overcome many problems associated with the biodiesel production. The wide range of application of natural heterogeneous catalyst is an attempt to find a way to produce an economical, high yield product, a novel green method, sustainable and reusability catalyst. Utilizing renewable feed stocks will reduce the environmental emissions and facilitate the nutrient recovery, wastewater reuse and recycling.

REFERENCES

Abbaszaadeh, A., Ghobadian, B., Omidkhah, M. R., & Najafi, G. (2012). Current Biodiesel Production Technologies: A Comparative Review. *Energy Conversion and Management*, *63*, 138–148. doi:10.1016/j.enconman.2012.02.027

Ashraful, A. M., Masjuki, H. H., Kalam, M. A., Rizwanul Fattah, I. M., Imtenan, S., Shahir, S. A., & Mobarak, H. M. (2014). Production and comparison of fuel properties, engine performance, and emission characteristics of biodiesel from various non-edible vegetable oils: A review. *Energy Conversion and Management*, *80*, 202–228. doi:10.1016/j.enconman.2014.01.037

Atadashi, I. M., Aroua, M. K., & Abdul Aziz, A. (2010). High Quality Biodiesel and Its Diesel Engine Application: A Review. *Renewable & Sustainable Energy Reviews*, *14*(7), 1999–2008. doi:10.1016/j.rser.2010.03.020

Atadashi, I. M., Aroua, M. K., Abdul Aziz, A. R., & Sulaiman, N. M. N. (2012). Production of Biodiesel Using High Free Fatty Acid Feedstocks. *Renewable & Sustainable Energy Reviews*, *16*(5), 3275–3285. doi:10.1016/j.rser.2012.02.063

Balat, M. (2012). Potential Alternatives to Edible Oils for Biodiesel Production – A Review of Current Work. *Energy Conversion and Management*, (52): 1479–1492.

Bálint, E., Tajti, Á., & Keglevich, G. (2019). Application of the Microwave Technique in Continuous Flow Processing of Organophosphorus Chemical Reactions. *Materials (Basel)*, *12*(5), 788. doi:10.3390/ma12050788 PubMed

Bernard, M. T., Leadbeater, N. E., Boucher, M. B., Stencel, L. M., & Wilhite, B. A. (2007). Continuous-flow preparation of biodiesel using microwave heating. *Energy & Fuels*, *21*(3), 1777–1781. doi:10.1021/ef0606207

Boey, P.-L., Maniam, G. P., & Hamid, S. A. (2009). Biodiesel production via transesterification of palm olein using waste mud crab (Scylla serrate) shell as a heterogeneous catalyst. *Bioresource Technology*, *100*(24), 6362–6368. doi:10.1016/j.biortech.2009.07.036 PubMed

Boey, P.-L., Maniam, G. P., Hamid, S. A., & Ali, D. M. H. (2011). Utilization oxidation stability waste cockle shell (Anadara granosa) in biodiesel production from palm olein: Optimization using response surface methodology. *Fuel*, *90*(7), 2353–2358. doi:10.1016/j.fuel.2011.03.002

Buasri, A., Chaiyut, N., Loryuenyong, V., Wongweang, C., & Khamsrisuk, S. (2013). Application of eggshell wastes as a heterogeneous catalyst for biodiesel production. *Sustain Energy*, *1*(2), 7–13. doi:10.12691/rse-1-2-1

Buasri, A., Chaiyut, N., Loryuenyong, V., Worawanitchaphong, P., & Trongyong, S. (2013). Calcium oxide derived from waste shells of mussel, cockle, and scallop as the heterogeneous catalyst for biodiesel production. *The Scientific World Journal*. doi:10.1155/2013/460923

Buasri, A., Rattanapan, T., Boonrin, C., Wechayan, C., & Loryuenyong, V. (2015). Oyster and Pyramidella shells as heterogeneous catalysts for the microwave-assisted biodiesel production from Jatropha curcas oil. *Journal of Chemistry*, 578625.

Chakraborty, R., Bepari, S., & Banerjee, A. (2011). Application of Calcined Waste Fish (Labeo Rohita) Scale as Low-Cost Heterogeneous Catalyst for Biodiesel Synthesis. *Bioresource Technology*, *102*(3), 3610–3618. doi:10.1016/j.biortech.2010.10.123 PubMed

Demirbas, A. (2005). Biodiesel production from vegetable oils via catalytic and non-catalytic supercritical methanol transesterification methods. *Progress in Energy and Combustion Science*, *31*(5-6), 466–487. doi:10.1016/j.pecs.2005.09.001

Demirbas, A. (2017). Tomorrow's biofuels: Goals and hopes. *Energy Sources. Part A, Recovery, Utilization, and Environmental Effects*, *39*(7), 673–679. doi:10.1080/15567036.2016.1252815

Divakara, B. N., Upadhyaya, H. D., Wani, S. P., & Gowda, C. L. L. (2010). Biology and genetic improvement of Jatropha curcas L.: A review. *Applied Energy*, *87*(3), 732–742. doi:10.1016/j.apenergy.2009.07.013

Encinar, J. M., Gonzalez, J. F., Martinez, H., Sanchez, N., & Pardal, A. (2012). Soybean oil transesterification by the use of microwave flow system. *Fuel*, (95): 383–393. doi:10.1016/j.fuel.2011.11.010

Fadhlullah, M., Widiyanto, S. N. B., & Restiawaty, E. (2015). The potential of nyamplung (Calophyllum inophyllum L.) seed oil as biodiesel feedstock: Effect of seed moisture content and particle size on oil yield. *Energy Procedia*, *68*, 177–185. doi:10.1016/j.egypro.2015.03.246

Fatah, M. A., Farah, H. A., & Ossman, M. (2012). Production of biodiesel from non-edible oil and effect of blending with diesel on fuel properties. *Engineering Science and Technology: An International Journal*, *2*(4), 2250–3498.

Gmunder, S., Singh, R., Pfister, S., Adheloya, A., & Zah, R. (2012). Environmental Impacts of Jatropha curcas Biodiesel in India. Journal of Biomedicine & Biotechnology, 62307. PubMed

Gude, V. G., Patil, P., Martinez-Guerra, E., Deng, S., & Nirmalakhandan, N. (2013). Microwave energy potential for biodiesel production. Sustainable Chemical Processes, 1(1), 5. doi:10.1186/2043-7129-1-5

Hara, M. (2009). Environmentally benign production of biodiesel using heterogeneous catalyst. *ChemSusChem*, *2*(2), 129–135. doi:10.1002/cssc.200800222 PubMed

Hernando, J., Leton, P., Matia, M. P., Novella, J. L., & Alvarez-Builla, J. (2007). Biodiesel and FAME synthesis assisted by microwaves: Homogeneous batch and flow processes. *Fuel*, *86*(10-11), 1641–1644. doi:10.1016/j.fuel.2006.11.003

Idris, A. M. (2016). The effects of alcohol to oil molar ratios and the type of alcohol on biodiesel production using transesterification process. *Egyptian Journal of Petroleum*, *25*(1), 21–31. doi:10.1016/j.ejpe.2015.06.007

Ikwuagwu, O. E., Ononogbu, I. C., & Njoku, O. U. (2000). Production of biodiesel using rubber [Hevea brasiliensis (Kunth. Muell.)] seed oil. *Industrial Crops and Products*, *12*(1), 57–62. doi:10.1016/S0926-6690(99)00068-0

Janaun, J., & Ellis, N. (2010). Perspectives on biodiesel as a sustainable fuel. *Renewable & Sustainable Energy Reviews*, *14*(4), 1312–1320. doi:10.1016/j.rser.2009.12.011

Kara, K., Ouanji, F., El Mahi, M., Lotfi, E. M., Kacimi, M., & Mahfoud, Z. (2019). Biodiesel synthesis from vegetable oil using eggshell waste as a heterogeneous catalyst. Biofuels, 1–7. doi:10.1080/17597269.2019.1580972

Khemthong, Luadthong, & Nualpaeng, Changsuwan, Tongprem, Viriya-empikul, & Faungnawakij. (2012). Industrial Eggshell Wastes as the Heterogeneous Catalysts for Microwave-Assisted Biodiesel Production. *Catalysis Today*, *90*(1), 112–116. doi:10.1016/j.cattod.2011.12.024

Kouzu & Hidaka. (2012). Transesterification of vegetable oil into biodiesel catalyzed by CaO: A review. *Fuel*, *93*, 1–12. doi:10.1016/j.fuel.2011.09.015

Lin, C.-Y., & Lin, Y.-W. (2012). Fuel Characteristics of Biodiesel Produced from a High-Acid Oil from Soybean Soapstock by Supercritical-Methanol Transesterification. *Energies*, *5*(5), 2370–2380. doi:10.3390/en5072370

Marchetti, J. M., Miguel, V. U., & Errazu, A. E. (2007). Possible methods for biodiesel production. *Renewable & Sustainable Energy Reviews*, *11*(6), 1300–1311. doi:10.1016/j.rser.2005.08.006

Mário, R. P. S., & Maria, J. M. S. (2016). Growth and development of Jatropha curcas seedlings using Terracotem soil conditioners under different irrigation levels. *Emirates Journal of Food and Agriculture*, *28*(5), 326–331. doi:10.9755/ejfa.2015-12-1135

Marvey, B. B. (2008). Sunflower-based Feedstocks in Nonfood Applications: Perspectives from Olefin Metathesis. *International Journal of Molecular Sciences*, *9*(8), 1393–1406. doi:10.3390/ijms9081393 PubMed

Marwaha, A., Rosha, P., Mohapatra, S. K., Mahla, S. K., & Dhir, A. (2018). Waste materials as potential catalysts for biodiesel production: Current state and future scope. *Fuel Processing Technology*, *181*, 175–186. doi:10.1016/j.fuproc.2018.09.011

Milano, J., Ong, H. C., Masjuki, H. H., Silitonga, A. S., Chen, W.-H., Kusumo, F., ... Sebayang, A. H. (2018). Optimization of biodiesel production by microwave irradiation-assisted transesterification for waste cooking oil- Calophyllum inophyllum oil via response surface methodology. *Energy Conversion and Management*, *158*, 400–415. doi:10.1016/j.enconman.2017.12.027

Motasemi, F., & Ani, F. N. (2012). A Review on Microwave-Assisted Production of Biodiesel. *Renewable & Sustainable Energy Reviews*, *16*(7), 4719–4733. doi:10.1016/j.rser.2012.03.069

Natural Rubber Statistic. (2018). Unit Kajian dan Kemajuan Pasaran, Bahagian Ekonomi dan Pasaran Getah, Lembaga Getah Malaysia. Retrieved from http://www.lgm.gov.my/nrstat/nrstats2018_(Jan-Mar)_nopword.pdf

Ngadi, Hamdan, Hassan, & Jaya. (2016). Production of Biodiesel from Palm Oil Using Egg Shell Waste as Heterogeneous Catalyst. Jurnal Teknologi (Sciences and Engineering), (78), 59–63.

Ngamcharussrivichai, C., Nunthasanti, P., Tanachai, S., & Bunyakiat, K. (2010). Biodiesel production through transesterification over natural calciums. *Fuel Processing Technology*, *91*(11), 1409–1415. doi:10.1016/j.fuproc.2010.05.014

Nisar, J., Razaq, R., Farooq, M., Iqbal, M., Khan, R. A., Sayed, M., . . . Rahman, I. (2017). Enhanced biodiesel production from Jatropha oil using calcined waste animal bones as catalyst. Renewable Energy, Elsevier., 101, 111–119. doi:10.1016/j.renene.2016.08.048

Atabani. Silitonga, Ong, Mahlia, Masjuki, Badruddin, & Fayaz. (2013). Non-edible vegetable oils: A critical evaluation of oil extraction, fatty acid compositions, biodiesel production, characteristics, engine performance and emissions production. Renewable & Sustainable Energy Reviews, 18(0), 211–245. doi:10.1016/j.rser.2012.10.013

Orwa, C., Mutua, A., Kindt, R., Jamnadass, R., & Anthony, S. (2009). Agroforestree database: A tree reference and selection guide version 4.0. Retrieved from http://www.worldagroforestrycentre.org/sea/products/afdbases/af/asp/SpeciesInfo.asp?SpID=272

Ramadhas, A., Jayaraj, S., & Muraleedharan, C. (2005). Biodiesel Production from High FFA Rubber Seed Oil. *Fuel*, *84*(4), 335–340. doi:10.1016/j.fuel.2004.09.016

Singh & Singh. (2010). Biodiesel Production through the Use of Different Sources and Characterization of Oils and Their Esters as the Substitute of Diesel: A Review. *Renewable & Sustainable Energy Reviews*, (14): 200–216.

Roschat, W., Siritanon, T., Yoosuk, B., Sudyoadsuk, T., & Promarak, V. (2017). Rubber seed oil as potential non-edible feedstock for biodiesel production using heterogeneous catalyst in Thailand. *Renewable Energy*, *101*, 937–944. doi:10.1016/j.renene.2016.09.057

Sai Bharadwaj, A. V. S. L., Singh, M., Niju, S., Begum, K. M., & Anantharaman, N. (2019). Biodiesel production from rubber seed oil using calcium oxide derived from eggshell as catalyst – optimization and modeling studies. Green Processing and Synthesis, 8(1), 430–442. doi:10.1515/gps-2019-0011

Sakai, T., Kawashima, A., & Koshikawa, T. (2009). Economic assessment of batch biodiesel production processes using homogeneous and heterogeneous alkali catalysts. *Bioresource Technology*, *100*(13), 3268–3276. doi:10.1016/j.biortech.2009.02.010 PubMed

Siriwardene, J. A., & Nugara, D. (1972). Metabolizable energy of rubber seed meal in poultry diets. *Ceylon Veterinary Journal*, *20*(3), 61–63.

Soni & Agarwal. (2013). Use of Ceramic Material (Cement Clinker) for the Production of Biodiesel. International Journal of Modern Physics: Conference Series, (22), 71–78.

Tan, K. T., & Lee, K. T. (2011). A review on supercritical fluids (SCF) technology in sustainable biodiesel production: Potential and challenges. *Renewable & Sustainable Energy Reviews*, *15*(5), 2452–2456. doi:10.1016/j.rser.2011.02.012

Tangy, A., Pulidindi, I. N., Perkas, N., & Gedenken, A. (2017). Continuous flow through a microwave oven for the large-scale production of biodiesel from waste cooking oil. *Bioresource Technology, 224*, 333–341. doi:10.1016/j.biortech.2016.10.068 PubMed

Tippayawong, N., & Sittisun, P. (2012). Continuous-Flow Transesterification of Crude Jatropha Oil with Microwave Irradiation. *Scientia Iranica, 19*(5), 1324–1328. doi:10.1016/j.scient.2012.08.004

Venkanna, B. K., & Venkataramana, R. C. (2009). Biodiesel production and optimization from Calophyllum inophyllum linn oil (honne oil)—A three stage method. *Bioresource Technology, 100*(21), 5122–5125. doi:10.1016/j.biortech.2009.05.023 PubMed

Vyas, A. P., Verma, J. L., & Subrahmanyam, N. (2010). A review on FAME production processes. *Fuel, 89*(1), 1–9. doi:10.1016/j.fuel.2009.08.014

Xie, J., Zheng, X., Dong, A., Xiao, Z., & Zhang, J. (2009). Biont shell catalyst for biodiesel production. *Green Chemistry, 11*(3), 255–264. doi:10.1039/B812139G

Yang, L., Zhang, A., & Zheng, X. (2009). Shrimp Shell catalyst for biodiesel production. Energy Fuel, (23), 3859-65.

Yusuff, A. S., Adeniyi, O. D., Olutoye, M. A., & Akpan, U. G. (2017). A review on application of heterogeneous catalyst in the production of biodiesel from vegetable oil. *J Appl Sci Process Eng, 2*(4), 142–157. doi:10.33736/jaspe.432.2017

Zahan, K. A., & Kano, M. (2018). Biodiesel production from palm oil, its by-products and mill effluent: A review. *Energies, 11*(8), 2132. doi:10.3390/en11082132

Zamberi, M. M., Ani, F. N., & Abdollah, M. F. (2016). Heterogeneous Transesterification of Rubber Seed Oil Biodiesel Production. *Jurnal Teknologi, 78*(6-10), 105–110.

Zamberi, M. M., Ayob, M. R., & Ibrahim, M. Z. (2011). Experimental Study on Low Cost Biodiesel Production Alkaline Based Catalysts by Using Frying Oil. *Journal of Engineering Technology, 2*, 55–66.

Chapter 6
Optimizing the Mercerisation Effect on the Mode I Fracture Toughness of Bambusa Vulgaris Bamboo Using Surface Response Method

Mohamad Zaki Hassan
https://orcid.org/0000-0002-2321-0604
Universiti Teknologi Malaysia, Malaysia

Siti Amni Roslan
Universiti Teknologi Malaysia, Malaysia

Zainudin A. Rasid
Universiti Teknologi Malaysia, Malaysia

Mohd Zuhri Mohamed Yusoff
Universiti Putra Malaysia, Malaysia

S. M. Sapuan
Universiti Putra Malaysia, Malaysia

Firdaus Muhammad-Sukki
https://orcid.org/0000-0002-5415-2259
Robert Gordon University, UK

DOI: 10.4018/978-1-7998-1374-3.ch006

ABSTRACT

Alkaline treatment is widely being promoted to treat natural fibres and improves the fibre bundle surface for better interlocking with the polymer matrix. The aim of this study is to optimize the merceration parameter including natrium hydroxide (NaOH) concentration, soaking and drying time for Bambusa Vulgaris bamboo using response surface methodology (RSM). Here, the treatment conditions were employed by the Box-Behnken design (BBD). The comparative study of the treated and untreated fibre on crack propagation behaviour, Mode I interlaminar fracture toughness (GIC) of the bamboo along the longitudinal direction test was carried out. Through the statistical analysis approach (ANOVA), it is suggested that bamboo treated with 1.5 wt.% concentration of NaOH is capable to reach the fracture toughness value up to 367.25 J/m2. It is also shown that all proposed variables for treatment in this study (i.e., the concentration of the NaOH is highly significant with the 2.85 hours of soaking and drying for 72.5 hours).

INTRODUCTION

Normally, fiber reinforced polymer materials are laminate composites consisting of high strength of synthetic fibers (i.e. glass, carbon and kevlar) reinforcement in polymeric matrices. The reinforcement by those fibers provides the polymer substantially enhanced mechanical properties and makes the fiber reinforced polymer composites suitable for a large number of diverse applications. Although this synthetic fiber reinforced polymer possess exclusive mechanical strength, they have also got some serious drawbacks such as high cost, poor recycling and non-biodegradable properties.

For these reasons, natural plant fibers reinforced polymer composites are increasingly gaining attention as viable alternative to replace the synthetic fiber reinforced polymer counterpart. These renewable fiber closely high specific strength to weight ratio and offering less abrasion as compared to glass fibers. Many researcher has studied the mechanical properties of these natural fibers such as bamboo, kenaf and jute (Chen et al., 2017; Bakhori et al., 2017; Bakhori, Zuikafly, Ahmad, & Hassan, 2017). Interestingly, fiber derived from bamboos have attracted more interests due to their offering higher performance and plenty sources for production (Yu, Jiang, Fei, Wang, & Wang, 2011).

In addition, an extensively study have been conducted on chemical treatment for natural fiber. This treatment is being suggested in order to decrease the incompatibility between natural fibre and polymer matrix due to hydrophilic and hydrophobic interaction. The common alkaline treatment (NaOH) on natural fibre has been considered as one of most popular and cost affective methods. The effect of chemical treatments of the bamboo fiber is reported by Yang et al. (2017). They immersed the bamboo fibers in a NaOH solution of 5, 10 and 25% concentration for further removing hemicellulose and heated at 75 °C for 2 h in a water bath. It is found that a higher concentration of NaOH degrades the long chain cellulose molecules at the fiber interface and consequently, weakens fiber load transfer. Reddy and Dhoria (2018) examined the effect of NaOH concentrations of 5% (w/t) on kenaf fibre reinforced composites. Mechanical testings including tensile, flexural and impact are obtained to determine tensile strength, bending and impact strength of the composites. It is found that the chemically treated fiber composites offer good mechanical properties compared to untreated fibers. Similarly, Mishra et al. (2003) reported that sisal fibre treated with 5% concentration of NaOH exhibited good tensile properties than those treated with 10% of alkaline concentration. They suggested that the excessive alkali concentration would cause delignification and weaken the sisal fibre. On the other hand, a comprehensive characterization of wettability and interfacial properties of kenaf fibers polyester composites fabricated by resin transfer molding is reported by Ariawan et al. (2017). Here, kenaf fibers are chemically modified by immersing in 6% NaOH concentration for 1 to 5 h to enhance the interaction between fiber and matrices. In order to investigate the effect of soaking time, surface energy and the interlaminar shear strength value of the composites are evaluated. They reported that 3 h treated of kenaf fiber composite enhanced the interface bonding characteristic of the composite laminate. Furthermore, the effect of alkali treatment under various conditions on physical properties of kenaf fiber is investigated by Khan et al. (Khan, Yousif, & Islam, 2017). In this study, kenaf fiber were treated at alkali concentrations 2, 6 and 10 (w/v%), immersion durations at 30, 240 and 480 minute and temperatures at 27, 60 and 100° C. Kenaf fiber weight loss and density value are decreased after alkali treatment compared to untreated kenaf fiber. Furthermore, Rao et al. (2010) reported that 1% concentration of NaOH is the optimum condition to treat bamboo polymeric composite. On the other hand, few studies claimed that a higher NaOH concentration is the best in treating natural fibre (Yan, Chouw, Huang, & Kasal, 2016) and capable of removing excess moisture thoroughly. In spite of that, the NaOH concentration is the most dominant factor that gives effect on the natural fibre and still remain inconsistency.

Various technique have been explored by researchers to accurately determine the optimum condition for alkaline treatment of the natural fiber. Singh, Mukhopadhyay, & Das (2017)used a three-factor and three-level Box-Behnken design method to optimize the mass fraction of fibres, percentage of crosslinking and plasticizing agent. Initialy, they treated the sugarcane baggase fiber with 6% of sodium hydroxide. Liu et al. (2018) suggested optimized solution using the Box-Benhken design for pretreatment of corn straw by the alkaline solution. Under the optimized conditions, the cellulose and hemicellulose of the corn straw increased and the lignin content was also reduced. Furthermore, a three-level Box–Behnken design, which is subset of the response surface methodology (RSM), has been investigated by Yaghoobi and Fereidoon (2018). The effect of three independent variables incuding kenaf fiber load, fiber length and polypropylene-grafted maleic anhydride (PP-g-MA) compatibilizer content have been investigated on the tensile strength and modulus of the biocomposite. The optimal tensile strength and tensile modulus are to be 32.70 MPa and 2,182.33 MPa, respectively; and achieved at 28.95 wt% of the kenaf fiber, fiber length of 6.22 mm and PP-g-MA content of 5 wt%.

In this study, the Box-Behnken design (BBD) of the response surface method is employed to obtain an experimental design of alkaline treatment conditions for the *bambusa vulgaris* bamboo. Three experimental conditions including the concentration of NaOH, soaking and drying time are considered. Mode I interlaminar fracture toughness (G_{IC}) test is conducted to investigate the effect of alkaline treatment conditions on crack propagation behaviour along the longitudinal direction of the bamboo.

EXPERIMENTAL METHOD

Box-Benhken Design Iteration

Table 1 shows the design points for low, middle and high level of each variable conditions. This input data then was randomized and modelled by BBD using the Design-expert (6.0.8) software. It offers a total number of 17 experimental runs that include of 12 runs and 5 replication runs of the centre point.

Material and Testing

The bamboo material used in this study belongs to *bambusa vulgaris* family. The raw bamboo is cut approximately 5 metres above the ground. The age of bamboo is approximately four years old and taken from Jeli, Kelantan at the north-eastern

Table 1. The initial setting of Box-Behnken design (BBD)

Factors / Independent Variables	Symbols	Coded and Actual Levels		
		Low (-1)	Middle (0)	High (+1)
Concentration of NaOH (%)	X_1	1	2	3
Soaking duration (hours)	X_2	3	6	9
Drying duration (hours)	X_3	2	48	72

state of Malaysia. The Mode I testing specimens are prepared as a double cantilever beam (DCB) according to the ASTM D5528 standard (ASTM D5528, 2001). The smaller parts of dried bamboo culm are trimmed into narrow strips and those strips were edge squared. The dimension of the DCB specimen is, longitudinal direction, $w = 200$ mm; tangential direction, $h = 20$ mm; and radial thickness, $b = 9$ mm with initial crack length, $\alpha_0 = 40$ mm. Initially, two holes of 5 mm in diameter are made at the point crack initiation as shown in Figure 1.

Following this, the specimens are chemically treated with an alkali solution. The concentration of this solution, sodium hydroxide (NaOH), is prepared by weight per volume (*w/v*) percentage. Here, 1 gram of NaOH pellets is diluted in 100 ml of distilled water to obtain 1 wt.% concentration of NaOH. After undergoing alkali treatment and drying process, an initial crack is cleaved along the middle-line of the bamboo DCB specimen parallel to grain by a stiff razor. The length of the initial crack (α_0) controlled at 40 mm from the centre of the loading holes. The purpose of this cut is to stimulate a naturally sharp crack during the test. Then, the crack tip is marked to obtain correct pre-crack length. Later, the bamboo DCB specimen is fitted to the U-shaped hook steel which connected to a 10 kN load cell on a Shidmazu universal testing machine as illustrated in Figure 2.

The test performed at a crosshead displacement rate of 1 mm/min with five specimens tested at each treatment conditions. During loading, the crack propagated in the bamboo DCB specimen is measured. The corresponding applied load versus the opening displacement (F-δ) is recorded by a computer. Once the load (F) suddenly dropped, the loading machine stopped. Here, the crack tip on the inner (α_{1inner}) and outer (α_{1outer}) parts of the specimen are marked and the average value is taken as the

Figure 1. Schematic diagram of DCB specimen

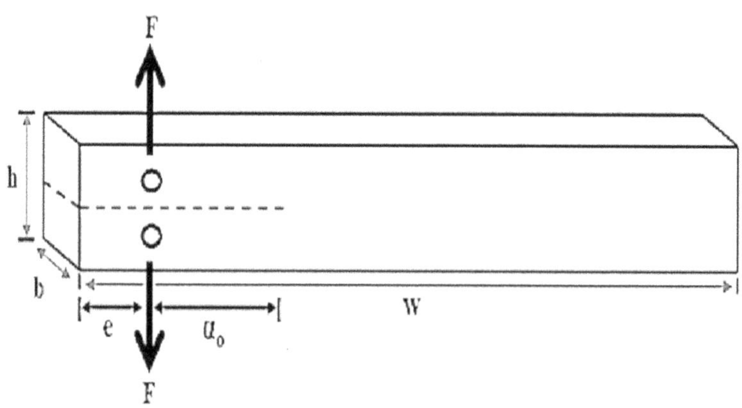

Figure 2. Schematic diagram of DCB specimen connected to U-hook

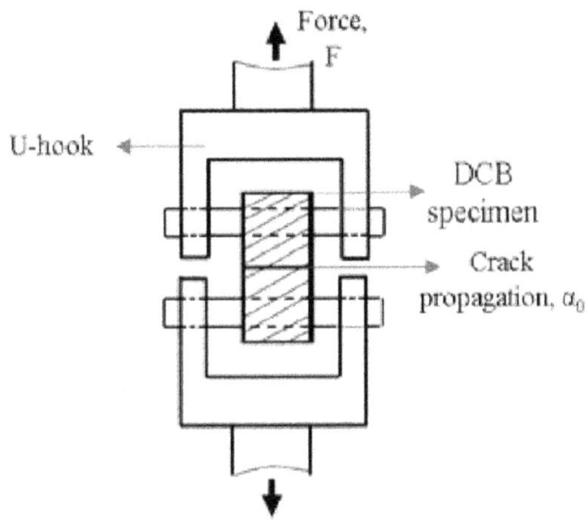

corresponding crack length (α_1). Afterwards, the specimen is unloaded and reloaded repeatedly until the specimen completely fractured. The equivalent crack length α_2, α_3, α_4… α_n can be obtained and recorded.

G_{IC} Analysis

The Mode I interlaminar fracture toughness (G_{IC}) is measured by symmetric bending tests using the DCB specimen. Hence, the energy release rate G of interlaminar fracture mode can be calculated according to Equation 1.

$$G = \frac{1}{b} . \frac{dU}{da} \tag{1}$$

where U is the total strain energy in the test specimen, b is the specimen width, and a is the delamination length. Then, U can also be written as

$$U = \frac{1}{2} F\delta = \frac{1}{2} F^2 C \tag{2}$$

where F is the load, δ is opening displacement of the DCB specimen and C is the DCB specimen compliance (δ/F) namely the reciprocal of the slope (k). It can be calculated based on two points from the F-δ traces obtained. The relationship between them can be explained in the following Equation 3.

$$C = \frac{1}{k} = \frac{\delta_2 - \delta_1}{F_2 - F_1} \quad (3)$$

So, the Mode I interlaminar fracture toughness (G_{IC}) is calculated using Equation 4.

$$G_{IC} = -\frac{1}{b} \cdot \frac{d(-\frac{1}{2}Fcr^2C)}{da} = \frac{Fcr^2}{2b} \cdot \frac{dc}{da} \quad (4)$$

where Fcr is the critical load. It is only necessary to measure the differential increase in compliance (dC) depending on the differential increase in the crack length (da).

According to the compliance method, the reciprocal of the slope is the corresponding compliance (Ci) of the DCB specimen with a certain crack length (α_i). The relationship between C and α can be described as stated in following Equation 5.

$$C = q\alpha\text{m} \quad (5)$$

where q and m are the fitting coefficients of the compliance curve of the DCB specimen. So, after taking a logarithm of Equation 5, the equation meets the linear model as stated in Equation 6.

$$\lg C = \lg q + m \lg \alpha \quad (6)$$

Following the Mode I test, the results are analysed and evaluated statistically using the Analysis of Variance (ANOVA).

ESULTS AND DISCUSSION

GI_C Characterization of Bamboo DCB Specimen

A typical force-displacement trace of an untreated bamboo specimen for a 42 mm of crack length (α_1) is shown in Figure 3. Initially, the trace increase linearly from the loading start to the maximum force. It begins to waver after a while due to the overbearing load capacity, which caused unstable crack propagation along the longitudinal direction of bamboo DCB specimen. Subsequently, the trace drops and the test suddenly stopped. At this point, maximum force is recorded as the critical force, F_{cr}, to indicate the decisive point of rapid cracking in the bamboo specimen.

Figure 4 shows an untreated bamboo DCB specimen subjected to the multi-point method of the Mode I fracture testing. The opening crack lengths for the outer layer of the bamboo that was recorded approximately 1 to 2 mm longer than those of the inner layer as can be seen in Figure 4. A similar observation also recorded on the treated bamboo DCB specimens. Shao, Fang, and Tian (2009) reported the resistance against crack propagation of bamboo is controlled by interlaminar strength between the bamboo fibres. As the amount of bamboo fibres on the outer layer is less than that of the inner layer, resistance against crack propagation exhibited by the outer layer is weaker. It indicates that the outer layer has lower interlaminar strength than the inner layer as described in (Shao, Fang, & Tian, 2009).

Table 2 summarises the average G_{IC} values for bamboo specimens with the corresponding treatment conditions subjected to Mode I testing. From the table, it can be seen that all of the treated specimens exhibited lower G_{IC} values compared to untreated specimens. The average of G_{IC} value for untreated bamboo specimens is recorded at 365.20 J/m^2. In contrast, the highest average G_{IC} value for the treated bamboo is shown by the treatment condition No.1, which is 362.85 J/m^2. The lowest G_{IC} value is found at the treatment condition No.13, exhibited more than half G_{IC} value of the treatment condition No.1. The lower G_{IC} value of treated bamboo specimens may be attributed to the softening effect caused by alkaline treatment which influences the stress transfer process along the specimen during the test is conducted (Shao, Fang, & Tian, 2009). It should be noted that even though the presence of alkaline lowers the value of the mean, G_{IC}, in the natural fibre composite application, the alkaline will improve the bonding between the fibre and the matrix and improve the mechanical properties of the composite.

Figure 3. The typical force-displacement trace of an untreated bamboo specimen

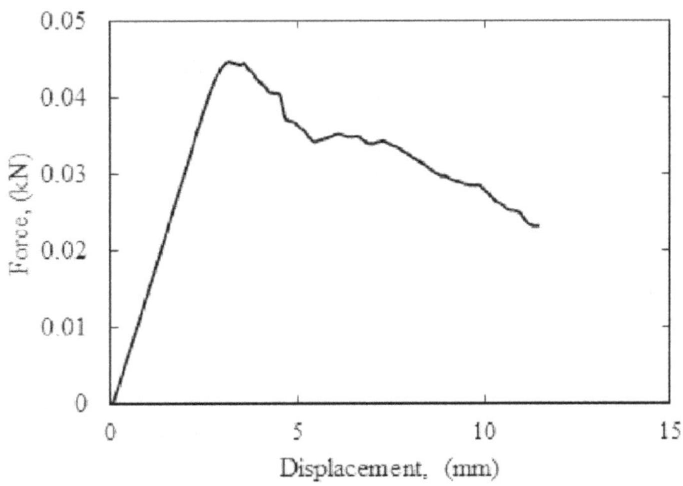

Figure 4. *The delamination of crack lengths with geometric views for (a) the outer and (b) the inner parts of an untreated bamboo DCB specimen*

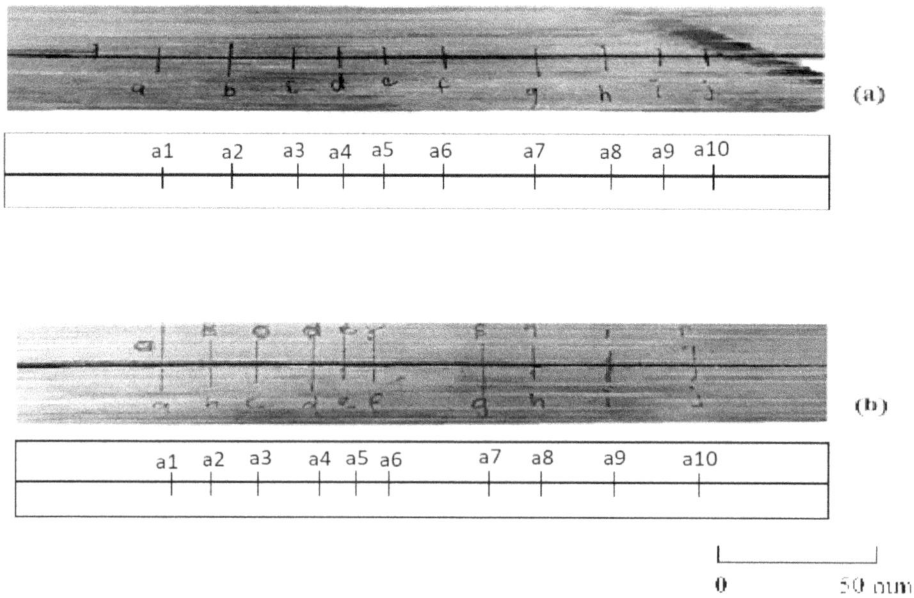

Quadratic Model Fitting and Analysis of Variance (ANOVA)

The G_{IC} results of treated bamboo are further investigated using analysis of variance (ANOVA) to determine which variables that significantly affect the G_{IC} value of treated bamboo. By applying a multiple regression analysis of the responses, the outcome proposed the highest order polynomial in which the additional terms are significant and the model is not aliased. Following that, backwards elimination method was applied to exclude insignificant terms automatically. The ANOVA for the reduced quadratic models that summarised in the G_{IC} value, is shown in Equation 7.

$$G_{IC} = 573.235 - 160.801\,X_1 - 14.743\,X_2 - 1.229\,X_3 + 27.696\,X_1^2 - 0.679\,X_2^2 + 0.033\,X_3^2 + 3.936\,X_1X_2 - 1.095\,X_1X_3 \quad (7)$$

Equation 7 represents the relationship between NaOH concentration (X_1), soaking time (X_2) and drying time (X_3) towards G_{IC} value of treated bamboo. This equation was also used to generate predictions of the response for a given level of each variable. Table 3 depicts the predicted G_{IC} values using Equation **7** and the experimental finding value. It is in good agreement between the yield predicted and the experimental responses, with a small difference in G_{IC} value are measured

Table 2. The effect of treatment conditions on G_{IC} values of bamboo

No.	NaOH concentration, X_1 (wt.%)	Soaking time, X_2 (hours)	Drying time, X_3 (hours)	Measurement numbers, k	Sample numbers, n	Mean G_{IC} (J/m²)
0	0	0	0	30	3	365.20
1	1	3	48	35	3	362.85
2	1	6	24	33	3	314.88
3	1	6	72	34	3	356.94
4	1	9	48	33	3	251.32
5	2	3	24	33	3	270.83
6	2	3	72	33	3	264.13
7	2	6	48	53	5	205.48
8	2	9	24	30	3	186.11
9	2	9	72	30	3	163.97
10	3	3	48	30	3	185.06
11	3	6	24	32	3	200.77
12	3	6	72	30	3	142.71
13	3	9	48	30	3	120.76

as can see in Table 3. The percentages of error are also calculated to determine the precision of calculations. Here, the error percentage of each run is less than 4%, which is considered effective (Montgomery, 2005). The bamboo treated with 1 wt.% concentration of NaOH offers the fracture toughness value up to 365.86 J/m², which 0.82% of error from the testing result. The G_{IC} value of treated bamboo specimens is declined with the increasing NaOH concentration and soaking time. Again, a higher NaOH concentration and soaking time could reduce the mechanical properties of the fibre due to the softening effect (Bledzki, Fink, & Specht, 2004). Longer drying time, however, has a positive correlation with the response. The model afterwards was inspected statistically using the F-test and regression coefficient, R^2 for validity purposes.

Model Accuracy Analysis Check

Table 4 lists the statistical data for analysis of variance of G_{IC} value. The coefficient of regression (R^2) value of 0.998 implies that the model competently represents the relationship between the significant terms. In fact, the closer R^2 value to 1, the higher reliability of the empirical model of the obtained data. A similar observation is noticed in the Adjusted R-squared of 0.996 where its value is almost 1, indicates

Table 3. The BBD analysis predicted and experimental G_{IC} values

Run	Independent Variables						G_{IC} (J/m²)			
	Coded values			Actual values			Experimental	Predicted	Residual	Error (%)
	X_1	X_2	X_3	X_1	X_2	X_3				
1	0	+1	-1	2	9	24	186.11	182.40	3.71	2.03
2	0	0	0	2	6	48	205.64	208.42	-2.78	1.33
3	0	-1	-1	2	3	24	270.83	272.58	-1.75	0.64
4	0	0	0	2	6	48	207.19	208.42	-1.23	0.59
5	0	+1	+1	2	9	72	163.97	169.94	-5.97	3.51
6	+1	-1	0	3	3	48	185.06	184.32	0.74	0.40
7	0	0	0	2	6	48	207.63	208.42	-0.79	0.38
8	0	-1	+1	2	3	72	264.13	260.12	4.01	1.54
9	-1	+1	0	1	9	48	251.32	252.06	-0.74	0.29
10	+1	0	+1	3	6	72	142.71	143.60	-0.89	0.62
11	0	0	0	2	6	48	208.73	208.42	0.31	0.15
12	-1	-1	0	1	3	48	362.85	365.86	-3.01	0.82
13	-1	0	+1	1	6	72	356.94	354.09	2.85	0.80
14	-1	0	-1	1	6	24	314.88	313.99	0.89	0.28
15	0	0	0	2	6	48	212.91	208.42	4.49	2.15
16	+1	+1	0	3	9	48	120.76	117.76	3.00	2.55
17	+1	0	-1	3	6	24	205.77	208.62	-2.85	1.37

that the suggested empirical model is significantly reliable. Adjusted *R*-squared of 0.996 is also in good agreement with the Predicted *R*-squared.

The coefficient of variation (C.V.) value measures the degree of precision with which treatments were executed and must be lower than 10% (Montgomery, 2005). At the same time, a lower C.V. value implied a higher reliability of the experiment. Therefore, the C.V. value of 1.81% indicates the experiment conducted in this study is reliable. Meanwhile, Adequate Precision value measures the signal to noise ratio were a ratio greater than 4 are desirable (Montgomery, 2005). The ratio for the studied model of 82.783 indicates an adequate signal, thus suggests the proposed model can be utilized to navigate the design space.

Response Analysis

The adequacy of the model is inspected to verify whether the suggested model provides sufficient approximation to the actual system (Shao, Fang, & Tian, 2009).

Table 4. The statistical data of ANOVA for G_{IC}

Source	Response value
R-Squared (R^2)	0.998
Adjusted R-Squared	0.996
Predicted R-Squared	0.987
Standard Deviation	4.120
Coefficient of Variation (C.V.) (%)	1.810
Adequate Precision	82.783
Mean	227.500

To evaluate the model satisfaction, the internal studentised residuals can be verified with the assumption of ANOVA. Here, the internally studentised residual is used to obtain the standard deviation between the experiment and predicted values. Figure 5 presents a relationship of normal probability distribution and the internally studentised residuals. All the residuals point fall closer to the line and fits with the model of the data as can be observed in Figure 5. It can be suggested that no response transformation is required. In addition, the examine residual is also having no apparent problem with the normality.

Figure 5. The normal probability plot of residuals

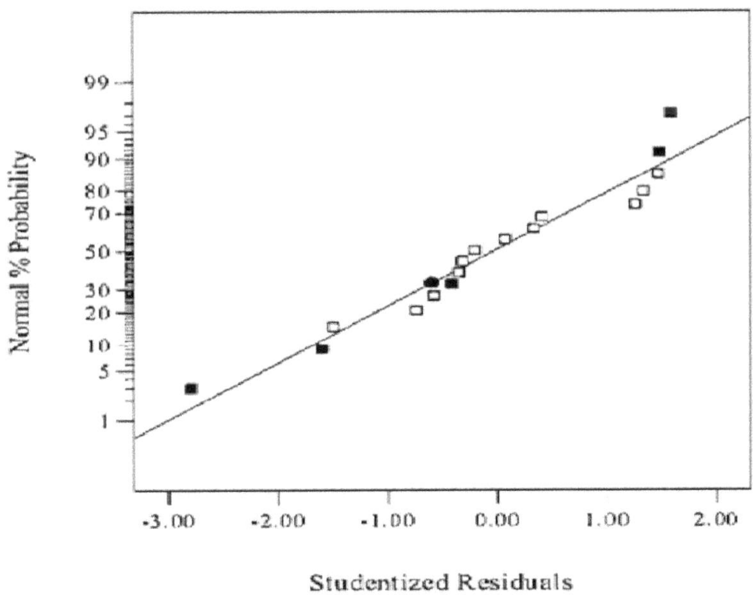

Figure 6 shows the relationship between studentized residuals and predicted response. In this figure, the studentized residuals was scattered randomly and spread in a constant range across the plot around the "0" line. It is indicated that no clear patterns in this plots and validating the initial assumption of constant variance competently. As can be seen from the Figure 6, the residuals and thus confirmed the adequacy of the model. Figure 7 presents a correlation between the experimental and predicted values of G_{IC}. Here, all scatter points are properly distributed close to the line, which suggests a high degree of correlation between the experiment and predicted values. It is therefore suggested that the closer data points to the reference line, the greater accuracy of the model (Montgomery, 2005). Thereby, all the fundamental analyses described are closely fitted, and the selected empirical model is acceptable in predicting the G_{IC} value.

In order to attain a better understanding of the results, the perturbation plot that provides a silhouette view of the response surface is presented in Figure 8. The perturbation plot demonstrates the response of a particular variable as it moves from the chosen reference point while the other variables are fixed to this point. In the Design-Expert software, the reference point is set at the centre of the design space that is the zero-coded level of each variable. So, the zero-coded level of each variable in this study is NaOH concentration (X_1) = 2.00, soaking time (X_2) = 6.00 and drying time (X_3) = 48.00. From Figure 8, the increasing variables X_1 and X_2 clearly

Figure 6. The plot of studentized residuals versus predicted response

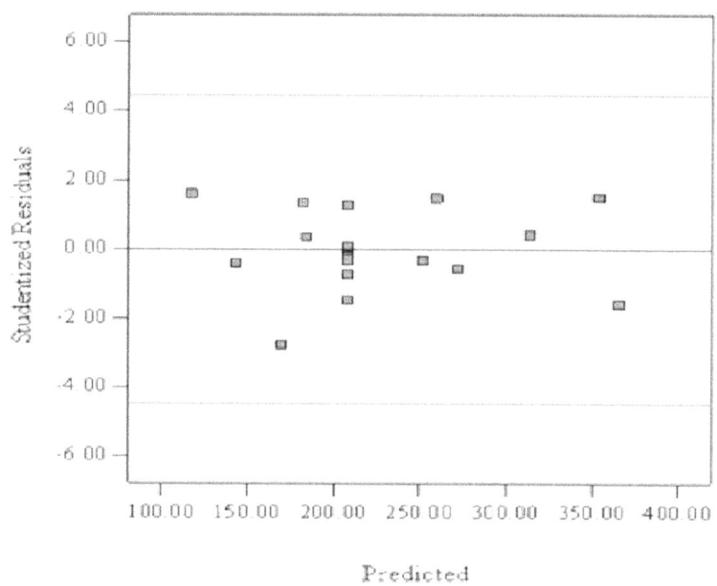

Figure 7. The plot of the predicted versus actual response

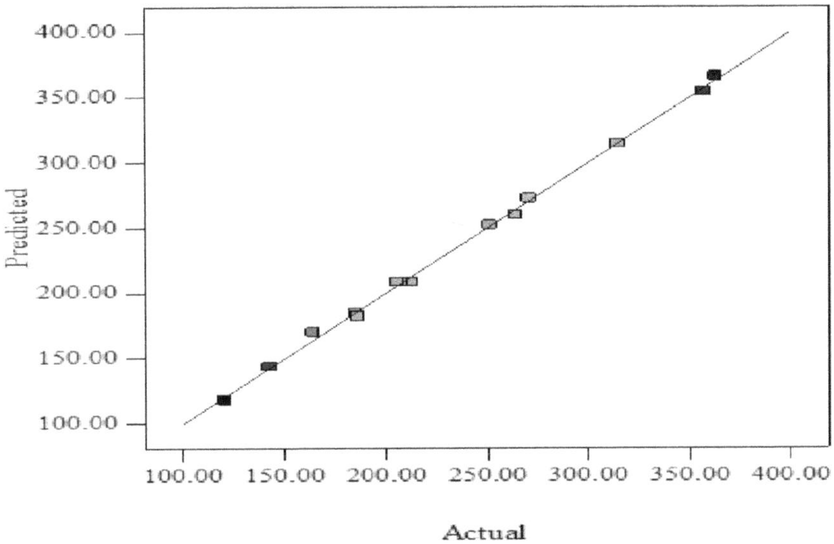

offer negative effect on G_{IC} values. Conversely, variable X_3 shows a slight positive effect on G_{IC} values of bamboo where the G_{IC} values increase with increasing X_3.

The contour plots and response surface for the interaction effect of NaOH concentration (X_1) and soaking time (X_2) on G_{IC} values of bamboo at 72 hours of drying time (X_3) are presented in Figure 9. The highest G_{IC} value of 365.86 J/m^2 was recorded at the lowest NaOH (1% wt) concentration and 3 hours of soaking time.

Validation of Optimization Process

Validation experiments are conducted to determine the performance of chemical treatment by evaluating the level of concentration, drying time and soaking duration studies at the optimum favorable conditions through Design Expert Software. The results recorded at (1.5% wt) NaOH concentration, 2.85 hours of soaking time and 72.5 hours of drying time (X_3) offers the highest G_{IC} value of 367.25 J/m^2. Three additional experiments are repeated and the average value for Mode 1 fracture is approximately 85%.

CONCLUSION

Based on the statistical analysis conducted using ANOVA, the proposed variables are found to significantly influence the G_{IC} values of treated bamboo. The analysis

Figure 8. The perturbation plot of the effect of all variables on G_{IC} value

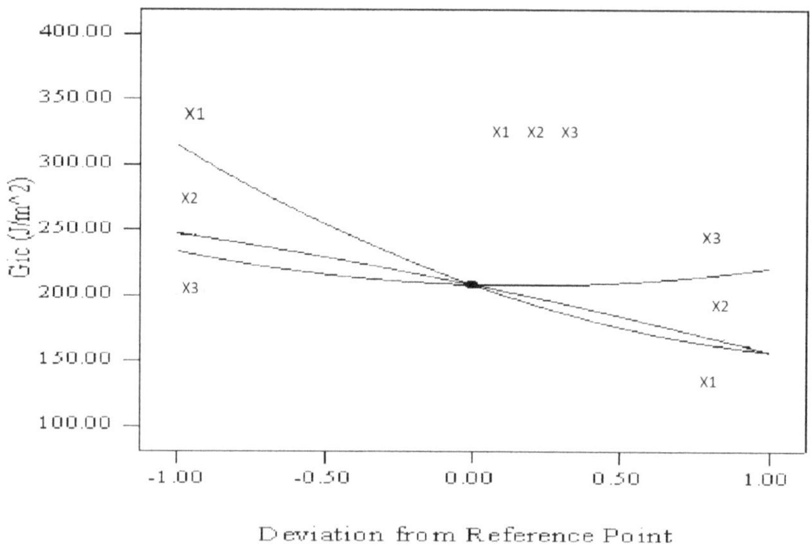

Figure 9. The effect of NaOH and drying time on G_{IC} value

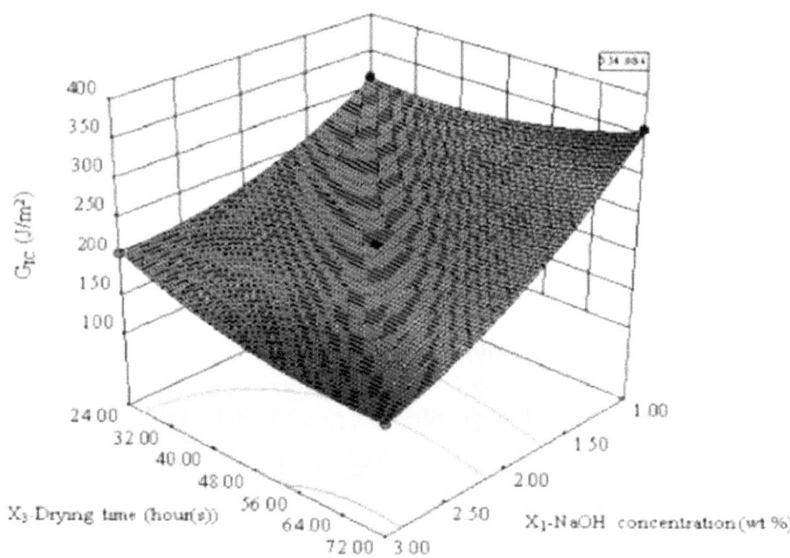

suggested that the bamboo needs to be treated with low NaOH concentration for a short soaking period and dried at a longer drying time, to obtain the optimum G_{IC} value of bamboo at room temperature. The highest value of G_{IC} for treated bamboo is evaluated at 367.25 J/m^2 when the bamboo treated with 1.5 wt.% of NaOH concentration for 2.85 hours and dried up to 72.5 hours at room temperature (1.5wt.%-2.85hours-72.5hours).

ACKNOWLEDGMENT

The appreciation is given to Universiti Teknologi Malaysia and Ministry of Higher Education under Fundamental Research Grant Scheme [R.K130000.7840.4F880] and "*Geran Universiti Penyelidik*' (GUP) Tier 1 [Q.K130000.2540.15H70] for financially support provided throughout the course of this research project.

REFERENCES

American Society for Testing and Materials (ASTM). (2001). *Standard test for Mode I interlaminar fracture toughness of unidirectional fibre-reinforced polymer matrix composites*. Philadelphia: D 5528-01.

Ariawan, D., Mohd Ishak, Z. A., Salim, M. S., Mat Taib, R., & Ahmad Thirmizir, M. Z. (2017). Wettability and interfacial characterization of alkaline treated kenaf fiber-unsaturated polyester composites fabricated by resin transfer molding. *Polymer Composites*, *38*(3), 507–515. doi:10.1002/pc.23609

Bakhori, S., Hassan, M. Z., Daud, Y., Sarip, S., Rahman, N., Ismail, Z., & Aziz, S. A. (2015). Mechanical behavior of Kenaf/Epoxy corrugated sandwich structures. *IOP Conference Series. Materials Science and Engineering*, *100*(1), 012011. doi:10.1088/1757-899X/100/1/012011

Bakhori, S. N. M., Zuikafly, S. N. F., Ahmad, F., & Hassan, M. Z. (2017). Tensile properties for MWCNT filled jute-Epoxy composites. *Journal of Advanced Research in Applied Mechanics*, *31*(1), 16–21.

Bledzki, A. K., Fink, H. P., & Specht, K. (2004). Unidirectional hemp and flax EP- and PP composites: Influence of defined fibre treatments. *Journal of Applied Polymer Science*, *93*(1), 2150–2156. doi:10.1002/app.20712

Chen, H., Yu, Y., Zhong, T., Wu, Y., Li, Y., Wu, Z., & Fei, B. (2017). Effect of alkali treatment on microstructure and mechanical properties of individual bamboo fibers. *Cellulose (London, England)*, *24*(1), 333–347. doi:10.100710570-016-1116-6

Islam, M. S. (2008). *The influence of fibre processing and treatments on hemp fibre/epoxy and hemp fibre/pla composites* (Unpublished doctoral dissertation). The University of Waikato, Hamilton, New Zealand.

Khan, Z., Yousif, B. F., & Islam, M. (2017). Fracture behaviour of bamboo fiber reinforced epoxy composites. *Composites. Part B, Engineering*, *116*, 186–199. doi:10.1016/j.compositesb.2017.02.015

Liu, Z., Li, L., Liu, C., & Xu, A. (2018). Pretreatment of corn straw using the alkaline solution of ionic liquids. *Bioresource Technology*, *260*, 417–420. doi:10.1016/j.biortech.2018.03.117 PMID:29631854

Mishra, S., Mohanty, A., Drzal, L., Misra, M., Parija, S., Nayak, K., & Tripathy, S. (2003). Studies on mechanical performance of biofibre/glass reinforced polyester hybrid composites. *Composites Science and Technology*, *63*(10), 1377–1385. doi:10.1016/S0266-3538(03)00084-8

Montgomery, D. C. (2005). *Design and Analysis of Experiments* (6th ed.). New York: John Wiley & Sons, Inc.

Rao, R., Varada, R., Ramachandra, R., & Hemachandra, K. (2010). Flexural and compressive properties of bamboo and glass fibre-reinforced epoxy hybrid composites. *Journal of Reinforced Plastics and Composites*, *29*(10), 1446–1450. doi:10.1177/0731684409105077

Reddy, B. R., & Dhoria, S. H. (2018). Effect of alkaline treatment on mechanical properties of kenaf fiber reinforced polyester composites. In AIP Conference Proceedings: Vol. 1943. *No.1* (p. 020097). AIP Publishing. doi:10.1063/1.5029673

Shao, Z., Fang, C., & Tian, G. (2009). Mode I interlaminar fracture property of moso bamboo. *Wood Science and Technology*, *43*(5), 527–536. doi:10.100700226-009-0265-2

Singh, V. K., Mukhopadhyay, S., & Das, D. (2017). Recipe optimisation for sugarcane bagasse fibre reinforced soy protein biocomposite. *Indian Journal of Fibre and Textile Research*, *42*(2), 132–137.

Yaghoobi, H., & Fereidoon, A. (2018). Modeling and optimization of tensile strength and modulus of polypropylene/kenaf fiber biocomposites using Box–Behnken response surface method. *Polymer Composites*, *39*, E463–E479. doi:10.1002/pc.24596

Yan, L., Chouw, N., Huang, L., & Kasal, B. (2016). Effect of alkali treatment on microstructure and mechanical properties of coir fibres, coir fibre reinforced-polymer composites and reinforced-cementitious composites. *Construction & Building Materials, 112*, 168–182. doi:10.1016/j.conbuildmat.2016.02.182

Yang, X., Shang, L., Xing'e, L., Shumin, Y., & Genlin, T. (2017). Changes in bamboo fiber subjected to different chemical treatments and freeze-drying as measured by nanoindentation. *Journal of Wood Science, 63*(1), 24–30. doi:10.100710086-016-1590-3

Yu, Y., Jiang, Z., Fei, B., Wang, G., & Wang, H. (2011). An improved microtensile technique for mechanical characterization of short plant fibers: A case study on bamboo fibers. *Journal of Materials Science, 46*(3), 739–746. doi:10.100710853-010-4806-8

Chapter 7
Investigation of Recycled Acrylonitrile Butadiene Styrene for Additive Manufacturing

Shajahan Bin Maidin
Universiti Teknikal Malaysia Melaka, Malaysia

Zulkeflee Abdullah
Universiti Teknikal Malaysia Melaka, Malaysia

Ting Kung Hieng
Universiti Teknikal Malaysia Melaka, Malaysia

ABSTRACT

One of the disadvantages of fused deposition modeling (FDM) is waste produced during the printing processes. This investigation focuses on using 100% recycled Acrylonitrile Butadiene Styrene (ABS) for the FDM process. The recycling begins with re-granule the waste ABS material and produces it into a new filament. The new recycled filament was used to print the test specimen. Investigation on the mechanical properties and the surface quality of the test specimen and comparison with standard ABS specimen was done. The result shows that the recycled ABS can be produced into filament with 335°C of extrusion temperature and 1.5 mm/s travel speed of the extruder conveyor. The surface roughness of recycled specimen is 6.94% higher than the standard ABS specimen. For ultimate tensile strength, there is a small difference in X and Y orientation between the standard and the recycled ABS specimen which are 22.93% and 19.98%, respectively. However, in Z orientation, it is 52.33% lower. This investigation proves that ABS can be recycled without significantly affecting its mechanical properties.

DOI: 10.4018/978-1-7998-1374-3.ch007

BACKGROUND_

Additive manufacturing (AM) is a process that creates a 3D object by using layer by layer approach to deposit material on the printer platform to form the desired 3D shape (Prakash et al., 2018). AM is able to provide the approachability to build, mass customization and has redefine the current production technology (Cam & Yanbing, 2015). The way to deposit the material and the printing process of the material will be manipulated by the information given by the computer-aided design (CAD) file and the process to convert it to stereolithography (STL) file. AM will have an important place in the future of manufacturing as numerous research and investigation need to be carried out to continuously improve the quality of 3D printed object (Wong & Hernandez, 2012). One of the AM systems is the fused deposition modeling (FDM). FDM 3D printing is a process which is used in the rapid prototyping and AM to create prototype and 3D product respectively. In FDM, the material which is in filament shape is melted when entering the heated nozzle and extruded out to stack a multiple layer of plastic on the printing platform. The material will solidify in a desired 3D shape by using the computer control process.

On the other hand, 3D product printed is not always 100% good in quality. Hence, the unqualified product or the fail printed part will be thrown away (Song & Telenko, 2016). The 3D printed object which is created by the Acrylonitrile Butadiene Styrene (ABS) filament will have a bad effect on the environment when it becomes redundant and as a waste from the lab or the industry due to failed print and support or extra material that need to be cut off from the product or prototype. Thus, a way to recycle the wasted ABS 3D printed product is needed to enhance the quality of environment and create an eco-3D printing in the rapid prototyping and AM process.

As plastic have highly beneficial material properties such as lightweight, high strength-to-weight ratio, low cost and easy to manufacture it has caused growth in global production which increases to 500% in the last three decades. The accelerated production and adoption of plastic has caused various extraordinary issue of disposal and recycling of huge amount of plastic. As plastic is produced from oil which is finite fossil-based resources and not biodegradable, so the recycling of plastic need to be investigated and increasingly important to environmental and resources issue. Thus, this paper produces the results on investigation of 100% recycled ABS printing quality and other material properties. Agreeably, plastic sustainable management needs to be solved as it becomes a serious problem in the modern day (Cruz Sanchez et al., 2017).

ABS filament in FDM process is made from the ABS material and is the most largely used and popular thermoplastics. ABS contains three monomers which are acrylonitrile, butadiene and styrene. Acrylonitrile is used as chemical and thermal enhancement, butadiene is used to improve the toughness and impact strength and the Styrene is make the surface of plastic become shiny and glossy. Although ABS is no biodegradable, it is able to be recycled. Hence, this investigation will be carried out for the method to recycle the ABS waste and compare the specimen quality that produced by the standard ABS filament and 100% recycled ABS filament.

AM aims to reduce material waste. However, it still creates a great deal of waste material due to printing error as most of the standard FDM 3D printer is used by the inexperienced user (Song & Telenko, 2016). According to Song & Telenko (2016), the amount of failed print is about 2.22 times from the predicted amount of waste produced. Hence, the first problem will be too much waste is create for the AM process and need to find a way to recycle the waste material. Furthermore, waste ABS material will cause negative effect on the quality of environment due to ABS material is non-biodegradable. In this situation, plastic residue will accumulate in environment and landfills due to a large volume of ABS material become redundant and take millions year to be degraded (Hopewell et al., 2009). Thus, the amount of waste material should be reduced by recycling it. Thus, FDM 3D printing process is still not an environment friendly process, so an eco-printing needs to be achieved and investigated.

As ABS filament for FDM 3D printer is made from oil which is a finite and limited resource and wasting of finite resource should be avoided. Hence, the oil stock resource cannot be wasted as it is keep decreasing and the recycled ABS material can be substitute to the standard ABS filament to decrease the rate of using oil to produce the standard ABS filament. The cost need to purchase the standard ABS filament is getting expensive as the ABS is made from oil. Hence, the cost of material for FDM 3D printing can be reduced by recycling the waste ABS material and take it as a substitute resource for the ABS 3D printing filament.

The literatures has less information on the process of recycling waste for AM process and as well as the mechanical properties particularly the recycled ABS filament for the FDM process. This research investigates the use of 100% recycled of ABS for FDM printing process. The investigation compares the quality and the mechanical properties of 3D object created by standard ABS filament with the recycled ABS material filament. The analysis includes the mechanical properties analysis which is tensile test by using the SHIMADZU AGS-X 20kN. The quality of the 3D printed product is measured by the surface roughness by using the MITUTOYO SJ-301. Besides, the optimum parameters related to the production of FDM filament were investigated which includes the size of recycled ABS pellet, filament diameter and extrusion temperature. The filament extruder used for this study is the Haake Rheomax OS and the FDM printer is Odyssey X2. Furthermore, this project focuses the study of the mechanical properties of specimen printed in X, Y and Z orientation. The parameter of 3D printing process is fixed for the bed temperature (90°C), print speed (60mm/s), layer thickness (0.2mm), raster angle (30°) and nozzle temperature (235°C). The result and data from the tensile test and surface roughness will indicate the feasibility of 100% recycled ABS filament.

LITERATURE REVIEW

Acrylonitrile Butadiene Styrene for the Fused Deposition Modeling 3D Printing Machine

ABS is an amorphous thermoplastic blend and is broadly used polymer in FDM 3D printing machine. ABS polymers are form by three monomers which are acrylonitrile (15-35%), butadiene (5 -30%), and styrene (40-60%) (Vishwakarma et al., 2017). Acrylonitrile enhances the chemical and heat resistance that allows FDM printed product taking the limited to extensive functional testing. Butadiene improves the impact strength and ductility that allow to produce 3D printed parts that are suitable for testing, prototyping and installation (Dandgaval & Bichkar, 2016). Styrene enhances the glossy surface and improves its machinability. Hence, ABS is a material that is easily molded, tough and resilient. Composition of ABS is (CH2-CH-C6H4).

Table 1 shows the mechanical properties for different type of ABS namely ABS, ABSi, ABS plus and ABS/PC. For ABSplus thermoplastic (ABS), it does not have significant shrinkage, warpage or moisture and it is environmentally stable. ABSplus is about 40% stronger than standard ABS material. For ABSi, it is a translucent material, good mechanical and appearance properties. ABSi is available in red, amber and natural color. For ABS/PC, it has better properties than the both PC and ABS material. ABS-PC has the good mechanical properties and resistance of heat of PC and good feature and aesthetic of ABS. It is good in strength of impact. After the ABS is recycled, the strength of impact is slightly reduced. This proves that ABS has good recyclability of mechanical. Besides, impact strength can be enhanced by adding the toughness agents (Hamad et al., 2013).

Recycling of Polymer for Additive Manufacturing

Waste of plastic is dramatically increasing which make it become a huge problem and reason to adopt recycling. Recycling of plastic is a process to collect the plastic waste, reprocess into a useful product. Polymer is widely used in AM. It will generate the polymer waste and released to the environment after the product is produced (Paris et al., 2016).

Table 1. Material for Fused Deposition Modeling, Source from: Ian Gibson et al. (2015)

Property	ABS	ABSi	ABSplus	ABS/PC
Tensile Strength (MPa)	22	37	36	34.8
Tensile modulus (MPa)	1627	1915	2265	1827
Elongation (%)	6	3.1	4	4.3
Flexural Strength (MPa)	41	61	52	50
Flexural Modulus (MPa)	1834	1820	2198	1863
IZOD Impact (J/m^2)	106.78	101.4	96	123
Heat Deflection at 66 psi (°C)	90	87	96	110
Heat Deflection at 264 psi (°C)	76	73	82	96
Thermal Expansion (in./in./F)	5.60E-05	6.7E-6	4.90E-0.5	4.10E-5
Specific Gravity	1.05	1.08	1.04	1.2

Although one of the aim of AM is to reduce waste of material, it still produce waste due to the support structure, human and machine error. Less research has been conducted to reduce the FDM printing waste (Jin et al. 2016). According to Cruz Sanchez et al., (2017), one of the polymer waste solution proposed is the adoption of open-source 3D printers with domestic waste plastic extruders. This approach can recycle wasted polymer to produce the 3D printer feedstock in order to reduce the waste. Hence, many benefits can be achieved if the wasted polymer from the AM process is recycled. However, there are insufficient knowledge and method to recycling process of polymer for the open-source 3D printing. For this reason, a study about the recycled material need to carry out to explore the potential method to recycle the material (Cruz Sanchez et al. 2017).

Process of Polymer Recycling

Polymer can be produced into the new product by reprocess and re-melt it. There are 3 types of method to recycling the polymer which are energy recovery, chemical recycling and mechanical recycling (Grigore, 2017). Method of mechanical recycling is better as chemical recycling as the waste is recycled by complex treatment of chemical (Hamad et al., 2013)

Mechanical Recycling

For mechanical recycling, polymer is not altered during the recycling process (Grigore, 2017). The first step of plastic recycling is to collect and sorting the polymer. Then the plastic is needed to be cleaned and dried to remove any contaminants. The third step is forming the plastic wastes into granulates, pellet or flakes by cutting, shredding or washing. These granule, pallet and flakes must have appropriate quality for manufacturing and needs to be tested, then the granule is ready to redesign into a new product. The granule can be used to make filament by using the filament extruder with suitable machine parameter (Dudek, 2013). Besides, the virgin material can be mixed with the reprocessed material to get better material properties. The quality of filament is based on its accuracy of dimension, mechanical properties and characteristics that are measureable. The controlled parameters of filament extruder are the screw rotating speed, nozzle temperature and the properties of raw material. In this way, amount of plastic waste can be decreased. The disadvantage of mechanical recycling is the solid waste heterogeneity and the of product's properties deterioration in each cycle which occurs due to the low molecular weight of the recycled resin. This happened because of containing the traces acidic impurities and water that causing the chain-scission reaction. Intensive drying is used to avoid lowering molecular weight, the use of reprocessing with vacuum degassing or chain extender compounds or can also solve this problem. Furthermore, mechanical recycling needs initial investment for further development of this method (Hopewell et al., 2009). Figure 1 shows the 6 steps in mechanical recycling.

Figure 1. The steps of mechanical recycling, source from (Grigore, 2017)

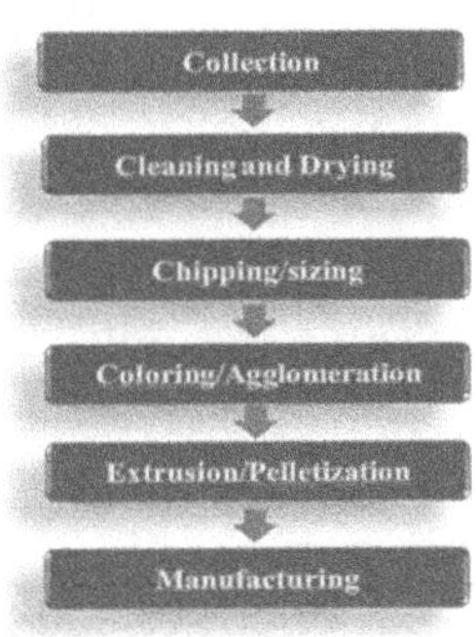

Chemical Recycling

Chemical recycling is one of the recycle methods for polymer. Chemical recycling is a recycling process that uses chemical reaction to partially depolymerize the monomer to oligomers which will change the structure of the polymer. The depolymerized monomer can be polymerized again to produce a new, original polymer product. Feedstock material that in the form of oligomer, monomer or mixture of hydrocarbon compound will suitable for this recycle method as chemical recycling is to change the plastic into smaller molecules (Grigore, 2017).

Energy Recovery

Energy recovery is a method to recycling the plastic by extracting out the energy content inside the plastic. Energy recovery is used to recover the energy content inside the plastic. Energy recovery is an effective method to decrease the amount of organic material that concern with energy recovery that called as incineration. It is a nice solution for generating energy from the polymers. However, energy recovery is not ecologically acceptable because the airborne toxic substance is not good for health such as dioxins (Grigore, 2017). Table 2 shows the comparison between the mechanical recycling, chemical recycling and energy recovery in term of advantages and challenges.

Advantages of Recycling Plastics

By recycling the wasted polymer, the cost for the AM process can be reduced. As from economically view, the cost of standard filament is 20 to 200 times more than the cost of raw plastic which is between \$18.86 and \$175.20 per kg (Pearce, 2014). From energy view, recycling wasted polymer can save the energy used for distributing recycling over a centralized recycling approach which is about 69% to

Table 2. The comparison of different technique to recycle plastic, Source from (Grigore, 2017)

Technique	Advantages	Challenges
Mechanical recycling	Cost-effective, efficiency, well-known	Deterioration of product's properties, pre-treatment
Chemical recycling	Operational for PET, simple technology	Mainly limited to condensation polymers
Energy recovery	Generates considerable energy from polymers	Not ecologically acceptable

82% embodied energy (Baechler et al., 2013). The saved energy can be used for other purpose in the economy. By recycling the plastic, it can provide a recycle material can be used as virgin material to the manufacturing industry. Moreover, as plastic can cause negative effect to the environment such as miss feed by animal, blocking drainage and pollution. Thus, reducing the plastic waste can reduce the diminishing of landfill site and the land can be used for the better purpose. It has been proved that the lifestyle can be improved by participate into the recycle business as it can also increase their daily income and better living environment. An example of organizations that creating filament from plastic waste is Dimension Polymers Inc. (USA).

FDM Printing Process Parameters

The FDM 3D printing process has several parameters that need to be control (Ali & Maharaj, 2014). The parameter needs to be controlled in order to get the optimum performance of the product with least material, better quality, safety, low manufacturing cost, short lead time and such parameter is called as optimum parameter. Optimized parameters of FDM for surface roughness is layer thickness, orientation angle and model temperature. The air gap and road width have least impact on the surface roughness (Singh Bual & Kumar, 2014).

According to Road (2017), there are also some other influence factor such as printing environment, variability of material properties and machine specification. These influence factors also need to be considered to get the best quality of 3D printed product. The strength and dimensional accuracy will be influence by the layer thickness and the printing orientation (X, Y and Z) (Farzadi et al., 2014). The accuracy of dimension and surface roughness will also effected by the orientation X, Y and Z (Rahman et al., 2018). So the orientation need to be controlled as the surface roughness will influence the mechanical coupling and functionality of a product (Boschett et al., 2016). Moreover, the printing parameter will also affect the surface roughness of the 3D printed product. Numerous researchers had investigated the relationship between the printing parameter with the surface roughness as the quality(Nourghassemi, 2011).

METHODOLOGY

The old prints or the wasted ABS material for this project was collected from the mechatronic and rapid prototyping lab at the Universiti Teknikal Malaysia Melaka. These waste ABS which is thrown, some are from the support materials or was redundant. Next, the process of separating the ABS material from the other type of plastic was carried out as sometime the ABS printed product will mix with PLA and TPU materials. Mixing of different plastic material will influence the parameter for extruding the filament later. To ensure that there will not have too much difference during comparison of the standard ABS filament and the recycled filament, the recycled ABS waste was taken from the same batch, same source of supplier and with the same grade.

For the granulation process, the wasted ABS plastics was crushed by using the bench vice and hammer and was cut into smaller pieces by using wire cutter. Then, it was feed into the polymer crusher TW-SC 400F to produce the ABS granules which shown in the Figure 2. This process was repeated 5 times to get the uniform pellet size. Next, 100% recycled ABS granules are used as input material to produce the 100% recycled ABS filament by using the filament extruder Haake Rheomex OS which shown in the Figure 3.

Figure 2. Polymer crusher TW-SC-400F

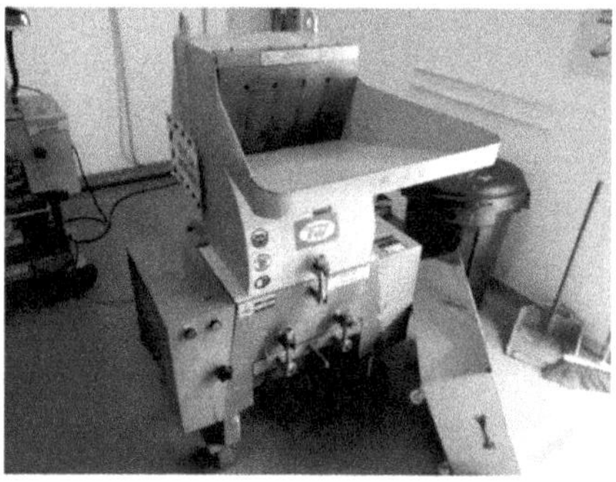

Figure 3. Haake rheomex OS filament extruder

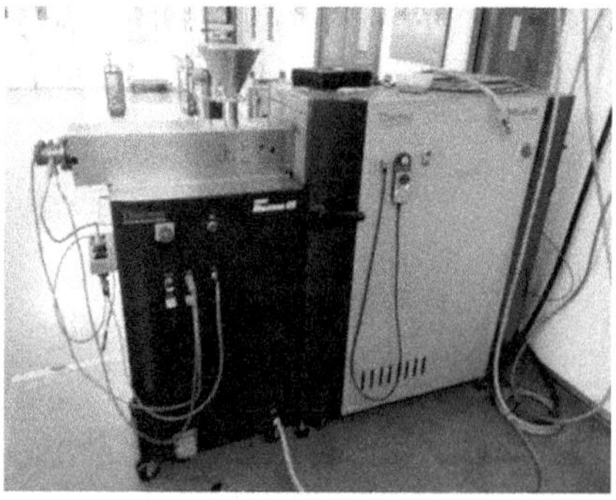

After the plastic granule is softened and melted, it was forced to flow through the heated nozzle which has fixed temperature. After the filament came out from the heated nozzle, it was laid on a conveyor belt that pull the filament to the desired diameter (1.75±0.1mm) as shown in Figure 4. The temperature of heated region will affect the viscosity of the melted ABS granule. The speed of screw rotation and the size of granule will affect the feed rate of granule into the heated region which means that it will also influence the pressure applied to molten plastic for extrusion. By controlling these parameters in best and constant condition, a good quality of filament can be produced that is suitable to be use as feed stock material for the FDM printer. The extrusion speed for filament extruder is fixed at 17.5 rpm and the nozzle temperature and the travel speed of conveyor is adjusted to obtain approximately 1.75mm diameter of the filament.

The FDM 3D printer (Odyssey X2) was used to print the testing specimen. To optimize the 3D printed product, some printing parameters of the 3D printer need to be controlled such as the extrusion temperature, feed rate and bed temperature. For this study, standard parameters of 3D printing were used for the standard ABS filaments is shown in Table 3. For the 100% recycled ABS filament, the bed temperature will remain the same unless the problem of adhesion between layers is occurred. If problem occur, the parameter was adjusted until optimum 3D printed part is produced with not gap and even layer adhesion. The specimen was printed in 3 different type of orientation in X, Y and Z which shown in the Figure 5. In this way the mechanical properties with different orientation can be tested and the comparison to investigate the best orientation for both standard ABS filament printed specimen and 100% recycled ABS printed specimen can be done.

To prove that the recycle ABS material can substitute the standard ABS material as another source for the FDM filament, the mechanical properties including the UTS, strain and modulus of elasticity need to be determined and compared. To test the mechanical properties, tensile test was used to determine the tensile strength

Figure 4. Filament pulled by conveyor

Table 3. The parameter of FDM 3D printing parameter (Vishwas & Basavaraj, 2017)

Type of Parameter	Level of Parameter
Nozzle temperature	Fixed 235?
Bed temperature	Fixed 105?
Printing speed	Fixed (60mm/s)
Orientation	3 Level (X,Y,Z)
Raster angle	Fixed 30°
Layer thickness	Fixed 0.2mm

Figure 5. Specimen in X, Y and Z orientation

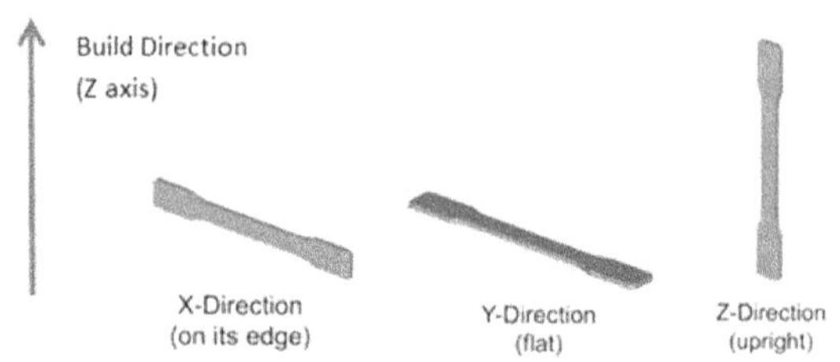

Figure 6. The shape and size of specimen

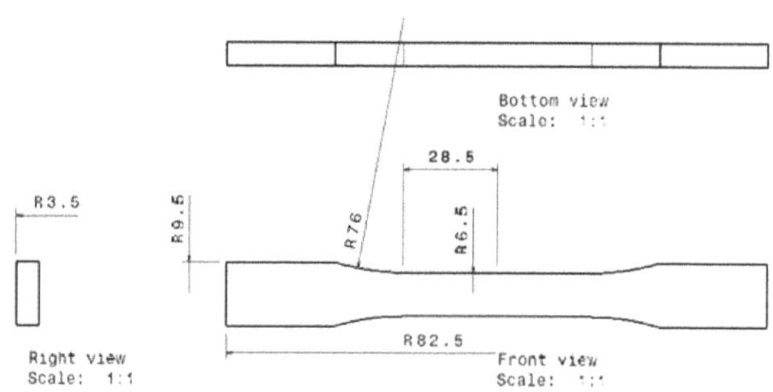

of the material. Specimens that printed for tensile tests follows the ASTM 638-02a type I standards for plastics which is shown in Figure 6. To get the better result of tensile test, type I standard of specimen which provide better result for the tensile test for rigid and semi rigid polymer was used. Hence, the dimension of specimen should follow the dimension state in the standard for specimen type 1. Each type of combination of parameter of specimen was tested 3 times for the tensile test and the average result was taken to ensure the accuracy.

Tensile test was performed using SHIMADZU AGS-X 20kN +500 tensile test system that is shown in Figure 7, where 3 repeat measurements were conducted. Thus, a total of 18 specimens were printed using the FDM 3D printing machine. The testing process has followed the standard testing procedure to ensure the result is correct. Besides the tensile test, different type of models was printed to determine the printability of the 100% recycled ABS filament such as passive and active model. Surface roughness testing was carried out to examine the quality of the 3D printed specimen. The surface roughness was measured by using the Mitutoyo SJ-301 which is shown in Figure 8. The surface roughness was carried out on the specimen after the tensile test. In this way, the surface conditions of specimen that was printed by the standard ABS filament and recycled ABS filament are able to be compared. During the surface roughness testing, the part was located on an even and clean surface and the surface of the specimen is also has to be clean to get an accurate result.

Figure 7. SHIMAZU tensile test machine

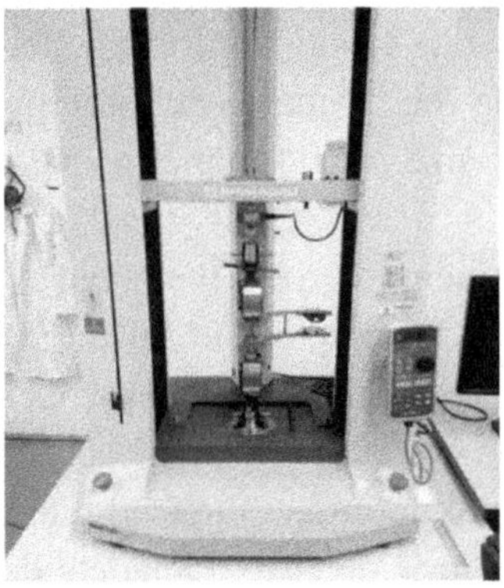

Figure 8. Mitutoyo SJ-301

RESULTS AND DISCUSSION

Mechanical Recycling Process

Referring to the process of producing the standard filament, the filament extrusion process used the uniform pelletised material in order to get a constant melt flow rate during the filament extrusion process. By using the mechanical recycling method, large ABS printed part was broken into smaller size by using hammer, wire cutter and bench vice. After that, the polymer crusher is used to crush the recycled material into a smaller and uniform pellet. The size of pellet can be even more uniform if it is repeatedly crushed by the polymer crusher for 5 times. The length, thickness and width of the pellet were controlled to be as same as the size of standard ABS pellet. The result of average size of pellet was taken randomly by picking 20 samples from the pallet that produced by using the mechanical recycling method and measuring each of them to get the length, width and the thickness of the pallet. Besides, the 100% recycled ABS pellet will also go through a sieve to filter some large pellet. In this way, the uniform pallet size of 100% recycled pellet can be produced. As one of the variables of filament extrusion is uniform pellet size, so this will not affect the feed rate of pellet during the filament extrusion process. The size of standard ABS pellet and the result of size of 20 samples of pellet are shown in Table 4 and Table 5 respectively.

With more advanced filament extruder, the range of the pellet size can be larger without effecting the uniform feedrate of the pellet. By using the polymer crusher, the consistency of the 100% recycled ABS pallet size can be improved. Hence, this research will still control the size of pallet to the acceptable size so that the size of pallet can be uniform and will not affect the feed rate of pallet and melt flow rate during the filament extrusion. If the size of pallets does not follow the range of acceptable size, potential thermal degradation of pallet will occur which will affect the result of mechanical test. Thus, the result obtain from the 20 samples proved that the mechanical recycling method is effective in creating pallet with acceptable size for filament extrusion.

Table 4. Dimension of standard ABS pellet

Size of Standard ABS Pellet	Length (mm)	Width (mm)	Thickness (mm)
	3.1	3.1	2.2

Table 5. Result of dimension of 20 sample of 100% recycled ABS pellet

No	Sample Size			No	Sample Size		
	Length (mm)	**Width (mm)**	**Thickness (mm)**		**Length (mm)**	**Width (mm)**	**Thickness (mm)**
1	3.3	3.0	2.4	11	3.1	3.1	2.1
2	3.5	3.2	2.6	12	3.1	3.2	2.3
3	3.3	3.1	2.1	13	3.3	3.2	2.1
4	2.9	2.7	2.3	14	2.8	3.0	2.4
5	3.1	2.9	2.3	15	3.0	2.7	2.2
6	2.8	3.1	2.4	16	3.1	2.9	2.3
7	3.3	3.5	2.2	17	3.3	3.1	2.1
8	3.6	3.1	1.9	18	2.9	3.1	2.2
9	3.4	3.1	2.4	19	3.1	3.2	2.3
10	3.3	2.9	2.3	20	3.4	3.2	2.1
Average dimension of 20 samples					3.18	3.07	2.25

Filament Extrusion Process

Surface Quality and Cross-sectional Shape of Filament

After the pallet of 100% recycled ABS material was poured into the hopper, the parameter setting of the filament extruder need to be controlled. The extrusion speed was fixed at 17 rpm, and the temperature of heating barrel was set at 4 different temperatures which are 185°C, 205°C, 225°C and 235°C. These temperature values are within the range of heating temperature of the ABS (180°C-240°C). The trial-and-error method was used to find the optimum parameter for the filament extrusion. The surface quality and the shape of the filament were determined by visual inspection initially. By using the correct parameter in the filament extruder, a filament with round shape from cross-sectional area can be produced. On the other hand, if extrusion temperature is too low, the produced filament will have defect like oval shape in cross-sectional area and containing sign of coarse material in the filament. The temperature of extrusion is one of the important variables. The extrusion temperature was controlled above the melting temperature of ABS. If the extrusion temperature of ABS is too low, the surface of the filament will be rough and will not be in round shape. If the extrusion temperature is within the acceptable range, a smooth surface and round filament can be produced.

Figure 9, 10, 11 and 12 shows the filament that was produced with different temperature. Figure 9 shows the filament produced with 185°C has melt fracture defect which means that the flow rate is too fast causing the tensile stress exceed the tensile strength of the material. The cross-sectional area is in oval shape and it is brittle. Figure 10 shows the filament that produce with 195°C which has a stick-slip defect. The condition of surface is getting better as the extrusion temperature is increased 10°C. The filament is still brittle. Remedy of this defect is to increase the extrusion temperature to get a better surface. Figure 11 shows the filament that produced with 215°C. The surface of the filament is still not smooth but better than the filament that produced with 195°C. The filament is not in round shape in cross-sectional area. Figure 12 shows that the filament produced with 235°C. The surface of the filament is smooth and the shape of cross-sectional area is round in shape. It also has better toughness as it will not easily break which is most suitable one between these temperatures. The summary of these results of 4 different temperatures is shown in Table 6.

Figure 9. Filament produced with 185°C

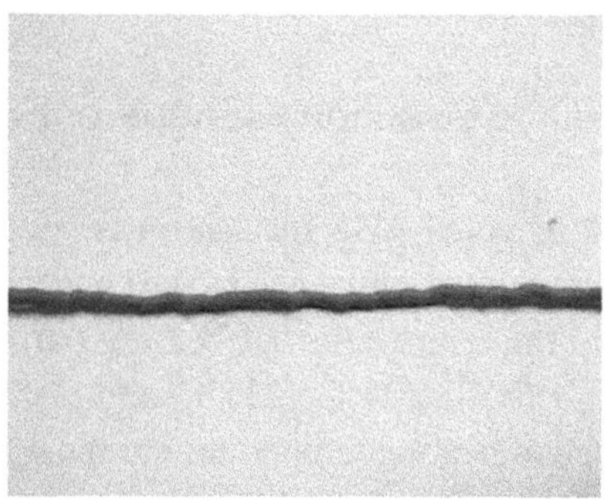

Figure 10. Filament produced with 195°C

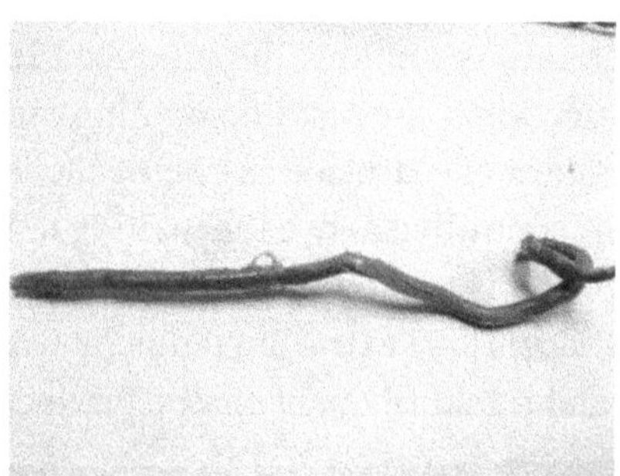

Figure 11. Filament produced with 215°C

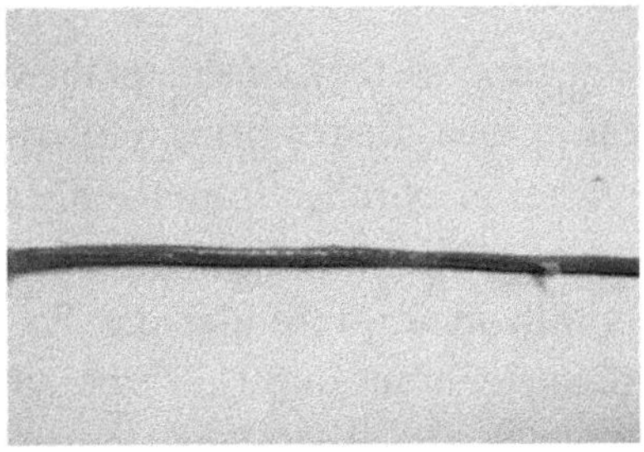

Figure 12. Filament produced with 235°C

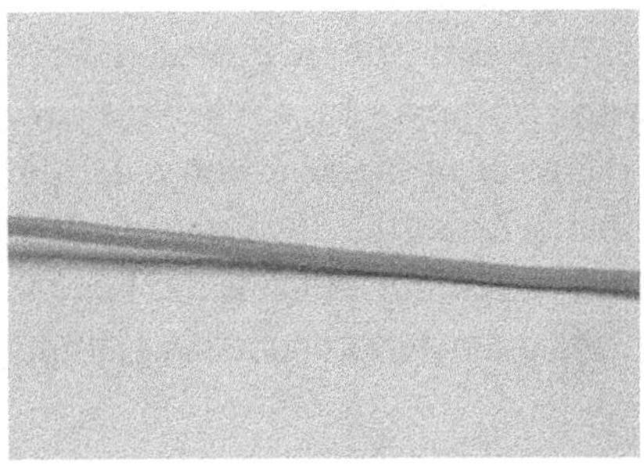

Table 6. Condition of filament produced with different temperature of filament extrusion

Temperature of Filament Extrusion Process (?)	Condition of the Filament
185	Rough, oval shape in cross-sectional area and brittle (melt fracture occurred)
205	Rough, oval shape in cross-sectional area and less brittle (stick-slip)
215-225	Less rough, oval shape in cross-sectional area and less brittle
235	Smooth, round shape in cross-sectional area and tough

Melt fracture occurred due to the tensile stress exceeds the tensile strength of melt which will cause irregular shape produced. According to Noroozi et al. (2012), melt fracture can be avoided when the heating temperature of extrusion is more than the melting temperature of ABS. By increase the heating temperature, it can increase the melt flow rate of the extruded material by reducing its viscosity. Hence, from the result obtained from these filament extrusions with different temperature, it was noticed that the 235°C is the most suitable extrusion temperature as it produces filament in good quality which has round shape in cross-sectional area, smooth surface and not easy to be broken.

Diameter of Extruded Filament

During the filament extrusion process, the diameter of the filament was determined. The diameter of filament can be measured by using the vernier caliper. This variable was controlled during the material extruded from the die and laid on the conveyor with certain travel speed which is shown in Figure 4. The faster the travel speed of the conveyor, the narrower the diameter of the filament. The diameter of the filament was not controlled by the diameter of the die. This is because the filament extruded will not follow the diameter of die as it will swell into large diameter. Before reaching the steady state of the filament extrusion, there was a measurable difference in the diameter of filament. Hence, the result are taken after it reached the steady state. The Table 7 shows the different diameter in different travel speed of conveyor.

Figure 13 shows that the speed of conveyor at 1.5mm/s is most suitable for producing the filament with 1.75mm (±1.0). When the conveyor speed is set as 0.5 mm/s, it shows that the diameter of the filament is 2.1mm which is too large for the FDM Odyssey X2 printer. Hence, the travel speed of conveyor was increased to 1.0 mm/s but the diameter is 1.915mm which is too large for FDM printer. For the travel speed pf 2.0 mm/s, the diameter produced is 1.56mm which is too small for

Table 7. Result of diameter of filament produced with different travel speed of conveyor

Travel Speed of Conveyer (cm/s)	Diameter of Filament (mm)
0.5	2.0-2.2
1.0	1.83-2.0
1.5	1.62-1.79
2.0	1.52-1.60
2.5	1.45-1.56

Figure 13. Graph of Diameter of filament (mm) against the speed of conveyor

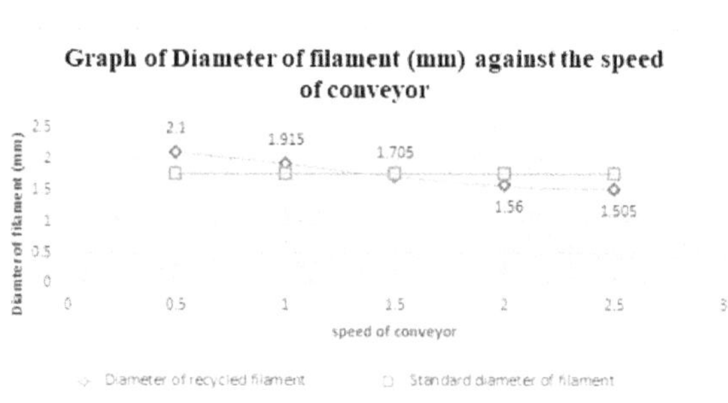

the FDM printer. Lastly, the 2.5 mm/s travel speed produce 1.505mm diameter of filament. Diameter of filament will affect the printing process of FDM printer. If the diameter of the filament is too large, it will inside the extrusion mechanism. When the filament is trapped, the extrusion mechanisms need to be separated to remove the trapped material. On the other hand, if the diameter is too small, the extrusion mechanism cannot pull the filament into the extrusion mechanism which will stop the melted material from coming out from the extrusion hole. After the 100% ABS recycled filament is produced, it will be used for the FDM 3D printer to print the test specimen for surface roughness testing and tensile test.

From the result that was taken for the surface roughness tester, it is noticed that there is no much differences between the standard ABS printed specimen and the recycled ABS printed specimen. The average surface roughness along the length of the standard ABS printed specimen is about 6.526µm and the average surface roughness along the length of recycled ABS printed object is 6.979µm which is shown in Figure 14. The value of surface roughness of recycled filament is higher than the standard filament in 0.453µm which is 6.94% higher than the standard ABS printed object.

For the result of strain failure that is shown in Figure 15, there is only a small difference in X and Y orientation between the standard ABS and the recycled ABS samples. On the other hand, in Z orientation there is a large difference. The ultimate tensile strength (UTS) of recycled ABS is 52.33% lower than the standard ABS. It has lowest difference in tensile test between the standard printed specimen and the 100% recycled ABS printed specimen which is 19.98% lower than the standard ABS specimen. Specimen in Y orientation has the best result of UTS. 100% recycled ABS printed specimen in X orientation is 22.93% lower than the standard ABS specimen in term of tensile strength. Specimen printed in Y orientation has the highest UTS, followed by the specimen printed in X orientation and Z orientation specimen has the lowest UTS.

Figure 14. Histogram of comparison of Surface roughness (µm) between standard filament printed object and recycled filament printed object

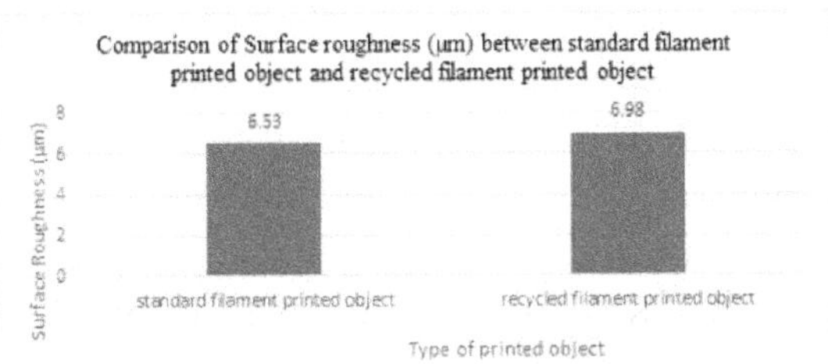

Figure 15. Histogram of Comparison of Ultimate Tensile Strength (MPa) between the standard ABS printed Specimen and 100% recycled ABS printed specimen

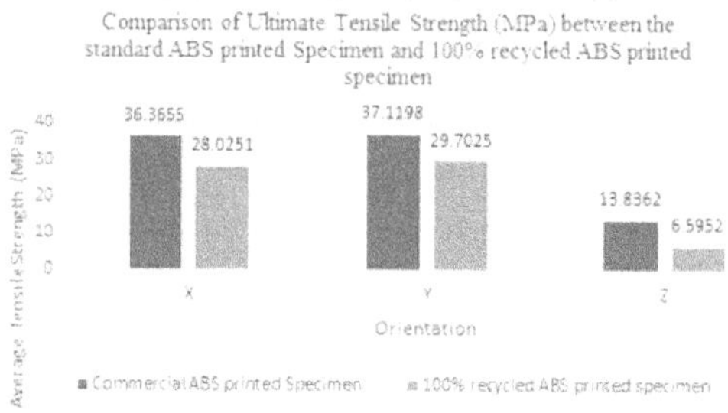

According to Carneiro et al. (2015), the load and the pulling force of the tensile test machine effects on the bonding of fibre instead of the bonding between the deposited layers when the specimen built in X and Y orientation. Thus, the build direction of the specimen is parallel to the pulling force (X and Y orientation) that needs more load or pulling force to pull to specimen until fracture. The weak material fusion between adjacent layers is due to the weak inter-layer bonding. Hence, the fracture path of specimen built in Z orientation will occur and progress at weak inter-layer bonding. Hence, the Z orientation specimen has lower UTS as the maximum strength of the inter-layer bonding is lower than the maximum strength of the material itself (Górski et al., 2015).

There is a small difference in terms of strain value between the standard ABS printed specimen and the 100% recycled ABS printed specimen which shown in the Figure 16. For the specimen printed in X orientation, the strain of recycled ABS specimen is 33.66% lower than the standard ABS specimen. For the specimen printed in Y orientation, the strain of recycled ABS specimen is 14.17% lower than standard ABS specimen. On the other hand, the specimens print in Z orientation have a large different in term of strain value. The strain of recycled ABS specimen that print in Z orientation is 59.76% lower than the standard ABS specimen.

The 100% recycled ABS printed specimen has shown a good result in term of modulus of elasticity which is shown in Figure 17. The recycled ABS only decrease 13.42%, 10.58% and 12.96% of modulus of elasticity in X, Y and Z orientation respectively. The lowest amount of decreasing in modulus of elasticity is in Y orientation which is 10.58%. The highest amount of decreasing in modulus elasticity is in X orientation which is 13.42%. The degradation of modulus of elasticity is still acceptable. Hence, the 100% recycled ABS filament can be accepted with respect to the modulus of elasticity as the result do not has a big different with the standard ABS printed specimen. For the modulus of elasticity, it shows the same result of the

Figure 16. Histogram of comparison of strain (%) between the standard ABS printed specimen and 100% recycled ABS printed specimen

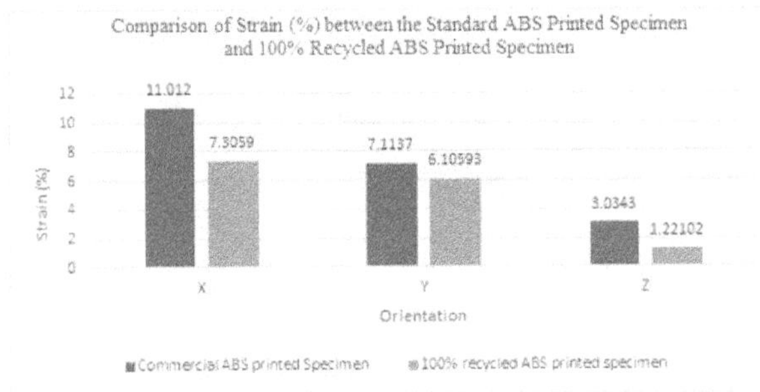

tensile strength that the Y orientation has the highest value which is 1032.997 MPa and 923.71 MPa for standard ABS printed specimen and 100% recycled ABS printed specimen respectively. For the X orientation, the tested specimens have second high of modulus of elasticity for both standard ABS printed specimen and 100% recycled ABS printed specimen which are 1032.997MPa and 923.71MPa respectively. For Z orientation, the tested specimens have the lowest value of modulus of elasticity for both standard ABS printed specimen and 100% recycled ABS printed specimen which are 793.881MPa and 690.961MPa. According to Gordon et al. (2016), the deformation of Z orientation specimen is delamination instead of deformation of fibre. Hence, it has lowest modulus of elasticity among these 3 orientations.

Mechanical properties need to be compared between the standard ABS filament and the 100% recycled ABS filament. Stress strain graph is needed to analyse the mechanical properties of both different type of material. The mechanical properties including the ultimate tensile strength, strain, modulus of elasticity and others.

Figure 17. Histogram of comparison of modulus of elasticity (MPa) between standard ABS printed specimen and 100% recycled ABS printed specimen

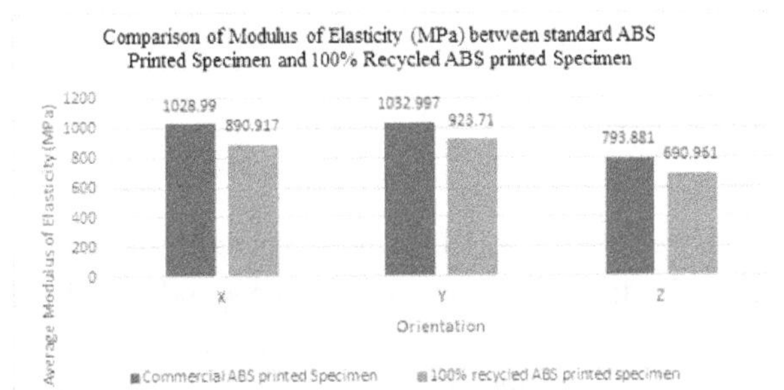

Figure 18. Stress-strain graph of 100% recycled ABS printed specimen in Y orientation

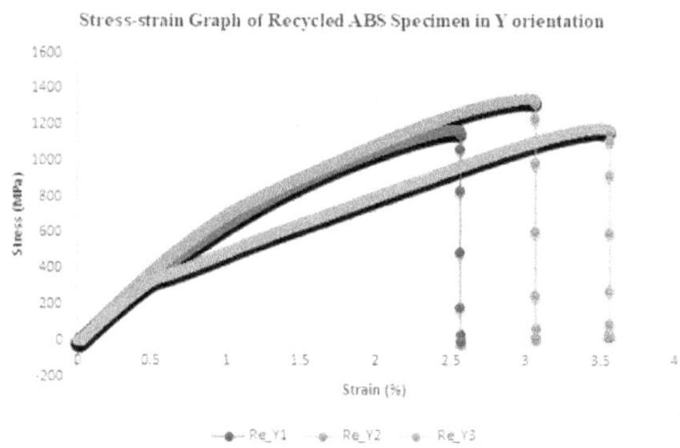

Figure 19. Stress-strain graph of standard ABS printed specimen in Y orientation

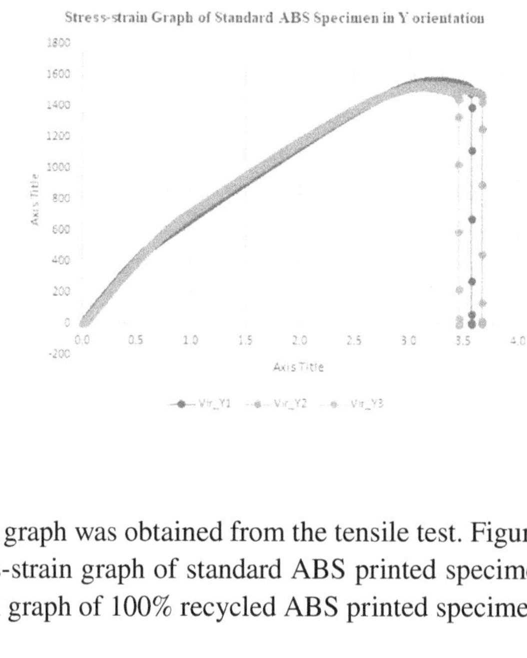

The stress strain graph was obtained from the tensile test. Figure 18 and Figure 19 shows the Stress-strain graph of standard ABS printed specimen in Y orientation and Stress-strain graph of 100% recycled ABS printed specimen in Y orientation.

CONCLUSION

For the granulation process of the ABS waste, the size of the pellets was controlled to 3.1x3.1x2.2mm that is similar with the size of the standard ABS pellet. By using the mechanical recycling method, the ABS waste pellet that was produced is 3.89% smaller than the pellet of standard ABS and has a good uniformity for the ABS waste pellets. The result obtained from the filament extrusion was the quality and the diameter of the filament. The best extrusion temperature obtained from the experiment was 235°C which can produce filament with smooth surface, round cross-sectional shape and without the sign of coarse material. The most suitable travel speed of conveyor is 1.5mm/s, which can produce the filament in the range of 1.62-1.79mm which is within the acceptable range of 1.75±0.1mm. For the surface roughness testing, the average result of standard filament printed specimen and the recycled filament printed specimen does not have large differences. The value of surface roughness of recycled object is 6.94% higher than the standard ABS object.

The result of tensile test shows that the specimen printed in Y orientation has the highest UTS, followed by the specimen printed in X orientation and Z orientation specimen has the lowest UTS. 100% recycled ABS specimen printed in Y orientation decrease in 19.98% if compare to standard ABS specimen. Besides, 100% recycled ABS printed specimen in X orientation is 22.93% lower than the standard ABS specimen in term of UTS. On the other hand, in Z orientation there has a large difference that the UTS of recycled ABS is 52.33% lower than the standard ABS.

For the strain, the 100% recycled ABS specimen printed in X orientation is 33.66% lower than the standard ABS specimen. For the specimen printed in Y orientation, the strain of recycled ABS specimen is 14.17% lower than standard ABS specimen. On the other hand, the strain of recycled ABS specimen that print in Z orientation is 59.76% lower than the standard ABS specimen.

For the modulus of elasticity, standard ABS printed specimen and the 100% recycled ABS printed specimen with Y orientation has the highest modulus of elasticity, then followed by the X orientation and the lowest one is Z orientation. After the ABS was recycled, the modulus of elasticity of recycled specimen in Y orientation is 10.58% lower than the modulus of elasticity of standard ABS printed specimen. Besides, the modulus of elasticity in X orientation is 13.42% lower than the standard ABS printed specimen. Lastly, the modulus of elasticity in Z orientation is 12.96% lower than standard ABS printed specimen.

Overall, the result of these experiments proved that there is some degradation of mechanical properties after the ABS waste was recycled. The degradation of mechanical properties was different when the specimen printed in different orientation. However, the percentage of degradation is within the range of 14.17% to 59.76% which means that the 100% recycled ABS filament can be use as another source for the FDM printer and could reduce waste produced.

ACKNOWLEDGMENT

We are grateful to all those who have assisted direct or indirectly to complete this study and appreciate financial support by Universiti Teknikal Malaysia Melaka.

REFERENCES

Ali, F., & Maharaj, J. (2014). Influence of some process parameters on build time, material consumption and surface roughness of FDM processed parts: Inferences based on the Taguchi design of experiments. *Proceedings of The 2014 IAJC/ISAM Joint International Conference*.

Baechler, C., DeVuono, M., & Pearce, J. M. (2013). Distributed recycling of waste polymer into RepRap feedstock. *Rapid Prototyping Journal*, *19*(2), 118–125. doi:10.1108/13552541311302978

Boschetto, A., Bottini, L., & Veniali, F. (2016). Integration of FDM surface quality modeling with process design. *Additive Manufacturing*, *12*, 334–344. doi:10.1016/j.addma.2016.05.008

Cam, C. A. D., & Yanbing, W. (2015). Application of Additive Manufacturing on Marine Diesel Industry. *Proceedings of PRADS2016*.

Carneiro, O. S., Silva, A. F., & Gomes, R. (2015). Materials & Design Fused deposition modeling with polypropylene. *Materials & Design*, *83*, 768–776. doi:10.1016/j.matdes.2015.06.053

Cruz Sanchez, F. A., Boudaoud, H., Hoppe, S., & Camargo, M. (2017). Polymer recycling in an open-source additive manufacturing context: Mechanical issues. *Additive Manufacturing*, *17*, 87–105. doi:10.1016/j.addma.2017.05.013

Dandgaval, O., & Bichkar, P. (2016). Rapid Prototyping Technology -Study Of Fused Deposition Modelling Technique. *International Journal of Mechanical And Production Engineering*, *4*(44), 2320–2092.

Dudek, P. (2013). FDM 3D printing technology in manufacturing composite elements. *Archives of Metallurgy and Materials*, *58*(4), 1415–1418. doi:10.2478/amm-2013-0186

Farzadi, A., Solati-Hashjin, M., Asadi-Eydivand, M., & Osman, N. A. A. (2014). Effect of layer thickness and printing orientation on mechanical properties and dimensional accuracy of 3D printed porous samples for bone tissue engineering. *PLoS One*, *9*(9), 1–14. doi:10.1371/journal.pone.0108252 PMID:25233468

Gordon, A. P., Torres, J., Cole, M., Owji, A., & Demastry, Z. (2016). An approach for mechanical property optimization of fused deposition modeling with polylactic acid via design of experiments. *Rapid Prototyping Journal*, *22*(2).

Górski, F., Wichniarek, R., Kuczko, W., Zawadzki, P., & Buń, P. (2015). *Strength Of Abs Parts Produced By Fused Deposition Modelling Technology – A Critical Orientation Problem*. Academic Press.

Grigore, M. (2017). Methods of Recycling, Properties and Applications of Recycled Thermoplastic Polymers. *Recycling*, *2*(4), 24. doi:10.3390/recycling2040024

Hamad, K., Kaseem, M., & Deri, F. (2013). Recycling of waste from polymer materials: An overview of the recent works. *Polymer Degradation & Stability*, *98*(12), 2801–2812. doi:10.1016/j.polymdegradstab.2013.09.025

Hopewell, J., Dvorak, R., & Kosior, E. (2009). Plastics recycling: Challenges and opportunities. *Philosophical Transactions of the Royal Society of London. Series B, Biological Sciences*, *364*(1526), 2115–2126. doi:10.1098/rstb.2008.0311 PMID:19528059

Mohamed, O. A., Masood, S. H., & Bhowmik, J. L. (2017). Process parameter optimization of viscoelastic properties of FDM manufactured parts using response surface methodology. *Materials Today: Proceedings*, *4*(8), 8250–8259.

Nourghassemi, B. (2011). *Surface Roughness Estimation for FDM Systems* (Thesis). Ryerson University.

Paris, H., Mokhtarian, H., Coatanéa, E., Museau, M., & Ituarte, I. F. (2016). Comparative environmental impacts of additive and subtractive manufacturing technologies. *CIRP Annals - Manufacturing Technology, 65*(1), 29–32.

Pearce, J. M. (2014). Life Cycle Analysis of Distributed Recycling of Post-consumer High Density Polyethylene for 3-D Printing Filament. *Journal of Cleaner Production, 2014*. doi:10.1016/j.jclepro.2014.02.009

Prakash, K. S., Nancharaih, T., & Rao, V. V. S. (2018). Additive Manufacturing Techniques in Manufacturing -An Overview. *Materials Today: Proceedings*, *5*(2), 3873–3882.

Rahman, H., John, T. D., Sivadasan, M., & Singh, N. K. (2018). Investigation on the Scale Factor applicable to ABS based FDM Additive Manufacturing. *Materials Today: Proceedings*, *5*(1), 1640–1648.

Singh Bual, G., & Kumar, P. (2014). Methods to Improve Surface Finish of Parts Produced by Fused Deposition Modeling. *Manufacturing Science and Technology*, *2*(3), 51–55.

Song, R., & Telenko, C. (2016). Material Waste of Commercial FDM Printers under Realstic Conditions. *Proceedings of the 27th Annual International Solid Freeform Fabrication Symposium*, 1217–1229.

Vishwakarma, S. K., Pandey, P., & Gupta, N. K. (2017). Characterization of ABS Material. *RE:view*, *3*(5), 13–16.

Vishwas, M., & Basavaraj, C. K. (2017). Studies on Optimizing Process Parameters of Fused Deposition Modelling Technology for ABS. *Materials Today: Proceedings*, *4*(10), 10994–11003.

Wong, K. V., & Hernandez, A. (2012). A Review of Additive Manufacturing. *ISRN Mechanical Engineering*, *2012*, 1–10. doi:10.5402/2012/208760

Related Readings

To continue IGI Global's long-standing tradition of advancing innovation through emerging research, please find below a compiled list of recommended IGI Global book chapters and journal articles in the areas of green cities, environmental management, and sustainable urban development. These related readings will provide additional information and guidance to further enrich your knowledge and assist you with your own research.

Ahuja, K., & Khosla, A. (2019). A Framework to Develop a Zero-Carbon Emission Sustainable Cognitive City. In K. Ahuja & A. Khosla (Eds.), *Driving the Development, Management, and Sustainability of Cognitive Cities* (pp. 1–26). Hershey, PA: IGI Global. doi:10.4018/978-1-5225-8085-0.ch001

Akkucuk, U. (2016). SCOR Model and the Green Supply Chain. In U. Akkucuk (Ed.), *Handbook of Research on Waste Management Techniques for Sustainability* (pp. 108–124). Hershey, PA: IGI Global. doi:10.4018/978-1-4666-9723-2.ch006

Albuquerque, C. M. (2017). Cities Really Smart and Inclusive: Possibilities and Limits for Social Inclusion and Participation. In L. Carvalho (Ed.), *Handbook of Research on Entrepreneurial Development and Innovation Within Smart Cities* (pp. 229–247). Hershey, PA: IGI Global. doi:10.4018/978-1-5225-1978-2.ch011

Angelucci, F., Di Girolamo, C., & Zazzero, E. (2018). New Designing Codes for Urban Infrastructures: A Hypothesis of a Transdisciplinary Approach. In G. Carlone, N. Martinelli, & F. Rotondo (Eds.), *Designing Grid Cities for Optimized Urban Development and Planning* (pp. 209–237). Hershey, PA: IGI Global. doi:10.4018/978-1-5225-3613-0.ch012

Avdimiotis, S., & Tilikidou, I. (2017). Smart Tourism Development: The Case of Halkidiki. In L. Carvalho (Ed.), *Handbook of Research on Entrepreneurial Development and Innovation Within Smart Cities* (pp. 491–513). Hershey, PA: IGI Global. doi:10.4018/978-1-5225-1978-2.ch021

Ayvaz, B., & Görener, A. (2016). Reverse Logistics in the Electronics Waste Industry. In U. Akkucuk (Ed.), *Handbook of Research on Waste Management Techniques for Sustainability* (pp. 155–171). Hershey, PA: IGI Global. doi:10.4018/978-1-4666-9723-2.ch008

Bagnato, V. P. (2018). Technology and Urban Structure: The Grid City Between Technological Innovation and New Public Space System. In G. Carlone, N. Martinelli, & F. Rotondo (Eds.), *Designing Grid Cities for Optimized Urban Development and Planning* (pp. 238–253). Hershey, PA: IGI Global. doi:10.4018/978-1-5225-3613-0.ch013

Bala, I., & Singh, G. (2019). Green Communication for Cognitive Cities. In K. Ahuja & A. Khosla (Eds.), *Driving the Development, Management, and Sustainability of Cognitive Cities* (pp. 87–110). Hershey, PA: IGI Global. doi:10.4018/978-1-5225-8085-0.ch004

Bernardino, S., & Santos, J. F. (2017). Building Smarter Cities through Social Entrepreneurship. In L. Carvalho (Ed.), *Handbook of Research on Entrepreneurial Development and Innovation Within Smart Cities* (pp. 327–362). Hershey, PA: IGI Global. doi:10.4018/978-1-5225-1978-2.ch015

Bernardo, M. D. (2017). Smart City Governance: From E-Government to Smart Governance. In L. Carvalho (Ed.), *Handbook of Research on Entrepreneurial Development and Innovation Within Smart Cities* (pp. 290–326). Hershey, PA: IGI Global. doi:10.4018/978-1-5225-1978-2.ch014

Bhatia, P., & Singh, P. (2019). Technological and Gamified Solutions for Pollution Control in Cognitive Cities. In K. Ahuja & A. Khosla (Eds.), *Driving the Development, Management, and Sustainability of Cognitive Cities* (pp. 234–249). Hershey, PA: IGI Global. doi:10.4018/978-1-5225-8085-0.ch010

Bhatt, R. (2017). Zero Tillage for Mitigating Global Warming Consequences and Improving Livelihoods in South Asia. In W. Ganpat & W. Isaac (Eds.), *Environmental Sustainability and Climate Change Adaptation Strategies* (pp. 126–161). Hershey, PA: IGI Global. doi:10.4018/978-1-5225-1607-1.ch005

Bolívar, M. P. (2018). Governance in Smart Cities: A Comparison of Practitioners' Perceptions and Prior Research. *International Journal of E-Planning Research*, *7*(2), 1–19. doi:10.4018/IJEPR.2018040101

Bostanci, S. H., & Albayrak, A. N. (2017). The Role of Eco-Municipalities in Climate Change for a Sustainable Future. In W. Ganpat & W. Isaac (Eds.), *Environmental Sustainability and Climate Change Adaptation Strategies* (pp. 213–231). Hershey, PA: IGI Global. doi:10.4018/978-1-5225-1607-1.ch008

Brkljačić, T., Majetić, F., & Tarabić, B. N. (2017). Smart Environment: Cyber Parks (Connecting Nature and Technology). In L. Carvalho (Ed.), *Handbook of Research on Entrepreneurial Development and Innovation Within Smart Cities* (pp. 150–172). Hershey, PA: IGI Global. doi:10.4018/978-1-5225-1978-2.ch008

Cai, Y. Z., Wu, F., Li, J., Wang, J., & Huang, M. (2019). Application of UAV Technology to Planning Study on Chinese Villages in Guanzhong. In J. Vargas-Hernández & J. Zdunek-Wielgołaska (Eds.), *Bioeconomical Solutions and Investments in Sustainable City Development* (pp. 180–195). Hershey, PA: IGI Global. doi:10.4018/978-1-5225-7958-8.ch008

Calisto, M. D., & Gonçalves, A. (2017). Smart Citizens, Wise Decisions: Sustainability-Driven Tourism Entrepreneurs. In L. Carvalho (Ed.), *Handbook of Research on Entrepreneurial Development and Innovation Within Smart Cities* (pp. 20–43). Hershey, PA: IGI Global. doi:10.4018/978-1-5225-1978-2.ch002

Capelo, C. (2017). Exploring the Dynamics of an Energy Service Venture. In L. Carvalho (Ed.), *Handbook of Research on Entrepreneurial Development and Innovation Within Smart Cities* (pp. 269–289). Hershey, PA: IGI Global. doi:10.4018/978-1-5225-1978-2.ch013

Caridi, G. (2018). Calabria 1783: The Orthogonal Grid as a Physical and Ideological Device of Reconstruction. In G. Carlone, N. Martinelli, & F. Rotondo (Eds.), *Designing Grid Cities for Optimized Urban Development and Planning* (pp. 176–187). Hershey, PA: IGI Global. doi:10.4018/978-1-5225-3613-0.ch010

Carlone, G. (2018). Cities and Extension Plans in the Kingdom of the Two Sicilies: Borgo Murattiano of Bari (1812-1859). In G. Carlone, N. Martinelli, & F. Rotondo (Eds.), *Designing Grid Cities for Optimized Urban Development and Planning* (pp. 1–18). Hershey, PA: IGI Global. doi:10.4018/978-1-5225-3613-0.ch001

Carvalho, L. C. (2017). Entrepreneurial Ecosystems: Lisbon as a Smart Start-Up City. In L. Carvalho (Ed.), *Handbook of Research on Entrepreneurial Development and Innovation Within Smart Cities* (pp. 1–19). Hershey, PA: IGI Global. doi:10.4018/978-1-5225-1978-2.ch001

Castagnolo, V. (2018). Analyzing, Classifying, Safeguarding: Drawing for the Borgo Murattiano Neighbourhood of Bari. In G. Carlone, N. Martinelli, & F. Rotondo (Eds.), *Designing Grid Cities for Optimized Urban Development and Planning* (pp. 93–108). Hershey, PA: IGI Global. doi:10.4018/978-1-5225-3613-0.ch006

Castilla-Polo, F., Gallardo-Vázquez, D., Sánchez-Hernández, M. I., & Ruiz-Rodríguez, M. D. (2017). Cooperatives as Responsible and Innovative Entrepreneurial Ecosystems in Smart Territories: The Olive Oil Industry in the South of Spain. In L. Carvalho (Ed.), *Handbook of Research on Entrepreneurial Development and Innovation Within Smart Cities* (pp. 459–490). Hershey, PA: IGI Global. doi:10.4018/978-1-5225-1978-2.ch020

Charalabidis, Y., & Theocharopoulou, C. (2019). A Participative Method for Prioritizing Smart City Interventions in Medium-Sized Municipalities. *International Journal of Public Administration in the Digital Age*, *6*(1), 41–63. doi:10.4018/IJPADA.2019010103

Chatsiwa, J., Mujere, N., & Maiyana, A. B. (2016). Municipal Solid Waste Management. In U. Akkucuk (Ed.), *Handbook of Research on Waste Management Techniques for Sustainability* (pp. 19–43). Hershey, PA: IGI Global. doi:10.4018/978-1-4666-9723-2.ch002

Chawla, R., Singhal, P., & Garg, A. K. (2019). Impact of Dust for Solar PV in Indian Scenario: Experimental Analysis. In K. Ahuja & A. Khosla (Eds.), *Driving the Development, Management, and Sustainability of Cognitive Cities* (pp. 111–138). Hershey, PA: IGI Global. doi:10.4018/978-1-5225-8085-0.ch005

Cılız, N., Yıldırım, H., & Temizel, Ş. (2016). Structure Development for Effective Medical Waste and Hazardous Waste Management System. In U. Akkucuk (Ed.), *Handbook of Research on Waste Management Techniques for Sustainability* (pp. 303–327). Hershey, PA: IGI Global. doi:10.4018/978-1-4666-9723-2.ch016

Collignon de Alba, C., Haberleithner, J., & López, M. M. (2017). Creative Industries in the Smart City: Overview of a Liability in Emerging Economies. In L. Carvalho (Ed.), *Handbook of Research on Entrepreneurial Development and Innovation Within Smart Cities* (pp. 107–126). Hershey, PA: IGI Global. doi:10.4018/978-1-5225-1978-2.ch006

Corum, A. (2016). Remanufacturing, an Added Value Product Recovery Strategy. In U. Akkucuk (Ed.), *Handbook of Research on Waste Management Techniques for Sustainability* (pp. 347–367). Hershey, PA: IGI Global. doi:10.4018/978-1-4666-9723-2.ch018

Crosas, C. (2018). Latin American Cities: Modern Grids From 1850s. In G. Carlone, N. Martinelli, & F. Rotondo (Eds.), *Designing Grid Cities for Optimized Urban Development and Planning* (pp. 39–51). Hershey, PA: IGI Global. doi:10.4018/978-1-5225-3613-0.ch003

D'Onofrio, S., Habenstein, A., & Portmann, E. (2019). Ontological Design for Cognitive Cities: The New Principle for Future Urban Management. In K. Ahuja & A. Khosla (Eds.), *Driving the Development, Management, and Sustainability of Cognitive Cities* (pp. 183–211). Hershey, PA: IGI Global. doi:10.4018/978-1-5225-8085-0.ch008

Dalzero, S. (2018). The Time of the Finished World Has Begun: A New Map of the World – National Borders Partially or Fully Fenced-Off. In G. Carlone, N. Martinelli, & F. Rotondo (Eds.), *Designing Grid Cities for Optimized Urban Development and Planning* (pp. 254–275). Hershey, PA: IGI Global. doi:10.4018/978-1-5225-3613-0.ch014

Damurski, L. (2016). Recent Progress in Online Communication Tools for Urban Planning: A Comparative Study of Polish and German Municipalities. *International Journal of E-Planning Research, 5*(1), 39–54. doi:10.4018/IJEPR.2016010103

Damurski, L. (2016). Smart City, Integrated Planning, and Multilevel Governance: A Conceptual Framework for e-Planning in Europe. *International Journal of E-Planning Research, 5*(4), 41–53. doi:10.4018/IJEPR.2016100103

Das, K. K., & Sharma, N. K. (2016). Post Disaster Housing Management for Sustainable Urban Development: A Review. *International Journal of Geotechnical Earthquake Engineering, 7*(1), 1–18. doi:10.4018/IJGEE.2016010101

Das, S., & Nayyar, A. (2019). Innovative Ideas to Manage Urban Traffic Congestion in Cognitive Cities. In K. Ahuja & A. Khosla (Eds.), *Driving the Development, Management, and Sustainability of Cognitive Cities* (pp. 139–162). Hershey, PA: IGI Global. doi:10.4018/978-1-5225-8085-0.ch006

Deakin, M., & Reid, A. (2017). The Embedded Intelligence of Smart Cities: Urban Life, Citizenship, and Community. *International Journal of Public Administration in the Digital Age, 4*(4), 62–74. doi:10.4018/IJPADA.2017100105

Djukic, A., Stupar, A., & Antonic, B. M. (2018). The Orthogonal Urban Matrix of the Towns in Vojvodina, Northern Serbia: Genesis and Transformation. In G. Carlone, N. Martinelli, & F. Rotondo (Eds.), *Designing Grid Cities for Optimized Urban Development and Planning* (pp. 128–156). Hershey, PA: IGI Global. doi:10.4018/978-1-5225-3613-0.ch008

Dolunay, O. (2016). A Paradigm Shift: Empowering Farmers to Eliminate the Waste in the Form of Fresh Water and Energy through the Implementation of 4R+T. In U. Akkucuk (Ed.), *Handbook of Research on Waste Management Techniques for Sustainability* (pp. 368–379). Hershey, PA: IGI Global. doi:10.4018/978-1-4666-9723-2.ch019

Dowling, C. M., Walsh, S. D., Purcell, S. M., Hynes, W. M., & Rhodes, M. L. (2017). Operationalising Sustainability within Smart Cities: Towards an Online Sustainability Indicator Tool. *International Journal of E-Planning Research*, *6*(4), 1–17. doi:10.4018/IJEPR.2017100101

Ekman, U. (2018). Smart City Planning: Complexity. *International Journal of E-Planning Research*, *7*(3), 1–21. doi:10.4018/IJEPR.2018070101

Erdogan, S. (2016). The Effect of Working Capital Management on Firm's Profitability: Evidence from Istanbul Stock Exchange. In U. Akkucuk (Ed.), *Handbook of Research on Waste Management Techniques for Sustainability* (pp. 244–261). Hershey, PA: IGI Global. doi:10.4018/978-1-4666-9723-2.ch013

Eudoxie, G., & Roopnarine, R. (2017). Climate Change Adaptation and Disaster Risk Management in the Caribbean. In W. Ganpat & W. Isaac (Eds.), *Environmental Sustainability and Climate Change Adaptation Strategies* (pp. 97–125). Hershey, PA: IGI Global. doi:10.4018/978-1-5225-1607-1.ch004

Fasolino, I. (2018). Rules for a New Town After a Disaster: The Gridded Schemes in the Plans. In G. Carlone, N. Martinelli, & F. Rotondo (Eds.), *Designing Grid Cities for Optimized Urban Development and Planning* (pp. 157–175). Hershey, PA: IGI Global. doi:10.4018/978-1-5225-3613-0.ch009

Fernandes, C. M., & Dias de Sousa, I. (2017). Digital Swarms: Social Interaction and Emergent Phenomena in Personal Communications Networks. In L. Carvalho (Ed.), *Handbook of Research on Entrepreneurial Development and Innovation Within Smart Cities* (pp. 44–59). Hershey, PA: IGI Global. doi:10.4018/978-1-5225-1978-2.ch003

Fernandes, V., Moreira, A., & Daniel, A. I. (2017). A Qualitative Analysis of Social Entrepreneurship Involving Social Innovation and Intervention. In L. Carvalho (Ed.), *Handbook of Research on Entrepreneurial Development and Innovation Within Smart Cities* (pp. 417–438). Hershey, PA: IGI Global. doi:10.4018/978-1-5225-1978-2.ch018

Gabriel, B. F., Valente, R. A., Dias-de-Oliveira, J., Neto, V. F., & Andrade-Campos, A. (2017). Methodologies for Engineering Learning and Teaching (MELT): An Overview of Engineering Education in Europe and a Novel Concept for Young Students. In L. Carvalho (Ed.), *Handbook of Research on Entrepreneurial Development and Innovation Within Smart Cities* (pp. 363–391). Hershey, PA: IGI Global. doi:10.4018/978-1-5225-1978-2.ch016

Garrick, T. A., & Liburd, O. E. (2017). Impact of Climate Change on a Key Agricultural Pest: Thrips. In W. Ganpat & W. Isaac (Eds.), *Environmental Sustainability and Climate Change Adaptation Strategies* (pp. 232–254). Hershey, PA: IGI Global. doi:10.4018/978-1-5225-1607-1.ch009

Gencer, Y. G. (2016). Mystery of Recycling: Glass and Aluminum Examples. In U. Akkucuk (Ed.), *Handbook of Research on Waste Management Techniques for Sustainability* (pp. 172–191). Hershey, PA: IGI Global. doi:10.4018/978-1-4666-9723-2.ch009

Gencer, Y. G., & Akkucuk, U. (2016). Reverse Logistics: Automobile Recalls and Other Conditions. In U. Akkucuk (Ed.), *Handbook of Research on Waste Management Techniques for Sustainability* (pp. 125–154). Hershey, PA: IGI Global. doi:10.4018/978-1-4666-9723-2.ch007

Gonçalves, J. M., Martins, T. G., & Vilhena da Cunha, I. B. (2017). Local Creative Ecosystems as a Strategy for the Development of Low-Density Urban Spaces. In L. Carvalho (Ed.), *Handbook of Research on Entrepreneurial Development and Innovation Within Smart Cities* (pp. 127–149). Hershey, PA: IGI Global. doi:10.4018/978-1-5225-1978-2.ch007

Goundar, S., & Appana, S. (2017). Mainstreaming Development Policies for Climate Change in Fiji: A Policy Gap Analysis and the Role of ICTs. In W. Ganpat & W. Isaac (Eds.), *Environmental Sustainability and Climate Change Adaptation Strategies* (pp. 1–31). Hershey, PA: IGI Global. doi:10.4018/978-1-5225-1607-1.ch001

Grochulska-Salak, M. (2019). Urban Farming in Sustainable City Development. In J. Vargas-Hernández & J. Zdunek-Wielgołaska (Eds.), *Bioeconomical Solutions and Investments in Sustainable City Development* (pp. 43–64). Hershey, PA: IGI Global. doi:10.4018/978-1-5225-7958-8.ch003

Gruszecka, K. (2019). Ecological Centre of Warsaw as a Development Path. In J. Vargas-Hernández & J. Zdunek-Wielgołaska (Eds.), *Bioeconomical Solutions and Investments in Sustainable City Development* (pp. 117–150). Hershey, PA: IGI Global. doi:10.4018/978-1-5225-7958-8.ch006

Hesapci-Sanaktekin, O., & Aslanbay, Y. (2016). The Networked Self: Collectivism Redefined in Civic Engagements through Social Media Causes. In U. Akkucuk (Ed.), *Handbook of Research on Waste Management Techniques for Sustainability* (pp. 262–276). Hershey, PA: IGI Global. doi:10.4018/978-1-4666-9723-2.ch014

Ikeda, M. (2019). Developing a Sustainable Eco-City in Pre-Olympic Tokyo: Potential of New Methods and Their Limits in an Urban Era. In J. Vargas-Hernández & J. Zdunek-Wielgołaska (Eds.), *Bioeconomical Solutions and Investments in Sustainable City Development* (pp. 196–223). Hershey, PA: IGI Global. doi:10.4018/978-1-5225-7958-8.ch009

İşcan, E. (2019). Strategies of Sustainable Bioeconomy in the Industry 4.0 Framework for Inclusive and Social Prosperity. In J. Vargas-Hernández & J. Zdunek-Wielgołaska (Eds.), *Bioeconomical Solutions and Investments in Sustainable City Development* (pp. 21–42). Hershey, PA: IGI Global. doi:10.4018/978-1-5225-7958-8.ch002

J., J., Samui, P., & Dixon, B. (2016). Determination of Rate of Medical Waste Generation Using RVM, MARS and MPMR. In U. Akkucuk (Ed.), *Handbook of Research on Waste Management Techniques for Sustainability* (pp. 1-18). Hershey, PA: IGI Global. doi:10.4018/978-1-4666-9723-2.ch001

Jiménez, R., Rodríguez, P. L., & Fernández, R. R. (2019). Green Spaces of the Metropolitan Area of Guadalajara. In J. Vargas-Hernández & J. Zdunek-Wielgołaska (Eds.), *Bioeconomical Solutions and Investments in Sustainable City Development* (pp. 151–179). Hershey, PA: IGI Global. doi:10.4018/978-1-5225-7958-8.ch007

Kais, S. M. (2017). Climate Change: Vulnerability and Resilience in Commercial Shrimp Aquaculture in Bangladesh. In W. Ganpat & W. Isaac (Eds.), *Environmental Sustainability and Climate Change Adaptation Strategies* (pp. 162–187). Hershey, PA: IGI Global. doi:10.4018/978-1-5225-1607-1.ch006

Khan, B. (2019). Bio-Economy: Visions, Strategies, and Policies. In J. Vargas-Hernández & J. Zdunek-Wielgołaska (Eds.), *Bioeconomical Solutions and Investments in Sustainable City Development* (pp. 1–20). Hershey, PA: IGI Global. doi:10.4018/978-1-5225-7958-8.ch001

Khanna, B. K. (2017). Indian National Strategy for Climate Change Adaptation and Mitigation. In W. Ganpat & W. Isaac (Eds.), *Environmental Sustainability and Climate Change Adaptation Strategies* (pp. 32–63). Hershey, PA: IGI Global. doi:10.4018/978-1-5225-1607-1.ch002

Khanna, B. K. (2017). Vulnerability of the Lakshadweep Coral Islands in India and Strategies for Mitigating Climate Change Impacts. In W. Ganpat & W. Isaac (Eds.), *Environmental Sustainability and Climate Change Adaptation Strategies* (pp. 64–96). Hershey, PA: IGI Global. doi:10.4018/978-1-5225-1607-1.ch003

Kocasoy, G. (2016). Economic Instruments for Sustainable Environmental Management. In U. Akkucuk (Ed.), *Handbook of Research on Waste Management Techniques for Sustainability* (pp. 192–211). Hershey, PA: IGI Global. doi:10.4018/978-1-4666-9723-2.ch010

Lallo, C. H., Smalling, S., Facey, A., & Hughes, M. (2017). The Impact of Climate Change on Small Ruminant Performance in Caribbean Communities. In W. Ganpat & W. Isaac (Eds.), *Environmental Sustainability and Climate Change Adaptation Strategies* (pp. 296–321). Hershey, PA: IGI Global. doi:10.4018/978-1-5225-1607-1.ch011

Li, Z., Wang, Y., & Chen, Q. (2017). Real-Time Monitoring of Intercity Passenger Flows Based on Big Data: A Decision Support Tool for Urban Sustainability. *International Journal of Strategic Decision Sciences*, *8*(4), 120–128. doi:10.4018/IJSDS.2017100106

Loi, N. K., Huyen, N. T., Tu, L. H., Tram, V. N., Liem, N. D., Dat, N. L., ... Minh, D. N. (2017). Sustainable Land Use and Watershed Management in Response to Climate Change Impacts: Case Study in Srepok Watershed, Central Highland of Vietnam. In W. Ganpat & W. Isaac (Eds.), *Environmental Sustainability and Climate Change Adaptation Strategies* (pp. 255–295). Hershey, PA: IGI Global. doi:10.4018/978-1-5225-1607-1.ch010

López-Arranz, M. A. (2017). The Role Corporate Social Responsibility Has in the Smart City Project in Spain. In L. Carvalho (Ed.), *Handbook of Research on Entrepreneurial Development and Innovation Within Smart Cities* (pp. 439–458). Hershey, PA: IGI Global. doi:10.4018/978-1-5225-1978-2.ch019

Lucas, M. R., Rego, C., Vieira, C., & Vieira, I. (2017). Proximity and Cooperation for Innovative Regional Development: The Case of the Science and Technology Park of Alentejo. In L. Carvalho (Ed.), *Handbook of Research on Entrepreneurial Development and Innovation Within Smart Cities* (pp. 199–228). Hershey, PA: IGI Global. doi:10.4018/978-1-5225-1978-2.ch010

Magrinho, A., Neves, J., & Silva, J. R. (2017). The Triple Helix Model: Evidence in the Internationalization of the Health Industry. In L. Carvalho (Ed.), *Handbook of Research on Entrepreneurial Development and Innovation Within Smart Cities* (pp. 60–79). Hershey, PA: IGI Global. doi:10.4018/978-1-5225-1978-2.ch004

Maiorano, A. C. (2018). Urban Fronts in Murattiano Neighbourhood of Bari: A Selective Survey of the Built Environment. In G. Carlone, N. Martinelli, & F. Rotondo (Eds.), *Designing Grid Cities for Optimized Urban Development and Planning* (pp. 78–92). Hershey, PA: IGI Global. doi:10.4018/978-1-5225-3613-0.ch005

Martinelli, N., & Mangialardi, G. (2018). Cities With Grid Layout: Ubiquitousness and Flexibility of an Urban Model. In G. Carlone, N. Martinelli, & F. Rotondo (Eds.), *Designing Grid Cities for Optimized Urban Development and Planning* (pp. 188–208). Hershey, PA: IGI Global. doi:10.4018/978-1-5225-3613-0.ch011

Mizutani, S., Liao, K., & Sasaki, T. G. (2019). Forest-River-Ocean Nexus-Based Education for Community Development: Aiming at Resilient Sustainable Society. In J. Vargas-Hernández & J. Zdunek-Wielgołaska (Eds.), *Bioeconomical Solutions and Investments in Sustainable City Development* (pp. 224–248). Hershey, PA: IGI Global. doi:10.4018/978-1-5225-7958-8.ch010

Moreira, A., & Ferreira, M. A. (2017). Strategic Challenges of the Portuguese Molds Industry: A Sectoral Innovation Perspective. In L. Carvalho (Ed.), *Handbook of Research on Entrepreneurial Development and Innovation Within Smart Cities* (pp. 534–560). Hershey, PA: IGI Global. doi:10.4018/978-1-5225-1978-2.ch023

Mujere, N., & Moyce, W. (2017). Climate Change Impacts on Surface Water Quality. In W. Ganpat & W. Isaac (Eds.), *Environmental Sustainability and Climate Change Adaptation Strategies* (pp. 322–340). Hershey, PA: IGI Global. doi:10.4018/978-1-5225-1607-1.ch012

Mundula, L., & Auci, S. (2017). Smartness, City Efficiency, and Entrepreneurship Milieu. In L. Carvalho (Ed.), *Handbook of Research on Entrepreneurial Development and Innovation Within Smart Cities* (pp. 173–198). Hershey, PA: IGI Global. doi:10.4018/978-1-5225-1978-2.ch009

Nath, R. (2019). Parametric Evaluation of Beam Deflection on Piezoelectric Material Using Implicit and Explicit Method Simulations: A Study in Energy Engineering. In J. Vargas-Hernández & J. Zdunek-Wielgołaska (Eds.), *Bioeconomical Solutions and Investments in Sustainable City Development* (pp. 65–87). Hershey, PA: IGI Global. doi:10.4018/978-1-5225-7958-8.ch004

Nayyar, A., Jain, R., Mahapatra, B., & Singh, A. (2019). Cyber Security Challenges for Smart Cities. In K. Ahuja & A. Khosla (Eds.), *Driving the Development, Management, and Sustainability of Cognitive Cities* (pp. 27–54). Hershey, PA: IGI Global. doi:10.4018/978-1-5225-8085-0.ch002

Odabasi, A., & Tiryaki, C. S. (2016). An Empirical Review of Long Term Electricity Demand Forecasts for Turkey. In U. Akkucuk (Ed.), *Handbook of Research on Waste Management Techniques for Sustainability* (pp. 227–243). Hershey, PA: IGI Global. doi:10.4018/978-1-4666-9723-2.ch012

Okay, E. (2016). Towards Smart Cities in Turkey?: Transitioning from Waste to Creative, Clean and Cheap Eco-Energy. In U. Akkucuk (Ed.), *Handbook of Research on Waste Management Techniques for Sustainability* (pp. 277–302). Hershey, PA: IGI Global. doi:10.4018/978-1-4666-9723-2.ch015

Pawlikowska-Piechotka, A., Łukasik, N., Ostrowska-Tryzno, A., & Sawicka, K. (2017). A Smart City Initiative: Urban Greens and Evaluation Method of the Sport and Recreation Potentials (SEM). In L. Carvalho (Ed.), *Handbook of Research on Entrepreneurial Development and Innovation Within Smart Cities* (pp. 561–583). Hershey, PA: IGI Global. doi:10.4018/978-1-5225-1978-2.ch024

Raisinghani, M. S., & Idemudia, E. C. (2016). Green Information Systems for Sustainability. In U. Akkucuk (Ed.), *Handbook of Research on Waste Management Techniques for Sustainability* (pp. 212–226). Hershey, PA: IGI Global. doi:10.4018/978-1-4666-9723-2.ch011

Rathee, D. S., Ahuja, K., & Hailu, T. (2019). Role of Electronics Devices for E-Health in Smart Cities. In K. Ahuja & A. Khosla (Eds.), *Driving the Development, Management, and Sustainability of Cognitive Cities* (pp. 212–233). Hershey, PA: IGI Global. doi:10.4018/978-1-5225-8085-0.ch009

Rotondo, F. (2018). The Grid Cities: Between Tradition and Innovation. In G. Carlone, N. Martinelli, & F. Rotondo (Eds.), *Designing Grid Cities for Optimized Urban Development and Planning* (pp. 109–127). Hershey, PA: IGI Global. doi:10.4018/978-1-5225-3613-0.ch007

Sánchez-Fernández, M. D., & Cardona, J. R. (2017). The Perception of the Effect of Tourism on the Local Community before the Ibiza Smart Island Project. In L. Carvalho (Ed.), *Handbook of Research on Entrepreneurial Development and Innovation Within Smart Cities* (pp. 392–416). Hershey, PA: IGI Global. doi:10.4018/978-1-5225-1978-2.ch017

Santos, B. (2017). Improving Urban Planning Information, Transparency and Participation in Public Administrations. *International Journal of E-Planning Research*, *6*(4), 58–75. doi:10.4018/IJEPR.2017100104

Sardà, J. F. (2018). Cerdà/Barcelona/Eixample: 1855-2017 … A Work in Progress. In G. Carlone, N. Martinelli, & F. Rotondo (Eds.), *Designing Grid Cities for Optimized Urban Development and Planning* (pp. 19–38). Hershey, PA: IGI Global. doi:10.4018/978-1-5225-3613-0.ch002

Selvi, M. S. (2016). Physical Distribution Problems of Textile Companies in Turkey. In U. Akkucuk (Ed.), *Handbook of Research on Waste Management Techniques for Sustainability* (pp. 328–346). Hershey, PA: IGI Global. doi:10.4018/978-1-4666-9723-2.ch017

Signorile, N. (2018). Pride and Prejudice: The Murattiano-Modernism. In G. Carlone, N. Martinelli, & F. Rotondo (Eds.), *Designing Grid Cities for Optimized Urban Development and Planning* (pp. 52–77). Hershey, PA: IGI Global. doi:10.4018/978-1-5225-3613-0.ch004

Singh, G., Kapoor, R., & Khosla, A. K. (2019). Intelligent Anomaly Detection Video Surveillance Systems for Smart Cities. In K. Ahuja & A. Khosla (Eds.), *Driving the Development, Management, and Sustainability of Cognitive Cities* (pp. 163–182). Hershey, PA: IGI Global. doi:10.4018/978-1-5225-8085-0.ch007

Singh-Ackbarali, D., & Maharaj, R. (2017). Mini Livestock Ranching: Solution to Reducing the Carbon Footprint and Negative Environmental Impacts of Agriculture. In W. Ganpat & W. Isaac (Eds.), *Environmental Sustainability and Climate Change Adaptation Strategies* (pp. 188–212). Hershey, PA: IGI Global. doi:10.4018/978-1-5225-1607-1.ch007

Sood, N., Saini, I., Awasthi, T., Saini, M. K., Bhoriwal, P., & Kaur, T. (2019). Fog Removal Algorithms for Real-Time Video Footage in Smart Cities for Safe Driving. In K. Ahuja & A. Khosla (Eds.), *Driving the Development, Management, and Sustainability of Cognitive Cities* (pp. 55–86). Hershey, PA: IGI Global. doi:10.4018/978-1-5225-8085-0.ch003

Stone, R. J. (2017). Modelling the Frequency of Tropical Cyclones in the Lower Caribbean Region. In W. Ganpat & W. Isaac (Eds.), *Environmental Sustainability and Climate Change Adaptation Strategies* (pp. 341–349). Hershey, PA: IGI Global. doi:10.4018/978-1-5225-1607-1.ch013

Stratigea, A., Leka, A., & Panagiotopoulou, M. (2017). In Search of Indicators for Assessing Smart and Sustainable Cities and Communities' Performance. *International Journal of E-Planning Research*, *6*(1), 43–73. doi:10.4018/IJEPR.2017010103

Touq, A. B., & Ijeh, A. (2018). Information Security and Ecosystems in Smart Cities: The Case of Dubai. *International Journal of Information Systems and Social Change*, *9*(2), 28–43. doi:10.4018/IJISSC.2018040103

Üç, M., & Elitaş, C. (2016). Life Cycle Costing for Sustainability. In U. Akkucuk (Ed.), *Handbook of Research on Waste Management Techniques for Sustainability* (pp. 96–107). Hershey, PA: IGI Global. doi:10.4018/978-1-4666-9723-2.ch005

Ulker-Demirel, E., & Demirel, E. (2016). Green Marketing and Stakeholder Perceptions. In U. Akkucuk (Ed.), *Handbook of Research on Waste Management Techniques for Sustainability* (pp. 75–95). Hershey, PA: IGI Global. doi:10.4018/978-1-4666-9723-2.ch004

Vargas-Hernández, J. G., López, J. J., & Zdunek-Wielgołaska, J. A. (2019). Entrepreneurial and Institutional Analysis of Biodiesel Companies in Mexico. In J. Vargas-Hernández & J. Zdunek-Wielgołaska (Eds.), *Bioeconomical Solutions and Investments in Sustainable City Development* (pp. 89–115). Hershey, PA: IGI Global. doi:10.4018/978-1-5225-7958-8.ch005

Vázquez, D. G., & Gil, M. T. (2017). Sustainability in Smart Cities: The Case of Vitoria-Gasteiz (Spain) – A Commitment to a New Urban Paradigm. In L. Carvalho (Ed.), *Handbook of Research on Entrepreneurial Development and Innovation Within Smart Cities* (pp. 248–268). Hershey, PA: IGI Global. doi:10.4018/978-1-5225-1978-2.ch012

Velozo, R. A., & Montanha, G. K. (2017). Evaluation of a Mobile Software Development Company. In L. Carvalho (Ed.), *Handbook of Research on Entrepreneurial Development and Innovation Within Smart Cities* (pp. 514–533). Hershey, PA: IGI Global. doi:10.4018/978-1-5225-1978-2.ch022

About the Contributors

Syahibudil Ikhwan Abdul Kudus has a PhD in design for digital fabrication from Loughborough University, UK. His area of interest includes additive manufacturing, consumer design and industrial design and actively published papers in related area.

Zulkeflee Abdullah is a senior lecturer in the Faculty of Manufacturing Engineering, Universiti Teknikal Malaysia Melaka (UTeM) with a PhD from the University of Melbourne. He joined academia after an early career as a process engineer in a multinational company in Malaysia. His research interests are principally associated with additive manufacturing, engineering design, design for manufacture, manufacturing sustainability, sustainable development, and total quality management.

Farid Nasir Ani is a professor in School of Mechanical Engineering in Universiti Teknologi Malaysia He has a Bachelor of Science in Engineering, University Glasgow, Scotland, 1982. Master of Science in Engineering, University Birmingham, England, 1985. Doctor of Philosophy, University Leeds, England, 1992.

M. N. M. Ansari is Associate Professor from the Institute of Power Engineering, Universiti Tenaga Nasional, 43000 Kajang, Selangor, Malaysia. He is working in the field of composites since 2000 and has 22 years of experience in teaching, research in the field of composites. His area of research interests includes advances in materials, natural fibre, biocomposite materials, polymer composite design and testing, composites modification and properties improvements, tissue engineering, hybrid composites and biomaterials.

A. Atiqah is currently working as a Postdoctoral Research Fellow Officer at the Institute of Power Engineering, Universiti Tenaga Nasional, 43000 Kajang, Selangor, Malaysia. She received her PhD from the Universiti Putra Malaysia, Selangor, Malaysia in Biocomposite Technology and Design in Institute of Tropical Forestry and Forest Products (INTROP). She is working in the composite field since 2013 and has five years of experience in teaching, research in the field of composites. Her area of research interests includes advances of materials, natural fibre, biocomposite materials, polymer composite design and testing, composites modification and properties improvements, plastic technology, hybrid composites and biomaterials.

Mohamad Zaki Bin Hassan began his career as a lecturer in the L& G Twintech in 1999 after graduating with a Bachelor's degree in Mechanical Engineering from Universiti Teknologi Malaysia (UTM). After spending almost a year there, in February 2000, he started his academic career as a tutor in UTM, Kuala Lumpur. From then, his journey into the scholarly world continued whereby he successfully completed his Masters in Mechanical Engineering at the UTM in 2003. In 2012, he went on to complete a Doctor of Philosophy (PhD) at the University of Liverpool. His PhD thesis was on "Low Velocity Impact Response of Sandwich Structures", where he modelled the polymeric sandwich structures using finite element analysis and mathematical approaches following quasi-static and dynamic loadings. This began of his journey as an academician, whereby he started publishing and establishing his expertise in polymer and sandwich structures. His expertise in polymer and composite structures has transcended from the industry to the academy and has proven to be extremely beneficial particularly in exploring new research opportunities in this field. He has an added advantage, as his research is focusing on an advanced material towards sustainability and green technology. In 2017, he was fortunate to study of recycling polymer that combining with low cost material in order for fabrication of the product. One of his research in this composite material focused on optimizing the amount of volume fraction between polymer and secondary material. Here, study also focused on the durability, life cycle and strength of the material due to humidity, UV and extreme weather conditions. This material has been employed as a breakwater structure that can be utilized to reduce wave strength in muddy and salty water condition along the coastline. This structure has installed at Pantai Haji Dorani. At the meantime, he also explored the possibility of recycle PET as a flood stick gauge. In addition, further study has also been conducting in use of this recycle polymer as a construction material including floor tile, roof tile, beam structure and brick. Since 2015, Dr Mohamad Zaki with his research students have been working to enhance the capabilities of selective laser melting (SLM) technique in order to manufacture a lightweight composite aerospace structure. This includes on the improvement of the fabrication time of the lattice structure in order to offer a higher strength to weight ratio of structural material. These sandwich structures have been verified using FEM and mathematical model. He also has been worked in enhancing the treatment parameter of natural fibres due to incompatibility with the matrixes. A major drawback from using natural fibre in polymeric composites is the incompatibility between natural fibre and polymer matrix due to hydrophilic and hydrophobic interaction. The high moisture content in natural fibre makes them difficult to bond properly with the polymer matrix and finally, degrading the mechanical properties of the structure laminate. Kenaf, jute, hemp and bamboo are mainly natural fibre that was obeyed in this work. The candidate has been involved by the university on numerous occasions for his knowledge and expertise in assisting UTM to move to a higher plane of excellence. His involvement in the College of Science & Technology from 2000 to 2007 in developing a curriculum for instituting collaboration academic programme where the academic programme was conducted as a joint programme. This was the first stage in offering a Vocational Diploma in a public university in Malaysia. The academic programme is recognized as Diploma of Technology. The collaborating institutions are MARA, IKTBN and Ministry of Defence Malaysia. At this time, he was also appointed as a coordinator of the Diploma of Technology (Tool and Die) specially offered by IKTBN, Sepang. With regards to curriculum development, he has committee in the development of Diploma of Technology (Tool and Die) (1999), Diploma of Technology (Automotive) (1999), Diploma of Technology (Military Vehicles) (2004), Diploma of Technology (Weaponry) (2004). He was appointed in various administrative posts in UTM including as the Head of Mechanical Engineering Panel, Fellow of Keloj Siswa Jaya, Head of Solid and Materials Lab, Fellow of PHB, Coordinator for Industrial Training (Industrial Design) and Head of Journal Club RSEAT. Dr Mohamad Zaki has been invited to lecture for short courses such as Life Cycle of Preventive Maintenance Using Weibull analysis conducted by SILK Hiway Sdn Bhd. He is also recognized by his peers abroad in his area of expertise where he was invited to become journals editor and reviewer for journal articles and research grant. He is also involved in consultation with government agency including Nahrim and Ministry of Defence Malaysia. In addition, he also have a contributed to consult with industry such as Intel, Petron Sdn Bhd, Voith Paper Fabrics Ipoh Sdn Bhd, Micromechanic Sdn Bhd, Bandus Jaya Enterprise and Babas Product (M) Sdn. Bhd. He has taught courses at undergraduate and postgraduate levels including subjects such as Solid Mechanics, Fluid Power, Manufacturing Processes, Production and Planning Control, Dynamic, Product Design, Mechanical Design, Static, Industrial Engineering, Machine Design, Engineering Economic. He also taught a subject for Masters and Doctoral Programmes including Research Methodology, Technology Management and Economic Analysis for Technology. Throughout his career as an academician, he has been instrumental in developing various curriculums and their content with input from various parties including industries. At the national level, he is a graduate member of the Institution of Engineers Malaysia (IEM). He also assisted in the preparation of MOU with Nahrim. He was honoured an Award of Excellence for Research Methodology and Technology Management taught classes.

In March 2017, Mohamad Zaki was awarded the Gold (Masters) for the invention of the ultrasonic horn for metal tube compression. He was also received bronze medal for invention of low cost micro hydro power generation: a cross –flow turbine conferred by Agensi Nuklear Malaysia.

Shajahan Bin Maidin is an Associates Professor at the Universiti Teknikal Malaysia Melaka specifically at the Faculty of Manufacturing Engineering. He is active in research, consultancy, teaching and learning at various undergraduate and postgraduates courses. His research interest and activity is related to additive manufacturing, product design, sustainable manufacturing. He graduated with a PhD from Loughborough University, UK in 2012. Prior to that he graduated from The University of Warwick specializing in Manufacturing Systems Engineering in 2002. and from the Portsmouth University also specializing in Manufacturing Systems Engineering in 1999.

Mohd Nurazzi Bin Norizan was born on 20th December 1988 in Jitra, Kedah. He completed his primary education at Sekolah Rendah Bukit Hampar, Segamat Johor in 2000. He completed his secondary education at Sekolah Teknik Melaka Tengah, Melaka (Civil Engineering program) in 2005. The author graduated from his Diploma in Polymer Technology, from Faculty of Applied Science, Universiti Teknologi MARA (UiTM) Arau in June 2009. Then he graduated in Bachelor of Science (Polymer Technology), from Faculty of Applied Science, Universiti Teknologi MARA (UiTM) Shah Alam in June 2011. In September 2011, he enrolled his Master program in the field of polymer synthesis under supervision of Associate Professor Dr. Rahmah Mohamed and graduated in Master of Science on November, 2014. On September 2015, he started his Ph.D. program in the field of Materials Engineering under Faculty of Engineering, Universiti Putra Malaysia (UPM) Serdang and graduated on October, 2018 under supervision of Associate Professor Dr Khalina Abdan.

Ridhwan Jumaidin has a PhD in Mechanical Engineering from Universiti Putra Malaysia. His area of interest includes natural fibre composites and bio-polymers and actively published papers in related area.

Yusriah Lazim is a Senior Lecturer at University Kuala Lumpur (UniKL MICET). Their research interests include: Characterization of materials, polymer and polymer composites, Natural fibre reinforced composites-physical, mechanical, thermal properties, Development and characterization of green bio-composites and hybrid composites, Polymer processing, Characterization and modification of thermoplastic starch.

Firdaus Muhammad-Sukki received his MEng degree in Electrical and Electronic Engineering from Imperial College, London in 2006 under the Yayasan TM scholarship. In 2009, he received studentship awards from Glasgow Caledonian University (GCU), Scottish Funding Council (SFC) and Yayasan TM to pursue his postgraduate study in Glasgow Caledonian University from which he obtained his PhD in 2013. Dr. Muhammad Sukki is a Chartered Engineer, a member of the Institution of Engineering and Technology (IET) and an associate of the City and Guilds, London (ACGI). Dr. Muhammad Sukki has been an active researcher and is currently a Lecturer in Robert Gordon University, United Kingdom. His research interest is in designing optical concentrators for use in many applications including solar technology, illumination, heating and cooling etc. On top of that, Dr. Muhammad Sukki also carried out a number of non-technical research including market trend and financial analysis related to renewable technologies for countries such as Malaysia, Japan, Nigeria, Cameroon and United Kingdom. He has excellent track record in collaborating with research universities around the world. He has published numerous articles in high impact factor journals (e.g. Nature, Renewable & Sustainable Energy Reviews, Applied Energy, Energy, Renewable Energy, Energy Policy etc.), as well as presenting in various conferences related to his area. Prior to joining the academia, he was a communication engineer in Malaysia's largest telecommunication company. He was in charge of the leased line servers for Malaysia's network and was involved in major projects related to the telecommunication while holding the post.

Nadlene Razali has a PhD in Mechanical Engineering from Universiti Putra Malaysia. Her area of interest includes natural fibre composites, and material characterization and actively published papers in related area.

Muda Siti Afifah is a lecturer in UniKL Micet and has a background in chemical engineering. Research interest on renewable energy, chemical extraction process and waste management. Currently pursuing a PhD in chemical engineering field focusing on carbon fibre production from biomass.

Mahanum Mohd Zamberi is currently a Mechanical Engineering Lecturer in Universiti Teknikal Malaysia Melaka since 2006. Received her Master in Combustion & Energy from University of Leeds in 2008 and in her final year of her phd in Universiti Teknologi Malaysia.

Index

3D printing 4-5, 9-16, 18, 131-134, 137, 139, 141

A

acrylonitrile butadiene styrene (ABS) 1-3, 5-6, 8, 14, 16-20, 22, 28-29, 34-35, 39-40, 42, 46-48, 56, 58-59, 63, 67, 69-70, 80, 83, 85-87, 92, 101-103, 106, 114-117, 122, 124-127, 130, 133-135, 138, 140-145, 147, 151-152
Additive manufacturing (AM) 131
ANOVA 113, 118, 120, 123, 127

B

bamboo 112-116, 118-121, 125, 127
bio-composites 1, 13, 16, 18, 28-30, 32-35, 38-41, 46
biodiesel 92-102, 104-105
biodiversity 92-93
biopolymer 10, 12-14, 30, 48
Box-Behnken design 112, 114-115
Box-Behnken design (BBD) 112, 115
build orientation 12

C

composites 1, 11, 14-18, 20, 22, 32-33, 38, 40-46, 48, 56-66, 68-70, 80, 113-114
computer-aided design (CAD) 3, 131
crosslinking 78, 81-84, 86-87, 114

D

Differential Scanning Calorimetry (DSC) 17

E

energy recovery 134, 136

F

fibre loading 59, 62-63, 65
fibre reinforcements 61
Free fatty acids (FFAs) 96
Fused Deposition Modeling (FDM) 130-134, 137, 139, 141, 147, 151
Fused Filament Fabrication (FFF) 1-8, 10, 12, 14-17, 19-22

G

geometry 2, 7-8, 59, 68
green composites 56-59, 80
green materials 1, 10, 13, 22, 56-58

H

heterogeneous catalyst 92-93, 102-103, 105
Hevea Brasiliensis (Rubber) 94-95, 104
high efficiency particulate air (HEPA) filter 9
hybrid composites 56, 64-65
hybridization 64-65

L

lignocellulosic 28, 32, 41-45
low density polyethylene (LDPE) 78

M

mechanical properties 4, 11-17, 30, 33-34, 38, 40-43, 58-61, 63-66, 68-69, 77, 81-82, 84, 86-87, 113-114, 120-121, 130, 132-133, 135, 139, 149, 151
mechanical recycling 134-136, 142, 150
microcellulose 32-34
microwave irradiation 92-93, 100-101, 104
Mode I Fracture Toughness 112

N

nanocellulose 28, 32-35, 46-47
natrium hydroxide (NaOH) 112
natural fibre 13-14, 56-58, 62, 64-67, 113-114, 120
non-edible oils 92-93, 96, 100

P

packaging 28, 35, 39, 46, 77-78, 80-82, 86
Polyethylene terephthalate (PET) 15
polylactic acid (PLA) 9
polypropylene (PP) 17, 78

R

recycling 8, 32, 102, 106, 113, 130-137, 142, 150
retrogradation 44, 77-78, 81-83, 87

S

sago starch 77-80, 82-84, 86-87
starch film 34-35, 38-39, 41, 44, 47

T

thermogravimetric analysis (TGA) 17, 84
Thermoplastic Starch (TPS) 30, 33-35, 40, 42, 44, 46-48, 77-78, 80-87

www.ingramcontent.com/pod-product-compliance
Lightning Source LLC
LaVergne TN
LVHW081321300926
849161LV00008B/1837
* 9 7 8 1 7 9 9 8 1 3 7 4 3 *